Lecture Notes in Computer Science 10820

Commenced Publication in 1973
Founding and Former Series Editors:
Gerhard Goos, Juris Hartmanis, and Jan van Leeuwen

More information about this series at http://www.springer.com/series/7410

Jesper Buus Nielsen · Vincent Rijmen (Eds.)

Advances in Cryptology – EUROCRYPT 2018

37th Annual International Conference on the Theory
and Applications of Cryptographic Techniques
Tel Aviv, Israel, April 29 – May 3, 2018
Proceedings, Part I

 Springer

Editors
Jesper Buus Nielsen
Aarhus University
Aarhus
Denmark

Vincent Rijmen
University of Leuven
Leuven
Belgium

ISSN 0302-9743   ISSN 1611-3349   (electronic)
Lecture Notes in Computer Science
ISBN 978-3-319-78380-2   ISBN 978-3-319-78381-9   (eBook)
https://doi.org/10.1007/978-3-319-78381-9

Library of Congress Control Number: 2018937382

LNCS Sublibrary: SL4 – Security and Cryptology

Printed on acid-free paper

This Springer imprint is published by the registered company Springer International Publishing AG part of Springer Nature
The registered company address is: Gewerbestrasse 11, 6330 Cham, Switzerland

Preface

Eurocrypt 2018, the 37th Annual International Conference on the Theory and Applications of Cryptographic Techniques, was held in Tel Aviv, Israel, from April 29 to May 3, 2018. The conference was sponsored by the International Association for Cryptologic Research (IACR). Orr Dunkelman (University of Haifa, Israel) was responsible for the local organization. He was supported by a local organizing team consisting of Technion's Hiroshi Fujiwara Cyber Security Research Center headed by Eli Biham, and most notably by Suzie Eid. We are deeply indebted to them for their support and smooth collaboration.

The conference program followed the now established parallel track system where the works of the authors were presented in two concurrently running tracks. Only the invited talks spanned over both tracks.

We received a total of 294 submissions. Each submission was anonymized for the reviewing process and was assigned to at least three of the 54 Program Committee members. Committee members were allowed to submit at most one paper, or two if both were co-authored. Submissions by committee members were held to a higher standard than normal submissions. The reviewing process included a rebuttal round for all submissions. After extensive deliberations, the Program Committee accepted 69 papers. The revised versions of these papers are included in these three-volume proceedings, organized topically within their respective track.

The committee decided to give the Best Paper Award to the papers "Simple Proofs of Sequential Work" by Bram Cohen and Krzysztof Pietrzak, "Two-Round Multiparty Secure Computation from Minimal Assumptions" by Sanjam Garg and Akshayaram Srinivasan, and "Two-Round MPC from Two-Round OT" by Fabrice Benhamouda and Huijia Lin. All three papers received invitations for the *Journal of Cryptology*.

The program also included invited talks by Anne Canteaut, titled "Desperately Seeking Sboxes", and Matthew Green, titled "Thirty Years of Digital Currency: From DigiCash to the Blockchain".

We would like to thank all the authors who submitted papers. We know that the Program Committee's decisions can be very disappointing, especially rejections of very good papers that did not find a slot in the sparse number of accepted papers. We sincerely hope that these works eventually get the attention they deserve.

We are also indebted to the members of the Program Committee and all external reviewers for their voluntary work. The Program Committee work is quite a workload. It has been an honor to work with everyone. The committee's work was tremendously simplified by Shai Halevi's submission software and his support, including running the service on IACR servers.

Finally, we thank everyone else — speakers, session chairs, and rump-session chairs — for their contribution to the program of Eurocrypt 2018. We would also like to thank the many sponsors for their generous support, including the Cryptography Research Fund that supported student speakers.

May 2018   Jesper Buus Nielsen
                                                                        Vincent Rijmen

Eurocrypt 2018

The 37th Annual International Conference on the Theory and Applications of Cryptographic Techniques

Sponsored by *the International Association for Cryptologic Research*

April 29 – May 3, 2018
Tel Aviv, Israel

General Chair

Orr Dunkelman — University of Haifa, Israel

Program Co-chairs

Jesper Buus Nielsen — Aarhus University, Denmark
Vincent Rijmen — University of Leuven, Belgium

Program Committee

Martin Albrecht	Royal Holloway, UK
Joël Alwen	IST Austria, Austria, and Wickr, USA
Gilles Van Assche	STMicroelectronics, Belgium
Paulo S. L. M. Barreto	University of Washington Tacoma, USA
Nir Bitansky	Tel Aviv University, Israel
Céline Blondeau	Aalto University, Finland
Andrey Bogdanov	DTU, Denmark
Chris Brzuska	TU Hamburg, Germany, and Aalto University, Finland
Jan Camenisch	IBM Research – Zurich, Switzerland
Ignacio Cascudo	Aalborg University, Denmark
Melissa Chase	Microsoft Research, USA
Alessandro Chiesa	UC Berkeley, USA
Joan Daemen	Radboud University, The Netherlands, and STMicroelectronics, Belgium
Yevgeniy Dodis	New York University, USA
Nico Döttling	Friedrich Alexander University Erlangen-Nürnberg, Germany
Sebastian Faust	TU Darmstadt, Germany
Serge Fehr	CWI Amsterdam, The Netherlands
Georg Fuchsbauer	Inria and ENS, France
Jens Groth	University College London, UK
Jian Guo	Nanyang Technological University, Singapore

Additional Reviewers

Daniel Apon
Gilad Asharov
Nuttapong Attrapadung
Benedikt Auerbach
Daniel Augot
Christian Badertscher
Saikrishna
   Badrinarayanan
Shi Bai
Josep Balasch
Marshall Ball
Valentina Banciu
Subhadeep Banik
Zhenzhen Bao
Gilles Barthe
Lejla Batina
Balthazar Bauer
Carsten Baum
Christof Beierle
Amos Beimel
Sonia Belaid
Aner Ben-Efraim
Fabrice Benhamouda
Iddo Bentov
Itay Berman
Kavun Elif Bilge
Olivier Blazy
Jeremiah Blocki
Andrey Bogdanov
Carl Bootland
Jonathan Bootle
Raphael Bost
Leif Both
Florian Bourse
Elette Boyle
Zvika Brakerski
Christian Cachin
Ran Canetti
Anne Canteaut
Brent Carmer
Wouter Castryck
Andrea Cerulli
André Chailloux
Avik Chakraborti
Yilei Chen
Ashish Choudhury

Chitchanok
   Chuengsatiansup
Michele Ciampi
Thomas De Cnudde
Ran Cohen
Sandro Coretti
Jean-Sebastien Coron
Henry Corrigan-Gibbs
Ana Costache
Geoffroy Couteau
Claude Crépeau
Ben Curtis
Dana Dachman-Soled
Yuanxi Dai
Bernardo David
Alex Davidson
Jean Paul Degabriele
Akshay Degwekar
Daniel Demmler
Amit Deo
Apoorvaa Deshpande
Itai Dinur
Christoph Dobraunig
Manu Drijvers
Maria Dubovitskaya
Léo Ducas
Yfke Dulek
Pierre-Alain Dupont
François Dupressoir
Avijit Dutta
Lisa Eckey
Maria Eichlseder
Maximilian Ernst
Mohammad Etemad
Antonio Faonio
Oriol Farràs
Pooya Farshim
Manuel Fersch
Dario Fiore
Viktor Fischer
Nils Fleischhacker
Christian Forler
Tommaso Gagliardoni
Chaya Ganesh
Juan Garay
Sanjam Garg

Romain Gay
Peter Gaži
Rosario Gennaro
Satrajit Ghosh
Irene Giacomelli
Federico Giacon
Benedikt Gierlichs
Junqing Gong
Dov Gordon
Divya Gupta
Lorenzo Grassi
Hannes Gross
Vincent Grosso
Paul Grubbs
Chun Guo
Siyao Guo
Mohammad Hajiabadi
Carmit Hazay
Gottfried Herold
Felix Heuer
Thang Hoang
Viet Tung Hoang
Akinori Hosoyamada
Kristina Hostáková
Andreas Hülsing
Ilia Iliashenko
Roi Inbar
Vincenzo Iovino
Tetsu Iwata
Abhishek Jain
Martin Jepsen
Daniel Jost
Chiraag Juvekar
Seny Kamara
Chethan Kamath
Bhavana Kanukurthi
Harish Karthikeyan
Suichi Katsumata
Jonathan Katz
John Kelsey
Dakshita Khurana
Eunkyung Kim
Taechan Kim
Elena Kirshanova
Ágnes Kiss
Susumu Kiyoshima

Ilya Kizhvatov
Alexander Koch
Konrad Kohbrok
Lisa Kohl
Stefan Kölbl
Ilan Komargodski
Yashvanth Kondi
Venkata Koppula
Thorsten Kranz
Hugo Krawczyk
Marie-Sarah Lacharite
Kim Laine
Virginie Lallemand
Gaëtan Leurent
Anthony Leverrier
Xin Li
Pierre-Yvan Liardet
Benoît Libert
Huijia Lin
Guozhen Liu
Jian Liu
Chen-Da Liu-Zhang
Alex Lombardi
Julian Loss
Steve Lu
Atul Luykx
Vadim Lyubashevsky
Saeed Mahloujifar
Hemanta Maji
Mary Maller
Umberto Martínez-Peñas
Daniel Masny
Takahiro Matsuda
Christian Matt
Patrick McCorry
Pierrick Méaux
Lauren De Meyer
Peihan Miao
Brice Minaud
Esfandiar Mohammadi
Ameer Mohammed
Maria Chiara Molteni
Tal Moran
Fabrice Mouhartem
Amir Moradi
Pratyay Mukherjee

Marta Mularczyk
Mridul Nandi
Ventzislav Nikov
Tobias Nilges
Ryo Nishimaki
Anca Nitulescu
Ariel Nof
Achiya Bar On
Claudio Orlandi
Michele Orrù
Clara Paglialonga
Giorgos Panagiotakos
Omer Paneth
Louiza Papachristodoulou
Kostas Papagiannopoulos
Sunoo Park
Anat Paskin-Cherniavsky
Alain Passelègue
Kenny Paterson
Michaël Peeters
Chris Peikert
Alice Pellet–Mary
Geovandro C. C. F.
    Pereira
Leo Perrin
Giuseppe Persiano
Thomas Peters
Krzysztof Pietrzak
Benny Pinkas
Oxana Poburinnaya
Bertram Poettering
Antigoni Polychroniadou
Christopher Portmann
Manoj Prabhakaran
Emmanuel Prouff
Carla Ràfols
Somindu C. Ramanna
Samuel Ranellucci
Shahram Rasoolzadeh
Divya Ravi
Ling Ren
Oscar Reparaz
Silas Richelson
Peter Rindal
Michal Rolinek
Miruna Rosca

Ron Rothblum
David Roubinet
Adeline Roux-Langlois
Vladimir Rozic
Andy Rupp
Yusuke Sakai
Simona Samardjiska
Niels Samwel
Olivier Sanders
Pratik Sarkar
Alessandra Scafuro
Martin Schläffer
Dominique Schröder
Sven Schäge
Adam Sealfon
Yannick Seurin
abhi shelat
Kazumasa Shinagawa
Luisa Siniscalchi
Maciej Skórski
Fang Song
Ling Song
Katerina Sotiraki
Florian Speelman
Gabriele Spini
Kannan Srinathan
Thomas Steinke
Uri Stemmer
Igors Stepanovs
Noah
    Stephens-Davidowitz
Alan Szepieniec
Seth Terashima
Cihangir Tezcan
Mehdi Tibouchi
Elmar Tischhauser
Radu Titiu
Yosuke Todo
Junichi Tomida
Patrick Towa
Boaz Tsaban
Daniel Tschudi
Thomas Unterluggauer
Margarita Vald
Kerem Varici
Prashant Vasudevan

Philip Vejre
Daniele Venturi
Benoît Viguier
Fernando Virdia
Damian Vizár
Alexandre Wallet
Michael Walter
Haoyang Wang
Qingju Wang

Hoeteck Wee
Felix Wegener
Christian Weinert
Erich Wenger
Daniel Wichs
Friedrich Wiemer
David Wu
Thomas Wunderer
Sophia Yakoubov

Shota Yamada
Takashi Yamakawa
Kan Yasuda
Attila Yavuz
Scott Yilek
Eylon Yogev
Greg Zaverucha
Mark Zhandry
Ren Zhang

Abstract of Invited Talks

Desperately Seeking Sboxes

Anne Canteaut

Inria, Paris, France
anne.canteaut@inria.fr

Abstract. Twenty-five years ago, the definition of security criteria associated to the resistance to linear and differential cryptanalysis has initiated a long line of research in the quest for Sboxes with optimal nonlinearity and differential uniformity. Although these optimal Sboxes have been studied by many cryptographers and mathematicians, many questions remain open. The most prominent open problem is probably the determination of the optimal values of the nonlinearity and of the differential uniformity for a permutation depending on an even number of variables.

Besides those classical properties, various attacks have motivated several other criteria. Higher-order differential attacks, cube distinguishers and the more recent division property exploit some specific properties of the representation of the whole cipher as a collection of multivariate polynomials, typically the fact that some given monomials do not appear in these polynomials. This type of property is often inherited from some algebraic property of the Sbox. Similarly, the invariant subspace attack and its nonlinear counterpart also originate from specific algebraic structure in the Sbox.

Thirty Years of Digital Currency:
From DigiCash to the Blockchain

Matthew Green

Johns Hopkins University
mgreen@cs.jhu.edu

Abstract. More than thirty years ago a researcher named David Chaum presented his vision for a cryptographic financial system. In the past ten years this vision has been realized. Yet despite a vast amount of popular excitement, it remains to be seen whether the development of cryptocurrencies (and their associated consensus technologies) will have a lasting positive impact—both on society and on our research community. In this talk I will examine that question. Specifically, I will review several important contributions that research cryptography has made to this field; survey the most promising deployed (or developing) technologies; and discuss the many challenges ahead.

Contents – Part I

Contents – Part II

Symmetric Cryptanalysis

Contents – Part III

Isogeny

Leakage

Key Exchange

Quantum

Non-maleable Codes

Provable Symmetric Cryptography

Foundations

On the Bit Security of Cryptographic Primitives

Daniele Micciancio[1] and Michael Walter[2(✉)]

[1] UC San Diego, San Diego, USA
daniele@cs.ucsd.edu
[2] IST Austria, Klosterneuburg, Austria
michael.walter@ist.ac.at

Abstract. We introduce a formal quantitative notion of "bit security" for a general type of cryptographic games (capturing both decision and search problems), aimed at capturing the intuition that a cryptographic primitive with k-bit security is as hard to break as an ideal cryptographic function requiring a brute force attack on a k-bit key space. Our new definition matches the notion of bit security commonly used by cryptographers and cryptanalysts when studying search (e.g., key recovery) problems, where the use of the traditional definition is well established. However, it produces a quantitatively different metric in the case of decision (indistinguishability) problems, where the use of (a straightforward generalization of) the traditional definition is more problematic and leads to a number of paradoxical situations or mismatches between theoretical/provable security and practical/common sense intuition. Key to our new definition is to consider adversaries that may explicitly declare failure of the attack. We support and justify the new definition by proving a number of technical results, including tight reductions between several standard cryptographic problems, a new hybrid theorem that preserves bit security, and an application to the security analysis of indistinguishability primitives making use of (approximate) floating point numbers. This is the first result showing that (standard precision) 53-bit floating point numbers can be used to achieve 100-bit security in the context of cryptographic primitives with general indistinguishability-based security definitions. Previous results of this type applied only to search problems, or special types of decision problems.

1 Introduction

It is common in cryptography to describe the level of security offered by a (concrete instantiation of a) cryptographic primitive P by saying that P provides a

Research supported in part by the Defense Advanced Research Projects Agency (DARPA) and the U.S. Army Research Office under the SafeWare program. Opinions, findings and conclusions or recommendations expressed in this material are those of the author(s) and do not necessarily reflect the views, position or policy of the Government.

J. B. Nielsen and V. Rijmen (Eds.): EUROCRYPT 2018, LNCS 10820, pp. 3–28, 2018.
https://doi.org/10.1007/978-3-319-78381-9_1

certain number of *bits of security*. E.g., one may say that AES offers 110-bits of security as a pseudorandom permutation [6], or that a certain lattice based digital signature scheme offers at least 160-bits of security for a given setting of the parameters. While there is no universally accepted, general, formal definition of bit security, in many cases cryptographers seem to have an intuitive (at least approximate) common understanding of what "n bits of security" means: any attacker that successfully breaks the cryptographic primitive must incur a cost[1] of at least $T > 2^n$, or, alternatively, any efficient attack achieves at most $\epsilon < 2^{-n}$ success probability, or, perhaps, a combination of these two conditions, i.e., for any attack with cost T and success probability ϵ, it must be $T/\epsilon > 2^n$. The intuition is that 2^n is the cost of running a brute force attack to retrieve an n-bit key, or the inverse success probability of a trivial attack that guesses the key at random. In other words, n bits of security means "as secure as an idealized perfect cryptographic primitive with an n-bit key".

The appeal and popularity of the notion of bit security (both in theory and in practice) rests on the fact that it nicely sits in between two extreme approaches:

- The *foundations of cryptography* asymptotic approach (e.g., see [9,10]) which identifies feasible adversaries with polynomial time computation, and successful attacks with breaking a system with non-negligible probability.
- The *concrete security* approach [3,5], which breaks the adversarial cost into a number of different components (running time, oracle queries, etc.), and expresses, precisely, how the adversary's advantage in breaking a cryptographic primitive depends on all of them.

The foundational/asymptotic approach has the indubious advantage of simplicity, but it only offers a *qualitative* classification of cryptographic functions into secure and insecure ones. In particular, it does not provide any guidance on choosing appropriate parameters and key sizes to achieve a desired level of security in practice. On the other hand, the concrete security treatment delivers (precise, but) substantially more complex security statements, and requires carefully tracking a number of different parameters through security reductions. In this respect, bit security offers a *quantitative*, yet simple, security metric, in the form of a single number: the *bit security* or *security level* of a primitive, typically understood as the logarithm (to the base 2) of the ratio T/ϵ between the cost T and advantage ϵ of the attack, minimized over all possible adversaries.

Capturing security level with a single number is certainly convenient and useful: it allows for direct comparison of the security level of different instances of the same primitive (or even between different primitives altogether), and it provides a basis for the study of *tight reductions*, i.e., constructions and reductions that approximately preserve the security level. Not surprisingly, bit security is

---

[1] For concreteness, the reader may think of the cost as the running time of the attack, but other cost measures are possible, and everything we say applies to any cost measure satisfying certain general closure properties, like the fact that the cost of repeating an attack k times is *at most* k times as large as the cost of a single execution.

widely used. However, there is no formal definition of this term at this point, but rather just an intuitive common understanding of what this quantity should capture. This understanding has led to some paradoxical situations that suggest that the current "definitions" might not capture exactly what they are meant to.

It has been noted that only considering the adversary's running time is a poor measure of the cost of an attack [7,8]. This is especially true if moving to the non-uniform setting, where an adversary may receive additional advice, and the question of identifying an appropriate cost measure has been studied before [6]. However, the paradoxical situations have not, to this day, been resolved to satisfaction, and it seems that considering only the adversary's resources is insufficient to address this issue.

In order to explain the problems with the current situation, we first distinguish between two types of primitives with respect to the type of game that defines their security (see Sect. 3 for a more formal definition): *search primitives* and *decision primitives*. Intuitively, the former are primitives where an adversary is trying to recover some secret information from a large search space, as in a key recovery attack. The latter are games where the adversary is trying to decide if a secret bit is 0 or 1, as in the indistinguishability games underlying the definition of pseudorandom generators or semantically secure encryption. For search games, the advantage of an adversary is usually understood to be the probability of finding said secret information, while for decision games it is usually considered to be the distinguishing advantage (which is equal to the probability that the output of the adversary is correct, over the trivial probability $\frac{1}{2}$ of a random guess).

The Peculiar Case of PRGs. Informally, a PRG is a function $f : \{0,1\}^n \mapsto \{0,1\}^m$, where $m > n$, such that its output under uniform input is indistinguishable from the uniform distribution over $\{0,1\}^m$. In the complexity community it is common knowledge according to [8] that a PRG with seed length n cannot provide more than $n/2$ bits of security under the current definition of security level. This is because there are non-uniform attacks that achieve distinguishing advantage $2^{-n/2}$ in very little time against any such function. Such attacks were generalized to yield other time-space-advantage trade-offs in [7]. This is very counter-intuitive, as the best generic seed recovery attacks do not prevent n-bit security (for appropriate cost measure), and thus one would expect n bits of security in such a case to be possible.

The Peculiar Case of Approximate Samplers. Many cryptographic schemes, in particular lattice based schemes, involve specific distributions that need to be sampled from during their execution. Furthermore, security reductions may assume that these distributions are sampled exactly. During the transition of such a cryptographic scheme from a theoretical construction to a practical implementation, the question arises as to how such a sampling algorithm should be implemented. In many cases, it is much more efficient or secure (against e.g. side channel attacks) or even only possible to approximate the corresponding distribution rather than generating it exactly. In such a case it is crucial to analyze how this approximation impacts the security of the scheme. Traditionally, statistical distance has been employed to quantify this trade-off between approximation and security guarantee, but it leads to the unfortunate situation where the 53-bit precision provided by floating point numbers (as implemented in hardware in commodity microprocessors) only puts a 2^{-53} bound on statistical distance, and results in a rather weak 53-bit security guarantee on the final application. Proving better security using statistical distance methods seems to require higher precision floating point numbers implemented in (substantially slower) software libraries. In recent years a number of generic results have shown improved analysis methods based on different divergences [2,15–17] and using the conventional definition of bit security. Surprisingly, all of them apply exclusively to search primitives (with the only exception of [2], which also considers decision primitives with a specific property). This has led to the unnatural situation where it seems that decision primitives, like encryption, require higher precision sampling than search primitives. This is counter-intuitive, because in search primitives, like signature schemes, the distribution is often used to hide a specific secret and a bad approximation may leak information about it. On the other hand, it is commonly believed within the research community that for encryption schemes the distribution does not necessarily have to be followed exactly, as long as it has sufficient entropy, since none of the cryptanalytic attacks seem to be able to take advantage of a bad approximation in this case [1]. However, a corresponding proof for generic decision primitives (e.g., supporting the use of hardware floating point numbers, while still targeting 100-bit or higher levels of security) has so far eluded the attempts of the research community.

1.1   Contribution and Techniques

We present a new notion of *bit security* associated to a general cryptographic game. Informally, we consider a game in which an adversary has to guess an n-bit secret string[2] x. This captures, in a unified setting, both decision/indistinguishability properties, when $n = 1$, and arbitrary search/unpredictability properties, for larger n. The definition of bit security is quite natural and intuitive, building on concepts from information theory, but we postpone its description to the end of this section. For now, what matters is that a distinguishing feature of our framework is to explicitly allow the adversary to output a special "don't know" symbol $\perp$, rather than a random guess. So, we can talk about the probability α that the adversary outputs something (other than $\perp$), and the (conditional) probability β that the output correctly identifies the secret. This makes little difference for search problems, but for decision problems it allows the adversary to express different degrees of confidence in its guess: admitting failure is more informative than a random guess. We proceed by specializing our notion of bit

---

[2] More generally, the adversary has to output a value satisfying a relation $R(x, a)$ which defines successful attacks. For simplicity, in this introduction, we assume R is the identity function. See Definition 5 for the actual definition.

security to the two important settings of search and decision problems and show that:

- For the case of search primitives (large secret size $n = |x|$), this yields the traditional notion of bit security, as the logarithm of the ratio T/ϵ between the attack cost T, and the success probability $\epsilon = \alpha\beta$. The fact that our definition is consistent with the current one in the case of search primitives gives us confidence in its validity, since in this case the traditional definition is very intuitive and there are no paradoxes casting doubts about it.
- Surprisingly, for decision primitives (i.e., for $n = 1$), our definition yields a different formula, which, instead of being linear the distinguishing advantage $\delta = 2\beta - 1$, is quadratic in δ. In other words, the bit security is the logarithm of $T/(\alpha\delta^2)$. This is not entirely new, as a similar proposal was already put forward in [11,14] in a more specific context, but has so far received very little attention.

One of the implications of our new definition is that it seemingly resolves the paradoxical situation about the bit security of pseudorandom generators (PRGs) described in [7]. (The significance of the nonuniform attacks to one-way functions described in [7] can already be addressed by an appropriate choice of cost measure.) For the PRG case, an attack achieving distinguishing advantage $\delta = 2^{-n/2}$ even in constant time does not necessarily contradict n-bit security. In fact, [7] shows that for any algorithm distinguishing the output of any function $f : \{0,1\}^n \mapsto \{0,1\}^{n+1}$ from uniform with distinguishing advantage δ must use at least $T = \Omega(\delta^2 2^n)$ resources (for a suitable definition of resources, similar to the one-way function case). So, this shows that by our definition, there exist PRGs with bit security $\log_2(T/\delta^2) = n$, as one would expect.

Of course, as definitions are arbitrary, it is not clear if changing a definition is really solving any real problem, and our definition of bit security needs to be properly supported and justified. Notice that a reduction $A \leq B$ showing that if A is n-bit secure, then B is $n/2$-bit secure, may be interpreted in different ways:

- Either the construction of B from A is not optimal/tight, i.c., it incurs an actual security degradation
- Or the construction is tight, but the reduction (i.e., the security proof) is not
- Or the definition of bit security is incorrect.

The last possibility is most delicate when reducing between different types of cryptographic primitives (e.g., from search to decision) where the definition of bit security may take different (and somehow arbitrary) forms. All these comments apply equally well to tight reductions, mapping n-bit security to n-bit security. We support and justify our definition by providing a collection of results (typically in the form of *tight reductions*[3] between different cryptographic primitives), which are the main technical contribution of this paper. For example,

---

[3] In the context of this work, "tight" means that bit security is (approximately) preserved, up to small additive logarithmic terms corresponding to the polynomial running time of an attack. More specifically, a reduction is tight if it maps a primitive providing n-bit security, to another with security level $n - O(\log n)$. For simplicity, we omit all the $O(\log n)$ in this introduction.

- We observe that the Goldreich-Levin hard-core predicate is tight according to our definition, i.e., if $f(x)$ is an n-bit secure one-way permutation,[4] then $G(r, x) = (r, f(x), \langle r, x \rangle)$ is an n-bit secure PRG.
- There is a simple reduction showing that if G is an n-bit secure PRG, then the same G (and also f) is an n-bit secure one-way function. (Interestingly, the reduction is not completely trivial, and makes critical use of the special symbol $\perp$ in our definition. See Theorem 4.)

Notice that, while both reductions are between different types of cryptographic primitives (search and decision, with different bit security formulas), combining them together gives a search-to-search reduction which uses the same security definition on both sides. Since it would be quite counterintuitive for such a simple reduction (from PRG to OWF) to increase the level of security from $n/2$ to n bits, this provides some confidence that our definition is on target, and the Goldreich-Levin PRG is indeed as secure as the underlying one-way function.

Other technical results presented in this paper include:

- Approximate samplers: we give a proof in Sect. 5.3 that shows for the first time that the sampling precision requirement is essentially the same for search and decision primitives to maintain security. We do this by extending a result from [15] for search primitives to decision primitives using our definition of bit security.
- Hybrid argument: since our new definition of advantage no longer matches the simple notion of statistical distance, the standard proof of the hybrid argument [12] (so ubiquitously used in cryptography and complexity) is no longer valid. While the proof in our setting becomes considerably more involved, we show (Theorem 7) that hybrid arguments are still valid.
- Additional examples about non-verifiable search problems (Theorem 5), and tight reductions for message-hiding encryption (Theorem 6), and multi-message security (Corollary 1).

Beside increasing our confidence in the validity of our new bit security notion, these reductions also start building a toolbox of techniques that can be used to fruitfully employ the new definition in the analysis of both old and new cryptographic primitives, and improve our theoretical understanding of the relation between different cryptographic primitives by means of tight reductions. Finally, they allow us to expand the use of divergence techniques [2, 15–17] to bound the floating point precision required to secure cryptographic primitives with indistinguishability security properties.

We conclude this section with an informal overview of the new bit security definition. As already mentioned, our definition is based on concepts from information theory. In a purely information theoretic setting, the advantage of an adversary A in discovering a secret X could be modeled by the mutual information $\epsilon = I(A, X)/H(X)$, normalized by the entropy of the secret $H(X)$ to ensure

---

[4] The actual reduction holds for any one-way functions. Here we focus on permutations just to emphasize the connection with PRGs. See Theorem 3.

$\epsilon \leq 1$. Of course, this approach completely fails in the computational setting, where the output of a one-way permutation $f(X)$ is perfectly correlated with the input X, but still we do not want to consider a trivial algorithm $A(f(X)) = f(X)$ as a successful attack (with advantage $\epsilon = I(A, X)/H(X) = 1$!) to the one-way permutation input recovery problem: what the adversary knows $(f(X))$ identifies the input X information theoretically, but it does not provide knowledge of it. We adapt this definition to the computational setting by replacing A with a different random variable Y which equals (1) the secret X when A is successful (i.e., $A = X$), and (2) an independent copy X' of the secret (conditioned on $X' \neq X$) when A failed to output X. We find this definition intuitively appealing, and we consider it the main conceptual contribution of this paper. But words are of limited value when arguing about the validity of a new definition. We view the technical results described above the most important evidence to support our definition, and the main technical contribution of this work.

1.2 Related Work

While the informal concept of bit security is widely used in cryptography, not many papers directly address the problem of its formal definition. Some of the works that are perhaps most directly related to our are [6–8], which pinpoint the shortcoming of the traditional definition. The work of Bernstein and Lange [6] provides an extensive survey of relevant literature, and attempts to provide a better definition. In [6, Appendix B] the authors analyze different measures to address the underlying problems, and show how each of them can be used to make partial progress towards a more sound definition of bit security, while pointing out that none of them seem to solve the problem entirely. In contrast, the definitions and results in this paper concern the definition of adversarial advantage, which we consider to be orthogonal to any of the ideas presented in [6]. So, we see our work as complementary to [6–8].

To the best of our knowledge there are only two works proposing an alternative definition of adversarial advantage for decision problems: the aforementioned works of Goldreich and Levin [11,14] and the infamous HILL paper [13]. The latter primarily works with the traditional definition of adversarial advantage, but presents the advantage function δ^2 (note the lack of α) as an alternative, observing that many of their reductions are much tighter in this case. Our work can be considered as a generalization of them, and supporting the definitional choices made in [11,14]. In the last years, bit security has been the focus on a body of work [2,15–17] aimed at optimizing the parameters and floating point precision requirements of lattice cryptography. Our work resolves the main problem left open in [15,17] of extending definitions and techniques from search to decision problems, and support the secure use of standard precision floating point numbers in the implementation of cryptographic primitives (like encryption) with indistinguishability security properties.

2   Preliminaries

Notation. We denote the integers by $\mathbb{Z}$ and the reals by $\mathbb{R}$. Roman and Greek letters can denote elements from either set, while bold letters denote vectors over them. Occasionally, we construct vectors on the fly using the notation $(\cdot)_{i \in S}$ for some set S (or in short $(\cdot)_i$ if the set S is clear from context), where $\cdot$ is a function of i. For a set S, we denote its complement by $\bar{S}$. We denote the logarithm to the base 2 by log and the one to the base e by ln.

Calligraphic letters are reserved for probability distributions and $x \leftarrow \mathcal{P}$ means that x is sampled from the distribution $\mathcal{P}$. For any x in the support of $\mathcal{P}$, we denote its probability under $\mathcal{P}$ by $\mathcal{P}(x)$. All distributions in this work are discrete, and $\mathcal{U}(S)$ is the uniform distribution over the support S. If S is clear from context, we simply write $\mathcal{U}$ instead of $\mathcal{U}(S)$. A probability ensemble $\{\mathcal{P}_\theta\}_\theta$ is a family of distributions indexed by a parameter θ (which may be a string or a vector). We extend any divergence δ between distributions to probability ensembles by $\delta(\{\mathcal{P}_\theta\}_\theta, \{\mathcal{Q}_\theta\}_\theta) = \max_\theta \delta(\mathcal{P}_\theta, \mathcal{Q}_\theta)$. For notational simplicity, we do not make a distinction between random variables, probability distributions, and probabilistic algorithms generating them.

Definition 1. *The* statistical distance *between two distributions $\mathcal{P}$ and $\mathcal{Q}$ over S is defined as* $\Delta_{SD}(\mathcal{P}, \mathcal{Q}) = \frac{1}{2} \sum_{x \in S} |\mathcal{P}(x) - \mathcal{Q}(x)|$.

2.1   Information Theory

For our definition, we need a few concepts from information theory.

Definition 2. *The* Shannon entropy *of a random variable X is given by*

$$H(X) = \mathbb{E}_X \left[\log \frac{1}{\Pr\{X\}} \right] = - \sum_x \Pr[X = x] \log \Pr[X = x].$$

Definition 3. *For two random variables X and Y, the* conditional entropy *of X given Y is*

$$H(X|Y) = \mathbb{E}_Y [H(X|Y)] = \sum_{x,y} \Pr[X = x, Y = y] \log \frac{\Pr[Y = y]}{\Pr[X = x, Y = y]}.$$

Definition 4. *The* mutual information *between two random variables X and Y is*

$$I(X;Y) = H(X) - H(X|Y).$$

3   Security Games

In this section we formally define the bit security of cryptographic primitives in a way that captures practical intuition and is theoretically sound. As the security of cryptographic primitives is commonly defined using games, we start by defining a general class of security games.

Definition 5. *An n-bit security game is played by an adversary A interacting with a challenger X. At the beginning of the game, the challenger chooses a secret x, represented by the random variable $X \in \{0,1\}^n$, from some distribution $\mathcal{D}_X$. At the end of the game, A outputs some value, which is represented by the random variable A. The goal of the adversary is to output a value a such that $R(x, a)$, where R is some relation. A may output a special symbol $\perp$ such that $R(x, \perp)$ and $\bar{R}(x, \perp)$ are both false.*

This definition is very general and covers a lot of standard games from the literature. Some illustrative examples are given in Table 1. But for the cryptographic primitives explicitly studied in this paper, it will be enough to consider the simplest version of the definition where $R = \{(x, x) | x \in X\}$ is the identity relation, i.e., the goal of the adversary is to guess the secret x. We formally define the indistinguishability game for two distributions because we refer to it extensively throughout this work.

Table 1. Typical instantiations of security games covered by Definition 5. The security parameter is denoted by κ. In the definition of digital signatures, the list Q of the adversary's queries are regarded as part of its output.

Game	R	n	$\mathcal{D}_X$
Uninvertibility of one-way permutations	$\{(x, y) \mid x = y\}$	$O(\kappa)$	$\mathcal{U}$
Uninvertibility of one-way functions f	$\{(x, y) \mid f(x) = f(y)\}$	$O(\kappa)$	$\mathcal{U}$
2nd pre-image resistance for hash functions h	$\{(x, y) \mid x \neq y, h(x) = h(y)\}$	$O(\kappa)$	$\mathcal{U}$
Indistinguishability of two distributions	$\{(x, y) \mid x = y\}$	1	$\mathcal{U}$
Unforgeability of signature scheme (K,S,V)	$\{(x, (m, \sigma, Q)) \mid (pk, sk) \leftarrow K(x), V(pk, m, \sigma) = 1, m \notin Q\}$	$O(\kappa)$	$K(\mathcal{U})$

Definition 6. *Let $\{\mathcal{D}_\theta^0\}_\theta$, $\{\mathcal{D}_\theta^1\}_\theta$ be two distribution ensembles. The indistinguishability game is defined as follows: the challenger C chooses $b \leftarrow \mathcal{U}(\{0,1\})$. At any time after that the adversary A may (adaptively) request samples by sending θ_i to C, upon which C draws samples $c_i \leftarrow \mathcal{D}_{\theta_i}^b$ and sends c_i to A. The goal of the adversary is to output $b' = b$.*

We loosely classify primitives into two categories according to their associated security games: we call primitives, where the associated security game is a 1-bit game ($O(\kappa)$-bit game), *decision primitives* (*search primitive*, respectively).

Note that we allow the adversary to always output $\perp$, which roughly means "I don't know", even for decision primitives. This is a crucial difference from previous definitions that force the distinguisher to always output a bit. The reason this is important is that in games, where the distinguisher is not able to check if it produced the correct result, it is more informative to admit defeat rather than guessing at random. In many cases this will allow for much tighter reductions (cf. Sect. 5.2). Such a definition in the context of indistinguishability games is not entirely new, as Goldreich and Levin [11,14] also allowed this type of flexibility for the distinguisher. To the best of our knowledge, this is the only place this has previously appeared in the cryptographic literature.

Now we are ready to define the advantage. The definition is trying to capture the amount of information that the adversary is able to learn about the secret. The reasoning is that the inverse of this advantage provides a lower bound on the number of times this adversary needs to be run in order to extract the entire secret. We use tools from information theory to quantify exactly this information, in particular the Shannon entropy. Other notions of entropy might be worth considering, but we focus on Shannon entropy as the most natural definition that captures information. A straight-forward definition could try to measure the mutual information between the random variables X (modeling the secret) and A (modeling the adversary output, cf. Definition 5). Unfortunately, the variable A might reveal X completely in an information theoretical sense, yet not anything in a computational sense. To break any computationally hidden connection between X and A, we introduce another random variable Y, which indicates, when A actually achieves its goal and otherwise does not reveal anything about the secret.

Definition 7. *For any security game with corresponding random variable X and $A(X)$, the adversary's advantage is*

$$\mathrm{adv}^A = \frac{I(X;Y)}{H(X)} = 1 - \frac{H(X|Y)}{H(X)}$$

where $I(\cdot;\cdot)$ is the mutual information, $H(\cdot)$ is the Shannon entropy, and $Y(X,A)$ is the random variable with marginal distributions $Y_{x,a} = \{Y \mid X = x, A = a\}$ defined as

1. $Y_{x,\perp} = \perp$, *for all x.*
2. $Y_{x,a} = x$, *for all $(x,a) \in R$.*
3. $Y_{x,a} = \{x' \leftarrow \mathcal{D}_X \mid x' \neq x\}$, *for all $(x,a) \in \bar{R}$.*

At first glance, the definition of Y might not be obviously intuitive, except for case 1. For case 2, note that x completely determines the set $R(x,\cdot)$ and if the adversary finds an element in it, then it wins the game. Therefore, one can think of $R(x,\cdot)$ as a secret set, and finding any element in it as completely breaking the scheme. Finally, the third case defines Y to follow the distribution of the secret, but is conditioned on the event that it is incorrect. The intuition here is that if an adversary outputs something, then his goal is to bias the secret distribution towards the correct one, i.e. it will allow us to quantify how much better A performs than random guessing.

With the definition of the advantage in place, the definition of bit security follows quite naturally.

Definition 8. *Let $T : \{A \mid A$ is any algorithm$\} \mapsto \mathbb{Z}_+$ be a measure of resources that is linear under repetition, i.e. $T(kA) = kT(A)$, where kA is the k time repetition of A. For any primitive, we define its bit security as $\min_A \log \frac{T(A)}{\mathrm{adv}^A}$.*

For convenience we will often write $T(A)$ as T^A or simply T if A is clear from context. Note that we leave out a concrete definition of the resources on purpose, since we focus on the advantage. Our definition can be used with many different measures, for example running time, space, advice, etc., or combinations of them.

4   The Adversary's Advantage

While the advantage as defined in the previous section captures the intuition about how well an adversary performs, it seems too complex to be handled in actual proofs or to be used in practice. A simple definition in terms of simple quantities related to the adversary would be much more desirable. We begin by defining the quantities of an adversary that we are interested in.

Definition 9. *For any adversary A playing a security game, we define its* output probability *as $\alpha^A = \Pr[A \neq \perp]$ and its* conditional success probability *as $\beta^A = \Pr[R(X, A)|A \neq \perp]$, where the probabilities are taken over the randomness of the entire security game (including the internal randomness of A). Finally, in the context of decision primitives, we also define A's conditional distinguishing* advantage *as $\delta^A = 2\beta^A - 1$. With all of these quantities, when the adversary A is clear from context, we drop the corresponding superscript.*

The goal of this section is to distill a simple definition of advantage in terms of α_A and β^A by considering a broad and natural class of adversaries and games.

Theorem 1. *For any n-bit security game with uniform secret distribution, let A be an adversary that for any secret $x \in \{0,1\}^n$ outputs $\perp$ with probability $1 - \alpha$, some value a such that $R(x, a)$ with probability $\beta\alpha$, and some value a such that $\bar{R}(x, a)$ with probability $(1 - \beta)\alpha$. Then*

$$\text{adv}^A = \alpha \left(1 - \frac{(1 - \beta)\log(2^n - 1) + H(\mathcal{B}_\beta)}{n} \right) \tag{1}$$

where $\mathcal{B}_\beta$ denotes the Bernoulli distribution with parameter β.

We defer the proof to Appendix A. Note that for large n we get $\text{adv}^A \approx \alpha^A \beta^A$, which is exactly A's success probability. Plugging this into Definition 8 matches the well-known definition of bit security for search primitives. On the other hand, for $n = 1$ this yields $\text{adv}^A = \alpha^A(1 - H(\mathcal{B}_{\beta^A})) = \alpha^A(\delta^A)^2/(2\ln 2) + O(\alpha^A(\delta^A)^4)$ by Taylor approximation, which, for our purposes, can be approximated by $\alpha^A(\delta^A)^2$. This matches the definition of Levin [14], who proposed this definition since it yields the inverse sample complexity of noticing the correlation between the adversary output and the secret. The fact that it can be derived from Definition 7 suggests that this is the "right" definition of the adversary's advantage.

We now redefine the adversary's advantage according to above observations, which, combined with Definition 8 yields the definition of bit security we actually put forward and will use throughout the rest of this work.

Definition 10. *For a search game, the advantage of the adversary A is*

$$\text{adv}^A = \alpha^A \beta^A$$

and for a decision game, it is

$$\text{adv}^A = \alpha^A (\delta^A)^2.$$

Note that assuming that Definition 10 is equivalent to 7 for all adversaries is quite a leap as we only proved it for a subclass of them, and in fact, it is not true at least for decision games. However, the following theorem shows that when used in the context of bit security (Definition 8) for decision games, Definitions 10 and 7 are in fact equivalent, since we are quantifying over all adversaries.

Theorem 2. *For any distinguisher D playing a decision game with $\text{adv}^D = \zeta$ according to Definition 7, there is a distinguisher D' such that $T^D = T^{D'}$ and $\alpha^{D'} (\delta^{D'})^2 \geq \zeta/9$ for the same game.*

Before we prove Theorem 2, we observe that the distinguisher D' that we construct from D will run D and decide on its output depending on the result. As such, D' is essentially a distinguisher for the indistinguishability game (restricted to one query) against the two distributions induced by the secret on D. We start with a simple lemma that analyzes how well such a simple distinguisher does in this game.

Lemma 1. *Let $\mathcal{D}_x$ for $x \in \{0,1\}$ be two distributions over the same support $\{a,b,c\}$ and denote their probabilities by $z_x = \mathcal{D}_x(z)$ for $z \in \{a,b,c\}$. Let D_z be a distinguisher for the indistinguishability game instantiated with $\mathcal{D}_x$ that on input z returns $\arg\max_x(z_x)$ and $\perp$ otherwise. Then,*

$$\alpha^{D_z} (\delta^{D_z})^2 = \frac{1}{2} \frac{(z_1 - z_0)^2}{z_1 + z_0}.$$

We now prove Theorem 2 by showing that for any distinguisher D there is an event $z \in \{\perp, 0, 1\}$ such that $\alpha^{D_z}(\delta^{D_z})^2 \approx \text{adv}^D$.

Proof (of Theorem 2). Since adv^D is independent of the support/domain of D (as long as it has size exactly 3), we identify $\{\perp, 0, 1\}$ with a, b, c to highlight this genericity.

With the same notation as in Lemma 1, we note that the conditional entropy of the secret X given Y is

$$H(X|Y) = \frac{1}{2} \left(H_1(a_0, a_1) + H_1(b_0, b_1) + H_1(c_0, c_1) \right)$$

where

$$H_1(z_0, z_1) = z_0 \log \frac{z_0 + z_1}{z_0} + z_1 \log \frac{z_0 + z_1}{z_1}$$
$$= ((z_0 + z_1) \log((z_0 + z_1) - z_0 \log z_0 - z_1 \log z_1.$$

Setting $\bar{z} = z_1 - z_0$, H_1 can be rewritten as

$$H_1(z_0, \bar{z}) = (2z_0 + \bar{z}) \log(2z_0 + \bar{z}) + z_0 \log z_0 + (z_0 + \bar{z}) \log(z_0 + \bar{z}).$$

We use the following bound on H_1:

$$H_1(z_0, \bar{z}) \geq 2z_0 \qquad\qquad\qquad\qquad \text{for } \bar{z} \geq 0 \qquad (2)$$

$$H_1(z_0, \bar{z}) \geq 2z_0 + \bar{z} - \frac{\bar{z}^2}{z_0} \qquad\qquad \text{for } |\bar{z}| \leq z_0 \qquad (3)$$

where (2) follows from monotonicity in $\bar{z}$ and (3) from Taylor approximation of order 2 in $\bar{z}$ at $\bar{z} = 0$. Since $\bar{z} > z_0$ implies that (2) is larger than (3), these bounds imply

$$H_1(z_0, \bar{z}) \geq \max\left(2z_0, 2z_0 + \bar{z} - \frac{\bar{z}^2}{z_0}\right) \qquad (4)$$

for all $\bar{z} \in [-z_0, 1 - z_0]$. In the following, we will apply the bound (3) for $\bar{z} \in [-z_0, 0]$ and (4) for $\bar{z} \in [0, 1 - z_0]$.

W.l.o.g. assume $\bar{a} \geq 0$, $\bar{b} \leq 0$ and $\bar{c} \leq 0$ (note that $\sum_{z \in \{a,b,c\}} \bar{z} = 0$). Using (3) and (4)

$$H(X|Y) \geq \frac{1}{2}\left[\max\left(2a_0, 2a_0 + \bar{a} - \frac{\bar{a}^2}{a_0}\right) + 2b_0 + \bar{b} - \frac{\bar{b}^2}{b_0} + 2c_0 + \bar{c} - \frac{\bar{c}^2}{c_0}\right]$$

$$= 1 + \frac{1}{2}\left[\max\left(-\bar{a}, -\frac{\bar{a}^2}{a_0}\right) - \frac{\bar{b}^2}{b_0} - \frac{\bar{c}^2}{c_0}\right]$$

which shows that

$$\mathrm{adv}^D \leq \frac{1}{2}\left[-\max\left(-\bar{a}, -\frac{\bar{a}^2}{a_0}\right) + \frac{\bar{b}^2}{b_0} + \frac{\bar{c}^2}{c_0}\right]$$

$$= \frac{1}{2}\left[\min\left(\bar{a}, \frac{\bar{a}^2}{a_0}\right) + \frac{\bar{b}^2}{b_0} + \frac{\bar{c}^2}{c_0}\right]$$

$$\leq \frac{3}{2}\max\left[\min\left(\bar{a}, \frac{\bar{a}^2}{a_0}\right), \frac{\bar{b}^2}{b_0}, \frac{\bar{c}^2}{c_0}\right].$$

Note that if the maximum is attained by one of the latter two terms, since $\bar{b}$ and $\bar{c}$ are negative, we have $\alpha^{D_b}(\delta^{D_b})^2 \geq \frac{\bar{b}^2}{4b_0}$ by Lemma 1 (and similarly for c). So $\mathrm{adv}^D \leq 6\alpha^{D_z}(\delta^{D_z})^2$ for one of $z \in \{b, c\}$.

Now assume the maximum is $\min(\bar{a}, \frac{\bar{a}^2}{a_0})$. If $\frac{\bar{a}^2}{a_0} \leq \bar{a}$, then $\bar{a} \leq a_0$ and so $a_0 + a_1 \leq 3a_0$. Again by Lemma 1, $\alpha^{D_a}(\delta^{D_a})^2 \geq \frac{\bar{a}^2}{6a_0}$. Finally, if $\bar{a} \leq \frac{\bar{a}^2}{a_0}$ then $a_0 \leq \bar{a}$, which means $a_0 + a_1 \leq 3\bar{a}$ and so by Lemma 1, $\alpha^{D_a}(\delta^{D_a})^2 \geq \frac{\bar{a}}{6}$. In both cases we have $\mathrm{adv}^D \leq 9\alpha^{D_a}(\delta^{D_a})^2$. □

5 Security Reductions

To argue that our definition is useful in a theoretical sense, we apply it to several natural reductions, which arise when constructing cryptographic primitives from other ones. As the novelty of our definition lies mostly with decision games, we will focus on decision primitives that are built from search primitives (cf. Sect. 5.1), search primitives that are built from decision primitives (cf. Sect. 5.2), and finally decision primitives that are built from other decision primitives (cf. Sect. 5.3).

Throughout this section we will refer to two distribution ensembles $\{\mathcal{D}_\theta^0\}_\theta$ and $\{\mathcal{D}_\theta^1\}_\theta$ as κ-bit indistinguishable, if the indistinguishability game from Definition 6 instantiated with $\{\mathcal{D}_\theta^0\}_\theta$ and $\{\mathcal{D}_\theta^1\}_\theta$ is κ-bit secure.

5.1 Search to Decision

A classical way to turn a search primitive into a decision primitive is the Goldreich-Levin hardcore bit [11].

Definition 11. *Let* $f : X \mapsto Y$ *be a function and* $b : X \mapsto \{0,1\}$ *be a predicate. The predicate* b *is a* κ-*bit secure hardcore bit for* f, *if the distributions* $(f(x), b(x))$ *and* $(f(x), \mathcal{U}(\{0,1\}))$, *where* $x \leftarrow \mathcal{U}(X)$, *are* κ-*bit indistinguishable.*

Goldreich and Levin showed a way to construct a function with a hardcore bit from any one-way function. In this setting, one would hope that if the one-way function is κ-bit secure then also the hardcore bit is close to κ bit secure. The next theorem due to Levin [14] establishes exactly such a connection.

Theorem 3 (adapted from [14]). *Let* $f : \{0,1\}^n \mapsto \{0,1\}^k$ *be a* κ-*bit secure one-way function. Then* $b(x,r) = \langle x,r \rangle \mod 2$ *is a* $(\kappa - O(\log n))$-*bit secure hardcore bit for* $g(x,r) = (f(x), r)$.

This theorem was proven in [14], and all we did was to adapt the statement from [14] to our notation/framework. So, we refer the reader to [14] for the proof details, and move on to make some general observations. The proof for this theorem assumes a distinguisher D for b and constructs from it an inverter A for f, where $\mathrm{adv}^D = \mathrm{adv}^A$ (and the running time is polynomially related). Such security preserving reductions are information theoretically only possible with a definition of advantage that is proportional to $(\delta^D)^2$ for decision primitives, if it is proportional to $\alpha^A \beta^A$ for search primitives. This is because any inverter querying a distinguisher with advantage δ^D and attempting to learn an $(\alpha^A \beta^A)$-fraction of a uniformly chosen n-bit secret, must make at least $\Omega(n\alpha^A\beta^A/(\delta^D)^2)$ queries. Denote the resources of D by T^D and note that $T^A \geq \Omega(\alpha^A\beta^A/(\delta^D)^2)T^D$ is a lower bound on the resources of A. The goal of the proof is to find an upper bound on $T^A/\mathrm{adv}^A = T^A/\alpha^A\beta^A \geq \Omega(T^D/(\delta^D)^2)$. This is only possible by assuming an upper bound on $T^D/(\delta^D)^2$. If only a bound on T^D/δ^D is assumed, then the upper bound on T^A/adv^A must contain a linear factor in $1/\delta^D$, which may be as large as $O(2^n)$ and thus result in a dramatic loss in (nominal) security.

5.2   Decision to Search

In the following subsections we show constructions and the corresponding reductions in the other direction. The first is just a straightforward converse to the Goldreich-Levin theorem, showing that any PRG is also a OWF for the same bit security. The second construction is presented as a very natural and straightforward way of turning a decision primitive into a search primitive. The third reduction is one that naturally arises in cryptographic applications, for example identification protocols.

PRGs Are One-Way Functions. While the following theorem is intuitively trivial (and technically simple), as explained in the introduction it serves to justify our definition of bit security. The proof also illustrates the subtle difference between an adversary that outputs $\bot$ and one that outputs a random guess.

Theorem 4. *If g is a PRG with κ-bit security, then it is also a $(\kappa-4)$-bit secure one-way function.*

Proof. Assume A is an attack to g as a one-way function with cost T, output probability α^A, and conditional success probability β^A. We turn A into an adversary D to g as a PRG by letting $D(y)$ output 1 if $G(A(y)) = y$ and $\bot$ otherwise. Assume that A has conditional success probability $\beta^A = 1$. This is without loss of generality because one-way function inversion is a verifiable search problem, and A can be modified (without affecting its advantage) to output $\bot$ when its answer is incorrect. So, A has advantage α^A, equal to its output probability. Notice that D is successful only when the indistinguishability game chooses the secret bit 1, and then A correctly inverts the PRG. So, the success probability of D is precisely $\alpha^D \beta^D = \alpha^A/2$. The output probability of D can be a bit higher, to take into account the possibility that on secret bit 0, the challenger picks a random string that belongs (by chance) to the image of the PRG, and A correctly inverts it. But, in any case, it always belongs to the interval $\alpha^D \in [1/2, 3/4] \cdot \alpha^A$. It follows that $\alpha^D \geq \alpha^A/2$ and $\beta^D = (\alpha^A/2)/\alpha^D \geq 2/3$. So, D has advantage at least $\alpha^D(\delta^D)^2 = \alpha^D(2\beta^D - 1)^2 \geq \alpha^A/9$. Since the two algorithms have essentially the same cost, they achieve the same level of bit security, up to a small constant additive term $\log 9 < 4$. $\square$

We remark that our proof differs from the standard text-book reduction that pseudorandom generators are one-way functions in a simple, but crucial way: when $A(y)$ fails to invert G, instead of outputting 0 as a "best guess" at the decision problem, it outputs $\bot$ to explicitly declare failure. The reader can easily check that the standard reduction has output probability $\alpha^D = 1$ and (conditional) success probability $\beta^D \leq (\alpha^A + 1)/2$. So, the advantage of the distinguisher in the standard proof is $\alpha^D(2\beta^D - 1)^2 = (\alpha^A)^2$, resulting in a substantial drop ($\log \alpha^A$) in the bit security proved by the reduction.

Secret Recovery. We proceed by giving a construction of a search primitive from two distributions. We are not aware of any immediate applications, but this simple example is supposed to serve as evidence that our definitions for search and decision primitives behave nicely under composition. It also provides an example of "non verifiable" search problem, i.e., a cryptographic problem with exponentially large secret space defined by a game at the end of which A cannot efficiently determine if the secret has been found. Differently from Theorem 4, this time one *cannot* assume without loss of generality that the (hypothetical) attacker to the search problem has conditional success probability $\beta = 1$.

Definition 12. *Let $\mathcal{D}_0, \mathcal{D}_1$ be two distributions. We define the n-bit secret recovery game as the following n-bit security game: the challenger X chooses an n-bit secret $x \leftarrow \mathcal{U}(\{0,1\}^n)$ and sends the vector $\mathbf{c} = (c_i \leftarrow \mathcal{D}_{x_i})_{i \leq n}$ to A. The adversary A attempts to guess x, i.e. R is the equality relation.*

The next theorem shows that when instantiating the game with two indistinguishable distributions, the secret recovery game enjoys essentially the same bit security.

Theorem 5. *If the κ-bit secret recovery game is instantiated with two κ-bit secure indistinguishable distributions $\mathcal{D}_0$ and $\mathcal{D}_1$, and $\mathcal{D}_0$ is publicly sampleable, then it is $(\kappa - 1)$-bit secure.*

Proof. Let A be an adversary against the secret recovery game that recovers x from the vector $\mathbf{c}$ with advantage $\mathrm{adv}^A = \alpha^A \beta^A$. We build a distinguisher D against the indistinguishability of $\mathcal{D}_0$ and $\mathcal{D}_1$ with essentially the same resources and advantage: D chooses a secret $x \in \{0,1\}^\kappa$ uniformly at random, which is non-zero with high probability (otherwise output $\bot$) and constructs the vector $\mathbf{c}$ by sampling $\mathcal{D}_0$ itself for every zero bit in x and querying its oracle for every 1 bit in x (which will return either samples from $\mathcal{D}_0$ or from $\mathcal{D}_1$). It sends $\mathbf{c}$ to A and returns 1 iff A returns x, otherwise it outputs $\bot$.

The resources of D are essentially the same as those of A, so we analyze its advantage $\mathrm{adv}^D = \alpha^D (\delta^D)^2$. The output probability of D, conditioned on $x \neq 0$, is almost exactly A's success probability, but note that A is only presented with the correct input distribution if D's challenger returns samples from $\mathcal{D}_1$, which is the case with probability $\frac{1}{2}$. So $\alpha^D \geq \frac{1-2^{-\kappa}}{2} \alpha^A \beta^A$. Furthermore, D's conditional distinguishing advantage is $\delta^D \geq 1 - 2^{-\kappa+1}$, because it only outputs the incorrect value if A returned x even though $\mathbf{c}$ consisted of samples only from $\mathcal{D}_0$. Note that in this case A has no information about x, which was chosen uniformly at random and thus the probability of this event is at most $2^{-\kappa}$. Accordingly, $\mathrm{adv}^D = \alpha_D (\delta^D)^2 \geq \frac{(1-2^{-\kappa+1})^2}{2} \alpha^A \beta^A \approx \mathrm{adv}_A / 2$. □

Indistinguishability Implies Message-Hiding. In our last example for this section we show that IND-CCA secure encryption schemes enjoy a message hiding property, which we first formally define.

Definition 13. *A private or public key encryption scheme is κ-bit message hiding, if the following security game is κ-bit secure: the challenger chooses a message $m \in \{0,1\}^n$ uniformly at random and sends its encryption to A. The adversary A attempts to guess m, while C provides it with encryption (in case of private key schemes) and decryption oracles.*

This property naturally arises in the context of constructions of identification protocols from encryption schemes (see e.g. [4]), where a random message is encrypted and identification relies on the fact that only the correct entity can decrypt it. While it seems intuitively obvious that breaking message hiding is no easier than distinguishing encrypted messages, showing that this is true in a quantifiable sense for specific definitions of bit security is not as obvious. The next theorem establishes this connection.

Theorem 6. *If a scheme with message space larger than 2^κ is κ-bit IND-CCA secure, it is κ-bit message hiding.*

Proof. Let A be an adversary that is able to extract a random message from an encryption scheme with advantage $\text{adv}^A = \alpha^A \beta^A$. We construct a IND-CCA distinguisher D against the scheme with essentially the same resources and advantage: D generates two messages $m_0, m_1 \leftarrow \{0,1\}^m$ uniformly at random, which are distinct with overwhelming probability (if not, output $\perp$). It sends them to the challenger, which encrypts one of them. Upon receiving the challenge cipher text c_b, D forwards it to A. Any queries to the encryption (in case of private key encryption) or decryption oracle are simply forwarded to D's own oracles. If A returns a message in $\{m_0, m_1\}$, D returns the corresponding bit. Otherwise, it outputs $\perp$.

The resources of D are essentially the same as for A, so we focus on its advantage. Note that conditioned on the event that $m_0 \neq m_1$, D's output probability α^D is at least as large as the success probability of A, so $\alpha^D \geq (1 - 2^{-\kappa})\alpha^A \beta^A$. The conditional distinguishing advantage of D is $\delta^D \geq 1 - 2^{-\kappa+1}$, since the only way D will guess incorrectly is when A somehow outputs the wrong message $m_{\bar{b}}$. Since A has no information about this message (which was chosen uniformly at random), the probability of this happening is at most $2^{-\kappa}$. This shows that D's advantage in the indistinguishability game is $\text{adv}^D = \alpha^D (\delta^D)^2 \geq (1 - 2^{-\kappa})\alpha^A \beta^A (1 - 2^{-\kappa+1})^2 \approx \alpha^A \beta^A = \text{adv}^A$, where the latter is A's advantage in the message hiding game. $\qquad\square$

5.3   Decision to Decision

Finally, we turn to reductions between decision primitives. The results in this section are very generic. The first establishes the validity of hybrid arguments when using our definition of advantage for decision primitives. Our second result extends a previous result for approximate sampling to any decision primitive fitting our definition.

The Hybrid Argument. This section is devoted to proving a general hybrid argument for indistinguishability games using our definition of advantage. Formally, we prove the following lemma.

Lemma 2. *Let $\mathcal{H}_i$ be k distributions and $G_{i,j}$ be the indistinguishability game instantiated with $\mathcal{H}_i$ and $\mathcal{H}_j$. Further, let $\epsilon_{i,j} = \max_A \mathrm{adv}^A$ over all T-bounded adversaries A against $G_{i,j}$. Then $\epsilon_{1,k} \leq 3k \sum_{i=1}^{k-1} \epsilon_{i,i+1}$.*

Applying the lemma to our definition of bit security, we immediately get the following theorem.

Theorem 7. *Let $\mathcal{H}_i$ be k distributions. If $\mathcal{H}_i$ and $\mathcal{H}_{i+1}$ are κ-bit indistinguishable for all i, then $\mathcal{H}_1$ and $\mathcal{H}_k$ are $(\kappa - 2(\log k + 1))$-bit indistinguishable.*

Proof. Let A be any adversary with resources T^A (when attacking $\mathcal{H}_1$ and $\mathcal{H}_k$). By assumption, $\epsilon_{i,i+1} \leq T^A/2^\kappa$ (where $\epsilon_{i,j}$ is defined as in Lemma 2) for all T^A-bounded adversaries against $\mathcal{H}_i$ and $\mathcal{H}_{i+1}$. By Lemma 2, $\epsilon_{i,k} \leq 3k^2 T^A/2^\kappa$ for all T^A-bounded adversaries, in particular A. □

As a simple application, we get the following corollary.

Corollary 1. *If a public key encryption scheme is κ-bit IND-CCA secure, then it is $(\kappa - 2(\log k + 1))$-bit IND-CCA secure in the k message setting.*

In contrast to the standard hybrid argument, which simply exploits the triangle inequality of statistical distance, we lose an additional factor of $3k$ in the advantage in Lemma 2. In particular, consider the case where the bounds $\epsilon_{i,i+1} = \epsilon$ are the same for all i. This means that $\epsilon_{1,k} \leq 3k^2\epsilon$. Note that this additional factor has only a minor impact on bit security. (See below for details.) Still, one may wonder if this additional factor is an artifact of a non-tight proof or if it is indeed necessary. Consider a distinguisher D that never outputs $\bot$ (i.e. $\alpha^D = 1$). Its distinguishing advantage $\delta_{i,j}^D$ in game $G_{i,j}$ is exactly the statistical distance between $D(\mathcal{H}_i)$ and $D(\mathcal{H}_j)$. Assume $\delta_{i,i+1}^D = \epsilon$ for all i, so D's advantage in the game $G_{i,j}$ according to Definition 10 is ϵ^2. The standard hybrid argument, or equivalently triangle inequality for statistical distance, implies that $\delta_{1,k}^D$ cannot be larger than – but may be as large as – $k\epsilon$. So, D's advantage in $G_{1,k}$ may be as large as $k^2\epsilon^2$, which is k^2 times as large as D's advantage against the individual hybrids. This seems to suggest that our argument is tight (up to the constant factor 3). Either way, as Theorem 7 and Corollary 1 demonstrate, this additional factor only affects the constant in front of the log term in the number of hybrids, so, we believe, it is only of secondary importance and we leave it as an open problem.

The rest of the subsection proves Lemma 2, where we make use of the following notation. For some distinguisher D, let $\alpha_{\mathcal{P},\mathcal{Q}}^D$ be its output probability, $\beta_{\mathcal{P},\mathcal{Q}}^D$ its conditional success probability, $\delta_{\mathcal{P},\mathcal{Q}}^D$ its conditional distinguishing advantage, and $\mathrm{adv}_{\mathcal{P},\mathcal{Q}}^D = \alpha_{\mathcal{P},\mathcal{Q}}^D(\delta_{\mathcal{P},\mathcal{Q}}^D)^2$ its advantage against the distributions $\mathcal{P}, \mathcal{Q}$.

Furthermore, let $\alpha_{\mathcal{P}}^{D} = \Pr[D(\mathcal{P}) \neq \perp]$ and $\gamma_{\mathcal{P}}^{D} = \Pr[D(\mathcal{P}) = 1]$ for any distribution $\mathcal{P}$. We can express the advantage of D against $\mathcal{P}$ and $\mathcal{Q}$ in terms of $\alpha_{\mathcal{P}}^{D}$, $\alpha_{\mathcal{Q}}^{D}$, $\gamma_{\mathcal{P}}^{D}$, $\gamma_{\mathcal{Q}}^{D}$:

$$\alpha_{\mathcal{P},\mathcal{Q}}^{D} = \frac{1}{2}(\alpha_{\mathcal{P}}^{D} + \alpha_{\mathcal{Q}}^{D})$$

$$\beta_{\mathcal{P},\mathcal{Q}}^{D} = \frac{\gamma_{\mathcal{P}}^{D} - \gamma_{\mathcal{Q}}^{D} + \alpha_{\mathcal{Q}}^{D}}{\alpha_{\mathcal{P}}^{D} + \alpha_{\mathcal{Q}}^{D}}$$

$$\delta_{\mathcal{P},\mathcal{Q}}^{D} = 2\beta_{\mathcal{P},\mathcal{Q}}^{D} - 1 = \frac{2(\gamma_{\mathcal{P}}^{D} - \gamma_{\mathcal{Q}}^{D}) + \alpha_{\mathcal{Q}}^{D} - \alpha_{\mathcal{P}}^{D}}{\alpha_{\mathcal{P}}^{D} + \alpha_{\mathcal{Q}}^{D}}$$

$$\text{adv}_{\mathcal{P},\mathcal{Q}}^{D} = \frac{(2(\gamma_{\mathcal{P}}^{D} - \gamma_{\mathcal{Q}}^{D}) + \alpha_{\mathcal{Q}}^{D} - \alpha_{\mathcal{P}}^{D})^2}{2(\alpha_{\mathcal{P}}^{D} + \alpha_{\mathcal{Q}}^{D})}. \tag{5}$$

We begin with the observation that for computationally indistuingishable distributions the output probabilities of any bounded distinguisher D cannot vary too much under the two distributions.

Lemma 3. *Let $\mathcal{P}, \mathcal{Q}$ be two distributions. If $\text{adv}_{\mathcal{P},\mathcal{Q}}^{D} \leq \epsilon$ for all T-bounded distinguishers, then we have $\alpha_{\mathcal{P}}^{D} \leq 2\alpha_{\mathcal{Q}}^{D} + 3\epsilon$ and $\alpha_{\mathcal{Q}}^{D} \leq 2\alpha_{\mathcal{P}}^{D} + 3\epsilon$ for any T bounded distinguisher.*

Proof. We prove the first claim. (The proof of the second claim is symmetrical.) Fix any distinguisher D. Assume $\alpha_{\mathcal{P}}^{D} \geq 2\alpha_{\mathcal{Q}}^{D}$, since otherwise we are done. Consider an alternative distinguisher D', which runs D and in the event that $D \neq \perp$, outputs 1 and otherwise $\perp$. Obviously, D' is also T-bounded, and (setting $\gamma_{\mathcal{P}}^{D'} = \alpha_{\mathcal{P}}^{D'}$, $\gamma_{\mathcal{Q}}^{D'} = \alpha_{\mathcal{Q}}^{D'}$ in (5)) we get

$$\begin{aligned}
\text{adv}_{\mathcal{P},\mathcal{Q}}^{D'} &= \frac{(\alpha_{\mathcal{P}}^{D} - \alpha_{\mathcal{Q}}^{D})^2}{2(\alpha_{\mathcal{P}}^{D} + \alpha_{\mathcal{Q}}^{D})} \\
&\geq \frac{(\alpha_{\mathcal{P}}^{D} - \alpha_{\mathcal{Q}}^{D})^2}{3\alpha_{\mathcal{P}}^{D}} \\
&= \frac{1}{3}\left(\alpha_{\mathcal{P}}^{D} - 2\alpha_{\mathcal{Q}}^{D} + \frac{(\alpha_{\mathcal{Q}}^{D})^2}{\alpha_{\mathcal{P}}^{D}}\right) \\
&\geq \frac{1}{3}\left(\alpha_{\mathcal{P}}^{D} - 2\alpha_{\mathcal{Q}}^{D}\right).
\end{aligned}$$

The first claim now follows from $\epsilon \geq \text{adv}_{\mathcal{P},\mathcal{Q}}^{D'}$. $\square$

Proof (of Lemma 2). We fix any distinguisher D and drop the superfix of α, γ, δ and adv for the rest of the proof. Furthermore, we will abbreviate $\mathcal{H}_i$ by i in the subfixes of α, γ, δ, and adv.

Using induction, one can prove

$$\sum_{i=1}^{k} \text{adv}_{i,i+1} \geq \frac{\alpha_1 + \alpha_k}{\alpha_1 + 2\sum_{i=2}^{k-1} \alpha_i + \alpha_k} \text{adv}_{1,k}$$

The proof proceeds by substituting in the definition of $\mathrm{adv}_{i,i+1}$ from (5), applying the induction hypothesis to the first $k-1$ terms of the sum, and then minimizing over γ_{k-1}. Details can be found in Appendix B.

It remains to show that

$$\frac{\alpha_1 + \alpha_k}{\alpha_1 + 2\sum_{i=2}^{k-1}\alpha_i + \alpha_k} \geq \frac{1}{3k}.$$

We again proceed by induction and can thus assume that $\mathrm{adv}_{1,i} \leq 3i\sum_{j=1}^{i-1}\epsilon_{j,j+1}$ for all $i < k$ and symmetrically $\mathrm{adv}_{i,k} \leq 3(k-i)\sum_{j=i}^{k-1}\epsilon_{j,j+1}$ for all $i > 1$. By Lemma 3, this means that $\alpha_i \leq 2\alpha_1 + 9i\sum_{j=1}^{i-1}\epsilon_{j,j+1}$ for all $i < k$ and again $\alpha_i \leq 2\alpha_k + 9(k-i)\sum_{j=i}^{k-1}\epsilon_{j,j+1}$ for all $i > 1$. We note that

$$\alpha_1 + 2\sum_{i=2}^{k-1}\alpha_i + \alpha_k = \alpha_1 + 2\sum_{i=2}^{\lfloor(k-1)/2\rfloor}\alpha_i + 2\sum_{\lfloor(k-1)/2\rfloor+1}^{k-1}\alpha_i + \alpha_k$$

and using the above inequalities, the two sums are bounded by

$$2\sum_{i=2}^{\lfloor(k-1)/2\rfloor}\alpha_i \leq 2(k-3)\alpha_1 + 3k^2\sum_{i=1}^{\lfloor(k-1)/2\rfloor}\epsilon_{i,i+1}$$

and

$$2\sum_{\lfloor(k-1)/2\rfloor+1}^{k-1}\alpha_i \leq 2(k-3)\alpha_k + 3k^2\sum_{\lfloor(k-1)/2\rfloor+1}^{k-1}\epsilon_{i,i+1}$$

respectively. This bounds the entire sum:

$$\alpha_1 + 2\sum_{i=2}^{k-1}\alpha_i + \alpha_k \leq 2k(\alpha_1 + \alpha_k) + 3k^2\sum_{i=1}^{k-1}\epsilon_{i,i+1}$$

This in turn leads to the lower bound

$$\frac{\alpha_1 + \alpha_k}{\alpha_1 + 2\sum_{i=2}^{k-1}\alpha_i + \alpha_k} \geq \frac{1}{2k + \frac{3k^2\sum_{i=1}^{k-1}\epsilon_{i,i+1}}{\alpha_1 + \alpha_k}}$$

The last step is noticing that we can assume that $(\alpha_1 + \alpha_k) \geq 6k\sum_{i=1}^{k-1}\epsilon_{i,i+1}$, because $(\alpha_1 + \alpha_k)/2 \geq \epsilon_{1,k}$ and otherwise we would be done. Using this assumption we have

$$\frac{\alpha_1 + \alpha_k}{\alpha_1 + 2\sum_{i=2}^{k-1}\alpha_i + \alpha_k} \geq \frac{1}{2k + \frac{3k^2}{6k}} \geq \frac{1}{3k}$$

as desired. □

Approximate Samplers. In this section we bridge the gap between search and decision primitives making use of approximate samplers, for the first time by extending a result from [15] to arbitrary decision primitives. It might be possible to extend other results from the literature [2, 16, 17] to decision primitives using our definition, but we leave that for future work. Our main result is given in Theorem 8. Combining it with results from [15] it implies that approximating a distribution with relative error bounded by $2^{-\kappa/2}$ (e.g., as provided by floating point numbers with $\kappa/2$-bit mantissa) allows to preserve almost all of κ bits of security.

Before introducing the result formally, we first need to cover some preliminaries from [15].

Background. Using the same terminology as [15], let $\delta(\mathcal{P}, \mathcal{Q})$ be some divergence on probability distributions. A λ-*efficient* divergence satisfies three properties:

1. *Sub-additivity for joint distributions:* if $(X_i)_i$ and $(Y_i)_i$ are two lists of discrete random variables over the support $\prod_i S_i$, then

$$\delta((X_i)_i, (Y_i)_i) \leq \sum_i \max_a \delta([X_i \mid X_{<i} = a], [Y_i \mid Y_{<i} = a]),$$

   where $X_{<i} = (X_1, \ldots, X_{i-1})$ (and similarly for $Y_{<i}$), and the maximum is taken over $a \in \prod_{j<i} S_j$.
2. *Data processing inequality:* $\delta(f(\mathcal{P}), f(\mathcal{Q})) \leq \delta(\mathcal{P}, \mathcal{Q})$ for any two distributions $\mathcal{P}$ and $\mathcal{Q}$ and (possibly randomized) algorithm $f(\cdot)$, i.e., the measure does not increase under function application.
3. *Pythagorean probability preservation* with parameter $\lambda \in \mathbb{R}$: if $(X_i)_i$ and $(Y_i)_i$ are two lists of discrete random variables over the support $\prod_i S_i$ and

$$\delta((X_i \mid X_{<i} = a_i), (Y_i \mid Y_{<i} = a_i)) \leq \lambda$$

for all i and $a_i \in \prod_{j<i} S_j$, then

$$\Delta_{SD}((X_i)_i, (Y_i)_i) \leq \left\| \left(\max_{a_i} \delta((X_i \mid X_{<i} = a_i), (Y_i \mid Y_{<i} = a_i)) \right)_i \right\|_2.$$

As an example, the max-log distance $\Delta_{ML}(\mathcal{P}, \mathcal{Q}) = \max|\log \mathcal{P}(x) - \log \mathcal{Q}(x)|$ is λ-efficient for any $\lambda \leq \frac{1}{3}$ [15].

Main Result for Approximate Samplers. The next theorem states the main result of this section. It shows that it suffices to approximate a distribution $\mathcal{P}$ up to distance $\delta(\mathcal{P}, \mathcal{Q}) \leq 2^{-\kappa/2}$ for an efficient divergence δ in order to maintain almost κ bits of security.

Theorem 8. *Let $S^{\mathcal{P}}$ be a 1-bit secrecy game with black-box access to a probability ensemble $(\mathcal{P}_\theta)_\theta$, and δ be a λ-efficient measure for any $\lambda \leq \frac{1}{4}$. If $S^{\mathcal{P}}$ is κ-bit secure and $\delta(\mathcal{P}_\theta, \mathcal{Q}_\theta) \leq 2^{-\kappa/2}$, then $S^{\mathcal{Q}}$ is $(\kappa - 8)$-bit secure.*

The remainder of this section is devoted to proving Theorem 8. We first rewrite a lemma from [15], which we will use in our proof.

Lemma 4 (adapted from [15]). *Let $S^{\mathcal{P}}$ be any security game with black-box access to a probability distribution ensemble $\mathcal{P}_\theta$. For any adversary A with resources T that plays $S^{\mathcal{P}}$ and event E over its output, denote $\gamma_{\mathcal{P}} = \Pr[A \in E]$. For the same event, denote by $\gamma_{\mathcal{Q}}$ the probability of E when A is playing $S^{\mathcal{Q}}$. If $\frac{T}{\gamma_{\mathcal{P}}} \geq 2^k$ and $\delta(\mathcal{P}_\theta, \mathcal{Q}_\theta) \leq 2^{-k/2}$ for any $2^{-k/2}$-efficient δ, then $\frac{T}{\gamma_{\mathcal{Q}}} \geq 2^{k-3}$.*

From Lemma 4 we can derive a bound on the output probability of an adversary when switching the distribution of the scheme.

Corollary 2. *For any adversary A with resources T attacking $S^{\mathcal{P}}$ and any event E over A's output, denote the probability of E by $\gamma_{\mathcal{P}}$. Denote the probability of E over A's output when attacking $S^{\mathcal{Q}}$ by $\gamma_{\mathcal{Q}}$. If δ is $\sqrt{\gamma_{\mathcal{Q}}/16T}$-efficient and $\delta(\mathcal{P}_\theta, \mathcal{Q}_\theta) \leq \sqrt{\gamma_{\mathcal{Q}}/16T}$, then $16\gamma_{\mathcal{P}} \geq \gamma_{\mathcal{Q}}$.*

Proof. We use Lemma 4 and set k such that $2^{k-4} = \frac{T}{\gamma_{\mathcal{Q}}}$. This implies that $\frac{T}{\gamma_{\mathcal{Q}}} \geq 2^{k-3}$ is false. Assuming towards a contradiction that $16\gamma_{\mathcal{P}} < \gamma_{\mathcal{Q}}$, we see that

$$2^{k-4} = \frac{T}{\gamma_{\mathcal{Q}}} \leq \frac{T}{16\gamma_{\mathcal{P}}}$$

contradicting Lemma 4.   □

With this bound in place, we are ready for the main proof.

Proof (of Theorem 8). Fix any T^A-bounded adversary A against $S^{\mathcal{P}}$, output probability $\alpha_{\mathcal{P}}^A$ and conditional success probability $\beta_{\mathcal{P}}^A$. By assumption we have $\alpha_{\mathcal{P}}^A(2\beta_{\mathcal{P}}^A - 1)^2 \leq T^A/2^\kappa$. Denote the output and conditional success probability of A against $S^{\mathcal{Q}}$ by $\alpha_{\mathcal{Q}}^A$ and $\beta_{\mathcal{Q}}^A$. Assume towards contradiction that $\alpha_{\mathcal{Q}}^A(2\beta_{\mathcal{Q}}^A - 1)^2 > T^A/2^{\kappa-8}$.

First we apply Corollary 2 to obtain $\alpha_{\mathcal{P}}^A \geq 2^{-4}\alpha_{\mathcal{Q}}^A$. Note that by assumption $\sqrt{\alpha_{\mathcal{Q}}^A/16T} > 2^{(-\kappa+4)/2} > 2^{-\kappa/2} \geq \delta(\mathcal{P}_\theta, \mathcal{Q}_\theta)$ and that trivially $\sqrt{\alpha_{\mathcal{Q}}^A/16T} \leq \frac{1}{4}$.

We now consider the hypothetical modified games $\hat{S}^{\mathcal{P}}$ and $\hat{S}^{\mathcal{Q}}$, which are the same as $S^{\mathcal{P}}$ and $S^{\mathcal{Q}}$ with the only difference that the adversary has the ability to restart the game with fresh randomness at any time. Consider the adversary B against $\hat{S}$ that simply runs A until $A \neq \perp$ (restarting the game if $A = \perp$) and outputs whatever A returns. Let $\alpha = \min(\alpha_{\mathcal{P}}^A, \alpha_{\mathcal{Q}}^A)$ and note that B's resources are $T^B < T^A/\alpha$, its output probability is 1 and the (conditional) success probability is $\beta_{\mathcal{P}}^B = \beta_{\mathcal{P}}^A$ (or $\beta_{\mathcal{Q}}^B = \beta_{\mathcal{Q}}^A$) if playing $\hat{S}^{\mathcal{P}}$ (or $\hat{S}^{\mathcal{Q}}$, respectively).

By the properties of δ and Δ_{SD}, we have $\beta_{\mathcal{P}}^B \geq \beta_{\mathcal{Q}}^B - \sqrt{T^B}\delta(\mathcal{P}_\theta, \mathcal{Q}_\theta)$ and so $2\beta_{\mathcal{P}}^B - 1 \geq 2\beta_{\mathcal{Q}}^B - 1 - 2\sqrt{T^B/2^\kappa}$. By assumption we also have that $2\beta_{\mathcal{P}}^A - 1 \leq \sqrt{T^A/\alpha_{\mathcal{P}}^A 2^\kappa}$, which yields

$$\sqrt{\frac{T^A}{\alpha 2^\kappa}} \geq \sqrt{\frac{T^A}{\alpha_{\mathcal{P}}^A 2^\kappa}} \geq 2\beta_{\mathcal{Q}}^B - 1 - 2\sqrt{\frac{T^A}{\alpha 2^\kappa}}$$

because $\beta_{\mathcal{P}}^B = \beta_{\mathcal{P}}^A$, and so

$$2\beta_{\mathcal{Q}}^A - 1 = 2\beta_{\mathcal{Q}}^B - 1 \leq 3\sqrt{\frac{T^A}{\alpha 2^{\kappa}}}.$$

If $\alpha_{\mathcal{Q}}^A \leq \alpha_{\mathcal{P}}^A$, then $\alpha = \alpha_{\mathcal{Q}}^A$ and the above inequality immediately yields the contradiction. Otherwise, we can derive an upper bound on $\alpha_{\mathcal{P}}^A$ from it:

$$\alpha_{\mathcal{P}}^A \leq \frac{9T^A}{2^{\kappa}(2\beta_{\mathcal{Q}}^A - 1)^2} < \frac{\alpha_{\mathcal{Q}}^A}{2^4}$$

where the latter inequality follows from the assumption. This contradicts our lower bound above. □

Acknowledgment. We would like to thank Krzysztof Pietrzak, Russell Impagliazzo, and Mihir Bellare for helpful discussions and pointers to relevant literature.

A   Proof of Theorem 1

Proof (of Theorem 1). From the definition of Y in Definition 7 we get for any $x, y \in \{0,1\}^n$ with $y \neq x$

- $\Pr[Y = \perp | X = x] = 1 - \alpha$
- $\Pr[Y = x | X = x] = \alpha\beta$
- $\Pr[Y = y | X = x] = \frac{\alpha(1-\beta)}{2^n - 1}$.

From this we compute

- $\Pr[Y = \perp] = 1 - \alpha$
- $\Pr[Y = y] = \Pr[Y = y | X = y]\Pr[X = y] + \Pr[Y = y | X \neq y]\Pr[X \neq y] = \frac{\alpha\beta}{2^n} + \frac{2^n - 1}{2^n}\frac{\alpha(1-\beta)}{2^n - 1} = \frac{\alpha}{2^n}$.

Now we calculate the conditional entropy

$$H(X|Y) = \sum_{x,y} \Pr[Y = y | X = x]\Pr[X = x] \log \frac{\Pr[Y = y]}{\Pr[Y = y | X = x]\Pr[X = x]}$$

$$= \sum_x \Pr[Y = \perp | X = x]\Pr[X = x] \log \frac{\Pr[Y = \perp]}{\Pr[Y = \perp | X = x]\Pr[X = x]}$$

$$+ \Pr[Y = x | X = x]\Pr[X = x] \log \frac{\Pr[Y = x]}{\Pr[Y = x | X = x]\Pr[X = x]}$$

$$+ \sum_{y \neq x \wedge y \neq \perp} \Pr[Y = y | X = x]\Pr[X = x] \log \frac{\Pr[Y = y]}{\Pr[Y = y | X = x]\Pr[X = x]}$$

$$= \sum_x \frac{1 - \alpha}{2^n} \log \frac{(1 - \alpha)2^n}{1 - \alpha} + \frac{\alpha\beta}{2^n} \log \frac{\alpha 2^n}{\alpha\beta 2^n}$$

$$+ (2^n - 1)\frac{\alpha(1-\beta)}{(2^n - 1)2^n} \log \frac{\alpha 2^n (2^n - 1)}{2^n \alpha (1 - \beta)}$$

$$= (1 - \alpha)n + \alpha\beta \log \frac{1}{\beta} + \alpha(1 - \beta) \log \frac{2^n - 1}{1 - \beta}$$

$$= (1 - \alpha)n + \alpha((1 - \beta)\log(2^n - 1) + H(\mathcal{B}_{\beta}))$$

Finally, we compute the advantage

$$\text{adv}^A = 1 - \frac{H(X|Y)}{n}$$

$$= 1 - (1 - \alpha) - \alpha \frac{(1 - \beta)\log(2^n - 1) + H(\mathcal{B}_\beta)}{n}$$

$$= \alpha \left(1 - \frac{(1 - \beta)\log(2^n - 1) + H(\mathcal{B}_\beta)}{n} \right).$$

□

B   Missing Details of Proof for Lemma 2

With the notation of Sect. 5.3, the goal of this section is to prove

$$\sum_{i=1}^{k} \text{adv}_{i,i+1} \geq \frac{\alpha_1 + \alpha_k}{\alpha_1 + 2\sum_{i=2}^{k-1} \alpha_i + \alpha_k} \text{adv}_{1,k}.$$

By Eq. (5)

$$\sum_{i=1}^{k} \text{adv}_{i,i+1} = \sum_{i=1}^{k} \frac{(2(\gamma_i - \gamma_{i+1}) + \alpha_{i+1} - \alpha_i)^2}{2(\alpha_i + \alpha_{i+1})}.$$

Applying the induction hypothesis, this is lower bounded by

$$f(\gamma_{k-1}) = \frac{(2(\gamma_1 - \gamma_{k-1}) + \alpha_{k-1} - \alpha_1)^2}{2(\alpha_1 + 2\sum_{i=2}^{k-2} \alpha_i + \alpha_{k-1})} + \frac{(2(\gamma_{k-1} - \gamma_k) + \alpha_k - \alpha_{k-1})^2}{2(\alpha_{k-1} + \alpha_k)}.$$

Taking f's derivative

$$f'(\gamma_{k-1}) = \frac{2(2(\gamma_{k-1} - \gamma_k) + \alpha_k - \alpha_{k-1})}{\alpha_{k-1} + \alpha_k} - \frac{2(2(\gamma_1 - \gamma_{k-1}) + \alpha_{k-1} - \alpha_1)}{\alpha_1 + 2\sum_{i=2}^{k-2} \alpha_i + \alpha_{k-1}}$$

Note that the second derivative is a positive constant, so if f has an extremum it must be a minimum, and since it is a quadratic function, it is a global minimum. Setting $f'(\gamma_{k-1}) = 0$ and solving for $2\gamma_{k-1}$, we get:

$$2\gamma_{k-1} \left(\alpha_1 + 2\sum_{i=2}^{k-1} \alpha_i + \alpha_k \right) = 2(\gamma_1 + \alpha_{k-1} - \alpha_1)(\alpha_{k-1} + \alpha_k)$$

$$+ 2(\gamma_k + \alpha_k - \alpha_{k-1}) \left(\alpha_1 + 2\sum_{i=2}^{k-2} \alpha_i + \alpha_k \right)$$

Plugging this into the terms of f:

$$(2(\gamma_1 - \gamma_{k-1}) + \alpha_{k-1} - \alpha_1) = \frac{2(\gamma_1 - \gamma_k) - \alpha_1 + \alpha_k \left(\alpha_1 + 2\sum_{i=2}^{k-2} \alpha_i + \alpha_k \right)}{\alpha_1 + 2\sum_{i=2}^{k-1} \alpha_i + \alpha_k}$$

and

$$(2(\gamma_{k-1} - \gamma_k) + \alpha_k - \alpha_{k-1}) = \frac{(2(\gamma_1 - \gamma_k) - \alpha_1 + \alpha_k)(\alpha_{k-1} + \alpha_k)}{\alpha_1 + 2\sum_{i=2}^{k-1} \alpha_i + \alpha_k}$$

which yields that

$$
\begin{aligned}
f(\gamma_{k-1}) &\geq \frac{(2(\gamma_1 - \gamma_k) - \alpha_1 + \alpha_k)^2 \left(\alpha_1 + 2\sum_{i=2}^{k-2} \alpha_i + \alpha_k\right)^2}{\left(\alpha_1 + 2\sum_{i=2}^{k-1} \alpha_i + \alpha_k\right)^2 \left(\alpha_1 + 2\sum_{i=2}^{k-2} \alpha_i + \alpha_k\right)} \\
&\quad + \frac{(2(\gamma_1 - \gamma_k) - \alpha_1 + \alpha_k)^2 (\alpha_{k-1} + \alpha_k)^2}{\left(\alpha_1 + 2\sum_{i=2}^{k-1} \alpha_i + \alpha_k\right)^2 (\alpha_{k-1} + \alpha_k)} \\
&= \frac{(2(\gamma_1 - \gamma_k) - \alpha_1 + \alpha_k)^2}{\left(\alpha_1 + 2\sum_{i=2}^{k-1} \alpha_i + \alpha_k\right)^2} \left(\alpha_1 + 2\sum_{i=2}^{k-1} \alpha_i + \alpha_k\right) \\
&= \frac{(2(\gamma_1 - \gamma_k) - \alpha_1 + \alpha_k)^2}{\left(\alpha_1 + 2\sum_{i=2}^{k-1} \alpha_i + \alpha_k\right)} \\
&= \frac{\alpha_1 + \alpha_k}{\left(\alpha_1 + 2\sum_{i=2}^{k-1} \alpha_i + \alpha_k\right)} \mathrm{adv}_{1,k}
\end{aligned}
$$

as desired.

References

1. Alkim, E., Ducas, L., Pöppelmann, T., Schwabe, P.: Post-quantum key exchange - a new hope. In: USENIX Security Symposium, pp. 327–343. USENIX Association (2016)
2. Bai, S., Langlois, A., Lepoint, T., Stehlé, D., Steinfeld, R.: Improved security proofs in lattice-based cryptography: using the Rényi divergence rather than the statistical distance. In: Iwata, T., Cheon, J.H. (eds.) ASIACRYPT 2015. LNCS, vol. 9452, pp. 3–24. Springer, Heidelberg (2015). https://doi.org/10.1007/978-3-662-48797-6_1
3. Bellare, M., Desai, A., Jokipii, E., Rogaway, P.: A concrete security treatment of symmetric encryption. In: 38th Annual Symposium on Foundations of Computer Science, pp. 394–403. IEEE Computer Society Press, October 1997
4. Bellare, M., Fischlin, M., Goldwasser, S., Micali, S.: Identification protocols secure against reset attacks. In: Pfitzmann, B. (ed.) EUROCRYPT 2001. LNCS, vol. 2045, pp. 495–511. Springer, Heidelberg (2001). https://doi.org/10.1007/3-540-44987-6_30
5. Bellare, M., Rogaway, P.: The exact security of digital signatures-how to sign with RSA and Rabin. In: Maurer, U. (ed.) EUROCRYPT 1996. LNCS, vol. 1070, pp. 399–416. Springer, Heidelberg (1996). https://doi.org/10.1007/3-540-68339-9_34
6. Bernstein, D.J., Lange, T.: Non-uniform cracks in the concrete: the power of free precomputation. In: Sako, K., Sarkar, P. (eds.) ASIACRYPT 2013. LNCS, vol. 8270, pp. 321–340. Springer, Heidelberg (2013). https://doi.org/10.1007/978-3-642-42045-0_17

7. De, A., Trevisan, L., Tulsiani, M.: Time space tradeoffs for attacks against one-way functions and PRGs. In: Rabin, T. (ed.) CRYPTO 2010. LNCS, vol. 6223, pp. 649–665. Springer, Heidelberg (2010). https://doi.org/10.1007/978-3-642-14623-7_35

8. Dodis, Y., Steinberger, J.P.: Message authentication codes from unpredictable block ciphers. In: Halevi, S. (ed.) CRYPTO 2009. LNCS, vol. 5677, pp. 267–285. Springer, Heidelberg (2009). https://doi.org/10.1007/978-3-642-03356-8_16

9. Goldreich, O.: Foundations of Cryptography: Basic Tools, vol. 1. Cambridge University Press, Cambridge (2001)

10. Goldreich, O.: Foundations of Cryptography: Basic Applications, vol. 2. Cambridge University Press, Cambridge (2004)

11. Goldreich, O., Levin, L.A.: A hard-core predicate for all one-way functions. In: 21st Annual ACM Symposium on Theory of Computing, pp. 25–32. ACM Press, May 1989

12. Goldwasser, S., Micali, S.: Probabilistic encryption. J. Comput. Syst. Sci. **28**(2), 270–299 (1984)

13. Håstad, J., Impagliazzo, R., Levin, L.A., Luby, M.: A pseudorandom generator from any one-way function. SIAM J. Comput. **28**(4), 1364–1396 (1999)

14. Levin, L.A.: Randomness and non-determinism. J. Symbol. Logic **58**, 1102–1103 (1993)

15. Micciancio, D., Walter, M.: Gaussian sampling over the integers: efficient, generic, constant-time. In: Katz, J., Shacham, H. (eds.) CRYPTO 2017. LNCS, vol. 10402, pp. 455–485. Springer, Cham (2017). https://doi.org/10.1007/978-3-319-63715-0_16

16. Pöppelmann, T., Ducas, L., Güneysu, T.: Enhanced lattice-based signatures on reconfigurable hardware. In: Batina, L., Robshaw, M. (eds.) CHES 2014. LNCS, vol. 8731, pp. 353–370. Springer, Heidelberg (2014). https://doi.org/10.1007/978-3-662-44709-3_20

17. Prest, T.: Sharper bounds in lattice-based cryptography using the Rényi divergence. In: Takagi, T., Peyrin, T. (eds.) ASIACRYPT 2017. LNCS, vol. 10624, pp. 347–374. Springer, Cham (2017). https://doi.org/10.1007/978-3-319-70694-8_13

On the Gold Standard for Security
of Universal Steganography

Sebastian Berndt[1](✉) and Maciej Liśkiewicz[2]

[1] Department of Computer Science, Kiel University, Kiel, Germany
seb@informatik.uni-kiel.de
[2] Institute for Theoretical Computer Science, University of Lübeck, Lübeck, Germany
liskiewi@tcs.uni-luebeck.de

Abstract. While symmetric-key steganography is quite well understood both in the information-theoretic and in the computational setting, many fundamental questions about its public-key counterpart resist persistent attempts to solve them. The computational model for public-key steganography was proposed by von Ahn and Hopper in EUROCRYPT 2004. At TCC 2005, Backes and Cachin gave the first universal public-key stegosystem – i.e. one that works on all channels – achieving security against replayable chosen-covertext attacks (SS-RCCA) and asked whether security against non-replayable chosen-covertext attacks (SS-CCA) is achievable. Later, Hopper (ICALP 2005) provided such a stegosystem for every efficiently sampleable channel, but did not achieve universality. He posed the question whether universality *and* SS-CCA-security can be achieved simultaneously. No progress on this question has been achieved since more than a decade. In our work we solve Hopper's problem in a somehow complete manner: As our main positive result we design an SS-CCA-secure stegosystem that works for *every* memoryless channel. On the other hand, we prove that this result is the best possible in the context of universal steganography. We provide a family of 0-*memoryless* channels – where the already sent documents have only marginal influence on the current distribution – and prove that no SS-CCA-secure steganography for this family exists in the standard non-look-ahead model.

1 Introduction

Steganography is the art of hiding the transmission of information to achieve secret communication without revealing its presence. In the basic setting, the aim of the steganographic encoder (often called Alice or the stegoencoder) is to hide a secret message in a document and to send it to the stegodecoder (Bob) via a public *channel* which is completely monitored by an *adversary* (Warden or steganalyst). The channel is modeled as a probability distribution of legal documents, called *covertexts*, and the adversary's task is to distinguish those from altered ones, called *stegotexts*. Although strongly connected with cryptographic encryption, steganography is not encryption: While encryption only tries to hide

© International Association for Cryptologic Research 2018
J. B. Nielsen and V. Rijmen (Eds.): EUROCRYPT 2018, LNCS 10820, pp. 29–60, 2018.
https://doi.org/10.1007/978-3-319-78381-9_2

the content of the transmitted message, steganography aims to hide both the message and the fact that a message was transmitted at all.

As in the cryptographic setting, the security of the stegosystems should only rely on the secrecy of the keys used by the system. *Symmetric-key* steganography, which assumes that Alice and Bob share a secret-key, has been a subject of intensive study both in an information-theoretic [7,36,40] and in a computational setting [13,22,23,25,26,30]. A drawback of such an approach is that the encoder and the decoder must have shared a key in a secure way. This may be unhandy, e.g. if the encoder communicates with several parties.

In order to avoid this problem in cryptography, Diffie and Hellman provided the notion of a *public-key scenario* in their groundbreaking work [15]. This idea has proved to be very useful and is currently used in nearly every cryptographic application. Over time, the notion of security against so-called *chosen ciphertext attacks* (chosen-ciphertext attack (CCA)-security) has established itself as the "gold standard" for security in the public-key scenario [20,27]. In this setting, an attacker has also access to a decoding oracle that decodes every ciphertext different from the challenge-text. Dolev et al. [16] proved that the simplest assumption for public-key cryptography – the existence of trapdoor permutations – is sufficient to construct a CCA-secure public key cryptosystem.

Somewhat in contrast to the research in cryptographic encryption, only very little studies in steganography have been concerned so far within the public-key setting. Von Ahn and Hopper [38,39] were the first to give a formal framework and to prove that secure public-key steganography exists. They formalized security against a *passive* adversary in which Warden is allowed to provide challenge-hiddentexts to Alice in hopes of distinguishing covertexts from stegotexts encoding the hiddentext of his choice. For a restricted model, they also defined security against an active adversary; It is assumed, however, that Bob must know the identity of Alice, which deviates from the common bare public-key scenario.

Importantly, the schemes provided in [38,39] are *universal* (called also *black-box* in the literature). This property guarantees that the systems are secure with respect not only to a concrete channel C but to a broad range of channels. The importance of universality is based on the fact that typically no good description of the distribution of a channel is known.

In [3], Backes and Cachin provided a notion of security for public-key steganography with *active* attacks, called *steganographic chosen-covertext attacks (SS-CCAs)*. In this scenario the warden may provide a challenge-hiddentext to Alice and enforce the stegoencoder to send stegotexts encoding the hiddentext of his choice. The warden may then insert documents into the channel between Alice and Bob and observe Bob's responses in hope of detecting the steganographic communication. This is the steganographic equivalent of a chosen ciphertext attack against encryption and it seems to be the most general type of security for public-key steganography with active attacks similar to CCA-security in encryption. Backes and Cachin also gave a universal public-key stegosystem which, although not secure in the general SS-CCA-setting, satisfies a relaxed notion called *steganographic security against publicly-detectable replayable adaptive chosen-covertext attacks* (steganographic replayable chosen-covertext attack (SS-RCCA)) inspired by the work of Canetti et al. [8]. In this relaxed setting, the warden may still provide a hiddentext to Alice and is allowed to insert documents into the channel between Alice and Bob but with the restriction that the warden's document does not encode the chosen hiddentext. Backes and Cachin left as an open problem if secure public-key steganography exists at all in the SS-CCA-framework.

This question was answered by Hopper [21] in the affirmative in case Alice and Bob communicate via an efficiently sampleable channel $\mathcal{C}$. He proved (under the assumption of a CCA-secure cryptosystem) that for every such channel $\mathcal{C}$ there is an SS-CCA-secure stegosystem $\mathsf{PKStS}_{\mathcal{C}}$ on $\mathcal{C}$. The system cleverly "derandomizes" sampling documents by using the sampling-algorithm of the channel and using a pseudorandom generator to deterministically embed the encrypted message. Hence, $\mathsf{PKStS}_{\mathcal{C}}$ is only secure on the single channel $\mathcal{C}$ and is thus not universal. Hopper [21] posed as a challenging open problem to show the (non)existence of a universal SS-CCA-secure stegosystem. Since more than a decade, public key steganography has been used as a tool in different contexts (e.g. broadcast steganography [17] and private computation [9,11]), but this fundamental question remained open.

We solve Hopper's problem in a complete manner by proving (under the assumption of the existence of doubly-enhanced trapdoor permutations and collision-resistant hash functions) the existence of an SS-CCA-secure public key stegosystem that works for every *memoryless* channel, i.e. such that the documents are independently distributed (for a formal definition see next section). On the other hand, we also prove that the influence of the history – the already sent documents – dramatically limits the security of stegosystems in the realistic non-look-ahead model: We show that no stegosystem can be SS-CCA-secure against all 0-memoryless channels in the non-look-ahead model. In these channels, the influence of the history is minimal. We thereby demonstrate a clear dichotomy result for universal public-key steganography: While memoryless channels do exhibit an SS-CCA-secure stegosystem, the introduction of the history prevents this kind of security.

Our Contribution. As noted above, the stegosystem of Backes and Cachin has the drawback that it achieves a weaker security than SS-CCA-security while it works on every channel [3]. On the other hand, the stegosystem of Hopper achieves SS-CCA-security but is specialized to a single channel [21]. We prove (under the assumption of the existence of doubly-enhanced trapdoor permutations and collision-resistant hash functions) that there is a stegosystem that is SS-CCA-secure on a large class of channels (namely the memoryless ones). The main technical novelty is a method to generate covertexts for the message m such that finding a second sequence of covertexts that encodes m is hard. Hopper achieves this at the cost of the universality of his system, while we still allow a very large class of channels. We thereby answer the question of Hopper in the affirmative, in case of memoryless channels. Note that before this work, it was not even known whether an SS-CCA-secure stegosystem exists that works for some class of channels (Hopper's system only works on a single channel that is hard-wired into the system). Furthermore, we prove that SS-CCA-security for memoryless channels is the best possible in a very natural model: If the history influences the channel distribution in a minor way, i.e. only by its length, we prove that SS-CCA-security is not achievable in the standard non-look-ahead model of von Ahn and Hopper. In Table 1, we compare our results with previous works.

Table 1. Comparison of the public-key stegosystems

Paper	Security	Channels	Applicability
von Ahn and Hopper [38]	Passive	Universal	Possible
Backes and Cachin [3]	SS-RCCA	Universal	Possible
Hopper [21]	SS-CCA	Single constr. channel	Possible
This work (Theorem 10)	SS-CCA	All memoryless channels	Possible
This work (Theorem 12)	SS-CCA	Universal	Impossible[a]

[a]In the non-look-ahead model against non-uniform wardens.

Related Results. Anderson and Petitcolas [1] and Craver [12], have both, even before the publication of the work by von Ahn and Hopper [38,39], described ideas for public-key steganography, however, with only heuristic arguments for security. Van Le and Kurosawa [28] showed that every efficiently sampleable channel has an SS-CCA-secure public-key stegosystem. A description of the channel is built into the stegosystem and it makes use of a pseudo-random generator G that encoder and decoder share. But the authors make a strong assumption concerning changes of internal states of G each time the embedding operation is performed, which does not fit into the usual models of cryptography and steganography. Lysyanskaya and Meyerovich [32] investigated the influence of the sampling oracle on the security of public key stegosystems with passive attackers. They prove that the stegosystem of von Ahn and Hopper [39] becomes insecure if the approximation of the channel distribution by the sampling oracle deviates only slightly from the correct distribution. They also construct a channel, where no incorrect approximation of the channel yields a secure stegosystem. This strengthens the need for universal stegosystems, as even tiny approximation errors of the channel distribution may lead to huge changes with regard to the security of the system. Fazio et al. [17] extended public-key steganography to the multi-recipient setting, where a single sender communicates with a dynamically set of receivers. Their system is designed such that no outside party and no unauthorized user is able to detect the presence of these broadcast communication. Cho et al. [11] upgraded the covert multi-party computation model of Chandran et al. [9] to the concurrent case and gave protocols for several fundamental operations, e.g. string equality and set intersection. Their steganographic (or *covert*) protocols are based upon the decisional Diffie-Hellman problem.

The paper is organized as follows. Section 2 contains the basic definitions and notations. In Sect. 3, we give an example attack on the stegosystem of Backes and Cachin to highlight the differences between SS-RCCA-security and SS-CCA-security. The following Sect. 4 contains a high-level view of our construction. Section 5 uses the results of [21] to prove that one can construct cryptosystems with ciphertexts that are indistinguishable from a distribution on bitstrings related to the hypergeometric distribution, which we will need later on. The main core of our protocol is an algorithm to order the documents in an undetectable way that still allows us to transfer information. This ordering is described in Sect. 6. Our results concerning the existence of SS-CCA-secure steganography for every memoryless channel are then presented and proved in Sect. 7. Finally, Sect. 8 contains the impossibility result for SS-CCA-secure stegosystems in the non-look-ahead model on 0-memoryless channels.

In order to improve the presentation, we moved proofs of some technical statements to the appendix.

2   Definitions and Notation

If S is a finite set, we write $x \leftarrow S$ to denote the *random* assignment of a uniformly chosen element of S to x. If A is a probability distribution or a randomized algorithm, we write $x \leftarrow A$ to denote the assignment of the output of A, taken over the internal coin-flips of A.

As our cryptographic and steganographic primitives will be parameterized by the key length κ, we want that the ability of any polynomial algorithm to attack this primitives is lower than the inverse of all polynomials in κ. This is modeled by the definition of a negligible function. A function $\mathsf{negl} \colon \mathbb{N} \to [0,1]$ is called *negligible*, if for every polynomial p, there is an $N_0 \in \mathbb{N}$ such that $\mathsf{negl}(N) < p(N)^{-1}$ for every $N \geq N_0$. For a probability distribution D on support X, the *min-entropy* $H_\infty(D)$ is defined as $\inf_{x \in X} \{ -\log D(x) \}$.

We also need the notion of a *strongly 2-universal hash function*, which is a set of functions G mapping bitstrings of length ℓ to bitstrings of length $\ell' < \ell$ such that for all $x, x' \in \{0,1\}^\ell$ with $x \neq x'$ and all (not necessarily different) $y, y' \in \{0,1\}^{\ell'}$, we have $|\{ f \in G \mid f(x) = y \wedge f(x') = y' \}| = \frac{|G|}{2^{2\ell'}}$. If $\ell/\ell' \in \mathbb{N}$, a typical example of such a family is the set of functions

$$\{ x \mapsto \left(\textstyle\sum_{i=1}^{\ell/\ell'} a_i x_i + b \right) \bmod 2^{\ell'} \mid a_1, \ldots, a_{\ell/\ell'}, b \in \{0, \ldots, 2^{\ell'} - 1\} \},$$

where x_i denotes the i-th block of length ℓ' of x and we implicitly use the canonical bijection between $\{0,1\}^n$ and the finite field $\{0, \ldots, 2^n - 1\}$. See e.g. the textbook of Mitzenmacher and Upfal [33] for more information on this. For two polynomials ℓ and ℓ', a *strongly 2-universal hash family* is a family $\mathcal{G} = \{ G_\kappa \}_{\kappa \in \mathbb{N}}$ such that every G_κ is a strongly 2-universal hash function mapping strings of length $\ell(\kappa)$ to strings of length $\ell'(\kappa)$.

Channels and Stegosystems. In order to be able to embed messages into unsuspicious communication, we first need to provide a definition for this. We model the communication as an unidirectional transfer of *documents* that we will treat as strings of length n over a constant-size alphabet Σ. The communication is defined via the concept of a *channel* $\mathcal{C}$ on Σ: A function, that maps, for every $n \in \mathbb{N}$, a *history* hist $\in (\Sigma^n)^*$ to a probability distribution on Σ^n. We denote this probability distribution by $\mathcal{C}_{\text{hist},n}$ and its *min-entropy* $H_\infty(\mathcal{C}, n)$ as $\min_{\text{hist}}\{H_\infty(\mathcal{C}_{\text{hist},n})\}$.

Definition 1. *We say that a channel* $\mathcal{C}$ *is* memoryless, *if* $\mathcal{C}_{\text{hist},n} = \mathcal{C}_{\text{hist}',n}$ *for all* hist, hist$'$, *i.e. if the history has no effect on the channel distribution.*

Note the difference between memoryless and 0-*memoryless* channels of Lysyanskaya and Meyerovich [32], where only the *length* of the history has an influence on the channel, since the channel distributions are described by the use of *memoryless Markov chains*:

Definition 2 ([32]). *A channel* $\mathcal{C}$ *is* 0-memoryless, *if* $\mathcal{C}_{\text{hist},n} = \mathcal{C}_{\text{hist}',n}$ *for all* hist, hist$'$ *such that* $|\text{hist}| = |\text{hist}'|$.

A stegosystem PKStS tries to embed messages of length PKStS.ml into PKStS.ol documents of the channel $\mathcal{C}$ that each have size PKStS.dl, such that this sequence is indistinguishable from a sequence of typical documents. A *public-key stegosystem* PKStS with message length PKStS.ml: $\mathbb{N} \to \mathbb{N}$, document length PKStS.dl: $\mathbb{N} \to \mathbb{N}$, and output length PKStS.ol: $\mathbb{N} \to \mathbb{N}$ (all functions of the security parameter κ) is a triple of polynomial probabilistic Turing machines (PPTMs) [PKStS.Gen, PKStS.Enc, PKStS.Dec][1] with the functionalities:

- The *key generation* Gen on input 1^κ produces a pair (pk, sk) consisting of a *public key* pk and a *secret key* sk (we assume that sk also fully contains pk).
- The *encoding* algorithm Enc takes as input the public key pk, a message $m \in \{0,1\}^{\text{ml}(\kappa)}$, a history hist $\in (\Sigma^{\text{dl}(\kappa)})^*$ and some state information $s \in \{0,1\}^*$ and produces a document $d \in \Sigma^{\text{dl}(\kappa)}$ and state information $s' \in \{0,1\}^*$ by being able to sample from $\mathcal{C}_{\text{hist},\text{dl}(\kappa)}$. By $\text{Enc}^\mathcal{C}(pk, m, \text{hist})$, we denote the complete output of $\text{ol}(\kappa)$ documents one by one. Note that generally, the encoder needs to decide upon document d_i before it is able to get samples for the $(i+1)$-th document, as in the secret-key model of Hopper et al. [23, Sect. 2, "channel access"] and the public-key model of von Ahn and Hopper [38, 39, Sect. 3]. This captures the notion that an attacker should have as much information as possible while the stegosystem is not able to look-ahead into the future. To highlight this restriction, we call this model the *non-look-ahead model*. Note that this is no restriction for memoryless channels.
- The *decoding* algorithm Dec takes as input the secret key sk, a sequence of documents $d_1, \ldots, d_{\text{ol}(\kappa)}$, history hist and outputs a message m'.

The following properties are essential for stegosystems PKStS with output length $\ell = \text{PKStS.ol}(\kappa)$. It is *universal* (*black box*), if it works on every channel without

---

[1] We will drop the prefix PKStS if the context is clear.

prior knowledge of the probability distribution of the channel. Clearly channels with too small min-entropy (such as deterministic channels) are not suitable for steganographic purposes. We thus concentrate only on channels with sufficiently large min-entropy.

The system is *reliable* if the probability that the decoding fails is bounded by a negligible function. Formally, the *unreliability* $\mathbf{UnRel}_{\mathsf{PKStS},\mathcal{C}}(\kappa)$ is defined as probability that the decoding fails, i.e.

$$\max_{m,\mathsf{hist}}\{\Pr_{(pk,sk)\leftarrow\mathsf{PKStS.Gen}(1^\kappa)}[\mathsf{PKStS.Dec}(sk,\mathsf{PKStS.Enc}^{\mathcal{C}}(pk,m,\mathsf{hist}),\mathsf{hist})\neq m]\}.$$

The system PKStS is *secure*, if every polynomial attacker W (the *warden*) has only negligible success probability. W works in two phases: In the first phase (called W.Find), the warden has access to the channel $\mathcal{C}$ and to a *decoding oracle* $\mathsf{Dec}_{sk}(\cdot)$, that returns upon input $d_1,\ldots,d_\ell$ and hist the same result as $\mathsf{PKStS.Dec}(sk,(d_1,\ldots,d_\ell),\mathsf{hist})$. At the end of the first phase, the warden chooses a message m^* and a history hist^*.

At the beginning of the second phase (called W.Guess), the warden gets a sequence of documents $\boldsymbol{d}^* = d_1^*,\ldots,d_\ell^*$, which is with probability of 50% the result of $\mathsf{PKStS.Enc}^{\mathcal{C}}(pk,m^*,\mathsf{hist}^*)$ and with probability of 50% just the result of sampling ℓ random documents from $\mathcal{C}_{\mathsf{hist}^*,n}$, which we denote as $\mathcal{C}_{\mathsf{hist}^*,n}^{\ell}$. Speaking more precisely, this sampling is done as follows: $d_i^* \leftarrow \mathcal{C}_{\mathsf{hist}^*||d_1^*||\ldots||d_{i-1}^*,n}$ for $i = 1,\ldots,\ell$. Next, the warden should distinguish between these two cases by having access to another *decoding oracle*. Depending on the oracle type, two definitions for security were proposed by Backes and Cachin in [3].

In the SS-CCA-model (chosen covertext attack), the decoding oracle, denoted as $\mathsf{Dec}_{sk,\boldsymbol{d}^*}(\cdot)$, works like $\mathsf{Dec}_{sk}(\cdot)$ on every input different from $\boldsymbol{d}^*$. If $\mathsf{Dec}_{sk,\boldsymbol{d}^*}(\cdot)$ is called with input $\boldsymbol{d}^*$, it simply returns $\perp$. In the weaker SS-RCCA-model (restricted chosen ciphertext attack), the decoding oracle, denoted as $\mathsf{Dec}_{sk,\boldsymbol{d}^*,m^*}(\cdot)$, works like $\mathsf{Dec}_{sk}(\cdot)$ on most inputs: If the input d equals $\boldsymbol{d}^*$ or is a valid encoding of m^* (a *replay* of $\boldsymbol{d}^*$), the oracle simply returns $\perp$.

Formally, SS-CCA-security is defined by the SS-CCA-security game given below and the advantage of W = [W.Find, W.Guess] is defined as

$$\mathbf{Adv}_{\mathsf{W,PKStS},\mathcal{C}}^{\mathrm{ss-cca}}(\kappa) = \left|\Pr[\text{SS-CCA-Dist}(\mathsf{W},\mathsf{PKStS},\mathcal{C},\kappa) = 1] - \frac{1}{2}\right|.$$

SS-CCA-security game: SS-CCA-Dist(W, PKStS, $\mathcal{C}$, κ)

Input: warden W, stegosystem PKStS, channel $\mathcal{C}$, security parameter κ
1: $(pk,sk) \leftarrow \mathsf{PKStS.Gen}(1^\kappa)$; $(m^*,\mathsf{hist}^*,s) \leftarrow \mathsf{W.Find}^{\mathsf{Dec}_{sk},\mathcal{C}}(pk)$
2: $b \leftarrow \{0,1\}$
3: **if** $b = 0$ **then** $\boldsymbol{d}^* \leftarrow \mathsf{PKStS.Enc}^{\mathcal{C}}(pk,m^*,\mathsf{hist}^*)$ **else** $\boldsymbol{d}^* \leftarrow \mathcal{C}_{\mathsf{hist}^*,n}^{\ell}$
4: $b' \leftarrow \mathsf{W.Guess}^{\mathsf{Dec}_{sk,\boldsymbol{d}^*},\mathcal{C}}(pk,m^*,\mathsf{hist}^*,s,\boldsymbol{d}^*)$
5: **if** $b' = b$ **then return** 1 **else return** 0

A stegosystem PKStS is called SS-CCA-secure against channel $\mathcal{C}$ if for some negligible function negl and all wardens W, we have $\mathbf{Adv}^{ss\text{-}cca}_{W,PKStS,\mathcal{C}}(\kappa) \leq negl(\kappa)$. We define SS-RCCA-security analogously, where the Guess phase uses Dec_{sk,d^*,m^*} as decoding oracle. Formally, a stegosystem is *universally SS-CCA-secure* (or just universal), if it is SS-CCA-secure against all channels of sufficiently large (i.e. super-logarithmic in κ) min-entropy.

Cryptographic Primitives. Due to space constraints, we only give informal definitions of the used cryptographic primitives and refer the reader to the textbook of Katz and Lindell [24] for complete definitions.

We will make use of different cryptographic primitives, namely hash functions, pseudorandom permutations and CCA-secure cryptosystems. A *collision-resistant hash function (CRHF)* H = (H.Gen, H.Eval) is a pair of PPTMs such that H.Gen upon input 1^κ produces a key $k \in \{0,1\}^\kappa$. The keyed function H.Eval takes the key $k \leftarrow H.Gen(1^\kappa)$ and a string $x \in \{0,1\}^{H.in(\kappa)}$ and produces a string $H.Eval_k(x)$ of length $H.out(\kappa) < H.in(\kappa)$. The probability of every PPTM Fi to find a collision – two strings $x \neq x'$ such that $H.Eval_k(x) = H.Eval_k(x')$ – upon random choice of k is negligible. For a set X, denote by $Perms(X)$ the set of all permutations on X. A *pseudorandom permutation (PRP)* P = (P.Gen, P.Eval) is a pair of PPTMs such that P.Gen upon input 1^κ produces a key $k \in \{0,1\}^\kappa$. The keyed function P.Eval takes the key $k \leftarrow P.Gen(1^\kappa)$ and is a permutation on the set $\{0,1\}^{P.in(\kappa)}$. An attacker Dist (the *distinguisher*) is given black-box access to $P \leftarrow Perms(\{0,1\}^{P.in(\kappa)})$ or to $P.Eval_k$ for a randomly chosen k and should distinguish between those scenarios. The success probability of every Dist is negligible. A *public key encryption scheme (PKES)* PKES = (PKES.Gen, PKES.Enc, PKES.Dec) is a triple of PPTMs such that PKES.Gen(1^κ) produces a pair of keys (pk, sk) with $|pk| = \kappa$ and $|sk| = \kappa$. The key pk is called the *public key* and the key sk is called the *secret key* (or *private key*). The *encryption algorithm* PKES.Enc takes as input pk and a plaintext $m \in \{0,1\}^{PKES.ml(\kappa)}$ of length PKES.ml(κ) and outputs a ciphertext $c \in \{0,1\}^{PKES.cl(\kappa)}$ of length PKES.cl(κ). The *decryption algorithm* PKES.Dec takes as input sk and the ciphertext c and produces a plaintext $m \in \{0,1\}^{PKES.ml(\kappa)}$. Informally, we will allow an attacker A to first choose a message m^* that should be encrypted and denote this by A.Find. In the next step (A.Guess), the attacker gets c^*, which is either $Enc(pk, m^*)$ or a random bitstring. He is allowed to decrypt ciphertexts different from c^* and his task is to distinguish between these two cases. This security notion is known as security against chosen-ciphertext\$ attack (CCA\$s). For an attacker A on cryptographic primitive $\Pi \in \{hash, prp, pkes\}$ with implementation X, we write $\mathbf{Adv}^{\Pi}_{A,X,\mathcal{C}}(\kappa)$ for the success probability of A against X relative to channel $\mathcal{C}$, i.e. the attacker A also has access to a sampling oracle of $\mathcal{C}$. In case of encryption schemes, the superscript cca\$ is used instead of pkes.

Due to the works [16,18,31,34] we know that CCA\$-secure cryptosystems and PRPs can be constructed from doubly-enhanced trapdoor permutations resp. one-way functions, while CRHFs can not be constructed from them in a *black-box* way, as Simon showed an oracle-separation in [37].

3   Detecting the Scheme of Backes and Cachin

In order to understand the difference between SS-CCA-security and the closely related, but weaker, SS-RCCA-security, we give a short presentation of the universal SS-RCCA-stegosystem of Backes and Cachin [3]. We also show that their system is not SS-CCA-secure, which was already noted by Hopper in [21]. The proof of insecurity nicely illustrates the difference between the security models. It also highlights the main difficulty of SS-CCA-security: One needs to prevent so called *replay attacks*, where the warden constructs upon stegotext c another stegotext c' – the *replay* of c – that embeds the same message as c.

Backes and Cachin [3] showed that there is a universal SS-RCCA-secure stegosystem under the assumption that a replayable chosen-covertext\$ (RCCA\$)-secure cryptosystem exists.[2] They make use of a technique called *rejection sampling*. Let $\{G_\kappa\}_{\kappa \in \mathbb{N}}$ be a strongly 2-universal hash function family, $f \in G_\kappa$ a function, $\mathcal{C}$ be a channel, hist be a history and $b \in \{0,1\}$ be a bit. The algorithm $\mathsf{rejsam}(f, \mathcal{C}, b, \mathsf{hist})$ samples documents $d \leftarrow \mathcal{C}_{\mathsf{hist},\mathsf{dl}(\kappa)}$ until it finds a document d^* such that $f(d^*) = b$ or until it has sampled κ documents. If PKES is an RCCA\$-secure cryptosystem, they define a stegosystem that computes $(b_1, \ldots, b_\ell) \leftarrow \mathsf{PKES.Enc}(pk, m)$ and then sends $d_1, d_2, \ldots, d_\ell$, where $d_i \leftarrow \mathsf{rejsam}(f, \mathcal{C}, b_i, \mathsf{hist}||d_1|| \ldots ||d_{i-1})$. The function $f \in G_\kappa$ is also part of the public key. The system is universal as it does not assume any knowledge on $\mathcal{C}$.

They then prove that this stegosystem is SS-RCCA-secure. And indeed, one can show that their stegosystem is not SS-CCA-secure by constructing a generic warden W that works as follows: The first phase W.Find chooses as message $m^* = 00 \cdots 0$ and as hist* the empty history $\varnothing$. The second phase W.Guess gets $\boldsymbol{d}^* = d_1^*, \ldots, d_\ell^*$ which is either a sequence of random documents or the output of the stegosystem on pk, m^*, and hist*. The warden W now computes another document d' via rejection sampling that embedds $f(d_\ell^*)$ (the *replay* of $\boldsymbol{d}^*$) and decodes $d_1^*, \ldots, d_{\ell-1}^*, d'$ via the decoder of the rejection sampling stegosystem. It then returns 0 if the returned message m' consists only of zeroes. If $\boldsymbol{d}^*$ was a sequence of random documents, it is highly unlikely that $\boldsymbol{d}^*$ decodes to a message that only consists of zeroes. If $\boldsymbol{d}^*$ was produced by the stegosystem, the decoder only returns something different from the all-zero-message if $d' = d_\ell^*$ which is highly unlikely. The warden W has advantage of $1 - \mathsf{negl}(\kappa)$ and the stegosystem is thus not SS-CCA-secure. Backes and Cachin posed the question whether a universal SS-CCA-secure stegosystem exists.

4   An High-Level View of Our Stegosystem

The stegosystem of Backes and Cachin only achieves SS-RCCA-security as a single ciphertext has many different possible encodings in terms of the documents used. Hopper achieves SS-CCA-security by limiting those encodings: Due to

---

[2] The definition of a RCCA\$-secure cryptosystem is analogous to SS-RCCA-security given in Sect. 2.

the sampleability of the channel, each ciphertext has exactly one deterministic encoding in terms of the documents. While Hopper achieves SS-CCA-security, he needs to give up the universality of the stegosystem, as a description of the channel is hard-wired into the stegosystem. In order to handle as many channels as possible, we will allow many different encodings of the same ciphertext, but make it hard to find them for anyone but the stegoencoder. To simplify the presentation, we focus on the case of embedding a single bit per document. Straightforward modifications allow embedding of $\log(\kappa)$ bits.

Our stegosystem, named PKStS* will use the following approach to encode a message m: It first samples, for sufficiently large N, a set D of N documents from the channel C and uses a strongly 2-universal hash function $f \in G_\kappa$ to split these documents into documents D_0 that encode bit 0 (i.e. $D_0 = \{d \in D \mid f(d) = 0\}$) and D_1 that encode bit 1 (i.e. $D_1 = \{d \in D \mid f(d) = 1\}$). Now we encrypt the message m via a certain public-key encryption system, named PKES* (described in the next section), and obtain a ciphertext $b = b_1, \ldots, b_L$ of length $L = \lfloor N/8 \rfloor$. Next our goal is to order the documents in D into a sequence $d = d_1, \ldots, d_N$ such that the first L documents $d_1, \ldots, d_L$ encode b (i.e. $f(d)_i = b_i$). This ordering is performed by the algorithm generate. However, the attacker still has several possibilities for a replay attack on this scheme, for example:

- He could exchange some document d_i by another document d_i' with $f(d_i) = f(d_i')$ (as f is publicly known) and the sequence $d_1, \ldots, d_{i-1}, d_i', d_{i+1}, \ldots, d_N$ would be a replay of d. Such attacks will be called *sampling attacks*. To prevent the attacker from exchanging a sampled document by a non-sampled one, we also encode a hash-value of all sampled documents D and transmit this hash value to Bob.
- The attacker can exchange documents d_i and d_j, with $i < j$ and $f(d_i) = f(d_j)$, and the resulting sequence $d_1, \ldots, d_{i-1}, d_j, d_{i+1}, \ldots, d_{j-1}, d_i, d_{j+1}, \ldots, d_N$ would be a replay of d. Such attacks will be called *ordering attacks*. We thus need to prevent the attacker from exchanging the positions of sampled documents. We achieve this by making sure that the ordering of the documents generated by generate is deterministic, i.e. for each set of documents D and each ciphertext b, the ordering d generated by generate is deterministic. This property is achieved by using PRPs to sort the sampled documents D. The corresponding keys of the PRPs are also transmitted to Bob and the stegodecoder can thus also compute this deterministic ordering.

In total, our stegoencoder PKStS*.Enc works on a secret message m and on a publicly known hash-function f as follows:

1. Sample N documents D from the channel;
2. Get a hash-key k_{H} and compute a hash-value $h = \mathsf{H.Eval}_{k_{\mathsf{H}}}(\mathrm{lex}(D))$ of the sampled documents, where $\mathrm{lex}(D)$ denotes the sequence of elements of D in lexicographic order. This prevents sampling attacks, where a sampled document is replaced by a non-sampled one;

3. Get two[3] PRP-keys k_P and k_P' that will be used to determine the unique ordering of the documents in D via generate. This prevents ordering attacks, where the order of the sampled documents is switched;
4. Encrypt the concatenation of m, k_H, k_P, k_P', h via a certain public key encryption scheme PKES* and obtain the ciphertext b of length $L = \lfloor N/8 \rfloor$. As long as PKES* is secure, the stegodecoder is thus able to verify whether all sampled documents were sent and can also verify the ordering of the documents.
5. Compute the ordering d of the documents D via generate that uses the PRP keys k_P and k_P' to determine the ordering of the documents. It also uses the ciphertext b to guarantee that the first L send documents encode the ciphertext b, i.e. $b_1 \ldots b_L = f(d_1) \ldots f(d_L)$;
6. Send the ordering of the documents d.

To decode a sequence of documents $d = d_1, \ldots, d_N$, the stegodecoder of PKStS* computes the ciphertext $b_1 = f(d_1), \ldots, b_L = f(d_L)$ encoded in the first L documents of d. It then decodes this ciphertext $b_1 \ldots b_L$ via PKES* to obtain the message m, the PRP keys k_P and k_P', the hash-key k_H and the hash-value h. First it verifies the hash-value by checking whether $\text{H.Eval}_{k_H}(\text{lex}(\{d_1, \ldots, d_N\}))$ equals the hash-value h to prevent sampling attacks. It then uses the PRP keys k_P and k_P', to compute an ordering of the received documents via generate to verify that no ordering attack was used. If these validations are successful, the decoder PKStS*.Dec returns m; Otherwise, it concludes that d is not a valid stegotext and returns $\bot$.

Intuitively, it is clear that a successful sampling attack on this scheme would break the collision-resistant hash function H, as it needs to create a collision of $\text{lex}(D)$ in order to pass the first verification step. Furthermore, a successful ordering attack would need manipulate the ciphertext b and thus break the security of the public key encryption scheme PKES*, as the PRP keys k_P and k_P' guarantee a deterministic ordering of the documents.

As explained above, our stegoencoder computes the ordering $d = d_1, \ldots, d_N$ of the documents $D = \{d_1, \ldots, d_N\}$ via the deterministic algorithm generate, that is given the following parameters: the set of documents D, the hash-function f and the ciphertext b to ensure that the first documents of the ordering encode b. It has furthermore access to the PRP keys k_P and k_P' that guarantee a deterministic ordering of the documents in D and thus prevents ordering attacks. As the ordering d produced by generate is sent by the stegoencoder, this ordering must be indistinguishable from a random permutation on D (which equals the channel distribution) in order to be undetectable. As $f(d_1) = b_1, \ldots, f(d_L) = b_L$, not every distribution upon the ciphertext b can be used to guarantee that d is indistinguishable from a uniformly random permutation. This indistinguishability is guaranteed by requiring that the ciphertext b is distributed according to a certain distribution corresponding to a random process modeled by drawing black and white balls from an urn without replacement. In our setting, the

---

[3] We believe that one permutation suffices. But in order to improve the readability of the proof for security, we use two permutations in our stegosystem.

documents in D will play the role of the balls and the coloring is given by the function f.

Section 5 describes this random process in detail and proves that we can indeed construct a public-key encryption system that produces ciphertexts that are indistinguishable from this process. Section 6 contains a formal description of generate, proves that no attacker can produce a replay of its output and shows that the generated permutation is indeed indistinguishable from a random permutation. Finally, Sect. 7 contains the complete description of the stegosystem.

5 Obtaining Biased Ciphertexts

We will now describe a probability distribution and show how one can derive a symmetric encryption scheme with ciphertexts that are indistinguishable from this distribution. In order to do this, we first define a channel that represents the required probability distribution together with appropriate parameters, use Theorem 3 to derive a stegosystem for this channel, and finally derive a cryptosystem from this stegosystem.

Based upon a CCA$-secure public-key cryptosystem PKES, Hopper [21] constructs for every efficiently sampleable channel C an SS-CCA-secure stegosystem PKStS$_C$ by "derandomizing" the rejection sampling algorithm. The only requirement upon the channel C is the existence of the efficient sampling algorithm and that the stegoencoder and the stegodecoder use the same sampling algorithm. Importantly, due to the efficient sampleability of C, the encoder of PKStS$_C$ does not need an access to the sample oracle. Thus, we get the following result.

Theorem 3 (Theorem 2 in [21]). *If C is an efficiently sampleable channel and* PKES *is a CCA$-secure public-key cryptosystem (which can be constructed from doubly enhanced trapdoor permutations[4]) then there is a stegosystem* PKStS$_C$ *(without an access to the sample oracle) such that for all wardens* W *there is a negligible function* negl *such that*

$$\mathbf{Adv}^{\text{ss-cca}}_{\text{W,PKStS}_C, C}(\kappa) \leq \text{negl}(\kappa) + 2^{-H_\infty(C,\kappa)/2}.$$

Note that the system PKStS$_C$ is guaranteed to be secure (under the assumption that CCA$-secure public-key cryptosystems exist), if the channel C is efficiently sampleable and has min-entropy $\omega(\log \kappa)$. We call such a channel *suitable*.

The probability distribution for the ciphertexts we are interested in is the distribution for the bitstrings b we announced in the previous section. As we will see later, the required probability can be described equivalently as follows:

– We are given N elements: N_0 of them are labeled with 0 and the remaining $N - N_0$ elements are labeled with 1.
– We draw randomly a sequence of K elements from the set (drawing without replacements) and look at the generated bitstring $b = b_1 \ldots b_K$ of length K determined by the labels of the elements.

---

[4] See e.g. the work [18] of Goldreich and Rothblum.

We will assume that there are enough elements of both types, i.e. that $N_0 \geq K$ and $N - N_0 \geq K$. The resulting probability distribution, denoted as $D^*_{(N,N_0,K)}$, upon bitstrings of length K is then given as

$$\Pr[D^*_{(N,N_0,K)} = b_1 \ldots b_K] = \frac{1}{\binom{K}{|b|_0}} \cdot \frac{\binom{N_0}{|b|_0} \cdot \binom{N-N_0}{K-|b|_0}}{\binom{N}{K}}$$

$$= \left(\prod_{j=0}^{K-1} \frac{1}{N-j} \right) \cdot \left(\prod_{j=0}^{|b|_0-1} [N_0 - j] \right) \cdot \left(\prod_{j=0}^{|b|_1-1} [N - N_0 - j] \right),$$

(1)

where $|b|_0$ denotes the number of zero bits in $b = b_1, \ldots, b_K$ and $|b|_1$ the number of one bits in b. Note that the distribution on the number of zeroes within such bitstrings is a hypergeometric distribution with parameters N, N_0, and K.

Now we will construct a channel C^* upon key parameter κ with document length $n = \mathsf{dl}(\kappa) = \kappa$. In the definition below, $\mathsf{bin}(x)_y$ denotes the binary representation of length exactly y for the integer x.

- For the empty history $\varnothing$, let $C^*_{\varnothing,\kappa}$ be the uniform distribution on all strings $\mathsf{bin}(N)_{\lceil \kappa/2 \rceil} \mathsf{bin}(N_0)_{\lfloor \kappa/2 \rfloor}$ that range over all positive integers $N, N_0 \leq 2^{\lfloor \kappa/2 \rfloor}$ such that $N \geq 8\kappa$ and $1/3 \leq N_0/N \leq 2/3$ (in our construction we need initially a stronger condition than just $N_0 \geq \kappa$ and $N - N_0 \geq \kappa$).
- If the history is of the form $\mathsf{hist}' = \mathsf{bin}(N)_{\lceil \kappa/2 \rceil} \mathsf{bin}(N_0)_{\lfloor \kappa/2 \rfloor} \mathsf{hist}$ for some $\mathsf{hist} \in \{0,1\}^*$ then we consider two cases: if $|\mathsf{hist}| \leq \frac{1}{8} N$ then the distribution $C^*_{\mathsf{hist}',\kappa}$ equals $D^*_{(N-|\mathsf{hist}|, N_0-|\mathsf{hist}|_0, \kappa)}$; Otherwise, i.e. if $|\mathsf{hist}| > \frac{1}{8} N$ then $C^*_{\mathsf{hist}',\kappa}$ equals the uniform distribution over $\{0,1\}^\kappa$.

It is easy to see that the min-entropy $H_\infty(C^*, n) = \min_{\mathsf{hist}'} \{H_\infty(C^*_{\mathsf{hist}',n})\}$ of the channel C^* is obtained for the history $\mathsf{hist}' = \mathsf{bin}(N)_{\lceil \kappa/2 \rceil} \mathsf{bin}(N_0)_{\lfloor \kappa/2 \rfloor} \mathsf{hist}$, with $8\kappa \leq N \leq 2^{\lfloor \kappa/2 \rfloor}$ and such that (i) $N_0 = \frac{1}{3} N$ and $\mathsf{hist} = 00 \ldots 0$ of length $\frac{1}{8} N - \kappa$ or (ii) $N_0 = \frac{2}{3} N$ and $\mathsf{hist} = 11 \ldots 1$ of length $\frac{1}{8} N - \kappa$. In the first case we get that the min-entropy of the distribution $C^*_{\mathsf{hist}',n}$ is achieved on the bitstring $11 \ldots 1$ of length κ and in the second case on $00 \ldots 0$ of length κ. By Eq. (1) the probabilities to get such strings are equal to each other and, since $\kappa \leq N/8$, they can be estimated as follows:

$$\prod_{j=0}^{\kappa-1} \frac{2N/3 - j}{7N/8 - \kappa - j} \leq \left(\frac{2N/3}{7N/8 - \kappa} \right)^\kappa \leq \left(\frac{2N/3}{6N/8} \right)^\kappa = (8/9)^\kappa.$$

Thus, we get that $H_\infty(C^*, n) \geq \kappa \log(9/8)$.

Moreover one can efficiently simulate the choice of N, N_0, the sampling process of $D^*_{(N,N_0,\kappa)}$ and the uniform sampling in $\{0,1\}^\kappa$. Therefore we can conclude

Lemma 4. *The channel C^* is suitable, i.e. it is efficiently sampleable and has min-entropy $\omega(\log \kappa)$. Furthermore, for history $\mathsf{hist} = \mathsf{bin}(N)_{\lceil \kappa/2 \rceil} \mathsf{bin}(N_0)_{\lfloor \kappa/2 \rfloor}$, with $8\kappa \leq N \leq 2^{\lceil \kappa/2 \rceil}$ and $1/3 \leq N_0/N \leq 2/3$, and for any integer $\ell \leq \frac{N}{8\kappa}$, the* bitstrings $\boldsymbol{b} = b_1 \ldots b_K$ of length $K = \kappa \cdot \ell \leq N/8$ obtained by the concatenation of ℓ consecutive documents sampled from the channel with history hist, i.e. $b_i \leftarrow \mathcal{C}^*_{\text{hist}b_1 \ldots b_{i-1}, n=\kappa}$, have distribution $D^*_{(N,N_0,K)}$.

A proof for the second statement of the lemma follows directly from the construction of the channel. Now, combining the first claim of the lemma with Theorem 3 we get the following corollary.

Corollary 5. *If doubly enhanced trapdoor permutations exists, there is a stegosystem* $\mathsf{PKStS}_{\mathcal{C}^*}$ *(without an access to the sample oracle) such that for all wardens* W *there is a negligible function* negl *such that* $\mathbf{Adv}^{\text{ss-cca}}_{\mathsf{W},\mathsf{PKStS}_{\mathcal{C}^*},\mathcal{C}^*}(\kappa) \leq \mathsf{negl}(\kappa)$.

Based upon this stegosystem $\mathsf{PKStS} = \mathsf{PKStS}_{\mathcal{C}^*}$, we construct a public-key cryptosystem PKES^*, with ciphertexts of length $\mathsf{PKES}^*.\mathsf{cl}(\kappa) = \kappa \cdot \mathsf{PKStS}.\mathsf{cl}(\kappa)$ such that PKES^* also has another algorithm, called $\mathsf{PKES}^*.\mathsf{Setup}$ that takes parameters: two integers N and N_0 which satisfy $8 \cdot \mathsf{PKES}^*.\mathsf{cl}(\kappa) \leq N \leq 2^{\lfloor \kappa/2 \rfloor}$ and $N_0/N \in [1/3, 2/3]$. Calling $\mathsf{PKES}^*.\mathsf{Setup}(N, N_0)$ stores the values N, N_0 such that $\mathsf{PKES}^*.\mathsf{Enc}$ and $\mathsf{PKES}^*.\mathsf{Dec}$ can use them.

- The key generation $\mathsf{PKES}^*.\mathsf{Gen}$ simply equals the key generation algorithm $\mathsf{PKStS}.\mathsf{Gen}$.
- The encoding algorithm $\mathsf{PKES}^*.\mathsf{Enc}$ takes as parameters the public key pk and a message m. It then simulates the encoder $\mathsf{PKStS}.\mathsf{Enc}$ on key pk, message m and history $\mathsf{hist} = \mathrm{bin}(N)_{\lceil \kappa/2 \rceil} \mathrm{bin}(N_0)_{\lfloor \kappa/2 \rfloor}$ and produces a bitstring of length $\mathsf{PKES}^*.\mathsf{cl}(\kappa) = \mathsf{PKStS}.\mathsf{ol}(\kappa) \cdot \kappa$.
- The decoder $\mathsf{PKES}^*.\mathsf{Dec}$ simply inverts this process by simulating the stegodecoder $\mathsf{PKStS}.\mathsf{Dec}$ on key sk and history $\mathsf{hist} = \mathrm{bin}(N)_{\lceil \kappa/2 \rceil} \mathrm{bin}(N_0)_{\lfloor \kappa/2 \rfloor}$.

Clearly, the ciphertexts of $\mathsf{PKES}^*.\mathsf{Enc}(pk, m)$ are indistinguishable from the distribution $D^*_{(N,N_0,\mathsf{PKES}^*.\mathsf{cl}(\kappa))}$ by the second statement of Lemma 4. This generalization of Theorem 3 yields the following corollary:

Corollary 6. *If doubly-enhanced trapdoor permutations exist, there is a secure public-key cryptosystem* PKES^*, *equipped with the algorithm* $\mathsf{PKES}^*.\mathsf{Setup}$ *that takes two parameters* N *and* N_0, *such that its ciphertexts are indistinguishable from the probability distribution* $D^*_{(N,N_0,\mathsf{PKES}^*.\mathsf{cl}(\kappa))}$ *whenever* N *and* N_0 *satisfy that* $8 \cdot \mathsf{PKES}^*.\mathsf{cl}(\kappa) \leq N \leq 2^{\lfloor \kappa/2 \rfloor}$ *and* $N_0/N \in [1/3, 2/3]$.

6   Ordering the Documents

As described before, to prevent replay attacks, we need to order the sampled documents. This is done via the algorithm generate described in this section. To improve the readability, we will abbreviate some terms and define $L = \mathsf{PKES}^*.\mathsf{cl}(\kappa)$ and $n = \mathsf{PKStS}^*.\mathsf{dl}(\kappa)$, where PKES^* is the public-key encryption scheme from the last section and PKStS^* is our target stegosystem that we will provide later on. We also define $N = 8L$.

To order the set of documents $D \subseteq \Sigma^n$, we use the algorithm **generate**, presented below. It takes the set of documents D with $|D| = N$, a hash function $f \colon \Sigma^n \to \{0,1\}$ from G_κ, a bitstring $b_1, \ldots, b_L$, and two keys $k_\mathsf{P}, k_\mathsf{P}'$ for PRPs. It then uses the PRPs to find the right order of the documents.

Algorithm: generate$(D, f, b_1, \ldots, b_L, k_\mathsf{P}, k_\mathsf{P}')$

Input: set D with $|D| = N$, hash function f, bits $b_1, \ldots, b_L$, PRP-keys $k_\mathsf{P}, k_\mathsf{P}'$
1: let $D_0 = \{d \in D \mid f(d) = 0\}$ and $D_1 = \{d \in D \mid f(d) = 1\}$ ▷ We assert that
$\quad$ $|D| = N$, and furthermore $|D_0| \in [N/3, 2N/3]$
2: **for** $i = 1$ to L **do**
3: $\quad$ $d_i := \arg\min_{d \in D_{b_i}}\{\mathsf{P.Eval}_{k_\mathsf{P}}(d)\}$; $D_{b_i} := D_{b_i} \setminus \{d_i\}$
4: let $D' = D_0 \cup D_1$ $\qquad\qquad\qquad\qquad\qquad$ ▷ collect remaining documents
5: **for** $i = L+1, \ldots, N$ **do**
6: $\quad$ $d_i := \arg\min_{d \in D'}\{\mathsf{P.Eval}_{k_\mathsf{P}'}(d)\}$; $D' := D' \setminus \{d_i\}$
7: **return** $d_1, d_2, \ldots, d_N$

Note that the permutation $\mathsf{P.Eval}_{k_\mathsf{P}}$ is a permutation upon the set $\{0,1\}^n$ (i.e. on the documents themselves) and the canonical ordering of $\{0,1\}^n$ thus implicitly gives us an ordering of the documents.

We note the following important property of **generate** that shows where the urn model of the previous section comes into play. For uniform random permutations P and P', we denote by $\mathsf{generate}(\cdots, P, P')$ the run of **generate**, where the use of $\mathsf{P.Eval}_{k_\mathsf{P}}$ is replaced by P and the use of $\mathsf{P.Eval}_{k_\mathsf{P}'}$ is replaced by P'. If the bits $\boldsymbol{b} = b_1, \ldots, b_L$ are distributed according to $D^*_{(N, |D_0|, L)}$, the resulting distribution on the documents then equals the channel distribution.

Lemma 7. *Let $\mathcal{C}$ be any memoryless channel, f be some hash function and D be a set of $N = 8L$ documents of $\mathcal{C}$ such that $N/3 \leq |D_0| \leq 2N/3$, where $D_0 = \{d \in D \mid f(d) = 0\}$. If the permutations P, P' are uniformly random and the bitstring $\boldsymbol{b} = b_1, \ldots, b_L$ is distributed according to $D^*_{(N, |D_0|, L)}$, the output of $\mathsf{generate}(D, f, \boldsymbol{b}, P, P')$ is a uniformly random permutation of D.*

Proof. Fix any document set D of size $N = 8L$ and a function f that splits D into $D_0 \dot\cup D_1$, with $|D_0| \geq N/3$ and $|D_1| \geq N/3$. Let $\hat{\boldsymbol{d}} = \hat{d}_1, \ldots, \hat{d}_N$ be any permutation on D. We will prove that the probability (upon bits $\boldsymbol{b}$ and permutations P, P') that $\hat{\boldsymbol{d}}$ is produced, is $1/N!$ and thus establish the result. Let $\boldsymbol{d} = d_1, \ldots, d_N$ be the random variables that denote the outcome of $\mathsf{generate}(D, f, b_1, \ldots, b_L, P, P')$.

Note that if $\boldsymbol{d}[i]$ (resp. $\hat{\boldsymbol{d}}[i]$) denotes the prefix of length i of $\boldsymbol{d}$ (resp. $\hat{\boldsymbol{d}}$), then using the chain rule formula we get

$$\Pr_{\boldsymbol{b}, P, P'}[d_1 d_2 \ldots d_N = \hat{d}_1 \hat{d}_2 \ldots \hat{d}_N] = \prod_{i=1}^{N} \Pr_{\boldsymbol{b}, P, P'}[d_i = \hat{d}_i \mid \boldsymbol{d}[i-1] = \hat{\boldsymbol{d}}[i-1]].$$

To estimate each of the factors of the product, we consider two cases:

– Case $i \leq L$: Let $\hat{\boldsymbol{b}} = \hat{b}_1, \ldots, \hat{b}_L$ be the bitstring such that $\hat{b}_i = f(\hat{d}_i)$ and let $\hat{\boldsymbol{b}}[i]$ be the prefix $\hat{b}_1, \ldots, \hat{b}_i$ of $\hat{\boldsymbol{b}}$ of length i. Clearly, for $i \leq L$ it holds that the event $d_i = \hat{d}_i$ under the condition $\boldsymbol{d}[i-1] = \hat{\boldsymbol{d}}[i-1]$ occurs iff (A) $d_i \in D_{\hat{b}_i}$ and (B) d_i is put on position $|\hat{\boldsymbol{b}}[i]|_{\hat{b}_i}$ by the permutation P with respect to $D_{\hat{b}_i}$. Due to the distribution of bit b_i in the random bits $\boldsymbol{b}$, the event $d_i \in D_{\hat{b}_i}$ occurs with probability $(|D_{\hat{b}_i}| - |\hat{\boldsymbol{b}}[i-1]|_{\hat{b}_i})/(N-i+1)$ (under the above condition). As $\boldsymbol{d}[i-1] = \hat{\boldsymbol{d}}[i-1]$ holds, exactly $|\hat{\boldsymbol{b}}[i-1]|_{\hat{b}_i}$ documents from $D_{\hat{b}_i}$ are already used in the output. As P is a uniform random permutation, the probability that d_i is put on position $|\hat{\boldsymbol{b}}[i]|_{\hat{b}_i}$ by the permutation P (with respect to $D_{\hat{b}_i}$) is thus $1/(|D_{\hat{b}_i}| - |\hat{\boldsymbol{b}}[i-1]|_{\hat{b}_i})$. Since (A) and (B) are independent, we conclude for $i \leq L$ that the probability $\mathrm{Pr}_{\boldsymbol{b},P,P'}[d_i = \hat{d}_i \mid \boldsymbol{d}[i-1] = \hat{\boldsymbol{d}}[i-1]]$ is equal to

$$\mathrm{Pr}_{\boldsymbol{b}}[d_i \in D_{\hat{b}_i} \mid \boldsymbol{d}[i-1] = \hat{\boldsymbol{d}}[i-1]] \times$$
$$\mathrm{Pr}_P[P \text{ puts } d_i \text{ on position } |\hat{\boldsymbol{b}}[i]|_{\hat{b}_i} \mid \boldsymbol{d}[i-1] = \hat{\boldsymbol{d}}[i-1]]$$
$$= \frac{|D_{\hat{b}_i}| - |\hat{\boldsymbol{b}}[i-1]|_{\hat{b}_i}}{N-i+1} \cdot \frac{1}{|D_{\hat{b}_i}| - |\hat{\boldsymbol{b}}[i-1]|_{\hat{b}_i}} = \frac{1}{N-i+1}.$$

– Case $i > L$: As the choice of P' is independent from the choice of P, the remaining $2L$ items are ordered completely random. Hence, for $i > L$ we also have

$$\mathrm{Pr}_{\boldsymbol{b},P,P'}[d_i = \hat{d}_i \mid \boldsymbol{d}[i-1] = \hat{\boldsymbol{d}}[i-1]] = \frac{1}{N-i+1}.$$

Putting it together, we get

$$\mathrm{Pr}_{\boldsymbol{b},P,P'}[d_1 d_2 \ldots d_N = \hat{d}_1 \hat{d}_2 \ldots \hat{d}_N] = \prod_{i=1}^{N} \frac{1}{N-i+1} = \frac{1}{N!}. \qquad \square$$

As explained above, a second property that we need is that no attacker should be able to produce a "replay" of the output of generate. Below, we formalize this notion and analyze the security of the algorithm. An attacker A on generate is a PPTM, that receives nearly the same input as generate: a set D of N documents, a hash function $f: \Sigma^n \to \{0,1\}$ from the family G_κ, a sequence $b_1, \ldots, b_L$ of L bits, and a key k_H for the CRHF H. Then A outputs a sequence $d'_1, \ldots, d'_N$ of documents. We say that the algorithm A is *successful* if

1. $f(d_i) = f(d'_i)$ for all $i = 1, \ldots, N$,
2. $d'_1, \ldots, d'_N = \mathsf{generate}(D', f, b_1, \ldots, b_L, k_\mathsf{P}, k'_\mathsf{P})$, and
3. $\mathsf{H.Eval}_{k_\mathsf{H}}(\mathrm{lex}(D')) = \mathsf{H.Eval}_{k_\mathsf{H}}(\mathrm{lex}(D))$.

where D' denotes the set $\{d'_1, \ldots, d'_N\}$ and, recall, $\mathrm{lex}(X)$ denotes the sequence of elements of set X in lexicographic order. We can then conclude the following lemma.

Lemma 8 (Informal). *Let $D \subseteq \Sigma^n$ be a set of documents with $|D| = N$, let $b_1, \ldots, b_L$ be a bitstring, and $f \in G_\kappa$. For every attacker A on generate, there is a collision finder Fi for the CRHF H such that the probability that A is successful on $D, f, b_1, \ldots, b_L, k_H$ is bounded by $\mathbf{Adv}_{Fi,H,C}^{hash}(\kappa)$.*

The formal definition of "A is successful" as well as a formal statement of the lemma can be found in Appendix A.

7   The Steganographic Protocol PKStS*

We now have all of the ingredients of our stegosystem, namely the CCA-secure cryptosystem PKES* from Sect. 5 and the ordering algorithm generate from Sect. 6. To improve the readability, we will abbreviate some terms and define $n = \mathsf{PKStS^*.dl}(\kappa)$, $\ell = \mathsf{PKStS^*.ol}(\kappa)$, and $L = \mathsf{PKES^*.cl}(\kappa)$, where PKES* is the public-key encryption scheme from Sect. 5 and PKStS* is the stegosystem that we will define in this section. We also let $N = 8L$.

In the following, let C be a memoryless channel, P be a PRP relative to C, H be a CRHF relative to C and $G = \{G_\kappa\}_{\kappa \in \mathbb{N}}$ be a strongly 2-universal hash family. Remember, that PKES* has the algorithm PKES*.Setup that takes the additional parameters $N, N_0 \leq 2^{\lceil \kappa/2 \rceil}$, such that if $N \geq 8 \cdot \mathsf{PKES^*.cl}(\kappa)$ and $N_0/N \in [1/3, 2/3]$ then the output of PKES*.Enc(pk, m) is indistinguishable from $D^*_{(N,N_0,\mathsf{PKES^*.cl}(\kappa))}$ (see Sect. 5 for a discussion). Furthermore, we assume that PKES* has very sparse support, i.e. the ratio of valid ciphertexts compared to $\{0,1\}^{\mathsf{PKES^*.cl}(\kappa)}$ is negligible: If PKES*.Enc(pk, m) is called, we first use some public key encryption scheme PKES with very sparse support to compute $c \leftarrow \mathsf{PKES.Enc}(pk, m)$ and then encrypt c via PKES*. This construction is due to Lindell [29] and also maintains the indistinguishability of the output of PKES*.Enc and the distribution D^*, as this properties hold for all fixed messages m. Now we are ready to provide our stegosystem named PKStS*. Its main core is the ordering algorithm generate.

- The key generating PKStS*.Gen queries PKES*.Gen for a key-pair (pk, sk) and chooses a hash-function $f \leftarrow G_\kappa$. The public key of the stegosystem will be $pk^* = (pk, f)$ and the secret key will be $sk^* = (sk, f)$.
- The encoding algorithm PKStS*.Enc presented below (as C_n is memoryless we skip hist in the description) works as described in Sect. 4: It chooses appropriate keys, samples documents D, computes a hash value of D, generates bitstring b via PKES*, and finally orders the documents via generate.[5]
- To decode a sequence of documents $d_1, \ldots, d_N$, the stegodecoder PKStS*.Dec first computes the bit string $b_1 = f(d_1), \ldots, b_N = f(d_N)$ and computes the number $N_0 = |\{d_i : f(d_i) = 0\}|$. In case $|\{d_1, \ldots, d_N\}| < N$ or $N_0/N \notin [1/3, 2/3]$, the decoder PKStS*.Dec returns $\perp$ and halts. Otherwise, using PKES*.Dec with sk and parameters N, N_0, it decrypts from the

---

[5] That the number of produced documents is always divisible by 8 does not hurt the security: The warden always gets the same number of documents, whether steganography is used or not.

ciphertext $b_1, b_2, \ldots, b_L$ the message m, the keys $k_{\mathsf{H}}, k_{\mathsf{P}}, k'_{\mathsf{P}}$ and the hash-value h. It then checks whether the hash-value h is correct and whether $d_1, \ldots, d_N = \mathsf{generate}(\{d_1, \ldots, d_N\}, f, b_1, \ldots, b_L, k_{\mathsf{P}}, k'_{\mathsf{P}})$. Only if this is the case, the message m is returned. Otherwise, $\mathsf{PKStS}^*.\mathsf{Dec}$ decides that it can not decode the documents and returns $\bot$.

The steganographic encoder: $\mathsf{PKStS}^*.\mathsf{Enc}(pk^*, m)$

Input: public key $pk^* = (pk, f)$, message m; access to channel $\mathcal{C}_n$
1: let $L = \mathsf{PKES}^*.\mathsf{cl}(\kappa)$ and $N = 8L$; let $D_0 := \emptyset$ and $D_1 := \emptyset$
2: **for** $j = 1$ to N **do**
3:     sample d_j from $\mathcal{C}_n$; let $D_{f(d_j)} := D_{f(d_j)} \cup \{d_j\}$
4: $N_0 = |D_0|$
5: **if** $|D_0 \cup D_1| < N$ or $N_0/N \notin [1/3, 2/3]$ **then return** $d_1, \ldots, d_N$ and **halt**
6: choose hash key $k_{\mathsf{H}} \leftarrow \mathsf{H.Gen}(1^\kappa)$
7: choose PRP keys $k_{\mathsf{P}}, k'_{\mathsf{P}} \leftarrow \mathsf{P.Gen}(1^\kappa)$
8: let $h := \mathsf{H.Eval}_{k_{\mathsf{H}}}(\mathrm{lex}(D_0 \cup D_1))$ ▷ compute hash
9: call $\mathsf{PKES}^*\mathsf{Setup}(N, N_0)$ ▷ setup N, N_0
10: let $b_1, b_2, \ldots, b_L \leftarrow \mathsf{PKES}^*.\mathsf{Enc}(pk, m \parallel k_{\mathsf{H}} \parallel k_{\mathsf{P}} \parallel k'_{\mathsf{P}} \parallel h)$
11: let $d := \mathsf{generate}(D_0 \cup D_1, f, b_1, \ldots, b_L, k_{\mathsf{P}}, k'_{\mathsf{P}})$
12: **return** d

Proofs of Reliability and Security. We will first concentrate on the reliability of the system PKStS^* and prove that its unreliability is negligible. This is due to the fact, that the decoding always works and the encoding can only fail if a document was drawn more than once or if the sampled documents are very imbalanced with regard to f.

Theorem 9. *The probability that a message is not correctly embedded by the encoder* $\mathsf{PKStS}^*.\mathsf{Enc}$ *is at most* $3N^2 \cdot 2^{-H_\infty(\mathcal{C},\kappa)} + 2\exp(-N/54)$.

If $1 < \lambda \leq \log(\kappa)$ bits per document are embedded, this probability is bounded by $2^{2\lambda} \cdot 3N^2 \cdot 2^{-H_\infty(\mathcal{C},\kappa)} + 2^{\lambda+1}\exp(-N/54)$, which is negligible in κ if $H_\infty(\mathcal{C},\kappa)$ sufficiently large. Now, it only remains to prove that our construction is secure. The proof proceeds similar to the security proof of Hopper [21]. But instead of showing that no other encoding of a message exists, we prove that finding any other encoding of the message is infeasible via Lemma 8.

Theorem 10. *Let* $\mathcal{C}$ *be a memoryless channel,* P *be a PRP relative to* $\mathcal{C}$, *the algorithm* H *be a CRHF relative to* $\mathcal{C}$, *the cryptosystem* PKES^* *be the cryptosystem designed in Sect. 5 with very sparse support relative to* $\mathcal{C}$, *and* $\mathcal{G}$ *be a strongly 2-universal hash family. The stegosystem* PKStS^* *is SS-CCA-secure against every memoryless channel.*

$H_1 = \mathcal{C}_n^N$	H_2

$H_1 = \mathcal{C}_n^N$ H_2

1:   $pk^* = (pk, f) \leftarrow \mathsf{PKStS}^*.\mathsf{Gen}(1^\kappa)$ $pk^* = (pk, f) \leftarrow \mathsf{PKStS}^*.\mathsf{Gen}(1^\kappa)$

2:   **for** $j := 1, 2, \ldots, N$:   Lines 1 to 4 in $\mathsf{PKStS}^*.\mathsf{Enc}$

3:     $d_j \leftarrow \mathcal{C}_{\mathsf{dl}(\kappa)}$ 5: $P \twoheadleftarrow \mathsf{Perms}$

4:   **return** $((d_1, \ldots, d_N), pk^*)$ 6:   **return** $((d_{P(1)}, \ldots, d_{P(N)}), pk^*)$

H_3

$pk^* = (pk, f) \leftarrow \mathsf{PKStS}^*.\mathsf{Gen}(1^\kappa)$

Lines 1 to 4 in $\mathsf{PKStS}^*.\mathsf{Enc}$

5:   $P \twoheadleftarrow \mathsf{Perms}; P' \twoheadleftarrow \mathsf{Perms}; k_\mathsf{H} \leftarrow \mathsf{H.Gen}(1^\kappa)$

6:   $b_1, b_2, \ldots, b_L \leftarrow D^*_{(N, N_0, L)}$

7:   **return** $(\mathbf{generate}(D_0 \cup D_1, f, b_1, \ldots, b_L, P, P'), pk^*)$

// $\mathsf{generate}(\ldots, P, P')$ uses the permutations P, P'

H_4

$pk^* = (pk, f) \leftarrow \mathsf{PKStS}^*.\mathsf{Gen}(1^\kappa)$

Lines 1 to 4 in $\mathsf{PKStS}^*.\mathsf{Enc}$

5:   $k_\mathsf{P} \leftarrow \mathsf{P.Gen}(1^\kappa); P' \twoheadleftarrow \mathsf{Perms}; k_\mathsf{H} \leftarrow \mathsf{H.Gen}(1^\kappa)$

6:   $b_1, b_2, \ldots, b_L \twoheadleftarrow D^*_{(N, N_0, L)}$

7:   **return** $(\mathbf{generate}(D_0 \cup D_1, f, b_1, \ldots, b_L, k_\mathsf{P}, P'), pk^*)$

// $\mathsf{generate}(\ldots, P')$ uses the permutation P'

H_5

$pk^* = (pk, f) \leftarrow \mathsf{PKStS}^*.\mathsf{Gen}(1^\kappa)$

Lines 1 to 4 in $\mathsf{PKStS}^*.\mathsf{Enc}$

5:   $k_\mathsf{P} \leftarrow \mathsf{P.Gen}(1^\kappa); k'_\mathsf{P} \leftarrow \mathsf{P.Gen}(1^\kappa); k_\mathsf{H} \leftarrow \mathsf{H.Gen}(1^\kappa)$

6:   $b_1, b_2, \ldots, b_L \twoheadleftarrow D^*_{(N, N_0, L)}$

7:   **return** $(\mathbf{generate}(D_0 \cup D_1, f, b_1, \ldots, b_L, k_\mathsf{P}, k'_\mathsf{P}), pk^*)$

$H_6 = \mathsf{PKStS}^*.\mathsf{Enc}$

$pk^* = (pk, f) \leftarrow \mathsf{PKStS}^*.\mathsf{Gen}(1^\kappa)$

Lines 1 to 4 in $\mathsf{PKStS}^*.\mathsf{Enc}$

5:   $k_\mathsf{P} \leftarrow \mathsf{P.Gen}(1^\kappa); k'_\mathsf{P} \leftarrow \mathsf{P.Gen}(1^\kappa); k_\mathsf{H} \leftarrow \mathsf{H.Gen}(1^\kappa)$

6:   $h := \mathsf{H.Eval}_{k_\mathsf{H}}(\mathrm{lex}(D_0 \cup D_1))$

7:   $\mathsf{PKES}^*.\mathsf{Setup}(N, N_0)$

8:   $b_1, b_2, \ldots, b_L \leftarrow \mathsf{PKES}^*.\mathsf{Enc}(pk, m \;\|\; k_\mathsf{H} \;\|\; k_\mathsf{P} \;\|\; k'_\mathsf{P} \;\|\; h)$

9:   **return** $(\mathbf{generate}(D_0 \cup D_1, f, b_1, \ldots, b_L, k_\mathsf{P}, k'_\mathsf{P}), pk^*)$

Fig. 1. An overview of hybrids H_1 and H_6 used in the proof of Theorem 10. Changes between the hybrids are marked as shadowed.

Proof (Proof sketch). We prove that the above construction is secure via a *hybrid argument*. We thus define six distributions $H_1, \ldots, H_6$ shown in Fig. 1.

We now proceed by proving that H_i and H_{i+1} are SS-CCA-indistinguishable (denoted by $H_i \sim H_{i+1}$). Informally, this means that we replace in SS-CCA-Dist the call to the stegosystem (if $b = 0$) by H_i and the call to the channel (if $b = 1$) by H_{i+1}. Denote by $\mathbf{Adv}_W^{(i)}(\kappa)$ the advantage of a warden W in this situation. Clearly, the SS-CCA-advantage of W is bounded as $\mathbf{Adv}_{W,\mathsf{PKStS}^*,\mathcal{C}}^{\text{ss-cca}}(\kappa) \leq$ $\mathbf{Adv}_W^{(1)}(\kappa) + \mathbf{Adv}_W^{(2)}(\kappa) + \mathbf{Adv}_W^{(3)}(\kappa) + \mathbf{Adv}_W^{(4)}(\kappa) + \mathbf{Adv}_W^{(5)}(\kappa)$. This implies the theorem, as H_1 simply describes the channel and H_6 describes the stegosystem. Informally, we argue that:

1. $H_1 \sim H_2$ because a uniform random permutation on a memoryless channel does not change any probabilities;
2. $H_2 \sim H_3$ because our choice of $b_1, \ldots, b_L$ and random permutations equal the channel by Lemma 7;
3. $H_3 \sim H_4$ because P is a PRP;
4. $H_4 \sim H_5$ because P is a PRP;
5. $H_5 \sim H_6$ because PKES* is secure due to Corollary 6 and because of Lemma 8. $\square$

8   An Impossibility Result

We first describe an argument for truly random channels using an infeasible assumption and then proceed to modify those channels to get rid of this. All channels will be 0-memoryless and we thus write $\mathcal{C}_{\eta,\mathsf{dl}}$ instead of $\mathcal{C}_{\mathsf{hist},\mathsf{dl}}$, if hist contains η document.

The main idea of our construction lies on the *unpredictability* of random channels. If $\mathcal{C}_\eta$ and $\mathcal{C}_{\eta+1}$ are independent and sufficiently random, we can not deduce anything about $\mathcal{C}_{\eta+1}$ before we have sampling access to it, which we only have *after* we sent the document of $\mathcal{C}_\eta$ in the standard non-look-ahead model. To be reliable, there must be enough documents in $\mathcal{C}_{\eta+1}$ continuing the already sent documents (call those documents *suitable*). To be SS-CCA-secure, the number of suitable documents in $\mathcal{C}_{\eta+1}$ must be very small to prevent replay attacks like those in Sect. 3. By replacing the random channels with pseudorandom ones, we can thus prove that *every* stegosystem is either unreliable or SS-CCA-insecure on one of those channels. To improve the readability, fix some stegosystem PKStS and let $n = \mathsf{PKStS.dl}(\kappa)$ and $\ell = \mathsf{PKStS.ol}(\kappa)$.

Lower Bound on Truly Random Channels. For $n \in \mathbb{N}$, we denote by $\mathcal{R}_n$ all subsets R of $\{0,1\}^n$ such that there is a negligible function negl with

- $|R| \geq \mathsf{negl}(n)^{-1}$ and
- $|R| \leq 2^{n/2}$.

This means each subset R has super-polynomial cardinality in n without being too large. For an infinite sequence $\boldsymbol{R} = R_0, R_1, \ldots$ with $R_i \in \mathcal{R}_n$, we construct a channel $\mathcal{C}(\boldsymbol{R})$ where the distribution $\mathcal{C}(\boldsymbol{R})_{i,n}$ is the uniform distribution on R_i. The family of all such channels is denoted by $\mathcal{F}(\mathcal{R}_n)$. We assume that a warden can test whether a document d belongs to the support of $\mathcal{C}(\boldsymbol{R})_{i,n}$ and denote this warden by $\mathsf{W}_{\boldsymbol{R}}$. In the next section, we replace the totally random channels by pseudorandom ones and will get rid of this infeasible assumption. For a stegosystem PKStS – like the system PKStS* from the last section – we are now interested in two possible events that may occur during the run of PKStS.Enc on a channel $\mathcal{C}(\boldsymbol{R})$. The first event, denoted by $\mathcal{E}_{\mathrm{Nq}}$ (for *Non queried*), happens if PKStS.Enc outputs a document it has not seen due to sampling. We are also interested in the case that PKStS.Enc outputs something in the support of the channel, denoted by $\mathcal{E}_{\mathrm{InS}}$ for *In Support*. Clearly, upon random choice of $\boldsymbol{R}$, η (the length of the history), m and pk we have

$$\Pr[\mathcal{E}_{\mathrm{InS}} \mid \mathcal{E}_{\mathrm{Nq}}] \leq \ell \cdot \frac{2^{n/2} - \mathsf{PKStS.query}(\kappa)}{2^n - \mathsf{PKStS.query}(\kappa)} \leq \ell \cdot 2^{-n/2},$$

where $\mathsf{PKStS.query}(\kappa)$ denotes the number of queries performed by PKStS. This is negligible in κ as n, query and ℓ are polynomials in κ. As warden $\mathsf{W}_{\boldsymbol{R}}$ can test whether a document belongs to the random sets, we have $\mathbf{Adv}^{\mathrm{ss\text{-}cca}}_{\mathsf{W}_{\boldsymbol{R}},\mathsf{PKStS},\mathcal{C}(\boldsymbol{R})}(\kappa) \geq \Pr[\overline{\mathcal{E}_{\mathrm{InS}}}]$. Clearly, since we can assume $\overline{\mathcal{E}_{\mathrm{InS}}} \subseteq \mathcal{E}_{\mathrm{Nq}}$ we thus obtain

$$\Pr[\mathcal{E}_{\mathrm{Nq}}] = \frac{\Pr[\overline{\mathcal{E}_{\mathrm{InS}}} \wedge \mathcal{E}_{\mathrm{Nq}}]}{\Pr[\overline{\mathcal{E}_{\mathrm{InS}}} \mid \mathcal{E}_{\mathrm{Nq}}]} \leq \frac{\mathbf{Adv}^{\mathrm{ss\text{-}cca}}_{\mathsf{W}_{\boldsymbol{R}},\mathsf{PKStS},\mathcal{C}(\boldsymbol{R})}(\kappa)}{1 - \ell \cdot 2^{-n/2}}.$$

Hence, if PKStS is SS-CCA-secure, the term $\Pr[\mathcal{E}_{\mathrm{Nq}}]$ must be negligible.

If PKStS is given a history of length η and it outputs documents $d_1, \ldots, d_\ell$, we note that PKStS.Enc only gets sampling access to $\mathcal{C}(\boldsymbol{R})_{\eta+\ell-1,n}$ after it sent $d_1, \ldots, d_{\ell-1}$ in the standard non-look-ahead model. Clearly, due to the random choice of $\boldsymbol{R}$, the set $R_{\eta+\ell}$ is independent of $R_\eta, R_{\eta+1}, \ldots, R_{\eta+\ell-1}$. The encoder PKStS.Enc thus needs to decide on the documents $d_1, \ldots, d_{\ell-1}$ without any knowledge of $R_{\eta+\ell}$. As PKStS.Enc draws a sample set D from $\mathcal{C}(\boldsymbol{R})_{\eta+\ell-1,n}$ with at most $q = \mathsf{PKStS.query}(\kappa)$ documents, we now look at the event $\mathcal{E}_{\mathrm{Nsui}}$ (for *Not suitable*) that none of the documents in D are suitable for the encoding, i.e. if the sequence $d_1, d_2, \ldots, d_{\ell-1}, d$ is not a suitable encoding of the message m for all $d \in D$. Denote the unreliability of the stegosystem by ρ. Clearly, if $\mathcal{E}_{\mathrm{Nsui}}$ occurs, there are two possibilities for the stegosystem: It either outputs something from D and thus increases the unreliability or it outputs something it has not queried. We thus have $\Pr[\mathcal{E}_{\mathrm{Nsui}}] \leq \max\{\rho, (1-\rho) \cdot \Pr[\mathcal{E}_{\mathrm{Nq}}]\}$. Note that ρ must be negligible if PKStS.Enc is reliable and, as discussed above, the term $\Pr[\mathcal{E}_{\mathrm{Nq}}]$ (and thus the term $(1-\rho) \cdot \Pr[\mathcal{E}_{\mathrm{Nq}}]$) must be negligible if PKStS.Enc is SS-CCA-secure. Hence, if PKStS.Enc is SS-CCA-secure and reliable, the probability $\Pr[\mathcal{E}_{\mathrm{Nsui}}]$ must be negligible. The insight, that $\Pr[\mathcal{E}_{\mathrm{Nsui}}]$ must be negligible directly leads us to the construction of a warden $\mathsf{W}_{\boldsymbol{R}}$ on the channel $\mathcal{C}(\boldsymbol{R})$. The warden chooses a random history of length η and a random message m and sends those to its challenging oracle. It then receives the document sequence $d_1, \ldots, d_\ell$. If $d_i \notin R_{\eta+i}$, the warden returns »Stego«. Else, it samples q documents D from $\mathcal{C}(R)_{\eta+\ell,n}$ and tests for all $d \in D$ via the decoding oracle $\mathsf{PKStS.Dec}_{sk}$ if the sequence $d_1, d_2, \ldots, d_{\ell-1}, d$ decodes to m. If we find such a d, return »Stego« and else return »Not Stego«. If the documents are randomly chosen from the channel, the probability to return »Stego« is at most $q/|2^{\mathsf{PKStS.ml}(\kappa)}|$, i.e. negligible. If the documents are chosen by the stegosystem, the probability of »Not Stego« is exactly $\Pr[\mathcal{E}_{\mathrm{Nsui}}]$. Hence, PKStS must be either unreliable or SS-CCA-insecure on some channel in $\mathcal{F}(\mathcal{R}_n)$.

Lower Bound on Pseudorandom Channels. To give a proof, we will replace the random channels $\mathcal{C}(R)$ by pseudorandom ones. The construction assumes existence of a CCA\$-secure cryptosystem PKES with $\mathsf{PKES.cl}(\kappa) \geq 2\,\mathsf{PKES.ml}(\kappa)$.

For $\omega = (pk, sk) \in \mathrm{supp}(\mathsf{PKES.Gen}(1^\kappa))$, let $\mathcal{C}(\omega)_{i,\mathsf{dl}(\kappa)}$ be the distribution $\mathsf{PKES.Enc}(pk, \mathrm{bin}(i)_{\mathsf{dl}(\kappa)})$, where $\mathrm{bin}(i)_{\mathsf{dl}(\kappa)}$ is the binary representation of the number i of length exactly $\mathsf{dl}(\kappa)$ modulo $2^{\mathsf{dl}(\kappa)}$. The family of channels $\mathcal{C}_{\mathsf{PKES}} = \{\mathcal{C}(\omega)\}_\omega$ thus has the following properties:

1. There is a negligible function negl such that for each ω and each i, we have $2^{\mathsf{PKES.ml}(\kappa)/2} \geq |\mathcal{C}(\omega)_{i,\mathsf{dl}(\kappa)}| \geq \mathsf{negl}(\kappa)^{-1}$ if PKES is CCA\$-secure. This follows easily from the CCA\$-security of PKES: If $|\mathcal{C}(\omega)_{i,\mathsf{dl}(\kappa)}|$ would be polynomial, an attacker could simply store all corresponding ciphertexts.
2. An algorithm with the knowledge of ω can test in polynomial time, whether $d \in \mathrm{supp}(\mathcal{C}(\omega)_{i,\mathsf{dl}(\kappa)})$, i.e. whether d belongs to the support by simply testing whether $\mathsf{PKES.Dec}(sk, d)$ equals $\mathrm{bin}(i)_{\mathsf{dl}(\kappa)}$.
3. Every algorithm Q that tries to distinguish $\mathcal{C}(\omega)$ from a random channel $\mathcal{C}(R)$ fails: For every polynomial algorithm Q, we have that the term

$$\left| \Pr_{R \leftarrow \mathcal{R}^*_{\mathsf{dl}(\kappa)}} [\mathsf{Q}^{\mathcal{C}(R)}(1^\kappa) = 1] - \Pr_{\omega \leftarrow \mathsf{PKES.Gen}(1^\kappa)} [\mathsf{Q}^{\mathcal{C}(\omega)}(1^\kappa) = 1] \right|$$

is negligible in κ if PKES is CCA\$-secure. This follows from the fact that no polynomial algorithm can distinguish $\mathcal{C}(R)$ upon random choice of R from the uniform distribution on $\{0,1\}^n$, as $|\mathcal{C}(R)_{i,n}|$ is super-polynomial. Furthermore, an attacker A on PKES can simulate Q for a successful attack.

Note that the third property directly implies that no polynomial algorithm can conclude anything about $\mathcal{C}(\omega)_{i,\mathsf{dl}(\kappa)}$ from samples of previous distributions $\mathcal{C}(\omega)_{0,\mathsf{dl}(\kappa)}, \ldots, \mathcal{C}(\omega)_{i-1,\mathsf{dl}(\kappa)}$, except for a negligible term. The second property directly implies that we can get rid of the infeasible assumption of the previous section concerning the ability of the warden to test whether a document belongs to the support of $\mathcal{C}(\omega)$: We simply equip the warden with the seed ω. Call the resulting warden W_ω. Note that this results in a non-uniform warden. As above, we are interested in the events that a stegosystem outputs a document that it has not seen $(\mathcal{E}_{\widehat{\text{Nq}}})$, that a document is outputted which does not belong to the support $(\mathcal{E}_{\widehat{\text{InS}}})$ and the event that a random set of q documents is not suitable to complete a given document prefix $d_1, d_2, \ldots, d_{\ell-1}$ $(\mathcal{E}_{\widehat{\text{Nsui}}})$.

As $\mathcal{E}_{\widehat{\text{InS}}}$ is a polynomially testable property (due to the second property of our construction), we can conclude a similar bound as above:

Lemma 11. *Let* PKStS *be an SS-CCA-secure universal stegosystem. For every warden* W *and every CCA\$-attacker* A, $\Pr[\mathcal{E}_{\widehat{\text{Nq}}}] \leq \frac{\mathbf{Adv}_{\text{W,PKStS},\mathcal{C}(\omega)}^{\text{ss-cca}}(\kappa)}{1-\ell\cdot 2^{-n/2}} + \mathbf{Adv}_{\text{A,PKES}}^{pkes}(\kappa).$

Hence, if the stegosystem PKStS is SS-CCA-secure and PKES is CCA\$-secure, the term $\Pr[\mathcal{E}_{\widehat{\text{Nq}}}]$ must be negligible. As above, we can conclude that $\Pr[\mathcal{E}_{\widehat{\text{Nsui}}}] \leq \max\{\rho, (1-\rho)\cdot\Pr[\mathcal{E}_{\widehat{\text{Nq}}}]\}$ for unreliability ρ. The warden W_ω similar to W_R from the preceding section thus succeeds with very high probability. Hence, no SS-CCA-secure and reliable stegosystem can exist for the family $\mathcal{C}_{\text{PKES}}$:

Theorem 12. *If doubly-enhanced trapdoor permutations exist, for every stegosystem* PKStS *in the non-look-ahead model there is a 0-memoryless channel* $\mathcal{C}$ *such that* PKStS *is either unreliable or it is not SS-CCA-secure on* $\mathcal{C}$ *against non-uniform wardens.*

9   Discussion

The work of Dedić et al. [13] shows that provable secure universal steganography needs a huge number of sample documents to embed long secret messages as high bandwidth stegosystems are needed for such messages. This limitation also transfers to the public-key scenario. However, such a limitation does not necessarily restrict applicability of steganography, especially in case of specific communication channels or if the length of secret messages is sufficiently short.

A prominent recent example for such applications is the use of steganography for channels determined by cryptographic primitives, like symmetric encryption scheme (SESs) or digital signature schemes. Bellare, Paterson, and Rogaway introduced in [5] so called *algorithm substitution attacks* against SESs, where an attacker replaces an honest implementation of the encryption algorithm by a modified version which allows to extract the secret key from the ciphertexts produced by the corrupted implementation. Several follow-up works have been done based on this paper, such as those by Bellare et al. [4], Ateniese et al. [2], or Degabriele et al. [14]. These works strengthened the model proposed in [5] and presented new attacks against SESs or against other cryptographic primitives, e.g. against signature schemes. Surprisingly, as we show in [6], all such algorithm substitution attacks can be analyzed in the framework of computational secret-key steganography and in consequence, the attackers can be identified as stegosystems on certain channels determined by the primitives. In such scenarios, the secret message embedded by the stegosystem corresponds to a secret key of the cryptographic algorithm.

A similar approach was used by Pasquini et al. [35] to show that so called *password decoy vaults* used for example by Chatterjee et al. [10] and Golla et al. [19] can also be interpreted as steganographic protocols.

A   Remaining Proofs

To improve the readability, we will abbreviate some terms and define $n = \text{PKStS}^*.\text{dl}(\kappa)$, $\ell = \text{PKStS}^*.\text{ol}(\kappa)$ and $L = \text{PKES}^*.\text{cl}(\kappa)$, where PKStS^* is our stegosystem constructed in Sect. 7 and PKES^* is the public-key cryptosystem constructed in Sect. 5. We also define $N = 8L$.

A.1   Formal Statement of Lemma 8 and its Proof

We start with a formal definition for "A is successful on $D, f, b_1, \ldots, b_L, k_H$".

Definition 13. *An attacker A on* generate *is a PPTM, that receives the following input:*

- *a sequence $d_1, \ldots, d_N$ of N pairwise different documents*
- *a hash function $f : \Sigma^n \to \{0, 1\}$ from the family $\mathcal{G} = \{G_\kappa\}_{\kappa \in \mathbb{N}}$,*
- *a sequence $b_1, \ldots, b_L$ of L bits, and*
- *a hash-key k_H for H.*

The attacker A then outputs a sequence $d'_1, \ldots, d'_N$ of documents. Note that the attacker knows the mapping function f and even the hash-key k_H for H.

We say that A is successful *if the experiment Sgen$(A, D, f, b_1, \ldots, b_L)$ returns value 1:*

Security of generate: Sgen$(A, D, f, b_1, \ldots, b_L)$

Input: Attacker A, set D, function f, bits $b_1, \ldots, b_L$

1: $k_P, k'_P \leftarrow \text{P.Gen}(1^\kappa)$
2: $k_H \leftarrow \text{H.Gen}(1^\kappa)$
3: $d_1, \ldots, d_N := \text{generate}(D, f, b_1, \ldots, b_L, k_P, k'_P)$
4: $d'_1, \ldots, d'_N \leftarrow A(d_1, \ldots, d_N, f, b_1, \ldots, b_L, k_H)$
5: **if** $f(d'_i) = b_i$ for every $i = 1, \ldots L$ **then**
6:     $D'_0 = \{d'_j \mid f(d'_j) = 0\}$; $D'_1 = \{d'_j \mid f(d'_j) = 1\}$
7:     **if** $d'_1, \ldots, d'_N = \text{generate}(D'_0 \cup D'_1, f, b_1, \ldots, b_L, k_P, k'_P)$ **then**
8:         **if** $\text{H.Eval}_{k_H}(\text{lex}(D'_0 \cup D'_1)) = \text{H.Eval}_{k_H}(\text{lex}(D_0 \cup D_1))$ **then**
9:             **if** $d'_1, \ldots, d'_N \neq d_1, \ldots, d_N$ **then**
10:                **return** 1 and **halt**
11: **return** 0

We are now ready to give the formal version of Lemma 8:

Lemma (formal version of Lemma 8). *Let $D \subseteq \Sigma^n$ be a set of documents, with $|D| = N$, let $b_1, \dots, b_L$ be a bitstring, and $f \in G_\kappa$. For every attacker A on generate, there is a collision finder Fi for the CRHF H such that*

$$\Pr[Sgen(\mathsf{A}, D, f, b_1, \dots, b_L) = 1] \leq \mathbf{Adv}^{hash}_{\mathsf{Fi},\mathsf{H},\mathcal{C}}(\kappa),$$

where the probability is taken over the random choices made in experiment Sgen.

Proof. Let A be an attacker on **generate** with maximal success probability. Let $D = D_0 \dot\cup D_1$ be the input to **generate**, the sequence $d_1, \dots, d_N$ its output and $d'_1, \dots, d'_N$ be the output of A. Furthermore, let $D'_b = \{d'_j \mid f(d'_j) = b\}$ and $D' = D'_0 \cup D'_1$. We now distinguish three cases of the relation between D and D'. If $D' \subsetneq D$, the sequence $d'_1, \dots, d'_N$ must contain the same element on at least two positions, but **generate** does only accept sets of size exactly N. Hence, A was not successful in this case. If $D' = D$ and A was successful, it holds that $d'_1, \dots, d'_N \neq d_1, \dots, d_N$. Hence, there must be positions $i < j$ and $j' < i'$ such that $d_i = d_{i'}$ and $d_j = d_{j'}$. As k_P and k'_P define a total order, the sequence $d'_1, \dots, d'_N$ could not be produced by **generate**. Thus, A can not be successful in this case.

The only remaining case is $D' \setminus D \neq \emptyset$. If A was successful, it holds that $\mathsf{H}_{k_\mathsf{H}}(\mathrm{lex}(D')) = \mathsf{H}_{k_\mathsf{H}}(\mathrm{lex}(D))$, i.e. this is a collision with regard to H. We will now construct a finder Fi for H, such that $\mathbf{Adv}^{hash}_{\mathsf{Fi},\mathsf{H},\mathcal{C}}(\kappa) \geq \Pr[\mathsf{A}\ \text{succeeds}]$. The finder Fi receives a hash key k_H. It then chooses $f \twoheadleftarrow G_\kappa$, samples D documents of cardinality $|D| = N$ via rejection sampling and PRP-keys $k_\mathsf{P}, k'_\mathsf{P}$. The finder simulates A and receives

$$d'_1, \dots, d'_N \leftarrow \mathsf{A}(\mathbf{generate}(D, f, b_1, \dots, b_L, k_\mathsf{P}, k'_\mathsf{P}), f, b_1, \dots, b_L, k_\mathsf{H}).$$

Then, it returns D and $D' = \{d'_1, \dots, d'_N\}$. Whenever A succeeds, we have $D \neq D'$ by the discussion above and thus also $\mathsf{H}_{k_\mathsf{H}}(\mathrm{lex}(D)) = \mathsf{H}_{k_\mathsf{H}}(\mathrm{lex}(D'))$. Hence, Fi has successfully found a collision. This implies that $\mathbf{Adv}^{hash}_{\mathsf{Fi},\mathsf{H},\mathcal{C}}(\kappa) \geq \Pr[\mathsf{A}\ \text{succeeds}]$. $\square$

A.2   Proof of Theorem 9

Recall the statement of the theorem:

Theorem (Theorem 9). *The probability that a message is not correctly embedded by* PKStS*.Enc *is at most* $3N^2 \cdot 2^{-H_\infty(\mathcal{C},\kappa)} + 2\exp(-N/54)$.

Proof. Note that PKStS*.Enc may not correctly embed a message m if (a) $|D_0 \cup D_1| < N$ i.e. a document sampled in line 3 was drawn twice, or (b) $N_0/N \notin [1/3, 2/3]$ i.e. the bias is too large, or (c) the number of elements of D_0 or D_1 is too small to embed $b_1, b_2, \dots, b_L$ by **generate**. The probability of (a) can be bounded similar to the birthday attack. It is roughly bounded by $3N^2 \cdot 2^{-H_\infty(C,\kappa)}$ as the probability of every document is bounded by $2^{-H_\infty(C,\kappa)}$.

A simple calculation shows that the probability of (b) and (c) is negligible. Note that the algorithm always correctly embeds a message, if $|D_0| \geq L$ and $|D_1| \geq L$. As $N_0/N = |D_0|/N$, this implies that $N_0/N \in [1/3, 2/3]$. We will thus estimate the probability for this. As f is drawn from a strongly 2-universal hash family, we note that the probability that a random document d is mapped to 1 is equal to $1/2$. For $i = 1, \ldots, N$, let X_i be the indicator variable such that X_i equals 1 if the i-th element drawn from the channel maps to 1. The random variable $X = \sum_{i=1}^{N} X_i$ thus has the size of D_1. Clearly, its expected value is $N/2$. The probability that $|X - N/2| > L$ (and thus $|D_1| < L$ or $|D_0| < L$) is hence bounded by

$$\Pr[|X - N/2| > L] \leq 2\exp(-\frac{L \cdot (1/3)^2}{3}) = 2\exp(-N/54)$$

using a Chernoff-like bound. The probability that the message m is incorrectly embedded is thus bounded by $2^{-H_\infty(C,\kappa)} + 2\exp(-N/54)$. □

A.3   Proof of Theorem 10

We recall:

Theorem (Theorem 10). *Let C be a memoryless channel, P be a PRP relative to C, the algorithm H be a CRHF relative to C, the cryptosystem PKES* be the cryptosystem designed in Sect. 5 with very sparse support relative to C, and G be a strongly 2-universal hash family. The stegosystem PKStS* is SS-CCA-secure against every memoryless channel.*

Proof. We prove that the above construction is secure via a *hybrid argument*. We thus define six distributions $H_1, \ldots, H_6$ shown in Fig. 1.

If P and Q are two probability distributions, denote by SS-CCA-Dist$_{P,Q}$ the modification of the game SS-CCA-Dist, where the call to the stegosystem (if $b = 0$) is replaced by a call to P and the call to the channel (if $b = 1$) is replaced by a call to Q. If W is some warden, denote by $\mathbf{Adv}_{W,P,Q}^{ss\text{-}cca}(\kappa)$ the winning probability of W in SS-CCA-Dist$_{P,Q}$. If $\mathbf{Adv}_{W,P,Q}^{ss\text{-}cca}(\kappa) \leq \mathsf{negl}(\kappa)$ for a negligible function negl, we denote this situation as $P \sim Q$ and say that P and Q are *indistinguishable* with respect to SS-CCA-Dist. Furthermore, we define $\mathbf{Adv}_W^{(i)}(\kappa) = \mathbf{Adv}_{W,H_i,H_{i+1}}^{ss\text{-}cca}(\kappa)$. As the term $\mathbf{Adv}_W^{(i)}(\kappa)$ can also be written as

$$\left|\Pr[\text{W.Guess outputs } b' = 0 \mid b = 0] - \Pr[\text{W.Guess outputs } b' = 0 \mid b = 1]\right|,$$

the triangle inequality implies that $\mathbf{Adv}_{W,\text{PKStS}^*,C}^{ss\text{-}cca}(\kappa) \leq \mathbf{Adv}_W^{(1)}(\kappa) + \mathbf{Adv}_W^{(2)}(\kappa) + \mathbf{Adv}_W^{(3)}(\kappa) + \mathbf{Adv}_W^{(4)}(\kappa) + \mathbf{Adv}_W^{(5)}(\kappa)$.

Informally, we argue that:

1. $H_1 = H_2 \implies H_1 \sim H_2$ because a uniform random permutation on a memoryless channel does not change any probabilities;

2. $H_2 = H_3 \implies H_2 \sim H_3$ because our choice of $b_1, \ldots, b_L$ and random permutations equal the channel by Lemma 7;
3. $H_3 \sim H_4$ because P is a PRP;
4. $H_4 \sim H_5$ because P is a PRP;
5. $H_5 \sim H_6$ PKES* is secure due to Corollary 6 and because of Lemma 8.

Distribution H_1 can be specified as follows:

$$H_1 = \mathcal{C}_n^N$$

---

1 : $\quad pk^* = (pk, f) \leftarrow \textsf{PKStS}^*.\textsf{Gen}(1^\kappa)$

2 : $\quad$ **for** $j := 1, 2, \ldots, N$:

3 : $\qquad d_j \leftarrow \mathcal{C}_{\textsf{dl}(\kappa)}$

4 : $\quad$ **return** $((d_1, \ldots, d_N), pk^*)$

Indistinguishability of H_1 and
$$H_2$$

$pk^* = (pk, f) \leftarrow \textsf{PKStS}^*.\textsf{Gen}(1^\kappa)$

Lines 1 to 4 in $\textsf{PKStS}^*.\textsf{Enc}$

5 : $\quad P \leftarrow \textsf{Perms}$

6 : $\quad$ **return** $((d_{P(1)}, \ldots, d_{P(N)}), pk^*)$

If $|D_0 \cup D_1| < N$, i.e. a document was sampled twice or $|D_0|/|D| \notin [1/3, 2/3]$, the system only outputs the sampled documents. Hence H_1 equals H_2 in this case. In the other case, we first permute the items before we output them. But, as P is a uniform random permutation and the documents are drawn independently from a memoryless channel, we have

$$\Pr_{H_1}[d_1, \ldots, d_N \text{ are drawn}] = \Pr_{H_1}[d_{P(1)}, \ldots, d_{P(N)} \text{ are drawn}].$$

As pk is not used in these hybrids, $H_1 = H_2$ follows.

Indistinguishability of H_2 and
$$H_3$$

$pk^* = (pk, f) \leftarrow \textsf{PKStS}^*.\textsf{Gen}(1^\kappa)$

Lines 1 to 4 in $\textsf{PKStS}^*.\textsf{Enc}$

5 : $\quad P \leftarrow \textsf{Perms}; \; P' \leftarrow \textsf{Perms}; \; k_\textsf{H} \leftarrow \textsf{H.Gen}(1^\kappa)$

6 : $\quad b_1, b_2, \ldots, b_L \leftarrow D_{(N, N_0, L)}^*$

7 : $\quad$ **return** $(\textsf{generate}(D_0 \cup D_1, f, b_1, \ldots, b_L, P, P'), pk^*)$

$/\!/$ generate$(\ldots, P, P')$ uses the permutations P, P'

If $|D_0 \cup D_1| < N$, i.e. a document was sampled twice or $|D_0|/|D| \notin [1/3, 2/3]$, the system only outputs the sampled documents. Hence H_2 equals H_3 in this case. If $|D_0 \cup D_1| = N$, Lemma 7 shows that H_2 equals H_3.

Indistinguishability of H_3 and H_4

$pk^* = (pk, f) \leftarrow \mathsf{PKStS}^*.\mathsf{Gen}(1^\kappa)$

Lines 1 to 4 in $\mathsf{PKStS}^*.\mathsf{Enc}$

5 : $k_\mathsf{P} \leftarrow \mathsf{P}.\mathsf{Gen}(1^\kappa); P' \twoheadleftarrow \mathsf{Perms}; k_\mathsf{H} \leftarrow \mathsf{H}.\mathsf{Gen}(1^\kappa)$

6 : $b_1, b_2, \ldots, b_L \twoheadleftarrow D^*_{(N, N_0, L)}$

7 : $\mathbf{return}\ (\mathsf{generate}(D_0 \cup D_1, f, b_1, \ldots, b_L, k_\mathsf{P}, P'), pk^*)$

$/\!/$ generate$(\ldots, P')$ uses the permutation P'

We will construct a distinguisher Dist on the PRP P with $\mathbf{Adv}^{\mathrm{prp}}_{\mathsf{Dist}, \mathsf{P}, \mathcal{C}}(\kappa) = \mathbf{Adv}^{(3)}_\mathsf{W}(\kappa)$. Note that such a distinguisher has access to an oracle that either corresponds to a truly random permutation or to $\mathsf{P}.\mathsf{Eval}_k$ for a key $k \leftarrow \mathsf{P}.\mathsf{Gen}(1^\kappa)$.

The PRP-distinguisher Dist simulates the run of W. It first chooses a key-pair $(pk, sk) \leftarrow \mathsf{PKStS}^*.\mathsf{Gen}(1^\kappa)$. It then simulates W. Whenever the warden W makes a call to its decoding-oracle $\mathsf{PKStS}^*.\mathsf{Dec}$, it computes $\mathsf{PKStS}^*.\mathsf{Dec}(sk, \cdot)$ (or $\perp$ if necessary). In order to generate the challenge sequence $\hat{d}$ upon the message m, it simulates the run of $\mathsf{PKStS}^*.\mathsf{Enc}$ and replaces every call to P or $\mathsf{P}.\mathsf{Eval}_{k_\mathsf{P}}$ by a call to its oracle. Similarly, the bits output by $\mathsf{PKES}^*.\mathsf{Enc}(pk, m)$ are ignored and replaced by truly random bits distributed according to $D^*_{(N, |D_0|, L)}$. If the oracle is a truly random permutation, the simulation yields exactly H_3 and if the oracle equals $\mathsf{P}.\mathsf{Eval}_k$ for a certain key k, the simulation yields H_4. The advantage of Dist is thus exactly $\mathbf{Adv}^{(3)}_\mathsf{W}(\kappa)$. As P is a secure PRP, this advantage is negligible and H_3 and H_4 are thus indistinguishable.

Indistinguishability of H_4 and H_5

$pk^* = (pk, f) \leftarrow \mathsf{PKStS}^*.\mathsf{Gen}(1^\kappa)$

Lines 1 to 4 in $\mathsf{PKStS}^*.\mathsf{Enc}$

5 : $k_\mathsf{P} \leftarrow \mathsf{P}.\mathsf{Gen}(1^\kappa); k'_\mathsf{P} \leftarrow \mathsf{P}.\mathsf{Gen}(1^\kappa); k_\mathsf{H} \leftarrow \mathsf{H}.\mathsf{Gen}(1^\kappa)$

6 : $b_1, b_2, \ldots, b_L \twoheadleftarrow D^*_{(N, N_0, L)}$

7 : $\mathbf{return}\ (\mathsf{generate}(D_0 \cup D_1, f, b_1, \ldots, b_L, k_\mathsf{P}, k'_\mathsf{P}), pk^*)$

We will construct a distinguisher Dist on the PRP P with $\mathbf{Adv}^{\mathrm{prp}}_{\mathsf{Dist}, \mathsf{P}, \mathcal{C}}(\kappa) = \mathbf{Adv}^{(4)}_\mathsf{W}(\kappa)$. Note that such a distinguisher has access to an oracle that either corresponds to a truly random permutation or to $\mathsf{P}.\mathsf{Eval}_k$ for a key $k \leftarrow \mathsf{P}.\mathsf{Gen}(1^\kappa)$.

The PRP-distinguisher Dist simulates the run of W. It first chooses a key-pair $(pk, sk) \leftarrow \mathsf{PKStS}^*.\mathsf{Gen}(1^\kappa)$ and a key $k_\mathsf{P} \leftarrow \mathsf{P}.\mathsf{Gen}(1^\kappa)$ for the PRP P. It then simulates W. Whenever the warden W makes a call to its decoding-oracle $\mathsf{PKStS}^*.\mathsf{Dec}$, it computes $\mathsf{PKStS}^*.\mathsf{Dec}(sk, \cdot)$ (or $\perp$ if necessary). In order to generate the challenge sequence $\hat{d}$ upon the message m, it simulates the run of $\mathsf{PKStS}^*.\mathsf{Enc}$ and replaces every call to P' or $\mathsf{P}.\mathsf{Eval}_{k_\mathsf{P}}$ by a call to its oracle.

Similarly, the bits output by PKES*.Enc(pk, m) are ignored and replaced by truly random bits distributed according to $D^*_{(N,|D_0|,L)}$. If the oracle is a truly random permutation, the simulation yields exactly H_4 and if the oracle equals P.Eval$_k$ for a certain key k, the simulation yields H_5. The advantage of Dist is thus exactly $\mathbf{Adv}^{(4)}_W(\kappa)$. As P is a secure PRP, this advantage is negligible and H_4 and H_5 are thus indistinguishable.

Indistinguishability of H_5 and
$$H_6 = \mathsf{PKStS}^*.\mathsf{Enc}$$

$pk^* = (pk, f) \leftarrow \mathsf{PKStS}^*.\mathsf{Gen}(1^\kappa)$

Lines 1 to 4 in PKStS*.Enc

5 : $k_\mathsf{P} \leftarrow \mathsf{P.Gen}(1^\kappa); k'_\mathsf{P} \leftarrow \mathsf{P.Gen}(1^\kappa); k_\mathsf{H} \leftarrow \mathsf{H.Gen}(1^\kappa)$

6 : $h := \mathsf{H.Eval}_{k_\mathsf{H}}(\mathrm{lex}(D_0 \cup D_1))$

7 : $\mathsf{PKES}^*.\mathsf{Setup}(N, N_0)$

8 : $b_1, b_2, \ldots, b_L \leftarrow \mathsf{PKES}^*.\mathsf{Enc}(pk, m \parallel k_\mathsf{H} \parallel k_\mathsf{P} \parallel k'_\mathsf{P} \parallel h)$

9 :   **return** (generate($D_0 \cup D_1, f, b_1, \ldots, b_L, k_\mathsf{P}, k'_\mathsf{P}$), pk^*)

We construct an attacker A on PKES* such that there is a negligible function negl with $\mathbf{Adv}^{cca}_{A,\mathsf{PKES}^*,\mathcal{C}}(\kappa) + \mathsf{negl}(\kappa) \geq \mathbf{Adv}^{(5)}_W(\kappa)$. Note that such an attacker A has access to the decryption-oracle PKES*.Dec$_{sk}(\cdot)$.

The attacker A simply simulates W. First, it chooses $f \leftarrow G_\kappa$. Whenever W uses its decryption-oracle to decrypt $d_1, \ldots, d_N$, the attacker A simulates PKStS*.Dec($d_1, \ldots, d_N$) and uses its own decryption-oracle PKES*.Dec$_{sk}(\cdot)$ in this. When W outputs the challenge m, the attacker A chooses all of the parameters $D_0, D_1, k_\mathsf{H}, k_\mathsf{P}, k'_\mathsf{P}$ as in PKStS*.Enc and chooses its own challenge $\widetilde{m} := m \parallel k_\mathsf{H} \parallel k_\mathsf{P} \parallel k'_\mathsf{P} \parallel h$, where $h = \mathsf{H.Eval}_{k_\mathsf{H}}(D_0 \cup D_1)$.

The attacker now either receives $b \leftarrow \mathsf{PKES}^*.\mathsf{Enc}(pk, \widetilde{m})$ or L random bits b from $D^*_{(N,|D_0|,L)}$ and computes

$$d_1, \ldots, d_N = \mathsf{generate}(D_0 \cup D_1, f, b_1, \ldots, b_L, k_\mathsf{P}, k'_\mathsf{P}).$$

If the bits correspond to PKES*.Enc($pk, \widetilde{m}$), this simulates the stegosystem and thus H_6 perfectly. If the bits are random, this equals H_5.

After the challenge is determined, A continues to simulate W. Whenever W uses its decryption-oracle to decrypt $d_1, \ldots, d_N$, it behaves as above. There is now a significant difference to the pre-challenge situation: The attacker A is not allowed to decrypt the bits $b = b_1, \ldots, b_L$. Hence, when W tries to decrypt documents $d_1, \ldots, d_N$ such that $f(d_i) = b_i$, it has no way to use its decryption-oracle and must simply return $\perp$. Suppose that this situation arises. Note that the decryption-oracle of W would only return a message not equal to $\perp$ then iff $d_1, \ldots, d_N = \mathsf{generate}(D_0 \cup D_1, f, b, k_\mathsf{P}, k'_\mathsf{P})$ and $\mathsf{H.Eval}_{k_\mathsf{H}}(\{d_1, \ldots, d_N\}) = h$. If b is a truly random string from $D^*_{(N,|D_0|,L)}$, the sparsity of PKES* implies that the probability that b is a valid encoding is negligible. Hence the probability that the decryption-oracle of W would return a message not equal to $\perp$ is negligible. It only remains to prove that the probability that the decryption-oracle of W returns a message not equal to $\perp$ is negligible if b is a valid encryption of a message. But Lemma 8 states just that. We thus have $\mathbf{Adv}^{cca}_{A,PKES^*,\mathcal{C}}(\kappa) + \mathsf{negl}(\kappa) \geq \mathbf{Adv}^{(5)}_W(\kappa)$. As the system PKES* is CCA-secure by Corollary 6, this advantage is negligible. Hence, H_5 and H_6 are indistinguishable.

Hence, the stegosystem PKStS* is SS-CCA-secure on $\mathcal{C}$.   □

References

1. Anderson, R.J., Petitcolas, F.A.P.: On the limits of steganography. IEEE J. Sel. Areas Commun. **16**(4), 474–481 (1998)
2. Ateniese, G., Magri, B., Venturi, D.: Subversion-resilient signature schemes. In: Proceedings of the CCS, pp. 364–375. ACM (2015)
3. Backes, M., Cachin, C.: Public-key steganography with active attacks. In: Kilian, J. (ed.) TCC 2005. LNCS, vol. 3378, pp. 210–226. Springer, Heidelberg (2005). https://doi.org/10.1007/978-3-540-30576-7_12
4. Bellare, M., Jaeger, J., Kane, D.: Mass-surveillance without the state: strongly undetectable algorithm-substitution attacks. In: Proceedings of the CCS 2015, pp. 1431–1440. ACM (2015)
5. Bellare, M., Paterson, K.G., Rogaway, P.: Security of symmetric encryption against mass surveillance. In: Garay, J.A., Gennaro, R. (eds.) CRYPTO 2014. LNCS, vol. 8616, pp. 1–19. Springer, Heidelberg (2014). https://doi.org/10.1007/978-3-662-44371-2_1
6. Berndt, S., Liśkiewicz, M.: Algorithm substitution attacks from a steganographic perspective. In: Proceedings of the CCS, pp. 1649–1660 (2017). https://doi.org/10.1145/3133956.3133981
7. Cachin, C.: An information-theoretic model for steganography. Inf. Comput. **192**(1), 41–56 (2004)
8. Canetti, R., Krawczyk, H., Nielsen, J.B.: Relaxing chosen-ciphertext security. In: Boneh, D. (ed.) CRYPTO 2003. LNCS, vol. 2729, pp. 565–582. Springer, Heidelberg (2003). https://doi.org/10.1007/978-3-540-45146-4_33
9. Chandran, N., Goyal, V., Ostrovsky, R., Sahai, A.: Covert multi-party computation. In: Proceedings of the FOCS, pp. 238–248. IEEE Computer Society (2007)
10. Chatterjee, R., Bonneau, J., Juels, A., Ristenpart, T.: Cracking-resistant password vaults using natural language encoders. In: Proceedings of the S&P, pp. 481–498 (2015). https://doi.org/10.1109/SP.2015.36
11. Cho, C., Dachman-Soled, D., Jarecki, S.: Efficient concurrent covert computation of string equality and set intersection. In: Sako, K. (ed.) CT-RSA 2016. LNCS, vol. 9610, pp. 164–179. Springer, Cham (2016). https://doi.org/10.1007/978-3-319-29485-8_10
12. Craver, S.: On public-key steganography in the presence of an active warden. In: Aucsmith, D. (ed.) IH 1998. LNCS, vol. 1525, pp. 355–368. Springer, Heidelberg (1998). https://doi.org/10.1007/3-540-49380-8_25
13. Dedić, N., Itkis, G., Reyzin, L., Russell, S.: Upper and lower bounds on black-box steganography. J. Cryptol. **22**(3), 365–394 (2009)

14. Degabriele, J.P., Farshim, P., Poettering, B.: A more cautious approach to security against mass surveillance. In: Leander, G. (ed.) FSE 2015. LNCS, vol. 9054, pp. 579–598. Springer, Heidelberg (2015). https://doi.org/10.1007/978-3-662-48116-5_28

15. Diffie, W., Hellman, M.E.: New directions in cryptography. IEEE Trans. Inf. Theory **22**(6), 644–654 (1976)

16. Dolev, D., Dwork, C., Naor, M.: Nonmalleable cryptography. SIAM J. Comput. **30**(2), 391–437 (2000)

17. Fazio, N., Nicolosi, A.R., Perera, I.M.: Broadcast steganography. In: Benaloh, J. (ed.) CT-RSA 2014. LNCS, vol. 8366, pp. 64–84. Springer, Cham (2014). https://doi.org/10.1007/978-3-319-04852-9_4

18. Goldreich, O., Rothblum, R.D.: Enhancements of trapdoor permutations. J. Cryptol. **26**(3), 484–512 (2013)

19. Golla, M., Beuscher, B., Dürmuth, M.: On the security of cracking-resistant password vaults. In: Proceedings of the CCS, pp. 1230–1241 (2016). https://doi.org/10.1145/2976749.2978416

20. Hofheinz, D., Rao, V., Wichs, D.: Standard security does not imply indistinguishability under selective opening. In: Hirt, M., Smith, A. (eds.) TCC 2016. LNCS, vol. 9986, pp. 121–145. Springer, Heidelberg (2016). https://doi.org/10.1007/978-3-662-53644-5_5

21. Hopper, N.: On steganographic chosen covertext security. In: Caires, L., Italiano, G.F., Monteiro, L., Palamidessi, C., Yung, M. (eds.) ICALP 2005. LNCS, vol. 3580, pp. 311–323. Springer, Heidelberg (2005). https://doi.org/10.1007/11523468_26

22. Hopper, N.J., Langford, J., von Ahn, L.: Provably secure steganography. In: Yung, M. (ed.) CRYPTO 2002. LNCS, vol. 2442, pp. 77–92. Springer, Heidelberg (2002). https://doi.org/10.1007/3-540-45708-9_6

23. Hopper, N.J., von Ahn, L., Langford, J.: Provably secure steganography. IEEE Trans. Comput. **58**(5), 662–676 (2009)

24. Katz, J., Lindell, Y.: Introduction to Modern Cryptography, 2nd edn. CRC Press, Boca Raton (2014)

25. Katzenbeisser, S., Petitcolas, F.A.P.: Defining security in steganographic systems. In: Proceedings of the Electronic Imaging, pp. 50–56. SPIE (2002)

26. Kiayias, A., Raekow, Y., Russell, A., Shashidhar, N.: A one-time stegosystem and applications to efficient covert communication. J. Cryptol. **27**(1), 23–44 (2014)

27. Kiltz, E., Mohassel, P., O'Neill, A.: Adaptive trapdoor functions and chosen-ciphertext security. In: Gilbert, H. (ed.) EUROCRYPT 2010. LNCS, vol. 6110, pp. 673–692. Springer, Heidelberg (2010). https://doi.org/10.1007/978-3-642-13190-5_34

28. Van Le, T., Kurosawa, K.: Bandwidth optimal steganography secure against adaptive chosen stegotext attacks. In: Camenisch, J.L., Collberg, C.S., Johnson, N.F., Sallee, P. (eds.) IH 2006. LNCS, vol. 4437, pp. 297–313. Springer, Heidelberg (2007). https://doi.org/10.1007/978-3-540-74124-4_20

29. Lindell, Y.: A simpler construction of CCA2-secure public-key encryption under general assumptions. In: Biham, E. (ed.) EUROCRYPT 2003. LNCS, vol. 2656, pp. 241–254. Springer, Heidelberg (2003). https://doi.org/10.1007/3-540-39200-9_15

30. Liśkiewicz, M., Reischuk, R., Wölfel, U.: Grey-box steganography. Theoret. Comput. Sci. **505**, 27–41 (2013)

31. Luby, M., Rackoff, C.: How to construct pseudo-random permutations from pseudo-random functions. In: Williams, H.C. (ed.) CRYPTO 1985. LNCS, vol. 218, p. 447. Springer, Heidelberg (1986). https://doi.org/10.1007/3-540-39799-X_34

32. Lysyanskaya, A., Meyerovich, M.: Provably secure steganography with imperfect sampling. In: Yung, M., Dodis, Y., Kiayias, A., Malkin, T. (eds.) PKC 2006. LNCS, vol. 3958, pp. 123–139. Springer, Heidelberg (2006). https://doi.org/10.1007/11745853_9
33. Mitzenmacher, M., Upfal, E.: Probability and Computing - Randomized Algorithms and Probabilistic Analysis. Cambridge University Press, Cambridge (2005)
34. Naor, M., Yung, M.: Universal one-way hash functions and their cryptographic applications. In: Proceedings of the STOC, pp. 33–43. ACM (1989)
35. Pasquini, C., Schöttle, P., Böhme, R.: Decoy password vaults: at least as hard as steganography? In: De Capitani di Vimercati, S., Martinelli, F. (eds.) SEC 2017. IAICT, vol. 502, pp. 356–370. Springer, Cham (2017). https://doi.org/10.1007/978-3-319-58469-0_24
36. Ryabko, B., Ryabko, D.: Constructing perfect steganographic systems. Inf. Comput. **209**(9), 1223–1230 (2011)
37. Simon, D.R.: Finding collisions on a one-way street: can secure hash functions be based on general assumptions? In: Nyberg, K. (ed.) EUROCRYPT 1998. LNCS, vol. 1403, pp. 334–345. Springer, Heidelberg (1998). https://doi.org/10.1007/BFb0054137
38. von Ahn, L., Hopper, N.J.: Public key steganography. IACR Cryptology ePrint Archive, 2003/233 (2003)
39. von Ahn, L., Hopper, N.J.: Public-key steganography. In: Cachin, C., Camenisch, J.L. (eds.) EUROCRYPT 2004. LNCS, vol. 3027, pp. 323–341. Springer, Heidelberg (2004). https://doi.org/10.1007/978-3-540-24676-3_20
40. Wang, Y., Moulin, P.: Perfectly secure steganography: capacity, error exponents, and code constructions. IEEE Trans. Inf. Theory **54**(6), 2706–2722 (2008)

Memory Lower Bounds of Reductions Revisited

Yuyu Wang[1,2,3](✉) , Takahiro Matsuda[2], Goichiro Hanaoka[2],
and Keisuke Tanaka[1]

[1] Tokyo Institute of Technology, Tokyo, Japan
wang.y.ar@m.titech.ac.jp, keisuke@is.titech.ac.jp
[2] National Institute of Advanced Industrial Science and Technology (AIST),
Tokyo, Japan
{t-matsuda,hanaoka-goichiro}@aist.go.jp
[3] IOHK, Tokyo, Japan

Abstract. In Crypto 2017, Auerbach et al. initiated the study on memory-tight reductions and proved two negative results on the memory-tightness of restricted black-box reductions from multi-challenge security to single-challenge security for signatures and an artificial hash function. In this paper, we revisit the results by Auerbach et al. and show that for a large class of reductions treating multi-challenge security, it is impossible to avoid loss of memory-tightness unless we sacrifice the efficiency of their running-time. Specifically, we show three lower bound results. Firstly, we show a memory lower bound of natural black-box reductions from the multi-challenge unforgeability of unique signatures to any computational assumption. Then we show a lower bound of restricted reductions from multi-challenge security to single-challenge security for a wide class of cryptographic primitives with unique keys in the multi-user setting. Finally, we extend the lower bound result shown by Auerbach et al. treating a hash function to one treating any hash function with a large domain.

Keywords: Memory · Tightness · Lower bound · Uniqueness
Black-box reduction

1 Introduction

1.1 Background

Security proofs for cryptographic primitives are typically supported by the black-box reduction paradigm. A black-box reduction $\mathcal{R}$, which is a probabilistic polynomial-time (PPT) algorithm, allows us to convert an adversary $\mathcal{A}$ against some security game (or we say problem) GM_1 into an algorithm $\mathcal{R}^{\mathcal{A}}$ against another security game GM_2. If breaking GM_2 is believed to be hard, then the existence of $\mathcal{R}$ implies the security of GM_1. The quality of $\mathcal{R}$ depends on its tightness, which measures how close the performances of $\mathcal{A}$ and $\mathcal{R}^{\mathcal{A}}$ are. The

© International Association for Cryptologic Research 2018
J. B. Nielsen and V. Rijmen (Eds.): EUROCRYPT 2018, LNCS 10820, pp. 61–90, 2018.
https://doi.org/10.1007/978-3-319-78381-9_3

tighter a reduction is, the larger class of adversaries can be ruled out. Tightness traditionally takes running-time and success probability into account. However, Auerbach et al. [1] observed that some types of reductions, which are tight in common sense, are memory-loose, meaning that they incur large increase in memory usage when converting adversaries. For example, suppose that $\mathcal{A}$ use t_1 time steps and m_1 memory units, and succeed with probability ϵ_1 in GM_1. Even if $\mathcal{R}^{\mathcal{A}}$ can succeed in GM_2 with probability $\epsilon_2 \approx \epsilon_1$ by using $t_2 \approx t_1$ time steps, it may use $m_2 \gg m_1$ memory units. If the security of GM_2 is memory-sensitive, i.e., it can be broken more quickly with large memory than small memory (when the running-time of $\mathcal{A}$ is reasonably long), then a memory-loose reduction does not rule out as many attacks as expected. Recall the instance about the learning parities with noise (LPN) problem in dimension 1024 and error rate 1/4 in [1]. A memory-loose reduction from some security game to this problem only ensures that adversaries running in time less than 2^{85} cannot succeed in the game. There are many memory-sensitive problems besides the LPN problem, such as factoring, discrete-logarithm in prime fields, learning with errors, approximate shortest vector problem, short integer solution, t-collision-resistance (CR_t) where $t > 2$, etc., as noted in [1]. When proving security of cryptographic primitives based on these problems, memory-tightness should be seriously taken into account.

Memory Lower Bound of Restricted Reductions. Auerbach et al. initiated the study on memory-tightness, and provided general techniques helping achieve memory-tight reductions. Surprisingly, as negative results, they showed a memory lower bound of reductions from multi-challenge unforgeability (mUF) to standard unforgeability (UF) for signatures. The former security notion is defined in exactly the same way as the latter except that it gives an adversary many chances to produce a valid forgery rather than one chance. Although it is trivial to reduce mUF security to UF security tightly in both running-time and success probability, Auerbach et al. showed that some class of reductions between these two security notions inherently and significantly increase memory usage, unless they sacrifice the efficiency of the running-time. Specifically, they proved that such a reduction must consume roughly $\Omega(q/(p+1))$ bits of memory, where $2q$ is the number of queries made by an adversary and p is the number of times an adversary is run. The class of black-box reductions they treated is restricted, in the sense that a reduction $\mathcal{R}$ only runs an adversary $\mathcal{A}$ sequentially from beginning to end, and is not allowed to rewind $\mathcal{A}$. Moreover, $\mathcal{R}$ only forwards the public keys and signing queries between its challenger and $\mathcal{A}$, and the forgery made by $\mathcal{R}$ should be amongst the ones generated by $\mathcal{A}$. This result implies that in practice, UF security and mUF security may not really be equivalent. As an open problem left by Auerbach et al., it is not clear whether this result holds when a reduction does not respect the restrictions. Moreover, this result does not rule out the possibility that there exists a memory-tight restricted reduction that directly derives mUF security from some memory-sensitive problem. Therefore, it is desirable to clarify whether there exists a memory lower bound of any natural reduction from mUF security to any common assumption.

Auerbach et al. also showed another similar lower bound of restricted reductions from multi-challenge t-collision-resistance (mCR_t) to standard CR_t security for an artificial hash function that truncates partial bits of its input. Here, both security notions prevent an adversary from finding a t-collision (i.e., outputting t distinct elements having the same hash value), while the mCR_t (respectively, CR_t) game allows an adversary to have many chances (respectively, only one chance) to find a t-collision. Since CR_t security is memory-sensitive, this result indicates that breaking mCR_t security might be much easier than breaking CR_t security in practice. However, since the hash function they considered is specific and not collision-resistant, it is still not clear whether this result holds for collision-resistant hash functions.

Finally, it is desirable to clarify whether there exist memory lower bounds for cryptographic primitives in other settings, which are potentially based on memory-sensitive problems.

1.2   Our Results

We revisit memory-tightness on black-box reductions, and show several lower bound results.

Lower Bound for Unique Signatures. In [6], Coron proved a tightness lower bound of black-box reductions from the security of unique signatures [10, 19, 20], in which there exists only one valid signature for each pair of public key (not necessarily output by the key generation algorithm) and message, to any non-interactive (computational) assumption. Later, Kakvi and Kiltz [15] and Bader et al. [4] respectively fixed a flaw in the proof and improved the bound. The reductions considered in these works are "natural" reductions, in the sense that they run adversaries only sequentially.

Although the study on the tightness of reductions for unique signatures has a long history, memory-tightness of such reductions has never been taken into account until [1], and it is still unclear, when considering natural reductions or reducing the security of unique signatures to common assumptions, whether memory-tightness is achievable. In our work, we focus on natural reductions for unique signatures from the angle of memory, and prove that loss of memory-tightness is inevitable when reducing their mUF security to computational assumptions. Specifically, we show the existence of a memory lower bound of any natural reduction from the mUF security of unique signatures to any computational assumption (rather than only UF security).[1] Here, a natural black-box reduction can interact with its challenger in any way it wants, and can adaptively rewind an adversary. We do not allow reductions to modify the internal state of an adversary, which is a very natural restriction. Similarly to [1], the bound is roughly $\Omega(q/(p+1))$ bits of memory, where $2q$ is the number of queries made by an adversary and p is the number of times an adversary is rewound. This result indicates that for a unique signature scheme, any natural reduction

---

[1] Note that all the memory-sensitive problems discussed in [1] fall under the notion of computational assumptions.

from its mUF security to a memory-sensitive problem may not rule out as many attacks as expected. Therefore, when using a unique signature scheme based on a memory-sensitive problem in practice, one should make its security parameter larger than indicated by traditional security proofs. As far as we know, this is the first negative result on memory-tight reductions to any computational assumptions, and also the first one treating memory-tightness of natural reductions.

Moreover, we give our result in a generalized way so that it also captures some other assumptions that do not fall under the definition of computational assumptions. By slightly modifying our proof, we can also show memory lower bounds for the notions of verifiable unpredictable functions (VUFs) and re-randomizable signatures, which are more general primitives and hence capture more instantiations (e.g., [13,19,20,23]).

Lower Bound for Unique-Key Primitives in the Multi-user Setting. Security notions of cryptographic primitives are usually considered in the single-user setting, where an adversary only sees one challenge public key. However, in practice, an attacker may see many public keys and adaptively corrupt secret keys. Hence, considering security of primitives in the multi-user setting [2,3] is necessary. In [4], Bader et al. showed that in this setting, it is impossible to avoid loss of tightness when deriving the security of unique-key primitives, in which there exists only one valid secret key for each public key, from non-interactive assumptions.

In this work, we give the first negative result on memory-tightness in the multi-user setting. Specifically, we show a memory lower bound of restricted black-box reductions from multi-challenge one-wayness in the multi-user setting (mU-mOW) to standard one-wayness in the multi-user setting (mU-OW) for unique-key relations. Compared with [1], the reductions we treat are less restricted. We only require them to forward the public keys and corruption queries between the challengers and adversaries, while they can forge secret keys in any way they want (i.e., a forgery is not necessarily amongst the ones output by an adversary). The bound is roughly $\Omega(\max\{q/(p+2), n/(p+2)\})$, where $2q$ is the number of queries, n is the number of users, and p is the number of rewinding procedures. Since unique-key relations are very fundamental primitives, from this result, we can easily derive lower bounds for a large class of primitives with unique keys (including public key encryption (PKE) schemes, signatures, trapdoor commitment schemes (with collision-resistance), etc.), which capture many constructions (e.g., [5,7,8,12,14,17,18,22,23]). These results imply that for unique-key primitives in the multi-user setting, the gaps between their multi-challenge security notions and single-challenge security notions might be wider than indicated by conventional security proofs via restricted reductions.

As a by-product result, our result can be extended for primitives with re-randomizable keys [4], where secret keys can be efficiently re-randomized and the distribution of a re-randomized key is uniform.

Lower Bound for Large-Domain Hash Functions. Finally, we revisit the memory lower bound of restricted reductions from mCR_t security to CR_t security for an artificial hash function shown in [1]. We firstly show a streaming lower bound for all the CR_t secure large-domain hash functions. Specifically, we show that determining whether there exists a t-collision in a data stream consumes large memory. Following from this fact, we extend the result in [1] to a lower bound for all the large-domain hash functions. Here, a hash function is said to have a large domain if its range is negligibly small compared with its domain (e.g., $\mathsf{H} : \{0,1\}^{2\lambda} \to \{0,1\}^{\lambda}$ where λ is the security parameter). It is a natural property satisfied by most practical hash functions. The bound is roughly $\Omega(\min\{(q - \kappa)/(p + 1)\})$ where q is the number of queries, κ is the length of the hash key, and p is the number of rewinding procedures. Since CR_t security (where $t > 2$) is memory-sensitive, this result implies that for any natural hash function, its mCR_t security directly derived from its CR_t security via restricted reductions does not rule out as many attacks as its CR_t security does in practice.

1.3   High-Level Ideas

Like in [1], our lower bound for unique signatures follows from a streaming lower bound result implying that determining the output of a specific function $G(\boldsymbol{y})$ consumes large memory. Here, $\boldsymbol{y}$ is a data stream that does not occupy local memory and can be accessed sequentially. We construct an inefficient adversary $\mathcal{A}_{\boldsymbol{y}}$ (storing $\boldsymbol{y}$) breaking the mUF security of any unique signature scheme iff $G(\boldsymbol{y}) = 1$. Let $\mathcal{R}$ be a black-box reduction from mUF security to a cryptographic game GM. $\mathcal{R}^{\mathcal{A}_{\boldsymbol{y}}}$ is likely to succeed in GM when $G(\boldsymbol{y}) = 1$. On the other hand, when $G(\boldsymbol{y}) = 0$, we use the meta-reduction method to show that $\mathcal{R}^{\mathcal{A}_{\boldsymbol{y}}}$ will fail. Roughly, we construct a PPT simulator $\mathcal{S}_{\boldsymbol{y}}$ that is indistinguishable from $\mathcal{A}_{\boldsymbol{y}}$ due to uniqueness. If $\mathcal{R}^{\mathcal{A}_{\boldsymbol{y}}}$ succeeds in GM, then the PPT algorithm $\mathcal{R}^{\mathcal{S}_{\boldsymbol{y}}}$ succeeds in GM as well, which gives us the conflict. As a result, we can obtain an algorithm that determines $G(\boldsymbol{y})$ with high probability by simulating the game GM and $\mathcal{R}^{\mathcal{A}_{\boldsymbol{y}}}$. Such an algorithm must consume large memory due to the streaming lower bound. Moreover, $\mathcal{A}_{\boldsymbol{y}}$ can be simulated by accessing the stream $\boldsymbol{y}$ with small memory usage. Therefore, $\mathcal{R}$ must use large memory if simulating the challenger in GM does not consume large memory. This is the case in most computational assumptions (including all the memory-sensitive problem noted in [1]), where the challenger saves an answer, which only occupies small memory, when sampling a challenge, and checks whether the final output of an adversary is equal to that answer.

The lower bound of restricted reductions from mU-mOW security to mU-OW security for unique-key primitives is shown in a similar way by constructing an inefficient adversary and its simulator in the mU-mOW game. However, in this case, we face a problem that it consumes large memory to store public keys of users when running the mU-OW and mU-mOW games. This spoils our result since the streaming lower bound does not imply that $\mathcal{R}$ consumes large memory any more. We deal with this problem by running a pseudorandom function (PRF) to simulate random coins used to generate public keys, which is similar to the technique used in [1] for achieving memory-tightness. Whenever a public key is needed, we only have to run the PRF to obtain the corresponding random coin and generate the key again, and hence there is no need to store public keys any more. Here, it might seem that outputs of PRF are not indistinguishable from real random coins since an inefficient adversary is involved in the interaction. However, we can show that the adversary can be simulated in polynomial-time (PT) due to the uniqueness of secret keys.

Extending the lower bound result for a specific hash function in [1] to all the large-domain hash functions satisfying CR_t security (where t is a constant) involves three steps. Firstly, we prove a theorem saying that for a large-domain hash function satisfying CR_t security, there exist many hash values with more than t pre-images. Intuitively, for a large-domain hash function (using a randomly chosen key), if there are few hash values with more than t pre-images, then there should exist some hash value with many pre-images. We prove that the set of all pre-images of such a hash value is so large that t randomly chosen inputs are very likely to fall into this set, which conflicts with CR_t security. Therefore, we conclude that a CR_t secure large-domain hash function should have many hash values with more than t pre-images. Then by exploiting this theorem and the technique used in previous works [1,16,21], we prove the existence of a memory lower bound for determining whether there exists a t-collision in a stream, based on the disjointness problem [16,21]. Following from this result, we achieve a memory lower bound of restricted reductions from mCR_t security to CR_t security for large-domain hash functions.

1.4   Outline of This Paper

In Sect. 2, we recall some notation and describe the computational model and data stream model. In Sect. 3, we show a lower bound of black-box reductions from the mUF security of unique signatures to cryptographic games. In Sect. 4, we show a lower bound of restricted reductions from mU-mOW security to mU-OW security for unique-key cryptographic primitives. In Sect. 5, we show a lower bound of restricted reductions from mCR_t security to CR_t security for large-domain hash functions.

2   Preliminaries

In this section, we give several terminologies that are necessary to describe our results, describe the computational model and data stream model, and recall the disjointness problem and a streaming lower bound.

2.1   Notation and Computational Model

In this paper, all algorithms are RAMs having access to memory and registers that each holds one word. Rewinding random bits used by RAMs is not permitted, so if an algorithm wants to access previously used random bits it must store them. If $\mathcal{A}$ is a deterministic (respectively, probabilistic) algorithm, then $y = \mathcal{A}(x)$ (respectively, $y \leftarrow \mathcal{A}(x)$) means that $\mathcal{A}$ takes as input x and outputs y. By $\mathcal{A}^{\mathcal{O}}$ we mean that $\mathcal{A}$ has access to an oracle $\mathcal{O}$. By $\mathcal{A}_z$ we mean that z is stored in the memory of $\mathcal{A}$. We denote the code and memory consumed by $\mathcal{A}$ (but not its oracle) in the worst case by **LocalMem**$(\mathcal{A})$, where the consumption is measured in bits. *negl* denotes an unspecified negligible function. If $\mathcal{Z}$ is a finite set, then $|\mathcal{Z}|$ denotes the number of (distinct) elements in $\mathcal{Z}$, and $z \leftarrow \mathcal{Z}$ denotes the process of sampling z at uniformly random from $\mathcal{Z}$.

2.2 Data Stream Model

Now we recall stream oracles. To a stream oracle, an algorithm is allowed to make queries to access a large stream of data sequentially, while the local memory consumption remains small. We adopt the notation in [1] to describe stream oracles as follows.

A stream oracle $\mathcal{O}_{\boldsymbol{y}}$ is parameterized by a vector $\boldsymbol{y} = (y_1, \cdots, y_n) \in \mathcal{U}^n$ where $\mathcal{U}$ is some finite set. Whenever receiving a query, $\mathcal{O}_{\boldsymbol{y}}$ runs $i = i+1 \mod n$ (where i is initialized with 0), and returns y_i. Let q be the total number of queries. *The number of passes* is defined as $p = \lceil q/n \rceil$.

2.3 Disjointness Problem and Streaming Lower Bound

Now we recall the disjointness problem, which derives streaming lower bounds.

Theorem 1 ([16,21]). *Let* $x_1, x_2 \in \{0,1\}^n$ *and* $\mathsf{DISJ}(x_1, x_2)$ *be defined by*

$$\mathsf{DISJ}(x_1, x_2) = \begin{cases} 1 & \text{if } \exists i : x_1[i] = x_2[i] = 1 \\ 0 & \text{otherwise} \end{cases},$$

where $x_b[j]$ *denotes the jth bit of x_b for $j \in \{1, \cdots, n\}$ and $b \in \{0,1\}$. Then any two-party protocol (P_1, P_2), such that $\Pr[\mathsf{DISJ}(x_1, x_2) \leftarrow (P_1(x_1) \leftrightharpoons P_2(x_2))] \geq c$ holds for some constant $c > 1/2$ and every $x_1, x_2 \in \{0,1\}^n$, must have communication $\Omega(n)$ in the worst case. Here, by $\mathsf{DISJ}(x_1, x_2) \leftarrow (P_1(x_1) \leftrightharpoons P_2(x_2))$ we mean that the interaction between P_1 and P_2 respectively on input x_1 and x_2 outputs $\mathsf{DISJ}(x_1, x_2)$.*

In [1], Auerbach et al. gave a streaming lower bound result, which is a corollary of prior works [16,21] based on the disjointness problem. It shows that determining whether the second half of a stream contains an element not in the first half requires large memory. We now follow [1] to define $G(\boldsymbol{y})$, where $\boldsymbol{y} = \boldsymbol{y}_1 \| \boldsymbol{y}_2$ and $\boldsymbol{y}_1, \boldsymbol{y}_2 \in \mathcal{U}^q$, and recall the streaming lower bound.

$$G(\boldsymbol{y}) = \begin{cases} 1 & \text{if } \exists j \; \forall i : \boldsymbol{y}_2[j] \neq \boldsymbol{y}_1[i] \\ 0 & \text{otherwise} \end{cases}.$$

Theorem 2. *Let $\mathcal{B}$ be a probabilistic algorithm and λ be a (sufficiently large) security parameter. Assuming that there exists some constant $c > 1/2$ such that $\Pr[\mathcal{B}^{\mathcal{O}_{\boldsymbol{y}}}(1^\lambda) = G(\boldsymbol{y})] \geq c$ holds for polynomials $q = q(\lambda)$ and $n = n(\lambda)$, and all $\boldsymbol{y} \in (\{0,1\}^n)^{2q}$ (respectively, $\boldsymbol{y} \in (\{i\}_{i=1}^n)^{2q}$). Then we have **LocalMem**$(\mathcal{B}) = \Omega(\min\{q/p, 2^n/p\})$ (respectively, **LocalMem**$(\mathcal{B}) = \Omega(\min\{q/p, n/p\}))$, where p is the number of passes $\mathcal{B}$ makes in the worst case.*

The above theorem is slightly different from the one in [1], in the sense that we let $y \in (\{0,1\}^n)^{2q}$ or $y \in (\{i\}_{i=1}^n)^{2q}$ (instead of $y \in \mathcal{U}^{2q}$ for all sufficiently large q and $|\mathcal{U}|$), and require q and n be polynomials in λ. However, the proof for the streaming lower bound in [1, Appendix A] can be directly applied to prove the above theorem. We refer the reader to [1, Appendix A] for details.

3   Lower Bound of Reductions from the mUF Security of Unique Signatures to Cryptographic Games

In this section, we show a memory lower bound of black-box reductions from the mUF security of unique signatures to assumptions captured by cryptographic games. We start by recalling the definition of unique signatures and mUF security, and then show the lower bound.

3.1   Unique Signatures and mUF Security

We now recall the definition of (digital) signatures.

Definition 1 (Digital signature). *A signature scheme consists of PT algorithms* (Gen, Sign, Verify). *(a)* Gen *is a probabilistic algorithm that takes as input* 1^λ, *and returns a public/secret key pair* (pk, sk). *(b)* Sign *is a probabilistic algorithm that takes as input a secret key* sk *and a message* $m \in \{0,1\}^\delta$ *where* $\delta = \delta(\lambda)$ *is some polynomial, and returns a signature* σ. *(c)* Verify *is a deterministic algorithm that takes as input a public key* pk, *a message* m, *and a signature* σ, *and returns* 1 *(accept) or* 0 *(reject).*

A signature scheme is required to satisfy correctness, which means that Verify$_{pk}(m, \sigma) = 1$ *holds for all* $\lambda \in \mathbb{N}$, *all* $(pk, sk) \leftarrow$ Gen(1^λ), *all* $m \in \{0,1\}^\delta$, *and all* $\sigma \leftarrow$ Sign$_{sk}(m)$.

Next we recall the definition of unique signatures, in which there exists only one valid signature for each pair of public key (not necessarily output by Gen(1^λ)) and message.

Definition 2 (Unique signature [19]). *A signature scheme* (Gen, Sign, Verify) *is said to be a* unique *signature scheme if for all* $\lambda \in \mathbb{N}$, *all* pk *(possibly outside the support of* Gen*), and all* $m \in \{0,1\}^\delta$, *there exists no pair* (σ, σ') *that simultaneously satisfies* $\sigma \neq \sigma'$ *and* Verify$_{pk}(m, \sigma) =$ Verify$_{pk}(m, \sigma') = 1$.

Now we recall mUF security. In the mUF game, an adversary has many chances to produce a valid forgery rather than one chance. Although mUF security can be tightly reduced to UF security straightforwardly in common sense, it is shown in [1] that restricted reductions between these two security notions inherently require increased memory usage.

Definition 3 (mUF [1]). *A signature scheme* (Gen, Sign, Verify) *is said to be* mUF *secure if for any PPT adversary* $\mathcal{A}$, *we have* $\mathbf{Adv}_{mUF}^{\mathcal{A}}(\lambda) =$ Pr[$\mathcal{CH}$ *outputs* 1] $\leq$ *negl*(λ) *in the following game.*

1. *The challenger $\mathcal{CH}$ sets $w = 0$ and $\mathcal{Q} = \emptyset$, samples $(pk, sk) \leftarrow \mathsf{Gen}(1^\lambda)$, and runs $\mathcal{A}$ on input $(1^\lambda, pk)$. $\mathcal{A}$ may make adaptive signing and verification queries to $\mathcal{CH}$, and $\mathcal{CH}$ responds as follows:*
   - *On receiving a signing query m, $\mathcal{CH}$ computes $\sigma \leftarrow \mathsf{Sign}_{sk}(m)$, adds m to $\mathcal{Q}$, and sends σ to $\mathcal{A}$.*
   - *On receiving a verification query (m^*, σ^*), if $\mathsf{Verify}_{pk}(m^*, \sigma^*) = 1$ and $m^* \notin \mathcal{Q}$, $\mathcal{CH}$ sets $w = 1$.*
2. *At some point, $\mathcal{A}$ makes a stopping query stp to $\mathcal{CH}$, and $\mathcal{CH}$ returns w.*

The definition of UF security is exactly the same as the above one except that $\mathcal{A}$ is allowed to make only one verification query and the advantage of $\mathcal{A}$ is denoted by $\mathbf{Adv}_{\mathsf{UF}}^{\mathcal{A}}(\lambda)$.

3.2 Lower Bound for Unique Signatures

Before giving the main theorem, we recall the definition of cryptographic games.

Definition 4 (Cryptographic game [11]). *A cryptographic game GM consists of a (possibly inefficient) random system (called the challenger) $\mathcal{CH}$ and a constant c. On input security parameter 1^λ, $\mathcal{CH}(1^\lambda)$ interacts with some adversary $\mathcal{A}(1^\lambda)$, and outputs a bit b. This interaction is denoted by $b \leftarrow (\mathcal{A}(1^\lambda) \leftrightharpoons \mathcal{CH}(1^\lambda))$, and the advantage of $\mathcal{A}$ in GM is $\mathbf{Adv}_{\mathsf{GM}}^{\mathcal{A}}(\lambda) = \Pr[1 \leftarrow (\mathcal{A}(1^\lambda) \leftrightharpoons \mathcal{CH}(1^\lambda))] - c$.*

  A cryptographic game $\mathsf{GM} = (\mathcal{CH}, c)$ is said to be secure if for any PPT adversary $\mathcal{A}$, we have $\mathbf{Adv}_{\mathsf{GM}}^{\mathcal{A}}(\lambda) \leq negl(\lambda)$.

All commonly used assumptions and most security games in cryptography fall under the framework of cryptographic games. We call a cryptographic game $\mathsf{GM} = (\mathcal{CH}, c)$ a *computational assumption* if $c = 0$.

Black-Box Reduction. Now we follow [1] to describe black-box reductions. Unlike in [1], we do not fix the random tape of an adversary, and do not give any restriction on the queries made by a reduction.[2]

  Let $\mathcal{R}$ be a black-box reduction from GM_1 to GM_2. We write $\mathcal{R}^{\mathcal{A}}$ to mean that $\mathcal{R}$ has oracle access to a (stateful) adversary $\mathcal{A}$ playing game GM_1. Whenever receiving a query from $\mathcal{R}$, $\mathcal{A}$ returns the "next" query to $\mathcal{R}$. $\mathcal{R}$ is not able to modify the current state of $\mathcal{A}$ (i.e., $\mathcal{A}$ runs sequentially), but is allowed to adaptively rewind $\mathcal{A}$ to previous states.

Definition 5 (c-black-box reduction). *Let GM_1 and GM_2 be cryptographic games and $c > 0$ be a constant. An oracle-access PPT machine $\mathcal{R}^{(\cdot)}$ is said to be a c-black-box reduction from GM_1 to GM_2, if for any (sufficiently large) security parameter λ and any (possibly inefficient) adversary $\mathcal{A}$, we have $\mathbf{Adv}_{\mathsf{GM}_2}^{\mathcal{R}^{\mathcal{A}}}(\lambda) \geq c \cdot \mathbf{Adv}_{\mathsf{GM}_1}^{\mathcal{A}}(\lambda)$.*

---

[2] Auerbach et al. requires a reduction to preserve the advantage of an adversary even if the random tape of the adversary is fixed. However, we observe that this restriction is not necessary in their work as well, which we will discuss after giving the proof.

Like many previous works (e.g., [1,4,6,15]), we do not consider reductions that can modify the current state of an adversary. This is a natural restriction, which is respected by most black-box reductions.

We now give a theorem showing a memory lower bound of c_r-black-box reductions from the mUF security of unique signatures to cryptographic games $\mathsf{GM} = (\mathcal{CH}, c_g)$, where $c_g < 1/2$ and $c_r + c_g > 1/2$. When $c_g = 0$, our result captures c_r-black-box reductions where $c_r > 1/2$ to any computational assumption. When $c_r = 1$, it captures 1-black-box reductions to any cryptographic game such that $c_g < 1/2$.[3]

Theorem 3. *Let λ be a (sufficiently large) security parameter, $\Sigma = (\mathsf{Gen}, \mathsf{Sign}, \mathsf{Verify})$ be a unique signature scheme with message length δ, $\mathsf{GM} = (\mathcal{CH}, c_g)$ be a secure cryptographic game, $\mathbf{LocalMem}(\mathcal{CH})$ be the amount of memory consumed by $\mathcal{CH}$, and $\mathcal{R}$ be a c_r-black box reduction from the mUF security of Σ to the security of GM. Let $q = q(\lambda)$ be the maximum numbers of signing queries and verification queries made by an adversary in the mUF game. If (a) $\mathcal{R}$ rewinds the adversary for at most $p = p(\lambda)$ times and (b) $c_g < 1/2$ and $c_r + c_g > 1/2$, then we have*

$$\mathbf{LocalMem}(\mathcal{R}) = \Omega(\min\{q/(p+1), 2^\delta/(p+1)\}) - O(\log q)$$
$$- \mathbf{LocalMem}(\mathcal{CH}) - \mathbf{LocalMem}(\mathsf{Verify}).$$

Roughly, this theorem implies that when the maximum number of signing queries made by an adversary in the mUF game is very large, $\mathcal{R}$ must consume large memory unless it rewinds $\mathcal{A}$ many times, which increases its running-time.

High-Level Idea. We firstly construct an inefficient adversary $\mathcal{A}_{\boldsymbol{y}}$ where $\boldsymbol{y} = (y_1, \cdots, y_{2q})$. $\mathcal{A}_{\boldsymbol{y}}$ makes signing queries $y_1, \cdots, y_q$, checks the validity of the answers, and then makes verification queries $(y_{q+1}, \sigma_1^*), \cdots, (y_{2q}, \sigma_q^*)$ which are generated by using brute force. Consider the interaction $\mathcal{R}^{\mathcal{A}_{\boldsymbol{y}}}(1^\lambda) \leftrightarrows \mathcal{CH}(1^\lambda)$. When $G(\boldsymbol{y}) = 1$ (see Sect. 2.3 for the definition of G), we have $\{y_{q+i}\}_{i=1}^q \not\subseteq \{y_i\}_{i=1}^q$, which means that $\mathcal{A}_{\boldsymbol{y}}$ is a deterministic algorithm breaking mUF security. Since $\mathcal{R}$ is a black-box reduction, $\mathcal{CH}$ is likely to output 1 in this case. When $G(\boldsymbol{y}) = 0$, we have $\{y_{q+i}\}_{i=1}^q \subseteq \{y_i\}_{i=1}^q$, in which case we can construct a PT algorithm $\mathcal{S}_{\boldsymbol{y}}$ running in the same way as $\mathcal{A}_{\boldsymbol{y}}$ does, except that $\mathcal{S}_{\boldsymbol{y}}$ uses the answers of signing queries as its forgeries instead of exploiting brute force. Due to uniqueness, $\mathcal{S}_{\boldsymbol{y}}$ perfectly simulates $\mathcal{A}_{\boldsymbol{y}}$. If $\mathcal{CH}$ outputs 1 with probability that is non-negligibly greater than c_g in the interaction with $\mathcal{R}^{\mathcal{A}_{\boldsymbol{y}}}$, then we have a PPT algorithm $\mathcal{R}^{\mathcal{S}_{\boldsymbol{y}}}$ breaking the security of GM, which gives us the conflict. Therefore, $\mathcal{CH}$ is likely to output 0 when $G(\boldsymbol{y}) = 0$.

Then we can construct an algorithm $\mathcal{B}$ with access to a stream $\boldsymbol{y}$ that simulates the interaction $\mathcal{R}^{\mathcal{A}_{\boldsymbol{y}}}(1^\lambda) \leftrightarrows \mathcal{CH}(1^\lambda)$ and outputs $G(\boldsymbol{y})$ with high probability. According to Theorem 2, the memory consumed by $\mathcal{B}$ is inherently large

---

[3] There are several typical cryptographic games with $0 < c_g < 1/2$, such as recipient-anonymity for IBE schemes [9] and one-wayness for encryption schemes with constantly large message spaces.

(to some extent). Moreover, although $\mathcal{A}_y$ consumes a large amount of memory to store y, $\mathcal{B}$ does not have to use large memory when simulating $\mathcal{A}_y$ by accessing its stream. As a result, if the memory consumed by $\mathcal{CH}$ is small (which is often the case in computational assumptions), then $\mathcal{R}$ must consume large memory.

Proof (of Theorem 3). Assuming the existence of the reduction $\mathcal{R}$ stated in Theorem 3, we show the existence of a probabilistic algorithm $\mathcal{B}$ such that $\Pr[G(y) \leftarrow \mathcal{B}^{\mathcal{O}_y}(1^\lambda)] \geq c$ for all $y = (y_1, \cdots, y_{2q}) \in (\{0,1\}^\delta)^{2q}$ and some constant $c > 1/2$ by giving hybrid games.

Game 0: In this game, $\mathcal{R}$ has access to an adversary $\mathcal{A}_y$ and interacts with $\mathcal{CH}$, where $\mathcal{A}_y$ runs as follows.

1. On receiving $(1^\lambda, pk)$, $\mathcal{A}_y$ stores $(1^\lambda, pk)$ and makes a signing query y_1.
2. For $i = 1, \cdots, q-1$, on receiving the answer σ_i to the ith signing query, if $\mathsf{Verify}_{pk}(y_i, \sigma_i) \neq 1$, $\mathcal{A}_y$ aborts. Otherwise, $\mathcal{A}_y$ makes a signing query y_{i+1}.
3. On receiving the answer σ_q to the qth signing query, if $\mathsf{Verify}_{pk}(y_q, \sigma_q) \neq 1$, $\mathcal{A}_y$ aborts. Otherwise, $\mathcal{A}_y$ exhaustively searches σ_1^* such that $\mathsf{Verify}_{pk}(y_{q+1}, \sigma_1^*) = 1$, and makes a verification query (y_{q+1}, σ_1^*).
4. For $i = 1, \cdots, q-1$, when invoked (with no input) for the ith time, $\mathcal{A}_y$ exhaustively searches σ_{i+1}^* such that $\mathsf{Verify}_{pk}(y_{q+i+1}, \sigma_{i+1}^*) = 1$, and makes a verification query $(y_{q+i+1}, \sigma_{i+1}^*)$.
5. When invoked (with no input) for the qth time, $\mathcal{A}_y$ makes a stopping query *stp*.

We now show the following lemma.

Lemma 1. $\Pr[G(y) \leftarrow (\mathcal{R}^{\mathcal{A}_y}(1^\lambda) \leftrightarrows \mathcal{CH}(1^\lambda))] \geq c$ *for all* $y \in (\{0,1\}^\delta)^{2q}$ *and some constant* $c > 1/2$ *in* **Game 0**.

Proof (of Lemma 1). Firstly, we show the existence of a PT algorithm $\mathcal{S}_y$ perfectly simulating $\mathcal{A}_y$ on condition that $G(y) = 0$. $\mathcal{S}_y$ runs in the same way as $\mathcal{A}_y$ except that it uses the answers of the signing queries as its verification queries. Formally, it runs as follows. (Below, the difference from $\mathcal{A}_y$ is *emphasized*.)

1. On receiving $(1^\lambda, pk)$, $\mathcal{S}_y$ stores $(1^\lambda, pk)$ and makes a signing query y_1.
2. For $i = 1, \cdots, q-1$, on receiving the answer σ_i to the ith signing query, if $\mathsf{Verify}_{pk}(y_i, \sigma_i) \neq 1$, $\mathcal{S}_y$ aborts. Otherwise, $\mathcal{S}_y$ *stores* (y_i, σ_i) *in its internal list* L *(initialized with* $\emptyset$*)*, and makes a signing query y_{i+1}.
3. On receiving the answer σ_q to the qth signing query, if $\mathsf{Verify}_{pk}(y_q, \sigma_q) \neq 1$, $\mathcal{S}_y$ aborts. Otherwise, $\mathcal{S}_y$ *stores* (y_q, σ_q) *in* L, *searches a pair* (m, σ) *in* L *such that* $m = y_{q+1}$, *and makes a verification query* (m, σ). *If the searching procedure fails,* $\mathcal{S}_y$ *aborts.*
4. For $i = 1, \cdots, q-1$, when invoked (with no input) for the ith time, $\mathcal{S}_y$ *searches a pair* (m, σ) *in* L *such that* $m = y_{q+i+1}$, *and makes a verification query* (m, σ). *If the searching procedure fails for some* i, $\mathcal{S}_y$ *aborts.*
5. When invoked (with no input) for the qth time, $\mathcal{S}_y$ makes a stopping query *stp*.

When $G(\boldsymbol{y}) = 0$, we have $\{y_{q+i}\}_{i=1}^{q} \subseteq \{y_i\}_{i=1}^{q}$, which means that the searching procedures executed by $\mathcal{S}_y$ (in Steps 3 and 4) will not fail. Moreover, due to the uniqueness of Σ, the verification queries made by $\mathcal{S}_y$ are exactly the same as those made by $\mathcal{A}_y$. Hence, $\mathcal{S}_y$ perfectly simulates $\mathcal{A}_y$ in the view of $\mathcal{R}$.

Due to the security of GM, we have $\mathbf{Adv}_{\mathsf{GM}}^{\mathcal{R}^{\mathcal{A}_y}}(\lambda) = \mathbf{Adv}_{\mathsf{GM}}^{\mathcal{R}^{\mathcal{S}_y}}(\lambda) \leq negl(\lambda)$ when $G(\boldsymbol{y}) = 0$, which implies $\Pr[1 \leftarrow (\mathcal{R}^{\mathcal{A}_y}(1^\lambda) \leftrightharpoons \mathcal{CH}(1^\lambda)) \mid G(\boldsymbol{y}) = 0] - c_g \leq negl(\lambda)$, i.e., $\Pr[0 \leftarrow (\mathcal{R}^{\mathcal{A}_y}(1^\lambda) \leftrightharpoons \mathcal{CH}(1^\lambda)) \mid G(\boldsymbol{y}) = 0] \geq 1 - c_g - negl(\lambda)$. On the other hand, when $G(\boldsymbol{y}) = 1$, there exists some $1 \leq j \leq q$ such that $y_{q+j} \notin \{y_i\}_{i=1}^{q}$, which implies $\mathbf{Adv}_{\mathsf{mUF}}^{\mathcal{A}_y}(\lambda) = 1$. Since $\mathcal{R}$ is a c_r-black-box reduction, we have $\Pr[1 \leftarrow (\mathcal{R}^{\mathcal{A}_y}(1^\lambda) \leftrightharpoons \mathcal{CH}(1^\lambda)) \mid G(\boldsymbol{y}) = 1] - c_g \geq c_r$. Since $c_g < 1/2$, $c_r + c_g > 1/2$, and λ *is sufficiently large*, there exists some constant $c > 1/2$ such that $\Pr[G(\boldsymbol{y}) \leftarrow (\mathcal{R}^{\mathcal{A}_y}(1^\lambda) \leftrightharpoons \mathcal{CH}(1^\lambda))] \geq c$ for all $\boldsymbol{y} \in (\{0,1\}^\delta)^{2q}$, completing the proof of Lemma 1. □

Game 1: This game is exactly the same as **Game 0** except that there exists an algorithm $\mathcal{A}'$ with access to the stream oracle $\mathcal{O}_y$ simulating $\mathcal{A}_y$ as follows. (Below, the difference from $\mathcal{A}_y$ is *emphasized*.)

1. On receiving $(1^\lambda, pk)$, $\mathcal{A}'$ stores $(1^\lambda, pk)$, *queries $\mathcal{O}_y$ to obtain y_1*, and makes a signing query y_1.
2. For $i = 1, \cdots, q - 1$, on receiving the answer σ_i to the ith signing query, if $\mathsf{Verify}_{pk}(y_i, \sigma_i) \neq 1$, $\mathcal{A}'$ aborts. Otherwise, $\mathcal{A}'$ *queries $\mathcal{O}_y$ to obtain y_{i+1}* and makes a signing query y_{i+1}.
3. On receiving the answer σ_q to the qth signing query, if $\mathsf{Verify}_{pk}(y_q, \sigma_q) \neq 1$, $\mathcal{A}'$ aborts. Otherwise, $\mathcal{A}'$ *queries $\mathcal{O}_y$ to obtain y_{q+1}*, exhaustively searches σ_1^* such that $\mathsf{Verify}_{pk}(y_{q+1}, \sigma_1^*) = 1$, and makes a verification query (y_{q+1}, σ_1^*).
4. For $i = 1, \cdots, q - 1$, when invoked (with no input) for the ith time, $\mathcal{A}'$ *queries $\mathcal{O}_y$ to obtain y_{q+i+1}*, exhaustively searches σ_{i+1}^* such that $\mathsf{Verify}_{pk}(y_{q+i+1}, \sigma_{i+1}^*) = 1$, and makes a verification query $(y_{q+i+1}, \sigma_{i+1}^*)$.
5. When invoked (with no input) for the qth time, $\mathcal{A}'$ makes a stopping query stp.

Whenever $\mathcal{R}$ executes a rewinding procedure, $\mathcal{A}'$ makes another pass on its stream so that it can access the message for the next signing or verification query. Since $\mathcal{A}'^{\mathcal{O}_y}$ perfectly simulates $\mathcal{A}_y$, we immediately obtain the following lemma.

Lemma 2. $\Pr[G(\boldsymbol{y}) \leftarrow (\mathcal{R}^{\mathcal{A}'^{\mathcal{O}_y}}(1^\lambda) \leftrightharpoons \mathcal{CH}(1^\lambda))] \geq c$ *for all* $\boldsymbol{y} \in (\{0,1\}^\delta)^{2q}$ *and some constant $c > 1/2$ in* **Game 1**.

Game 2: This game is exactly the same as **Game 1** except that there exists a stream-access algorithm $\mathcal{B}^{\mathcal{O}_y}$ that simulates $\mathcal{CH}$, $\mathcal{R}$, and $\mathcal{A}'^{\mathcal{O}_y}$ and returns the output of $\mathcal{CH}$. Since the view of $\mathcal{R}$ does not change at all, we have the following lemma.

Lemma 3. $\Pr[G(\boldsymbol{y}) \leftarrow \mathcal{B}^{\mathcal{O}_y}(1^\lambda)] \geq c$ *for all* $\boldsymbol{y} \in (\{0,1\}^\delta)^{2q}$ *and some constant $c > 1/2$ in* **Game 2**.

Since $\mathcal{B}$ makes $p + 1$ passes on its stream in total, according to Theorem 2 and Lemma 3, we have

$$\mathbf{LocalMem}(\mathcal{B}) = \Omega(\min\{q/(p+1)\}, 2^\delta/(p+1)).$$

Furthermore, the memory used to simulate $\mathcal{CH}$ and $\mathcal{A}'$ is $O(\log q) + \mathbf{LocalMem}(\mathcal{CH}) + \mathbf{LocalMem}(\mathsf{Verify})$, where $O(\log q)$ is the amount of memory used to record q and the index of the next query $\mathcal{A}'$ will make. Therefore, we have

$$\mathbf{LocalMem}(\mathcal{B}) = O(\mathbf{LocalMem}(\mathcal{R})) + O(\log q)$$
$$+ \mathbf{LocalMem}(\mathcal{CH}) + \mathbf{LocalMem}(\mathsf{Verify}).$$

Combining the above two bounds completes the proof of Theorem 3. $\qquad\square$

Remark on Security Parameter. Theorem 3 holds only when the security parameter λ is sufficiently large, while one may wonder why memory-tightness makes sense when λ is already required to be very large. In fact, λ only has to be large enough to ensure $c_g + \mathbf{Adv}_{\mathsf{GM}}^{\mathcal{R}^{S_y}}(\lambda) < 1/2$ in the proof of Lemma 1. When c_g is small (e.g., $c_g = 1/4$), it is obvious that $c_g + \mathbf{Adv}_{\mathsf{GM}}^{\mathcal{R}^{S_y}}(\lambda) < 1/2$ should hold even if λ is small (to some extent) and $\mathcal{R}^{S_y}$ may consume large memory, due to the security of GM. Therefore, λ is not necessarily very large unless c_g is very close to $1/2$.

Remark on Advantage-Preserving Reductions. In [1], it is required that the black-box reductions are advantage-preserving, which means that they should work well for adversaries with fixed random tapes. However, we observe that this restriction is not necessary. The reason is that we can treat adversaries with fixed random tapes as deterministic ones, for which any black-box reduction should work well. Furthermore, although a deterministic adversary consumes large memory in this case (compared with an adversary with fixed random tape), simulating it with stream does not, hence our result is not spoiled. The same argument is made for our results in other sections.

Remark on Reductions to UF Security. Auerbach et al. [1] showed a lower bound on the memory usage of restricted reductions from mUF security to UF security. A restricted reduction forwards the public keys generated by $\mathcal{CH}$ to $\mathcal{A}_y$, and forwards the signing queries y and one of the forgery made by $\mathcal{A}_y$ to the challenger $\mathcal{CH}$ in the UF game. One can see that $\mathcal{CH}$ uses large memory to store y so that it can check whether $\mathcal{R}^{\mathcal{A}}$ succeeds later. Since $\mathbf{LocalMem}(\mathcal{CH})$ is very large in this case, the result in [1] is not directly captured by Theorem 3. However, one can easily modify our proof by letting $\mathcal{CH}$ in **Game 2** access to the stream y instead of storing y. By doing this, $\mathbf{LocalMem}(\mathcal{CH})$ can remain small when $\mathcal{R}$ forwards signing queries from $\mathcal{A}$ to $\mathcal{CH}$, and hence, the lower bound in [1] or ones in other similar cases can be derived from our result (when treating unique signatures). We do not take this into account in our formal proof only for simplicity.

Re-randomizable Signatures and VUFs. If we give an additional restriction that a reduction does not control the random tape of an adversary, i.e., an adversary uses real random coins (but not ones from the reduction), then by slightly modifying our proof, we can also show a memory lower bound for re-randomizable signatures [13,23], where signatures can be efficiently re-randomized (we refer the reader to [13] for the formal definition). In this case, we only have to let both the inefficient adversary and the simulator re-randomize the forged signatures so that $\mathcal{R}$ cannot distinguish them.

We can also extend our result for the notion of VUFs [19,20], which is exactly the same as the notion of unique signatures except that a proof (which is not necessarily unique) is needed when verifying the validity of a signature. We omit the details since the extension is straightforward.

4   Lower Bound of Restricted Reductions from mU-mOW to mU-OW for Unique-Key Cryptographic Primitives

In this section, we give a memory lower bound of restricted reductions from mU-mOW security to mU-OW security for unique-key one-way primitives. For simplicity, we treat a basic primitive called *unique-key relation* [24] and argue that this result can be easily extended for other unique-key primitives. We start by recalling the definition of unique-key relations and their security in the multi-user setting, and then show the lower bound.

4.1   Unique-Key Relations

We now recall the definition of a unique-key relation. In a unique-key relation, there exists at most one valid secret key for every public key in the support of the key generation algorithm.[4]

Definition 6 (Unique-key relation). *A* unique-key relation *consists of PT algorithms* (Gen, Check). *(a)* Gen *is a probabilistic algorithm that takes as input* 1^λ, *and returns a public/secret key pair* (pk, sk). *(b)* Check *is a deterministic algorithm that takes as input a public/secret key pair* (pk, sk), *and returns 1 (accept) or 0 (reject).*

A unique-key relation *is required to satisfy* correctness *and* uniqueness. *Correctness is satisfied if* Check$(pk, sk) = 1$ *holds for all* $\lambda \in \mathbb{N}$ *and all* $(pk, sk) \leftarrow$ Gen(1^λ). *Uniqueness is satisfied if for all* $\lambda \in \mathbb{N}$ *and all pk in the support of* Gen(1^λ), *there exists no pair* (sk, sk') *that simultaneously satisfies* $sk \neq sk'$ *and* Check$(pk, sk) =$ Check$(pk, sk') = 1$.

Now we give the definitions of the mU-mOW and mU-OW security of unique-key relations [2,3]. In these security games, an adversary sees many public keys and can adaptively corrupt the secret keys. It succeeds if it outputs a valid secret key that is not corrupted.

---

[4] Unlike the definition of unique signatures, here we do not require uniqueness for public keys outside the support of the key generation algorithm.

Definition 7 (mU-mOW). *A unique-key relation* (Gen, Check) *is said to be* mU-mOW *secure if for any PPT adversary* $\mathcal{A}$, *we have* $\mathbf{Adv}^{\mathcal{A}}_{\text{mU-mOW}}(\lambda) = \Pr[\mathcal{CH} \text{ outputs } 1] \leq negl(\lambda)$ *in the following game.*

1. *The challenger* $\mathcal{CH}$ *sets* $w = 0$ *and* $\mathcal{Q} = \emptyset$, *and runs* $\mathcal{A}$ *on input* 1^{λ}. *Then* $\mathcal{A}$ *may make sampling queries to* $\mathcal{CH}$, *and* $\mathcal{CH}$ *responds as follows.*
   - *On receiving the ith sampling query sp,* $\mathcal{CH}$ *samples* $(pk_i, sk_i) \leftarrow \text{Gen}(1^{\lambda})$ *and sends* pk_i *to* $\mathcal{A}$.
2. *Then* $\mathcal{A}$ *may make adaptive corruption and verification queries to* $\mathcal{CH}$, *and* $\mathcal{CH}$ *responds as follows:*
   - *On receiving a corruption query* i, $\mathcal{CH}$ *adds* i *to* $\mathcal{Q}$, *and sends* sk_i *to* $\mathcal{A}$.
   - *On receiving a verification query* (i^*, sk^*), *if* $\text{Check}(pk_{i^*}, sk^*) = 1$ *and* $i^* \notin \mathcal{Q}$, $\mathcal{CH}$ *sets* $w = 1$.
3. *At some point,* $\mathcal{A}$ *makes a stopping query stp to* $\mathcal{CH}$, *and* $\mathcal{CH}$ *returns* w.

Definition 8 (mU-OW). mU-OW *security is defined in exactly the same way as* mU-mOW *security except that* $\mathcal{A}$ *is allowed to make only one verification query and the advantage of* $\mathcal{A}$ *is denoted by* $\mathbf{Adv}^{\mathcal{A}}_{\text{mU-OW}}(\lambda)$.

4.2   Lower Bound for Unique-Key Relations

In this section, we define restricted reductions from the mU-mOW security to mU-OW security of unique-key relations and show a memory lower bound of such reductions.

Restricted Black-Box Reductions from mU-mOW **to** mU-OW. Let $\mathcal{R}$ be a black-box reduction from mU-mOW security to mU-OW security. As before, we write $\mathcal{R}^{\mathcal{A}}$ to mean that $\mathcal{R}$ has oracle access to a (stateful) adversary $\mathcal{A}$ playing the mU-mOW game. Whenever receiving a query from $\mathcal{R}$, $\mathcal{A}$ returns the "next" query to $\mathcal{R}$. $\mathcal{R}$ is not able to modify the current state of $\mathcal{A}$ (i.e., $\mathcal{A}$ runs sequentially), but is allowed to adaptively rewind $\mathcal{A}$ to previous states.

Definition 9 (c-restricted black-box reduction from mU-mOW **to** mU-OW). *Let* $c > 0$ *be a constant. An oracle-access PPT machine* $\mathcal{R}^{(\cdot)}$ *is said to be a c-restricted black-box reduction from the* mU-mOW *security to the* mU-OW *security of a unique-key relation, if for any (possibly inefficient) adversary* $\mathcal{A}$, *we have* $\mathbf{Adv}^{\mathcal{R}^{\mathcal{A}}}_{\text{mU-OW}}(\lambda) \geq c \cdot \mathbf{Adv}^{\mathcal{A}}_{\text{mU-mOW}}(\lambda)$, *and* $\mathcal{R}$ *respects the following restriction.*

- *The public keys* $(pk_1, \cdots, pk_n)$ *that* $\mathcal{R}$ *sends to* $\mathcal{A}$ *are the ones generated by the challenger and given to* $\mathcal{R}$ *in the* mU-OW *game.*
- *The set of corruption queries* $\{y_1, \cdots, y_q\}$ *made by* $\mathcal{R}$ *is the same as the set of all corruption queries made by* $\mathcal{A}$.

Before showing the lower bound, we recall the definition of PRFs which will be exploited in our proof.

Definition 10 (Pseudorandom function (PRF)). F : $\{0,1\}^{\kappa(\lambda)} \times \{0,1\}^{\delta(\lambda)} \to \{0,1\}^{\rho(\lambda)}$, where $\kappa = \kappa(\lambda)$, $\delta = \delta(\lambda)$, and $\rho = (\lambda)$ are polynomials, is said to be a pseudorandom function, if for any PPT adversary $\mathcal{A}$, we have

$$\mathbf{Adv}_{\mathsf{PR}}^{\mathcal{A}}(\lambda) = |\Pr[1 \leftarrow \mathcal{A}^{\mathcal{O}_k}(1^\lambda) \mid k \leftarrow \{0,1\}^\kappa] - \Pr[1 \leftarrow \mathcal{A}^{\mathcal{O}}(1^\lambda)]| \le negl(\lambda).$$

Here, $\mathcal{O}_k(i)$ returns $\mathsf{F}(k,i)$. $\mathcal{O}(i)$ returns r if there exists (i,r) in its internal list (initiated with $\emptyset$). Otherwise, $\mathcal{O}(i)$ returns $r \leftarrow \{0,1\}^\rho$ and adds (i,r) to its list.

The main theorem is as follows.

Theorem 4. Let λ be a (sufficiently large) security parameter, $\Phi = (\mathsf{Gen}, \mathsf{Check})$, where the internal randomness space of Gen is $\{0,1\}^\rho$, be a $\mathsf{mU}\text{-}\mathsf{OW}$ secure unique-key relation, $\mathsf{F} : \{0,1\}^\kappa \times \{0,1\}^\lambda \to \{0,1\}^\rho$ be a PRF, and $\mathcal{R}$ be a c_r-restricted black-box reduction from the $\mathsf{mU}\text{-}\mathsf{mOW}$ security to the $\mathsf{mU}\text{-}\mathsf{OW}$ security of Φ. Let $n = n(\lambda)$ be the maximum number of sampling queries and $q = q(\lambda)$ be the maximum numbers of corruption and verification queries made by an adversary in the $\mathsf{mU}\text{-}\mathsf{mOW}$ game, and $\mathcal{U} = \{i\}_{i=1}^n$. If (a) $\mathcal{R}$ rewinds the adversary for at most $p = p(\lambda)$ times and (b) $c_r > 1/2$, then we have

$$\mathbf{LocalMem}(\mathcal{R}) = \Omega(\max\{q/(p+2), n/(p+2)\}) - O(\log q) - O(\log n) - \kappa$$
$$- \max\{\mathbf{LocalMem}(\mathsf{Gen}), \mathbf{LocalMem}(\mathsf{Check}), \mathbf{LocalMem}(\mathsf{F})\}.$$

Roughly, this theorem implies that when the maximum number of users and that of corruption queries made by an adversary in the $\mathsf{mU}\text{-}\mathsf{mOW}$ game are very large, $\mathcal{R}$ must consume large memory unless it rewinds $\mathcal{A}$ many times, which increases its running-time.

High-Level Idea of the Proof. We firstly construct an inefficient adversary $\mathcal{A}_y$ where $y = (y_1, \cdots, y_{2q})$. $\mathcal{A}_y$ takes as input and stores public keys $pk_1, \cdots, pk_n$, makes corruption queries $y_1, \cdots, y_q$, checks the validity of the answers, and then makes verification queries $(pk_{y_{q+1}}, sk_1^*), \cdots, (pk_{y_{2q}}, sk_q^*)$ generated by using brute force. When $G(y) = 1$, $\mathcal{R}^{\mathcal{A}_y}$ is likely to succeed in the $\mathsf{mU}\text{-}\mathsf{OW}$ game, since $\mathcal{R}$ is a black-box reduction and $\mathcal{A}_y$ is a deterministic algorithm breaking $\mathsf{mU}\text{-}\mathsf{mOW}$ security. When $G(y) = 0$, we can construct a PT algorithm $\mathcal{S}_y$, which runs in the same way as $\mathcal{A}_y$ does except that $\mathcal{S}_y$ uses the answers of corruption queries to make verification queries. Due to uniqueness, $\mathcal{S}_y$ perfectly simulates $\mathcal{A}_y$. Since the PPT algorithm $\mathcal{R}^{\mathcal{S}_y}$ is likely to fail in the $\mathsf{mU}\text{-}\mathsf{OW}$ game, $\mathcal{R}^{\mathcal{A}_y}$ is likely to fail as well.

Then, similarly to the proof of Theorem 3, we can construct an algorithm $\mathcal{B}$ with access to a stream y that simulates the $\mathsf{mU}\text{-}\mathsf{OW}$ game with $\mathcal{R}^{\mathcal{A}_y}$ and outputs $G(y)$ with high probability. Therefore, we can show the lower bound on memory consumed by $\mathcal{R}$ since the memory consumed by $\mathcal{B}$ is inherently large, due to Theorem 2. However, one may notice that $\mathcal{B}$ uses a large amount of memory to store $pk_1, \cdots, pk_n$, which spoils our result since $\mathcal{B}$ using large memory does not imply $\mathcal{R}$ using large memory any more. We deal with this problem by using a PRF to simulate random coins used by the challenger and running the PRF to output the corresponding random coin used to generate a public key when the key is needed. In this way, $\mathcal{B}$ does not store the public keys anymore. Here, there is a point that $\mathcal{B}$ can simulate $\mathcal{A}_y$ efficiently by using secret keys generated by the challenger in the mU-OW game, so that the whole interaction $\mathcal{B}$ simulates only runs in polynomial-time and cannot distinguish outputs of the PRF with real random coins.

Proof (of Theorem 4). Assuming the existence of the reduction $\mathcal{R}$ stated in Theorem 4, we show the existence of a probabilistic algorithm $\mathcal{B}$ such that $\Pr[G(\boldsymbol{y}) \leftarrow \mathcal{B}^{\mathcal{O}_y}(1^\lambda)] \geq c$ for all $\boldsymbol{y} = (y_1, \cdots, y_{2q}) \in \mathcal{U}^{2q}$ and some constant $c > 1/2$ by giving hybrid games.

Game 0: In this game, $\mathcal{R}$ has access to an adversary $\mathcal{A}_y$ and interacts with the challenger $\mathcal{CH}$ in the mU-OW game. $\mathcal{A}_y$ runs as follows.

1. On receiving 1^λ, $\mathcal{A}_y$ makes a sampling query sp.
2. For $i = 1, \cdots, n - 1$, on receiving pk_i, $\mathcal{A}_y$ stores pk_i and makes a sampling query sp.
3. On receiving pk_n, $\mathcal{A}_y$ stores pk_n and makes a corruption query y_1.
4. For $i = 1, \cdots, q - 1$, on receiving the answer sk'_i to the ith corruption query, if $\mathsf{Check}(pk_{y_i}, sk'_i) \neq 1$, $\mathcal{A}_y$ aborts. Otherwise, $\mathcal{A}_y$ makes a corruption query y_{i+1}.
5. On receiving the answer sk'_q to the qth corruption query, if $\mathsf{Check}(pk_{y_q}, sk'_q) \neq 1$, $\mathcal{A}_y$ aborts. Otherwise, $\mathcal{A}_y$ exhaustively searches sk_1^* such that $\mathsf{Verify}(pk_{y_{q+1}}, sk_1^*) = 1$, and makes a verification query (y_{q+1}, sk_1^*).
6. For $i = 1, \cdots, q - 1$, when invoked (with no input) for the ith time, $\mathcal{A}_y$ exhaustively searches sk_{i+1}^* such that $\mathsf{Check}(pk_{y_{q+i+1}}, sk_{i+1}^*) = 1$, and makes a verification query (y_{q+i+1}, sk_{i+1}^*).
7. When invoked (with no input) for the qth time, $\mathcal{A}_y$ makes a stopping query stp.

We now show the following lemma.

Lemma 4. $\Pr[G(\boldsymbol{y}) \leftarrow (\mathcal{R}^{\mathcal{A}_y}(1^\lambda) \leftrightharpoons \mathcal{CH}(1^\lambda))] \geq c_r$ *for all* $\boldsymbol{y} \in \mathcal{U}^{2q}$ *in* **Game 0**.

Proof (of Lemma 4). Firstly, we show the existence of a PT algorithm $\mathcal{S}_y$ perfectly simulating $\mathcal{A}_y$ on condition that $G(\boldsymbol{y}) = 0$. $\mathcal{S}_y$ runs as follows. (Below, the difference from $\mathcal{A}_y$ is *emphasized*.)

1. On receiving 1^λ, $\mathcal{S}_y$ makes a sampling query sp.
2. For $i = 1, \cdots, n - 1$, on receiving pk_i, $\mathcal{S}_y$ stores pk_i and makes a sampling query sp.
3. On receiving pk_n, $\mathcal{S}_y$ stores pk_n and makes a corruption query y_1.
4. For $i = 1, \cdots, q-1$, on receiving the answer sk'_i to the ith corruption query, if $\mathsf{Check}(pk_{y_i}, sk'_i) \neq 1$, $\mathcal{S}_y$ aborts. Otherwise, $\mathcal{S}_y$ *stores* (y_i, sk'_i) *in its internal list* L *(initialized with* $\emptyset$*)*, and makes a corruption query y_{i+1}.
5. On receiving the answer sk'_q to the qth corruption query, if $\mathsf{Check}(pk_{y_q}, sk'_q) \neq 1$, $\mathcal{S}_y$ aborts. Otherwise, $\mathcal{S}_y$ *stores* (y_q, sk'_q) *in* L*, searches a pair* (i^*, sk) *in* L *such that* $i^* = y_{q+1}$*, and makes a verification query* (y_{q+1}, sk)*. If the searching procedure fails,* $\mathcal{S}_y$ *aborts.*

6. For $i = 1, \cdots, q - 1$, when invoked (with no input) for the ith time, $\mathcal{S}_y$ searches a pair (i^*, sk) in L such that $i^* = y_{q+i+1}$, and makes a verification query (y_{q+i+1}, sk). If the searching procedure fails for some i, $\mathcal{S}_y$ aborts.
7. When invoked (with no input) for the qth time, $\mathcal{S}_y$ makes a stopping query stp.

When $G(\boldsymbol{y}) = 0$, we have $\{y_{q+i}\}_{i=1}^q \subseteq \{y_i\}_{i=1}^q$, which means that the searching procedures executed by $\mathcal{S}_y$ (in Steps 5 and 6) will not fail. Moreover, due to the uniqueness of Φ, the verification queries made by $\mathcal{S}_y$ are exactly the same as those made by $\mathcal{A}_y$. Hence, $\mathcal{S}_y$ perfectly simulates $\mathcal{A}_y$ in the view of $\mathcal{R}$.

Due to the mU-OW security of Φ, we have $\mathbf{Adv}_{\mathsf{mU\text{-}OW}}^{\mathcal{R}^{\mathcal{A}_y}}(\lambda) = \mathbf{Adv}_{\mathsf{mU\text{-}OW}}^{\mathcal{R}^{\mathcal{S}_y}}(\lambda) \leq negl(\lambda)$ when $G(\boldsymbol{y}) = 0$, which implies $\Pr[1 \leftarrow (\mathcal{R}^{\mathcal{A}_y}(1^\lambda) \leftrightharpoons \mathcal{CH}(1^\lambda)) \mid G(\boldsymbol{y}) = 0] \leq negl(\lambda)$, i.e., $\Pr[0 \leftarrow (\mathcal{R}^{\mathcal{A}_y}(1^\lambda) \leftrightharpoons \mathcal{CH}(1^\lambda)) \mid G(\boldsymbol{y}) = 0] \geq 1 - negl(\lambda)$. On the other hand, when $G(\boldsymbol{y}) = 1$, there exists some $1 \leq j \leq q$ such that $y_{q+j} \notin \{y_i\}_{i=1}^q$, which implies $\mathbf{Adv}_{\mathsf{mU\text{-}mOW}}^{\mathcal{A}_y}(\lambda) = 1$. Since $\mathcal{R}$ is a c_r-restricted black-box reduction, we have $\Pr[1 \leftarrow (\mathcal{R}^{\mathcal{A}_y}(1^\lambda) \leftrightharpoons \mathcal{CH}(1^\lambda)) \mid G(\boldsymbol{y}) = 1] \geq c_r$. Since $c_r > 1/2$ and λ is sufficiently large, we have $\Pr[G(\boldsymbol{y}) \leftarrow (\mathcal{R}^{\mathcal{A}_y}(1^\lambda) \leftrightharpoons \mathcal{CH}(1^\lambda))] \geq c_r$, completing the proof of Lemma 4. $\square$

Game 1: This game is exactly the same as **Game 0** except that for each i, $\mathcal{CH}$ generates the ith key pair by computing $(pk_i, sk_i) \leftarrow \mathsf{Gen}(1^\lambda; \mathsf{F}(k, i))$ where k is randomly chosen from $\{0, 1\}^\kappa$ at the beginning of the game.

Lemma 5. $\Pr[G(\boldsymbol{y}) \leftarrow (\mathcal{R}^{\mathcal{A}_y}(1^\lambda) \leftrightharpoons \mathcal{CH}(1^\lambda))] \geq c$ for all $\boldsymbol{y} \in \mathcal{U}^{2q}$ and some constant $c > 1/2$ in **Game 1**.

Proof (of Lemma 5). Let $\Pr[G(\boldsymbol{y}) \leftarrow (\mathcal{R}^{\mathcal{A}_y}(1^\lambda) \leftrightharpoons \mathcal{CH}(1^\lambda))]$ be $c_0^{\boldsymbol{y}}$ (respectively, $c_1^{\boldsymbol{y}}$) in **Game 0** (respectively, **Game 1**). For any $\boldsymbol{y}$, we can construct a PPT adversary $\mathcal{D}$ breaking the pseudorandom property of F with advantage $\mathbf{Adv}_{\mathsf{PR}}^{\mathcal{D}}(\lambda) = |c_0^{\boldsymbol{y}} - c_1^{\boldsymbol{y}}|$ as follows.

$\mathcal{D}$ has access to an oracle $\mathcal{O}_k$ parameterized by $k \leftarrow \{0, 1\}^\kappa$ or an oracle $\mathcal{O}$ (see Definition 10 for the descriptions of $\mathcal{O}_k$ and $\mathcal{O}$). $\mathcal{D}$ runs $\mathcal{R}^{\mathcal{A}_y}(1^\lambda) \leftrightharpoons \mathcal{CH}(1^\lambda)$ in exactly the same way as in **Game 0**, except that $\mathcal{A}_y$ receives secret keys generated by $\mathcal{CH}$ from $\mathcal{D}$ to make verification queries, instead of using brute force to recover them. This is possible due to the restriction that all the public keys $\mathcal{R}$ sends to $\mathcal{A}$ are generated by $\mathcal{CH}$. Furthermore, when $\mathcal{CH}$ requires the ith random coin, $\mathcal{D}$ makes a query to its oracle and sends the answer of the query back. If $\mathcal{CH}$ outputs $G(\boldsymbol{y})$, $\mathcal{D}$ outputs 1. Otherwise, $\mathcal{D}$ outputs 0.

When the oracle is $\mathcal{O}$ (respectively, $\mathcal{O}_k$), the view of $\mathcal{CH}$ is exactly the same as its view in **Game 0** (respectively, **Game 1**) due to the unique key property. Therefore, we have $\mathbf{Adv}_{\mathsf{PR}}^{\mathcal{D}}(\lambda) = |c_0^{\boldsymbol{y}} - c_1^{\boldsymbol{y}}|$. Due to the pseudorandom property of F, we have $|c_0^{\boldsymbol{y}} - c_1^{\boldsymbol{y}}| \leq negl(\lambda)$. Since λ is sufficiently large, combining this bound with Lemma 4 completes the proof of Lemma 5. $\square$

Game 2: This game is exactly the same as **Game 1** except that there exists an algorithm $\mathcal{A}'$ with access to the stream oracle $\mathcal{O}_y$ simulating $\mathcal{A}_y$ as follows. (Below, the difference from $\mathcal{A}_y$ is *emphasized*.)

1. On receiving 1^λ, $\mathcal{A}'$ makes a sampling query sp.
2. For $i = 1, \cdots, n - 1$, on receiving pk_i, $\mathcal{A}'$ stores pk_i and makes a sampling query sp.
3. On receiving pk_n, $\mathcal{A}'$ stores pk_n, queries $\mathcal{O}_y$ to obtain y_1, and makes a corruption query y_1.
4. For $i = 1, \cdots, q - 1$, on receiving the answer sk_i' to the ith corruption query, if $\mathsf{Check}(pk_{y_i}, sk_i') \neq 1$, $\mathcal{A}'$ aborts. Otherwise, $\mathcal{A}'$ queries $\mathcal{O}_y$ to obtain y_{i+1}, and makes a corruption query y_{i+1}.
5. On receiving the answer sk_q' to the qth corruption query, if $\mathsf{Check}(pk_{y_q}, sk_q') \neq 1$, $\mathcal{A}'$ aborts. Otherwise, $\mathcal{A}'$ queries $\mathcal{O}_y$ to obtain y_{q+1}, exhaustively searches sk_1^* such that $\mathsf{Check}(pk_{y_{q+1}}, sk_1^*) = 1$, and makes a verification query (y_{q+1}, sk_1^*).
6. For $i = 1, \cdots, q - 1$, when invoked (with no input), $\mathcal{A}'$ queries $\mathcal{O}_y$ to obtain y_{q+i+1}, exhaustively searches sk_{i+1}^* such that $\mathsf{Check}(pk_{y_{q+i+1}}, sk_{i+1}^*) = 1$, and makes a verification query (y_{q+i+1}, sk_{i+1}^*).
7. When invoked (without input) for the qth time, $\mathcal{A}'$ makes a stopping query stp.

Whenever $\mathcal{R}$ executes a rewinding procedure, $\mathcal{A}'$ makes another pass on its stream to obtain the index for the next corruption or verification query. Since $\mathcal{A}'^{\mathcal{O}_y}$ perfectly simulates $\mathcal{A}_y$, we have the following lemma.

Lemma 6. $\Pr[G(\boldsymbol{y}) \leftarrow (\mathcal{R}^{\mathcal{A}'^{\mathcal{O}_y}}(1^\lambda) \leftrightarrows \mathcal{CH}(1^\lambda))] \geq c$ *for all* $\boldsymbol{y} \in \mathcal{U}^{2q}$ *and some constant* $c > 1/2$ *in* **Game 2**.

Game 3: This game is the same as **Game 2** except that there exists a stream access algorithm $\mathcal{A}''^{\mathcal{O}_y}$ that runs $k \leftarrow \{0,1\}^\kappa$, stores k, simulates $\mathcal{CH}$, $\mathcal{R}$, and $\mathcal{A}'^{\mathcal{O}_y}$, and generates the ith key pair by computing $(pk_i, sk_i) \leftarrow \mathsf{Gen}(1^\lambda; \mathsf{F}(k, i))$. When $\mathcal{R}$ makes a verification query (i, sk^*), $\mathcal{CH}$ makes another pass on the stream $\boldsymbol{y}$ through $\mathcal{A}''$, and checks whether $i \in \{y_1, \cdots, y_q\}$ and $\mathsf{Check}(pk_i, sk^*) = 1$.[5] If the check works, $\mathcal{CH}$ outputs 1. Otherwise, $\mathcal{CH}$ outputs 0. Then $\mathcal{A}''$ returns the output of $\mathcal{CH}$. Since the view of $\mathcal{CH}$ in this game is identical to its view in **Game 2**, we have the following lemma.

Lemma 7. $\Pr[G(\boldsymbol{y}) \leftarrow \mathcal{A}''^{\mathcal{O}_y}(1^\lambda)] \geq c$ *for all* $\boldsymbol{y} \in \mathcal{U}^{2q}$ *and some constant* $c > 1/2$ *in* **Game 3**.

Game 4: In this game, there exists an algorithm $\mathcal{B}^{\mathcal{O}_y}$ which runs in exactly the same way as $\mathcal{A}''^{\mathcal{O}_y}$ except that it does not store $(pk_i)_{i=1}^n$ generated by $\mathcal{CH}$. Instead, whenever $\mathcal{B}^{\mathcal{O}_y}$ needs to see pk_i, $\mathcal{B}^{\mathcal{O}_y}$ computes $(pk_i, sk_i) \leftarrow \mathsf{Gen}(1^\lambda; \mathsf{F}(k, i))$. Since the view of $\mathcal{CH}$ in **Game 4** is identical to its view in **Game 3**, we have the following lemma.

Lemma 8. $\Pr[G(\boldsymbol{y}) \leftarrow \mathcal{B}^{\mathcal{O}_y}(1^\lambda)] \geq c$ *for all* $\boldsymbol{y} \in \mathcal{U}^{2q}$ *and some constant* $c > 1/2$ *in* **Game 4**.

---

[5] According to the second restriction in Definition 9, the corruption queries $\mathcal{R}$ has made are $\{y_1, \cdots, y_q\}$.

Since $\mathcal{B}$ makes $p + 2$ passes on its stream in total, according to Theorem 2 and Lemma 8, we have

$$\mathbf{LocalMem}(\mathcal{B}) = \Omega(\min\{q/(p+2), n/(p+2)\}).$$

Furthermore, the memory used to simulate $\mathcal{CH}$, $\mathcal{A}''$, and random coins is $O(\log q) + O(\log n) + \max\{\mathbf{LocalMem}(\mathsf{Gen}), \mathbf{LocalMem}(\mathsf{Check}), \mathbf{LocalMem}(\mathsf{F})\} + \kappa$, where $O(\log q) + O(\log n)$ is the amount of memory used to record q, n, and the index of the next query $\mathcal{A}''$ will make. Therefore we have

$$\mathbf{LocalMem}(\mathcal{B}) = O(\mathbf{LocalMem}(\mathcal{R})) + O(\log q) + O(\log n) + \kappa$$
$$+ \max\{\mathbf{LocalMem}(\mathsf{Gen}), \mathbf{LocalMem}(\mathsf{Check}), \mathbf{LocalMem}(\mathsf{F})\}.$$

Combining the above two bounds completes Theorem 4. □

Remark on Security Parameter. Similarly to the case of Theorem 3, Theorem 4 holds only when the security parameter λ is sufficiently large, while one may wonder why memory-tightness makes sense when λ is already required to be large. In fact, λ only has to be large enough to ensure $1 - \mathbf{Adv}_{\mathsf{mU\text{-}OW}}^{\mathcal{R}^{\mathcal{S}_y}}(\lambda) - \mathbf{Adv}_{\mathsf{PR}}^{\mathcal{D}}(\lambda) > 1/2$ and $c_r - \mathbf{Adv}_{\mathsf{PR}}^{\mathcal{D}}(\lambda) > 1/2$ in the proofs of Lemmas 4 and 5. When c_r is large, these two inequations should hold even if λ is small (to some extent) and $\mathcal{R}^{\mathcal{S}_y}$ and $\mathcal{D}$ may consume large memory, due to mU-mOW security and pseudorandomness. Therefore, λ is not necessarily very large unless c_r is very close to $1/2$.

Lower Bound for Other Unique-Key and Re-randomizable Primitives. It is not hard to see that the above result can be easily extended to lower bound results for (one-way secure) PKE schemes, signatures, and many other primitives in the multi-user setting, in which key pairs satisfy unique-key relations. Since unique-key primitives capture many existing natural constructions [5,7,8,12, 14,17,18,22,23], a very wide class of memory lower bounds in the multi-user setting can be directly derived from our result stated in Theorem 4. For ease of understanding, we take unique-key PKE schemes and unique-key signatures as examples in Appendix A. Concretely, we give the definitions of unique-key PKE schemes and unique-key signatures and their security notions in the multi-user setting. Then we give two corollaries showing that in this setting, reductions from multi-challenge security to single-challenge security for these two types of primitives must consume large memory unless they increase running-time.

Similarly to the case of unique signatures, this result can also be extended for primitives with key re-randomization [4] if reductions do not control random tapes of adversaries.[6]

5   Lower Bound of Restricted Reductions from mCR_t to CR_t for Large-Domain Hash Functions

In [1], Auerbach et al. showed a memory lower bound of restricted reductions from mCR_t security to CR_t security for an artificial function which just truncates

---

[6] Similarly to the definition of unique-key relations, we do not require re-randomization for public keys outside the support of the key generation algorithm.

last λ bits of its input, while such a function does not satisfy CR_t security itself. In this section, we extend this result to a lower bound for all the large-domain hash functions satisfying CR_t security (where t is a constant). To achieve the goal, we prove a streaming lower bound with respect to hash functions.

5.1 Hash Functions

In this section, we define large-domain hash functions, recall mCR_t security and CR_t security, and show a theorem for large-domain hash functions.

Definition 11 (Large-domain hash function). *A hash function* $H : \{0,1\}^\kappa \times \{0,1\}^\delta \to \{0,1\}^\rho$, *where* $\kappa = \kappa(\lambda)$, $\delta = \delta(\lambda)$, *and* $\rho = \rho(\lambda)$ *are polynomials, is said to have a large domain if* $2^{\rho-\delta} \le negl(\lambda)$.

Definition 12 (mCR_t [1]). *A hash function* $H : \{0,1\}^\kappa \times \{0,1\}^\delta \to \{0,1\}^\rho$ *is said to satisfy* mCR_t *security (where t is some constant independent of the security parameter λ), if for any PPT adversary $\mathcal{A}$, we have* $\mathbf{Adv}^{\mathcal{A}}_{mCR_t}(\lambda) = \Pr[\mathcal{CH}$ *outputs* $1] \le negl(\lambda)$ *in the following game.*

1. *The challenger $\mathcal{CH}$ sets $\mathcal{Q} = \emptyset$, randomly chooses $k \leftarrow \{0,1\}^\kappa$, and runs $\mathcal{A}$ on input $(1^\lambda, k)$. $\mathcal{A}$ may make adaptive input queries to $\mathcal{CH}$. Every time on receiving a query $m \in \{0,1\}^\delta$ from $\mathcal{A}$, $\mathcal{CH}$ adds m to $\mathcal{Q}$.*
2. *At some point, $\mathcal{A}$ makes a stopping query stp to $\mathcal{CH}$. If there exists $\{m_i^*\}_{i=1}^t \subseteq \mathcal{Q}$ such that $H_k(m_1^*) = \cdots = H_k(m_t^*)$ and $|\{m_i^*\}_{i=1}^t| = t$, $\mathcal{CH}$ outputs 1. Otherwise, $\mathcal{CH}$ outputs 0.*

Definition 13 (CR_t [1]). CR_t *security is defined in exactly the same way as that of mCR_t security, except that $\mathcal{A}$ is allowed to make at most t input queries and the advantage of $\mathcal{A}$ is denoted by* $\mathbf{Adv}^{\mathcal{A}}_{CR_t}(\lambda)$.

Next we give a theorem that will be used to prove a streaming lower bound later. This theorem shows that for a large-domain hash function satisfying CR_t security, there exist "many" hash values with more than t pre-images. Intuitively, if this theorem does not hold, then there will be some hash value with many pre-images, so that t randomly chosen inputs are likely to fall into the class of these pre-images, which breaks its CR_t security.

Theorem 5. *Let λ be a (sufficiently large) security parameter, $H : \{0,1\}^\kappa \times \{0,1\}^\delta \to \{0,1\}^\rho$ be a large-domain hash function satisfying CR_t security, and $n = n(\lambda)$ be any polynomial in λ. For $k \leftarrow \{0,1\}^\kappa$, the probability that there exist more than n elements in $\{0,1\}^\rho$ with more than t pre-images (with respect to H_k) is $1 - negl(\lambda)$.*

Proof (of Theorem 5). Let $n' \le n$ and E be the event that the number of elements in $\{0,1\}^\rho$ that have more than t pre-images is exactly n'. To prove Theorem 5, we just need to prove $\Pr[E] \le negl(\lambda)$.

Let $k \leftarrow \{0,1\}^\kappa$, $m \leftarrow \{0,1\}^\delta$, and E_0 be the event that the number of pre-images of $H_k(m)$ is more than t. Since the number of elements in $\{0,1\}^\delta$, the hash values of which respectively have less than t pre-image, is at most 2^ρ, we have $\Pr[\overline{E_0} \mid E] \leq t \cdot 2^\rho/2^\delta$, i.e., $\Pr[E_0 \mid E] \geq 1 - t \cdot 2^\rho/2^\delta$. Let E_i be the event that $\mathsf{H}_k(m)$ is the ith lexicographically smallest value in $\{0,1\}^\rho$ with more than t pre-images. Since $\Pr[E_1 \lor \cdots \lor E_{n'} \mid E] = \Pr[E_0 \mid E]$, there must exist some $i^* \in \{1, \cdots, n'\}$ such that

$$\Pr[E_{i^*} \mid E] \geq 1/n' \cdot \Pr[E_0 \mid E] \geq (1/n) \cdot (1 - t \cdot 2^\rho/2^\delta).$$

Now we construct a PPT adversary $\mathcal{A}$ in the CR_t game of H. On receiving $k \leftarrow \{0,1\}^\kappa$, $\mathcal{A}$ randomly chooses $m_1, \cdots, m_t \leftarrow \{0,1\}^\delta$, and uses them as input queries. Let E' be the event that there exist some $i, j \in \{1, \cdots, t\}$ such that $m_i = m_j$. We have $\Pr[E'] \leq 1 - (1 - (t-1)/2^\delta)^t \leq O(t^2/2^\delta)$. Therefore, we have

$$\begin{aligned}
\mathbf{Adv}_{CR_t}^{\mathcal{A}}(\lambda) &= \Pr[\overline{E'} \land \mathsf{H}_k(m_1) = \cdots = \mathsf{H}_k(m_t)] \\
&= \Pr[\mathsf{H}_k(m_1) = \cdots = \mathsf{H}_k(m_t)] - \Pr[E' \land \mathsf{H}_k(m_1) = \cdots = \mathsf{H}_k(m_t)] \\
&\geq \Pr[\mathsf{H}_k(m_1) = \cdots = \mathsf{H}_k(m_t) \land E] - \Pr[E'] \\
&= \Pr[\mathsf{H}_k(m_1) = \cdots = \mathsf{H}_k(m_t) \mid E] \cdot \Pr[E] - \Pr[E'] \\
&\geq \Pr[E_{i^*}|E]^t \cdot \Pr[E] - O(t^2/2^\delta) \\
&\geq (1/n \cdot (1 - t \cdot 2^\rho/2^\delta))^t \cdot \Pr[E] - O(t^2/2^\delta).
\end{aligned}$$

As a result, the probability that $\mathcal{A}$ breaks CR_t security is larger than $(1/n \cdot (1 - t \cdot 2^\rho/2^\delta))^t \cdot \Pr[E] - O(t^2/2^\delta)$, where t is some constant. Since $(1/n \cdot (1 - t \cdot 2^\rho/2^\delta))^t \geq 1/n^t - negl(\lambda)$ and $O(t^2/2^\delta) \leq negl(\lambda)$, we have $\Pr[E] \leq negl(\lambda)$, completing the proof of Theorem 5. □

5.2 Streaming Lower Bound for Hash Functions

In this section, we give a theorem, which is another corollary of prior works [16, 21] based on the disjointness problem. It is also a variant of a streaming lower bound shown in [1]. It shows the existence of a memory lower bound for determining whether there exists a t-collision, with respect to a CR_t secure large-domain hash function, in a data stream. Before giving the main theorem, we define the function $F_{\mathsf{H},t}(\boldsymbol{y})$ as follows.

Let $\boldsymbol{y} \in (\{0,1\}^\delta)^q$, $F_{\mathsf{H}}(\boldsymbol{y})$ be defined as $F_{\mathsf{H}}(\boldsymbol{y}) = \max_{s \in \{0,1\}^\rho} |\{y_i : \mathsf{H}(y_i) = s\}|$, and $F_{\mathsf{H},t}(\boldsymbol{y})$ be defined as

$$F_{\mathsf{H},t}(\boldsymbol{y}) = \begin{cases} 1 & \text{if } F_{\mathsf{H}}(\boldsymbol{y}) \geq t \\ 0 & \text{otherwise} \end{cases}.$$

Theorem 6. *Let $\mathcal{B}$ be a probabilistic algorithm, λ be a (sufficiently large) security parameter, and $\mathsf{H} : \{0,1\}^\kappa \times \{0,1\}^\delta \to \{0,1\}^\rho$ be a large-domain hash function satisfying CR_t security. Assuming that there exists some constant $c > 1/2$ such that $\Pr[\mathcal{B}^{\mathcal{O}_{\boldsymbol{y}}}(1^\lambda, k) = F_{\mathsf{H}_k,t}(\boldsymbol{y}) \mid k \leftarrow \{0,1\}^\kappa] \geq c$ holds for polynomial $q = q(\lambda)$ and all $\boldsymbol{y} \in (\{0,1\}^\delta)^q$. Then $\mathbf{LocalMem}(\mathcal{B}) = \Omega((q - \kappa)/p)$, where p is the number of passes $\mathcal{B}$ makes in the worst case.*

Proof (of Theorem 6). Let $n = \lfloor q/t \rfloor$. We now construct a two-party protocol (P_1, P_2) by using $\mathcal{B}$ as follows. Taking as input $x_1 \in \{0,1\}^n$, P_1 samples $k \leftarrow \{0,1\}^\kappa$ and sends k to P_2. If there do not exist n elements in $\{0,1\}^\rho$ with more than t pre-images for the hash function $\mathsf{H}_k(\cdot)$, P_1 aborts. Let h_i be the ith lexicographically smallest element in $\{0,1\}^\rho$ with more than t pre-images, m_{ij} be the jth smallest pre-image of h_i, and $\overline{m}$ be an element in $\{0,1\}^\delta$ such that $\mathsf{H}_k(\overline{m}) \notin \{h_i\}_{i=1}^n$. For $i = 1, \cdots, n$, if the ith bit of x_1 is 1, P_1 adds $(m_{ij})_{j=1}^{\lfloor t/2 \rfloor}$ to $\boldsymbol{y}_1$. Taking as input $x_2 \in \{0,1\}^n$, for $i = 1, \cdots, n$, if the ith bit of x_2 is 1, P_2 adds $(m_{ij})_{j=\lceil t/2 \rceil}^t$ to $\boldsymbol{y}_2$. Then P_1 and P_2 respectively pad $\boldsymbol{y}_1$ and $\boldsymbol{y}_2$ with $\overline{m}$ so that $\boldsymbol{y} = \boldsymbol{y}_1 \| \boldsymbol{y}_2$ consists of q elements in total.

Then (P_1, P_2) starts to run $\mathcal{B}(1^\lambda, k)$ in multiple rounds until $\mathcal{B}$ stops and returns $b \in \{0,1\}$. More specifically, in each round, P_1 runs $\mathcal{B}(1^\lambda, k)$, answers queries from $\mathcal{B}$ to the stream $\boldsymbol{y}_1$, and sends the local memory state of $\mathcal{B}$ denoted by s to P_2 after all the elements in $\boldsymbol{y}_1$ having been queried by $\mathcal{B}$. P_2 runs $\mathcal{B}(1^\lambda, k)$ starting from state s, answers queries from $\mathcal{B}$ to the stream $\boldsymbol{y}_2$, and then sends the local memory state of $\mathcal{B}$ back to P_1 after all the elements in $\boldsymbol{y}_2$ having been queried. The final output of (P_1, P_2) is $\mathcal{B}$'s output b.

Since the probability that there exist more than n elements with more than t pre-images is $1 - negl(\lambda)$, we have $\Pr[\mathsf{DISJ}(\boldsymbol{x}_1, \boldsymbol{x}_2) = F_{\mathsf{H}_k, t}(\boldsymbol{y}) \mid k \leftarrow \{0,1\}^\kappa] \geq 1 - negl(\lambda)$ and $\Pr[\mathcal{B}^{\mathcal{O}_{\boldsymbol{y}}}(1^\lambda, k) = F_{\mathsf{H}_k, t}(\boldsymbol{y}) \mid k \leftarrow \{0,1\}^\kappa] \geq c$. As a result, we have $\Pr[\mathcal{B}^{\mathcal{O}_{\boldsymbol{y}}}(1^\lambda, k) = \mathsf{DISJ}(\boldsymbol{x}_1, \boldsymbol{x}_2) \mid k \leftarrow \{0,1\}^\kappa] \geq c - negl(\lambda)$, which implies $\Pr[P_1(\boldsymbol{x}_1) \leftrightarrow P_2(\boldsymbol{x}_2) = \mathsf{DISJ}(\boldsymbol{x}_1, \boldsymbol{x}_2)] \geq c - negl(\lambda)$. Since $c > 1/2$, there must exist some constant $c' > 1/2$ such that $c - negl(\lambda) > c'$ for a *sufficiently large* λ. Therefore, (P_1, P_2) solves the disjointness problem.

Since P_1 and P_2 have communication $\kappa + O(p \cdot \mathbf{LocalMem}(\mathcal{B}))$, and Theorem 1 implies that the communication must be $\Omega(n) = \Omega(\lfloor q/t \rfloor) = \Omega(q)$ (in the worst case), we have $\mathbf{LocalMem}(\mathcal{B}) = \Omega((q - \kappa)/p)$, completing the proof. $\qquad\square$

5.3   Lower Bound for Large-Domain Hash Functions

In this section, we recall the definition of restricted reductions from mCR_t security to CR_t security and show a memory lower bound of these reductions.

Restricted Black-Box Reductions from mCR_t to CR_t. Let $\mathcal{R}$ be a black-box reduction from mCR_t security to CR_t security. As before, we write $\mathcal{R}^\mathcal{A}$ to mean that $\mathcal{R}$ has oracle access to a (stateful) adversary $\mathcal{A}$ playing the mCR_t game. Whenever receiving a query from $\mathcal{R}$, $\mathcal{A}$ returns the "next" query to $\mathcal{R}$. $\mathcal{R}$ is not able to modify the current state of $\mathcal{A}$ (i.e., $\mathcal{A}$ runs sequentially), but is allowed to adaptively rewind $\mathcal{A}$ to previous states.

Definition 14 (c-restricted black-box reduction from mCR_t to CR_t [1]). *Let $c > 0$ be a constant. An oracle-access PPT machine $\mathcal{R}^{(\cdot)}$ is said to be a c-restricted black-box reduction from the mCR_t security to the CR_t security of a hash function, if for any (possibly inefficient) adversary $\mathcal{A}$, we have* $\mathbf{Adv}_{\mathsf{CR}_t}^{\mathcal{R}^\mathcal{A}}(\lambda) \geq c \cdot \mathbf{Adv}_{\mathsf{mCR}_t}^{\mathcal{A}}(\lambda)$, *and $\mathcal{R}$ respects the following restrictions.*

- *The key k that $\mathcal{R}$ sends to $\mathcal{A}$ is the one generated by the challenger and given to $\mathcal{R}$ in the CR_t game.*
- *The queries made by $\mathcal{R}$ are amongst the queries made by $\mathcal{A}$.*

Theorem 7. *Let λ be a (sufficiently large) security parameter, $\mathsf{H} : \{0,1\}^\kappa \times \{0,1\}^\delta \to \{0,1\}^\rho$ be a large-domain hash function satisfying CR_t security, and $\mathcal{R}$ be a c_r-restricted black-box reduction from the mCR_t security to the CR_t security of H. Let $q = q(\lambda)$ be the maximum numbers of input queries made by an adversary in the mCR_t game. If (a) $\mathcal{R}$ rewinds the adversary for at most $p = p(\lambda)$ times and (b) $c_r > 1/2$, then we have*

$$\textbf{LocalMem}(\mathcal{R}) = \Omega(\min\{(q-\kappa)/(p+1)\}) - O(\log q) - \textbf{LocalMem}(\mathsf{H}).$$

Similarly to before, this theorem implies that when the maximum number of input queries made by an adversary in the mCR_t game is very large, $\mathcal{R}$ must consume large memory unless it rewinds $\mathcal{A}$ many times, which increases its running-time.

Proof (of Theorem 7). Assuming the existence of the reduction $\mathcal{R}$ stated in Theorem 7, we show the existence of a probabilistic algorithm $\mathcal{B}$ such that $\Pr[\mathcal{B}^{\mathcal{O}_y}(1^\lambda, k) = F_{\mathsf{H}_k,t}(\boldsymbol{y}) \mid k \leftarrow \{0,1\}^\kappa] \geq c_r > 1/2$ for all $\boldsymbol{y} = (y_1, \cdots, y_q) \in (\{0,1\}^\delta)^q$.

Game 0: In this game, $\mathcal{R}$ has access to an adversary $\mathcal{A}_{\boldsymbol{y}}$, and interacts with the challenger $\mathcal{CH}$ in the mCR_t game. $\mathcal{A}_{\boldsymbol{y}}$ runs as follows.

- On receiving $(1^\lambda, k)$, $\mathcal{A}_{\boldsymbol{y}}$ makes an input query y_1.
- For $i = 1, \cdots, q - 1$, when invoked (with empty input) for the ith time, $\mathcal{A}_{\boldsymbol{y}}$ makes an input query y_{i+1}.
- When invoked (with empty input) for the qth time, $\mathcal{A}_{\boldsymbol{y}}$ makes a stopping query stp.

We now show the following lemma.

Lemma 9. $\Pr[F_{\mathsf{H}_k,t}(\boldsymbol{y}) \leftarrow (\mathcal{R}^{\mathcal{A}_{\boldsymbol{y}}}(1^\lambda) \leftrightharpoons \mathcal{CH}(1^\lambda))] \geq c_r$ *for all* $\boldsymbol{y} = (y_1, \cdots, y_q) \in (\{0,1\}^\delta)^q$ *in* **Game 0**.

Proof (of Lemma 9). If $F_{\mathsf{H}_k,t}(\boldsymbol{y}) = 0$, then $\mathcal{CH}$ will output 0. This is due to the restriction that all the input queries made by $\mathcal{R}$ are amongst the elements in $\boldsymbol{y}$. On the other hand, one can see that $\textbf{Adv}_{\mathsf{mCR}_t}^{\mathcal{A}_{\boldsymbol{y}}}(\lambda) = 1$ when $F_{\mathsf{H}_k,t}(\boldsymbol{y}) = 1$. Since $\mathcal{R}$ is a c_r-restricted black-box reduction, we have $\textbf{Adv}_{\mathsf{CR}_t}^{\mathcal{R}^{\mathcal{A}_{\boldsymbol{y}}}}(\lambda) \geq c_r \cdot \textbf{Adv}_{\mathsf{mCR}_t}^{\mathcal{A}_{\boldsymbol{y}}}(\lambda) = c_r$, i.e., $\Pr[F_{\mathsf{H}_k,t}(\boldsymbol{y}) \leftarrow (\mathcal{R}^{\mathcal{A}_{\boldsymbol{y}}}(1^\lambda) \leftrightharpoons \mathcal{CH}(1^\lambda))] \geq c_r$, completing the proof of Lemma 9. ☐

Game 1: This game is exactly the same as **Game 0** except that there exists an algorithm $\mathcal{B}$ with access to the stream $\boldsymbol{y}$ that simulates $\mathcal{A}_{\boldsymbol{y}}$, $\mathcal{CH}$, and $\mathcal{R}$. Here, $\mathcal{B}$ takes as input $k \leftarrow \{0,1\}^\kappa$ from an external party and uses it as the hash key generated by $\mathcal{CH}$. Moreover, $\mathcal{B}$ does not store $\boldsymbol{y}$ in its local memory but queries

$\mathcal{O}_y$ to obtain the ith input query y_i for $i = 1, \cdots, q$. Whenever $\mathcal{R}$ executes a rewinding procedure, $\mathcal{B}^{\mathcal{O}_y}$ makes another pass on its stream so that it can access its next input query to $\mathcal{R}$. Since $\mathcal{B}^{\mathcal{O}_y}$ perfectly simulates $\mathcal{A}_y$, we immediately obtain the following lemma.

Lemma 10. $\Pr[F_{\mathsf{H}_k,t}(\boldsymbol{y}) \leftarrow \mathcal{B}^{\mathcal{O}_y}(1^\lambda, k)] \geq c_r$ *for all* $\boldsymbol{y} \in (\{0,1\}^\delta)^q$ *in* **Game 1**.

Since $c_r > 1/2$ and $\mathcal{B}$ makes $p + 1$ passes on its stream in total, according to Theorem 6 and Lemma 10, we have

$$\mathbf{LocalMem}(\mathcal{B}) = \Omega(\min\{(q - \kappa)/(p + 1)\}).$$

Furthermore, the memory used to simulate $\mathcal{CH}$ and $\mathcal{A}_y$ is $O(\log q) +$ **LocalMem**(H), where $O(\log q)$ is the amount of memory used to record q and the index of the next input query $\mathcal{A}_y$ will make. Therefore, we have

$$\mathbf{LocalMem}(\mathcal{B}) = \mathbf{LocalMem}(\mathcal{R}) + O(\log q) + \mathbf{LocalMem}(\mathsf{H}).$$

Combining the two bounds completes Theorem 7. $\qquad\qquad\qquad\qquad\qquad\square$

Remark on Security Parameter. Similarly to the case of Theorems 3 and 4, Theorem 7 holds only when the security parameter λ is sufficiently large, while one may wonder why memory-tightness makes sense when λ is already required to be large. Notice that λ is required to be large only when c (in Theorem 6) is very close to $1/2$. However, it is not hard to see that when c_r (which is the parameter of the reduction in Theorem 7) is not close to $1/2$, c (in Theorem 6) is not necessarily close to $1/2$. Hence, λ is not necessarily very large, unless c_r (in Theorem 7) is very close to $1/2$.

6   Open Problem

The lower bound results shown in Sect. 4 and Sect. 5 only treat reductions which respect restrictions on their queries. It is desirable to clarify whether memory lower bounds of natural black-box reductions exist with respect to those security games. Showing some novel streaming lower bounds based on other problems about parity learning might be a promising way. It is also desirable to know whether there exist memory lower bounds of reductions for the multi-challenge security of other class of cryptographic primitives, and whether it is possible to unify these bounds. Finally, it would be interesting to find memory lower bounds in the random oracle model.

Acknowledgement. A part of this work was supported by Input Output Cryptocurrency Collaborative Research Chair funded by IOHK, Nomura Research Institute, NTT Secure Platform Laboratories, Mitsubishi Electric, I-System, JSPS Fellowship for Young Scientists, JST CREST JPMJCR14D6 and JPMJCR1688, JST OPERA, JSPS KAKENHI JP16H01705, 16J10697, JP17H01695.

A   Lower Bounds for Unique-Key PKE and Signature Schemes in the Multi-user Setting

In this section, we give the security notions of unique-key signatures and encryption schemes in the multi-user setting. Then we show two memory lower bounds of restricted reductions, which are extensions of the result in Sect. 4.

A.1   Unique-Key PKE Schemes and Signatures

A cryptographic primitive (which can be PKE scheme, signature scheme, trapdoor commitment scheme (with collision resistance), etc.) with key generation algorithm Gen is called a *unique-key primitive* if there exists some algorithm Check such that (Gen, Check) forms a unique-key relation (see Definition 7). We now recall the definition of PKE schemes and define unique-key signatures and unique-key PKE schemes as follows.

Definition 15 (Public key encryption (PKE)). *A PKE scheme consists of the PT algorithms* (Gen, Enc, Dec). *(a)* Gen *is a probabilistic algorithm that takes as input* 1^λ, *and returns a public/secret key pair* (pk, sk). *(b)* Enc *is a probabilistic algorithm that takes as input a public key pk and a message* $m \in \{0,1\}^\delta$, *and returns a ciphertext ct. (c)* Dec *is a deterministic algorithm that takes as input a secret key sk and a ciphertext ct, and returns a message* $m \in \{0,1\}^\delta$ *or* $\perp$.

A PKE scheme is required to satisfy correctness, *which means that* $\mathsf{Dec}_{sk}(ct)$ *= m holds for all* $\lambda \in \mathbb{N}$, *all* $(pk, sk) \leftarrow \mathsf{Gen}(1^\lambda)$, *all* $m \in \{0,1\}^\delta$, *and all* $ct \leftarrow \mathsf{Enc}_{pk}(m)$.

Definition 16 (Unique-key signature and PKE). *A signature (respectively, PKE) scheme* (Gen, Sign, Verify) *(respectively,* (Gen, Enc, Dec)*) is said to have the unique-key property if there exists a deterministic PT algorithm* Check *such that* (Gen, Check) *is a unique-key relation.*

Now we define the security notions for unique-key signatures and PKE schemes. We denote mUF security and UF security in the multi-user setting by mU-mUF and mU-UF respectively. Moreover, we overload the notions mU-mOW and mU-OW (defined for unique-key relations) so that they apply to PKE schemes.

Definition 17 (mU-mUF). *A unique-key signature scheme* (Gen, Check, Sign, Verify) *is said to be* mU-mUF *secure if for any PPT adversary* $\mathcal{A}$, *we have* $\mathbf{Adv}^{\mathcal{A}}_{\mathsf{mU-mUF}}(\lambda) = \Pr[\mathcal{CH} \text{ outputs } 1] \leq negl(\lambda)$ *in the following game.*

1. *The challenger* $\mathcal{CH}$ *sets* $w = 0$, $\mathcal{Q} = \emptyset$, *and* $\mathcal{Q}_s = \emptyset$, *and runs* $\mathcal{A}$ *on input* 1^λ. *Then* $\mathcal{A}$ *may make sampling queries to* $\mathcal{CH}$, *and* $\mathcal{CH}$ *responds as follows.*
   – *On receiving the ith sampling query sp,* $\mathcal{CH}$ *samples* $(pk_i, sk_i) \leftarrow \mathsf{Gen}(1^\lambda)$ *and sends* pk_i *to* $\mathcal{A}$.
   Then $\mathcal{A}$ *may make adaptive corruption, signing, and verification queries to* $\mathcal{CH}$, *and* $\mathcal{CH}$ *responds as follows:*

    – *On receiving a corruption query i, $\mathcal{CH}$ adds i to $\mathcal{Q}$, and sends sk_i to $\mathcal{A}$.*

    – *On receiving a signing query (i, m), $\mathcal{CH}$ computes $\sigma \leftarrow \mathsf{Sign}_{sk_i}(m)$, adds (i, m) to $\mathcal{Q}_s$, and sends σ to $\mathcal{A}$.*

    – *On receiving a verification query (i^*, m^*, σ^*), if $\mathsf{Verify}_{pk_{i^*}}(m^*, \sigma^*) = 1$, $i^* \notin \mathcal{Q}$, and $(i^*, m^*) \notin \mathcal{Q}_s$, $\mathcal{CH}$ sets $w = 1$.*

2. *At some point, $\mathcal{A}$ makes a stopping query stp to $\mathcal{CH}$, and $\mathcal{CH}$ returns w.*

Definition 18 (mU-mOW (for PKE)). *A unique-key PKE scheme* (Gen, Check, Enc, Dec) *is said to be* mU-mOW *secure if for any PPT adversary $\mathcal{A}$, we have* $\mathbf{Adv}^{\mathcal{A}}_{\mathsf{mU\text{-}mOW}}(\lambda) = \Pr[\mathcal{CH}$ *outputs* $1] \le negl(\lambda)$ *in the following game.*

1. *The challenger $\mathcal{CH}$ sets $w = 0$, $\mathcal{Q} = \emptyset$, and $\mathcal{Q}_m = \emptyset$, and runs $\mathcal{A}$ on input 1^λ. Then $\mathcal{A}$ may make sampling queries to $\mathcal{CH}$, and $\mathcal{CH}$ responds as follows.*

    – *On receiving the ith sampling query sp, $\mathcal{CH}$ samples $(pk_i, sk_i) \leftarrow \mathsf{Gen}(1^\lambda)$ and sends pk_i to $\mathcal{A}$.*

2. *$\mathcal{A}$ may make adaptive corruption and challenge queries to $\mathcal{CH}$, and $\mathcal{CH}$ responds as follows:*

    – *On receiving a corruption query i, $\mathcal{CH}$ adds i to $\mathcal{Q}$, and sends sk_i to $\mathcal{A}$.*

    – *On receiving a challenge query i, $\mathcal{CH}$ searches $(i, m) \in \mathcal{Q}_m$. If the searching procedure fails, $\mathcal{CH}$ runs $m \leftarrow \{0, 1\}^\delta$ and adds (i, m) to $\mathcal{Q}_m$. Then it returns $ct \leftarrow \mathsf{Enc}_{pk_i}(m)$ to $\mathcal{A}$.*

    – *On receiving a verification query (i^*, m') from $\mathcal{A}$. If $i^* \notin \mathcal{Q}$ and $(i^*, m') \in \mathcal{Q}_m$, $\mathcal{CH}$ sets $w = 1$.*

3. *At some point, $\mathcal{A}$ makes a stopping query stp to $\mathcal{CH}$, and $\mathcal{CH}$ returns w.*

Definition 19 (mU-UF and mU-OW (for PKE)). mU-UF *security (respectively, mU-OW security for PKE) is defined in exactly the same way as mU-mUF security (respectively, mU-mOW security for PKE) except that $\mathcal{A}$ is allowed to make only one verification query and the advantage of $\mathcal{A}$ is denoted by* $\mathbf{Adv}^{\mathcal{A}}_{\mathsf{mU\text{-}UF}}(\lambda)$ *(respectively,* $\mathbf{Adv}^{\mathcal{A}}_{\mathsf{mU\text{-}OW}}(\lambda)$*).*

A.2 Lower Bounds for Unique-Key PKE Schemes and Signatures

We now show two memory lower bounds for restricted reductions respectively from mU-mUF security to mU-UF security and mU-mOW security to mU-OW security. The definition of the latter type of restricted reductions is exactly the same as Definition 9. The definition of the former type is also the same as Definition 9 except that the following restriction is additionally required.

– The set of signing queries made by $\mathcal{R}$ is the same as the set of all signing queries made by $\mathcal{A}$.[7]

---

[7] This restriction is made due to the fact that if the signing queries are chosen by $\mathcal{R}$, then the challenger may consume large memory to store them, which spoils our result. When considering random message attacks, this restriction can be removed. Also, this restriction is not made for other primitives such as PKE schemes, trapdoor commitment, and chameleon hash function schemes (with collision resistance).

These two by-product results can be treated as two examples of memory lower bounds derived from our lower bound result for unique-key relations stated in Theorem 4.

Corollary 1. *Let λ be a (sufficiently large) security parameter, $\Sigma = $ (Gen, Check, Sign, Verify), where the internal randomness space of Gen is $\{0,1\}^{\rho}$, be a mU-UF secure unique-key signature scheme, F : $\{0,1\}^{\kappa} \times \{0,1\}^{\lambda} \to \{0,1\}^{\rho}$ be a PRF, and $\mathcal{R}$ be a c_r-restricted black-box reduction from the mU-mUF security to the mU-UF security of Σ. Let $n = n(\lambda)$ be the maximum number of sampling queries and $q = q(\lambda)$ be the maximum numbers of corruption and verification queries made by an adversary in the mU-mUF game, and $\mathcal{U} = \{i\}_{i=1}^{n}$. If (a) $\mathcal{R}$ rewinds the adversary for at most $p = p(\lambda)$ times and (b) $c_r > 1/2$, then we have*

$$\textbf{LocalMem}(\mathcal{R}) = \Omega(\max\{q/(p+2), n/(p+2)\}) - O(\log q) - O(\log n) - \kappa$$
$$- \max\{\textbf{LocalMem}(\text{Gen}), \textbf{LocalMem}(\text{Check}),$$
$$\textbf{LocalMem}(\text{Sign}), \textbf{LocalMem}(\text{Verify}), \textbf{LocalMem}(\text{F})\}.$$

Corollary 2. *Let λ be a (sufficiently large) security parameter, $\Pi = $ (Gen, Check, Enc, Dec) with message space $\mathcal{M}$, where the internal randomness space of Gen is $\{0,1\}^{\rho}$, be a mU-OW secure unique-key PKE scheme, F : $\{0,1\}^{\kappa} \times \{0,1\}^{\lambda} \to \{0,1\}^{\rho}$ and F' : $\{0,1\}^{\kappa} \times \{0,1\}^{\lambda} \to \mathcal{M}$ be PRFs, and $\mathcal{R}$ be a c_r-restricted black-box reduction from the mU-mOW security to the mU-OW security of Π. Let $n = n(\lambda)$ be the maximum number of sampling queries and $q = q(\lambda)$ be the maximum numbers of corruption, challenge, and verification queries made by an adversary in the mU-mOW game, and $\mathcal{U} = \{i\}_{i=1}^{n}$. If (a) $\mathcal{R}$ rewinds the adversary for at most $p = p(\lambda)$ times and (b) $c_r > 1/2$, then we have*

$$\textbf{LocalMem}(\mathcal{R}) = \Omega(\max\{q/(p+2), n/(p+2)\}) - O(\log q) - O(\log n) - 2\kappa$$
$$- \max\{\textbf{LocalMem}(\text{Gen}), \textbf{LocalMem}(\text{Check}), \textbf{LocalMem}(\text{Enc}),$$
$$\textbf{LocalMem}(\text{Dec}), \textbf{LocalMem}(\text{F}), \textbf{LocalMem}(\text{F}')\}.$$

We omit the proofs of the above two corollaries since they are very similar to the proof of Theorem 4. The main difference is that instead of directly giving secret keys to $\mathcal{R}$ as verification queries, the adversary playing the mU-mUF or mU-mOW game uses the secret keys to forge signatures or decrypt challenge ciphertexts. Moreover, the mU-OW challenger uses F' to simulate the random messages chosen for challenge queries so that it does not have to consume large memory to store the list $\mathcal{Q}_m$.

References

1. Auerbach, B., Cash, D., Fersch, M., Kiltz, E.: Memory-tight reductions. In: Katz, J., Shacham, H. (eds.) CRYPTO 2017, Part I. LNCS, vol. 10401, pp. 101–132. Springer, Cham (2017). https://doi.org/10.1007/978-3-319-63688-7_4
2. Bader, C.: Efficient signatures with tight real world security in the random-oracle model. In: Gritzalis, D., Kiayias, A., Askoxylakis, I. (eds.) CANS 2014. LNCS, vol. 8813, pp. 370–383. Springer, Cham (2014). https://doi.org/10.1007/978-3-319-12280-9_24
3. Bader, C., Hofheinz, D., Jager, T., Kiltz, E., Li, Y.: Tightly-secure authenticated key exchange. In: Dodis, Y., Nielsen, J.B. (eds.) TCC 2015, Part I. LNCS, vol. 9014, pp. 629–658. Springer, Heidelberg (2015). https://doi.org/10.1007/978-3-662-46494-6_26
4. Bader, C., Jager, T., Li, Y., Schäge, S.: On the impossibility of tight cryptographic reductions. In: Fischlin, M., Coron, J.-S. (eds.) EUROCRYPT 2016, Part II. LNCS, vol. 9666, pp. 273–304. Springer, Heidelberg (2016). https://doi.org/10.1007/978-3-662-49896-5_10
5. Boyen, X., Mei, Q., Waters, B.: Direct chosen ciphertext security from identity-based techniques. In: ACM CCS 2005, pp. 320–329 (2005)
6. Coron, J.-S.: Optimal security proofs for PSS and other signature schemes. In: Knudsen, L.R. (ed.) EUROCRYPT 2002. LNCS, vol. 2332, pp. 272–287. Springer, Heidelberg (2002). https://doi.org/10.1007/3-540-46035-7_18
7. Cramer, R., Shoup, V.: Signature schemes based on the strong RSA assumption. ACM Trans. Inf. Syst. Secur. 3(3), 161–185 (2000)
8. Gamal, T.E.: A public key cryptosystem and a signature scheme based on discrete logarithms. IEEE Trans. Inf. Theory 31(4), 469–472 (1985)
9. Gentry, C.: Practical identity-based encryption without random oracles. In: Vaudenay, S. (ed.) EUROCRYPT 2006. LNCS, vol. 4004, pp. 445–464. Springer, Heidelberg (2006). https://doi.org/10.1007/11761679_27
10. Goldwasser, S., Ostrovsky, R.: *Invariant* signatures and non-interactive zero-knowledge proofs are equivalent (extended abstract). In: Brickell, E.F. (ed.) CRYPTO 1992. LNCS, vol. 740, pp. 228–245. Springer, Heidelberg (1993). https://doi.org/10.1007/3-540-48071-4_16
11. Haitner, I., Holenstein, T.: On the (Im)possibility of key dependent encryption. In: Reingold, O. (ed.) TCC 2009. LNCS, vol. 5444, pp. 202–219. Springer, Heidelberg (2009). https://doi.org/10.1007/978-3-642-00457-5_13
12. Haralambiev, K., Jager, T., Kiltz, E., Shoup, V.: Simple and efficient public-key encryption from computational diffie-hellman in the standard model. In: Nguyen, P.Q., Pointcheval, D. (eds.) PKC 2010. LNCS, vol. 6056, pp. 1–18. Springer, Heidelberg (2010). https://doi.org/10.1007/978-3-642-13013-7_1
13. Hofheinz, D., Jager, T., Knapp, E.: Waters signatures with optimal security reduction. In: Fischlin, M., Buchmann, J., Manulis, M. (eds.) PKC 2012. LNCS, vol. 7293, pp. 66–83. Springer, Heidelberg (2012). https://doi.org/10.1007/978-3-642-30057-8_5
14. Hohenberger, S., Waters, B.: Short and stateless signatures from the RSA assumption. In: Halevi, S. (ed.) CRYPTO 2009. LNCS, vol. 5677, pp. 654–670. Springer, Heidelberg (2009). https://doi.org/10.1007/978-3-642-03356-8_38
15. Kakvi, S.A., Kiltz, E.: Optimal security proofs for full domain hash, revisited. In: Pointcheval, D., Johansson, T. (eds.) EUROCRYPT 2012. LNCS, vol. 7237, pp. 537–553. Springer, Heidelberg (2012). https://doi.org/10.1007/978-3-642-29011-4_32

16. Kalyanasundaram, B., Schnitger, G.: The probabilistic communication complexity of set intersection. SIAM J. Discrete Math. **5**(4), 545–557 (1992)
17. Kravitz, D.W.: Digital signature algorithm. US Patent 5,231,668, 27 July 1993
18. Krawczyk, H., Rabin, T.: Chameleon signatures. In: NDSS. The Internet Society (2000)
19. Lysyanskaya, A.: Unique signatures and verifiable random functions from the DH-DDH separation. In: Yung, M. (ed.) CRYPTO 2002. LNCS, vol. 2442, pp. 597–612. Springer, Heidelberg (2002). https://doi.org/10.1007/3-540-45708-9_38
20. Micali, S., Rabin, M.O., Vadhan, S.P.: Verifiable random functions. In: FOCS, pp. 120–130. IEEE Computer Society (1999)
21. Razborov, A.A.: On the distributional complexity of disjointness. Theor. Comput. Sci. **106**(2), 385–390 (1992)
22. Rivest, R.L., Shamir, A., Adleman, L.M.: A method for obtaining digital signatures and public-key cryptosystems (reprint). Commun. ACM **26**(1), 96–99 (1983)
23. Waters, B.: Efficient identity-based encryption without random oracles. In: Cramer, R. (ed.) EUROCRYPT 2005. LNCS, vol. 3494, pp. 114–127. Springer, Heidelberg (2005). https://doi.org/10.1007/11426639_7
24. Wichs, D.: Barriers in cryptography with weak, correlated and leaky sources. In: ITCS, pp. 111–126. ACM (2013)

Fiat-Shamir and Correlation Intractability from Strong KDM-Secure Encryption

Ran Canetti[1,2]($\boxtimes$), Yilei Chen[1], Leonid Reyzin[1], and Ron D. Rothblum[3,4]

[1] Boston University, Boston, USA
{canetti,chenyl,reyzin}@bu.edu
[2] Tel Aviv University, Tel Aviv, Israel
canetti@tau.ac.il
[3] MIT, Cambridge, USA
ronr@csail.mit.edu
[4] Northeastern University, Boston, USA
r.rothblum@northeastern.edu

Abstract. A hash function family is called correlation intractable if for all sparse relations, it is hard to find, given a random function from the family, an input-output pair that satisfies the relation (Canetti et al., STOC 1998). Correlation intractability (CI) captures a strong Random-Oracle-like property of hash functions. In particular, when security holds for all sparse relations, CI suffices for guaranteeing the soundness of the Fiat-Shamir transformation from any constant round, statistically sound interactive proof to a non-interactive argument. However, to date, the only CI hash function for all sparse relations (Kalai et al., Crypto 2017) is based on general program obfuscation with exponential hardness properties.

We construct a simple CI hash function for arbitrary sparse relations, from any symmetric encryption scheme that satisfies some natural structural properties, and in addition guarantees that key recovery attacks mounted by polynomial-time adversaries have only exponentially small success probability - even in the context of key-dependent messages (KDM). We then provide parameter settings where ElGamal encryption and Regev encryption plausibly satisfy the needed properties. Our techniques are based on those of Kalai et al., with the main contribution being substituting a statistical argument for the use of obfuscation, therefore greatly simplifying the construction and basing security on better-understood intractability assumptions.

In addition, we extend the definition of correlation intractability to handle moderately sparse relations so as to capture the properties required in proof-of-work applications (e.g. Bitcoin). We also discuss the applicability of our constructions and analyses in that regime.

The full version [25] is available at https://eprint.iacr.org/2018/131.

J. B. Nielsen and V. Rijmen (Eds.): EUROCRYPT 2018, LNCS 10820, pp. 91–122, 2018.
https://doi.org/10.1007/978-3-319-78381-9_4

1 Introduction

The random oracle methodology [12,39] models cryptographic hash functions as completely random functions. The model yields simple constructions of cryptographic primitives both in theory and practice, but is known to be inherently unsound in principle [26,32,44,51,68]. A natural alternative is to formalize concrete "random-oracle-like" properties of hash functions, and then (a) construct hash functions that provably demonstrate these properties based on established hardness assumptions, and (b) show how security of applications follow from these properties. Indeed, a number of such notions have been proposed and used in the literature, with multiple applications e.g. [10,11,18,23,26,29,47,52,57].

Correlation intractability. We focus on one of such notion called *correlation intractability*, defined by Canetti et al. [26]. Correlation intractability attempts to capture the following property of random functions. Consider a random function O from $\{0,1\}^n$ to $\{0,1\}^m$, along with some fixed binary relation $R : \{0,1\}^n \times \{0,1\}^m \to \{0,1\}$ such that for any $x \in \{0,1\}^n$, the fraction of $y \in \{0,1\}^m$ such that $R(x,y)$ holds is at most μ. Then, the best possible way to find x such that $R(x, O(x))$ holds is to randomly try different x's. The probability of success after t attempts is at most $t\mu$. A function family is *correlation intractable (CI)* if it behaves similarly against polytime algorithms. Specifically, a function family H is correlation intractable if, for any relation R with negligible density μ, no polytime adversary can, given the description of a function $h : \{0,1\}^n \to \{0,1\}^m$ chosen randomly from H, find x such that $R(x, h(x))$ holds, except with negligible probability. Note that there are no secrets here: The adversary sees the entire description of h, which succinctly encodes the values $h(x)$ for all possible values of x.

Correlation intractability captures a large class of natural properties of random functions. For example, the infeasibility of finding preimages of any *fixed* value c in the range can be formalized as correlation intractability w.r.t. any constant relations $R_c = \{(x, c) \mid \forall x$ in the domain$\}$. The "fixed output value" in the example can be extended to "a sufficiently long fixed prefix", e.g. sufficiently many leading 0s. Indeed, correlation intractability (in its quantitative form) is the natural formalization of the requirements expected from the hash function used for mining chaining values in the Bitcoin protocol [66] and other applications relied on proof-of-work [35]. We further discuss these application later on.

Another natural and prominent application of correlation intractable hash functions is their use for sound realization of the Fiat-Shamir (FS) heuristic [39]. Recall that, originally, the idea of Fiat and Shamir was to transform a three-message, public coin identification scheme to a signature scheme by having the signer first generate the first prover message α of the identification scheme (incorporating the message-to-be-signed in the identity), then computing the verifier message as $\beta = h(\alpha)$ for some public hash function h, and then having the signature consist of (α, γ), where γ is the corresponding third message of the identification scheme. Verification first reconstructs $\beta = h(\alpha)$ and then verifies the identification. It can be seen that if h is modeled as a random function, then the resulting signature scheme is unforgeable [1]. In fact, the same transform can be used to build a non-interactive argument from any public-coin interactive proof (even multi-round ones), as long as the initial proof is *resettably sound* (see e.g. [13]).[1] Furthermore, if the original proof is honest-verifier zero-knowledge, then the resulting non-interactive protocol (in the random oracle model) is a non-interactive zero-knowledge argument [12,39].

It has been demonstrated that CI families that withstand arbitrary binary relations suffice for realizing the Fiat-Shamir heuristic in the case of constant-round proofs. That is, if the initial interactive proof is constant-round and is statistically sound, then computational soundness of the resulting non-interactive protocol holds even when the random oracle is replaced by a CI hash function family that withstands arbitrary binary relations (the only difference from the original Fiat-Shamir heuristic is that now the resulting protocol has an initial verifier message that determines the actual function h in the CI family.) Indeed, CI families that withstand arbitrary binary relations are *entropy preserving* [24], and entropy preserving families suffice for the soundness of the Fiat-Shamir heuristic for constant-round proofs [10]. A direct proof is also implicit in [59, Sect. 4]. (We note that soundness for the case of three-message proofs was observed already in [36,49].)

Constructing correlation intractable hash functions. Canetti et al. [26] show that there do not exist CI function families where the key is shorter than the input, but leave open the possibility of CI functions with longer keys. Still no construction of CI functions, even for restricted cases, was known until very recently. Furthermore, over the years evidence accumulated that coming up with CI functions, and in particular a sound instantiation of the FS paradigm, would not be easy. Goldwasser and Kalai [44] construct a public coin interactive *argument* (i.e. a protocol that is only computationally sound) that becomes unsound if it is turned into an non-interactive argument by applying the Fiat-Shamir transformation with any function. Bitansky et al. show that it is impossible to prove soundness of the FS paradigm using a black-box reduction to falsifiable assumptions [14].

Recently, two papers independently suggested using an obfuscated puncturable pseudorandom function family as a CI family. Canetti et al. [24] show that this construction is CI for relations that are computable by circuits of a priori bounded polynomial size, assuming sub-exponentially secure puncturable pseudorandom functions and indistinguishability obfuscation, and in addition, input hiding obfuscation for evasive functions. Kalai et al. [59] show that the same construction is CI for arbitrary relations, assuming sub-exponentially secure puncturable pseudorandom functions and indistinguishability obfuscation, plus exponentially secure point obfuscation. In particular, the latter result implies that this function family suffices for sound realization of the Fiat-Shamir heuristic (when applied to constant-round interactive proofs).

---

[1] In particular, every *constant-round* interactive proof with negligible soundness, is resettably sound.

1.1 Our Results

We provide new constructions of CI function families for arbitrary binary relations. Compared to [24,59], our constructions are dramatically more efficient, and are based on better-understood assumptions. Furthermore, while sampling a hash function from the family of obfuscated puncturable PRFs involves secret randomness, we present candidates where the sampling can be done with only public randomness.

The main tool (or, abstraction) we use is symmetric encryption with the following two properties: First, the scheme should guarantee that polynomial time key-recovery attacks have only exponentially small success probability even after seeing encryptions of key-dependent messages (KDM). That is, for any super-polynomial function s, for an arbitrary key-dependency function f (not necessarily computable in polynomial time), any polynomial time adversary that obtains $c = \mathsf{Enc}(k, f(k))$ outputs k with probability no more than $\frac{s(\lambda)}{2^\lambda}$, where λ is the key length.

The second property, which we refer to as *universal ciphertexts*, is statistical. Loosely speaking, it requires that any ciphertext is "decryptable" under any key. More precisely, the requirement is that (a) for every key, random ciphertexts decrypt to random messages; (b) for every key k and message m, the encryption algorithm generates ciphertexts that are uniformly sampled from the space of ciphertexts that are decrypted to m with key k. (The actual definition includes also public parameters, which are omitted here for simplicity.) Given an encryption scheme that satisfies the above requirements, we obtain the following result:

Theorem 1 (Informally stated). *Assuming the existence of encryption schemes that have universal ciphertexts and that are exponentially KDM-secure against polytime key-recovery attacks, there exist:*

- *Correlation intractable hash functions for arbitrary binary sparse relations.*
- *Hash functions that guarantee soundness of the Fiat-Shamir transformation, when applied to interactive proof-systems.*
- *Non-interactive, publicly verifiable arguments for all languages computable in polynomial-time and* bounded *polynomial space (in particular, the class* SC*).*

The last bullet follows by applying the Fiat-Shamir transformation to the recent public-coin, constant-round interactive proof-system of Reingold et al. [74].

Our second main contribution is in providing concrete instantiations of Theorem 1. Specifically, we show that variants of El-Gamal encryption [37] and Regev encryption [72] satisfy the universal ciphertext property and *plausibly* satisfy the foregoing exponential security against KDM key recovery.

Non-interactive zero-knowledge. As an additional result, we show that if the Fiat-Shamir transformation is applied to a three-round honest-verifier zero-knowledge proof, and the CI function family in use is *programmable*, then the resulting protocol is a Non-Interactive Zero-Knowledge (NIZK) argument, with the description of the hash function serving as a common reference string. (Here programmability means that, given random values a, b from the family's domain and range, respectively, it is possible to efficiently sample a random function h from the family such that $h(a) = b$.) We also observe that the CI functions we construct are programmable. Furthermore, if the initial three-round protocol is delayed-input (as in, e.g., [38]), then the resulting NIZK argument is both adaptive ZK and adaptively sound. We thus have:

Theorem 2 (Informally stated). *Assuming the existence of encryption schemes that have universal ciphertexts and that are exponentially KDM-secure against polytime key-recovery attacks, there exist* NIZK *arguments for all of* NP. *Furthermore, these* NIZK*s have adaptive soundness and zero-knowledge.*

We note that, prior to this work, NIZK arguments for NP were not known based on any variant of the Diffie-Hellman assumption in groups that do not admit bilinear pairings, nor any variant of the LWE assumption—even exponentially strong ones. Also, for the NIZK application we only need the CI family to withstand relations that are exponentially sparse, which somewhat relaxes the assumption. For example, if the soundness of the interactive proof system is $2^{-\lambda^\epsilon}$, then we can tolerate encryption schemes where the success probability of polytime key-recovery attack is $\frac{\text{superpoly}(\lambda)}{2^{\lambda - \lambda^\epsilon}}$.

Quantitative correlation intractability and its connection to the Bitcoin protocol. A central component in the Bitcoin protocol [66] is a probabilistic mechanism for guaranteeing that the amount of influence participants have on the process of producing the public ledger is proportional to their computing power. The idea here is that since each individual entity has only a fraction of the overall computing power, the influence of each entity is limited. Indeed, the core validity of the currency (i.e., the mechanism for preventing double spending) hinges upon that guarantee.

The Bitcoin mechanism for limiting influence was sketched earlier in the introduction: In order to incorporate a block of new transactions in the public registry, the individual ("miner") is asked to present a value x such that the pair $(x, h(x))$ satisfies some known relation R_w, where h is a hash function defined by the protocol, and w is determined by the current state of the system, the new block, and the miner's identity. R_w is set so that it is "moderately sparse". That is, for any x, w the fraction of values y such that $R_w(x, y)$ holds is small, but not too small.

Clearly, if h were a random function then this mechanism would work well: Given w, the best way to find x such that $R_w(x, h(x))$ holds is to keep guessing random x's until one is found. This means that the probability of success is proportional to the number of guesses, which is correlated to the computational power of the miner. However, when h is an explicit function with a succinct description, it is not clear how to provide rigorous guarantees regarding the amount of time needed to find a "wining x" given w. Indeed, "shortcut attacks" on the Bitcoin mechanism have been reported, e.g. [53].

We argue that correlation intractability, or more precisely a quantitative variant of the notion, captures the properties needed from the underlying hash function so as to guarantee the soundness of the Bitcoin mechanism for limiting influence. Specifically, say that a binary relation $R : \{0,1\}^n \times \{0,1\}^m \to \{0,1\}$ is μ-sparse if for any $x \in \{0,1\}^n$, the fraction of $y \in \{0,1\}^m$ such that $R(x,y)$ holds is at most μ. A family H of functions $h : \{0,1\}^n \to \{0,1\}^m$ is f-**correlation intractable** if for any binary μ-sparse relation R and for any adversary Adv that runs in time t, the probability that Adv, given a random function h in H, outputs x such that $R(x, h(x))$ holds is at most $f(t, \mu)$. The smaller f grows the better the guarantee. Clearly we must have $f(t, \mu) \geq t\mu$. A good "fudge function" f will not grow much faster than that.

It should also be stressed that the quantitative correlation intractability, as presented here, only bounds the success probability in solving a *single challenge*. Asserting the overall stability of the protocol would require bounding the aggregate success probability over multiple related challenges. Formalizing a set of properties for concrete, non-idealized hash functions, that would suffice for the security of Bitcoin-like applications, as well as proposing constructions with rigorous analyses is a fascinating research direction.

1.2 Our Techniques

The construction of our CI hash function is simple. Let $(\mathsf{Enc}, \mathsf{Dec})$ be an encryption scheme with key space K, message space M and ciphertext space C. The constructed hash function family $H = \{h_c\}_{c \in C}$, where $h_c : K \to M$, is defined by $h_c(k) = \mathsf{Dec}_k(c)$. That is, a function h_c in the family is defined via a ciphertext $c \in C$. Given an input k, the function h_c decrypts c using key k and returns the decrypted plaintext.

In general, key generation (i.e., choosing a random $c \in C$) is done by encrypting a random message with a random key. We note however that for both of our specific candidates, choosing a random ciphertext can be done obliviously and publicly without any secret randomness.

A high level rationale for the construction may be the following: Consider a ciphertext $c = \mathsf{Enc}(k, m)$ where both k and m are random. If the encryption scheme is good, then it should be guaranteed that, when trying to decrypt c with any key $k' \neq k$, then the result should be completely "random looking". Intuitively, this means that finding a key k' such that $\mathsf{Dec}(k', c) = m'$ for some target m' should be hard. The universal ciphertexts property guarantees that a random ciphertext looks like the result of encrypting a random message with a random key. KDM security guarantees that the above intuition applies even when considering relations that look at both the key and the corresponding message together (as is indeed the case for correlation intractability).

Indeed, the crux of the proof is in translating correlation intractability, which is a requirement on the (in)ability of polynomial time adversaries to find structure in a succinctly represented public function (namely the decryption algorithm along with a random ciphertext), to a *secrecy* requirement on the corresponding encryption process.

The actual proof is strongly inspired by that of [59]. In fact, we follow essentially the same sequence of logical steps. However, the argumentation used to move from one step to the next is different in some key places. Specifically, our goal is to turn an adversary A that breaks correlation intractability of the hash function into an adversary that breaks KDM security of the underlying encryption scheme. Following [59], we start by considering a conditional experiment where we fix some random value k^*, and consider only the probability that A, given the hash key c, outputs a key k such that the correlation $R(k, \mathsf{Dec}(k, c))$ holds, *and in addition $k = k^*$*. While this probability is very small, it allows us to move (with some loss) to a different experiment where the value c that A sees is the result of encrypting $f(k^*)$ with key k^*, where f is a function related to R. We now observe that recovering the right k^* corresponds to breaking the KDM security of the scheme.

As in [59], the price of this analytical approach is an exponential loss in security against guessing attacks. On the other hand, in the case of the [59] scheme and analysis, the critical switch from one conditional experiment to another relies on sub-exponentially secure indistinguishability obfuscation. Here, in contrast, the move is purely statistical.

1.3   A Closer Look at the Hardness Assumptions

We sketch the assumptions we use and briefly discuss their plausibility.

The scheme based on ElGamal encryption. We first consider the ElGamal based scheme. For simplicity, we discuss a restricted case where both the key and the message are represented by group elements. (See Sect. 6 for a more general construction and the associated assumption.) Assuming there exists a family of groups $\mathbb{G}(\lambda)$ of sizes $N(\lambda) \approx 2^\lambda$, with a generator g and efficient group operations, such that for any super-polynomial function s, any (not necessarily efficiently computable) function $f : [N] \to [N]$, and any polynomial time adversary A:

$$\Pr_{k, a \leftarrow [N]} \left[A\left(g^a, g^{ak+f(k)} \right) = k \right] \leq \frac{s(\lambda)}{2^\lambda}$$

We discuss the plausibility of this assumption. For the discrete-log problem over $\mathbb{F}_q^*$, there are well-known sub-exponential time algorithms with constant success probability [2, 30]. However, a 2^t-time algorithm with constant success probability does not necessary imply a polynomial time algorithms with success probability 2^{-t}. For example, Pollard's rho algorithm [70] runs in time $O(2^{\lambda/2})$ and achieves constant success probability. But its polynomial time version only gives polynomial advantage over simply guessing. In fact, Shoup [77] shows that any generic algorithm (like Pollard's rho algorithm) cannot achieve success probability better than $O(T^2/2^\lambda)$ if it only makes T oracle queries.

However, the index-calculus algorithm does achieve a $2^{-\lambda/c}$ success probability if it is allowed to have a super-polynomial preprocessing phase, keep advices of polynomial size, and run a polynomial time online phase. We leave the algorithm and analysis in Appendix A. Although it is not a complete polynomial time algorithm (i.e. without a super-polynomial preprocessing phase) with non-trivial success probability, it suggests that the extra structure of the group $\mathbb{F}_q^*$ can be utilized even if the algorithm is restricted in polynomial time in a meaningful model.

Still, the above assumption is plausible for the discrete-log problem over elliptic curve groups (ECDLP), especially for those defined over prime fields. Over decades, ECDLP algorithms only out-perform generic algorithms for specific families of curves (e.g. [42, 63]). Useful factor bases for index calculus algorithms were not known for the elliptic curve groups, until the work of Semaev [76] which proposes the use of summation polynomials, later developed by Gaudry [41] and Diem [31]. But so far they are only known to out-perform Pollard's rho algorithm for elliptic curve groups defined over $\mathbb{F}_{q^n}$ when certain relations of q and n hold. For elliptic curve groups defined over prime fields, the recent attempts by [69] and others provide plausible factor bases. Still, no algorithms are known to achieve non-negligible success probability with less than $O(2^{\lambda/2})$ running time. See [40] for a survey of the recent progress on ECDLP.

To conclude, based on the current understanding ECDLP for curves defined over prime fields, polytime algorithms that perform super-polynomially better than guessing appear to be out of reach. In particular, any such algorithm must exploit more structures in the elliptic curve groups than in generic groups [77].

The scheme based on Regev encryption. Consider the Regev scheme [73] with an even polynomial modulus $q(\lambda) \in \text{poly}(\lambda)$, and key space $\{0, ..., B-1\}^\ell$ where $B^\ell \in [2^{\lambda - \log(\lambda)}, 2^{\lambda + \log(\lambda)}]$ and $B \leq q$. The message space is $\{0, 1\}^w$ where $w(\lambda) \in \text{poly}(\lambda)$. For the security of this scheme we make the following assumption: for any (not necessarily efficiently computable) function $f : \{0, ..., B-1\}^\ell \to \{0, 1\}^w$, any super-polynomial function s, and any polynomial time adversary A:

$$\Pr_{\substack{\mathbf{k} \in_R \{0,...,B-1\}^\ell \\ \{\mathbf{a}_j \in_R \mathbb{Z}_q^{1 \times \ell}, e_j \in_R [0, q/2) \cap \mathbb{Z}\}}} \left[A\left(\{\mathbf{a}_j, \mathbf{a}_j \cdot \mathbf{k} + e_j + f_j(\mathbf{k}) \cdot q/2\}_{j \in [w]}\right) = \mathbf{k} \right] \leq \frac{s(\lambda)}{2^\lambda}$$

where $f_j(\mathbf{k})$ denotes the jth bit of $f(\mathbf{k})$.

Note that super-polynomial algorithms that break LWE with constant success probability are known (e.g. [8, 16, 60, 61, 75], see the analyses and surveys of [4, 55, 62, 65, 67]). Still, within this setting of parameters, especially given a polynomial size modulus q and high noise magnitude $q/2$, we are not aware of any polynomial time algorithms that succeed in guessing the key super-polynomially better than a random guess.

Possible relaxations on the assumptions of success probability. The restriction on the success probability (smaller than $\frac{s(\lambda)}{2^\lambda}$ for any super-polynomial s) mentioned in the foregoing paragraphs suffices for implying correlation intractability for *all* negligible sparse relations under *any* given input and output length parameters. We note that even if there are polynomial time algorithms that achieve better success probability for these problems, our result may still apply to correlation intractability for *certain* classes of relations. For example, if a polynomial time algorithm were found for LWE that succeeds with probability $2^{-\lambda/3}$, then the Regev-based hash function may still be secure for Fiat-Shamir transformation applied on a 3-round proof system where the length of the first message is λ, the length of the second message is $2\lambda/3$, and the soundness of the protocol is $2^{-2\lambda/3}$.

On the quantitative hardness of our assumptions. One may wonder if the ElGamal or Regev-like hash functions were used for proof-of-work, what are the precise bounds of the "fudge function" f we can guarantee. For the ElGamal-based function, as we mentioned before, the Pollard's rho algorithm achieves success probability $O(T^2/2^\lambda)$ in T steps for any group of size $\approx 2^\lambda$. So the smallest possible f is $O(T^2 \cdot \mu)$, which is far from the dream bound $T \cdot \mu$. For LWE, when T is relatively small (say a small polynomial), the success probabilities of LWE solvers are typically tiny and less studied, so the precise bound is unclear to us. We leave to future work any additional quantitative analysis of the possible values for f for the concrete functions.

1.4   Additional Related Works

Notions related to Fiat-Shamir paradigm. Hada, Tanaka [49] and Dwork et al. [36] show that the existence of correlation intractable functions implies the soundness of Fiat-Shamir paradigm for proofs, which in turn rules out the possibility of constant-round public-coin zero-knowledge proofs for languages beyond BPP. This means that, assuming KDM-secure encryption as defined above, there do not exist constant-round public-coin zero-knowledge protocols with negligible soundness error for languages beyond BPP.

Among the attempts to better capture the property of a hash function suitable for the Fiat-Shamir paradigm, Barak et al. define *entropy-preserving hashing* and show it is sufficient for Fiat-Shamir [10]. Dodis et al. then provide a property of condensers that is necessary for entropy-preserving hashing [33]. It is shown by Canetti et al. that entropy-preservation is implied by correlation intractability w.r.t. sparse relations whose memberships are not efficiently checkable [24].

A different way of reducing rounds in interactive proofs was shown by Kalai and Raz [58]. However, in contrast to the Fiat-Shamir paradigm, the Kalai-Raz transform inherently yields a *private-coin* argument-system (and in particular does not yield NIZK proof-systems).

Background on KDM. The potential security risk of encrypting one's own key was noted already in the seminal work of Goldwasser and Micali [45]. Potential applications and suitable formalizations were provided by Camenisch and Lysyanskaya [22] and Black, Rogaway and Shrimpton [15]. More recently, Gentry's breakthrough construction of fully homomorphic encryption also utilizes KDM security in a fundamental way for the "bootstrapping" process (transforming somewhat homomorphic schemes to *fully* homomorphic ones) [43].

Encryption schemes that are KDM secure[2] with respect to the class of affine functions were constructed by Boneh et al. [19], Applebaum et al. [6] and Brakerski and Goldwasser [20]. Using techniques developed in [5, 9, 21] the foregoing schemes can be amplified to provide security for the class of KDM functions computable by *polynomial-size* circuits. Canetti et al. [28] construct strong KDM-secure encryption from multi-bit point obfuscation. However, their construction inherently does *not* have the universal ciphertexts property. We also note that fully-homomorphic encryption schemes that are KDM secure w.r.t. the identity function are automatically KDM secure for arbitrary polynomial functions [9]. However achieving KDM secure FHE w.r.t. the identity function from standard assumptions is an open problem.

Haitner and Holenstein [50] showed limitations to the possibility of constructing KDM secure encryption schemes via blackbox techniques. They first show that there is no fully blackbox reduction from the KDM security of an encryption scheme (with respect to a certain class of functions) to the existence of one-way permutations. More relevant for us is their second result, which shows that there is no reduction from the KDM security of an encryption scheme to "essentially any cryptographic assumption" if the adversary can obtain an encryption of an arbitrary function g of the key, and the reduction treats both the adversary and the function g as black boxes. A significant difference from our notion of KDM security with respect to all functions is that [50] assume that the adversary also obtains oracle access to the function g, which is not the case in our setting. Namely, we only provide the adversary with an encryption of $g(k)$, where k is the key, but no additional access to g. Indeed, the oracle constructed by Haitner and Holenstein becomes useless in this setting.

The works of Halevi, Krawczyk [51] and Hofheinz, Unruh [56] construct several variants of KDM symmetric encryption assuming only pseudorandom functions. However these schemes don't achieve the level of security we require (exponentially small probability of key-recovery) and we were unable to extend them to schemes that do.

Relation to Extremely Lossy Functions (ELFs). Our work bears a high-level similarity to the work of Zhandry [79] in terms of the motivation, constructions and assumptions. However, the actual contributions are very different.

In terms of the motivation, both papers attempt to capture the properties of random oracles. Our paper focuses on correlation intractability and its implication to Fiat-Shamir, whereas [79] defines the notion of k-ary output-intractability, where the relation checks k output values and an additional auxiliary input w. Indeed, as was mentioned in [79], k-ary output-intractability roughly corresponds to a special case of k-ary correlation intractability (namely, correlation intractability where the relation R takes k pairs of values (x, y).) However, k-ary output-intractability is interesting only for $k > 1$. For $k = 1$, output intractability is trivially satisfiable. In contrast, in this work we concentrate on correlation intractability with $k = 1$.

---

[2] More precisely, the KDM security of these scheme reduces to their plain (i.e., non key dependent) semantic security.

In terms of constructions and assumptions, both papers make exponential hardness assumptions on discrete-log or DDH type problems. However the precise ways of making the assumptions are different. [79] assumes that for DDH over group size $B(\lambda) \approx 2^\lambda$, the best attack takes time $B(\lambda)^c$ for some constant c. Whereas we assume (modulo KDM) that all the polynomial time algorithm solves discrete-log problem with success probability less than $\frac{\text{superpoly}(\lambda)}{B(\lambda)}$.

1.5 Organization

In Sect. 2 we provide standard notations and definitions that will be used throughout this work. In Sect. 3 we give an overview of our construction, focusing on the discrete-log based construction as a warm-up. In Sect. 4 we formally define our notion of "universal ciphertexts" and strong KDM security. In Sect. 5 we show how to use encryption schemes satisfying the foregoing properties to construct correlation intractable functions. In Sect. 6 we describe parameter settings where the variants of ElGamal and Regev encryption schemes plausibly satisfy these properties. Finally, in Sect. 7 we show how to construct NIZKs for NP from our correlation intractable functions.

2   Preliminaries

Notations and terminology. Denote $\mathbb{R}$, $\mathbb{Z}$, $\mathbb{N}$ as the set of reals, integers and natural numbers. Let $\mathbb{Z}_q$ denote $\mathbb{Z}/(q\mathbb{Z})$. For $n \in \mathbb{N}$, let $[n]$ denote $\{1, 2, ..., n\}$. The rounding operation $\lfloor a \rceil : \mathbb{Z}_q \to \mathbb{Z}_p$ is defined as multiplying a by p/q and rounding the result to the nearest integer.

In cryptography, the security parameter (denoted as λ) is a variable that is used to parameterize the computational complexity of the cryptographic algorithm or protocol, and the adversary's probability of breaking security. An algorithm is "efficient" if it runs in (probabilistic) polynomial time over λ.

For any definition based on computational hardness, we refer the relevant security level to the success probability of any efficient adversary. For example, a security notion is *subexponential* if for every efficient adversary there exists $\epsilon > 0$ such that the adversary's advantage is less or equal to $2^{-\lambda^\epsilon}$.

Many experiments and probability statements in this paper contain randomized algorithms. When a variable v is drawn uniformly random from the set S we denote as $v \in_R S$ or $v \leftarrow U(S)$, sometimes abbreviated as v when the context is clear. Distributions written in multiple lines under Pr means they are sampled in sequence.

A function ensemble $\mathcal{F}$ has a key generation function $g : \mathcal{S} \to \mathcal{K}$; on a seed $s \in \mathcal{S}(\lambda)$, g produces a key $k \in \mathcal{K}(\lambda)$ for a function with domain $\mathcal{D}(\lambda)$ and range $\mathcal{C}(\lambda)$:

$$\mathcal{F} = \{f_k : \mathcal{D}(\lambda) \to \mathcal{C}(\lambda), k = g(s), s \in \mathcal{S}(\lambda)\}_{\lambda \in \mathbb{N}}$$

The bit-lengths of the seed, key, input and output are denoted as σ, κ, ℓ and w, unless specified otherwise.

The main object studied in this article is families of public key hash functions. We assume the key k is public. For certain key generation algorithm g, publishing k implies publishing s (e.g. when g is the identity function). We call such functions *public-coin*. By default we treat the bit-length of its input as being equal to the security parameter, i.e. $|\mathcal{D}(\lambda)| = 2^{\lambda}$.

2.1   Correlation Intractability

We recall the definition of correlation intractability [27].

Definition 1 (Density of a binary relations). *A binary relation $R = R(\lambda) \subseteq \{ (x,y) \mid x \in \mathcal{D}(\lambda), y \in \mathcal{C}(\lambda) \}$ has density $\mu = \mu(\lambda)$ if for every $x \in \mathcal{D}(\lambda)$ it holds that $\Pr_{y \in \mathcal{C}(\lambda)}[(x,y) \in R(\lambda)] < \mu(\lambda)$. A relation R is* sparse *if it has negligible density.*

Definition 2 (Correlation intractability w.r.t. binary sparse relations [27]). *A family of functions $\mathcal{H} = \{H_k : \mathcal{D}(\lambda) \to \mathcal{C}(\lambda)\}_{\lambda \in \mathbb{N}}$ is correlation intractable w.r.t. binary sparse relations if for every polynomial-size adversary A and every sparse relation R, there is a negligible function $\mathsf{negl}(\cdot)$ such that:*

$$\Pr_{\substack{k, \\ x \leftarrow A(H_k)}} \left[(x, H_k(x)) \in R \right] \leq \mathsf{negl}(\lambda).$$

We introduce a quantitative generalization of correlation intractability.

Definition 3 (f-correlation intractability). *A family of functions $\mathcal{H} = \{H_k : \mathcal{D}(\lambda) \to \mathcal{C}(\lambda)\}_{\lambda \in \mathbb{N}}$ is f-correlation intractable w.r.t. a function $f : \mathbb{N} \times [0,1] \to [0,1]$ if for all time function $T(\cdot)$, for all density function $\mu(\cdot)$, for every adversary A of running time $T(\lambda)$, and every relation R with density $\mu(\lambda)$, it holds that*

$$\Pr_{\substack{k, \\ x \leftarrow A(H_k)}} \left[(x, H_k(x)) \in R \right] \leq f(T, \mu).$$

For example, random oracles satisfy f-correlation intractability for $f(T,\mu) = T \cdot \mu$. Definition 2 can be viewed as f-correlation intractability w.r.t. $f(T,\mu) = T \cdot \mu$, for all polynomial $T(\cdot)$, and all negligible $\mu(\cdot)$. In the rest of the paper, "correlation intractability" refers to Definition 2 unless explicitly stated otherwise.

Survey of impossible parameters for correlation intractability. For some parameters relevant to the length of seed, key, input and output of the function, correlation intractability w.r.t. binary sparse relations is impossible to achieve. We survey some of the results.

[27] shows that a function family cannot be correlation intractable when the bit-length of the key $\kappa(\lambda)$ of the function is short compared to the bit-length of the input $\ell(\lambda)$:

Claim 1 ([27]). *$\mathcal{H}_\lambda$ is not correlation intractable w.r.t. efficiently checkable relations when $\kappa(\lambda) \leq \ell(\lambda)$.*

Proof. Consider the diagonalization relation $R_{\mathsf{diag}} = \{(k, h_k(k)) | k \in \mathcal{K}(\lambda)\}$. The attacker outputs k. □

The impossibility result generalizes to keys that are slightly larger than the bit-length of the input, but still smaller than the sum of the bit-length of input plus output $\ell(\lambda) + w(\lambda)$. The idea is to consider an extension of the diagonalization relation s.t. the relation checks a prefix of k—as long as the key is not too long, the relation is still sparse, albeit not necessarily efficient checkable.

Claim 2 ([27]). *$\mathcal{H}_\lambda$ is not correlation intractable w.r.t. possibly inefficiently checkable relations when $\kappa(\lambda) \leq \ell(\lambda) + w(\lambda) - \omega(\log(\lambda))$.*

We also observe when the "family size" of the function is relatively small, precisely, when the seed length is small w.r.t. the output length, then the function family is not correlation intractable w.r.t. possibly inefficiently checkable relations. This case is not ruled out by Claim 2 when the key is potentially long but derived from a short seed (e.g. from applying a PRG on a short seed).

Claim 3. *$\mathcal{H}_\lambda$ is not correlation intractable when the seed space $\mathcal{S}(\lambda)$ and the range $\mathcal{C}(\lambda)$ satisfies $|\mathcal{S}(\lambda)| \leq \mathsf{negl}(\lambda) \cdot |\mathcal{C}(\lambda)|$.*

Proof. Fix the hash function family $\mathcal{H}_\lambda$, consider the relation $R_\mathcal{H}$ that collects every functions in the function family $R_\mathcal{H} = \{(x, h_k(x)) \mid s \in \mathcal{S}, \ k = g(s), \ x \in \mathcal{D}(\lambda)\}$. The density of the relation less or equal to $|\mathcal{S}(\lambda)|/|\mathcal{C}(\lambda)| \leq \mathsf{negl}(\lambda)$. The attacker simply outputs any input. □

For the discussions of the other impossibility results, we refer the readers to [27] for the details.

2.2 Fiat-Shamir Heuristics

Definition 4 (Interactive proof-systems [46]). *An interactive proof-system for a language L is a protocol between a prover P and a verifier V. The prover's runtime is unbounded. The verifier runs in probabilistic polynomial time. The protocol satisfies*

- **Completeness:** *For every $x \in L$, the verifier V accepts with probability 1 after interacting with P on common input x.*
- **Soundness:** *For every $x \notin L$ and every cheating prover P^*, the verifier accepts with negligible probability after interacting with P^* on common input x.*

An interactive protocol is called an *argument-system* if it satisfies Definition 4 except that the prover is restricted to run in (non-uniform) polynomial time. An interactive proof or argument is called *public-coin* if the verifier's messages are random coins.

Correlation intractability and public-coin interactive proofs. Consider a language L and a 3-round public-coin interactive proof Π for L. Let α, β, γ be the 3 messages in the protocol (α and γ are sent by the prover P, β is sent by the verifier V). The relation $R_{\notin L, \Pi}$ is defined by

$$R_{\notin L, \Pi} = \left\{ ((x, \alpha), \beta) : x \notin L \text{ and } \exists \gamma \text{ s.t. } V(x, \alpha, \beta, \gamma) = \mathsf{Accept} \right\}. \tag{1}$$

Observe that the relation $R_{\notin L, \Pi}$ is sparse due to the statistical soundness of the underlying proof, i.e. the density of $R_{\notin L, \Pi}$ is equal to the soundness error of Π.

Interestingly, correlation intractability can also capture a stronger notion of soundness called *adaptive soundness*. We say that a 3 message interactive proof-system as above has adaptive soundness, if the message α sent by the honest prover does not depend on x, and soundness is guaranteed even if the adversary may choose the input $x \notin L$ on which to cheat *after* seeing β. For such protocols we define the relation $R_{\notin L, \Pi}$ as

$$R_{\notin L, \Pi} = \left\{ (\alpha, \beta) : \exists x, \gamma \text{ s.t. } x \notin L \land V(x, \alpha, \beta, \gamma) = \mathsf{Accept} \right\} \tag{2}$$

Again, the relation $R_{\notin L, \Pi}$ is sparse due to the adaptive soundness of Π.

Correlation intractability also implies the soundness of Fiat-Shamir for general constant-round public-coin interactive proof-systems. Without loss of generality assuming the number of rounds in the starting proof-system is $2c$ for a constant c. In the resulting 2-message argument, the verifier samples c independent correlation intractable hash functions. For $i \in \{1, 2, ..., c\}$, the prover applies the ith hash function on $(\alpha_1 || \beta_1 || ... || \alpha_{i-1} || \beta_{i-1} || \alpha_i)$ to generate β_i, where α_i is the ith message from the prover in the starting proof-system. The message from the prover in the resulting 2-message argument is then $(\alpha_1 || \beta_1 || ... || \alpha_c || \beta_c)$.

It is shown that the transformation above yields a sound 2-message argument if the hash functions are *entropy preserving* [10]. Given that CI families that withstand arbitrary binary relations are entropy preserving [24], we have

Lemma 1 ([10, 24, 36, 49]). *Assuming correlation intractable function family w.r.t. all binary sparse relations exists, then the Fiat-Shamir transformation is sound when applied on any constant-round public-coin interactive proof-systems.*

3   A Warm-Up Construction from Discrete Logarithm

We present a simple construction based on the discrete-log program as a warm-up to the general scheme. Along the way we will give the rationale of the proof strategy adapted from the work of Kalai et al. [59], and explain the level of KDM security we need for the underlying discrete-log problem.

Let $\mathbb{G}$ be a cyclic group where the discrete-log problem is hard. Assume the size of $\mathbb{G}$ is roughly 2^λ where λ is the security parameter. Let g be a generator of $\mathbb{G}$, $A = g^a, B = g^b$ be two random elements in $\mathbb{G}$. Consider the following length preserving function $H : \{1, ..., |\mathbb{G}|\} \to \mathbb{G}$

$$H_{A,B}(x) := A^x \cdot B = g^{ax+b} \in \mathbb{G}. \tag{3}$$

Theorem 4. *Given $\mathbb{G}(\lambda)$ of sizes $N(\lambda) \approx 2^\lambda$, with a generator g and efficient group operations, such that for any super-polynomial function s, any (not necessarily efficiently computable) function $f : [N] \to [N]$, and any polynomial time adversary A:*

$$\Pr_{k,a \leftarrow [N]} \left[A\left(g^a, g^{ak+f(k)}\right) = k \right] \leq \frac{s(\lambda)}{2^\lambda}.$$

Then $H_{A,B}$ is correlation intractable w.r.t. all sparse relations.

Towards a contradiction, let R be any sparse relation with negligible density $\mu(\lambda)$. Suppose there exists an efficient adversary Adv that breaks correlation intractability w.r.t. R with non-negligible probability ν:

$$\Pr_{A,B} \left[\left(\mathsf{Adv}(H_{A,B}) \to x\right) \wedge \left((x, H_{A,B}(x)) \in R\right) \right] \geq \nu, \qquad (4)$$

where the notation $\mathsf{Adv}(H_{A,B}) \to x$ simply means that we use x to refer to the string that $\mathsf{Adv}(H_{A,B})$ outputs.

In the first step, we translate the probability of outputting some x to the probability of outputting a particular x^*. For a random x^* from the domain, the probability that the adversary outputs x^* as the answer is greater or equal to ν divided by the domain size

$$\Pr_{\substack{x^* \in_R \{0,1\}^\lambda \\ A,B}} \left[\left(\mathsf{Adv}(H_{A,B}) \to x'\right) \wedge \left(x' = x^*\right) \wedge \left((x^*, H_{A,B}(x^*)) \in R\right) \right] \geq \nu/2^\lambda. \qquad (5)$$

Focusing on a single x^* costs a huge loss in the success probability. The readers may wonder what is the motivation of doing so. The purpose of fixing an input x^* is to prepare for replacing the winning condition $\left(x^*, H_{A,B}(x^*)\right) \in R$ by another condition that is "key independent". Towards this goal, consider the following sampling procedure: first sample a random y^* from the range, then sample the key (A', B') randomly under the condition $H_{A',B'}(x^*) = y^*$. Now we use the fact that H is a "one-universal" function, which means that for a fixed input, a uniformly random key projects the input to a uniformly random output. In turn, a uniformly random output corresponds to a uniformly random key. Therefore the key (A', B') obtained from reverse sampling distributes the same as before. Hence we have

$$\Pr_{\substack{x^* \in_R \{0,1\}^\lambda \\ y^* \in_R \mathbb{G}, \\ A',B' \text{ s.t. } H_{A',B'}(x^*)=y^*}} \left[\left(\mathsf{Adv}(H_{A',B'}) = x'\right) \wedge \left(x' = x^*\right) \wedge \left((x^*, H_{A',B'}(x^*)) \in R\right) \right] \geq \nu/2^\lambda. \qquad (6)$$

Given that $y^* = H_{A',B'}(x^*)$, we can change the winning condition in Eq. (6) into one which is independent from the function $H_{A',B'}$:

$$\Pr_{\substack{x^* \in_R \{0,1\}^\lambda \\ y^* \in_R \mathbb{G} \\ A',B' \text{ s.t. } H_{A',B'}(x^*)=y^*}} \left[(\mathsf{Adv}(H_{A',B'}) = x') \wedge (x' = x^*) \wedge ((x^*, y^*) \in R) \right] \geq \nu/2^\lambda. \qquad (7)$$

Separating the winning condition $(x^*, y^*) \in R$ from the hash key paves the way for connecting correlation intractability to a property that is only about hiding one specific point in the key (instead of hiding a bunch of potential input-output pairs in the relation). In the next statement, the first equality follows by the definition of conditional probability. The inequality follows from Eq. (7) together with the fact that R is μ sparse:

$$\Pr_{\substack{x^*, y^* \text{ s.t. } (x^*, y^*) \in R, \\ A', B' \text{ s.t. } H_{A', B'}(x^*) = y^*}} \left[(\mathsf{Adv}(H_{A', B'}) \to x') \wedge (x' = x^*) \right]$$

$$= \frac{\Pr_{\substack{x^* \in_R \{0,1\}^\lambda \\ y^* \in_R \mathbb{G} \\ A', B' \text{ s.t. } H_{A', B'}(x^*) = y^*}} \left[\begin{array}{c} \mathsf{Adv}(H_{A', B'}) = x' \\ x' = x^* \\ (x^*, y^*) \in R \end{array} \right]}{\Pr_{\substack{x^* \in_R \{0,1\}^\lambda \\ y^* \in_R \mathbb{G}}} \left[(x^*, y^*) \in R \right]} \quad (8)$$

$$\geq \frac{\nu}{2^\lambda \cdot \mu(\lambda)}$$

The LHS of Eq. (8) spells out as an efficient adversary's success probability of finding the input x^* embedded in A', B', where the key A', B' is sampled conditioned on mapping some input-output pair in the relation $(x^*, y^*) \in R$. Let's examine A', B', and for simplicity consider only the constant relations $R_c = \{(x, c) \mid \forall x \in \{0,1\}^\lambda\}$. Fix a $c^* \in \mathbb{G}$, a random input-output pair from R_{c^*} distributes as (x^*, c^*), where x^* is uniformly random from $\{0,1\}^\lambda$. For $A' = g^{a'}$, $B = g^{b'}$ sampled randomly from the set $\{g^{a'}, g^{b'} \mid g^{z^*} := c^* = g^{a'x^*+b'}\}$, where z^* is explicitly defined as the discrete-log of c^* over base g for the convenience of explanation. Observe that the marginal distribution of a' is uniform, and b' equals to $z^* - a'x^*$. In other words, the adversary is asked to find x^* given $A' = g^{a'}$, $B' = g^{z^* - a'x^*}$ where z^* is fixed. The hardness of this problem follows directly from the hardness of the discrete-log problem.

What is the hardness required for the underlying discrete-log problem in order to form a contradiction? For the probability in the hypothesis $\frac{\nu(\lambda)}{2^\lambda \cdot \mu(\lambda)}$, where ν is a non-negligible function; μ, the density of a sparse relation, is an arbitrary negligible function. We can form a contradiction by assuming that every polynomial time algorithm for the discrete-log problem over $\mathbb{G}$ succeeds with probability less than $s(\lambda)/2^\lambda$ for any super-polynomial function s.

What happens when we consider all sparse relations instead of only the constant relations? For a general sparse relation, sampling a random pair (x^*, y^*) from the relation may result into an output y^* that is correlated to the input x^*. Take the "fixed point" relation $R_{x=y} := \{(x, y) \mid x = y\}$ as an example. A random input-output pair from $R_{x=y}$ distributes as (x^*, x^*), where x^* is uniformly random. For $A' = g^{a'}$, $B = g^{b'}$ sampled randomly from the set $\{g^{a'}, g^{b'} \mid g^{z^*(x^*)} := x^* = g^{a'x^*+b'}\}$, where $z^*(x^*)$ is the discrete-log of x^* over base g (unlike in the previous example, now z^* depends on the input x^*). The marginal distribution of a' is still uniform, and b' equals to $z^*(x^*) - a'x^*$. In other words, the adversary is asked to find x^* given $A' = g^{a'}$, $B' = g^{z^*(x^*)-a'x^*}$ where $z^*(\cdot)$ is a function on x^*, a' is independent from x^* and uniform. The latter corresponds to the hardness of finding the decryption key x^* given a ciphertext of ElGamal encryption with uniform randomness a', and key-dependent message $z^*(x^*)$.

To summarize, the proof strategy translates the hardness of finding any solution in a sparse relation to the hardness of finding the key from the encryption of possibly key-dependent messages. The translation is purely statistical, but it results into a significant cost in the final computational assumption—the success probability for any polytime attacker has to be extremely small. To capture arbitrary relations, arbitrary key dependency functions are considered.

4   Encryption Scheme with Universal Ciphertext and KDM Security

Let $\mathcal{M} = \{\mathcal{M}_\lambda\}_{\lambda \in \mathbb{N}}$ be an ensemble of message spaces (i.e., $\mathcal{M}_\lambda$ is the message space with respect to security parameter $\lambda \in \mathbb{N}$). An *encryption scheme*, with respect to the message space $\mathcal{M}$, consists of three probabilistic polynomial-time algorithm PP-Gen, Enc and Dec. The public-parameter generation algorithm PP-Gen gets as input 1^λ and outputs some public-parameters pp (without loss of generality we assume that pp contains λ). Given the public-parameters pp, a key $k \in \{0,1\}^\lambda$ and a message $m \in \mathcal{M}_\lambda$ the encryption algorithm Enc outputs a ciphertext c. The decryption algorithm Dec gets as input the public-parameters pp, a key k as well as a ciphertext c and outputs a message in $\mathcal{M}_\lambda$. We require that (with probability 1), for every setting of the public-parameters pp, message $m \in \mathcal{M}_\lambda$ and key $k \in \{0,1\}^\lambda$ it holds that $\mathsf{Dec}(pp, k, \mathsf{Enc}(pp, k, m)) = m$.

In many encryption schemes each ciphertext is associated with some particular key. We will be interested in schemes where this is not the case. Namely, ciphertexts are not associated with a specific key, but rather "make sense" under any possible key. We denote by $\mathcal{C}_{pp}$ the distribution obtained by encrypting a random message using a random key. Namely, the distribution $\mathsf{Enc}(pp, k, m)$ where $k \in_R \{0,1\}^\lambda$ and $m \in_R \mathcal{M}_\lambda$.

Definition 5 (Universal Ciphertexts). *We say that an encryption scheme* (PP-Gen, Enc, Dec) *with respect to message space* $\mathcal{M} = \{\mathcal{M}_\lambda\}_{\lambda \in \mathbb{N}}$ *has* universal ciphertexts *if the following two conditions hold for all constant* $\eta > 0$*, for all (sufficiently large)* $\lambda \in \mathbb{N}$ *and public parameters* $pp \in \mathsf{PP\text{-}Gen}(1^\lambda)$*:*

1. *For every fixed key* $k^* \in \{0,1\}^\lambda$*, a random ciphertext decrypts to a random message. Namely, the distribution* $m \leftarrow \mathsf{Dec}(pp, k^*, c)$*, where* $c \leftarrow \mathcal{C}_{pp}$*, is* $2^{-(1+\eta)\lambda}$*-statistically close to uniform.*
2. *For all* $k^* \in \{0,1\}^\lambda$ *and* $m^* \in \mathcal{M}_\lambda$*, the following distributions are* $2^{-(1+\eta)\lambda}$*- statistically close*
   - $c \leftarrow \mathcal{C}_{pp}$ *conditioned on* $\mathsf{Dec}(pp, k^*, c) = m^*$*.*
   - *c is sampled from* $c \leftarrow \mathsf{Enc}(pp, k^*, m^*)$ *(i.e., a fresh encryption of* m^* *under public parameters pp and key k^*).*

Definition 6 (ϵ-KDM Security). *Let $\epsilon = \epsilon(\lambda) \in [0,1]$. We say that an encryption scheme* (PP-Gen, Enc, Dec) *is ϵ-KDM secure, if for every polynomial-time adversary $\mathcal{A}$, for all sufficiently large values of λ and any (possibly inefficient) function $f : \{0,1\}^\lambda \to \mathcal{M}_\lambda$ it holds that:*

$$\Pr_{\substack{pp \leftarrow \text{PP-Gen}(1^\lambda) \\ k \in_R \{0,1\}^\lambda}} \left[\mathcal{A}(pp, \text{Enc}(pp, k, f(k))) = k \right] < \epsilon.$$

5   Correlation Intractability from Universal-Ciphertexts KDM Encryption

Let PP-Gen, Enc, Dec be an encryption scheme with respect to an ensemble of message spaces $\mathcal{M} = \{\mathcal{M}_\lambda\}_{\lambda \in \mathbb{N}}$, as defined in Sect. 4. For public parameters pp recall that we denote by $\mathcal{C}_{pp}$ the distribution obtained by encrypting a random message using a random key (with respect to public parameters pp).

Construction 5. *We construct a hash function family $\mathcal{H} = \{\mathcal{H}_\lambda : \{0,1\}^\lambda \to \mathcal{M}_\lambda\}_{\lambda \in \mathbb{N}}$ as follows.*

The key generation algorithm of the hash function takes input 1^λ, samples public parameters pp of the encryption scheme and a random ciphertext $c \leftarrow \mathcal{C}_{pp}$. The hash key is $hk = (pp, c)$. On input the key (pp, c) and a message to be hashed $\alpha \in \{0,1\}^\lambda$, the hashing algorithm views α as a key of the encryption scheme and outputs $\text{Dec}(pp, \alpha, c)$.

The main result that we prove in this section is if the encryption scheme has *universal ciphertexts* (as per Definition 5) and is *ε-KDM secure* (as per Definition 6), for sufficiently small $\varepsilon = \varepsilon(\lambda) > 0$, then Construction 5 forms a correlation intractable hash function family.

Theorem 6. *If there exists an encryption scheme with universal ciphertexts that is ε-KDM secure for $\varepsilon \leq \left(\text{poly}(\lambda) \cdot 2^\lambda \cdot \mu(\lambda) \right)^{-1}$, then Construction 5 is correlation intractable for all sparse relations with negligible density $\mu(\lambda)$.*

5.1   Proof of Theorem 6

Let R be any sparse relation with negligible density $\mu = \mu(\lambda)$. Suppose toward a contradiction that there exists a probabilistic polynomial-time adversary Adv that breaks the correlation intractability of Construction 5 with non-negligible probability $\nu = \nu(\lambda)$. Namely,

$$\Pr_{hk} \left[\text{Adv}(H_{hk}) \text{ outputs some } \alpha \wedge (\alpha, H_{hk}(\alpha)) \in R \right] \geq \nu(\lambda).$$

Thus, by construction of our hash function it holds that:

$$\Pr_{\substack{pp \\ c \leftarrow \mathcal{C}_{pp}}} \Big[\mathsf{Adv}(pp, c) \text{ outputs some } \alpha \text{ s.t. } \big(\alpha, \mathsf{Dec}(pp, \alpha, c)\big) \in R \Big] \geq \nu(\lambda), \qquad (9)$$

where here and below we use pp to denote public parameters sampled from PP-Gen(1^λ).

For the analysis, we consider a relaxed relation R' where $(\alpha, \beta) \in R'$ if $(\alpha, \beta) \in R$ or if the first $\lfloor \log(\nu/2\mu) \rfloor$ bits of β are all 0. The density of R' is bounded by $\mu' \leq 4\mu/\nu$, which is negligible when μ is negligible. Looking ahead, the purpose of "padding" R is so that the marginal distribution of α^*, obtained from jointly sampling a pair (α^*, β^*) randomly from R', is close to uniform. More specifically, following [59, Proposition 3.4] we can bound the point-wise *multiplicative* difference (or ratio) between these distributions:

Fact 7. *For all $\alpha' \in \{0,1\}^\lambda$, $\beta' \in \mathcal{M}_\lambda$,*

$$\Pr_{\substack{\alpha^* \\ \beta^* \text{ s.t } (\alpha^*,\beta^*) \in R'}} \big[\alpha^* = \alpha', \beta^* = \beta'\big] \geq \frac{1}{4} \cdot \Pr_{\alpha^*,\beta^* \text{ s.t } (\alpha^*,\beta^*) \in R'} \big[\alpha^* = \alpha', \beta^* = \beta'\big]$$

$$(10)$$

Since $R \subseteq R'$, Eq. (9) implies that:

$$\Pr_{\substack{pp \leftarrow \text{PP-Gen}(1^\lambda), \\ c \leftarrow \mathcal{C}_{pp}}} \Big[\mathsf{Adv}(pp, c) \text{ outputs } \alpha \text{ s.t. } \big(\alpha, \mathsf{Dec}(pp, \alpha, c)\big) \in R' \Big] \geq \nu(\lambda). \quad (11)$$

We will use Eq. (11) to show that Adv breaks the KDM security of our encryption scheme, with respect to the randomized KDM function f that given a key α^*, outputs a random β^* such that $(\alpha^*, \beta^*) \in R'$.

We now fix some setting of the public parameters pp. Using the structure of R', and the fact that our encryption scheme has universal ciphertexts (property 2 of Definition 5), it holds that:

$$\Pr_{\substack{\alpha^* \\ \beta^* \text{ s.t } (\alpha^*,\beta^*) \in R' \\ c \leftarrow \mathsf{Enc}(pp,\alpha^*,\beta^*)}} \Big[\mathsf{Adv}(pp, c) \text{ outputs } \alpha^* \Big]$$

$$\geq (1/4) \cdot \Pr_{\substack{\alpha^*,\beta^* \text{ s.t } (\alpha^*,\beta^*) \in R' \\ c \leftarrow \mathsf{Enc}(pp,\alpha^*,\beta^*)}} \Big[\mathsf{Adv}(pp, c) \text{ outputs } \alpha^* \Big]$$

$$\geq (1/4) \cdot \left(\Pr_{\substack{\alpha^*,\beta^* \text{ s.t } (\alpha^*,\beta^*) \in R' \\ c \leftarrow \mathcal{C}_{pp} \text{ s.t. } \mathsf{Dec}(pp,\alpha^*,c)=\beta^*}} \Big[\mathsf{Adv}(pp, c) \text{ outputs } \alpha^* \Big] - 2^{-(1+\eta)\lambda} \right) \quad (12)$$

where the first inequality is due to Fact 7; the second is due to the universal ciphertexts property.

Our key step is captured by the following proposition, which relates the adversary's advantage of recovering the *specific* key α^* in a ciphertext encrypting possibly key-dependent messages, to the advantage of outputting *any* α

that breaks correlation intractability. While the winning probability in the key-recovery game is exponentially small, it is lower bounded by a function of the success probability of breaking correlation intractability.

Proposition 1. *For every setting of the public-parameters pp it holds that:*

$$\Pr_{\substack{\alpha^*,\beta^* \text{ s.t } (\alpha^*,\beta^*)\in R' \\ c \text{ s.t. } \mathsf{Dec}(pp,\alpha^*,c)=\beta^*}} \left[\mathsf{Adv}(pp,c) \text{ outputs } \alpha^*\right]$$

$$\geq \frac{2^{-\lambda}}{\mu'} \cdot \left(\Pr_c\left[\begin{array}{c}\mathsf{Adv}(pp,c) \text{ outputs } \alpha \text{ s.t.} \\ (\alpha, \mathsf{Dec}(pp,\alpha,c)) \in R'\end{array}\right] - 2^{-\eta\lambda}\right),$$

Proof. Fix the public parameters pp. By the fact that the random variables (α^*, β^*) and c are independent, it holds that:

$$\Pr_{\substack{\alpha^*,\beta^* \text{ s.t } (\alpha^*,\beta^*)\in R' \\ c \text{ s.t. } \mathsf{Dec}(pp,\alpha^*,c)=\beta^*}} \left[\mathsf{Adv}(pp,c) \text{ outputs } \alpha^*\right] \tag{13}$$

$$= \Pr_{\substack{\alpha^*,\beta^* \\ c \text{ s.t. } \mathsf{Dec}(pp,\alpha^*,c)=\beta^*}} \left[\mathsf{Adv}(pp,c) \text{ outputs } \alpha^* \mid (\alpha^*,\beta^*) \in R'\right].$$

By definition of conditional probability, it holds that:

$$\Pr_{\substack{\alpha^*,\beta^* \\ c \text{ s.t. } \mathsf{Dec}(pp,\alpha^*,c)=\beta^*}} \left[\mathsf{Adv}(pp,c) \text{ outputs } \alpha^* \mid (\alpha^*,\beta^*) \in R'\right]$$

$$= \frac{\Pr_{\substack{\alpha^*,\beta^* \\ c \text{ s.t. } \mathsf{Dec}(pp,\alpha^*,c)=\beta^*}} \left[\begin{array}{c}\mathsf{Adv}(pp,c) \text{ outputs } \alpha^* \\ (\alpha^*,\beta^*) \in R'\end{array}\right]}{\Pr_{\alpha^*,\beta^*}\left[(\alpha^*,\beta^*) \in R'\right]}$$

$$\geq (1/\mu') \cdot \Pr_{\substack{\alpha^*,\beta^* \\ c \text{ s.t. } \mathsf{Dec}(pp,\alpha^*,c)=\beta^*}} \left[\begin{array}{c}\mathsf{Adv}(pp,c) \text{ outputs } \alpha^* \\ (\alpha^*, \mathsf{Dec}(pp,\alpha^*,c)) \in R'\end{array}\right], \tag{14}$$

where the inequality follows from the density of R'.

Claim 8. *The following two distributions are $2^{-(1+\eta)\lambda}$-close:*

1. (α^*,c): *such that $\alpha^* \in_R \{0,1\}^\lambda$, $\beta^* \in_R \mathcal{M}_\lambda$ and $c \leftarrow \mathcal{C}_{pp}$ conditioned on $\mathsf{Dec}(pp,\alpha^*,c) = \beta^*$.*
2. (α^*,c'): *such that $\alpha^* \in_R \{0,1\}^\lambda$ and $c' \leftarrow \mathcal{C}_{pp}$.*

Proof. A different way to sample the exact same distribution as in item (2) is to first sample $\alpha^* \in_R \{0,1\}^\lambda$, then $c'' \leftarrow \mathcal{C}_{pp}$ and finally $c' \leftarrow \mathcal{C}_{pp}$ *conditioned on* $\mathsf{Dec}(pp,\alpha^*,c') = \mathsf{Dec}(pp,\alpha^*,c'')$.

By the universal ciphertext property 5(1) of the encryption scheme, the distribution $\mathsf{Dec}(pp,\alpha^*,c'')$ is $2^{-(1+\eta)\lambda}$ close to the uniform distribution over $\mathcal{M}_\lambda$. The claim follows. □

Combining Claim 8 together with Eqs. (13) and (14) yields that:

$$\Pr_{\substack{\alpha^*,\beta^* \text{ s.t } (\alpha^*,\beta^*)\in R' \\ c \text{ s.t. } \mathsf{Dec}(pp,\alpha^*,c)=\beta^*}} \Big[\mathsf{Adv}(pp,c) \text{ outputs } \alpha^*\Big]$$

$$\geq (1/\mu') \cdot \left(\Pr_{\alpha^*,c}\left[\frac{\mathsf{Adv}(pp,c) \text{ outputs } \alpha^*}{(\alpha^*,\mathsf{Dec}(pp,\alpha^*,c)) \in R'}\right] - 2^{-(1+\eta)\lambda}\right)$$

$$= (1/\mu') \cdot \left(2^{-\lambda} \cdot \Pr_{c}\left[\begin{matrix}\mathsf{Adv}(pp,c) \text{ outputs } \alpha \text{ s.t.} \\ (\alpha,\mathsf{Dec}(pp,\alpha,c)) \in R'\end{matrix}\right] - 2^{-(1+\eta)\lambda}\right)$$

$$= (2^{-\lambda}/\mu') \cdot \left(\Pr_{c}\left[\begin{matrix}\mathsf{Adv}(pp,c) \text{ outputs } \alpha \text{ s.t.} \\ (\alpha,\mathsf{Dec}(pp,\alpha,c)) \in R'\end{matrix}\right] - 2^{-\eta\lambda}\right) \tag{15}$$

This concludes the proof of Proposition 1. $\qquad\square$

Using Proposition 1 and Eq. (12) we obtain that:

$$\Pr_{\substack{pp \\ \alpha^* \\ \beta^* \text{ s.t } (\alpha^*,\beta^*)\in R' \\ c\leftarrow\mathsf{Enc}(pp,\alpha^*,\beta^*)}} \Big[\mathsf{Adv}(pp,c) \text{ outputs } \alpha^*\Big]$$

$$= \mathop{\mathbf{E}}_{pp}\left[\Pr_{\substack{\alpha^* \\ \beta^* \text{ s.t } (\alpha^*,\beta^*)\in R' \\ c\leftarrow\mathsf{Enc}(pp,\alpha^*,\beta^*)}} \Big[\mathsf{Adv}(pp,c) \text{ outputs } \alpha^*\Big]\right]$$

$$\geq 1/4 \cdot \mathop{\mathbf{E}}_{pp}\left[\Pr_{\substack{\alpha^*,\beta^* \text{ s.t } (\alpha^*,\beta^*)\in R' \\ c\leftarrow\mathcal{C}_{pp} \text{ s.t. } \mathsf{Dec}(pp,\alpha^*,c)=\beta^*}} \Big[\mathsf{Adv}(pp,c) \text{ outputs } \alpha^*\Big]\right] - 2^{-(1+\eta)\lambda}$$

$$\geq \frac{1}{4 \cdot 2^{\lambda} \cdot \mu'} \cdot \mathop{\mathbf{E}}_{pp}\left[\Pr_{c}\left[\begin{matrix}\mathsf{Adv}(pp,c) \text{ outputs } \alpha \text{ s.t.} \\ (\alpha,\mathsf{Dec}(pp,\alpha,c))\in R'\end{matrix}\right] - 2^{-\eta\lambda}\right] - 2^{-(1+\eta)\lambda}$$

$$= \frac{1}{4 \cdot 2^{\lambda} \cdot \mu'} \cdot \left(\Pr_{pp,c}\left[\begin{matrix}\mathsf{Adv}(pp,c) \text{ outputs } \alpha \text{ s.t.} \\ (\alpha,\mathsf{Dec}(pp,\alpha,c))\in R'\end{matrix}\right] - 2^{-\eta\lambda}\right) - 2^{-(1+\eta)\lambda}$$

$$\geq \frac{\nu}{8 \cdot 2^{\lambda} \cdot \mu'}$$

$$= \omega\left(\frac{\mathsf{poly}(\lambda)}{2^{\lambda}}\right).$$

Thus, Adv breaks KDM security with probability $\varepsilon \geq (1/\mathsf{negl}) \cdot 2^{-\lambda}$, in contradiction to our assumption.

6   Candidate KDM Encryption with Universal Ciphertexts

We present two encryption schemes that satisfy the ciphertext universality (Definition 5), and plausibly satisfy ϵ-KDM security (Definition 6) for exponentially small ϵ.

6.1   Discrete-Log Based

We first present the encryption scheme based on a generic multiplicative group, and then specify its instantiation over the elliptic curve groups. The scheme can be viewed as a bit-encryption variant of ElGamal.

Construction 9. *Fix a small constant $\eta > 0$ (e.g. $\eta = 0.01$). Let the message space be $\mathcal{M} = \{\mathcal{M}_\lambda\}_{\lambda \in \mathbb{N}}$, where $\mathcal{M}_\lambda = \{0,1\}^{w(\lambda)}$ and $w = w(\lambda) \in \mathbb{N}$. We construct an encryption scheme as follows.*

- *Public parameters Generation $\mathsf{PP\text{-}Gen}(1^\lambda)$: the key-generation algorithm selects a prime $N = N(\lambda) \geq 2^{(1+2\eta)\lambda}$, a group $\mathbb{G} = \mathbb{G}(\lambda)$ of size N, and a generator g (the exact algorithm for determining these depends on the specific group family we use - see instantiations below).*
  Let $\mathsf{ext} : \mathbb{G} \to \{0,1\}$ be a deterministic efficiently computable function that outputs 0 on $\lceil N/2 \rceil$ of the group elements, and 1 on the remaining $\lfloor N/2 \rfloor$ elements.
  The public-parameters pp include a concise[3] description of the group G, generator g, and function ext.
- *Encrypt $\mathsf{Enc}(pp, k, y)$: We view k as an integer in $[2^\lambda]$. Let $y_1 \ldots y_w \in \{0,1\}$ be the bit decomposition of y.*
  For each $j \in [w]$, sample $a_j \in_R \{0, 1, ..., N-1\}$ and let $A_j := g^{a_j}$. Sample C_j uniformly from $\mathsf{ext}^{-1}(y_j)$ and let $B_j = C_j \cdot A_j^k$. Output $c = (A_j, B_j)_{j \in [w]}$ as the ciphertext.
- *Decrypt $\mathsf{Dec}(pp, k, c)$: Decompose the ciphertext c as $(A_j, B_j)_{j \in [w]}$. For $j \in [w]$, let $C_j = B_j / A_j^k$ and let the jth output bit be $\mathsf{ext}(C_j)$.*

Remark 1. To ensure the KDM problem is as hard as possible, the group order is set to be a prime so that not only the discrete-log problem but also the decisional Diffie-Hellman problem is plausibly hard.

Since the group order is a prime, a deterministic function that extracts a bit from the group cannot be perfectly balanced. So we set the group order to be slightly larger than $2^{(1+\eta)\lambda}$ in order to allow $2^{-(1+\eta)\lambda}$-statistical distance for the statistical properties.

We first show that the scheme satisfies the *universal ciphertext* requirement (see Definition 5).

Proposition 2. *The encryption scheme of Construction 9 has universal ciphertexts.*

Proof. The first condition in Definition 5 follows from the fact that for a fixed encryption key k, and random ciphertext $(A_j, B_j)_{j \in [w]}$, it holds that each $C_j = B_j / A_j^k$ is uniformly distributed and so we only need to account for the deviation from ext. Overall we get that the output is at most $2^{-(1+\eta)\lambda}$-close to uniform.

---

[3] By concise description of the group, we mean a description of length $\mathsf{poly}(\lambda)$ that allows performing group operations such as multiplication, inversion, equality testing and sampling random elements.

The second condition in Definition 5 can be verified as follows. For every $j \in [w]$ and every possible value of A_j, there are exactly $|\text{ext}^{-1}(y_j)|$ possible values B_j that Enc can output, and each of them is equally likely. Therefore, each pair (A_j, B_j) subject to the condition $\text{ext}(B_j \cdot A_j^k) = w_j$ is equally likely to be output by Enc, and thus the distribution output by Enc is identical to a random ciphertext for the given plaintext. $\qquad\qquad\square$

As noted above, we need to assume that Construction 9 is exponentially KDM secure.

Assumption 10 (KDM security for the discrete-log based encryption).
Let $\lambda \in \mathbb{N}$, $w(\lambda) \in \text{poly}(\lambda)$. There exists a family of groups $\mathbb{G}(\lambda)$ (of efficiently computable sizes $N(\lambda)$, with efficiently computable generators, efficient group operations, and efficient $\text{ext} : \mathbb{G} \to \{0,1\}$) such that for all function $f : \{1,\ldots,2^\lambda\} \to \{0,1\}^w$ (including those that are not efficiently computable), the following holds. For any polynomial-time adversary Adv, for a uniformly random $k \in \{1,\ldots,2^\lambda\}$; for each $j \in [w]$, sample $a_j \in_R \{0,1,\ldots,N\}$, $C_j \in_R \text{ext}^{-1}(f(k)_j)$. The probability that adversary outputs k on input $(A_j = g^{a_j}, B_j = g^{a_j k} \cdot C_j)_{j \in [w]}$ is smaller than $\frac{1}{2^\lambda \cdot \text{negl}(\lambda)}$, i.e.

$$\Pr_{\substack{k \in_R \{1,\ldots,2^\lambda\} \\ \{a_j \in_R \{0,1,\ldots,N\}, C_j \in_R \text{ext}^{-1}(f(k)_j)\}_{j \in [w]} \\ \{A_j = g^{a_j}, B_j = g^{a_j k} \cdot C_j\}_{j \in [w]}}} \left[\text{Adv}(\{A_j, B_j\}_{j \in [w]}) = k \right] \leq \frac{1}{2^\lambda \cdot \text{negl}(\lambda)}$$

Thus, using Theorem 6, we obtain the following corollary.

Corollary 1. *Suppose that Assumption 10 holds. Then, there exists correlation intractable function for all sparse relations.*

Remark 2. In Assumption 10, if the function f is a constant (i.e. is independent of the key), the problem can be reduced from the discrete-log problem over $\mathbb{G}$ with the key restricted to $\{1,\ldots,2^\lambda\}$, i.e. computing $k \in \{1,\ldots,2^\lambda\}$ given g, $g^k \in \mathbb{G}$. In the reduction, the discrete-log attacker, given g, g^k, and f, can sample $(A_j, B_j)_{j \in [w]}$ from the correct distribution, send over to the adversary in Assumption 10.

Remark 3. We chose bit encryption for simplicity of notation. Instead of representing messages as bits, we can represent them in any base b, as long as there is an efficient and nearly-regular map ext from $\mathbb{G}$ to $\{0,\ldots,b-1\}$. The regularity requirement, however, is quite strong: because of the first requirement in Definition 5, the preimage size of every digit under ext must be very close to the average, so that the statistical distance between $\text{ext}(\mathbb{G})$ and uniform is $2^{-(1+2\eta)\lambda}$.

We can use seeded extractors and put the seed in the public parameters. Specifically, if we choose N to be at least $2^{2(1+2\eta)\lambda} \cdot b$ and $\text{ext} : \mathbb{G} \to [b]$ to be a pairwise independent hash function, then for the average seed, by the leftover hash lemma [54, Lemma 4.8], the output will be $\sqrt{|\mathbb{G}|/b} = 2^{-(1+2\eta)\lambda}$-close to uniform. This ensures that a good seed exists (nonconstructively). If want to make sure the average seed is good with all but exponential probability, we can choose N to be at least $2^{4(1+2\eta)\lambda} \cdot b$ instead. Then for the average seed, the output will be $\sqrt{|\mathbb{G}|/b} = 2^{-2(1+2\eta)\lambda}$-close to uniform, and therefore for all but a $1 - 2^{-(1+2\eta)\lambda}$ fraction of the seeds, it will be at least $2^{-(1+2\eta)\lambda}$-close to uniform, as required.

An instantiation over elliptic curves groups. The group $\mathbb{G}$ and the extraction function ext are chosen such that they avoid the known weakness instances of the underlying ECDLP, and at the same time enjoy the statistical properties.

An elliptic curve group $E(\mathbb{F}_q)$ is represent by the curve E (in the short Weierstrass form) over finite field $\mathbb{F}_q$: $E(\mathbb{F}_q) = \{ (x, y) \mid y^2 = x^3 + ax + b \bmod q \} \cup \mathcal{O}$. Choose the curve (namely, choose a, b and q) so that q is an odd prime, the order of the group $\#E(\mathbb{F}_q)$ is a prime $N > 2^{(1+2\eta)\lambda}$.

In the short Weierstrass form, if $(x, y) \in E(\mathbb{F}_q)$, then $(x, -y) \in E(\mathbb{F}_q)$. Any point P whose y-coordinate is zero does not exist in a prime order group, since $P = (x, 0)$ implies the order of P is 2. So one option of the extraction function ext : $E(\mathbb{F}_q) \to \{0, 1\}$ is to take the sign of the y-coordinate of a point $P = (x, y) \in E(\mathbb{F}_q)$. To be precise, if $y \in \{1, ..., (q-1)/2\}$, output 1; if $y \in \{(q+1)/2, ..., q-1\}$, output 0. As an exception, if $P = \mathcal{O}$, output 0.

6.2   LWE Based

The LWE based encryption scheme is a variant of Regev's scheme [73]. We remark that the hash function obtained by applying Construction 5 on Construction 11 yields a variant of Ajtai's hash function [3], in the sense that we apply rounding on the output vector.

Construction 11. *The message space is* $\mathcal{M} = \{\mathcal{M}_\lambda\}_{\lambda \in \mathbb{N}}$, *where* $\mathcal{M}_\lambda = \{0, 1\}^{w(\lambda)}$ *and* $w = w(\lambda) \in \mathbb{N}$. *We construct an encryption scheme as follows.*

- **Public parameters generation** PP-Gen(1^λ): *Fix an even number* $q(\lambda)$ *as the modulus. Select* $B(\lambda), \ell(\lambda) \in \mathbb{N}$ *such that* $B^\ell \in [2^{\lambda - \log(\lambda)}, 2^{\lambda + \log(\lambda)}]$ *and* $B \leq q$. *The public-parameters pp are* (B, q, ℓ).
- **Representation of the secret key:** *we view the secret key* $k \in \{0, 1\}^\lambda$ *as a vector* $\mathbf{k} \in \{0, ..., B(\lambda) - 1\}^{\ell(\lambda)}$, *written as a column vector.*
- **Encryption** Enc($pp, \mathbf{k}, y$): *Given a message* $y \in \{0, 1\}^w$. *For* $j \in [w]$, *sample* $\mathbf{a}_j \in_R \mathbb{Z}_q^{1 \times \ell}$. *compute* $b_j = y_j \cdot \frac{q}{2} + e_j - \mathbf{a}_j \cdot \mathbf{k} \pmod{q}$, *where* $e_j \leftarrow U([0, q/2) \cap \mathbb{Z})$. *Output* $c = (\mathbf{a}_j, b_j)_{j \in [w]}$ *as the ciphertext.*
- **Decryption** Dec($pp, \mathbf{k}, c$): *View c as* $(\mathbf{a}_j, b_j)_{j \in [w]}$. *For* $j \in [w]$, *let the* j^{th} *output bit be* $\lfloor b_j + \mathbf{a}_j \cdot \mathbf{k} \bmod q \rceil_2$, *where* $\lfloor \cdot \rceil_2 : \mathbb{Z}_q \to \{0, 1\}$ *outputs 0 if the input is from* $[0, q/2)$, *1 if the input is from* $[q/2, q - 1]$.

The parameters are set according to the following constraints to minimize the adversary's advantage on the KDM problem, and to guarantee the statistical properties. The choices of parameters are guided by the reductions from the worst case problems, as well as the known attacks (e.g. [7,8,16,60,61,75]), even though some of the attacks were designed to achieve non-trivial (sub)exponential running time and do not clearly achieving non-trivial success probability when running in polynomial time.

1. q is even so that we can get perfect ciphertext-universality.
2. The error term e is sampled uniformly from $[0, q/2) \cap \mathbb{Z}$, differing from the typical setting of discrete Gaussian distribution. Noise sampled uniformly from a sufficiently large range is as good as Gaussian for some parameter settings [34,64]. In particular, $q/2$ is sufficiently large, even larger than the typical settings of the norm of the noise.
3. B, ℓ, q are selected so that each coordinate of the secret vector has enough entropy (i.e. $B > \sqrt{n}$), the vector dimension ℓ is sufficiently close to λ, B/q is not too small (i.e. $q/B \in \mathsf{poly}(\lambda)$). One way of setting the parameter is to let $q = O(\lambda^3)$, $B(\lambda) = 2^{\lfloor \log \lambda \rfloor}$, $\ell(\lambda) = \left\lfloor \frac{\lambda}{\lceil \log \lambda \rceil} \right\rfloor$.

We first show that the scheme satisfies the *universal ciphertext* requirement (see Definition 5).

Proposition 3. *The encryption scheme of Construction 11 has universal ciphertexts.*

Proof. The first property (as per Definition 5(1)) follows immediately from the perfect 1-universality of the decryption function.

The second property (as per Definition 5(2)) can be verified as follows. For $j \in [w]$, the randomness in the encryption includes $\mathbf{a}_j \in \mathbb{Z}_q^{1 \times \ell}$ and the error term $e_j \in \mathbb{Z}_q$. For all $y_j^* \in \{0, 1\}$ and $\mathbf{k}^* \in \{0, ..., B-1\}^\ell$, $(b_j, \mathbf{a}_j) \in \mathbb{Z}_q \times \mathbb{Z}_q^n$ is sampled uniformly random conditioned on $b_j + \mathbf{a}_j \cdot \mathbf{k}^* \bmod q \in y_j^* \cdot \frac{q}{2} + [0, q/2) \cap \mathbb{Z}$. Viewing the equality as a 1-universal function $\mathbf{a}_j \cdot \mathbf{k}^* \bmod q \in y_j^* \cdot \frac{q}{2} + [0, q/2) \cap \mathbb{Z} - b_j$ with key $\mathbf{a}_j$, the marginal distribution of $\mathbf{a}_j$ is uniform over $\mathbb{Z}_q^{1 \times \ell}$. Then, $e_j = b_j - y_j^* \cdot \frac{q}{2} + \mathbf{a}_j \cdot \mathbf{k}^*$ is distributed uniformly over $[0, q/2) \cap \mathbb{Z}$.   □

Assumption 12 (KDM security for LWE-based encryption). *Let $\lambda \in \mathbb{N}$, $w(\lambda) \in \mathsf{poly}(\lambda)$. For all functions $f : \{0, ..., B-1\}^\ell \to \{0,1\}^w$ (including those who are not efficiently computable). The probably that any polynomial time adversary* Adv, *given* $\{\mathbf{a}_j, \mathbf{a}_j \cdot \mathbf{k} + e_j + f_j(\mathbf{k}) \cdot q/2\}_{j \in [w]}$ *where* $\mathbf{k} \in_R \{0, ..., B-1\}^\ell$, $\mathbf{a}_j \in_R \mathbb{Z}_q^{1 \times \ell}$, $e_j \in_R [0, q/2) \cap \mathbb{Z}$, *outputs* $\mathbf{k}$ *is smaller than* $\frac{1}{2^\lambda \cdot \mathsf{negl}(\lambda)}$, *i.e.*

$$\Pr_{\substack{\mathbf{k} \in_R \{0,...,B-1\}^\ell \\ \{\mathbf{a}_j \in_R \mathbb{Z}_q^{1 \times \ell}, e_j \in_R [0, q/2) \cap \mathbb{Z}, \\ b_j = \mathbf{a}_j \cdot \mathbf{k} + e_j + f_j(\mathbf{k}) \cdot q/2\}_{j \in [w]}}} \left[\mathsf{Adv}(\{\mathbf{a}_j, b_j\}_{j \in [w]}) = \mathbf{k} \right] \leq \frac{1}{2^\lambda \cdot \mathsf{negl}(\lambda)}$$

Thus, using Theorem 6, we obtain the following corollary.

Corollary 2. *Suppose that Assumption 12 holds. Then, there exists correlation intractable function for all sparse relations.*

Remark 4. In Assumption 12, if the function f is a constant (i.e. is independent of the key), then the problem is equivalent to search-LWE (for the same distributions of secret, noise, and public matrices, and the same requirement on the success probability as described in Assumption 12).

7 NIZK from Fiat-Shamir

In this section we show how to use our hash functions to construct non-interactive zero-knowledge (NIZK) arguments for NP. We follow the folklore approach of applying the Fiat-Shamir transformation to a constant-round public-coin honest-verifier zero-knowledge proof-system. The point however is that we can establish soundness based on a concrete assumption (with a meaningful security reduction) rather than just heuristically assuming that the Fiat-Shamir transformation preserves soundness. Further, we show that if we start from an interactive proof with *adaptive soundness* (where the instance x can be chosen adaptively in the last message), as in [38]; then in the resulting NIZK, the soundness and zero-knowledge properties hold even if the instance is chosen adaptively given the CRS.

We remark that for this result to go through we require an additional property from the hash function family that we use, beyond correlation intractability. Namely, that it is possible to efficiently sample a uniformly random hash function h from the family, *conditioned* on $h(a) = b$, for some arbitrary fixed values a and b. We refer to this property as "programmability".

Definition 7 (Programmability of hash function). *A hash function ensemble* $\mathcal{H} = \{h_k : \mathcal{D}(\lambda) \to \mathcal{C}(\lambda)\}_{\lambda \in \mathbb{N}}$ *is called* programmable *if there exists an efficient algorithm* M *that given* $x \in \mathcal{D}(\lambda)$ *and* $y \in \mathcal{C}(\lambda)$, *outputs a uniformly random hash function* h_k *from the family such that* $h_k(x) = y$.

Translating the requirement to the hash function instantiated using our KDM-secure encryption scheme, it means the encryption algorithm given a key a and message b outputs the ciphertext efficiently.

We recall the definition of NIZK with adaptive soundness and zero-knowledge.

Definition 8 (NIZK with adaptive soundness and ZK [17,38]). *Let* $\lambda \in \mathbb{N}$ *be the security parameter. A non-interactive (computational) zero-knowledge argument system (NIZK) for an NP language* $\mathcal{L} \in$ NP, *with witness relation* $R_{\mathcal{L}}$, *is a pair of probabilistic polynomial-time algorithms* (P, V) *such that:*

- **Completeness:** *For every* $x \in \mathcal{L}$ *and witness* w *for* x *(i.e.,* $(x, w \in R_{\mathcal{L}})$*), for all* $\sigma \in \{0,1\}^{\mathsf{poly}(\lambda)}$,
$$V\big(\sigma, x, P(x, \sigma, w)\big) = 1.$$

- **Adaptive Soundness:** *For every polynomial-size cheating prover* P^*, *we have*
$$\Pr_{\substack{\sigma \in_R \{0,1\}^{\mathsf{poly}(\lambda)} \\ (x,a) \leftarrow P^*(\sigma)}} \Big[\big(V(x, \sigma, a) = 1\big) \wedge \big(x \notin \mathcal{L}\big) \Big] < \mathsf{negl}(\lambda).$$

- **Adaptive Zero-Knowledge:** *There exists a probabilistic polynomial-time simulator* $S = (S_1, S_2)$ *such that for every polynomial time adversary* $A = (A_1, A_2)$,

$$\left| \Pr_{\substack{\sigma \in_R \{0,1\}^{\mathrm{poly}(\lambda)} \\ (x,w,\zeta) \leftarrow A_1(\sigma) \\ \pi \leftarrow P(x,\sigma,w)}} \left[(A_2(\sigma,x,\pi,\zeta) = 1) \ \wedge \ (x \in \mathcal{L}) \right] \right.$$

$$\left. - \Pr_{\substack{\sigma,\tau \leftarrow S_1(1^\lambda) \\ (x,w,\zeta) \leftarrow A_1(\sigma) \\ \pi \leftarrow S_2(\tau,x,\sigma)}} \left[(A_2(\sigma,x,\pi,\zeta) = 1) \ \wedge \ (x \in \mathcal{L}) \right] \right| \leq \mathsf{negl}(\lambda),$$

where ζ (resp., τ) denote an internal state of the adversary (resp., simulator).

The random string σ received by both P and V is referred to as the *common random string* or CRS.

We establish the following result.

Theorem 13. *Assume there exists one-way functions and a programmable correlation intractable function ensemble for all sparse relations. Then, any language in NP has a non-interactive zero-knowledge argument-system with adaptive soundness and adaptive zero-knowledge.*

As a corollary of Theorem 13 and the results obtained in the previous sections, we obtain that:

Corollary 3. *If either Assumption 10 or Assumption 12 holds, then any language in NP has a non-interactive zero-knowledge argument-system with adaptive soundness and adaptive zero-knowledge.*

The readers are referred to the full version [25] for the Proof of Theorem 13.

Acknowledgments. We thank the anonymous reviewers for their helpful comments.

R.C. is a member of the Check Point Institute for Information Security, and is supported by ISF grant 1523/14. R.C. and Y.C. are supported by the NSF MACS project. L.R. is supported by NSF grant 1422965. R.D.R is supported by DARPA and the U.S. Army Office under contract numbers W911NF-15-C-0226 and W911NF-15-C-0236, a SIMONS Investigator Award Agreement Dated 6-5-12, and the Cybersecurity and Privacy Institute at Northeastern University.

Appendices

A Success Probability of Polynomial Time Algorithms on Discrete-Log Problem

The discrete-log problem over $\mathbb{F}_q^*$ can be solved by the index calculus algorithms in heuristic subexponential time $\exp(C(\log q)^{1/3}(\log \log q)^{2/3})$.

We consider a (commonly used) variant of the index calculus algorithm with an online phase and an offline phase. The offline (preprocessing) phase only gets the modulus q and the generator g, the online phase gets the challenge $h \equiv g^x \bmod q$, computes x. The offline part calculates the discrete log of $\log_g(2)$, $\log_g(3)$, ..., $\log_g(B)$. The online phase picks a random r, try to factorize $g^r \cdot h \equiv g^{r+x} \bmod q$ in $\mathbb{Z}$, see if all the factors are smaller or equal to a prescribed prime bound B. If $g^r \cdot h = 2^{x_2} \cdot 3^{x_3} \cdot ... \cdot B^{x_B}$, then $r + x \equiv \log_g(2) \cdot x_2 + \log_g(3) \cdot x_3 + ... + \log_g(B) \cdot x_B \bmod \phi(q)$.

The algorithm achieves $O(2^{-\frac{\lambda}{c}})$ success probability even if the online phase is only allowed to run in polynomial time, and the preprocessing phase is allowed to spend super-polynomial running time, but restricted to leave polynomially many bits as the advice for the online phase. The analysis of the success probability relies on the estimation of the number of smooth integers $\Psi(q, B)$, which stands for the number of integers in the range $[1, q]$ whose factors are all under B. Since the online phase is forced to receive only polynomial size advice and run in polynomial time, B will be chosen as a polynomial, whereas $q \approx 2^\lambda$.

The smooth integer bound can be derived from Rankin [71] (see the survey of [48]) that for any $A > 1$, $\Psi(q, \log(q)^A) = q^{1-1/A+O\left(\frac{1}{\log\log q}\right)}$. This means the probability of a number within $[1, 2^\lambda]$ to be $O(\lambda^c)$ smooth is $2^{-\frac{\lambda}{c}+O\left(\frac{\lambda}{\log \lambda}\right)}$.

References

1. Abdalla, M., An, J.H., Bellare, M., Namprempre, C.: From identification to signatures via the Fiat-Shamir transform: minimizing assumptions for security and forward-security. In: Knudsen, L.R. (ed.) EUROCRYPT 2002. LNCS, vol. 2332, pp. 418–433. Springer, Heidelberg (2002). https://doi.org/10.1007/3-540-46035-7_28
2. Adleman, L.: A subexponential algorithm for the discrete logarithm problem with applications to cryptography. In: 1979 20th Annual Symposiumon Foundations of Computer Science, pp. 55–60. IEEE (1979)
3. Ajtai, M.: Generating hard instances of lattice problems (extended abstract). In: STOC, pp. 99–108 (1996)
4. Albrecht, M.R., Player, R., Scott, S.: On the concrete hardness of learning with errors. J. Math. Cryptol. 9(3), 169–203 (2015)
5. Applebaum, B.: Key-dependent message security: generic amplification and completeness. J. Cryptol. 27(3), 429–451 (2014)
6. Applebaum, B., Cash, D., Peikert, C., Sahai, A.: Fast cryptographic primitives and circular-secure encryption based on hard learning problems. In: Halevi, S. (ed.) CRYPTO 2009. LNCS, vol. 5677, pp. 595–618. Springer, Heidelberg (2009). https://doi.org/10.1007/978-3-642-03356-8_35
7. Arora, S., Ge, R.: New algorithms for learning in presence of errors. In: Aceto, L., Henzinger, M., Sgall, J. (eds.) ICALP 2011. LNCS, vol. 6755, pp. 403–415. Springer, Heidelberg (2011). https://doi.org/10.1007/978-3-642-22006-7_34
8. Babai, L.: On Lovász' lattice reduction and the nearest lattice point problem. Combinatorica 6(1), 1–13 (1986)
9. Barak, B., Haitner, I., Hofheinz, D., Ishai, Y.: Bounded key-dependent message security. In: Gilbert, H. (ed.) EUROCRYPT 2010. LNCS, vol. 6110, pp. 423–444. Springer, Heidelberg (2010). https://doi.org/10.1007/978-3-642-13190-5_22
10. Barak, B., Lindell, Y., Vadhan, S.P.: Lower bounds for non-black-box zero knowledge. J. Comput. Syst. Sci. 72(2), 321–391 (2006)

11. Bellare, M., Hoang, V.T., Keelveedhi, S.: Instantiating random oracles via UCEs. In: Canetti, R., Garay, J.A. (eds.) CRYPTO 2013. LNCS, vol. 8043, pp. 398–415. Springer, Heidelberg (2013). https://doi.org/10.1007/978-3-642-40084-1_23

12. Bellare, M., Rogaway, P.: Random oracles are practical: a paradigm for designing efficient protocols. In: ACM Conference on Computer and Communications Security, pp. 62–73 (1993)

13. Ben-Sasson, E., Chiesa, A., Spooner, N.: Interactive oracle proofs. In: Hirt, M., Smith, A. (eds.) TCC 2016. LNCS, vol. 9986, pp. 31–60. Springer, Heidelberg (2016). https://doi.org/10.1007/978-3-662-53644-5_2

14. Bitansky, N., Dachman-Soled, D., Garg, S., Jain, A., TaumanKalai, Y., López-Alt, A., Wichs, D.: Why "fiat-shamir for proofs" lacks a proof. In: TCC, pp. 182–201 (2013)

15. Black, J., Rogaway, P., Shrimpton. T.: Encryption-scheme security in the presence of key-dependent messages. In: Selected Areas in Cryptography, pp. 62–75 (2002)

16. Blum, A., Kalai, A., Wasserman, H.: Noise-tolerant learning, the parity problem, and the statistical query model. J. ACM (JACM) 50(4), 506–519 (2003)

17. Blum, M., Feldman, P., Micali, S.: Non-interactive zero-knowledge and its applications (extended abstract). In: Proceedings of the 20th Annual ACM Symposium on Theory of Computing, 2–4 May 1988, Chicago, Illinois, USA, pp. 103–112 (1988)

18. Boldyreva, A., Cash, D., Fischlin, M., Warinschi, B.: Foundations of non-malleable hash and one-way functions. In: Matsui, M. (ed.) ASIACRYPT 2009. LNCS, vol. 5912, pp. 524–541. Springer, Heidelberg (2009). https://doi.org/10.1007/978-3-642-10366-7_31

19. Boneh, D., Halevi, S., Hamburg, M., Ostrovsky, R.: Circular-secure encryption from decision Diffie-Hellman. In: Wagner, D. (ed.) CRYPTO 2008. LNCS, vol. 5157, pp. 108–125. Springer, Heidelberg (2008). https://doi.org/10.1007/978-3-540-85174-5_7

20. Brakerski, Z., Goldwasser, S.: Circular and leakage resilient public-key encryption under subgroup indistinguishability. In: Rabin, T. (ed.) CRYPTO 2010. LNCS, vol. 6223, pp. 1–20. Springer, Heidelberg (2010). https://doi.org/10.1007/978-3-642-14623-7_1

21. Brakerski, Z., Goldwasser, S., Kalai, Y.T.: Black-Box circular-secure encryption beyond affine functions. In: Ishai, Y. (ed.) TCC 2011. LNCS, vol. 6597, pp. 201–218. Springer, Heidelberg (2011). https://doi.org/10.1007/978-3-642-19571-6_13

22. Camenisch, J., Lysyanskaya, A.: An efficient system for non-transferable anonymous credentials with optional anonymity revocation. In: EUROCRYPT: Advances in Cryptology: Proceedings of EUROCRYPT (2001)

23. Canetti, R.: Towards realizing random oracles: hash functions that hide all partial information. In: Kaliski, B.S. (ed.) CRYPTO 1997. LNCS, vol. 1294, pp. 455–469. Springer, Heidelberg (1997). https://doi.org/10.1007/BFb0052255

24. Canetti, R., Chen, Y., Reyzin, L.: On the correlation intractability of obfuscated pseudorandom functions. In: Kushilevitz, E., Malkin, T. (eds.) TCC 2016. LNCS, vol. 9562, pp. 389–415. Springer, Heidelberg (2016). https://doi.org/10.1007/978-3-662-49096-9_17

25. Canetti, R., Chen, Y., Reyzin, L., Rothblum, R.D.: Fiat-Shamir and correlation intractability from strong KDM-secure encryption. Cryptology ePrint Archive, Report 2018/131 (2018)

26. Canetti, R., Goldreich, O., Halevi, S.: The random oracle methodology, revisited (preliminary version). In: [78], pp. 209–218. ACM (1998)

27. Canetti, R., Goldreich, O., Halevi, S.: The random oracle methodology, revisited. J. ACM 51(4), 557–594 (2004)

28. Canetti, R., Tauman Kalai, Y., Varia, M., Wichs, D.: On symmetric encryption and point obfuscation. In: Micciancio, D. (ed.) TCC 2010. LNCS, vol. 5978, pp. 52–71. Springer, Heidelberg (2010). https://doi.org/10.1007/978-3-642-11799-2_4
29. Canetti, R., Micciancio, D., Reingold, O.: Perfectly one-way probabilistic hash functions (preliminary version). In: [78], pp. 131–140. ACM (1998)
30. Coppersmith, D., Odlyzko, A.M., Schroeppel, R.: Discrete logarithms in GF(p). Algorithmica 1(1), 1–15 (1986)
31. Diem, C.: On the discrete logarithm problem in elliptic curves. Compositio Mathematica 147(1), 75–104 (2011)
32. Dodis, Y., Oliveira, R., Pietrzak, K.: On the generic insecurity of the full domain hash. In: Shoup, V. (ed.) CRYPTO 2005. LNCS, vol. 3621, pp. 449–466. Springer, Heidelberg (2005). https://doi.org/10.1007/11535218_27
33. Dodis, Y., Ristenpart, T., Vadhan, S.P.: Randomness condensers for efficiently samplable, seed-dependent sources. In: TCC, pp. 618–635 (2012)
34. Döttling, N., Müller-Quade, J.: Lossy codes and a new variant of the learning-with-errors problem. In: Johansson, T., Nguyen, P.Q. (eds.) EUROCRYPT 2013. LNCS, vol. 7881, pp. 18–34. Springer, Heidelberg (2013). https://doi.org/10.1007/978-3-642-38348-9_2
35. Dwork, C., Naor, M.: Pricing via processing or combatting junk mail. In: Brickell, E.F. (ed.) CRYPTO 1992. LNCS, vol. 740, pp. 139–147. Springer, Heidelberg (1993). https://doi.org/10.1007/3-540-48071-4_10
36. Dwork, C., Naor, M., Reingold, O., Stockmeyer, L.J.: Magic functions. J. ACM 50(6), 852–921 (2003)
37. ElGamal, T.: A public key cryptosystem and a signature scheme based on discrete logarithms. IEEE Trans. Inf. Theor. 31(4), 469–472 (1985)
38. Feige, U., Lapidot, D., Shamir, A.: Multiple noninteractive zero knowledge proofs under general assumptions. SIAM J. Comput. 29, 1–28 (1999)
39. Fiat, A., Shamir, A.: How to prove yourself: practical solutions to identification and signature problems. In: Odlyzko, A.M. (ed.) CRYPTO 1986. LNCS, vol. 263, pp. 186–194. Springer, Heidelberg (1987). https://doi.org/10.1007/3-540-47721-7_12
40. Galbraith, S.D., Gaudry, P.: Recent progress on the elliptic curve discrete logarithm problem. Des. Codes Crypt. 78(1), 51–72 (2016)
41. Gaudry, P.: Index calculus for abelian varieties of small dimension and the elliptic curve discrete logarithm problem. J. Symb. Comput. 44(12), 1690–1702 (2009)
42. Gaudry, P., Hess, F., Smart, N.P.: Constructive and destructive facets of weil descent on elliptic curves. J. Cryptology 15(1), 19–46 (2002)
43. Gentry, C.: Fully homomorphic encryption using ideal lattices. In: Mitzenmacher, M. (ed.) STOC, pp. 169–178. ACM (2009)
44. Goldwasser, S., Kalai, Y.T.: On the (in)security of the Fiat-Shamir paradigm. In: FOCS, pp. 102–113 (2003)
45. Goldwasser, S., Micali, S.: Probabilistic encryption. J. Comput. Syst. Sci. 28(2), 270–299 (1984)
46. Goldwasser, S., Micali, S., Rackoff, C.: The knowledge complexity of interactive proof-systems (extended abstract). In: STOC, pp. 291–304 (1985)
47. Goyal, V., O'Neill, A., Rao, V.: Correlated-input secure hash functions. In: Ishai, Y. (ed.) TCC 2011. LNCS, vol. 6597, pp. 182–200. Springer, Heidelberg (2011). https://doi.org/10.1007/978-3-642-19571-6_12
48. Granville, A.: Smooth numbers: computational number theory and beyond, pp. 267–323 (2008)
49. Hada, S., Tanaka, T.: Zero-knowledge and correlation intractability. IEICE Trans. 89–A(10), 2894–2905 (2006)

50. Haitner, I., Holenstein, T.: On the (im)possibility of key dependent encryption. In: Reingold, O. (ed.) TCC 2009. LNCS, vol. 5444, pp. 202–219. Springer, Heidelberg (2009). https://doi.org/10.1007/978-3-642-00457-5_13
51. Halevi, S., Krawczyk, H.: Security under key-dependent inputs. In: Ning, P., De Capitani di Vimercati, S., Syverson, P.F. (eds.) Proceedings of the 2007 ACM Conference on Computer and Communications Security, CCS 2007, Alexandria, Virginia, USA, 28–31, October 2007, pp. 466–475. ACM (2007)
52. Halevi, S., Myers, S., Rackoff, C.: On seed-incompressible functions. In: Canetti, R. (ed.) TCC 2008. LNCS, vol. 4948, pp. 19–36. Springer, Heidelberg (2008). https://doi.org/10.1007/978-3-540-78524-8_2
53. Hanke, T.: Asicboost - a speedup for bitcoin mining. CoRR abs/1604.00575 (2016)
54. Håstad, J., Impagliazzo, R., Levin, L.A., Luby, M.: A pseudorandom generator from any one-way function. SIAM J. Comput. **28**(4), 1364–1396 (1999)
55. Herold, G., Kirshanova, E., May, A.: On the asymptotic complexity of solving lWE. Des. Codes Crypt., 1–29 (2015)
56. Hofheinz, D., Unruh, D.: Towards key-dependent message security in the standard model. In: Smart, N. (ed.) EUROCRYPT 2008. LNCS, vol. 4965, pp. 108–126. Springer, Heidelberg (2008). https://doi.org/10.1007/978-3-540-78967-3_7
57. Ishai, Y., Kilian, J., Nissim, K., Petrank, E.: Extending oblivious transfers efficiently. In: Boneh, D. (ed.) CRYPTO 2003. LNCS, vol. 2729, pp. 145–161. Springer, Heidelberg (2003). https://doi.org/10.1007/978-3-540-45146-4_9
58. Kalai, Y.T., Raz, R.: Probabilistically checkable arguments. In: Halevi, S. (ed.) CRYPTO 2009. LNCS, vol. 5677, pp. 143–159. Springer, Heidelberg (2009). https://doi.org/10.1007/978-3-642-03356-8_9
59. Kalai, Y.T., Rothblum, G.N., Rothblum, R.D.: From obfuscation to the security of Fiat-Shamir for proofs. In: Katz, J., Shacham, H. (eds.) CRYPTO 2017. LNCS, vol. 10402, pp. 224–251. Springer, Cham (2017). https://doi.org/10.1007/978-3-319-63715-0_8
60. Kirchner, P., Fouque, P.-A.: An improved BKW algorithm for LWE with applications to cryptography and lattices. In: Gennaro, R., Robshaw, M. (eds.) CRYPTO 2015. LNCS, vol. 9215, pp. 43–62. Springer, Heidelberg (2015). https://doi.org/10.1007/978-3-662-47989-6_3
61. Lenstra, A.K., Lenstra, H.W., Lovász, L.: Factoring polynomials with rational coefficients. Math. Ann. **261**(4), 515–534 (1982)
62. Lindner, R., Peikert, C.: Better key sizes (and attacks) for LWE-based encryption. In: Kiayias, A. (ed.) CT-RSA 2011. LNCS, vol. 6558, pp. 319–339. Springer, Heidelberg (2011). https://doi.org/10.1007/978-3-642-19074-2_21
63. Menezes, A., Okamoto, T., Vanstone, S.A.: Reducing elliptic curve logarithms to logarithms in a finite field. IEEE Trans. Inf. Theor. **39**(5), 1639–1646 (1993)
64. Micciancio, D., Peikert, C.: Hardness of SIS and LWE with small parameters. In: Canetti, R., Garay, J.A. (eds.) CRYPTO 2013. LNCS, vol. 8042, pp. 21–39. Springer, Heidelberg (2013). https://doi.org/10.1007/978-3-642-40041-4_2
65. Micciancio, D., Regev, O.: Lattice-based cryptography. In: Bernstein, D.J., Buchmann, J., Dahmen, E. (eds.) Post-Quantum Cryptography, pp. 147–191. Springer, Heidelberg (2009). https://doi.org/10.1007/978-3-540-88702-7_5
66. Nakamoto, S.: Bitcoin: a peer-to-peer electronic cash system (2008). Accessed 28 Jan 2018
67. Nguyen, P.Q., Stehlé, D.: LLL on the average. In: Hess, F., Pauli, S., Pohst, M. (eds.) ANTS 2006. LNCS, vol. 4076, pp. 238–256. Springer, Heidelberg (2006). https://doi.org/10.1007/11792086_18

68. Nielsen, J.B.: Separating random oracle proofs from complexity theoretic proofs: the non-committing encryption case. In: Yung, M. (ed.) CRYPTO 2002. LNCS, vol. 2442, pp. 111–126. Springer, Heidelberg (2002). https://doi.org/10.1007/3-540-45708-9_8

69. Petit, C., Kosters, M., Messeng, A.: Algebraic approaches for the elliptic curve discrete logarithm problem over prime fields. In: Cheng, C.-M., Chung, K.-M., Persiano, G., Yang, B.-Y. (eds.) PKC 2016. LNCS, vol. 9615, pp. 3–18. Springer, Heidelberg (2016). https://doi.org/10.1007/978-3-662-49387-8_1

70. Pollard, J.M.: A Monte Carlo method for factorization. BIT Numer. Math. 15(3), 331–334 (1975)

71. Rankin, R.A.: The difference between consecutive prime numbers. J. London Math. Soc. 1(4), 242–247 (1938)

72. Regev, O.: On lattices, learning with errors, random linear codes, and cryptography. In: Gabow, H.N., Fagin, R. (eds.) Proceedings of the 37th Annual ACM Symposium on Theory of Computing, Baltimore, MD, USA, 22–24 May 2005, pp. 84–93. ACM (2005)

73. Regev, O.: On lattices, learning with errors, random linear codes, and cryptography. J. ACM 56(6), 84–93 (2009)

74. Reingold, O., Rothblum, G.N.., Rothblum, R.D.: Constant-round interactive proofs for delegating computation. In: STOC, pp. 49–62. ACM (2016)

75. Schnorr, C.-P.: A hierarchy of polynomial time lattice basis reduction algorithms. Theor. Comput. Sci. 53, 201–224 (1987)

76. Semaev, I.A.: Summation polynomials and the discrete logarithm problem on elliptic curves (2004)

77. Shoup, V.: Lower bounds for discrete logarithms and related problems. In: Fumy, W. (ed.) EUROCRYPT 1997. LNCS, vol. 1233, pp. 256–266. Springer, Heidelberg (1997). https://doi.org/10.1007/3-540-69053-0_18

78. Vitter, J.S. (eds.): Proceedings of the Thirtieth Annual ACM Symposium on the Theory of Computing, Dallas, Texas, USA, 23–26 May 1998. ACM (1998)

79. Zhandry, M.: The magic of ELFs. In: Robshaw, M., Katz, J. (eds.) CRYPTO 2016. LNCS, vol. 9814, pp. 479–508. Springer, Heidelberg (2016). https://doi.org/10.1007/978-3-662-53018-4_18

Lattices

Shortest Vector from Lattice Sieving: A Few Dimensions for Free

Léo Ducas[✉]

Cryptology Group, CWI, Amsterdam, The Netherlands
ducas@cwi.nl

Abstract. Asymptotically, the best known algorithms for solving the Shortest Vector Problem (SVP) in a lattice of dimension n are sieve algorithms, which have heuristic complexity estimates ranging from $(4/3)^{n+o(n)}$ down to $(3/2)^{n/2+o(n)}$ when Locality Sensitive Hashing techniques are used. Sieve algorithms are however outperformed by pruned enumeration algorithms in practice by several orders of magnitude, despite the larger super-exponential asymptotical complexity $2^{\Theta(n\log n)}$ of the latter.

In this work, we show a concrete improvement of sieve-type algorithms. Precisely, we show that a few calls to the sieve algorithm in lattices of dimension less than $n - d$ solves SVP in dimension n, where $d = \Theta(n/\log n)$.

Although our improvement is only sub-exponential, its practical effect in relevant dimensions is quite significant. We implemented it over a simple sieve algorithm with $(4/3)^{n+o(n)}$ complexity, and it outperforms the best sieve algorithms from the literature by a factor of 10 in dimensions 70–80. It performs less than an order of magnitude slower than pruned enumeration in the same range.

By design, this improvement can also be applied to most other variants of sieve algorithms, including LSH sieve algorithms and tuple-sieve algorithms. In this light, we may expect sieve-techniques to outperform pruned enumeration in practice in the near future.

Keywords: Cryptanalysis · Lattice · Sieving · Nearest-Plane

1 Introduction

The concrete hardness of the Shortest Vector Problem (SVP) is at the core of the cost estimates of attacks against lattice-based cryptosystems. While those schemes may use various underlying problems (NTRU [HPS98], SIS [Ajt99], LWE [Reg05]) their cryptanalysis boils down to solving large instances of the Shortest Vector Problem inside BKZ-type algorithms. There are two classes of algorithms for SVP: enumeration algorithms and sieve algorithms.

Supported by a Veni Innovational Research Grant from NWO under project number 639.021.645.

J. B. Nielsen and V. Rijmen (Eds.): EUROCRYPT 2018, LNCS 10820, pp. 125–145, 2018.
https://doi.org/10.1007/978-3-319-78381-9_5

The first class of algorithms (enumeration) was initiated by Pohst [Poh81]. Kannan [Kan83, HS07, MW15] proved that with appropriate pre-processing, the shortest vector could be found in time $2^{\Theta(n \log n)}$. This algorithm only requires a polynomial amount of memory. These algorithms can be made much faster in practice using some heuristic techniques, in particular the pruning technique [SE94, SH95, GNR10, Che13].

The second class of algorithms (sieving) started with Ajtai et al. [AKS01], and requires single exponential time and memory. Variants were heuristically analyzed [NV08, MV10], giving a $(4/3)^{n+o(n)}$ time complexity and a $(4/3)^{n/2+o(n)}$ memory complexity. A long line of work, including [BGJ13, Laa15a, Laa15b, BDGL16] decrease this time complexity down to $(3/2)^{n/2+o(n)}$ at the cost of more memory. Other variants (tuple-sieving) are designed to lower the memory complexity [BLS16, HK17].

The situation is rather paradoxical: asymptotically, sieving algorithms should outperform enumeration algorithms, yet in practice, sieving remains several orders of magnitude slower. This situation makes security estimates delicate, requiring both algorithms to be considered. In that respect, one would much prefer enumeration to become irrelevant, as the heuristics used in this algorithm makes prediction of its practical cost tedious and maybe inaccurate.

To this end, an important goal is to improve not only the asymptotic complexity of sieving, but also its practical complexity. Indeed, much can been gained from asymptotically negligible tricks, fine-tuning of the parameters, and optimized implementation effort [FBB+15, BNvdP14, MLB17].

This work. We propose a new practical improvement for sieve algorithms. In theory, we can heuristically show that it contributes a sub-exponential gain in the running time and the memory consumption. In practice, our implementation outperforms all sieving implementations of the literature by a factor of 10 in dimensions 70–80, despite the fact that we did not implement some known improvements [BDGL16, MLB17]. Our improved sieving algorithm performs reasonably close to pruned enumeration; more precisely, within less than an order of magnitude of the optimized pruned enumeration implementation in fplll's library [Ste10, FPL16b, FPL16a].[1]

In brief, the main idea behind our improvement is exploiting the fact that sieving produces many short vectors, rather than only one. We use this fact to our advantage by solving SVP in lattices of dimension n while running a sieve algorithm in projected sub-lattices of dimension smaller than $n - d$. Using an appropriate pre-processing, we show that one may choose d as large as $\Theta(n/\log n)$. Heuristic arguments lead to a concrete prediction of $d \approx \frac{n \ln(4/3)}{\ln(n/2\pi e)}$. This prediction is corroborated by our experiments.

At last, we argue that, when combined with the LSH techniques [BDGL16, MLB17], our new technique should lead to a sieve algorithm that outperforms

---

[1] Please note that this library was not so fast for SVP and BKZ a few years ago and it recently caught up with the state of the art with the addition of a pruner module [FPL16b], and of an external Strategizer [FPL16a].

enumeration in practice, for dimensions maybe as low as $n = 90$. We also suggest four approaches to further improve sieving, including amortization inside BKZ.

Outline. We shall start with preliminaries in Sect. 2, including a generic presentation of sieve algorithms in Sect. 2.3. Our main contribution is presented in Sect. 3. In Sect. 4, we present details of our implementation, including other algorithmic tricks. In Sect. 5 we report on the experimental behavior of our algorithm, and compare its performances to the literature. We conclude with a discussion in Sect. 6, on combining our improvement with the LSH techniques [Laa15a, BDGL16, MLB17], and suggest further improvements.

2   Preliminaries

2.1   Notations and Basic Definitions

All vectors are denoted by bold lower case letters and are to be read as column-vectors. Matrices are denoted by bold capital letters. We write a matrix $\mathbf{B}$ as $\mathbf{B} = (\mathbf{b}_0, \cdots, \mathbf{b}_{n-1})$ where $\mathbf{b}_i$ is the i-th column vector of $\mathbf{B}$. If $\mathbf{B} \in \mathbb{R}^{m \times n}$ has full-column rank n, the lattice $\mathcal{L}$ generated by the basis $\mathbf{B}$ is denoted by $\mathcal{L}(\mathbf{B}) = \{\mathbf{B}\mathbf{x} \mid \mathbf{x} \in \mathbb{Z}^n\}$. We denote by $(\mathbf{b}_0^*, \cdots, \mathbf{b}_{n-1}^*)$ the Gram-Schmidt orthogonalization of the matrix $(\mathbf{b}_0, \cdots, \mathbf{b}_{n-1})$. For $i \in \{0, \cdots, n-1\}$, we denote the orthogonal projection to the span of $(\mathbf{b}_0, \cdots, \mathbf{b}_{i-1})$ by π_i. For $0 \leq i < j \leq n$, we denote by $\mathbf{B}_{[i,j]}$ the local projected block $(\pi_i(\mathbf{b}_i), \cdots, \pi_i(\mathbf{b}_{j-1}))$, and when the basis is clear from context, by $\mathcal{L}_{[i,j]}$ the lattice generated by $\mathbf{B}_{[i,j]}$. We use $\mathbf{B}_i$ and $\mathcal{L}_i$ as shorthands for $\mathbf{B}_{[i,n]}$ and $\mathcal{L}_{[i,n]}$.

The Euclidean norm of a vector $\mathbf{v}$ is denoted by $\|\mathbf{v}\|$. The volume of a lattice $\mathcal{L}(\mathbf{B})$ is $\mathrm{Vol}(\mathcal{L}(\mathbf{B})) = \prod_i \|\mathbf{b}_i^*\|$, that is an invariant of the lattice. The first minimum of a lattice $\mathcal{L}$ is the length of a shortest non-zero vector, denoted by $\lambda_1(\mathcal{L})$. We use the abbreviations $\mathrm{Vol}(\mathbf{B}) = \mathrm{Vol}(\mathcal{L}(\mathbf{B}))$ and $\lambda_1(\mathbf{B}) = \lambda_1(\mathcal{L}(\mathbf{B}))$.

2.2   Lattice Reduction

The Gaussian Heuristic predicts that the number $|\mathcal{L} \cap \mathcal{B}|$ lattice of points inside a measurable body $\mathcal{B} \subset \mathbb{R}^n$ is approximately equal to $\mathrm{Vol}(\mathcal{B})/\mathrm{Vol}(\mathcal{L})$. Applied to Euclidean n-balls, it leads to the following prediction of the length of a shortest non-zero vector in a lattice.

Definition 1 (Gaussian Heuristic). *We denote by* $\mathrm{gh}(\mathcal{L})$ *the expected first minimum of a lattice* $\mathcal{L}$ *according to the Gaussian Heuristic. For a full rank lattice* $\mathcal{L} \subset \mathbb{R}^n$, *it is given by:*

$$\mathrm{gh}(\mathcal{L}) = \sqrt{n/2\pi e} \cdot \mathrm{Vol}(\mathcal{L})^{1/n}.$$

We also denote $\mathrm{gh}(n)$ *for* $\mathrm{gh}(\mathcal{L})$ *of any* n-*dimensional lattice* $\mathcal{L}$ *of volume 1:* $\mathrm{gh}(n) = \sqrt{n/2\pi e}$.

Definition 2 (Hermite-Korkine-Zolotarev and Block-Korkine-Zolotarev reductions [Ngu09]). *The basis* $\mathbf{B} = (\mathbf{b}_0, \ldots, \mathbf{b}_{n-1})$ *of a lattice* $\mathcal{L}$ *is said to be HKZ reduced if* $\|\mathbf{b}_i^*\| = \lambda_1(\mathcal{L}(\mathbf{B}_i))$ *for all* $i < n$. *It is said BKZ reduced with block-size* b *(for short BKZ-b reduced)* $\|\mathbf{b}_i^*\| = \lambda_1(\mathcal{L}(\mathbf{B}_{[i:\max(i+b,n)]}))$ *for all* $i < n$.[2]

Under the Gaussian Heuristic, we can predict the shape $\ell_0 \ldots \ell_{n-1}$ of an HKZ reduced basis, i.e., the sequence of expected norms for the vectors $\mathbf{b}_i^*$. The sequence is inductively defined as follows:

Definition 3. *The HKZ-shape of dimension* n *is defined by the following sequence:*

$$\ell_0 = \mathrm{gh}(n) \quad \text{and} \quad \ell_i = \mathrm{gh}(n-i) \cdot \Big(\prod_{j<i} \ell_j\Big)^{-\frac{1}{n-i}}.$$

Note that the Gaussian Heuristic is known to be violated in small dimensions [CN11], fortunately we only rely on the above prediction for $i \ll n$.

Definition 4 (Geometric Series Assumption). *Let* $\mathbf{B}$ *be a BKZ-b reduced basis of a lattice of volume* 1. *The Geometric Series Assumption states that:*

$$\|\mathbf{b}_i^*\| = \alpha_b^{\frac{n-1}{2}-i}.$$

where $\alpha_b = \mathrm{gh}(b)^{2/b}$.

This model is reasonably accurate in practice for $b > 50$ and $b \ll n$. For further discussion on this model and its accuracy, the reader may refer to [CN11, Che13, YD16].

2.3 Sieve Algorithms

There are several variants of sieving algorithms, even among the restricted class of Sieving algorithms having asymptotic complexity $(4/3)^{n+o(n)}$ [NV08, MV10]. Its generic form is given below.

Algorithm 1. Sieve($\mathcal{L}$)

Require: The basis $\mathbf{B}$ of a lattice $\mathcal{L}$ of dimension n
Ensure: A list L of vectors
    $L \leftarrow$ a set of N random vectors (of length at most $2^n \cdot \mathrm{Vol}(\mathcal{L})^{1/n}$) from $\mathcal{L}$ where $N = (4/3)^{n/2+o(n)}$.
    while $\exists(\mathbf{v}, \mathbf{w}) \in L^2$ such that $\|\mathbf{v} - \mathbf{w}\| < \|\mathbf{v}\|$ **do**
        $\mathbf{v} \leftarrow \mathbf{v} - \mathbf{w}$
    end while
    return L

---

[2] The notion of BKZ-reduction is typically slightly relaxed for algorithmic purposes, see [HPS11].

The initialization of the list L can be performed by first computing an LLL-reduced basis of the lattice [LLL82], and taking small random linear combinations of that basis.

Using heuristic arguments, one can show [NV08] that this algorithm will terminate in time $N^2 \cdot \text{poly}(n)$, and that the output list contains a shortest vector of the lattice. The used heuristic reasoning might fail in some special lattices, such as $\mathbb{Z}^n$. However, nearly all lattices occurring in a cryptographic context are random-looking lattices, for which these heuristics have been confirmed extensively.

Many tricks can be implemented to improve the hidden polynomial factors. The most obvious one consists of working modulo negation of vectors (halving the list size), and to exploit the identity $\|\mathbf{v} \pm \mathbf{w}\|^2 = \|\mathbf{v}\|^2 + \|\mathbf{w}\|^2 \pm 2\langle \mathbf{v}, \mathbf{w} \rangle$: two reductions can be tested for the price of one inner product.

More substantial algorithmic improvements have been proposed in [MV10]: sorting the list by Euclidean length to make early reduction more likely, having the list size be adaptive, and having a queue of updated vectors to avoid considering the same pair several times. Another natural idea used in [MLB17] consists of strengthening the LLL-reduction to a BKZ-reduction with medium block-size, so as to decrease the length of the initial random vectors.

One particularly cute low-level trick proposed by Fitzpatrick et al. [FBB+15] consists of quickly rejecting pairs of vectors depending on the Hamming weight of the XOR of their bit signs. We shall re-use (a variant of) this trick in our implementation. This technique is in fact well known in the Nearest-Neighbor-Search (NNS) literature [Cha02], and sometimes referred to as SimHash.

The N^2 factor may also be improved to a sub-quadratic factor N^c, $1 < c < 2$ using advanced NNS data-structures [Laa15a,Laa15b,BDGL16]. While improving the exponential term, those techniques introduce extra hidden sub-exponential factors, and typically require more memory.[3] In practice these improvements remain substantial [MLB17]. Yet, as the new improvements presented in this paper are orthogonal, we leave it to the interested reader to consult this literature.

3   The SubSieve Algorithm and its Analysis

3.1   Approach

Our improvements rely on the remark that the output of the sieve contains much more information than a shortest vector of $\mathcal{L}$. Indeed, the analysis of [NV08, MV10] suggests that the outputted list contains the N shortest vector of the lattice, namely, all the vectors of the lattice of length less than $\sqrt{4/3} \cdot \text{gh}(\mathcal{L})$.

We proceed to exploit this extra information by solving SVP in a lattice of larger dimension. Let us choose an index d, and run the sieve in the projected sub-lattice $\mathcal{L}_d$, of dimension $n - d$. We obtain the list:

$$L := \text{Sieve}(\mathcal{L}_d) = \{\mathbf{x} \in \mathcal{L}_d \setminus \{\mathbf{0}\} \mid \|\mathbf{x}\| \leq \sqrt{4/3} \cdot \text{gh}(\mathcal{L}_d)\}. \tag{1}$$

---

[3] Becker *et al.* [BGJ15] proposed a way to not require extra memory, yet it may hide an extra polynomial factor on time.

Our hope is that the desired shortest non-zero vector $\mathbf{s}$ (of expected length $\mathrm{gh}(\mathcal{L})$) of the full lattice $\mathcal{L}$ projects to a vector contained in L, i.e. $\pi_d(\mathbf{s}) \in L$ or equivalently by Eq. (1), that $\|\pi_d(\mathbf{s})\| \leq \sqrt{4/3}\,\mathrm{gh}(\mathcal{L}_d)$. Because $\|\pi_d(\mathbf{s})\| \leq \|\mathbf{s}\| = \mathrm{gh}(\mathcal{L})$, it is sufficient that:

$$\mathrm{gh}(\mathcal{L}) \leq \sqrt{4/3} \cdot \mathrm{gh}(\mathcal{L}_d). \tag{2}$$

In fact, we may relax this condition, as we rather expect the projection to be shorter: $\|\pi_d(\mathbf{s})\| \approx \sqrt{(n-d)/n}\|\mathbf{s}\|$ assuming the direction of $\mathbf{s}$ is uniform and independent of the basis $\mathbf{B}$. More precisely, it will happen with constant probability that $\|\pi_d(\mathbf{s})\| \leq \sqrt{(n-d)/n}\|\mathbf{s}\|$. Instead we may therefore optimistically require:

$$\sqrt{\frac{n-d}{n}} \cdot \mathrm{gh}(\mathcal{L}) \leq \sqrt{4/3} \cdot \mathrm{gh}(\mathcal{L}_d). \tag{3}$$

We are now searching for a vector $\mathbf{s} \in L$ such that $\|\mathbf{s}\| \approx \mathrm{gh}(\mathcal{L})$, and such that $\mathbf{s}_d := \pi_d(\mathbf{s}) \in L$. By exhaustive search over the list L, let us assume we know $\mathbf{s}_d$; we now need to recover the full vector $\mathbf{s}$. We write $\mathbf{s} = \mathbf{B}\mathbf{x}$ and split $\mathbf{x} = (\mathbf{x}', \mathbf{x}'')$ where $\mathbf{x}' \in \mathbb{Z}^d$ and $\mathbf{x}'' \in \mathbb{Z}^{n-d}$. Note that $\mathbf{s}_d = \pi_d(\mathbf{B}\mathbf{x}) = \mathbf{B}_d\mathbf{x}''$, so we may recover $\mathbf{x}''$ from $\mathbf{s}_d$.

We are left with the problem of recovering $\mathbf{x}' \in \mathbb{Z}^d$ such that $\mathbf{B}'\mathbf{x}' + \mathbf{B}''\mathbf{x}''$ is small where $[\mathbf{B}'|\mathbf{B}''] = \mathbf{B}$, i.e., finding the short vector $\mathbf{s}$ in the lattice coset $\mathcal{L}(\mathbf{B}') - \mathbf{B}''\mathbf{x}$.

For appropriate parameters, this is an easy BDD instance over the d-dimensional lattice spanned by $\mathbf{B}'$. More precisely, a sufficient condition to solve this problem using Babai's Nearest-Plane algorithm [Bab86] is that $|\langle \mathbf{b}_i^*, \mathbf{s}\rangle| \leq \frac{1}{2}\|\mathbf{b}_i^*\|^2$ for all $i < d$. A sufficient condition is that:

$$\mathrm{gh}(\mathcal{L}) \leq \frac{1}{2} \min_{i<d} \|\mathbf{b}_i^*\|. \tag{4}$$

This conditions is far from tight, and in practice should not be a serious issue. Indeed, even for a strongly reduced basis, the d first Gram-Schmidt lengths won't be much smaller than $\mathrm{gh}(\mathcal{L})$, say by more than a factor 2. On the other hand assuming $\mathbf{s}$ has a random direction we expect $|\langle \mathbf{b}_i^*, \mathbf{s}\rangle| \leq \omega(\ln n)/\sqrt{n} \cdot \|\mathbf{b}_i^*\| \cdot \|\mathbf{s}\|$ except with super-polynomially small probability. We will check this condition in the complexity analysis below (Sect. 3.2), and will simply ignore it in the rest of this paper.

Algorithm 2. SubSieve($\mathcal{L}, d$)

Require: The basis $\mathbf{B} = [\mathbf{B}'|\mathbf{B}'']$ of a lattice $\mathcal{L}$ of dimension n
Ensure: A short vector of $\mathcal{L}$
    $L \leftarrow \mathsf{Sieve}(\mathcal{L}_d)$
    for each $\mathbf{w}_i \in L$ **do**
        Compute $\mathbf{x}_i''$ such that $\mathbf{B}_d \cdot \mathbf{x}_i'' = \mathbf{w}_i$
        $\mathbf{t}_i = \mathbf{B}'' \cdot \mathbf{x}''$
        $\mathbf{s}_i \leftarrow \mathsf{Babai}(\mathbf{B}', \mathbf{t}_i) + \mathbf{t}_i$
    end for
    return the shortest $\mathbf{s}_i$

Heuristic Claim 1. *For a random lattice, and under conditions* (2) *and* (4), SubSieve($\mathcal{L}, d$) *outputs the shortest vector of* $\mathcal{L}$, *and its complexity is dominated by the cost* $N^2 \cdot \text{poly}(n)$ *of* Sieve($\mathcal{L}_d$), *with an additive overhead of* $n^2 \cdot N$ *real arithmetic operations.*

We note that the success of our approach depends crucially on the length of the Gram-Schmidt norms $\|\mathbf{b}_i^*\|$ (indeed for a fixed d, gh($\mathcal{L}_d$) depends only of $\prod_{i \geq d} \|\mathbf{b}_i^*\|$). In the following Sect. 3.2, we will argue that our approach can be successfully instantiated with $d = \Theta(n/\ln n)$ using an appropriate pre-processing of negligible cost.

3.2 Complexity Analysis

Assume that our lattice $\mathcal{L}$ has volume 1 (without loss of generality by scaling), and that its given basis $\mathbf{B}$ is BKZ-b reduced. Using the Geometric Series Assumption (Definition 4) we calculate the volume of $\mathcal{L}_d$:

$$\text{Vol}(\mathcal{L}_d) = \prod_{i=d}^{n-1} \|\mathbf{b}_i^*\| = \prod_{i=d}^{n-1} \alpha_b^{\frac{n-1}{2}-i} = \alpha_b^{d(d-n)/2}.$$

Recalling that for a k-dimensional lattice we have gh($\mathcal{L}$) $\approx \text{Vol}(\mathcal{L})^{1/k}\sqrt{k/(2\pi e)}$, condition (2) is rewritten to

$$\sqrt{\frac{n}{2\pi e}} \leq \sqrt{\frac{4}{3}} \cdot \sqrt{\frac{n-d}{2\pi e}} \cdot \alpha_b^{-d/2}.$$

Taking logarithms, we rewrite the above condition as

$$d \ln \alpha_b \leq \ln(4/3) + \ln(1 - d/n).$$

We (arbitrarily) choose $b = n/2$ which ensures that the cost of the BKZ-preprocessing is negligible compared to the cost of sieving in dimension $n - o(n)$. Unrolling the definitions, we notice that $\ln \alpha_b = \Theta((\ln b)/b) = \Theta((\ln n)/n)$. We conclude that condition (2) is satisfied for some $d = \Theta(n/\ln n)$.

The second condition (4) for the correctness of Babai lifting is easily satisfied: for $i < d = o(n)$ we have $\|\mathbf{b}_i^*\| = \text{gh}(b)^{(n-o(n))/b} = \text{gh}(b)^{2-o(1)} = n^{1-o(1)}$, while gh($n$) $= \Theta(n^{1/2})$. This concludes our argument of the following claim.

Heuristic Claim 2. *Having preprocessed the basis* $\mathbf{B}$ *of* $\mathcal{L}$ *with the BKZ algorithm with blocksize* $b = n/2$—*for a cost of at most* poly(n) *time the cost of* Sieve *in dimension* $n/2$—*our* SubSieve($\mathcal{L}, d$) *algorithm will find the shortest vector of* $\mathcal{L}$ *for some* $d = \Theta(n/\ln n)$.

In particular, SubSieve($\mathcal{L}, d$) *is faster than* Sieve($\mathcal{L}$) *by a sub-exponential factor* $2^{\Theta(n/\ln n)}$.

The fact that BKZ-b requires only poly(n) calls to an SVP oracle in dimension b is justified in [HPS11].

3.3   (Progressive) Iteration as Pre-processing

We now propose an alternative approach to provide pre-processing in our context. It consists of applying an extension of the SubSieve algorithm iteratively from a weakly reduced basis to a strongly reduced one. To proceed, we first need to slightly extend our algorithm, to not only provide one short vector, but a partial basis $\mathbf{V} = [\mathbf{v}_0 | \dots | \mathbf{v}_m]$ of rank m, such that their Gram-Schmidt lengths are as short as possible. In other words, the algorithm now attempts to provide the first vectors of an HKZ-reduced basis. For all practical purpose, $m = n/2$ is sufficiently large. This extension comes at a negligible additional cost of $O(n^3) \cdot N$ compared to the sieve of complexity $\text{poly}(n) \cdot N^2$.

Algorithm 3. SubSieve$^+(\mathcal{L}, d)$

Require: The basis $\mathbf{B} = [\mathbf{B}'|\mathbf{B}'']$ of a lattice $\mathcal{L}$ of dimension n
Ensure: A short vector of $\mathcal{L}$
    $L \leftarrow \text{Sieve}(\mathcal{L}_d)$
    for each $\mathbf{w}_i \in L$ **do**
       Compute $\mathbf{x}_i''$ such that $\mathbf{B}_d \cdot \mathbf{x}_i'' = \mathbf{w}_i$
       $\mathbf{t}_i = \mathbf{B}'' \cdot \mathbf{x}''$
       $\mathbf{s}_i \leftarrow \text{Babai}(\mathbf{B}', \mathbf{t}_i) + \mathbf{t}_i$
    end for
    for $j = 0 \dots n/2 - 1$ **do**
       Set $\mathbf{v}_j$ to be the $\mathbf{s}_i$ vector minimizing $\|\pi_{(\mathbf{v}_0 \dots \mathbf{v}_{j-1})^\perp}(\mathbf{s}_i)\|$ such that $\mathbf{s} \notin$ Span$(\mathbf{v}_0 \dots \mathbf{v}_{j-1})$
    end for
    return $(\mathbf{v}_0 \dots \mathbf{v}_{n/2-1})$

Then, the iteration consists of completing $\mathbf{V}$ into a basis of $\mathcal{L}$, and to use it as our new input basis $\mathbf{B}$.[4]

Additionally, as conditions (2) or even its optimistic variant (3) are not necessary conditions, we may hope that a larger value of d may probabilistically lead faster to the shortest vector. In fact, hoping to obtain the shortest vector with d larger than required by the pessimistic condition (2) can be interpreted in the pruning framework of [GNR10, Che13]; this will be discussed in Sect. 6.2.

For this work, we proceed with a simple strategy, namely we iterate starting with a large value of d (say $n/4$) and decrease d by 1 until the shortest vector (or a vector of the desired length) is found. This way, the failed attempts with too small d nevertheless contribute to the approximate HKZ-reduction, improving the basis for the next attempt.

---

[4] This can be done by applying LLL [LLL82] on the matrix $[\mathbf{V}|\mathbf{B}]$, which eliminates linear dependencies. As LLL can only decrease partial determinants, the volume of the first d-vectors after this process can only be smaller than the volume of $\mathbf{V}$: this does not affect condition (2) and (3).

The author admit to have no theoretical arguments (or even heuristic) to justify that this iterating approach should be more efficient than the preprocessing approach presented in Sect. 3.2. Yet, as we shall see, this method works quite well in practice, and has the advantage of being much simpler to implement.

Remark. One natural tweak is to also consider the vectors in $\mathbf{B}'$ when constructing the new partial basis $\mathbf{V}$ so as to ensure that the iteration never introduces a regression. Yet, as the optimistic condition is probabilistic, we may get stuck with an unlucky partial basis, and prefer to change it at each iteration. This is reminiscent of the rerandomization of the basis in the extreme pruning technique of Gama et al. [GNR10]. It is therefore not entirely clear if this tweak should be applied. In practice, we noted that applying this trick made the running time of the algorithm much more erratic, making it hard to determine if it should be better on average. For the sake of this initial study, we prefer to stick with the more stable version of the algorithm.

3.4 Tentative Prediction of d on Quasi-HKZ Reduced Basis

We now attempt to estimate the concrete maximal value d allowing our algorithm to succeed. We nevertheless warn the reader against strong conclusions on the concrete hardness of SVP from the analysis below. Indeed, it does not capture some practical phenomena, such as the fact that (1) is not strictly true in practice,[5] or more subtly that the directions of the vectors of $\mathbf{B}$ are not independent of the direction of the shortest vector $\mathbf{s}$ when $\mathbf{B}$ is so strongly reduced. Additionally, we identify in Sect. 6.2 avenues for improvements that could make this analysis obsolete.

We work under the heuristic assumption that the iterations up to $d_{\text{last}} - 1$ have almost produced an HKZ-reduced basis: $\|\mathbf{b}_i^*\| \approx \ell_i$ where ℓ_i follows the HKZ-shape of dimension n (Definition 3). From there, we determine whether the last iteration with $d = d_{\text{last}}$ should produce the shortest vector according to both the pessimistic and optimistic condition. For $i \ll n$ we use the first order approximation $\ln \ell_i \approx \ln \ell_0 - i \cdot \ln \ell_0 / \ell_1$ and obtain

$$\ln \ell_i \approx \ln \ell_0 - i \cdot \frac{\ln(n/2\pi)}{2n}.$$

The pessimistic condition (2) and the optimistic condition (3) respectively rewrite as:

$$\ln \ell_0 \leq \ln \sqrt{4/3} + \ln \ell_d \qquad \text{and} \qquad \ln \sqrt{\frac{n-d}{n}} + \ln \ell_0 \leq \ln \sqrt{4/3} + \ln \ell_d.$$

With a bit of rewriting, we arrive at the following maximal value of d respectively under the following pessimistic and optimistic conditions:

$$d \approx \frac{n \ln 4/3}{\ln(n/2\pi)} \qquad \text{and} \qquad d \approx \frac{n \ln 4/3}{\ln(n/2\pi e)}.$$

---

[5] Some vectors below the $\sqrt{4/3} \cdot \text{gh}(\mathcal{L}_d)$ bound may be missing, while other vectors above this bound may be included.

We can also numerically simulate more precisely the maximal value of d using the exact values of the ℓ_i. All four predictions are depicted on Fig. 1. Our plots start at dimension 50, the conventional cut-off for the validity of the Gaussian Heuristic [GN08, Che13]. We note that the approximated predictions are accurate, up to an additive term 2 over the value of d for relevant dimensions $n \leq 250$. We also note that in this range the dimension gain d looks very much linear: for all practical concerns, our improvement should appear essentially exponential.

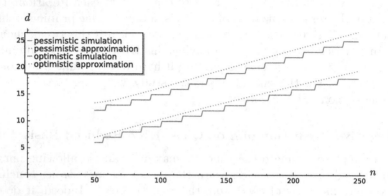

Fig. 1. Predictions of the maximal successful choice of d, under various methods and conditions.

4   Other Optimizations and Implementation Details

In this section, we describe a baseline sieve algorithm and two additional tricks to improve its practical efficiency. So as to later report the improvement brought by each trick and by our main contribution, we shall refer to 4 versions of our algorithm, activating one feature at the time:

- V0: GaussSieve baseline implementation
- V1: GaussSieve with XOR-POPCNT trick
- V2: GaussSieve with XOR-POPCNT trick and progressive sieving
- V3: Iterated SubSieve$^+$ with XOR-POPCNT trick and progressive sieving.

4.1   Baseline Implementation

As a baseline algorithm, we essentially use the Gauss-Sieve algorithm of [MV10], with the following tweaks.

First, we do not resort to Gaussian Sampling [Kle00] for the construction of the list L as the sphericity of the initial list does not seem so crucial in practice, and leads to starting the sieve with vectors longer than necessary. Instead, we choose vectors by sampling their $n/4$ last coordinates in base $\mathbf{B}$ uniformly in $\{0, \pm 1, \pm 2\}$, and choose the remaining coordinates deterministically using the Babai Nearest-Plane algorithm [Bab86].

Secondly, we do not maintain the list perfectly sorted, but only re-sort it periodically. This makes the implementation somewhat easier[6] and does not affect performances noticeably. Similarly, fresh random vectors are not inserted in L one by one, but in batches.

Thirdly, we use a hash table to prevent collisions: if $\mathbf{v} \pm \mathbf{w}$ is already in the list, then we cancel the reduction $\mathbf{v} \leftarrow \mathbf{v} \pm \mathbf{w}$. Our hash function is defined as random linear function $h : \mathbb{Z}^n \rightarrow \mathbb{Z}/2^{64}\mathbb{Z}$ tweaked so that $h(\mathbf{x}) = h(-\mathbf{x})$; hashing is fast, and false collisions should be very rare. This function is applied to the integer coordinates of the vector in base $\mathbf{B}$.

At last, the termination condition is as follows: the algorithm terminates when no pairs can be reduced, and when the ball of radius $\sqrt{4/3}\,\mathrm{gh}(\mathcal{L})$ is half-saturated according to the Gaussian Heuristic, $i.e.$ when the list L contains at least $\frac{1}{2}\sqrt{4/3}^n$ vectors of length less than $\sqrt{4/3}\,\mathrm{gh}(\mathcal{L})$.

At the implementation level, and contrary to most implementations of the literature, our implementation works by representing vectors in bases $\mathbf{B}$ and $\mathbf{B}^*$ rather than in the canonical basis of $\mathbb{R}^n$. It makes application of Babai's algorithm [Bab86] more idiomatic, and should be a crucial feature to use it as an SVP solver inside BKZ.

4.2   The XOR-POPCNT Trick (a.k.a. SimHash)

This trick—which can be traced back to [Cha02]—was developed for sieving in [FBB+15]. It consists of compressing vectors to a short binary representation that still carries some geometrical information: it allows for a quick approximation of inner-products. In more detail, they choose to represent a real vector $\mathbf{v} \in \mathbb{R}^n$ by the binary vector $\tilde{\mathbf{v}} \in \mathbb{Z}_2^n$ of it signs, and compute the Hamming weight $H = |\tilde{\mathbf{w}} \oplus \tilde{\mathbf{v}}|$ to determine whether $\langle \mathbf{v}, \mathbf{w} \rangle$ is expected to be small or large (which in turn informs us about the length $\|\mathbf{v} - \mathbf{w}\|^2 = \|\mathbf{v}\|^2 + \|\mathbf{w}\|^2 - 2\langle \mathbf{v}, \mathbf{w} \rangle$). If H is small enough then the exact length is computed, otherwise the pair is directly rejected.

This trick greatly decreases the practical computational cost and the memory bandwidth of the algorithm, in particular by exploiting the native POPCNT instruction available on most modern CPUs.

Following the original idea [Cha02], we use a generalized version of this trick, allowing the length of the compressed representation to differ from the lattice dimension. Indeed, we can for example choose $c \neq n$ vectors $\mathbf{r}_1, \ldots, \mathbf{r}_c$, and compress $\mathbf{v}$ as $\tilde{\mathbf{v}} \in \mathbb{Z}_2^c$ where $\tilde{v}_i = \mathrm{sign}(\langle \mathbf{v}, \mathbf{r}_i \rangle)$. This allows not only to align c to machine-word size, but also to tune the cost and the fidelity of this compressed representation.

In practice we choose $c = 128$ (2 machine words), and set the $\mathbf{r}_i$'s to be sparse random ternary vectors. We set the acceptance threshold to $|\tilde{\mathbf{w}} \oplus \tilde{\mathbf{v}}| < 47$,[7] having

---

[6] It avoids resorting to non-contiguous containers, following the nomenclature of c++ standard library.

[7] Of course, we also test whether $|\tilde{\mathbf{w}} \oplus \tilde{\mathbf{v}}| > 128 - 47$ in which case we attempt the reduction $\mathbf{v} \leftarrow \mathbf{v} + \mathbf{w}$ instead of $\mathbf{v} \leftarrow \mathbf{v} - \mathbf{w}$.

optimized this threshold by trial and error. Experimentally, the overall positive rate of this test is of about 2%, with a false negative rate of less than 30%. The sieve algorithm automatically compensates for false-negatives by increasing the list size.

4.3   Progressive Sieving

The trick described in this section was independently invented by Laarhoven and Mariano in [LM18]; and their work provides a much more thorough investigation of it. It consists of progressively increasing the dimension, first running the sieve in sublattices $\mathcal{L}_{[0,i]}$ for i increasing from (say) $n/2$ to n.[8]

It allows us to obtain an initial small pool of rather short vectors for a much cheaper cost. In turn, when we increase the dimension and insert new fresh vectors, the long fresh vectors get shorter noticeably faster thanks to this initial pool. We use the same terminating condition over $\mathcal{L}_{[0,i]}$ to decide when to increase i than the one described over the full lattice in Sect. 4.1.

4.4   Implementation Details

The core of the our Sieving implementation is written in c++ and the high level algorithm in python. It relies mostly on the fpylll [FPL16c] python wrapper for the fplll [FPL16b] library, used for calls to floating-point LLL [Ste10] and providing the Gram-Schmidt orthogonalization. Our code is not templated by the dimensions, doing so could improve the performance substantially by allowing the compiler to unroll and vectorize the inner-product loop.

Our implementation is open source, available at https://github.com/lducas/SubSieve.

5   Experiments and Performances

In this section, we report on the behavior in practice of our algorithm and the performances of our implementation. All experiments were ran on a single core (Intel Core i7-4790 @3.60 GHz).

For these experiments, we use the Darmstadt lattice challenges [SG10]. We make a first run of fplll's pruned enumeration (repeating it until 99% success probability) to determine the exact shortest vector.[9] Then, for our experiments, we stop our iteration of the SubSieve+ algorithm when it returns a vector of the same length.

---

[8] Note that unlike in our main algorithm SubSieve, the sublattices considered here are not projected sublattices, but simply the lattice spanned by the first basis vectors.

[9] Which is significantly harder than finding the approximation required by [SG10] to enter in the hall of fame.

5.1   The Dimension Gain d in Practice

In Fig. 2, we compare the experimental value of d to the predictions of Sect. 3.4. The area of each disc at position (n, d) is proportional the number of experiments that succeeded with $d_{last} = d$. We repeated the experiment 20 times for each dimension n.

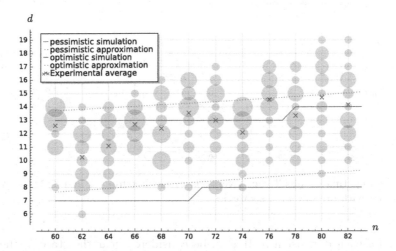

Fig. 2. Comparison between experimental value of d with the prediction of Sect. 3.4.

We note that the average d_{last} fits reasonably well with the simulated optimistic prediction. Also, in the worst case, it is never lower than the simulated pessimistic prediction, except for one outlier in dimension 62.

Remark. The apparent erratic behavior of the average for varying n is most likely due to the fact that our experiments are only randomized over the input basis, and not over the lattice itself. Indeed the actual length of the shortest vectors vary a bit around the Gaussian Heuristic, and it seems that the shorter it actually is, the easier it is to find with our algorithm.

5.2   Performances

We present in Fig. 3 the perfomances of the 4 versions of our implementation and of fplll's pruned enumeration with precomputed strategies [FPL16a].

Remark. In fplll, a strategy consists of the choice of a pre-processing blocksize b and of pruning parameters for the enumeration, as an attempt to reconstruct the BKZ 2.0 algorithm of Chen and Nguyen [CN11].

The external program Strategizer [FPL16a] first applies various descent techniques to optimize the pruning parameters, following the analysis of [GNR10, CN11, Che13], and iterates over all (reasonable) choices of b, to return the best

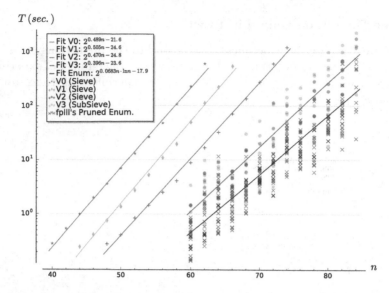

Fig. 3. Runing time T of all the 4 versions of sieving from 4 and `fplll`'s pruned enumeration with precomputed strategies.

strategy for each dimension n. It may be considered near the state of the art, at least for the dimensions at hand. Unfortunately, we are unaware of timing reports for exact-SVP in this range of dimensions for other implementations.

It would also be adequate to compare ourselves to the recent discrete-pruning techniques of Fukase and Kashiwabara [FK15, AN17], but again, we lack matching data. We note that neither the analysis of [AN17] nor the experiments of [TKH18] provide evidences that this new method is significantly more efficient than the method of [GNR10].

For a fair comparison with SubSieve, we stop repeating the pruned enumeration as soon as it finds the shortest vector, without imposing a minimal success probability (unlike the first run used to determine the of length shortest vectors). We also inform the enumerator of the exact length of that shortest vector, making its task somehow easier: without this information, it would enumerate at a larger radius.

As Algorithms V0, V1 and V2 have a rather deterministic running time depending only on the dimension, we only provide one sample. For V3 and enumeration, we provide 20 samples. To compute the fits, we first averaged the running times for each dimension n, and then computed the least-square linear fit of their logarithms (computing directly an exponential least-square fit leads to a fit only capturing the two last dimensions).

The given fits are only indicative and we warn against extrapolations. In particular, we note that the linear fit of V3 is below the heuristic asymptotic estimate of $(4/3)^{n+o(n)}$.

We conclude that our main contribution alone contributes a speed-up of more than an order of magnitude in the dimensions ≥ 70 (V3 versus V2), and that all the tricks taken together provide a speed-up of more than two orders of magnitudes (V3 versus V0). It performs within less than an order of magnitude of enumeration (V3 versus Pruned Enum).

5.3 Performance Comparison to the Literature

The literature on lattice sieving algorithms is vast [NV08, MV10, BGJ13, Laa15a, Laa15b, BDGL16, BLS16, HK17], and many papers do report implementation timings. We compare ourselves to four of them, namely a baseline implementation [MV10], and three advanced sieve implementations [FBB+15, MLB17, HK17], which represent (to the best of our knowledge) the state of the art in three different directions. This is given in Table 1.

Accounting for the CPU frequencies, we conclude that the implementation of our algorithm is more than 10 times faster than the current fastest sieve, namely the implementation of the Becker et al. algorithm [BDGL16] from Mariano et al. [MLB17].[10]

Remark. While we can hardly compare to this computation considering the lack of documentation, we note that T. Kleinjung holds the record for the shortest

Table 1. Comparison with other Sieve implementations.

Features	Algorithms							
	V0	V1	V2	V3	[MV10][a]	[FBB+15]	[MLB17]	[HK17]
XOR-POPCNT trick		x	x	x		x		
Progressive sieving			x	x				
SubSieve				x				
LSH (more mem.)							x	
tuple (less mem.)								x
Dimension	Running times							
$n = 60$	227 s	49s	8 s	.9 s	464 s	79 s	13 s	1080 s
$n = 70$	-	-	276 s	10 s	23933 s	4500 s	≈250 s [b]	33000 s
$n = 80$	-	-	-	234 s	-	-	4320 s	94700 s
CPU frequency (GHz)	3.6	3.6	3.6	3.6	4.0	4.0	2.3	2.3

[a] As reported by [FBB+15].
[b] This value is not given in [MLB17] as their implementation only handles dimensions that are multiples of 4. We estimated it from the given values for $n = 68$ (169 s) and $n = 72$ (418 s).

---

[10] The CPU frequency may not be the only property of the machines to take account of for a perfect comparison: memory access delay, memory bandwidth and cache sizes may have noticeable impacts.

vector found in Darmstadt Lattice challenge [SG10] of dimension 116 (seed 0), since May 2014, and reported having used *a* sieve algorithm. According to Herold and Kirshanova [HK17, Acknowledgments], the algorithm used by Kleinjung is similar to theirs.

Another Sieving record was achieved by Bos et al. [BNvdP14], for an ideal lattice of dimension 128, exploiting symmetries of ideal lattices to improve time and memory substantially. The computation ran over 1024 cores for 9 days. Similar computation have been run on GPU's [YKYC17], using 8 GPU's for about 35 days.

6   Conclusion

6.1   Sieve will Outperform Enumeration

While this statement is asymptotically true, it was a bit unclear where the cross-over should be, and therefore whether sieving algorithms have any practical relevance for concrete security levels. For example, it is argued in [MW16] that the cross-over would happen somewhere between $n = 745$ and $n = 1895$.

Our new results suggest otherwise. We do refrain from computing a cross-over dimension from the fits of Fig. 3 which are far from reliable enough for such an extrapolation; our prediction is of a different nature.

Our prediction is that—unless new enumerations techniques are discovered—further improvements of sieving techniques and implementations will outperform enumeration for exact-SVP in practice, for reachable dimensions, maybe even as low as $n = 90$. This, we believe, would constitute a landmark result. This prediction is backed by the following guesstimates, but also by the belief that fine-tuning, low-level optimizations and new ideas should further improve the state of the art. Some avenues for further improvements are discussed in Sect. 6.2.

Guesstimates. We can try to guesstimate how our improvements would combine with other techniques, in particular with List-Decoding Sieve [BDGL16]. The exact conclusion could be affected by many technical details, and is mostly meant to motivate further research and implementation effort.

Mariano et al. [MLB17] report a running time of $1850s$ for LDSieve [BDGL16] in dimension $n = 76$. First, the XOR-POPCNT trick is not orthogonal to LSH techniques, so we shall omit it.[11] The progressive sieving trick provides a speed up of about 4 in the relevant dimensions (V1 vs V2). Then, our main contribution offers 14 dimensions "for free", ($n = 90$, $d_{last} = 14$). More accurately, the iteration for increasing d would come at cost a factor $\sum_{i \geq 0} (\frac{3}{2})^{-i/2} \approx 5.5$. Overall we may expect to solve exact-SVP 90 in time $\approx 5.5 \cdot 1850/4 \approx 2500\,s$. In comparison, fpylll's implementation of BKZ 2.0 [CN11] solved exact-SVP in average time $2612\,s$ over Darmstadt lattice challenge 90 (seed 0) over 20 samples on our machine. For a fairer comparison across different machines, this Enumeration timing could be scaled up by $3.6\,\text{GHz}/2.3\,\text{GHz} \approx 1.5$.

---

[11] It could still be that, with proper tuning, combining them gives an extra speed-up.

6.2   Avenues for Further Improvements

Pruning in SubSieve. As we mentioned in Sect. 3.3, our optimistic condition (3) can be viewed as a form of pruning: this condition corresponds in the framework of [GNR10, Che13] to a pruning vector of the form $(1, 1, \ldots, 1, \gamma, \ldots \gamma) \in \mathbb{R}^n$ with d many 1's, and $\gamma = (n - d)/n$. A natural idea is to attempt running SubSieve using $\gamma < (n - d)/n$, *i.e.* being even more optimistic than condition (3). Indeed, rather than cluelessly increasing d at each iteration, we could compute for each d the success probability, and choose the value of d giving the optimal cost over success probability ratio.

Walking beyond $\sqrt{4/3}{\cdot}\mathbf{gh}(\mathcal{L_d})$. Noting $m = n - d$, another idea could consist of trying to get more vectors than the $\sqrt{4/3}^m$ shortest for a similar or slightly higher cost than the initial sieve, as this would allow d to increase a little bit. For example, we can extract the sublist A of all the vectors of length less than $\alpha \cdot \mathrm{gh}(\mathcal{L_d})$ where $\alpha \leq \sqrt{4/3}$ from the initial sieve, and use them to walk inside the ball of radius $\beta \cdot \mathrm{gh}(\mathcal{L_d}) \geq \sqrt{4/3}$ where $\frac{\alpha}{\beta}\sqrt{\beta^2 - \alpha^2/4} = 1$. Indeed, one can show that the volume of $(\mathbf{v} + \alpha\mathcal{B}) \cap (\beta\mathcal{B}) = \Omega(n^c)$ for some constant c, where $\|\mathbf{v}\| = \beta$. According to the Gaussian Heuristic, this means that from any lattice point in the ball of radius $\beta + \epsilon$, there exists a step in the list A that leads to another lattice point in the ball of radius $\beta + \epsilon$, for some $\epsilon = o(1)$. This kind of variation have already been considered in the Sieving literature [BGJ13, Laa16].

Each step of this walk would cost α^m and there are $\beta^{m+o(m)}$ many points to visit. Note that in our context, this walk can be done without extra memory, by instantly applying Babai lifting and keeping only interesting lifted vectors. We suspect that this approach could be beneficial in practice for $\beta = \sqrt{4/3} + o(1)$, if not for the running time, at least for the memory complexity.

Amortization Inside BKZ. We now consider two potential amortizations inside BKZ. Both ideas are not orthogonal to each others (yet may not be incompatible). If our SubSieve algorithm is to be used inside BKZ, we suggest fixing d_{last} (say, using the optimistic simulation), and to accept that we may not always solve SVP exactly; this is already the case when using pruned enumeration.

Already pre-processed. One notes that SubSieve$^+$ does more than ensure the shortness of the first vector, and in fact attempts a partial HKZ reduction. This means that the second block inside the BKZ loop is already quite reduced when we are over with the first one. One could therefore hope that directly starting the iteration of Sect. 3.3 at $d = d_{\text{last}}$ could be sufficient for the second block, and so forth.

Optimistically, this would lead to an amortization factor f of $f = \sum_{i \geq 0} (\frac{4}{3})^{-i} = 4$, or even $f = \sum_{i \geq 0} (\frac{3}{2})^{-i/2} \approx 5.5$ depending on which sieve is used. In practice, it may be preferable to start at $d = d_{\text{last}} - 1$ for example.
5 blocks for the price of 9/4. A second type of amortization consists of overshooting the blocksize by an additive term k, so as to SVP-reduce $k + 1$ consecutive blocks of dimension b for the price of one sieving in dimension $b + k$.

Indeed, an HKZ-reduction of size $b+k$ as attempted by SubSieve$^+$ directly guarentees the BKZ-b reduction of the first $k + 1$ blocks: we may jump directly by $k + 1$ blocks. This overshoot costs a factor $(3/2)^{k/2}$ using the List-Decoding-Sieve [BDGL16]. We therefore expect to gain a factor $f = (k + 1)/(3/2)^{k/2}$, which is maximal at $k = 4$, with $f = 20/9 \approx 2.2$.

Further, we note that the obtained basis could be better than a usual BKZ-b reduced basis, maybe even as good as a BKZ-$(b + \frac{k-1}{2})$ reduced basis. If so, the gain may be as large as $f' = (k + 1)/(3/2)^{(k+1)/4}$, which is maximal at $k = 9$, with $f' \approx 3.6$.

Acknowledgments. The author wishes to thank Koen de Boer, Gottfried Herold, Pierre Karman, Elena Kirshanova, Thijs Laarhoven, Marc Stevens and Eamonn Postlethwaite for enlightening conversations on this topic. The author is also extremely grateful to Martin Albrecht and the FPLLL development team for their thorough work on the fplll and fpylll libraries. This work was supported by a Veni Innovational Research Grant from NWO under project number 639.021.645.

References

[Ajt99] Ajtai, M.: Generating hard instances of the short basis problem. In: Wiedermann, J., van Emde Boas, P., Nielsen, M. (eds.) ICALP 1999. LNCS, vol. 1644, pp. 1–9. Springer, Heidelberg (1999). https://doi.org/10.1007/3-540-48523-6_1

[AKS01] Ajtai, M., Kumar, R., Sivakumar, D.: A sieve algorithm for the shortest lattice vector problem. In: 33rd Annual ACM Symposium on Theory of Computing, pp. 601–610. ACM Press, July 2001

[AN17] Aono, Y., Nguyen, P.Q.: Random sampling revisited: lattice enumeration with discrete pruning. In: Coron, J.-S., Nielsen, J.B. (eds.) EUROCRYPT 2017. LNCS, vol. 10211, pp. 65–102. Springer, Cham (2017). https://doi.org/10.1007/978-3-319-56614-6_3

[Bab86] Babai, L.: On Lovász' lattice reduction and the nearest lattice point problem. Combinatorica 6(1), 1–13 (1986)

[BDGL16] Becker, A., Ducas, L., Gama, N., Laarhoven, T.: New directions in nearest neighbor searching with applications to lattice sieving. In: Krauthgamer, R. (ed.) 27th Annual ACM-SIAM Symposium on Discrete Algorithms, pp. 10–24. ACM-SIAM, January 2016

[BGJ13] Becker, A., Gama, N., Joux, A.: Solving shortest and closest vector problems: The decomposition approach. Cryptology ePrint Archive, Report 2013/685 (2013). http://eprint.iacr.org/2013/685

[BGJ15] Becker, A., Gama, N., Joux, A.: Speeding-up lattice sieving without increasing the memory, using sub-quadratic nearest neighbor search. Cryptology ePrint Archive, Report 2015/522 (2015). http://eprint.iacr.org/2015/522

[BLS16] Bai, S., Laarhoven, T., Stehlé, D.: Tuple lattice sieving. LMS J. Comput. Math. 19(A), 146–162 (2016)

[BNvdP14] Bos, J.W., Naehrig, M., van de Pol, J.: Sieving for shortest vectors in ideal lattices: a practical perspective. Cryptology ePrint Archive, Report 2014/880 (2014). http://eprint.iacr.org/2014/880

[Cha02]   Charikar, M.: Similarity estimation techniques from rounding algorithms. In: 34th Annual ACM Symposium on Theory of Computing, pp. 380–388. ACM Press, May 2002

[Che13]   Chen, Y.: Réduction de réseau et sécurité concrète du chiffrement complètement homomorphe. Ph. D. thesis (2013)

[CN11]   Chen, Y., Nguyen, P.Q.: BKZ 2.0: better lattice security estimates. In: Lee, D.H., Wang, X. (eds.) ASIACRYPT 2011. LNCS, vol. 7073, pp. 1–20. Springer, Heidelberg (2011). https://doi.org/10.1007/978-3-642-25385-0_1

[FBB+15] Fitzpatrick, R., Bischof, C., Buchmann, J., Dagdelen, Ö., Göpfert, F., Mariano, A., Yang, B.-Y.: Tuning GaussSieve for speed. In: Aranha, D.F., Menezes, A. (eds.) LATINCRYPT 2014. LNCS, vol. 8895, pp. 288–305. Springer, Cham (2015). https://doi.org/10.1007/978-3-319-16295-9_16

[FK15]   Fukase, M., Kashiwabara, K.: An accelerated algorithm for solving svp based on statistical analysis. J. Inf. Process. 23(1), 67–80 (2015)

[FPL16a] FPLLL development team. Strategizer, an optimizer for pruned enumeration (2016). https://github.com/fplll/fpylll

[FPL16b] FPLLL development team. fplll, a lattice reduction library (2016). https://github.com/fplll/fplll

[FPL16c] FPLLL development team. fpylll, a python interface for fplll (2016). https://github.com/fplll/fpylll

[GN08]   Gama, N., Nguyen, P.Q.: Predicting lattice reduction. In: Smart, N. (ed.) EUROCRYPT 2008. LNCS, vol. 4965, pp. 31–51. Springer, Heidelberg (2008). https://doi.org/10.1007/978-3-540-78967-3_3

[GNR10]   Gama, N., Nguyen, P.Q., Regev, O.: Lattice enumeration using extreme pruning. In: Gilbert, H. (ed.) EUROCRYPT 2010. LNCS, vol. 6110, pp. 257–278. Springer, Heidelberg (2010). https://doi.org/10.1007/978-3-642-13190-5_13

[HK17]   Herold, G., Kirshanova, E.: Improved algorithms for the approximate k-list problem in euclidean norm. In: Fehr, S. (ed.) PKC 2017. LNCS, vol. 10174, pp. 16–40. Springer, Heidelberg (2017). https://doi.org/10.1007/978-3-662-54365-8_2

[HPS98]   Hoffstein, J., Pipher, J., Silverman, J.H.: NTRU: a ring-based public key cryptosystem. In: Buhler, J.P. (ed.) ANTS 1998. LNCS, vol. 1423, pp. 267–288. Springer, Heidelberg (1998). https://doi.org/10.1007/BFb0054868

[HPS11]   Hanrot, G., Pujol, X., Stehlé, D.: Analyzing blockwise lattice algorithms using dynamical systems. In: Rogaway, P. (ed.) CRYPTO 2011. LNCS, vol. 6841, pp. 447–464. Springer, Heidelberg (2011). https://doi.org/10.1007/978-3-642-22792-9_25

[HS07]   Hanrot, G., Stehlé, D.: Improved analysis of kannan's shortest lattice vector algorithm. In: Menezes, A. (ed.) CRYPTO 2007. LNCS, vol. 4622, pp. 170–186. Springer, Heidelberg (2007). https://doi.org/10.1007/978-3-540-74143-5_10

[Kan83]   Kannan, R.: Improved algorithms for integer programming and related lattice problems. In: 15th Annual ACM Symposium on Theory of Computing, pp. 193–206. ACM Press, April 1983

[Kle00]   Klein, P.N.: Finding the closest lattice vector when it's unusually close. In: Shmoys, D.B. (eds.) 11th Annual ACM-SIAM Symposium on Discrete Algorithms, pp. 937–941. ACM-SIAM, January 2000

[Laa15a] Laarhoven, T.: Search problems in cryptography (2015). http://thijs. com/docs/phd-final.pdf

[Laa15b] Laarhoven, T.: Sieving for shortest vectors in lattices using angular locality-sensitive hashing. In: Gennaro, R., Robshaw, M. (eds.) CRYPTO 2015. LNCS, vol. 9215, pp. 3–22. Springer, Heidelberg (2015). https://doi.org/10.1007/978-3-662-47989-6_1

[Laa16] Laarhoven, T.: Randomized lattice sieving for the closest vector problem (with preprocessing). Cryptology ePrint Archive, Report 2016/888 (2016). http://eprint.iacr.org/2016/888

[LLL82] Lenstra, A.K., Lenstra, H.W., Lovász, L.: Factoring polynomials with rational coefficients. Math. Ann. **261**(4), 515–534 (1982)

[LM18] Laarhoven, T., Mariano, A.: Progressive lattice sieving. In: PQcrypto 2018, Cryptology ePrint Archive, Report 2018/079 (2018, to appear). https://eprint.iacr.org/2018/079

[MLB17] Mariano, A., Laarhoven, T., Bischof, C.: A parallel variant of LDSIEVE for the SVP on lattices. In: 2017 25th Euromicro International Conference on Parallel, Distributed and Network-based Processing (PDP), pp. 23–30. IEEE (2017)

[MV10] Micciancio, D., Voulgaris, P.: Faster exponential time algorithms for the shortest vector problem. In: Charika, M. (ed.) 21st Annual ACM-SIAM Symposium on Discrete Algorithms, pp. 1468–1480. ACM-SIAM, January 2010

[MW15] Micciancio, D., Walter, M.: Fast lattice point enumeration with minimal overhead. In: Indyk, P. (ed.) 26th Annual ACM-SIAM Symposium on Discrete Algorithms, pp. 276–294. ACM-SIAM, January 2015

[MW16] Micciancio, D., Walter, M.: Practical, predictable lattice basis reduction. In: Fischlin, M., Coron, J.-S. (eds.) EUROCRYPT 2016. LNCS, vol. 9665, pp. 820–849. Springer, Heidelberg (2016). https://doi.org/ 10.1007/978-3-662-49890-3_31

[Ngu09] Nguyen, P.Q.: Hermite's constant and lattice algorithms. In: Nguyen, P., Vallée, B. (eds.) The LLL Algorithm. Information Security and Cryptography. Springer, Heidelberg (2009). https://doi.org/10.1007/ 978-3-642-02295-1_2

[NV08] Nguyen, P.Q., Vidick, T.: Sieve algorithms for the shortest vector problem are practical. J. Math. Cryptol. **2**(2), 181–207 (2008)

[Poh81] Pohst, M.: On the computation of lattice vectors of minimal length, successive minima and reduced bases with applications. ACM SIGSAM Bull. **15**(1), 37–44 (1981)

[Reg05] Regev, O.: On lattices, learning with errors, random linear codes, and cryptography. In: Gabow, H.N., Fagin, R. (eds.) 37th Annual ACM Symposium on Theory of Computing, pp. 84–93. ACM Press, May 2005

[SE94] Schnorr, C.-P., Euchner, M.: Lattice basis reduction: improved practical algorithms and solving subset sum problems. Math. Program. **66**(1–3), 181–199 (1994)

[SG10] Schneider, M., Gama, N.: SVP Challenge (2010). https:// latticechallenge.org/svp-challenge

[SH95] Schnorr, C.P., Hörner, H.H.: Attacking the chor-rivest cryptosystem by improved lattice reduction. In: Guillou, L.C., Quisquater, J.-J. (eds.) EUROCRYPT 1995. LNCS, vol. 921, pp. 1–12. Springer, Heidelberg (1995). https://doi.org/10.1007/3-540-49264-X_1

[Ste10] Stehlé, D.: Floating-point LLL: theoretical and practical aspects. In: Nguyen, P., Vallée, B. (eds.) The LLL Algorithm. Information Security and Cryptography, pp. 179–213. Springer, Heidelberg (2010). https://doi.org/10.1007/978-3-642-02295-1_5

[TKH18] Teruya, T., Kashiwabara, K., Hanaoka, G.: Fast lattice basis reduction suitable for massive parallelization and its application to the shortest vector problem. Cryptology ePrint Archive, Report 2018/044 (2018, to appear). https://eprint.iacr.org/2018/044

[YD16] Yu, Y., Ducas, L.: Second order statistical behavior of LLL and BKZ. In: Adams, C., Camenisch, J. (eds.) SAC 2017. LNCS, vol. 10719, pp. 3–22. Springer, Cham (2018). https://doi.org/10.1007/978-3-319-72565-9_1

[YKYC17] Yang, S.-Y., Kuo, P.-C., Yang, B.-Y., Cheng, C.-M.: Gauss Sieve Algorithm on GPUs. In: Handschuh, H. (ed.) CT-RSA 2017. LNCS, vol. 10159, pp. 39–57. Springer, Cham (2017). https://doi.org/10.1007/978-3-319-52153-4_3

On the Ring-LWE and Polynomial-LWE Problems

Miruna Rosca[1,2], Damien Stehlé[1], and Alexandre Wallet[1(✉)]

[1] ENS de Lyon, Laboratoire LIP (U. Lyon, CNRS, ENSL, INRIA, UCBL),
Lyon, France
damien.stehle@gmail.com, wallet.alexandre@gmail.com
[2] Bitdefender, Bucharest, Romania

Abstract. The Ring Learning With Errors problem (RLWE) comes in various forms. Vanilla RLWE is the decision dual-RLWE variant, consisting in distinguishing from uniform a distribution depending on a secret belonging to the dual $\mathcal{O}_K^\vee$ of the ring of integers $\mathcal{O}_K$ of a specified number field K. In primal-RLWE, the secret instead belongs to $\mathcal{O}_K$. Both decision dual-RLWE and primal-RLWE enjoy search counterparts. Also widely used is (search/decision) Polynomial Learning With Errors (PLWE), which is not defined using a ring of integers $\mathcal{O}_K$ of a number field K but a polynomial ring $\mathbb{Z}[x]/f$ for a monic irreducible $f \in \mathbb{Z}[x]$. We show that there exist reductions between all of these six problems that incur limited parameter losses. More precisely: we prove that the (decision/search) dual to primal reduction from Lyubashevsky *et al.* [EUROCRYPT 2010] and Peikert [SCN 2016] can be implemented with a small error rate growth for all rings (the resulting reduction is non-uniform polynomial time); we extend it to polynomial-time reductions between (decision/search) primal RLWE and PLWE that work for a family of polynomials f that is exponentially large as a function of $\deg f$ (the resulting reduction is also non-uniform polynomial time); and we exploit the recent technique from Peikert *et al.* [STOC 2017] to obtain a search to decision reduction for RLWE for arbitrary number fields. The reductions incur error rate increases that depend on intrinsic quantities related to K and f.

1 Introduction

DIFFERENT SHADES OF RLWE. Ring Learning With Errors (RLWE) was introduced by Lyubashevsky *et al.* in [LPR10], as a means of speeding up cryptographic constructions based on the Learning With Errors problem (LWE) [Reg05]. Let K be a number field, $\mathcal{O}_K$ its ring of integers and $q \geq 2$ a rational integer. The search variant of RLWE with parameters K and q consists in recovering a secret $s \in \mathcal{O}_K^\vee / q\mathcal{O}_K^\vee$ with $\mathcal{O}_K^\vee$ denoting the dual of $\mathcal{O}_K$, from arbitrarily many samples $(a_i, a_i \cdot s + e_i)$. Here each a_i is uniformly sampled in $\mathcal{O}_K / q\mathcal{O}_K$ and each e_i is a small random element of $K_{\mathbb{R}} := K \otimes_{\mathbb{Q}} \mathbb{R}$. The noise term e_i is sampled such that its Minkowski embedding vector follows a Gaussian distribution with a small

© International Association for Cryptologic Research 2018
J. B. Nielsen and V. Rijmen (Eds.): EUROCRYPT 2018, LNCS 10820, pp. 146–173, 2018.
https://doi.org/10.1007/978-3-319-78381-9_6

covariance matrix (relative to $q\mathcal{O}_K^\vee$). The decision variant consists in distinguishing arbitrarily many such pairs for a common s chosen uniformly in $\mathcal{O}_K^\vee/q\mathcal{O}_K^\vee$, from uniform samples in $\mathcal{O}_K/q\mathcal{O}_K \times K_\mathbb{R}/q\mathcal{O}_K^\vee$. More formal definitions are provided in Sect. 2, but these suffice for describing our contributions.

Lyubashevsky *et al.* backed in [LPR10] the conjectured hardness of RLWE with a quantum polynomial-time reduction from the (worst-case) Approximate Shortest Vector Problem (ApproxSVP) restricted to the class of Euclidean lattices corresponding to ideals of $\mathcal{O}_K$, with geometry inherited from the Minkowski embeddings. They showed its usefulness by describing a public-key encryption with quasi-optimal efficiency: the bit-sizes of the keys and the run-times of all involved algorithms are quasi-linear in the security parameter. A central technical contribution was a reduction from search RLWE to decision RLWE, when K is cyclotomic, and decision RLWE for cyclotomic fields is now pervasive in lattice-based cryptography, including in practice [ADPS16,BDK+18,DLL+18]. The search-to-decision reduction from [LPR10] was later extended to the case of general Galois rings in [EHL14,CLS17].

Prior to RLWE, Stehlé *et al.* [SSTX09] introduced what is now referred to as Polynomial Ring Learning With Errors (PLWE), for cyclotomic polynomials of degree a power of 2. PLWE is parametrized by a monic irreducible $f \in \mathbb{Z}[x]$ and an integer $q \geq 2$, and consists in recovering a secret $s \in \mathbb{Z}_q[x]/f$ from arbitrarily many samples $(a_i, a_i \cdot s + e_i)$ where each a_i is uniformly sampled in $\mathbb{Z}_q[x]/f$ and each e_i is a small random element of $\mathbb{R}[x]/f$. The decision variant consists in distinguishing arbitrarily many such samples for a common s sampled uniformly in $\mathbb{Z}_q[x]/f$, from uniform samples. Here the noise term e_i is sampled such that its coefficient vector follows a Gaussian distribution with a small covariance matrix. Stehlé *et al.* gave a reduction from the restriction of ApproxSVP to the class of lattices corresponding to ideals of $\mathbb{Z}[x]/f$, to search PLWE, for f a power-of-2 cyclotomic polynomial.

Finally, a variant of RLWE with $s \in \mathcal{O}_K/q\mathcal{O}_K$ rather than $\mathcal{O}_K^\vee/q\mathcal{O}_K^\vee$ was also considered (see, e.g., [DD12] among others), to avoid the complication of having to deal with the dual $\mathcal{O}_K^\vee$ of $\mathcal{O}_K$. In the rest of this paper, we will refer to the latter as primal-RLWE and to standard RLWE as dual-RLWE.

THE CASE OF CYCLOTOMICS. Even though [LPR10] defined RLWE for arbitrary number fields, the problem was mostly studied in the literature for K cyclotomic. This specialization had three justifications:

- it leads to very efficient cryptographic primitives, in particular if q totally splits over K;
- the hardness result from [LPR10] holds for cyclotomics;
- no particular weakness was known for these fields.

Among cyclotomics, those of order a power of 2 are a popular choice. In the case of a field K defined by the cyclotomic polynomial f, we have that $\mathcal{O}_K = \mathbb{Z}[\alpha]$ for α a root of f. Further, in the case of power-of-2 cyclotomics, mapping the coefficient vector of a polynomial in $\mathbb{Z}[x]/f$ to its Minkowski embedding is a scaled isometry. This makes primal-RLWE and PLWE collapse into a single problem. Still in the case of power-of-2 cyclotomics, the dual $\mathcal{O}_K^\vee$ is a scaling of $\mathcal{O}_K$, implying that dual and primal-RLWE are equivalent. Apart from the monogenicity property, these facts do not hold for all cyclotomics. Nevertheless, Ducas and Durmus [DD12] showed it is still possible to reduce dual-RLWE to primal-RLWE.

LOOKING AT OTHER FIELDS. The RLWE hardness proof holds with respect to a fixed field: the reduction in [LPR10] maps ApproxSVP for lattices corresponding to $\mathcal{O}_K$-ideals with small approximation factors, to decision/search dual-RLWE on K. Apart from the very specific case of field extensions [GHPS12], hardness on K seems unrelated to hardness on another field K'. One may then wonder if RLWE is easier for some fields. The attacks presented in [EHL14, ELOS15, CLS17, CLS16] were used to identify weak generating polynomials f of a number field K, but they only work for error distributions with small width relative to the geometry of the corresponding ring [CIV16b, CIV16a, Pei16]. At this occasion, the relationships between the RLWE and PLWE variants were more closely investigated.

Building upon [CGS14, CDPR16], Cramer *et al.* [CDW17] gave a quantum polynomial-time ApproxSVP algorithm for ideals of $\mathcal{O}_K$ when K is a cyclotomic field of prime-power conductor, when the ApproxSVP approximation factor is $2^{\tilde{O}(\sqrt{\deg K})}$. For general lattices, the best known algorithm [SE94] runs in time $2^{\tilde{O}(\sqrt{n})}$ for such an approximation factor, where n is the lattice dimension (here $n = \deg K$). We note that the result from [CGS14, CDPR16] was partly extended in [BBdV+17] to principal ideals generated by a short element in a completely different family of fields. These results show that all fields are not equal in terms of ApproxSVP hardness (unless they turn out to be all weak!). So far, there is no such result for RLWE.

On the constructive front, Bernstein *et al.* [BCLvV16] showed that some non-cyclotomic polynomials f also enjoy practical arithmetic over $\mathbb{Z}_q[x]/f$ and lead to efficient cryptographic design (though the concrete scheme relies on the presumed hardness of another problem than RLWE).

HEDGING AGAINST THE WEAK FIELD RISK. Two recent works propose complementary approaches to hedge against the risk of a weakness of RLWE for specific fields. First, in [PRS17], Peikert *et al.* give a new (quantum) reduction from ApproxSVP for $\mathcal{O}_K$-ideals to decision dual-RLWE for the corresponding field K. All fields support a (quantum) reduction from ApproxSVP, and hence, from this respect, one is not restricted to cyclotomics. Second, following an analogous result by Lyubashevsky for the Small Integer Solution problem [Lyu16], Roşca *et al.* [RSSS17] introduced the Middle-Product LWE problem and showed that it is at least as hard as PLWE for any f in an exponentially large family of f's (as a function of their degree). Neither result is fully satisfactory. In the first case, it could be that ApproxSVP is easy for lattices corresponding to ideals of $\mathcal{O}_K$ for any K: this would make the result vacuous. In the second case, the result of [RSSS17] focuses on PLWE rather than the more studied RLWE problem.

OUR RESULTS. The focus on the RLWE hardness for non-cyclotomic fields makes the discrepancies between the RLWE and PLWE variants more critical. In this article, we show that the six problems considered above — dual-RLWE, primal-RLWE and PLWE, all in both decision and search forms — reduce to one another in polynomial time with limited error rate increases, for huge classes of rings. More precisely, these reductions are obtained with the following three results.

- We show that for every field K, it is possible to implement the reduction from decision (resp. search) dual-RLWE to decision (resp. search) primal-RLWE from [LPR10, Le. 2.15] and [Pei16, Se. 2.3.2], with a limited error growth. Note that there exists a trivial converse reduction from primal-RLWE to dual-RLWE.
- We show that the reduction mentioned above can be extended to a reduction from decision (resp. search) primal-RLWE in K to decision (resp. search) PLWE for f, where K is the field generated by the polynomial f. The analysis is significantly more involved. It requires the introduction of the so-called conductor ideal, to handle the transformation from the ideal $\mathcal{O}_K$ to the order $\mathbb{Z}[x]/f$, and upper bounds on the condition number of the map that sends the coefficient embeddings to the Minkowski embeddings, to show that the noise increases are limited. Our conditioning upper bound is polynomial in n only for limited (but still huge) classes of polynomials that include those of the form $x^n + x \cdot P(x) - a$, with $\deg P < n/2$ and a prime that is $\geq 25 \cdot \|P\|_1^2$ and $\leq \mathrm{poly}(n)$. A trivial converse reduction goes through for the same f's.
- We exploit the recent technique from [PRS17] to obtain a search to decision reduction for dual-RLWE.

Concretely, the error rate increases are polynomial in $n = \deg K$, the root discriminant $|\Delta_K|^{1/n}$ and, for the reduction to PLWE, in the root algebraic norm $\mathcal{N}(\mathcal{C}_{\mathbb{Z}[\alpha]})^{1/n}$ of the conductor ideal $\mathcal{C}_{\mathbb{Z}[\alpha]}$ of $\mathbb{Z}[\alpha]$, where α is a root of f defining K. We note that in many cases of interest, all these quantities are polynomially bounded in n. To enjoy these limited error rate growths, the first two reductions require knowledge of specific data related to K, namely, a short element (with respect to the Minkowski embeddings) in the different ideal $(\mathcal{O}_K^\vee)^{-1}$ and a short element in $\mathcal{C}_{\mathbb{Z}[\alpha]}$. In general, these are hard to compute.

TECHNIQUES. The first reduction is derived from [LPR10, Le. 2.15] and [Pei16, Se. 2.3.2]: if it satisfies some arithmetic properties, a multiplication by an element $t \in \mathcal{O}_K$ induces an $\mathcal{O}_K$-module isomorphism from $\mathcal{O}_K^\vee/q\mathcal{O}_K^\vee$ to $\mathcal{O}_K/q\mathcal{O}_K$. For the reduction to be meaningful, we need t to have small Minkowski embeddings. We prove the existence of such a small t satisfying the appropriate arithmetic conditions, by generalizing the inclusion-exclusion technique developed in [SS13] to study the key generation algorithm of the NTRU signature scheme [HHPW10].

The Lyubashevsky et al. bijection works with $\mathcal{O}_K^\vee$ and $\mathcal{O}_K$ replaced by arbitrary ideals of K, but this does not provide a bijection from $\mathcal{O}_K/q\mathcal{O}_K$ to $\mathbb{Z}[\alpha]/q\mathbb{Z}[\alpha]$, as $\mathbb{Z}[\alpha]$ may only be an order of $\mathcal{O}_K$ (and not necessarily an ideal). We circumvent this difficulty by using the conductor ideal of $\mathbb{Z}[\alpha]$. Intuitively, the conductor ideal describes the relationship between $\mathcal{O}_K$ and $\mathbb{Z}[\alpha]$. As far as we are aware, this is the first time the conductor ideal is used in the RLWE context. This bijection and the existence of an appropriate multiplier t as above provide a (non-uniform) reduction from primal-RLWE to a variant of PLWE for which the noise terms have small Minkowski embeddings (instead of small polynomial coefficients).

We show that for many number fields, the linear map between polynomial coefficients and Minkowski embeddings has a condition number that is polynomially bounded in n, i.e., the map has bounded distortion and behaves not too noticeably differently from a scaling. This implies that the latter reduction is also a reduction from primal-RLWE to standard PLWE for these rings. We were able to show condition number bounds that are polynomial in n only for restricted families of polynomials f, yet exponentially large as n increases. These include in particular those of the form mentioned above. Note that the primality condition on the constant coefficient is used only to ensure that f is irreducible and hence defines a number field. For these f's, we use Rouché's theorem to prove that the roots are close to the scaled n-th roots of unity $(a^{1/n} \cdot \alpha_n^k)_{0 \leq k < n}$, and then that f "behaves" as $x^n - a$ in terms of geometric distortion.

Our search-to-decision reduction for dual-RLWE relies on techniques developed in [PRS17]. In that article, Peikert *et al.* consider the following 'oracle hidden center' problem (OHCP). In this problem, we are given access to an oracle $\mathcal{O}$ taking as inputs a vector $z \in \mathbb{R}^k$ and a scalar $t \in \mathbb{R}^{\geq 0}$, and outputting a bit. The probability that the oracle outputs 1 (over its internal randomness) is assumed to depend only on $\exp(t) \cdot \|z - x\|$, for some vector x. The goal is to recover $\mathcal{O}$'s center x. On the one hand, Peikert *et al.* give a polynomial-time algorithm for this problem, assuming the oracle is 'well-behaved' ([PRS17, Prop. 4.4]). On the other hand, they show how to map a Bounded Distance Decoding (BDD) instance to such an OHCP instance if they have access to Gaussian samples in the dual of the BDD lattice, where the engine of the oracle is the decision dual-RLWE oracle [PRS17, Se. 6.1]. We construct the OHCP instance from the decision RLWE oracle in a different manner. We use our input search dual-RLWE samples and take small Gaussian combinations of them. By re-randomizing the secret and adding some noise, we can obtain arbitrarily many dual-RLWE samples. Subtracting from the input samples well-chosen z_i's in $K_\mathbb{R}$ and setting the standard deviation of the Gaussian combination appropriately leads to a valid OHCP instance. The main technical hurdle is to show that a Gaussian combination of elements of $\mathcal{O}_K^\vee / q \mathcal{O}_K^\vee$ is close to uniform. For this, we generalize a ring Leftover Hash Lemma proved for specific pairs $(\mathcal{O}_K, q)$ in [SS11].

RELATED WORKS. The reductions studied in this work can be combined with those from ApproxSVP for $\mathcal{O}_K$-ideals to dual-RLWE [LPR10, PRS17]. Recently, Albrecht and Deo [AD17] built upon [BLP+13] to obtain a reduction from Module-LWE to RLWE. This can be both combined with our reductions and the quantum reductions from ApproxSVP for $\mathcal{O}_K$-modules to Module-LWE[1] [LS15, PRS17]. Downstream, the reductions can be combined with the

---

[1] The reduction from [LS15] is limited to cyclotomic fields, but [PRS17] readily extends to module lattices.

reduction from PLWE to Middle-Product LWE from [RSSS17]. The latter was showed to involve an error rate growth that is linearly bounded by the so-called the expansion factor of f: it turns out that those f's for which we could bound the condition number of the Minkowski map by a polynomial function of $\deg f$ also have polynomially bounded expansion factor. These reductions and those considered in the present work are pictorially described in Fig. 1.

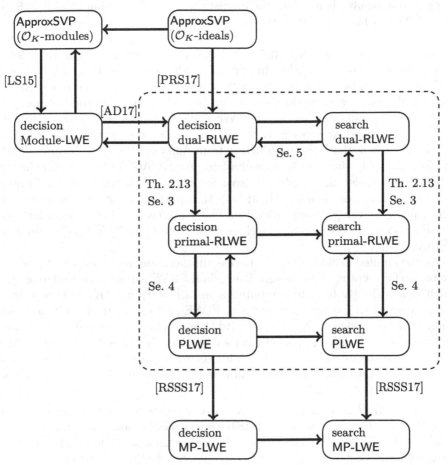

Fig. 1. Relationships between variants of RLWE and PLWE. The dotted box contains the problems studied in this work. Each arrow may hide a noise rate degradation (and module rank - modulus magnitude transfer in the case of [AD17]). The top to bottom arrows in the dotted box correspond to non-uniform reductions. The reductions involving PLWE are analyzed for limited family of defining polynomials. The arrows without references correspond to trivial reductions.

The ideal-changing scaling element t and the distortion of the Minkowski map were closely studied in [CIV16b, CIV16a, Pei16] for a few precise polynomials and fields. We use the same objects, but provide bounds that work for all (or many) fields.

IMPACT. As it is standard for the hardness foundations of lattice-based cryptography, our reductions *should not* be considered for setting practical parameters. They should rather be viewed as a strong evidence that the six problems under scope are essentially equivalent and do not suffer from a design flaw (unless they all do). We hope they will prove useful towards understanding the plausibility of weak fields for RLWE.

Our first result shows that there exists a way of reducing dual-RLWE to primal-RLWE while controlling the noise growth. Even though the reduction is non-uniform, it gives evidence that these problems are qualitatively equivalent. Our second result shows that RLWE and PLWE are essentially equivalent for a large class of polynomials/fields. In particular, the transformation map between the Minkowski embeddings and the coefficient embeddings has a bounded distortion. Finally, our search to decision fills an important gap. On the one hand, it precludes the possibility that search RLWE could be harder than decision RLWE. On the other hand, it gives further evidence of the decision RLWE hardness. In [PRS17], the authors give a reduction from ApproxSVP for $\mathcal{O}_K$-ideals to decision RLWE. But in the current state of affairs, ApproxSVP for this special class of lattices seems easier than RLWE, at least for some parameters. Indeed, Cramer *et al.* [CDW17] gave quantum algorithms that outperform generic lattice algorithms for some range of approximation factors in the context of ideal lattices. On the opposite, RLWE is qualitatively equivalent to ApproxSVP for $\mathcal{O}_K$-modules [LS15, AD17].

As the studied problems reduce to one another, one may then wonder which one to use for cryptographic design. Using dual-RLWE requires knowledge of $\mathcal{O}_K$, which is notoriously hard to compute for an arbitrary field K. This may look as an incentive to use the corresponding PLWE problem instead, as it does not require the knowledge of $\mathcal{O}_K$. Yet, for it to be useful in cryptographic design, one must be able to decode the noise from its representative modulo a scaled version of the lattice corresponding to $\mathbb{Z}[\alpha]$. This seems to require the knowledge of a good basis of that lattice, which may not be easy to obtain either, depending on the considered polynomial f.

NOTATIONS. If D is a distribution, we write $x \hookleftarrow D$ to say that we sample x from D. If D_1, D_2 are continuous distributions over the same measurable set Ω, we let $\Delta(D_1, D_2) = \int_\Omega |D_1(x) - D_2(x)| \mathrm{d}x$ denote their statistical distance. Similarly, we let $R(D_1\|D_2) = \int_\Omega D_1(x)^2/D_2(x)\mathrm{d}x$ denote their Rényi divergence. If E is a set of finite measure, we let $U(E)$ denote the uniform distribution over E. For a matrix $V = (v_{ij})$, we let $\|V\| = \sqrt{\sum_{1 \le i,j \le n} |v_{ij}|^2}$ denote its Frobenius norm.

This is the proceedings' version. The full version contains additional appendices and it is available on the IACR eprint archive.

2   Preliminaries

In this section, we give some background on algebraic number theory used in lattice-based cryptography, recall properties of Euclidean lattices, and state the precise definitions of the RLWE variants we will consider. More details on standard tools of algebraic number theory can be found in the full version. Useful references include [Ste17, Cona].

2.1 Some Algebraic Number Theory

RINGS AND IDEALS IN NUMBER FIELDS. In this article, we call any subring of K a number ring. For a number ring R, an (integral) R-ideal is an additive subgroup $I \subseteq R$ which is closed by multiplication in R, i.e., such that $IR = I$. A more compact definition is to say that I is an R-module. If $a_1, \ldots, a_k$ are elements in R, we let $\langle a_1, \ldots, a_k \rangle = a_1 R + \ldots + a_k R$ and call it the ideal generated by the a_i's. The product of two ideals I, J is the ideal generated by all elements xy with $x \in I$ and $y \in J$. The sum, product and intersection of two R-ideals are again R-ideals.

Two integral R-ideals I, J are said to be coprime if $I + J = R$, and, in this case, we have $I \cap J = IJ$. Any non-zero ideal in a number ring has finite index, i.e., the quotient ring R/I is always finite when I is a non-zero R-ideal. An R-ideal $\mathfrak{p}$ is said to be prime if whenever $\mathfrak{p} = IJ$ for some R-ideals I, J, then either $I = \mathfrak{p}$ or $J = \mathfrak{p}$. In a number ring, any prime ideal $\mathfrak{p}$ is maximal [Ste17, p. 19], i.e., R is the only R-ideal containing it. It also means that the quotient ring $R/\mathfrak{p}$ is a finite field. It is well-known that any $\mathcal{O}_K$-ideal admits a unique factorization into prime $\mathcal{O}_K$-ideals, i.e., it can be written $I = \mathfrak{p}_1^{e_1} \ldots \mathfrak{p}_k^{e_k}$ with all $\mathfrak{p}_i$'s distinct prime ideals. It fails to hold in general number rings and orders, but we describe later in Lemma 2.1 how the result can be extended in certain cases.

A fractional R-ideal I is an R-module such that $xI \subseteq R$ for some $x \in K^\times$. An integral ideal is a fractional ideal, and so are the sum, the product and the intersection of two fractional ideals. A fractional R-ideal I is said to be invertible if there exists a fractional R-ideal J such that $IJ = R$. In this case, the (unique) inverse is the integral ideal $I^{-1} = \{x \in K : xI \subseteq R\}$. Any $\mathcal{O}_K$-ideal is invertible, but it is again false for a general number ring.

The algebraic norm of a non-zero integral R-ideal I is defined as $\mathcal{N}_R(I) = |R/I|$, and we will omit the subscript when $R = \mathcal{O}_K$. It satisfies $\mathcal{N}_R(IJ) = \mathcal{N}_R(I)\mathcal{N}_R(J)$ for every R-ideals I, J.

The dual of a fractional R-ideal I is $I^\vee = \{\alpha \in K : \mathrm{Tr}(\alpha I) \subseteq \mathbb{Z}\}$, which is also a fractional R-ideal. We always have $II^\vee = R^\vee$, so that $I^\vee = I^{-1}R^\vee$ when I is invertible. We also have $I^{\vee\vee} = I$ for any R-ideal I.

A particularly interesting dual is $\mathcal{O}_K^\vee$, whose inverse $(\mathcal{O}_K^\vee)^{-1}$ is called the different ideal. The different ideal is an integral ideal, whose norm $\Delta_K = \mathcal{N}((\mathcal{O}_K^\vee)^{-1})$ is called the discriminant of the number field. We note that, for every f defining K, the field discriminant Δ_K is a factor of the discriminant of f. The latter is denoted Δ_f and is defined as $\Delta_f = \prod_{i \neq j}(\alpha_i - \alpha_j)$, where $\alpha_1, \ldots, \alpha_n$ are the roots of f. This provides an upper bound on Δ_K in terms of the defining polynomial f.

ORDERS IN NUMBER FIELDS. An order $\mathcal{O}$ in K is a number ring which is a finite index subring of $\mathcal{O}_K$. In particular, the ring of integers $\mathcal{O}_K$ is the maximal order in K. Number rings such as $\mathbb{Z}[\alpha]$, with α a root of a defining polynomial f, are of particular interest. The conductor of an order $\mathcal{O}$ is defined as the set $C_{\mathcal{O}} = \{x \in K : x\mathcal{O}_K \subseteq \mathcal{O}\}$. It is contained in $\mathcal{O}$, and it is both an $\mathcal{O}$-ideal and an $\mathcal{O}_K$-ideal: it is in fact the largest ideal with this property. It is never empty, as it contains the index $[\mathcal{O}_K : \mathcal{O}]$.

If it is coprime with the conductor, an ideal in $\mathcal{O}_K$ can be naturally considered as an ideal in $\mathcal{O}$, and reciprocally. This is made precise in the following lemma.

Lemma 2.1 ([Cona, **Th. 3.8**]). *Let $\mathcal{O}$ be an order in K.*

1. *Let I be an $\mathcal{O}_K$-ideal coprime to $C_{\mathcal{O}}$. Then $I \cap \mathcal{O}$ is an $\mathcal{O}$-ideal coprime to $C_{\mathcal{O}}$ and the natural map $\mathcal{O}/I \cap \mathcal{O} \longrightarrow \mathcal{O}_K/I$ is a ring isomorphism.*
2. *Let J be an $\mathcal{O}$-ideal coprime to $C_{\mathcal{O}}$. Then $J\mathcal{O}_K$ is an $\mathcal{O}_K$-ideal coprime to $C_{\mathcal{O}}$ and the natural map $\mathcal{O}/J \longrightarrow \mathcal{O}_K/J\mathcal{O}_K$ is a ring isomorphism.*
3. *The set of $\mathcal{O}_K$-ideals coprime to $C_{\mathcal{O}}$ and the set of $\mathcal{O}$-ideals coprime to $C_{\mathcal{O}}$ are in multiplicative bijection by $I \longmapsto I \cap \mathcal{O}$ and $J \longmapsto J\mathcal{O}_K$.*

The above description does not tell how to "invert" the isomorphisms. This can be done by a combination of the following lemmas and passing through the conductor, as we will show in the next section.

Lemma 2.2. *Let $\mathcal{O}$ be an order in K and I an $\mathcal{O}_K$-ideal coprime to the conductor $C_{\mathcal{O}}$. Then the inclusions $C_{\mathcal{O}} \subseteq \mathcal{O}$ and $C_{\mathcal{O}} \subseteq \mathcal{O}_K$ induce isomorphisms $C_{\mathcal{O}}/I \cap C_{\mathcal{O}} \simeq \mathcal{O}/I \cap \mathcal{O}$ and $C_{\mathcal{O}}/I \cap C_{\mathcal{O}} \simeq \mathcal{O}_K/I$.*

Proof. By assumption we have $C_{\mathcal{O}} + I = \mathcal{O}_K$, so that the homomorphism $C_{\mathcal{O}} \to \mathcal{O}_K/I$ is surjective. By Lemma 2.1, the set $I \cap \mathcal{O}$ is an $\mathcal{O}$-ideal coprime to $C_{\mathcal{O}}$ so that $C_{\mathcal{O}} + I \cap \mathcal{O} = \mathcal{O}$. This implies that the homomorphism $C_{\mathcal{O}} \to \mathcal{O}/I \cap \mathcal{O}$ is surjective too. Both homomorphisms have kernel $I \cap C_{\mathcal{O}}$. $\square$

Lemma 2.3 ([Cona, **Cor. 3.10**]). *Let $\mathcal{O}$ be an order in K and $\beta \in \mathcal{O}$ such that $\beta\mathcal{O}_K$ is coprime to $C_{\mathcal{O}}$. Then $\beta\mathcal{O}_K \cap \mathcal{O} = \beta\mathcal{O}$.*

QUOTIENTS OF IDEALS. We will use the following result.

Lemma 2.4 ([LPR10, **Le. 2.14**]). *Let I and J two $\mathcal{O}_K$-ideals. Let $t \in I$ such that the ideals $t \cdot I^{-1}$ and J are coprime and let $\mathcal{M}$ be any fractional $\mathcal{O}_K$-ideal. Then the function $\theta_t : \mathcal{M} \to \mathcal{M}$ defined as $\theta_t(x) = t \cdot x$ induces an $\mathcal{O}_K$-module isomorphism from $\mathcal{M}/J\mathcal{M}$ to $I\mathcal{M}/IJ\mathcal{M}$.*

The authors of [LPR10] also gave an explicit way to obtain a suitable t by solving a set of conditions stemming from the Chinese Remainder Theorem. However, this construction does not give good control on the magnitudes of the Minkowski embeddings of t.

2.2 Lattices

For the remainder of this article, a lattice is defined as a full-rank discrete additive subgroup of an $\mathbb{R}$-vector space V which is a Cartesian power H^m (for $m \geq 1$) of $H := \{x \in \mathbb{R}^{s_1} \times \mathbb{C}^{2s_2} : \forall i \leq s_2 : x_{s_1+s_2+i} = \overline{x_{s_1+i}}\}$. This space H is sometimes called the "canonical" space. A given lattice $\mathcal{L}$ can be thought as the set of $\mathbb{Z}$-linear combinations $(b_i)_i$ of some linearly independent vectors of V. These vectors are said to form a lattice basis, and we define the lattice determinant as $\det \mathcal{L} = (\det(\langle b_i, b_j \rangle)_{i,j})^{1/2}$ (it does not depend on the choice of the basis of $\mathcal{L}$). For $v \in V$, let $\|v\| = (\sum_{i \leq \dim V} |v_i|^2)^{1/2}$ denote the standard Hermitian norm on V and $\|v\|_\infty = \max_{i \leq \dim V} |v_i|$ denote the infinity norm. The minimum $\lambda_1(\mathcal{L})$ is the Hermitian norm of a shortest non-zero element in $\mathcal{L}$. We define $\lambda_1^\infty(\mathcal{L})$ similarly. If $\mathcal{L}$ is a lattice, then we define its dual as $\mathcal{L}^* = \{y \in V : y^T \mathcal{L} \subseteq \mathbb{Z}\}$.

IDEAL LATTICES. While it is possible to associate lattices with fractional ideals of a number ring, we will not need it. Any fractional $\mathcal{O}_K$-ideal I is a free $\mathbb{Z}$-module of rank $n = \deg(K)$, i.e., it can be written as $\mathbb{Z}u_1 + \cdots + \mathbb{Z}u_n$ for some u_i's in K. Its canonical embedding $\sigma(I)$ is a lattice of dimension n in the $\mathbb{R}$-vector space $H \subseteq \mathbb{R}^{s_1} \times \mathbb{C}^{2s_2}$. Such a lattice is called an ideal lattice (for $\mathcal{O}_K$). For the sake of readability, we will abuse notations and often identify I and $\sigma(I)$. It is possible to look at the coefficient embedding of such lattices as well, but we will not need it in this work. The lattice corresponding to $I^\vee$ is $\overline{I}^*$. The discriminant of K satisfies $\Delta_K = (\det \mathcal{O}_K)^2$. In the following lemma, the upper bounds follow from Minkowski's theorem whereas the lower bounds are a consequence of the algebraic structure underlying ideal lattices.

Lemma 2.5 (Adapted from [PR07, Se. 6.1]). *Let K be a number field of degree n. For any fractional $\mathcal{O}_K$-ideal I, we have:*

$$\sqrt{n} \cdot \mathcal{N}(I)^{1/n} \leq \lambda_1(I) \leq \sqrt{n} \cdot (\mathcal{N}(I)\sqrt{\Delta_K})^{1/n},$$
$$\mathcal{N}(I)^{1/n} \leq \lambda_1^\infty(I) \leq (\mathcal{N}(I)\sqrt{\Delta_K})^{1/n}.$$

GAUSSIANS. It is standard practice in the RLWE setting to consider Gaussian distributions with diagonal covariance matrices. In this work, we will be interested in the behavior of samples after linear transformations that are not necessarily diagonal. As the resulting covariance matrix may not be diagonal, we adopt a more general framework. Let $\Sigma \succ 0$, i.e., a symmetric positive definite matrix. We define the Gaussian function on $\mathbb{R}^n$ of covariance matrix Σ as $\rho_\Sigma(x) := \exp(-\pi \cdot x^T \Sigma^{-1} x)$ for every vector $x \in \mathbb{R}^n$. The Gaussian distribution D_Σ is the probability distribution whose density is proportional to ρ_Σ. When $\Sigma = \mathrm{diag}(r_i^2)_i$ for some $r \in \mathbb{R}^n$, we write ρ_r and D_r, respectively.

Let $(e_i)_{i \leq n}$ be the canonical basis of $\mathbb{C}^n$. We define $h_i = e_i$ for $i \leq s_1$, and $h_{s_1+i} = (e_{s_1+i} + e_{s_1+s_2+i})/\sqrt{2}$ and $h_{s_1+s_2+i} = (e_{s_1+i} - e_{s_1+s_2+i})/\sqrt{-2}$ for $i \leq s_2$. The h_i's form an orthonormal $\mathbb{R}$-basis of H. We define the Gaussian distribution D_Σ^H as the distribution obtained by sampling $x \hookleftarrow D_\Sigma$ and returning $\sum_i x_i h_i$. We will repeatedly use the observation that if x is sampled from D_Σ^H and t belongs to $K_\mathbb{R}$, then $t \cdot x$ is distributed as $D_{\Sigma'}^H$ with $\Sigma' = \mathrm{diag}(|\sigma_i(t)|) \cdot \Sigma \cdot \mathrm{diag}(|\sigma_i(t)|)$.

For a lattice $\mathcal{L}$ over $V = H^m$ (for some $m \geq 1$) and a coset $\boldsymbol{c} \in V/\mathcal{L}$, we let $D_{\mathcal{L}+\boldsymbol{c},r}$ denote the discretization of $D_{r\mathbf{I}}^H$ over $\mathcal{L} + \boldsymbol{c}$ (we omit the subscript for $D_{\mathcal{L}+\boldsymbol{c},r}$ as all our lattices are over Cartesian powers of H). For $\varepsilon > 0$, we define the smoothing parameter $\eta_\varepsilon(\mathcal{L})$ as the smallest $r > 0$ such that $\rho_{(1/r)\mathbf{I}}(\mathcal{L}^* \setminus \boldsymbol{0}) \leq \varepsilon$. We have the following upper bounds.

Lemma 2.6 ([MR04, Le. 3.3]). *For any lattice $\mathcal{L}$ over H^m and $\varepsilon \in (0,1)$, we have $\eta_\varepsilon(\mathcal{L}) \leq \sqrt{\log(2mn(1+1/\varepsilon))/\pi}/\lambda_1^\infty(\mathcal{L}^*)$.*

Lemma 2.7 (Adapted from [PR07, Le. 6.5]). *For any $\mathcal{O}_K$-ideal I and $\varepsilon \in (0,1)$, we have $\eta_\varepsilon(I) \leq \sqrt{\log(2n(1+1/\varepsilon))/(\pi n)} \cdot (\mathcal{N}(I)\Delta_K)^{1/n}$.*

The following are standard applications of the smoothing parameter.

Lemma 2.8 ([GPV08, Cor. 2.8]). *Let $\mathcal{L}' \subseteq \mathcal{L}$ be full-rank lattices, $\varepsilon \in (0, 1/2)$ and $r \geq \eta_\varepsilon(\mathcal{L}')$. Then $\Delta(D_{\mathcal{L},r} \bmod \mathcal{L}', U(\mathcal{L}/\mathcal{L}')) \leq 2\varepsilon$.*

Lemma 2.9 ([PR06, Le. 2.11]). *Let $\mathcal{L}$ be an n-dimensional lattice, $\varepsilon \in (0,1/3)$ and $r \geq 4\eta_\varepsilon(\mathcal{L})$. Then $D_{\mathcal{L},r}(\boldsymbol{0}) \leq 2^{-2n+1}$.*

Lemma 2.10 (Adapted from [MR04, Le. 4.4]). *Let $\mathcal{L}$ be an n-dimensional lattice, $\varepsilon \in (0,1/3)$ and $r \geq \eta_\varepsilon(\mathcal{L})$. Then $\mathrm{Pr}_{\boldsymbol{x} \hookleftarrow D_{\mathcal{L},r}}[\|\boldsymbol{x}\| \geq 2r\sqrt{n}] \leq 2^{-2n}$.*

2.3   Computational Problems

We now formally define the computational problems we will study.

Definition 2.11 (RLWE and PLWE distributions). *Let K a degree n number field defined by f, $\mathcal{O}_K$ its ring of integers, $\Sigma \succ 0$ and $q \geq 2$.*

For $s \in \mathcal{O}_K^\vee/q\mathcal{O}_K^\vee$, we define the dual-RLWE distribution $\mathcal{A}_{s,\Sigma}^\vee$ as the distribution over $\mathcal{O}_K/q\mathcal{O}_K \times K_\mathbb{R}/q\mathcal{O}_K^\vee$ obtained by sampling $a \hookleftarrow U(\mathcal{O}_K/q\mathcal{O}_K)$, $e \hookleftarrow D_\Sigma^H$ and returning the pair $(a, a \cdot s + e)$.

For $s \in \mathcal{O}_K/q\mathcal{O}_K$, we define the primal-RLWE distribution $\mathcal{A}_{s,\Sigma}$ as the distribution over $\mathcal{O}_K/q\mathcal{O}_K \times K_\mathbb{R}/q\mathcal{O}_K$ obtained by sampling $a \hookleftarrow U(\mathcal{O}_K/q\mathcal{O}_K)$, $e \hookleftarrow D_\Sigma^H$ and returning the pair $(a, a \cdot s + e)$.

For $s \in \mathbb{Z}_q[x]/f$, we define the PLWE distribution $\mathcal{B}_{s,\Sigma}$ as the distribution over $\mathbb{Z}_q[x]/f \times \mathbb{R}_q[x]/f$ obtained by sampling $a \hookleftarrow U(\mathbb{Z}_q[x]/f)$, $e \hookleftarrow D_\Sigma$ and returning the pair $(a, a \cdot s + e)$ (with $\mathbb{R}_q = \mathbb{R}/q\mathbb{Z}$).

In the definition above, we identified the support H of D_Σ^H with $K_\mathbb{R}$, and the support $\mathbb{R}^n$ of D_Σ with $\mathbb{R}[x]/f$. Note that sampling from $\mathcal{A}_{s,\Sigma}^\vee$ and $\mathcal{A}_{s,\Sigma}$ seems to require the knowledge of a basis of $\mathcal{O}_K$. It is not known to be computable in polynomial-time from a defining polynomial f of an arbitrary K. In this article, we assume that a basis of $\mathcal{O}_K$ is known.

Definition 2.12 (The RLWE and PLWE problems). *We use the same nota-tions as above. Further, we let $\mathcal{E}_\succ$ be a subset of $\Sigma \succ 0$ and $D_\succ$ be a distribution over $\Sigma \succ 0$.*

Search dual-RLWE$_{q, \mathcal{E}_\succ}$ (resp. primal-RLWE and PLWE) consists in finding s from a sampler from $\mathcal{A}^\vee_{s, \Sigma}$ (resp. $\mathcal{A}_{s,\Sigma}$ and $\mathcal{B}_{s,\Sigma}$), where $s \in \mathcal{O}^\vee_K / q\mathcal{O}^\vee_K$ (resp. $s \in \mathcal{O}_K / q\mathcal{O}_K$ and $s \in \mathbb{Z}_q[x]/f$) and $\Sigma \in \mathcal{E}_\succ$ are arbitrary.

Decision dual-RLWE$_{q, D_\succ}$ (resp. primal-RLWE and PLWE) consists in distin-guishing between a sampler from $\mathcal{A}^\vee_{s, \Sigma}$ (resp. $\mathcal{A}_{s,\Sigma}$ and $\mathcal{B}_{s,\Sigma}$) and a uniform sampler over $\mathcal{O}_K / q\mathcal{O}_K \times K_\mathbb{R}/q\mathcal{O}^\vee_K$ (resp. $\mathcal{O}_K / q\mathcal{O}_K \times K_\mathbb{R}/q\mathcal{O}_K$ and $\mathbb{Z}_q[x]/f \times \mathbb{R}_q[x]/f$), with non-negligible probability over $s \hookleftarrow \mathcal{O}^\vee_K / q\mathcal{O}^\vee_K$ (resp. $s \in \mathcal{O}_K / q\mathcal{O}_K$ and $s \in \mathbb{Z}_q[x]/f$) and $\Sigma \hookleftarrow D_\succ$.

The problems above are in fact defined for sequences of number fields of growing degrees n such that the bit-size of the problem description grows at most polynomially in n. The run-times, success probabilities and distinguishing advantages of the algorithms solving the problems are considered asymptotically as functions of n.

The following reduction from dual-RLWE to primal-RLWE is a consequence of Lemma 2.4. A proof is given in the full version.

Theorem 2.13 (Adapted from [Pei16, Se. 2.3.2]). *Let $\Sigma \succ 0$ and $s \in \mathcal{O}^\vee_K / q\mathcal{O}^\vee_K$. Let $t \in (\mathcal{O}^\vee_K)^{-1}$ such that $t(\mathcal{O}^\vee_K) + q\mathcal{O}_K = \mathcal{O}_K$. Then the map $(a, b) \mapsto (a, t \cdot b)$ transforms $\mathcal{A}^\vee_{s, \Sigma}$ to $\mathcal{A}_{t \cdot s, \Sigma'}$ and $U(\mathcal{O}_K / q\mathcal{O}_K \times K_\mathbb{R}/q\mathcal{O}^\vee_K)$ into $U(\mathcal{O}_K / q\mathcal{O}_K \times K_\mathbb{R}/q\mathcal{O}_K)$, with $\Sigma' = \operatorname{diag}(|\sigma_i(t)|) \cdot \Sigma \cdot \operatorname{diag}(|\sigma_i(t)|)$. The natural inclusion $\mathcal{O}_K \to \mathcal{O}^\vee_K$ induces a map that transforms $U(\mathcal{O}_K / q\mathcal{O}_K \times K_\mathbb{R}/q\mathcal{O}_K)$ to $U(\mathcal{O}_K / q\mathcal{O}_K \times K_\mathbb{R}/q\mathcal{O}^\vee_K)$, and $\mathcal{A}_{s, \Sigma}$ to $\mathcal{A}^\vee_{s, \Sigma}$.*

We will consider variants of the decision problems for which the distinguishing must occur for all $s \in \mathcal{O}^\vee_K / q\mathcal{O}^\vee_K$ (resp. $s \in \mathcal{O}_K / q\mathcal{O}_K$ and $s \in \mathbb{Z}_q[x]/f$) and all $\Sigma \succ 0$ rather than with non-negligible probability over s. We call this variant worst-case decision dual-RLWE (resp. primal-RLWE and PLWE). Under some conditions on $D_\succ$ and $\mathcal{E}_\succ$, these variants are computationally equivalent.

Lemma 2.14 (Adapted from [LPR10, Se. 5.2]). *We use the same nota-tions as above. If $\Pr_{\Sigma \hookleftarrow D_\succ}[\Sigma \notin \mathcal{E}_\succ] \leq 2^{-n}$, then decision dual-RLWE$_{q, D_\succ}$ (resp. primal-RLWE and PLWE) reduces to worst-case decision dual-RLWE$_{q, \mathcal{E}_\succ}$ (resp. primal-RLWE and PLWE).*

Assume further that $D_\succ$ can be sampled from in polynomial-time. If $\max_{\Sigma \in \mathcal{E}_\succ} R(D_\succ \| D_\succ + \Sigma) \leq \operatorname{poly}(n)$, then worst-case decision dual-RLWE$_{q, \mathcal{E}_\succ}$ (resp. primal-RLWE and PLWE) reduces to decision dual-RLWE$_{q, D_\succ}$ (resp. primal-RLWE and PLWE).

Note that it is permissible to use the Rényi divergence here even though we are considering decision problems. Indeed, the argument is applied to the random choice of the noise distribution and not to the distinguishing advantage. The same argument has been previously used in [LPR10, Se. 5.2].

Proof. The first statement is direct. We prove the second statement only for dual-RLWE, as the proofs for primal-RLWE and PLWE are direct adaptations. Assume we are given a sampler that outputs (a_i, b_i) with $a_i \hookleftarrow U(\mathcal{O}_K/q\mathcal{O}_K)$ and b_i either uniform in $K_\mathbb{R}/q\mathcal{O}_K^\vee$ or of the form $b_i = a_i s + e_i$ with $s \in \mathcal{O}_K^\vee/q\mathcal{O}_K^\vee$ and $e_i \hookleftarrow D_\Sigma^H$. The reduction proceeds by sampling $s' \hookleftarrow U(\mathcal{O}_K^\vee/q\mathcal{O}_K^\vee)$ and $\Sigma' \hookleftarrow D_\succ$, and mapping all input (a_i, b_i)'s to $(a_i', b_i') = (a_i, b_i + a_i s' + e_i')$ with $e_i' \hookleftarrow D_{\Sigma'}^H$. This transformation maps the uniform distribution to itself, and $\mathcal{A}_{s,\Sigma}^\vee$ to $\mathcal{A}_{s+s',\Sigma''}^\vee$ with $\Sigma_{ij}'' = \Sigma_{ij} + \Sigma_{ij}'$ for all i, j. If the success probability (success being enjoying a non-negligible distinguishing advantage) over the error parameter sampled from $D_\succ$ is non-negligible, then so is it for the error parameter sampled $D_\succ + \Sigma$, as, by assumption, the Rényi divergence $R(D_\succ \| D_\succ + \Sigma)$ is polynomially bounded. $\qquad\qquad\square$

Many choices of $D_\succ$ and $\mathcal{E}_\succ$ satisfy the conditions of Lemma 2.14. The following is inspired from [LPR10, Se. 5.2]. We define the distribution $\mathcal{E}_\succ$ as follows, for an arbitrary r: Let $s_{ij} = r^2(1 + nx_{ij})$ for all $i > j$, $s_{ii} = r^2(1 + n^3 x_{ii})$ for all i and $s_{ij} = s_{ji}$ for all $i < j$, where the x_{ij}'s are independent samples from the $\Gamma(2, 1)$ distribution (of density function $x \mapsto x \exp(-x)$); the output matrix is $(s_{ij})_{ij}$. Note that it is symmetric and strictly diagonally dominant (and hence $\succ 0$) with probability $1 - 2^{-\Omega(n)}$. Then the set of all $\Sigma \succ 0$ with coefficients of magnitudes $\leq r^2 n^4$ satisfies the first condition of Lemma 2.14, and the set of all $\Sigma \succ 0$ with coefficients of magnitudes $\leq r^2$ satisfies the second condition of Lemma 2.14. We can hence switch from one variant to the other while incurring an error rate increase that is $\leq \mathrm{poly}(n)$.

3   Controlling Noise Growth in Dual to Primal Reduction

The reduction of Theorem 2.13 is built upon the existence of t as in Lemma 2.4. While this existence is guaranteed constructively by [LPR10], the size is not controlled by the construction. Another t that satisfies the conditions is $t = f'(\alpha)$, where f' is the derivative of f defining $K = \mathbb{Q}[\alpha]$. Indeed, from [Conb, Rem. 4.5], we know that $f'(\alpha) \in (\mathcal{O}_K^\vee)^{-1}$. However, the noise growth incurred by multiplication by $f'(\alpha)$ may be rather large in general: we have $N(f'(\alpha)) = \Delta_f = [\mathcal{O}_K : \mathbb{Z}[\alpha]]^2 \cdot \mathcal{N}((\mathcal{O}_K^\vee)^{-1})$.

In this section, we give a probabilistic proof that adequate t's with controlled size can be found by Gaussian sampling.

Let I and J be integral ideals of $\mathcal{O}_K$. Theorem 3.1 below states that a Gaussian sample t in I is such that $t \cdot I^{-1} + J = \mathcal{O}_K$ with non-negligible probability. The main technical hurdle is to show that the sample is not trapped in IJ' with J' a non-trivial factor of J. We handle this probability in different ways depending on the algebraic norm of J', extending an idea used in [SS13, Se. 4].

- For small-norm factors J' of J, the Gaussian folded modulo IJ' is essentially uniform over I/IJ', by Lemma 2.8. This requires the standard deviation parameter s to be above the smoothing parameter of IJ'. We use the smoothing parameter bound from Lemma 2.7.

- For large-norm factors J', we argue that the non-zero points of IJ' are very unlikely to be hit, thanks to the Gaussian tail bound given in Lemma 2.10 and the fact that the lattice minimum of IJ' is large, by Lemma 2.5.
- For middle-norm factors J', neither of the arguments above applies. Instead, we bound the probability that t belongs to IJ' by the probability that t belongs to IJ'', where J'' is a non-trivial factor of J', and use the first argument above. The factor J'' must be significantly denser than J' so that we have smoothing. But it should also be significantly sparser than $\mathcal{O}_K$ so that the upper bound is not too large.

Setting the standard deviation parameter of the discrete Gaussian so that at least one of the three arguments above applies is non-trivial. In particular, this highly depends on how the ideal J factors into primes (whether the pieces are numerous, balanced, unbalanced, etc.). The choice we make below works in all cases while still providing a reasonably readable proof and still being sufficient for our needs, from an asymptotic perspective. In many cases, better choices can be made. If J is prime, we can take a very small s and use only the second argument. If all factors of J are small, there is good enough 'granularity' in the factorization to use the third argument, and again s can be chosen very small.

Theorem 3.1. *Let I and J be integral $\mathcal{O}_K$-ideals, and write $J = \mathfrak{p}_1^{e_1} \ldots \mathfrak{p}_k^{e_k}$ for some prime ideals $\mathfrak{p}_i$. We sort the $\mathfrak{p}_i$'s by non-decreasing algebraic norms. Assume that we can take $\delta \in [\frac{4n + \log_2 \Delta_K}{\log_2 \mathcal{N}(J)}, 1]$.[2] We define:*

$$
s = \begin{cases} \left(\mathcal{N}(J)^{1/2} \mathcal{N}(I) \Delta_K \right)^{1/n} & \text{if } \mathcal{N}(\mathfrak{p}_k) \geq \mathcal{N}(J)^{1/2+\delta}, \\ \left(\mathcal{N}(J)^{1/2+2\delta} \mathcal{N}(I) \Delta_K \right)^{1/n} & \text{else.} \end{cases}
$$

Then we have

$$
\Pr_{t \hookleftarrow D_{I,s}} [tI^{-1} + J = \mathcal{O}_K] \geq 1 - \frac{k}{\mathcal{N}(\mathfrak{p}_1)} - 2^{-n+4}.
$$

Proof. We bound the probability P of the negation, from above. We have

$$
P = \Pr_{t \hookleftarrow D_{I,s}} [t \in \bigcup_{i \in [k]} I\mathfrak{p}_i] = \sum_{S \subseteq [k], S \neq \emptyset} (-1)^{|S|+1} \cdot \Pr_{t \hookleftarrow D_{I,s}} [t \in I \cdot \prod_{i \in S} \mathfrak{p}_i].
$$

We rewrite it as $P = P_1 + P_2$ with

$$
P_1 = \sum_{S \subseteq [k], S \neq \emptyset} (-1)^{|S|+1} \frac{1}{\prod_{i \in S} \mathcal{N}(\mathfrak{p}_i)} = 1 - \prod_{i \in [k]} \left(1 - \frac{1}{\mathcal{N}(\mathfrak{p}_i)} \right),
$$

$$
P_2 = \sum_{S \subseteq [k], S \neq \emptyset} (-1)^{|S|+1} \left(\Pr_{t \hookleftarrow D_{I,s}} [t \in I \cdot \prod_{i \in S} \mathfrak{p}_i] - \prod_{i \in S} \frac{1}{\mathcal{N}(\mathfrak{p}_i)} \right).
$$

---

[2] The parameter δ should be thought as near 0. It can actually be chosen such if $\mathcal{N}(J)$ is sufficiently large.

We have $P_1 \leq 1 - (1 - 1/\mathcal{N}(\mathfrak{p}_1))^k \leq k/\mathcal{N}(\mathfrak{p}_1)$. Our task is now to bound P_2.

Assume first that $\mathcal{N}(\mathfrak{p}_k) \geq \mathcal{N}(J)^{1/2+\delta}$. This implies that $\prod_{i \in S} \mathcal{N}(\mathfrak{p}_i) \leq \mathcal{N}(J)^{1/2-\delta}$ for all $S \subseteq [k]$ not containing k. By Lemma 2.7, we have $s \geq \eta_\varepsilon(I \prod_{i \in S} \mathfrak{p}_i)$ for all such S's, with $\varepsilon = 2^{-2n}$. We "smooth" out those ideals, i.e., we use Lemma 2.8 to obtain, for all $S \subseteq [k] \setminus \{k\}$:

$$\left| \Pr_{t \leftarrow D_{I,s}} [t \in I \cdot \prod_{i \in S} \mathfrak{p}_i] - \prod_{i \in S} \frac{1}{\mathcal{N}(\mathfrak{p}_i)} \right| \leq 2\varepsilon.$$

Now if S is a subset containing k, then we have $\mathcal{N}(\prod_{i \in S} \mathfrak{p}_i) \geq \mathcal{N}(J)^{1/2+\delta}$. By Lemma 2.5, we have $\lambda_1(I \prod_{i \in S} \mathfrak{p}_i) \geq \sqrt{n} \cdot \mathcal{N}(I)^{1/n} \mathcal{N}(J)^{(1/2+\delta)/n}$. On the other hand, by Lemma 2.10, we have $\Pr_{t \leftarrow D_{I,s}}[\|t\| \geq 2s\sqrt{n}] \leq 2^{-2n}$. Thanks to our choice of s, the assumption on δ and Lemma 2.9, we obtain

$$\Pr_{t \leftarrow D_{I,s}} [t \in I \prod_{i \in S} \mathfrak{p}_i] \leq \Pr_{t \leftarrow D_{I,s}} [t = 0] + 2^{-2n} \leq 2^{-2n+2}.$$

This allows us to bound P_2 as follows:

$$P_2 \leq 2^k \cdot \left(\varepsilon + 2^{-2n+2} + \mathcal{N}(J)^{-(1/2+\delta)} \right).$$

By assumption on δ, we have $\mathcal{N}(J) \geq 2^{2n}$ and $P_2 \leq 2^{-n+3}$. This completes the proof for the large $\mathcal{N}(\mathfrak{p}_k)$ case.

Now, assume that $\mathcal{N}(\mathfrak{p}_k) < \mathcal{N}(J)^{1/2+2\delta}$. Then, as above, the definition of s implies that, for any $S \subseteq [k]$ with $\mathcal{N}(\prod_{i \in S} \mathfrak{p}_i) \leq \mathcal{N}(J)^{1/2+\delta}$, we have $|\Pr[t \in I \prod_{i \in S} \mathfrak{p}_i] - 1/\prod_{i \in S} \mathcal{N}(\mathfrak{p}_i)| \leq 2^{-2n+1}$. Also as above, if we have $\mathcal{N}(\prod_{i \in S} \mathfrak{p}_i) \geq \mathcal{N}(J)^{1/2+3\delta}$, then $\lambda_1(I \prod_{i \in S} \mathfrak{p}_i)$ is too large for a non-zero element of $I \prod_{i \in S} \mathfrak{p}_i$ to be hit with significant probability. Assume finally that

$$\mathcal{N}(J)^{1/2+2\delta} \leq \mathcal{N}(\prod_{i \in S} \mathfrak{p}_i) \leq \mathcal{N}(J)^{1/2+3\delta}.$$

As $\mathcal{N}(\mathfrak{p}_k) < \mathcal{N}(J)^{1/2+\delta}$, there exists $S' \subseteq S$ such that

$$\mathcal{N}(J)^{\delta} \leq \mathcal{N}(\prod_{i \in S'} \mathfrak{p}_i) \leq \mathcal{N}(J)^{1/2+2\delta}.$$

By inclusion, we have that $\Pr[t \in I \prod_{i \in S} \mathfrak{p}_i] \leq \Pr[t \in I \prod_{i \in S'} \mathfrak{p}_i]$. Now, as the norm of $\prod_{i \in S'} \mathfrak{p}_i$ is small enough, we can use the smoothing argument above to claim that

$$\Pr_{t \leftarrow D_{I,s}} [t \in I \prod_{i \in S'} \mathfrak{p}_i] \leq 2^{-2n+1} + \frac{1}{\mathcal{N}(\prod_{i \in S'} \mathfrak{p}_i)} \leq 2^{-2n+1} + \frac{1}{\mathcal{N}(J)^{\delta}}.$$

By assumption on δ, the latter is $\leq 2^{-n+2}$. Collecting terms allows to complete the proof. $\qquad\square$

The next corollary shows that the needed t can be found with non-negligible probability.

Corollary 3.2. *Let I be an integral $\mathcal{O}_K$-ideal. Let $q \geq \max(2n, 2^{16} \cdot \Delta_K^{8/n})$ be a prime rational integer and $\mathfrak{p}_k$ a prime factor of $q\mathcal{O}_K$ with largest norm. We define:*

$$s = \begin{cases} q^{1/2} \cdot (\mathcal{N}(I)\Delta_K)^{1/n} & \text{if } \mathcal{N}(\mathfrak{p}_k) \geq q^{(5/8)\cdot n}, \\ q^{3/4} \cdot (\mathcal{N}(I)\Delta_K)^{1/n} & \text{else.} \end{cases}$$

Then, for sufficiently large n, we have

$$\Pr_{t \hookleftarrow D_{I,s}} [tI^{-1} + q\mathcal{O}_K = \mathcal{O}_K] \geq 1/2.$$

Proof. The result follows from applying Theorem 3.1 with $J = q\mathcal{O}_K$ and $\delta = 1/8$. The first lower bound on q ensures that $k/\mathcal{N}(\mathfrak{p}_1) \leq 1/2$, where $k \leq n$ denotes the number of prime factors of $q\mathcal{O}_K$ and $\mathfrak{p}_1$ denotes a factor with smallest algebraic norm. The second lower bound on q ensures that we can indeed set $\delta = 1/8$. □

We insist again on the fact that the required lower bounds on s can be much improved under specific assumptions on the factorization of q. For example, one could choose a q such that all the factors of $q\mathcal{O}_K$ have large norms, by sampling q randomly and checking its primality and the factorization of the defining polynomial f modulo q. In that case, the factors $q^{1/2}$ and $q^{3/4}$ can be decreased drastically.

We note that if the noise increase incurred by a reduction from an LWE-type problem to another is bounded as $n_1^c \cdot q_2^c$ for some $c_1 < 1$ and some $c_2 > 0$, then one may set the working modulus q so that the starting LWE problem has a sufficient amount of noise to not be trivially easy to solve, and the ending LWE problem has not enough noise to be information-theoretically impossible to solve (else the reduction would be vacuous). Indeed, it suffices to set q sufficiently larger than $n^{c_1/(1-c_2)}$.

4   From Primal-RLWE to PLWE

The goal of this section is to describe a reduction from primal-RLWE to PLWE. As an intermediate step, we first consider a reduction from primal-RLWE to a variant PLWE$^\sigma$ of PLWE where the noise is small with respect to the Minkowski embedding rather than the coefficient embedding. Then, we assess the noise distortion when looking at its Minkowski embedding versus its coefficient embedding.

If $K = \mathbb{Q}[x]/f$ for some $f = \prod_{j \leq n}(x - \alpha_j)$, the associated Vandermonde matrix V_f has jth row $(1, \alpha_j, \ldots, \alpha_j^{n-1})$ and corresponds to the linear map between the coefficient and Minkowski embedding spaces. Thus a good approximation of the distortion is given by the condition number $\text{Cond}(V_f) = s_n/s_1$, where the s_i's refer to the largest/smallest singular values of V_f. As we also have $\text{Cond}(V_f) = \|V_f\| \cdot \|V_f^{-1}\|$, these matrix norms also quantify how much V_f distorts the space. For a restricted, yet exponentially large, family of polynomials defining number fields, we show that both $\|V_f\|$ and $\|V_f^{-1}\|$ are polynomially bounded.

To do this, we start from $f_{n,a} = x^n - a$ whose distortion is easily computable. Then we add a "small perturbation" to this polynomial. Intuitively, the roots of the resulting polynomial should not move much, so that the norms of the "perturbed" Vandermonde matrices should be essentially the same. We formalize this intuition in Sect. 4.2 and locate the roots of the perturbed polynomial using Rouché's theorem.

Mapping a sample of PLWE$^\sigma$ to a sample of the corresponding PLWE simply consists in changing the geometry of the noise distribution. A noise distribution with covariance matrix Σ in the Minkowski embedding corresponds to a noise distribution of covariance matrice $(V_f^{-1})^T \Sigma V_f^{-1}$ in the coefficient space. The converse is also true, replacing V_f^{-1} by V_f. Moreover, the noise growths incurred by the reductions remain limited whenever $\|V_f\|$ and $\|V_f^{-1}\|$ are small.

Overall, reductions between primal-RLWE to PLWE can be obtained by combining Theorems 4.2 and 4.7 below (with Lemma 2.14 to randomize the noise distributions).

4.1 Reducing Primal-RLWE to PLWE$^\sigma$

We keep the notations of the previous section, and let $\mathbb{Z}[x]/(f) = \mathcal{O}$.

Definition 4.1 (The PLWE$^\sigma$ problem). *Let also Σ be a positive definite matrix, and $q \geq 2$. For $s \in \mathcal{O}/q\mathcal{O}$, we define the PLWE$^\sigma$ distribution $\mathcal{B}_{s,\Sigma}^\sigma$ as the distribution over $\mathcal{O}/q\mathcal{O} \times K_\mathbb{R}/q\mathcal{O}$ obtained by sampling $a \hookleftarrow U(\mathcal{O}/q\mathcal{O})$, $e \hookleftarrow D_\Sigma^H$ and returning the pair $(a, a \cdot s + e)$*

Let $D_\succ$ be a distribution over $\Sigma \succ 0$. Decision PLWE$^\sigma$ consists in distinguishing between a sampler from $\mathcal{B}_{s,\Sigma}^\sigma$ and a uniform sampler over $\mathcal{O}/q\mathcal{O} \times K_\mathbb{R}/q\mathcal{O}$, with non-negligible probability over $s \hookleftarrow \mathcal{O}/q\mathcal{O}$ and $\Sigma \hookleftarrow D_\succ$.

Theorem 4.2. *Assume that $q\mathcal{O}_K + \mathcal{C}_\mathcal{O} = \mathcal{O}_K$. Let Σ be a positive definite matrix and $s \in \mathcal{O}_K/q\mathcal{O}_K$. Let $t \in \mathcal{C}_\mathcal{O}$ such that $t\mathcal{C}_\mathcal{O}^{-1} + q\mathcal{O}_K = \mathcal{O}_K$. Then the map $(a, b) \mapsto (t \cdot a, t^2 \cdot b)$ transforms $U(\mathcal{O}_K/q\mathcal{O}_K \times K_\mathbb{R}/q\mathcal{O}_K)$ to $U(\mathcal{O}/q\mathcal{O} \times K_\mathbb{R}/q\mathcal{O})$ and $\mathcal{A}_{s,\Sigma}$ to $\mathcal{B}_{t \cdot s, \Sigma'}^\sigma$, where the new covariance is $\Sigma' = \mathrm{diag}(|\sigma(t_i)|^2) \cdot \Sigma \cdot \mathrm{diag}(|\sigma_i(t)|^2)$.*

Let $\mathcal{B}_{s,\Sigma}^\sigma$ be a PLWE$^\sigma$ distribution. The natural inclusion $\mathcal{O} \to \mathcal{O}_K$ induces a map that transforms $U(\mathcal{O}/q\mathcal{O} \times K_\mathbb{R}/q\mathcal{O})$ to $U(\mathcal{O}_K/q\mathcal{O}_K \times K_\mathbb{R}/q\mathcal{O}_K)$ and $\mathcal{B}_{s,\Sigma}^\sigma$ to $\mathcal{A}_{s,\Sigma}$.

Proof. Let $(a, b = a \cdot s + e)$ be distributed as $\mathcal{A}_{s,\Sigma}$. Let $a' = t \cdot a$ and $b' = t^2 \cdot b = a' \cdot (t \cdot s) + e'$, with $e' = t^2 \cdot e$. Then a' is uniformly distributed in $\mathcal{C}_\mathcal{O}/q\mathcal{C}_\mathcal{O}$ by applying Lemma 2.4 for $I = \mathcal{C}_\mathcal{O}$, $J = q\mathcal{O}_K$ and $\mathcal{M} = \mathcal{O}_K$. It is also uniformly distributed in $\mathcal{O}/q\mathcal{O}$ by combining Lemmas 2.2 and 2.3. The noise follows the claimed distribution, see the observation in Sect. 2.2. The fact that $t \cdot s \in \mathcal{O}/q\mathcal{O}$ completes the proof that $\mathcal{A}_{s,\Sigma}$ is mapped to $\mathcal{B}_{t \cdot s, \Sigma'}^\sigma$.

Now, let (a, b) be uniform in $\mathcal{O}_K/q\mathcal{O}_K \times K_\mathbb{R}/q\mathcal{O}_K$. We already know that a' is uniformly distributed in $\mathcal{O}/q\mathcal{O}$. Let us now consider the distribution of b'. Thanks to the assumption on $q\mathcal{O}_K$, we also have $t^2\mathcal{C}_\mathcal{O}^{-1} + q\mathcal{O}_K = \mathcal{O}_K$. Therefore, by Lemma 2.4, multiplication by t^2 induces an isomorphism $\mathcal{O}_K/q\mathcal{O}_K \simeq \mathcal{C}_\mathcal{O}/q\mathcal{C}_\mathcal{O}$, and hence, by Lemmas 2.2 and 2.3, an isomorphism $\mathcal{O}_K/q\mathcal{O}_K \simeq \mathcal{O}/q\mathcal{O}$. This gives the first reduction.

We now turn to the converse reduction. By coprimality and Lemmas 2.2 and 2.4, we have $|\mathcal{O}/q\mathcal{O}| = |\mathcal{O}_K/q\mathcal{O}_K|$. This implies that any (a,b) uniform in $\mathcal{O}/q\mathcal{O} \times K_\mathbb{R}/q\mathcal{O}$ is also uniform in $\mathcal{O}_K/q\mathcal{O}_K \times K_\mathbb{R}/q\mathcal{O}_K$. The inclusion $\mathcal{O} \subseteq \mathcal{O}_K$ allows to conclude. □

As Theorem 2.13, Theorem 4.2 relies on a the existence of a good multiplier. Writing $K = \mathbb{Q}[x]/(f) = \mathbb{Q}[\alpha]$ and $\mathcal{O} = \mathbb{Z}[\alpha]$, the element $f'(\alpha)$ again satisfies the constraints. Indeed, we know that $\mathcal{O}^\vee = \frac{1}{f'(\alpha)}\mathcal{O}$ (see [Conb, Th. 3.7]), and we have the inclusion $\mathcal{O}_K \subseteq \mathcal{O}^\vee$. Multiplying by $f'(\alpha)$, we obtain $f'(\alpha)\mathcal{O}_K \subseteq \mathcal{O}$. By definition, this means that $f'(\alpha) \in \mathcal{C}_\mathcal{O}$, as claimed. While a large $f'(\alpha)$ would mean a large noise growth in the primal-RLWE to PLWE$^\sigma$ reduction, we described in Sect. 3 how to find a smaller adequate multiplier if needed.

We have $\mathcal{N}(f'(\alpha)) = [\mathcal{O}_K : \mathbb{Z}[\alpha]]^2 \cdot \Delta_K$, and, from [Ste17, p. 48], the prime factors of $[\mathcal{O}_K : \mathbb{Z}[\alpha]]$ are exactly those of $\mathcal{N}(\mathcal{C}_\mathcal{O})$. Provided the valuations are not too high, there should be smaller elements in $\mathcal{C}_\mathcal{O}$ than $f'(\alpha)$. We provide in the full version concrete examples of number fields with defining polynomials f such that the norm of $f'(\alpha)$ is considerably larger than both the norms of $\mathcal{C}_\mathcal{O}$ and $(\mathcal{O}_K^\vee)^{-1}$.

4.2 Distortion Between Embeddings

To bound the norms of a Vandermonde matrix associated to a polynomial and its inverse, we study the magnitude of the roots and their pairwise distances. It is known that $\|V\|^2 = \text{Tr}(V^*V)$, where $*$ denotes the transpose-conjugate operator. For Vandermonde matrices, this gives

$$\|V_f\|^2 = \sum_{j \in [n]} \sum_{k \in [n]} |\alpha_j|^{2(k-1)}, \tag{1}$$

which can be handled when the magnitudes of the α_j's are known. The entries of $V_f^{-1} = (w_{ij})$ have well-known expressions as:

$$w_{ij} = (-1)^{n-i} \frac{e_{n-i}(\overline{\alpha}^j)}{\prod_{k \neq j} (\alpha_j - \alpha_k)}, \tag{2}$$

where $e_0 = 1$, e_j for $j > 0$ stands for the elementary symmetric polynomial of total degree j in $n-1$ variables, and $\overline{\alpha}^j = (\alpha_1, \ldots, \alpha_{j-1}, \alpha_{j+1}, \ldots, \alpha_n)$, the vector of all roots but α_j. We have the following useful relations with the symmetric functions E_i of all the roots (for all j):

$$\begin{cases} E_1(\boldsymbol{\alpha}) = & \alpha_j + e_1(\overline{\alpha}^j), \\ E_i(\boldsymbol{\alpha}) = & \alpha_j e_{i-1}(\overline{\alpha}^j) + e_i(\overline{\alpha}^j) \text{ for } 2 \leq i \leq n-1, \\ E_n(\boldsymbol{\alpha}) = & \alpha_j e_{n-1}(\overline{\alpha}^j). \end{cases} \tag{3}$$

Combining (3) with Vieta's formulas, bounds on the magnitudes of the roots leads to bounds on the numerators of the w_{ij}'s. The denominators encode the separation of the roots, and deriving a precise lower bound turns out to be the main difficulty. By differentiating $f(x) = \prod_{j \in [n]} (x - \alpha_j)$, we note that $\prod_{k \neq j} |\alpha_j - \alpha_k| = |f'(\alpha_j)|$.

A FAMILY OF POLYNOMIALS WITH EASILY COMPUTABLE DISTORTION. We first introduce a family of polynomials for which $\|V_f\|$ and $\|V_f^{-1}\|$ are both simple to estimate. For $n \geq 2$ and $a \geq 1$, we define $f_{n,a} = x^n - a$. The roots can be written[3] as $\alpha_j = a^{1/n} e^{2i\pi \frac{j}{n}}$, for $0 \leq j < n$. As these are scalings of the roots of unity, both their magnitude and separation are well-understood. With (1), we obtain $\|V_{f_{n,a}}\| \leq n a^{\frac{n-1}{n}} \leq na$.

For any j, we readily compute $|f'_{n,a}(\alpha_j)| = n a^{\frac{n-1}{n}}$. Using (3), we observe that $|e_i(\overline{\alpha}^j)| = |\alpha_j|^i$ for $1 \leq i < n$. We obtain that the row norm of $V_{f_{n,a}}^{-1}$ is given by its first row as

$$\sum_{j \in [n]} |w_{1j}| = \frac{1}{n a^{\frac{n-1}{n}}} \cdot \sum_{j \in [n]} |\alpha_j|^{n-1} = 1,$$

from which it follows that $\|V_{f_{n,a}}^{-1}\| \leq \sqrt{n}$.

SMALL PERTURBATIONS OF $f_{n,a}$. Let $P(x) = \sum_{1 \leq j \leq \rho \cdot n} p_j x^j$ for some constant $\rho \in (0, 1)$, where the p_j's are a priori complex numbers. Locating the roots of $g_{n,a} = f_{n,a} + P$ is our first step towards estimating $\|V_{g_{n,a}}\|$ and $\|V_{g_{n,a}}^{-1}\|$. We will use the following version of Rouché's theorem.

Theorem 4.3 (Rouché, adapted from [Con95, pp. 125–126]). *Let f, P be complex polynomials, and let D be a disk in the complex plane. If for any z on the boundary ∂D we have $|P(z)| < |f(z)|$, then f and $f + P$ have the same number of zeros inside D, where each zero is counted as many times as its multiplicity.*

The lemma below allows to determine sufficient conditions on the parameters such that the assumptions of Theorem 4.3 hold. We consider small disks $D_k = D(\alpha_k, 1/n)$ of radius $1/n$ around the roots $\alpha_1, \ldots, \alpha_n$ of $f_{n,a}$, and we let ∂D_k denote their respective boundaries. We let $\|P\|_1 = \sum_j |p_j|$ denote the 1-norm of P.

Lemma 4.4. *We have, for all $k \leq n$ and $z \in \partial D_k$:*

$$|P(z)| \leq (ae)^\rho \cdot \|P\|_1 \quad and \quad |f_{n,a}(z)| \geq a \left(1 - \cos(a^{-1/n}) - \frac{2e^{a^{-1/n}}}{na^{2/n}} \right).$$

Proof. Write $z = \alpha_k + \frac{e^{it}}{n}$ for some $t \in [0, 2\pi)$. We have $|z| \leq a^{1/n} + 1/n$, and hence $|z|^{\rho n} \leq a^\rho \left(1 + \frac{1}{na^{1/n}} \right)^{\rho n}$. The first claim follows from the inequality $|P(z)| \leq \max(1, |z|^{\rho n}) \cdot \|P\|_1$.

---

[3] For the rest of this section, 'i' will refer to the imaginary unit.

Next, we have $|f_{n,a}(z)| = a|(1 + \frac{e^{it'}}{na^{1/n}})^n - 1|$, where $t' = t - 2k\pi/n$. W.l.o.g., we assume that $k = 0$. Let Log denote the complex logarithm, defined on $\mathbb{C} \backslash \mathbb{R}^-$. Since the power series $\sum_{k \geq 1} (-1)^{k-1} u^k/k$ converges to $\mathrm{Log}(1 + u)$ on the unit disk, we have $\mathrm{Log}(1 + \frac{e^{it}}{na^{1/n}}) = \frac{e^{it}}{na^{1/n}} + \delta$, for some δ satisfying $|\delta| \leq |u| \cdot \sum_{k \geq 1} |u|^k/(k+1) \leq |u|^2$ for $u = \frac{e^{it}}{na^{1/n}}$ (note that it has modulus $\leq 1/n \leq 1/2$). Similarly, we can write $\exp(n\delta) = 1 + \varepsilon$ for some ε satisfying $|\varepsilon| \leq 2n|\delta| \leq 2/(na^{2/n})$. We hence have:

$$|f_{n,a}(z)| = a \cdot |A \cdot (1 + \varepsilon) - 1| \geq a \cdot ||A - 1| - |\varepsilon \cdot A||,$$

with $A = \exp(e^{it} a^{-1/n})$. Elementary calculus leads to the inequalities $|A - 1| > 1 - \cos(a^{-1/n})$ and $|A| \leq e^{a^{-1/n}}$ for all $t \in [0, 2\pi)$. Details can be found in the full version. The second claim follows. $\qquad\square$

We note that when $a = 2^{o(n)}$ and n is sufficiently large, then the lower bound on $|f_{n,a}(z)|$ may be replaced by $|f_{n,a}(z)| > a/3$. To use Rouché's theorem, it is then enough that a, ρ and $\|P\|_1$ satisfy $a > (3e^\rho \|P\|_1)^{\frac{1}{1-\rho}}$. We can now derive upper bounds on the norms of $V_{g_{n,a}}$ and its inverse.

Lemma 4.5. *For any* $a > (\|P\|_1 \cdot C^{-1} \cdot e^\rho)^{\frac{1}{1-\rho}}$ *with* $C = |1 - \cos(a^{-1/n}) - \frac{2e^{a^{-1/n}}}{na^{2/n}}|$, *we have:*

$$\|V_{g_{n,a}}\| \leq ane \quad and \quad \|V_{g_{n,a}}^{-1}\| \leq n^{5/2}(\|P\|_1 + 1)a^{1/n}e^2.$$

Proof. Let $\alpha_j = a^{1/n} e^{2i\pi j/n}$ be the roots of $f_{n,a}$ (for $0 \leq j < n$). Thanks to the assumptions and Lemma 4.4, Theorem 4.3 allows us to locate the roots $(\beta_j)_{0 \leq j < n}$ of $g_{n,a}$ within distance $1/n$ from the α_j's. Up to renumbering, we have $|\alpha_j - \beta_j| \leq 1/n$ for all j. In particular, this implies that $|\beta_j| \leq a^{1/n} + 1/n$ for all j. The first claim follows from (1).

Another consequence is that any power less than n of any $|\beta_j|$ is $\leq ae$. We start the estimation of $\|V_{g_{n,a}}^{-1}\|$ by considering the numerators in (2). Let $k_0 = 1 + \lfloor n(1 - \rho) \rfloor$. For any $k < k_0$, we know that $E_k(\beta) = 0$. Using (3), we obtain $|e_k(\overline{\beta}^j)| = |\beta_j|^k \leq ae$ for $k < k_0$ and that $e_{k_0-1}(\overline{\beta}^j) = (-1)^{k_0-1}\beta_j^{k_0-1}$. Then (3) gives $E_{k_0}(\beta) = (-1)^{k_0} p_{n-k_0} = (-1)^{k_0-1}\beta_j^{k_0} + e_{k_0}(\overline{\beta}^j)$, which implies that $|e_{k_0}(\overline{\beta}^j)| \leq |\beta_j|^{k_0} + |p_{n-k_0}|$. By induction, we obtain, for all $k < n - k_0$:

$$|e_{k_0+k}(\overline{\beta}^j)| \leq |p_{n-k_0-k}| + |p_{n-k_0-k+1}\beta_j| + \cdots + |p_{n-k_0}\beta_j^k| + |\beta_j|^{k_0+k}$$
$$\leq (\|P\|_1 + 1)\max(1, |\beta_j|^n),$$

so that $|e_k(\overline{\beta}^j)| \leq (\|P\|_1 + 1)ae$ for $k \geq 1$.

We now derive a lower bound on the denominators in (2). The separation of the β_j's is close to that of the α_j's. Concretely: $|\beta_j - \beta_k| \geq |\alpha_j - \alpha_k| - 2/n$ for all j, k. Therefore, we have $\prod_{k \neq j} |\beta_j - \beta_k| \geq \prod_{k \neq j}(|\alpha_j - \alpha_k| - 2/n)$. Using the identity $|\alpha_j - \alpha_k| = 2a^{1/n} \sin(|k - j|\pi/n)$ and elementary calculus, we obtain $\prod_{k \neq j} |\beta_j - \beta_k| \geq a^{\frac{n-1}{n}}/(ne)$. Details can be found in the full version. Thus any coefficient w_{ij} of $V_{g_{n,a}}^{-1}$ satisfies $|w_{ij}| \leq n(\|P\|_1 + 1)a^{1/n}e^2$. The claim follows from equivalence between the row and Frobenius norms.   □

We now assume that the p_j's and a are integers. The following lemma states that, for a prime and sufficiently large, the polynomial $g_{n,a}$ is irreducible, and thus defines a number field.

Lemma 4.6. *Assume that P is an integer polynomial. For any prime $a >$ $\|P\|_1 + 1$, the polynomial $g_{n,a}$ is irreducible over $\mathbb{Q}$.*

Proof. Let β be a root of $g_{n,a}$. Then we have $a = |\beta^n + P(\beta)| \leq |\beta|^n + \|P\|_1 \max(1, |\beta|^n)$. The assumption on a implies that $|\beta| > 1$. In other words, all the roots of $g_{n,a}$ have a magnitude >1. Now, assume by contradiction that $g_{n,a} = h_1 h_2$ for some rational polynomials h_1, h_2. Since $g_{n,a}$ is monic, it is primitive and we can choose h_1, h_2 as integer polynomials. The product of their constant coefficients is then the prime a. Hence the constant coefficient of h_1 or h_2 is ± 1, which contradicts the fact that the roots of $g_{n,a}$ have magnitude >1.   □

Overall, we have proved the following result.

Theorem 4.7. *Let $\rho \in (0,1)$ and $p_j \in \mathbb{Z}$ for $1 \leq j \leq \rho \cdot n$. Then for $a \geq (3e^\rho \|P\|_1)^{1/(1-\rho)}$ smaller than $2^{o(n)}$ and prime, and n sufficiently large, the polynomial $g_{n,a} = x^n + \sum_{1 \leq j \leq \rho \cdot n} p_j x^j + a$ is irreducible over $\mathbb{Q}$ and satisfies:*

$$\|V_{g_{n,a}}\| \leq ane \quad and \quad \|V_{g_{n,a}}^{-1}\| \leq n^{5/2}(\|P\|_1 + 1)a^{1/n}e^2.$$

In particular, if a and $\|P\|_1$ are polynomial in n, then both $\|V_{g_{n,a}}\|$ and $\|V_{g_{n,a}}^{-1}\|$ are polynomial in n.

In the full version of this article, we give another family of well-behaved polynomials.

5   Search to Decision Dual-RLWE

The reduction relies on the recent technique of [PRS17]. To leverage it, we use a generalized Leftover Hash Lemma over rings. The proof generalizes a technique used in [SS11] to the case where the irreducible factors of the defining polynomial (of K) reduced modulo q do not share the same degree. Alternatively, a generalization of the regularity lemma from [LPR13, Se. 7] to arbitrary number fields could be used. Such a generalization may go through and improve our results a little.

5.1   A Ring-Based Leftover Hash Lemma

Let $m \geq 2$. We identify any rank m $\mathcal{O}_K$-module $M \subseteq K^m$ with the lattice $\sigma(M) \subseteq H^m$. For such modules, the dual may be defined as

$$\widehat{M} = \{\mathbf{t} \in K^m : \forall \mathbf{x} \in M, \mathrm{Tr}(\langle \mathbf{t}, \mathbf{x} \rangle) \in \mathbb{Z}\}.$$

Here $\langle \cdot, \cdot \rangle$ is the K-bilinear map defined by $\langle \mathbf{x}, \mathbf{y} \rangle = \sum_{i=1}^{m} x_i y_i$. We have $\sigma(\widehat{M}) = \overline{\sigma(M)^*}$ in H^m. For some $q \geq 2$ and a fixed $\mathbf{a} \in (\mathcal{O}_K/q\mathcal{O}_K)^m$, we focus on the modules:

$$L(\mathbf{a}) = \frac{\mathbf{a}}{q}\mathcal{O}_K^\vee + (\mathcal{O}_K^\vee)^m \quad \text{and} \quad \mathbf{a}^\perp = \{\mathbf{t} \in \mathcal{O}_K^m : \langle \mathbf{t}, \mathbf{a} \rangle = 0 \bmod q\mathcal{O}_K\}.$$

To prove our Leftover Hash Lemma variant, the main argument relies on an estimation of $\lambda_1^\infty(\widehat{\mathbf{a}^\perp})$, which is obtained by combining the following two lemmas. The first one was stated in [LS15, Se. 5] without a proof, for the case of cyclotomic fields (this restriction is unnecessary). We give a proof of the general case in the full version of this article.

Lemma 5.1. *Let $q \geq 2$ and $\mathbf{a} \in (\mathcal{O}_K/q\mathcal{O}_K)^m$. Then we have $\widehat{\mathbf{a}^\perp} = L(\mathbf{a})$.*

We now obtain a probabilistic lower bound on $\lambda_1^\infty(\widehat{\mathbf{a}^\perp}) = \lambda_1^\infty(L(\mathbf{a}))$. In full generality, it should depend on the ramification of the selected prime integer q, i.e., the exponents appearing in the factorization of $q\mathcal{O}_K$ in prime ideals. It is a classical fact that the ramified prime integers are exactly the primes dividing the discriminant of the field, so that there are only finitely many such q's. Moreover, it is always possible to use modulus switching techniques [BLP+13,LS15] if q ramifies. Therefore, we consider only the non-ramified case.

Lemma 5.2. *Let $q \geq 2$ a prime that does not divide Δ_K. For any $m \geq 2$ and $\delta > 0$, and except with a probability $\leq 2^{3n(m+1)} q^{-mn\delta}$ over the uniform choice of $\mathbf{a} \in ((\mathcal{O}_K/q\mathcal{O}_K)^\times)^m$, we have:*

$$\lambda_1^\infty(L(\mathbf{a})) \geq \Delta_K^{-1/n} \cdot q^{-\frac{1}{m} - \delta}.$$

Proof. Thanks to the assumption on q, we can write $q\mathcal{O}_K = \mathfrak{p}_1 \ldots \mathfrak{p}_k$ for distinct prime ideals $\mathfrak{p}_i$. By Lemma 2.4 and the Chinese Remainder Theorem, we have $\mathcal{O}_K^\vee/q\mathcal{O}_K^\vee \simeq \mathcal{O}_K/q\mathcal{O}_K \simeq \bigoplus_{i=1}^{k} \mathbb{F}_{q^{d_i}}$, where $q^{d_i} = \mathcal{N}(\mathfrak{p}_i)$.

Let $\mathbf{a} = (a_1, \ldots, a_m)$ sampled uniformly in $((\mathcal{O}_K/q\mathcal{O}_K)^\times)^m$. Fix some bound $B > 0$ and let p_B be the probability that $qL(\mathbf{a}) = \mathbf{a}\mathcal{O}_K^\vee + q(\mathcal{O}_K^\vee)^m$ contains a $\mathbf{t} = (t_1, \ldots, t_m)$ such that $0 < \|\mathbf{t}\|_\infty < B$. Our goal is to bound p_B from above. By the union bound, we have that

$$p_B \leq \sum_{\substack{s \in \mathcal{O}_K^\vee/q\mathcal{O}_K^\vee}} \sum_{\substack{\mathbf{t} \in (\mathcal{O}_K^\vee/q\mathcal{O}_K^\vee)^m \\ 0 < \|\mathbf{t}\|_\infty < B}} p(\mathbf{t}, s),$$

with $p(\mathbf{t}, s) = \mathrm{Pr}_{\mathbf{a}}[\forall j, t_j = a_j s \bmod q\mathcal{O}_K^\vee]$ for any s and $\mathbf{t}$ over $\mathcal{O}_K^\vee/q\mathcal{O}_K^\vee$. By independance of the a_j's, we can write $p(\mathbf{t}, s) = \prod_{j \in [m]} p(t_j, s)$ with $p(t_j, s) = \mathrm{Pr}_{a_j}[t_j = a_j s \bmod q\mathcal{O}_K^\vee]$. As $\mathcal{O}_K^\vee/q\mathcal{O}_K^\vee$ and $\mathcal{O}_K/q\mathcal{O}_K$ are isomorphic, estimating this probability amounts to studying the solutions in $(\mathcal{O}_K/q\mathcal{O}_K)^\times$ of the equation $t = as \bmod q\mathcal{O}_K$, for all $t, s \in \mathcal{O}_K/q\mathcal{O}_K$.

Note that if there is an i such that $t = 0 \bmod \mathfrak{p}_i$ and $s \neq 0 \bmod \mathfrak{p}_i$, or vice-versa, then there is no solution, so that $p(t, s) = 0$. Now, assume that s and t are 0 modulo the same $\mathfrak{p}_i$'s. Let $S \subseteq [k]$ denote the set of their indices, and let d_S be such that $q^{d_S} = \mathcal{N}(\prod_{i \in S} \mathfrak{p}_i)$. On the one hand, for all $i \in [k] \setminus S$, both t and s are invertible modulo $\mathfrak{p}_i$ so there is exactly one solution modulo those i's. On the other hand, for all $i \in S$, all the elements of $\mathbb{F}_{q^{d_i}}^\times$ are solutions. This gives $\prod_{i \in S}(q^{d_i} - 1)$ possibilities out of the $\prod_i (q^{d_i} - 1)$ elements of $(\mathcal{O}_K/q\mathcal{O}_K)^\times$. Overall, we obtain that $p(t, s) = \prod_{i \in [k] \setminus S}(q^{d_i} - 1)^{-1}$. Hence, for $\mathbf{t}$ with coordinates t_j such that s and all t_j's are 0 modulo the same $\mathfrak{p}_i$'s, we have:

$$p(\mathbf{t}, s) = q^{-m(n - d_S)} \prod_{i \in [k] \setminus S} (1 - \frac{1}{q^{d_i}})^{-m} \leq q^{-m(n - d_S)} \cdot 2^{mk},$$

the last inequality coming from the fact that $1 - 1/q^{d_i} \geq 1/2$ for all i.

Let τ denote the isomorphism mapping $\mathcal{O}_K^\vee/q\mathcal{O}_K^\vee$ to $\mathcal{O}_K/q\mathcal{O}_K$. The probability to bound is now

$$p_B \leq 2^{mk} \cdot \sum_{S \subseteq [k]} \sum_{\substack{\tau(s) \in \mathcal{O}_K/q\mathcal{O}_K \\ \forall i \in S: \mathfrak{p}_i \,|\, \tau(s)}} \sum_{\substack{\tau(\mathbf{t}) \in (\mathcal{O}_K/q\mathcal{O}_K)^m \\ 0 < \|\mathbf{t}\|_\infty < B \\ \forall j, \forall i \in S: \mathfrak{p}_i \,|\, \tau(t_j)}} q^{-m(n - d_S)}.$$

For any $r > 0$, we let $\mathcal{B}(r)$ denote the (open) ball in H of center 0 and radius r, with respect to the infinity norm. Such a ball has a volume $\mathrm{Vol}(\mathcal{B}(r)) = (2r)^n$. For any $S \subseteq [k]$, we define $N(B, S) = |\mathcal{B}(B) \cap \mathcal{L}(\tau^{-1}(\prod_{i \in S} \mathfrak{p}_i))| - 1$. Since there are 2^k subsets in $[k]$ and $q^{n - d_S}$ elements $\tau(s) \in \mathcal{O}_K/q\mathcal{O}_K$ such that $\mathfrak{p}_i | s$ for all $i \in S$, we have

$$p_B \leq 2^{k(m+1)} \cdot \max_{S \subseteq [k]} \frac{N(B, S)^m}{q^{(n - d_S)(m - 1)}}. \tag{4}$$

We now give an upper bound for $N(B, S)$, from which we will obtain the result. Let $I_S = \prod_{i \in S} \mathfrak{p}_i$ and $\lambda_S = \lambda_1^\infty(\tau^{-1}(I_S))$. Observe that any two distinct balls of radius $\lambda_S/2$ and centered around elements of $\mathcal{B}(B) \cap \mathcal{L}(\tau^{-1}(I_S))$ do not intersect. Moreover, all of them are contained in $\mathcal{B}(B + \lambda_S/2)$. This implies that

$$N(B, S) \leq \frac{\mathrm{Vol}(\mathcal{B}(B + \lambda_S/2))}{\mathrm{Vol}(\mathcal{B}(\lambda_S/2))} = \left(\frac{2B}{\lambda_S} + 1\right)^n.$$

It remains to give a lower bound on λ_S. As $\tau^{-1}(I_S) = I_S \mathcal{O}_K^\vee$, we have $\mathcal{N}(\tau^{-1}(I_S)) = q^{d_S}/\Delta_K$. With Lemma 2.5, this gives $\Delta_K^{-1/n} q^{d_S/n} \leq \lambda_S$. If we set $B = \Delta_K^{-1/n} q^\beta$, then $n\beta < d_S$ leads to $N(B, S) = 0$ and $n\beta \geq d_S$ implies the upper bound $N(B, S) \leq 2^{2n} q^{n\beta - d_S}$. With (4), this gives

$$p_B \leq 2^{(m+1)(k+2n)} \cdot \max_{\substack{S \subseteq [k] \\ d_S \leq n\beta}} q^{m(\beta-1)n + (n-d_S)}.$$

The maximum is reached for $d_S = 0$ (i.e., when $S = \emptyset$). In this case, the exponent of q is $-mn\delta$ for $\beta = 1 - \frac{1}{m} - \delta$. We obtain that $\lambda_1^\infty(qL(\mathbf{a})) \geq \Delta_K^{-1/n} q^{1 - \frac{1}{m} - \delta}$ except with probability $\leq 2^{3n(m+1)} q^{-mn\delta}$.   □

We are now ready to state the variant of the Leftover Hash Lemma.

Theorem 5.3. *Let $q \geq 2$ prime that does not divide Δ_K. Let $\delta > 0, \varepsilon \in (0, 1/2)$ and $m \geq 2$. For a given $\mathbf{a}$ in $((\mathcal{O}_K/q\mathcal{O}_K)^\times)^m$, let $U_{\mathbf{a}}$ be the distribution of $\sum_{i \leq m} t_i a_i$ where the vector $\mathbf{t} = (t_1, \ldots, t_m)$ is sampled from $D_{\mathcal{O}_K, s}$ with $s \geq \sqrt{\log(2mn(1 + 1/\varepsilon))/\pi} \cdot \Delta_K^{1/n} q^{1/m + \delta}$. Then, except for $\leq 2^{3n(m+1)} q^{-mn\delta}$ of $\mathbf{a}$'s, the distance to uniformity of $U_{\mathbf{a}}$ is $\leq 2\varepsilon$.*

Proof. First we note that the map $\mathbf{t} \mapsto \sum_{i \leq m} t_i a_i$ is a well-defined surjective $\mathcal{O}_K$-module homomorphism from $\mathcal{O}_K^m$ to $\mathcal{O}_K/q\mathcal{O}_K$, with kernel $\mathbf{a}^\perp$. The distance to uniformity of $U_{\mathbf{a}}$ is hence the same as the distance to uniformity of $\mathbf{t} \bmod \mathbf{a}^\perp$. By Lemma 2.8, the claim follows whenever $s \geq \eta_\varepsilon(\mathbf{a}^\perp)$. By Lemma 2.6, t it suffices to find an appropriate lower bound on $\lambda_1^\infty(L(\mathbf{a}))$. Lemma 5.2 allows to complete the proof.   □

Corollary 5.4 (Leftover Hash lemma). *If $\mathbf{t}$ is sampled from $D_{\mathcal{O}_K, s}$ with $s \geq \sqrt{\log(2mn(1 + 1/\epsilon))/\pi} \cdot \Delta_K^{1/n} q^{1/m + \delta}$, and the a_i's are sampled from $U((\mathcal{O}_K/q\mathcal{O}_K)^\times)$, then:*

$$\Delta \left[\left(a_1, \ldots, a_m, \sum_{i \leq m} t_i a_i \right), U\left(((\mathcal{O}_K/q\mathcal{O}_K)^\times)^m \times \mathcal{O}_K/q\mathcal{O}_K \right) \right]$$
$$\leq 2\varepsilon + 2^{3n(m+1)} \cdot q^{-mn\delta}.$$

5.2 Search RLWE to Decision RLWE

We now give the reduction from search to decision. As all proofs can be done similarly, we focus on the dual-RLWE version of the problems. For the sake of simplicity, we consider only the case of diagonal covariance matrices. The proof readily extends to general covariance matrices. To obtain the reduction, we need to generate suitable new samples from a starting set of samples from search dual-RLWE.

The lemma below is adapted from [LS15, Le. 4.15]. We will use it to analyze the error distribution we get when generating new samples.

Lemma 5.5. *Let $\alpha > 0$, $\mathcal{L}$ a rank-m $\mathcal{O}_K$-module, $\varepsilon \in (0, 1/2)$, a vector $\mathbf{t} \in D_{\mathcal{L}+\mathbf{c}, \mathbf{r}}$ for some $\mathbf{c} \in H^m$, and $e' \in K_\mathbb{R}$ chosen according to D_α^H. If $r_i \geq \eta_\varepsilon(\mathcal{L})$ and $\frac{\alpha}{\delta_i} \geq \eta_\varepsilon(\mathcal{L})$ for all i, then $\Delta(\langle \mathbf{t}, \mathbf{e} \rangle + e', D_\mathbf{x}^H) \leq 4\varepsilon$ with $x_i = \sqrt{(r_i \delta_i)^2 + \alpha^2}$ and $\delta_i = (\sum_{k \in [m]} |\sigma_i(e_k)|^2)^{1/2}$ for all i.*

We can now give a reduction from search dual-RLWE to worst-case decision dual-RLWE. It may be combined with the worst-case decision dual-RLWE to decision dual-RLWE from Lemma 2.14.

Theorem 5.6. *Let* $r \in (\mathbb{R}^{\geq 0})^n$ *be such that* $r_i = r_{i+s_2}$ *for any* $i > s_1$ *and* $r_i \leq r$ *for some* $r > 0$. *Let* $d = \sqrt{n} \cdot \Delta_K^{1/n} q^{1/m+1/n}$, *and consider* $\Sigma = \{r' : r'_i \leq \sqrt{d^2 \cdot r^2 \cdot m + d^2}\}$. *Then there exists a probabilistic polynomial-time reduction from search dual-*RLWE$_{q,D_r}$ *with* $m \leq q/(2n)$ *input samples to worst-case decision dual-*RLWE$_{q,\Sigma}$.

Proof. We have m samples $(a_i, b_i = a_i s + e_i) \in \mathcal{O}_K/q\mathcal{O}_K \times K_\mathbb{R}/q\mathcal{O}_K^\vee$ from the dual-RLWE distribution $\mathcal{A}_{s,r}^\vee$, for a uniform $s \in \mathcal{O}_K^\vee/q\mathcal{O}_K^\vee$ that we want to find. This is equivalent to finding the error term $\mathbf{e} = (e_1, \ldots, e_m)$. By assumption on m, the a_i's are all invertible with non-negligible probability. If it is not the case, the reduction aborts. From now on, we hence assume that they are uniformly distributed in $(\mathcal{O}_K/q\mathcal{O}_K)^\times$.

We use the same technique as in [PRS17], in that we find the ith embeddings $\sigma_i(e_1), \ldots, \sigma_i(e_m)$ of the error terms by constructing an m-dimensional instance of the Oracle Hidden Center Problem (OHCP). The only difference consists in the way we create the samples that we give to the decision oracle. The reduction uses the dual-RLWE decision oracle to build the oracles $\mathcal{O}_i : \mathbb{R}^m \times \mathbb{R}^{\geq 0} \to \{0,1\}$ for $i \leq s_1$ and $\mathcal{O}_i : \mathbb{C}^m \times \mathbb{R}^{\geq 0} \to \{0,1\}$ for $s_1 < i \leq s_1 + s_2$.

For $i \leq s_1$, we define $k_i : \mathbb{R} \to K_\mathbb{R}$ as $k_i(x) = \sigma^{-1}(x \cdot \mathbf{v}_i)$ and for $s_1 < i \leq s_1 + s_2$, we define $k_i : \mathbb{C} \to K_\mathbb{R}$ as $k_i(x) = \sigma^{-1}(x \cdot \mathbf{v}_i + \overline{x} \cdot \mathbf{v}_{i+s_2})$, where the $\mathbf{v}_i$'s form the canonical basis of H.

On input $(z_1, \ldots, z_m, \alpha)$, oracle $\mathcal{O}_i$ will output 1 with probability depending on $\exp(\alpha)\|\mathbf{e} - \overline{\mathbf{z}}\|$, where $\overline{\mathbf{z}} = (k_i(z_1), \ldots, k_i(z_m))$. It works as follows. It first chooses a uniform $s' \in \mathcal{O}_K^\vee/q\mathcal{O}_K^\vee$. On input $(z_1, \ldots, z_m, \alpha)$, it samples $\mathbf{t} = (t_1, \ldots, t_m) \in \mathcal{O}_K^m$ Gaussian with parameter $\exp(\alpha) \cdot \sqrt{n} \cdot \Delta_K^{1/n} q^{1/m+1/n}$ and some e' from D_d. The oracle then creates $(a', b') = (\langle \mathbf{t}, \mathbf{a} \rangle, \langle \mathbf{t}, \mathbf{b} - \overline{\mathbf{z}} \rangle + a's' + e')$, where $\mathbf{b} = (b_1, \ldots, b_m)$.

By Corollary 5.4, the distribution of $(\mathbf{a}, \langle \mathbf{t}, \mathbf{a} \rangle)$ is exponentially close to $U(((\mathcal{O}_K/q\mathcal{O}_K)^\times)^m \times \mathcal{O}_K/q\mathcal{O}_K)$. Since $b_j = a_j s + e_j$ for all j, we get $b' = a'(s + s') + \langle \mathbf{t}, \mathbf{e} - \overline{\mathbf{z}} \rangle + e'$, so oracle $\mathcal{O}_i$ creates RLWE samples for a uniformly distributed $s + s'$, provided the error term follows a suitable distribution. We let $\delta_\ell = (\sum_{j \in [m]} \sigma_\ell(e_j - k_i(z_j))|^2)^{1/2}$ for $\ell \leq n$. In particular, we have $\delta_i = \|\sigma_i(e_1) - z_1, \ldots, \sigma_i(e_m) - z_m\|$. Let us now study the distribution of the error term $\langle \mathbf{t}, \mathbf{e} - \overline{\mathbf{z}} \rangle + e'$. We can see that once the value of $\langle \mathbf{t}, \mathbf{a} \rangle = c$ and the a_i's are known, one can write $\mathbf{t} = (ca_1^{-1}, 0, \ldots, 0) + (-a_1^{-1}\sum_{i \geq 2} t_i a_i, t_2, \ldots, t_m)$, where the second vector belongs to $\mathbf{a}^\perp$. This means that the actual support of $\mathbf{t}$ is a shift of the $\mathbf{a}^\perp$ lattice by the vector $(ca_1^{-1}, 0, \ldots, 0)$. Using Lemma 5.5, we get that the distribution of the error is $D_\mathbf{x}^H$ where $x_j = \sqrt{\exp^2(\alpha) \cdot d^2 \cdot \delta_j^2 + d^2}$.

Let $\mathcal{S}_{i,(z_1,\ldots,z_m,\alpha)}$ be the samples obtained by applying the procedure above many times. Oracle $\mathcal{O}_i$ calls the dual-RLWE decision oracle with these and outputs 1 if and only if the latter accepts. With non-negligible probability over the choice of the initial errors, the distribution of the samples we get when we call the oracle $\mathcal{O}_i$ on $(\mathbf{0}, 0)$ belongs to the set Σ. One can now show that using the same technique as in [PRS17], it is possible to recover good approximations of the vector $(\sigma_i(e_1), \ldots, \sigma_i(e_m))$. By substracting them from the initial search samples, rounding and then taking the inverses of the a_i's, we obtain s.   □

Acknowledgments. We thank Karim Belabas, Guillaume Hanrot, Alice Pellet-- Mary, Bruno Salvy and Elias Tsigaridas for helpful discussions. This work has been supported in part by ERC Starting Grant ERC-2013-StG-335086-LATTAC, by the European Union PROMETHEUS project (Horizon 2020 Research and Innovation Program, grant 780701) and by BPI-France in the context of the national project RISQ (P141580).

References

[AD17] Albrecht, M.R., Deo, A.: Large modulus ring-LWE $\geq$ module-LWE. In: Takagi, T., Peyrin, T. (eds.) ASIACRYPT 2017. LNCS, vol. 10624, pp. 267–296. Springer, Cham (2017). https://doi.org/10.1007/978-3-319-70694-8_10

[ADPS16] Alkim, E., Ducas, L., Pöppelmann, T., Schwabe, P.: Post-quantum key exchange - a new hope. In: USENIX (2016)

[BBdV+17] Bauch, J., Bernstein, D.J., de Valence, H., Lange, T., van Vredendaal, C.: Short generators without quantum computers: the case of multiquadratics. In: Coron, J.-S., Nielsen, J.B. (eds.) EUROCRYPT 2017. LNCS, vol. 10210, pp. 27–59. Springer, Cham (2017). https://doi.org/10.1007/978-3-319-56620-7_2

[BCLvV16] Bernstein, D.J., Chuengsatiansup, C., Lange, T., van Vredendaal, C.: NTRU Prime (2016). http://eprint.iacr.org/2016/461

[BDK+18] Bos, J.W., Ducas, L., Kiltz, E., Lepoint, T., Lyubashevsky, V., Schanck, J.M., Schwabe, P., Stehlé, D.: CRYSTALS - Kyber: a CCA-secure module-lattice-based KEM. In: EuroS&P (2018)

[BLP+13] Brakerski, Z., Langlois, A., Peikert, C., Regev, O., Stehlé, D.: Classical hardness of learning with errors. In: STOC (2013)

[CDPR16] Cramer, R., Ducas, L., Peikert, C., Regev, O.: Recovering short generators of principal ideals in cyclotomic rings. In: Fischlin, M., Coron, J.-S. (eds.) EUROCRYPT 2016. LNCS, vol. 9666, pp. 559–585. Springer, Heidelberg (2016). https://doi.org/10.1007/978-3-662-49896-5_20

[CDW17] Cramer, R., Ducas, L., Wesolowski, B.: Short stickelberger class relations and application to ideal-SVP. In: Coron, J.-S., Nielsen, J.B. (eds.) EUROCRYPT 2017. LNCS, vol. 10210, pp. 324–348. Springer, Cham (2017). https://doi.org/10.1007/978-3-319-56620-7_12

[CGS14] Campbell, P., Groves, M., Shepherd, D.: Soliloquy: a cautionary tale. In: ETSI 2nd Quantum-Safe Crypto Workshop (2014). http://docbox. etsi.org/Workshop/2014/201410_CRYPTO/S07_Systems_and_Attacks/ S07_Groves_Annex.pdf

[CIV16a] Castryck, W., Iliashenko, I., Vercauteren, F.: On the tightness of the error bound in Ring-LWE. LMS J. Comput. Math. 130–145 (2016)

[CIV16b] Castryck, W., Iliashenko, I., Vercauteren, F.: Provably weak instances of ring-LWE revisited. In: Fischlin, M., Coron, J.-S. (eds.) EUROCRYPT 2016. LNCS, vol. 9665, pp. 147–167. Springer, Heidelberg (2016). https://doi.org/10.1007/978-3-662-49890-3_6

[CLS17] Chen, H., Lauter, K., Stange, K.E.: Attacks on search RLWE. SIAM J. Appl. Algebra Geom. (SIAGA) 1, 665–682 (2017)

[CLS16] Chen, H., Lauter, K., Stange, K.E.: Vulnerable Galois RLWE families and improved attacks. In: Proceedings of SAC. Springer (2016)

[Cona] Conrad, K.: The conductor ideal. http://www.math.uconn.edu/~kconrad/blurbs/gradnumthy/conductor.pdf

[Conb] Conrad, K.: The different ideal. http://www.math.uconn.edu/~kconrad/blurbs/gradnumthy/different.pdf

[Con95] Conway, J.B.: Functions of One Complex Variable. Springer, New York (1995). https://doi.org/10.1007/978-1-4612-6313-5

[DD12] Ducas, L., Durmus, A.: Ring-LWE in polynomial rings. In: Fischlin, M., Buchmann, J., Manulis, M. (eds.) PKC 2012. LNCS, vol. 7293, pp. 34–51. Springer, Heidelberg (2012). https://doi.org/10.1007/978-3-642-30057-8_3

[DLL+18] Ducas, L., Lepoint, T., Lyubashevsky, V., Schwabe, P., Seiler, G., Stehlé, D.: CRYSTALS - Dilithium: digital signatures from module lattices. In: TCHES (2018)

[EHL14] Eisenträger, K., Hallgren, S., Lauter, K.: Weak instances of PLWE. In: SAC (2014)

[ELOS15] Elias, Y., Lauter, K.E., Ozman, E., Stange, K.E.: Provably weak instances of ring-LWE. In: Gennaro, R., Robshaw, M. (eds.) CRYPTO 2015. LNCS, vol. 9215, pp. 63–92. Springer, Heidelberg (2015). https://doi.org/10.1007/978-3-662-47989-6_4

[GHPS12] Gentry, C., Halevi, S., Peikert, C., Smart, N.P.: Ring switching in BGV-style homomorphic encryption. In: Visconti, I., De Prisco, R. (eds.) SCN 2012. LNCS, vol. 7485, pp. 19–37. Springer, Heidelberg (2012). https://doi.org/10.1007/978-3-642-32928-9_2

[GPV08] Gentry, C., Peikert, C., Vaikuntanathan, V.: Trapdoors for hard lattices and new cryptographic constructions. In: STOC (2008)

[HHPW10] Hoffstein, J., Howgrave-Graham, N., Pipher, J., Whyte, W.: Practical lattice-based cryptography: NTRUEncrypt and NTRUSign. In: Nguyen, P., Vallée, B. (eds.) The LLL Algorithm. Information Security and Cryptography, pp. 349–390. Springer, Heidelberg (2010). https://doi.org/10.1007/978-3-642-02295-1_11

[LPR10] Lyubashevsky, V., Peikert, C., Regev, O.: On ideal lattices and learning with errors over rings. JACM 60(6), 43 (2013)

[LPR13] Lyubashevsky, V., Peikert, C., Regev, O.: A toolkit for ring-LWE cryptography. In: Johansson, T., Nguyen, P.Q. (eds.) EUROCRYPT 2013. LNCS, vol. 7881, pp. 35–54. Springer, Heidelberg (2013). https://doi.org/10.1007/978-3-642-38348-9_3

[LS15] Langlois, A., Stehlé, D.: Worst-case to average-case reductions for module lattices. Des. Codes Cryptogr. 75(3), 565–599 (2015)

[Lyu16] Lyubashevsky, V.: Digital signatures based on the hardness of ideal lattice problems in all rings. In: Cheon, J.H., Takagi, T. (eds.) ASIACRYPT 2016. LNCS, vol. 10032, pp. 196–214. Springer, Heidelberg (2016). https://doi.org/10.1007/978-3-662-53890-6_7

[MR04] Micciancio, D., Regev, O.: Worst-case to average-case reductions based on Gaussian measure. In: Proceedings of FOCS, pp. 371–381. IEEE (2004)

[Pei16] Peikert, C.: How (not) to instantiate ring-LWE. In: Zikas, V., De Prisco, R. (eds.) SCN 2016. LNCS, vol. 9841, pp. 411–430. Springer, Cham (2016). https://doi.org/10.1007/978-3-319-44618-9_22

[PR06] Peikert, C., Rosen, A.: Efficient collision-resistant hashing from worst-case assumptions on cyclic lattices. In: Halevi, S., Rabin, T. (eds.) TCC 2006. LNCS, vol. 3876, pp. 145–166. Springer, Heidelberg (2006). https://doi.org/10.1007/11681878_8

[PR07] Peikert, C., Rosen, A.: Lattices that admit logarithmic worst-case to average-case connection factors. In: STOC (2007)

[PRS17] Peikert, C., Regev, O., Stephens-Davidowitz, N.: Pseudorandomness of Ring-LWE for any ring and modulus. In: STOC (2017)

[Reg05] Regev, O.: On lattices, learning with errors, random linear codes, and cryptography. J. ACM 56(6), 1–40 (2009)

[RSSS17] Roşca, M., Sakzad, A., Stehlé, D., Steinfeld, R.: Middle-product learning with errors. In: Katz, J., Shacham, H. (eds.) CRYPTO 2017. LNCS, vol. 10403, pp. 283–297. Springer, Cham (2017). https://doi.org/10.1007/978-3-319-63697-9_10

[SE94] Schnorr, C.-P., Euchner, M.: Lattice basis reduction: improved practical algorithms and solving subset sum problems. Math. Program. 66, 181–199 (1994)

[SS11] Stehlé, D., Steinfeld, R.: Making NTRU as secure as worst-case problems over ideal lattices. In: Paterson, K.G. (ed.) EUROCRYPT 2011. LNCS, vol. 6632, pp. 27–47. Springer, Heidelberg (2011). https://doi.org/10.1007/978-3-642-20465-4_4

[SS13] Stehlé, D., Steinfeld, R.: Making NTRUEncrypt and NTRUSign as secure standard worst-case problems over ideal lattices (2013). http://perso.ens-lyon.fr/damien.stehle/NTRU.html

[SSTX09] Stehlé, D., Steinfeld, R., Tanaka, K., Xagawa, K.: Efficient public key encryption based on ideal lattices. In: Matsui, M. (ed.) ASIACRYPT 2009. LNCS, vol. 5912, pp. 617–635. Springer, Heidelberg (2009). https://doi.org/10.1007/978-3-642-10366-7_36

[Ste17] Stevenhagen, P.: Lecture notes on number rings (2017). http://websites.math.leidenuniv.nl/algebra/ant.pdf

Faster Gaussian Sampling for Trapdoor Lattices with Arbitrary Modulus

Nicholas Genise$^{(\boxtimes)}$ and Daniele Micciancio

University of California, San Diego, USA
{ngenise,daniele}@eng.ucsd.edu

Abstract. We present improved algorithms for gaussian preimage sampling using the lattice trapdoors of (Micciancio and Peikert, CRYPTO 2012). The MP12 work only offered a highly optimized algorithm for the on-line stage of the computation in the special case when the lattice modulus q is a power of two. For arbitrary modulus q, the MP12 preimage sampling procedure resorted to general lattice algorithms with complexity cubic in the bitsize of the modulus (or quadratic, but with substantial preprocessing and storage overheads). Our new preimage sampling algorithm (for any modulus q) achieves linear complexity with very modest storage requirements, and experimentally outperforms the generic method of MP12 already for small values of q. As an additional contribution, we give a new, quasi-linear time algorithm for the off-line perturbation sampling phase of MP12 in the ring setting. Our algorithm is based on a variant of the Fast Fourier Orthogonalization (FFO) algorithm of (Ducas and Prest, ISSAC 2016), but avoids the need to precompute and store the FFO matrix by a careful rearrangement of the operations. All our algorithms are fairly simple, with small hidden constants, and offer a practical alternative to use the MP12 trapdoor lattices in a broad range of cryptographic applications.

1 Introduction

Lattice cryptography provides powerful techniques to build a wide range of advanced cryptographic primitives, like identity based encryption [2–4,11,23,31], attribute based encryption [14,16,17,19,33], some types of fully homomorphic encryption and signatures [12,13,24,32,35], group signatures [21,36,44,45,54] and much more (e.g., see [6,10,34,46,51,56,57,60]). Most of the advanced applications of lattice cryptography rely on a notion of strong lattice trapdoor, introduced in [31], which allows to sample points from an n-dimensional lattice L with a gaussian-like distribution. This gaussian sampling operation is often the main bottleneck in the implementation of advanced cryptographic functions that make use of strong lattice trapdoors, and improving the methods to generate and use lattice trapdoors has been the subject of several investigations [5,7,31,55].

Research supported in part by the DARPA SafeWare program. Opinions, findings and conclusions or recommendations expressed in this material are those of the author(s) and do not necessarily reflect the views of DARPA.

J. B. Nielsen and V. Rijmen (Eds.): EUROCRYPT 2018, LNCS 10820, pp. 174–203, 2018.
https://doi.org/10.1007/978-3-319-78381-9_7

The current state of the art in lattice trapdoor generation and sampling is given by the work of Micciancio and Peikert [51], which introduces a new notion of lattice trapdoor, specialized to the type of q-ary lattices used in cryptography, i.e., integer lattices $L \subseteq \mathbb{Z}^n$ that are periodic modulo $q \cdot \mathbb{Z}^n$. The trapdoor is then used to efficiently sample lattice points with gaussian distribution around a given target. Building on techniques from [55], the sampling algorithm of [51] includes both an on-line and an off-line stage, and [51] focuses on improving the complexity of the on-line stage, which is far more critical in applications. Unfortunately, the most efficient algorithms proposed in [51] for (the on-line stage of) preimage sampling only apply to lattices with modulus $q = 2^k$ equal to a power of 2 (or, more generally, the power $q = p^k$ of a small prime p,) which is not compatible with the functional or efficiency requirements of many applications. Moreover, only the on-line stage of [51] takes full advantage of the structure of algebraic lattices [48–50] typically employed in the efficient instantiation of lattice cryptography, and essential to reduce the running time of lattice operations from quadratic (in the lattice dimension) to quasi-linear. A straightforward implementation of the off-line stage (e.g., using a generic Cholesky decomposition algorithm) completely destroys the algebraic structure, and degrades the running time of the (off-line) algorithm from quasi-linear to quadratic or worse. For lattices over "power-of-two" cyclotomic rings (the most popular class of algebraic lattices used in cryptography), a much faster algorithm for the off-line stage was proposed by Ducas and Nguyen in [27, Sect. 6], and subsequently simplified, improved and extended to a more general class of cyclotomic rings by the *Fast Fourier Orthogonalization (FFO)* of Ducas and Prest [28].

Our Contribution. We present improved algorithms for gaussian preimage sampling using the lattice trapdoors of [51]. Specifically, we present a new algorithm (for the on-line stage) capable of handling any modulus q (including the large prime moduli required by some applications) and still achieve the same level of performance of the specialized algorithm of [51] for power-of-two modulus $q = 2^k$. This improves the running time of [51] for arbitrary modulus from cubic $\log^3 q$ (or quadratic $\log^2 q$, using precomputation and a substantial amount of storage) to just linear in $\log q$ and with minimal storage requirements.

As an additional contribution, we present an improved algorithm for the off-line perturbation generation problem which takes full advantage of the algebraic structure of ring lattices. We remark that this problem can already be solved (in quasilinear time $\tilde{O}(n)$) using the (FFO) algorithm of [28], which first produces a compact representation of the orthogonalized lattice basis (or covariance matrix), and then uses it to quickly generate lattice samples. We improve on the algorithm of [28] on two fronts. First, the FFO algorithm is quasi-linear in the ring dimension, but quadratic in the module dimension (which, in our application, is $\log q$). We combine [28] with the "sparse matrix" optimization of [9] to yield an algorithm that is linear both in the ring dimension and $\log q$. Moreover, we provide a variant of the FFO algorithm that performs essentially the same operations as [28], but without requiring the precomputation and storage of the

Table 1. Running time and storage of the (G-sampling) algorithm. G-Sampling running times are scaled by a factor n to take into account that each sample requires n independent calls to the underlying G-sampling operation.

	MP12	MP12	This work
Modulus q	2^k	any	any
G-sampling precomp.	—	$O(\log^3 q)$	—
G-sampling space	$O(\log q)$	$O(\log^2 q)$	$O(\log q)$
G-sampling time	$O(n \log q)$	$O(n \log^2 q)$	$O(n \log q)$

FFO (structured) matrix, thereby simplifying the implementation and improving the space complexity of [28].

The G-sampling improvements are summarized in Table 1. The improvements are not just asymptotic: our new algorithms are fairly simple, with small hidden constants, and include a careful choice of the parameters that allows to implement most steps using only integer arithmetic on very small numbers. In Sect. 3.3, we provide an experimental comparison showing that the new algorithm outperforms the generic method of [51] already for small values of the moduli, making it an attractive choice for implementations even in applications where the modulus $q = n^{O(1)}$ has logarithmic bit-size. For applications using an exponentially large $q = \exp(n)$, the projected performance improvements are dramatic. The concrete efficiency of our algorithms in the context of full blown cryptographic applications, has been recently confirmed by independent implementation efforts [25,37,38].

Technical details. In order to describe our techniques, we need first to provide more details on the lattice trapdoor sampling problem. Given a lattice L and a target point $\mathbf{t}$, the lattice gaussian sampling problem asks to generate (possibly with the help of some trapdoor information) a random lattice point $\mathbf{v} \in L$ with probability proportional to $\exp(-c\|\mathbf{v} - \mathbf{t}\|^2)$. Building on techniques from [55], this problem is solved in [51] by mapping L to a fixed (key independent) lattice G^n, generating a gaussian sample in G^n, and then mapping the result back to L. (The linear function T mapping G^n to L serves as the trapdoor.) Without further adjustments, this produces a lattice point in L with ellipsoidal gaussian distribution, with covariance which depends on the linear transformation T. In order to produce spherical samples (as required[1] by applications), [51] employs a perturbation technique of Peikert [55] which adds some noise (with complementary covariance) to the target $\mathbf{t}$, before using it as a center for the G^n-lattice sampling operation. In summary, the sampling algorithm of [51,55] consists of two stages:

– an off-line (target independent) stage, which generates perturbation vectors with covariance matrix defined by the trapdoor transformation T, and
– an on-line (target dependent) stage which generates gaussian samples from an (easy to sample) lattice G^n.

---

[1] More generally, applications require samples to be generated according to a distribution that does not depend on the trapdoor/secret key.

Not much attention is paid in [51] to the perturbation generation, as it does not depend on the target vector $\mathbf{t}$, and it is far less time critical in applications.[2] As for the on-line stage, one of the properties that make the lattice G^n easy to sample is that it is the orthogonal sum of n copies of a $(\log q)$-dimensional lattice G. So, even using generic algorithms with quadratic running time, G sampling takes a total of $O(n \log^2 q)$ operations. For moduli $q = n^{O(1)}$ polynomial in the lattice dimension n, this results in quasilinear running time $O(n \log^2 n)$. However, since the G-sampling operation directly affects the on-line running time of the signing algorithm, even a polylogarithmic term $\log^2 q$ can be highly undesirable. To this end, [51] gives a particularly efficient (and easy to implement) algorithm for G-lattice sampling when the lattice modulus $q = 2^k$ is a power of 2 (or more generally, a power $q = p^k$ of a small prime p.) The running time of this specialized G-sampling algorithm is $\log q$, just linear in the lattice dimension, and has minimal (constant) storage requirements. Thanks to its simplicity and efficiency, this algorithm has quickly found its way in concrete implementations of lattice based cryptographic primitives (e.g., see [9]), largely solving the problem of efficient lattice sampling for $q = 2^k$. However, setting q to a power of 2 (or more generally, the power of a small prime), may be incompatible with applications and other techniques used in lattice cryptography, like *attribute based encryption (ABE)* schemes [14] and fast implementation via the *number theoretic transform* [47,49]. For arbitrary modulus q, [51] falls back to generic algorithms (for arbitrary lattices) with quadratic complexity. This may still be acceptable when the modulus q is relatively small. But it is nevertheless undesirable, as even polylogarithmic factors have a significant impact on the practical performance of cryptographic functions (easily increasing running times by an order of magnitude), and can make applications completely unusable when the modulus $q = \exp(n)$ is exponentially large. The concrete example best well illustrates the limitations of [51] is the recent conjunction obfuscator of [20], which requires the modulus q to be prime with bitsize $\log(q) = O(n)$ linear in the security parameter. In this setting, the specialized algorithm of [51] (for $q = 2^k$) is not applicable, and using a generic algorithm slows down the on-line stage by a factor $O(n)$, or, more concretely, various orders of magnitude for typical parameter settings. Another, less drastic, example is the arithmetic circuit ABE scheme of [14] where q is $O(2^{n^\epsilon})$ for some fixed $0 < \epsilon < 1/2$. Here the slow down is asymptotically smaller, n^ϵ, but still polynomial in the security parameter n.

Unfortunately, the specialized algorithm from [51] makes critical use of the structure of the G-basis when $q = 2^k$, and is not easily adapted to other moduli. (See Sect. 3 for details). In order to solve this problem we resort to the same approach used in [51,55] to generate samples from arbitrary lattices: we map G to an even simpler lattice D using an easy to compute linear transformation T', perform the gaussian sampling in D, and map the result back to G. As usual, the error shape is corrected by including a perturbation term with appropriate covariance matrix. The main technical problem to be solved is to find a suitable

---

E.g., in lattice based digital signature schemes [31,51], the off-line computation depends only on the secret key, and can be performed in advance without knowing the message to be signed.

linear transformation T' such that D can be efficiently sampled and perturbation terms can be easily generated. In Sect. 3 we demonstrate a choice of transformation T' with all these desirable properties. In particular, using a carefully chosen transformation T', we obtain lattices D and perturbation matrices that are triangular, sparse, and whose entries admit a simple (and efficiently computable) closed formula expression. So, there is not even a need to store these sparse matrices explicitly, as their entries can be easily computed on the fly. This results in a G-sampling algorithm with linear running time, and minimal (constant) space requirements, beyond the space necessary to store the input, output and randomness of the algorithm.

Next, in Sect. 4, we turn to the problem of efficiently generating the perturbations of the off-line stage. Notice that generating these perturbations is a much harder problem than the one faced when mapping G to D (via T'). The difference is that while G, D, T' are fixed (sparse, carefully designed) matrices, the transformation T is a randomly chosen matrix that is used as secret key. In this setting, there is no hope to reduce the computation time to linear in the lattice dimension, because even reading/writing the matrix T can in general take quadratic time. Still, when using algebraic lattices, matrix T admits a compact (linear size) representation, and one can reasonably hope for faster perturbation generation algorithms. As already noted, this can be achieved using the Fast Fourier Orthogonalization algorithm of Ducas and Prest [28], which has running time quasilinear in the ring dimension, but quadratic in the dimension (over the ring) of the matrix T, which is $O(\log q)$ in our setting. As usual, while for polynomial moduli $q = n^{O(1)}$, this is only a polylogarithmic slow down, it can be quite significant in practice [9]. We improve on a direct application of the FFO algorithm by first employing an optimization of Bansarkhani and Buchmann [9] to exploit the sparsity of T. (This corresponds to the top level function SAMPLEPZ in Fig. 4.) This optimization makes the computation linear in the dimension of T ($\log q$ in our setting), while keeping the quasilinear dependency on the ring dimension n from [28]. We further improve this combined algorithm by presenting a variant of FFO (described by the two mutually recursive functions SAMPLEFZ/SAMPLE2Z in Fig. 4) that does not require the precomputation and storage of the FFO matrix.

Comparison with FFO. Since our SAMPLEPZ function (Fig. 4) uses a subprocedure SAMPLEFZ which is closely related to the FFO algorithm [28], we provide a detailed comparison between the two. We recall that FFO works by first computing a *binary tree* data structure [28, Algorithm 3], where the root node is labeled by an n-dimensional vector, its two children are labeled by $(n/2)$-dimensional vectors, and so on, all the way down to n leaves which are labeled with 1-dimensional vectors. Then, [28, Algorithm 4] uses this binary tree data structure within a block/recursive variant of Babai's nearest plane algorithm.[3] Our SAMPLEFZ is

---

[3] Technically, [28, Algorithm 4] deterministically rounds a target point to a point in the lattice, rather than producing a probability distribution. But, as observed in [28], the algorithm is easily adapted to perform gaussian sampling by replacing deterministic rounding operations with probabilistic gaussian rounding.

based on the observation that one can blend/interleave the computation of [28, Algorithm 3] and [28, Algorithm 4], leading to a substantial (asymptotic) memory saving. Specifically, combining the two algorithms avoids the need to precompute and store the FFO binary tree data structure altogether, which is now implicitly generated, on the fly, one node/vector at a time, and discarding each node/vector as soon as possible in a way similar to a depth-first tree traversal. The resulting reduction in space complexity is easily estimated. The original FFO builds a tree with $\log n$ levels, where level l stores 2^l vectors in dimension $n/2^l$. So, the total storage requirement for each level is n, giving an overall space complexity of $n \log n$. Our FFO variant only stores one node/vector per level, and has space complexity $\sum_l (n/2^l) = 2n$, a $O(\log n)$ improvement over the space of original FFO algorithm. Moreover, the nodes/vectors are stored implicitly in the execution stack of the program, rather than an explicitly constructed binary tree data structure, yielding lower overhead and an algorithm that is easier to implement. For simplicity we specialized our presentation to power-of-two cyclotomics, which are the most commonly used in lattice cryptography, but everything works equally well for the larger class of cyclotomic rings, in the canonical embedding, considered in [28].

2   Preliminaries

We denote the complex numbers as $\mathbb{C}$, the real numbers as $\mathbb{R}$, the rational numbers as $\mathbb{Q}$, and the integers as $\mathbb{Z}$. A number is denoted by a lower case letter, $z \in \mathbb{Z}$ for example. We denote the conjugate of a complex number y as y^*. When q is a positive integer, $\log q$ is short for its rounded up logarithm in base two, $\lceil \log_2 q \rceil$. A floating point number with mantissa length m representing $x \in \mathbb{R}$ is denoted as $\bar{x}$. The index set of the first n natural numbers is $[n] = \{1, \ldots, n\}$. Vectors are denoted by bold lower case letters, $\mathbf{v}$, and are in column form ($\mathbf{v}^T$ is a row vector) unless stated otherwise. The inner product of two vectors is $\langle \mathbf{x}, \mathbf{y} \rangle = \mathbf{x}^T \mathbf{y}$. We denote matrices with bold upper case letters $\mathbf{B}$ or with upper case Greek letters (for positive-definite matrices). The transpose of a matrix is $\mathbf{B}^T$. The entry of $\mathbf{B}$ in row i and column j is denoted $B_{i,j}$. Unless otherwise stated, the norm of a vector is the ℓ_2 norm. The norm of a matrix $\|\mathbf{B}\| = \max_i \|\mathbf{b}_i\|$ is the maximum norm of its column vectors. Given two probability distributions over a countable domain D, the statistical distance between them is $\Delta_{\mathrm{SD}}(X, Y) = \frac{1}{2} \sum_{\omega \in D} |X(\omega) - Y(\omega)|$. In order to avoid tracing irrelevant terms in our statistical distance computations, we define $\hat{\epsilon} = \epsilon + O(\epsilon^2)$.

We denote a random variable x sampled from a distribution $\mathcal{D}$ as $x \leftarrow \mathcal{D}$. A random variable distributed as $\mathcal{D}$ is denoted $x \sim \mathcal{D}$. We denote an algorithm $\mathcal{A}$ with oracle access to another algorithm $\mathcal{B}$ (distribution $\mathcal{D}$) as $\mathcal{A}^{\mathcal{B}}$ ($\mathcal{A}^{\mathcal{D}}$).

The max-log, or ML, distance between two distributions was recently introduced by [53] in order to prove tighter bounds for concrete security. The *ML distance* between two discrete distributions over the same support, S, as

$$\Delta_{\mathrm{ML}}(\mathcal{P}, \mathcal{Q}) = \max_{x \in S} |\ln \mathcal{Q}(x) - \ln \mathcal{P}(x)|.$$

Let $\mathcal{P}, \mathcal{Q}$ be distributions over a countable domain again and let S be the support of $\mathcal{P}$.

The *Rényi divergence* of order infinity of $\mathcal{Q}$ from $\mathcal{P}$ is

$$R_\infty(\mathcal{P}\|\mathcal{Q}) = \max_{x \in S} \frac{\mathcal{P}(x)}{\mathcal{Q}(x)}.$$

Rényi divergence is used in [8] to yield a tighter security analysis than one using statistical distance.

2.1   Linear Algebra

The (foreward) *Gram-Schmidt orthogonalization* of an ordered set of linearly independent vectors $\mathbf{B} = \{\mathbf{b}_1, \ldots, \mathbf{b}_k\}$ is $\widetilde{\mathbf{B}} = \{\widetilde{\mathbf{b}}_1, \ldots, \widetilde{\mathbf{b}}_k\}$ where each $\widetilde{\mathbf{b}}_i$ is the component of $\mathbf{b}_i$ orthogonal to $\mathrm{span}(\mathbf{b}_1, \ldots, \mathbf{b}_{i-1})$ (and the backward GSO is defined as $\mathbf{b}_i^\dagger = \mathbf{b}_i \perp \mathrm{span}(\mathbf{b}_{i+1}, \ldots, \mathbf{b}_n)$). An *anti-cylic* matrix is an $n \times n$ matrix of the form

$$\begin{bmatrix} a_0 & -a_{n-1} & \cdots & -a_1 \\ a_1 & a_0 & \cdots & -a_2 \\ \vdots & \vdots & \ddots & \vdots \\ a_{n-1} & a_{n-2} & \cdots & a_0 \end{bmatrix}.$$

For any two (symmetric) matrices $\Sigma, \Gamma \in \mathbb{R}^{n \times n}$, we write $\Sigma \succeq \Gamma$ if $\mathbf{x}^T(\Sigma - \Gamma)\mathbf{x} \geq 0$ for all (nonzero) vectors $\mathbf{x} \in \mathbb{R}^n$, and $\Sigma \succ \Gamma$ if $\mathbf{x}^T(\Sigma - \Gamma)\mathbf{x} > 0$. It is easy to check that $\succeq$ is a partial order relation. Relations $\preceq$ and $\prec$ are defined symmetrically. When one of the two matrices $\Gamma = s\mathbf{I}$ is scalar, we simply write $\Sigma \succeq s$ or $\Sigma \preceq s$. A symmetric matrix $\Sigma \in \mathbb{R}^{n \times n}$ is called *positive definite* if $\Sigma \succ 0$, and *positive semidefinite* if $\Sigma \succeq 0$. Equivalently, Σ is positive semidefinite if and only if it can be written as $\Sigma = \mathbf{B}\mathbf{B}^T$ for some (square) matrix $\mathbf{B}$, called a *square root* of Σ and denoted $\mathbf{B} = \sqrt{\Sigma}$. (Notice that any $\Sigma \succ 0$ has infinitely many square roots $\mathbf{B} = \sqrt{\Sigma}$.) Σ is positive definite if and only if its square root $\mathbf{B}$ is a square nonsingular matrix. When $\mathbf{B}$ is upper (resp. lower) triangular, the factorization $\Sigma = \mathbf{B}\mathbf{B}^T$ is called the upper (resp. lower) triangular *Cholesky decomposition* of Σ. The Cholesky decomposition of any positive definite $\Sigma \in \mathbb{R}^{n \times n}$ can be computed with $O(n^3)$ floating point arithmetic operations. For any scalar s, $\Sigma \succ s$ if and only if all eigenvalues of Σ are strictly greater than s. In particular, positive definite matrices are nonsingular.

For any $n \times n$ matrix $\mathbf{S}$ and non-empty index sets $I, J \subseteq \{1, \ldots, n\}$, we write $\mathbf{S}[I, J]$ for the $|I| \times |J|$ matrix obtained by selecting the elements at positions $(i, j) \in I \times J$ from $\mathbf{S}$. When $I = J$, we write $\mathbf{S}[I]$ as a shorthand for $\mathbf{S}[I, I]$. For any nonsingular matrix $\mathbf{S} \in \mathbb{R}^{n \times n}$ and index partition $I \cup \bar{I} = \{1, \ldots, n\}$, $I \cap \bar{I} = \emptyset$, the $I \times I$ matrix

$$\mathbf{S}/I = \mathbf{S}[I] - \mathbf{S}[I, \bar{I}] \cdot \mathbf{S}[\bar{I}]^{-1} \cdot \mathbf{S}[\bar{I}, I]$$

is called the *Schur complement* of $\mathbf{S}[\bar{I}]$, often denoted by $\mathbf{S}/\mathbf{S}[\bar{I}] = \mathbf{S}/I$. In particular, if $\mathbf{S} = \begin{bmatrix} \mathbf{A} & \mathbf{B} \\ \mathbf{B}^T & \mathbf{D} \end{bmatrix}$ then the Schur complement of $\mathbf{A}$ is the matrix

$\mathbf{S}/\mathbf{A} = \mathbf{D} - \mathbf{B}^T\mathbf{A}^{-1}\mathbf{B}$. For any index set I, a symmetric matrix $\mathbf{S}$ is positive definite if and only if both $\mathbf{S}[I]$ and its Schur's complement $\mathbf{S}/\mathbf{S}[I]$ are positive definite.

Let $\Sigma = \begin{bmatrix} \mathbf{A} & \mathbf{B} \\ \mathbf{B}^T & \mathbf{D} \end{bmatrix} \succ 0$. We can factor Σ in terms of a principal submatrix, say $\mathbf{D}$, and its Schur complement, $\Sigma/\mathbf{D} = \mathbf{A} - \mathbf{B}\mathbf{D}^{-1}\mathbf{B}^T$, as follows:

$$\Sigma = \begin{bmatrix} \mathbf{I} & \mathbf{B}\mathbf{D}^{-1} \\ \mathbf{0} & \mathbf{I} \end{bmatrix} \begin{bmatrix} \Sigma/\mathbf{D} & \mathbf{0} \\ \mathbf{0} & \mathbf{D} \end{bmatrix} \begin{bmatrix} \mathbf{I} & \mathbf{0} \\ \mathbf{D}^{-1}\mathbf{B}^T & \mathbf{I} \end{bmatrix}.$$

The next two theorems regarding the spectra of principal submatrices and Schur complements of positive definite matrices are used in Sect. 4. In both theorems, λ_i is the ith (in non-increasing order, with multiplicity) eigenvalue of a symmetric matrix.

Theorem 1 (Cauchy). *For any symmetric matrix* $\mathbf{S} \in \mathbb{R}^{n \times n}$, $I \subseteq \{1, \dots, n\}$ *and* $1 \leq i \leq |I|$

$$\lambda_i(\mathbf{S}) \geq \lambda_i(\mathbf{S}[I]) \geq \lambda_{i+n-|I|}(\mathbf{S}).$$

Theorem 2 ([61, Corollary 2.3]). *For any positive definite* $\Sigma \in \mathbb{R}^{n \times n}$, $I \subseteq \{1, \dots, n\}$ *and* $1 \leq i \leq |I|$

$$\lambda_i(\Sigma) \geq \lambda_i(\Sigma/I) \geq \lambda_{i+n-|I|}(\Sigma).$$

In other words, the eigenvalues of principal submatrices and Schur complements of a positive definite matrix are bounded from below and above by the smallest and largest eigenvalues of the original matrix.

2.2 Gaussians and Lattices

A *lattice* $\Lambda \subset \mathbb{R}^n$ is a discrete subgroup of $\mathbb{R}^n$. Specifically, a lattice of *rank* k is the integer span $\mathcal{L}(\mathbf{B}) = \{z_1\mathbf{b}_1 + \cdots + z_k\mathbf{b}_k \mid z_i \in \mathbb{Z}\}$ of a basis $\mathbf{B} = \{\mathbf{b}_1, \dots, \mathbf{b}_k\} \subset \mathbb{R}^n$ ($k \leq n$). There are infinitely many bases for a given lattice since right-multiplying a basis by a unimodular transformation gives another basis. The *dual lattice* of Λ, denoted by Λ^*, is the lattice $\{\mathbf{x} \in \text{span}(\Lambda) \mid \langle \mathbf{x}, \Lambda \rangle \subseteq \mathbb{Z}\}$. It is easy to see that $\mathbf{B}^{-T}$ is a basis for $\mathcal{L}(\mathbf{B})^*$ for a full rank lattice ($n = k$).

The n-dimensional *gaussian* function $\rho : \mathbb{R}^n \to (0,1]$ is defined as $\rho(\mathbf{x}) := \exp(-\pi\|\mathbf{x}\|^2)$. Applying an invertible linear transformation $\mathbf{B}$ to the gaussian function yields

$$\rho_{\mathbf{B}}(\mathbf{x}) = \rho(\mathbf{B}^{-1}\mathbf{x}) = \exp(-\pi \cdot \mathbf{x}^T \Sigma^{-1}\mathbf{x})$$

with $\Sigma = \mathbf{B}\mathbf{B}^T \succ 0$. For any $\mathbf{c} \in \text{span}(\mathbf{B}) = \text{span}(\Sigma)$, we also define the shifted gaussian function (centered at $\mathbf{c}$) as $\rho_{\sqrt{\Sigma},\mathbf{c}}(\mathbf{x}) = \rho_{\sqrt{\Sigma}}(\mathbf{x} - \mathbf{c})$. Normalizing the function $\rho_{\mathbf{B},\mathbf{c}}(\mathbf{x})$ by the measure of $\rho_{\mathbf{B},\mathbf{c}}$ over the span of $\mathbf{B}$ gives the *continuous gaussian distribution* with covariance $\Sigma/(2\pi)$, denoted by $D_{\sqrt{\Sigma},\mathbf{c}}$. Let $S \subset \mathbb{R}^n$ be any discrete set in $\mathbb{R}^n$, then $\rho_{\sqrt{\Sigma}}(S) = \sum_{\mathbf{s} \in S} \rho_{\sqrt{\Sigma}}(\mathbf{s})$. The *discrete gaussian* distribution over a lattice Λ, denoted by $D_{\Lambda,\sqrt{\Sigma},\mathbf{c}}$, is defined by restricting the support of the distribution to Λ. Specifically, a sample $\mathbf{y} \leftarrow D_{\Lambda,\sqrt{\Sigma},\mathbf{c}}$

has probability mass function $\rho_{\sqrt{\Sigma},\mathbf{c}}(\mathbf{x})/\rho_{\sqrt{\Sigma},\mathbf{c}}(\Lambda)$ for all $\mathbf{x} \in \Lambda$. Discrete gaussians on lattice cosets $\Lambda + \mathbf{c}$, for $\mathbf{c} \in \text{span}(\Lambda)$, are defined similarly setting $\Pr\{\mathbf{y} \leftarrow D_{\Lambda+\mathbf{c},\sqrt{\Sigma},\mathbf{p}}\} = \rho_{\sqrt{\Sigma},\mathbf{p}}(\mathbf{y})/\rho_{\sqrt{\Sigma},\mathbf{p}}(\Lambda+\mathbf{c})$ for all $\mathbf{y} \in \Lambda+\mathbf{c}$. For brevity we let $D_{\Lambda+\mathbf{c},\sqrt{\Sigma},\mathbf{p}}(\mathbf{y}) := \Pr\{\mathbf{y} \leftarrow D_{\Lambda+\mathbf{c},\sqrt{\Sigma},\mathbf{p}}\}$.

For a lattice Λ and any (typically small) positive $\epsilon > 0$, the *smoothing parameter* $\eta_\epsilon(\Lambda)$ [52] is the smallest $s > 0$ such that $\rho(s \cdot \Lambda^*) \leq 1 + \epsilon$. A one-dimensional discrete gaussian with a tail-cut, t, is a discrete gaussian $D_{\mathbb{Z},c,s}$ restricted to a support of $\mathbb{Z} \cap [c - t \cdot s, c + t \cdot s]$. We denote this truncated discrete gaussian as $D_{\mathbb{Z},c,s}^t$. In order to use the ML distance in Sect. 3, we will restrict all tail-cut discrete gaussians to a universal support of $\mathbb{Z} \cap [c - t \cdot s_{max}, c + t \cdot s_{max}]$ for some s_{max}.

Lemma 1 ([31, **Lemma 4.2**]). *For any $\epsilon > 0$, any $s \geq \eta_\epsilon(\mathbb{Z})$, and any $t > 0$,*

$$\Pr_{x \leftarrow D_{\mathbb{Z},s,c}} [|x - c| \geq t \cdot s] \leq 2e^{-\pi t^2} \cdot \frac{1 + \epsilon}{1 - \epsilon}.$$

More generally, for any positive definite matrix Σ and lattice $\Lambda \subset \text{span}(\Sigma)$, we write $\sqrt{\Sigma} \geq \eta_\epsilon(\Lambda)$, or $\Sigma \succeq \eta_\epsilon^2(\Lambda)$, if $\rho(\sqrt{\Sigma}^T \cdot \Lambda^*) \leq 1 + \epsilon$. The reader is referred to [31,52,55] for additional information on the smoothing parameter.

Here we recall two bounds and a discrete gaussian convolution theorem to be used later.

Lemma 2 ([31, **Lemma 3.1**]). *Let $\Lambda \subset \mathbb{R}^n$ be a lattice with basis $\mathbf{B}$, and let $\epsilon > 0$. Then,*

$$\eta_\epsilon(\Lambda) \leq \|\widetilde{\mathbf{B}}\| \sqrt{\log(2n(1 + 1/\epsilon))/\pi}.$$

Lemma 3 ([55, **Lemma 2.5**]). *For any full rank n-dimensional lattice Λ, vector $\mathbf{c} \in \mathbb{R}^n$, real $\epsilon \in (0, 1)$, and positive definite $\Sigma \succeq \eta_\epsilon^2(\Lambda)$,*

$$\rho_{\sqrt{\Sigma}}(\Lambda + \mathbf{c}) \in \left[\frac{1 - \epsilon}{1 + \epsilon}, 1\right] \cdot \rho_{\sqrt{\Sigma}}(\Lambda).$$

Theorem 3 ([55, **Theorem 3.1**]). *For any vectors $\mathbf{c}_1, \mathbf{c}_2 \in \mathbb{R}^n$, lattices $\Lambda_1, \Lambda_2 \subset \mathbb{R}^n$, and positive definite matrices $\Sigma_1, \Sigma_2 \succ 0$, $\Sigma = \Sigma_1 + \Sigma_2 \succ 0$, $\Sigma_3^{-1} = \Sigma_1^{-1} + \Sigma_2^{-1} \succ 0$, if $\sqrt{\Sigma_1} \succeq \eta_\epsilon(\Lambda_1)$ and $\sqrt{\Sigma_3} \succeq \eta_\epsilon(\Lambda_2)$ for some $0 < \epsilon \leq 1/2$, then the distribution*

$$X = \{\mathbf{x} \mid \mathbf{p} \leftarrow D_{\Lambda_2+\mathbf{c}_2,\sqrt{\Sigma_2}}, \mathbf{x} \leftarrow D_{\Lambda_1+\mathbf{c}_1,\sqrt{\Sigma_1},\mathbf{p}}\}$$

is within statistical distance $\Delta(X, Y) \leq 8\epsilon$ from the discrete gaussian $Y = D_{\Lambda_1+\mathbf{c}_1,\sqrt{\Sigma}}$.

Below we have the correctness theorem for the standard, randomized version of Babai's nearest plane algorithm. The term *statistically close* is the standard cryptographic notion of negligible statistical distance. Precisely, a function $f : \mathbb{N} \to \mathbb{R}_{\geq 0}$ is negligible if for every $c > 1$ there exists an N such that for all $n > N$, $f(n) < n^{-c}$. We emphasize that the algorithm reduces to sampling $D_{\mathbb{Z},s,c}$.

Theorem 4 ([31, **Theorem 4.1**]). *Given a full-rank lattice basis $\mathbf{B} \in \mathbb{R}^{n \times n}$, a parameter $s \geq \|\widetilde{\mathbf{B}}\| \omega(\sqrt{\log n})$, and a center $\mathbf{c} \in \mathbb{R}^n$, there is an $O(n^2)$-time, with a $O(n^3)$-time preprocessing, probabilistic algorithm whose output is statistically close to $D_{\mathcal{L}(\mathbf{B}),s,\mathbf{c}}$.*

2.3 Cyclotomic Fields

Let n be a positive integer. The *n-th cyclotomic field over* $\mathbb{Q}$ is the number field $\mathcal{K}_n = \mathbb{Q}[x]/(\Phi_n(x)) \cong \mathbb{Q}(\zeta)$ where ζ is an *n-th primitive root of unity* and $\Phi_n(x)$ is the minimal polynomial of ζ over $\mathbb{Q}$. The nth *cyclotomic ring* is $\mathcal{O}_n = \mathbb{Z}[x]/(\Phi_n(x))$. Let $\varphi(n)$ be Euler's totient function. $\mathcal{K}_n$ is a $\varphi(n)$-dimensional $\mathbb{Q}$-vector space, and we can view $\mathcal{K}_n$ as a subset of $\mathbb{C}$ by viewing ζ as a complex primitive n-th root of unity.

Multiplication by a fixed element f, $g \mapsto f \cdot g$, is a linear transformation on $\mathcal{K}_n$ as a $\mathbb{Q}$-vector space. We will often view field elements as $\varphi(n)$-dimensional rational vectors via the *coefficient embedding*. This is defined by $f(x) = \sum_{i=0}^{\varphi(n)-1} f_i x^i \mapsto (f_0, \cdots, f_{\varphi(n)-1})^T$ mapping a field element to its vector of coefficients under the *power basis* $\{1, x, \cdots, x^{\varphi(n)-1}\}$ (or equivalently $\{1, \zeta, \cdots, \zeta^{\varphi(n)-1}\}$). We can represent a field element as the matrix in $\mathbb{Q}^{\varphi(n) \times \varphi(n)}$ that represents the linear transformation by its multiplication in the coefficient embedding. This matrix is called a field element's coefficient *multiplication* matrix. When n is a power of two, an element's coefficient multiplication matrix is anti-cyclic.

An *isomorphism* from the field F to the field K is a bijection $\theta : F \to K$ such that $\theta(fg) = \theta(f)\theta(g)$, and $\theta(f + g) = \theta(f) + \theta(g)$ for all $f, g \in F$. An *automorphism* is an isomorphism from a field to itself. For example, if we view the cyclotomic field $\mathcal{K}_n$ as a subset of the complex numbers, then the *conjugation* map $f(\zeta) \mapsto f(\zeta)^* = f(\zeta^*)$ is an automorphism and can be computed in linear time $O(n)$. In power-of-two cyclotomic fields, the conjugation of a field element corresponds to the matrix transpose of an element's anti-cyclic multiplication matrix.

Another embedding is the *canonical* embedding which maps an element $f \in \mathcal{K}_n$ to the vector of evaluations of f, as a polynomial, at each root of $\Phi_n(x)$. When n is a power of two, the linear transformation between the coefficient embedding and the canonical embedding is a scaled isometry.

Let n be a power of two, then the field $\mathcal{K}_{2n}$ is a two-dimensional $\mathcal{K}_n$-vector space as see by splitting a polynomial $f(x) \in \mathcal{K}_{2n}$ into $f(x) = f_0(x^2) + x \cdot f_1(x^2)$ for $f_i \in \mathcal{K}_n$. Now, we can view the linear transformation given by multiplication by some $f \in \mathcal{K}_{2n}$ as a linear transformation over $\mathcal{K}_n \oplus \mathcal{K}_n \cong \mathcal{K}_{2n}$. Let $\phi_{2n} : \mathcal{K}_{2n} \to \mathbb{Q}^{n \times n}$ be the injective ring homomorphism from the field to an element's anti-cyclic matrix. Then, we have the following relationship where $\mathbf{P}$ below is a simple re-indexing matrix known as a stride permutation (increasing evens followed by increasing odds in $\{0, 1, \ldots, n-1\}$),

$$\mathbf{P}\phi_n(f)\mathbf{P}^T = \begin{bmatrix} \phi_{n/2}(f_0) & \phi_{n/2}(x \cdot f_1) \\ \phi_{n/2}(f_1) & \phi_{n/2}(f_0) \end{bmatrix}.$$

3 Sampling G-Lattices

For any positive integers $b \geq 2$, $k \geq 1$ and non-negative integer $u < b^k$, we write $[u]_b^k$ for the base-b expansion of u, i.e., the unique vector $(u_0, \ldots, u_{k-1})$ with entries $0 \leq u_i < b$ such that $u = \sum_i u_i b^i$. Typically, $b = 2$ and $[u]_2^k$ is just the k-digits binary representation of u, but larger values of b may be used to obtain interesting efficiency trade-offs. Throughout this section, we consider the values of b and k as fixed, and all definitions and algorithms are implicitly parametrized by them.

In this section we study the so-called G-lattice sampling problem, i.e., the problem of sampling the discrete Gaussian distribution on a lattice coset

$$\Lambda_u^\perp(\mathbf{g}^T) = \{\mathbf{z} \in \mathbb{Z}^k : \mathbf{g}^T \mathbf{z} = u \bmod q\}$$

where $q \leq b^k$, $u \in \mathbb{Z}_q$, $k = \lceil \log_b q \rceil$, and $\mathbf{g} = (1, b, \ldots, b^{k-1})$. G-lattice sampling is used in many lattice schemes employing a trapdoor. Both schemes with polynomial modulus, like IBE [2,4,11,18], group signatures [36,44,45,54], and others (double authentication preventing and predicate authentication preventing signatures, constraint-hiding PRFs) [15,22], and schemes with super-polynomial modulus [1,17,19,20,33,35,42] (ABE, obfuscation, watermarking, etc.), as well as [39], use G-lattice sampling.

A very efficient algorithm to solve this problem is given in [51] for the special case when $q = b^k$ is a power of the base b. The algorithm, shown in Fig. 1, is very simple. This algorithm reduces the problem of sampling the k-dimensional lattice coset $\Lambda_u^\perp(\mathbf{g}^T)$ for $u \in \mathbb{Z}_q$ to the much simpler problem of sampling the *one-dimensional* lattice cosets $u + b\mathbb{Z}$ for $u \in \mathbb{Z}_b$. The simplicity of the algorithm is due to the fact that, when $q = b^k$ is an exact power of b, the lattice $\Lambda^\perp(\mathbf{g}^T)$ has a very special basis

$$\mathbf{B}_{b^k} = \begin{bmatrix} b & & & & \\ -1 & b & & & \\ & -1 & \ddots & & \\ & & \ddots & b & \\ & & & -1 & b \end{bmatrix}$$

which is sparse, triangular, and with small integer entries. (In particular, its Gram-Schmidt orthogonalization $\widetilde{\mathbf{B}}_{b^k} = b\mathbf{I}$ is a scalar matrix.) As a result, the general lattice sampling algorithm of [31,43] (which typically requires $O(k^3)$-time preprocessing, and $O(k^2)$ storage and online running time) can be specialized to the much simpler algorithm in Fig. 1 that runs in linear time $O(k)$, with minimal memory requirements and no preprocessing at all.

We give a specialized algorithm to solve the same sampling problem when $q < b^k$ is an arbitrary modulus. This is needed in many cryptographic applications

$$\text{SAMPLEG}(q = b^k, s, u)$$
$$\textbf{for } i = 0, \dots, k - 1 :$$
$$x_i \leftarrow \mathcal{D}_{b\mathbb{Z}+u,s}$$
$$u := (u - x_i)/b \in \mathbb{Z}.$$
$$\textbf{return } (x_0, \dots, x_{k-1}).$$

Fig. 1. A sampling algorithm for G-lattices when the modulus q is a perfect power of the base b. The algorithm is implicitly parametrized by a base b and dimension k.

where the modulus q is typically a prime. As already observed in [51] the lattice $\Lambda^{\perp}(\mathbf{g}^T)$ still has a fairly simple and sparse basis matrix

$$\mathbf{B}_q = \begin{bmatrix} b & & & & q_0 \\ -1 & b & & & q_1 \\ & -1 & \ddots & & \vdots \\ & & \ddots & b & q_{k-2} \\ & & & -1 & q_{k-1} \end{bmatrix}$$

where $(q_0, \dots, q_{k-1}) = [q]_b^k = \mathbf{q}$ is the base-b representation of the modulus q. This basis still has good geometric properties, as all vectors in its (left-to-right) Gram-Schmidt orthogonalization have length at most $O(b)$. So, it can be used with the algorithm of [31,43] to generate good-quality gaussian samples on the lattice cosets with small standard deviation. However, since the basis is no longer triangular, its Gram-Schmidt orthogonalization is not sparse anymore, and the algorithm of [31,43] can no longer be optimized to run in linear time as in Fig. 1. In applications where $q = n^{O(1)}$ is polynomial in the security parameter n, the matrix dimension $k = O(\log n)$ is relatively small, and the general sampling algorithm (with $O(k^2)$ storage and running time) can still be used with an acceptable (albeit significant) performance degradation. However, for larger q this becomes prohibitive in practice. Moreover, even for small q, it would be nice to have an optimal sampling algorithm with $O(k)$ running time, linear in the matrix dimension, as for the exact power case. Here we give such an algorithm, based on the convolution methods of [55], but specialized with a number of concrete technical choices that result in a simple and very fast implementation, comparable to the specialized algorithm of [51] for the exact power case.

The reader may notice that the alternating columns of $\mathbf{B}_q$, $\mathbf{b}_1, \mathbf{b}_3, \dots$ and $\mathbf{b}_2, \mathbf{b}_4, \dots$, are pair-wise orthogonal. Let us call these sets $\mathbf{B}_1$ and $\mathbf{B}_2$, respectively. Then, another basis for $\Lambda^{\perp}(\mathbf{g}^T)$ is $(\mathbf{B}_1, \mathbf{B}_2, \mathbf{q})$ and this might suggest that the GSO of this basis is sparse. Unfortunately, this leads to a GSO of $(\mathbf{B}_1, \mathbf{B}_2^*, \mathbf{q}^*)$ where $\mathbf{B}_2^*$ is a dense, upper triangular block. Let $\mathbf{b}$ be the $i - th$ vector in $\mathbf{B}_2$. Then, there are $2 + i - 1$ non-orthogonal vectors to $\mathbf{b}$ preceding it in $\mathbf{B}_1$ and $\mathbf{B}_2^*$, filling in the upper portion of $\tilde{\mathbf{b}}$.

Overview. The idea is the following. Instead of sampling $\Lambda_u^{\perp}(\mathbf{g}^T)$ directly, we express the lattice basis $\mathbf{B}_q = \mathbf{TD}$ as the image (under a linear transformation $\mathbf{T}$) of some other matrix $\mathbf{D}$ with very simple (sparse, triangular) structure. Next, we sample the discrete gaussian distribution (say, with variance σ^2) on an appropriate coset of $\mathcal{L}(\mathbf{D})$. Finally, we map the result back to the original lattice applying the linear transformation $\mathbf{T}$ to it. Notice that, even if $\mathcal{L}(\mathbf{D})$ is sampled according to a spherical gaussian distribution, the resulting distribution is no longer spherical. Rather, it follows an ellipsoidal gaussian distribution with (scaled) covariance $\sigma^2 \mathbf{T}\mathbf{T}^T$. This problem is solved using the convolution method of [55], i.e., initially adding a perturbation with complementary covariance $s^2\mathbf{I} - \sigma^2 \mathbf{T}\mathbf{T}^T$ to the target, so that the final output has covariance $\sigma^2 \mathbf{T}\mathbf{T}^T + (s^2\mathbf{I} - \sigma^2 \mathbf{T}\mathbf{T}^T) = s^2\mathbf{I}$. In summary, at a very high level, the algorithm performs (at least implicitly) the following steps:

1. Compute the covariance matrix $\Sigma_1 = \mathbf{T}\mathbf{T}^T$ and an upper bound r on the spectral norm of $\mathbf{T}\mathbf{T}^T$
2. Compute the complementary covariance matrix $\Sigma_2 = r^2\mathbf{I} - \Sigma_1$
3. Sample $\mathbf{p} \leftarrow D_{\Lambda_1, \sigma\sqrt{\Sigma_2}}$, from some convenient lattice Λ_1 using the Cholesky decomposition of Σ_2
4. Compute the preimage $\mathbf{c} = \mathbf{T}^{-1}(\mathbf{u} - \mathbf{p})$
5. Sample $\mathbf{z} \leftarrow D_{\mathcal{L}(\mathbf{D}), -\mathbf{c}, \sigma}$
6. Output $\mathbf{u} + \mathbf{T}\mathbf{z}$

The technical challenge is to find appropriate matrices $\mathbf{T}$ and $\mathbf{D}$ that lead to a very efficient implementation of all the steps. In particular, we would like $\mathbf{T}$ to be a very simple matrix (say, sparse, triangular, and with small integer entries) so that $\mathbf{T}$ has small spectral norm, and both linear transformations $\mathbf{T}$ and $\mathbf{T}^{-1}$ can be computed efficiently. The matrix $\mathbf{D}$ (which is uniquely determined by $\mathbf{B}$ and $\mathbf{T}$) should also be sparse and triangular, so that the discrete gaussian distribution on the cosets of $\mathcal{L}(\mathbf{D})$ can be efficiently sampled. Finally (and this is the trickiest part in obtaining an efficient instantiation) the complementary covariance matrix $\Sigma_2 = r^2\mathbf{I} - \Sigma_1$ should also have a simple Cholesky decomposition $\Sigma_2 = \mathbf{L}\mathbf{L}^T$ where $\mathbf{L}$ is triangular, sparse and with small entries, so that perturbations can be generated efficiently. Ideally, all matrices should also have a simple, regular structure, so that they do not need to be stored explicitly, and can be computed on the fly with minimal overhead.

In the next subsection we provide an instantiation that satisfies all of these properties. Next, in Subsect. 3.2 we describe the specialized sampling algorithm resulting from the instantiation, and analyze its correctness and efficiency properties.

3.1   Instantiation

In this subsection, we describe a specific choice of linear transformations and matrix decompositions that satisfies all our desiderata, and results in a very efficient instantiation of the convolution sampling algorithm on G-lattices.

A tempting idea may be to map the lattice basis $\mathbf{B}_q$ to the basis $\mathbf{B}_{b^k}$, and then use the efficient sampling algorithm from Fig. 1. However, this does not quite work because it results in a pretty bad transformation $\mathbf{T}$ which has both poor geometrical properties and a dense matrix representation. It turns out that a very good choice for a linear transformation $\mathbf{T}$ is given precisely by the matrix $\mathbf{T} = \mathbf{B}_{b^k}$ describing the basis when q is a power of b. We remark that $\mathbf{T}$ is used as a linear transformation, rather than a lattice basis. So, the fact that it equals $\mathbf{B}_{b^k}$ does not seem to carry any special geometric meaning, it just works! In particular, what we do here should not be confused with mapping $\mathbf{B}_q$ to $\mathbf{B}_{b^k}$. The resulting factorization is

$$
\mathbf{B}_q =
\begin{bmatrix}
b & & & & q_0 \\
-1 & b & & & q_1 \\
 & -1 & \ddots & & \vdots \\
 & & \ddots & b & q_{k-2} \\
 & & & -1 & q_{k-1}
\end{bmatrix}
=
\begin{bmatrix}
b & & & & \\
-1 & b & & & \\
 & -1 & \ddots & & \\
 & & \ddots & b & \\
 & & & -1 & b
\end{bmatrix}
\begin{bmatrix}
1 & & & & d_0 \\
 & 1 & & & d_1 \\
 & & \ddots & & \vdots \\
 & & & 1 & d_{k-2} \\
 & & & & d_{k-1}
\end{bmatrix}
= \mathbf{B}_{b^k} \mathbf{D}
$$

where the entries of the last column of $\mathbf{D}$ are defined by the recurrence $d_i = \frac{d_{i-1}+q_i}{b}$ with initial condition $d_{-1} = 0$. Notice that all the d_i are in the range $[0, 1)$, and $b^{i+1} \cdot d_i$ is always an integer. In some sense, sampling from $\mathcal{L}(\mathbf{D})$ is even easier than sampling from $\mathcal{L}(\mathbf{B}_{b^k})$ because the first $k - 1$ columns of $\mathbf{D}$ are orthogonal and the corresponding coordinates can be sampled independently in parallel. (This should be contrasted with the sequential algorithm in Fig. 1).

We now look at the geometry and algorithmic complexity of generating perturbations. The covariance matrix of $\mathbf{T} = \mathbf{B}_{b^k}$ is given by

$$
\Sigma_1 = \mathbf{B}_{b^k} \mathbf{B}_{b^k}^T =
\begin{bmatrix}
b^2 & -b & & & \\
-b & (b^2+1) & -b & & \\
 & \ddots & \ddots & \ddots & \\
 & & -b & (b^2+1) & -b \\
 & & & -b & (b^2+1)
\end{bmatrix}.
$$

The next step is to find an upper bound r^2 on the spectral norm of Σ_2, and compute the Cholesky decomposition $\mathbf{LL}^T$ of the complementary covariance matrix $\Sigma_2 = r^2\mathbf{I} - \Sigma_1$. By the Gershgorin circle theorem, all eigenvalues of Σ_1 are in the range $(b \pm 1)^2$. So, we may set $r = b + 1$. Numerical computations also suggest that this choice of r is optimal, in the sense that the spectral norm of Σ_1 approaches $b + 1$ as k tends to infinity. The Cholesky decomposition is customarily defined by taking $\mathbf{L}$ to be a *lower* triangular matrix. However, for sampling purposes, an upper triangular $\mathbf{L}$ works just as well. It turns out that using an upper triangular $\mathbf{L}$ in the decomposition process leads to a much simpler solution, where all (squared) entries have a simple, closed form expression, and can be easily computed on-line without requiring any preprocessing computation or storage. (By contrast, numerical computations suggest that the standard Cholesky decomposition with lower triangular $\mathbf{L}$ is far less regular, and even precomputing it requires exponentially higher precision arithmetic than our upper triangular solution.) So, we let $\mathbf{L}$ be an upper triangular matrix, and set $r = b+1$.

For any r, the perturbation's covariance matrix $\Sigma_2 = r^2 \mathbf{I} - \Sigma_1$ has Cholesky decomposition $\Sigma_2 = \mathbf{L} \cdot \mathbf{L}^T$ where $\mathbf{L}$ is the sparse upper triangular matrix defined by the following equations:

$$\mathbf{L} = \begin{bmatrix} l_0 & h_1 & & & \\ & l_1 & h_2 & & \\ & & \ddots & \ddots & \\ & & & & h_{k-1} \\ & & & & l_{k-1} \end{bmatrix} \quad \text{where} \quad \begin{aligned} l_0^2 + h_1^2 &= r^2 - b^2 \\ l_i^2 + h_{i+1}^2 &= r^2 - (b^2 + 1) \ (i = 1, \ldots, k-2) \\ l_{k-1}^2 &= r^2 - (b^2 + 1) \\ l_i h_i &= b \qquad\qquad (i = 1, \ldots, k-1) \end{aligned}$$

It can be easily verified that these equations have the following simple closed form solution:

$$r = b + 1, \quad l_0^2 = b\left(1 + \frac{1}{k}\right) + 1, \quad l_i^2 = b\left(1 + \frac{1}{k-i}\right), \quad h_{i+1}^2 = b\left(1 - \frac{1}{k-i}\right) \tag{1}$$

We observe that also the inverse transformation $\mathbf{B}_{b^k}^{-1}$ has a simple, closed-form solution: the ith column of $\mathbf{B}_{b^k}^{-1}$ equals $(0, \cdots, 0, \frac{1}{b}, \ldots, (\frac{1}{b})^{k-i})$. Notice that this matrix is not sparse, as it has $O(k^2)$ nonzero entries. However, there is no need to store it and the associated transformation can still be computed in linear time by solving the sparse triangular system $\mathbf{T}x = \mathbf{b}$ by back-substitution.

3.2   The Algorithm

The sampling algorithm, SAMPLEG, is shown in Fig. 2. It takes as input a modulus q, an integer variance s, a coset u of $\Lambda^{\perp}(\mathbf{g}^T)$, and outputs a sample statistically close to $D_{\Lambda_u^{\perp}(\mathbf{g}^T),s}$. SAMPLEG relies on subroutines PERTURB and SAMPLED where PERTURB(σ) returns a perturbation, $\mathbf{p}$, statistically close to $D_{\mathcal{L}(\Sigma_2),\sigma \cdot \sqrt{\Sigma_2}}$, and SAMPLED$(\sigma, \mathbf{c})$ returns a sample $\mathbf{z}$ such that $\mathbf{Dz}$ is statistically close to $D_{\mathcal{L}(\mathbf{D}),-\mathbf{c},\sigma}$.

Both PERTURB and SAMPLED are instantiations of the randomized nearest plane algorithm [31,43]. Consequently, both algorithms rely on a subroutine SAMPLEZ$_t(\sigma, c, \sigma_{max})$ which returns a sample statistically close to one-dimensional discrete gaussian with it a tail-cut t, $D_{\mathbb{Z},\sigma,c}^t$ over the *fixed* support of $\mathbb{Z} \cap [c - t \cdot \sigma_{max}, c + t \cdot \sigma_{max}]$. We fix the support of all one dimensional discrete gaussians for compatibility with ML distance. In addition, we only feed SAMPLEZ centers $c \in [0, 1)$ since we can always shift by an integer.

Storage. The scalars c_i in SAMPLEG, representing $\mathbf{c} = \mathbf{B}_{b^k}^{-1}(\mathbf{u} - \mathbf{p})$, and d_i in SAMPLED, representing the last column of $\mathbf{D}$, are rational numbers of the form x/b^i for a small integer x and $i \in [k]$. The numbers l_i, h_i are positive numbers of magnitude less than $\sqrt{2b+1}$.

A naive implementation of the algorithms store floating point numbers c_i, d_i, h_i, and l_i for a total storage of $4k$ floating point numbers. However, this can be adapted to constant time storage since they are determined by simple recurrence relations (c_i, d_i) or simple formulas (h_i, l_i).

SAMPLEG$(s, \mathbf{u} = [u]_b^k, \mathbf{q} = [q]_b^k)$

  $\sigma := s/(b+1)$

  $\mathbf{p} \leftarrow$ PERTURB(σ)

  for $i = 0, \ldots, k-1$:

    $c_i := (c_{i-1} + u_i - p_i)/b$

  $\mathbf{z} \leftarrow$ SAMPLED$(\sigma, \mathbf{c})$

  for $i = 0, \ldots, k-2$:

    $t_i := b \cdot z_i - z_{i-1} + q_i \cdot z_{k-1} + u_i$

  $t_{k-1} := q_{k-1} \cdot z_{k-1} - z_{k-2} + u_{k-1}$

  return t

PERTURB(σ)

  $\beta := 0$

  for $i = 0, \ldots, k-1$:

    $c_i := \beta/l_i$, and $\sigma_i := \sigma/l_i$

    $z_i \leftarrow \lfloor c_i \rfloor +$ SAMPLEZ$_t(\sigma_i, \lfloor c_i \rceil_{[0,1)}, s)$

    $\beta := -z_i h_i$

  $p_0 := (2b+1)z_0 + bz_1$

  for $i := 1, \ldots, k-1$:

    $p_i := b(z_{i-1} + 2z_i + z_{i+1})$

  return p

SAMPLED$(\sigma, \mathbf{c})$

  $z_{k-1} \leftarrow \lfloor -c_{k-1}/d_{k-1} \rfloor$

  $z_{k-1} \leftarrow z_{k-1} +$ SAMPLEZ$_t(\sigma/d_{k-1}, \lfloor -c_{k-1}/d_{k-1} \rceil_{[0,1)}, s)$

  $\mathbf{c} := \mathbf{c} - z_{k-1}\mathbf{d}$

  for $i \in \{0, \ldots, k-2\}$:

    $z_i \leftarrow \lfloor -c_i \rfloor +$ SAMPLEZ$_t(\sigma, \lfloor -c_i \rceil_{[0,1)}, s)$

  return z

Fig. 2. Sampling algorithm for G-lattices for any modulus $q < b^k$. The algorithms take b and k as implicit parameters, and SAMPLEG outputs a sample with distribution statistically close to $D_{A_u^\perp(\mathbf{g}^T),s}$. Any scalar with an index out of range is 0, i.e. $c_{-1} = z_{-1} = z_k = 0$. SAMPLEZ$_t(\sigma, c, \sigma_{max})$ is any algorithm that samples from a discrete gaussian over $\mathbb{Z}$ exactly or approximately with centers in $[0,1)$ and a fixed truncated support $\mathbb{Z} \cap [c - t \cdot \sigma_{max}, c + t \cdot \sigma_{max}]$ (t is the tail-cut parameter). We denote $x - \lfloor x \rfloor$ as $\lfloor x \rceil_{[0,1)}$.

Time Complexity. Assuming constant time sampling for SAMPLEZ and scalar arithmetic, SAMPLEG runs in time $O(k)$. Now let us consider all operations: there are $6k$ integer additions/subtractions, $3k + 2$ integer multiplications, $3(k + 1)$ floating point divisions, $2k$ floating point multiplications, and $2k$ floating point additions. The analysis below shows we can use double precision floating point numbers for most applications.

Statistical Analysis and Floating Point Precision. We now perform a statistical analysis on SAMPLEG with a perfect one-dimensional sampler (and no tail-bound), then with a tail-bounded imperfect sampler in terms of ML distance. This allows us to measure loss in concrete security. We direct the reader to [53, Sect. 3] for more details on the ML distance and a complete concrete security analysis.

The following lemma is needed in order to make sense of the "Σ_3 condition" in Theorem 3.

Lemma 4. *Let Σ_3 be defined by $\Sigma_3^{-1} = \frac{(b+1)^2}{s^2}[\Sigma_1^{-1} + [(b+1)^2\mathbf{I} - \Sigma]^{-1}]$, then its eigenvalues are $\Theta(s^2/b)$. Moreover, if λ_i is the i−th eigenvalue of Σ_1, then the i−th eigenvalue of Σ_3 is $(s/[b+1])^2 \cdot \frac{\lambda_i[(b+1)^2 - \lambda_i]}{(b+1)^2}$.*

Proof. Let $\Sigma_1 = \mathbf{Q}^T \mathbf{D} \mathbf{Q}$ be its diagonalization. Then, $\Sigma_1^{-1} = \mathbf{Q}^T \mathbf{D}^{-1} \mathbf{Q}$ and the rest follows from algebraic manipulations of the individual eigenvalues along with the Gershgorin circle theorem on Σ_1. $\qquad\square$

Let $C_{\epsilon,k} = \sqrt{\log(2k(1+1/\epsilon))/\pi}$. Now we can easily bound s from below. We need the following three conditions for s: $s \geq (b+1)\eta_\epsilon(\mathbf{D})$, $\sqrt{\Sigma_3} \geq \eta_\epsilon(\Sigma_2)$, and $s \geq (b+1)\eta_\epsilon(\mathbf{L})$. The middle condition determines s with a lower bound of $s \geq \sqrt{2b} \cdot (2b+1) \cdot C_{\epsilon,k}$ (the last two conditions both have $s = \Omega(b^{1.5} \cdot C_{\epsilon,k})$).

Corollary 1. *Fix $0 < \epsilon \leq 1/2$ and let $s \geq \sqrt{2b} \cdot (2b+1) \cdot C_{\epsilon,k}$. Then,* SAMPLEG *returns a perturbation within a statistical distance $\Theta(k\hat{\epsilon})$ from $D_{\Lambda_u^\perp(\mathbf{g}^T),s}$ for any $q < b^k$ when* PERTURB *and* SAMPLED *use a perfect one-dimensional sampler,* SAMPLEZ. *In addition, the Rényi divergence of order infinity of $D_{\Lambda_u^\perp(\mathbf{g}^T),s}$ from* SAMPLEG *with a perfect one-dimensional sampler is less than or equal to $1 + \Theta(k\hat{\epsilon})$.*

The statistical distance bound of $\Theta(k\hat{\epsilon})$ results in about a loss of $\log\log q$ bits in security if $\epsilon = 2^{-\kappa}$ for a security parameter κ by [53, Lemma 3.1]. (The multiplicative factor of k comes from the randomized nearest plane algorithm's analysis: see [31, Theorem 4.1].)

Next, we turn to the ML distance for a tighter analysis on the bits of security lost in using SAMPLEG with an imperfect one-dimensional sampler. Since the centers, c, and variances, s, given to SAMPLEZ are computed from two or three floating point computations, we assume both $\bar{c}$ and $\bar{s}$ are within a relative error of 2^{-m} of c and s.

Proposition 1. *Fix an $\epsilon > 0$ and let $s \geq (b+1) \cdot \eta_\epsilon(\mathbb{Z})$. For any one-dimensional sampler* SAMPLEZ$_t(\bar{\sigma}, \bar{c}, s)$ *that takes as inputs approximated centers $\bar{c} \in [0,1)$ and variances $\bar{\sigma} \in [s/(b+1), s \cdot b/(b+1)]$ represented as floating point numbers with mantissa length m, $\Delta_{\mathrm{ML}}(\mathrm{SAMPLEG}^{D_{\mathbb{Z},\sigma,c}^t}, \mathrm{SAMPLEG}^{\mathrm{SAMPLEZ}_t(\bar{\sigma},\bar{c})}) \leq 2k[O(b^2 t^2 2^{-m}) + \max_{\bar{\sigma},\bar{c}} \Delta_{\mathrm{ML}}(\mathrm{SAMPLEZ}_t(\bar{\sigma}, \bar{c}, s), D_{\mathbb{Z},\bar{\sigma},\bar{c}}^t)]$.*

Assuming a cryptosystem using a perfect sampler for $D_{\Lambda_u^\perp(g^T),s}$ has κ bits of security, we can combine the results of Corollary 1, Proposition 1, and [53, Lemma 3.3] to conclude that swapping $D_{\Lambda_u^\perp(g^T),s}$ with SAMPLEG yields about $\kappa - 2\log(tb^2) - 3\log\log q - 5$ bits of security when $m = \kappa/2$, $\Delta_{\mathrm{ML}}(\mathrm{SAMPLEZ}_t(\bar{s}, \bar{c}), D_{\mathbb{Z},\bar{s},\bar{c}}^t) < 2^{-\kappa/2}$, and $\epsilon = 2^{-\kappa}$.

3.3   Implementation and Comparison

In this subsection, we compare simple implementations of both SAMPLEG and the generic randomized nearest plane algorithm [31, Sect. 4] used in the G-lattice setting. The implementations were carried out in C++ with double precision floating point numbers for non-integers on an Intel i7-2600 3.4 GHz CPU. Clock cycles were measured with the "time.h" library and the results are charted in Fig. 3.

The one-dimensional sampler, SAMPLEZ, was an instantiation of a discrete version of Karney's sampler [41], which is a modified rejection sampler. The moduli q were chosen from the common parameters subsection of [40, Sect. 4.2],

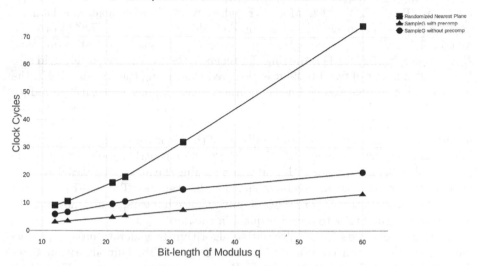

Fig. 3. Measured clock cycles with $q = \{4093, 12289, 1676083, 8383498, 4295967357, \approx 9 \cdot 10^{18}\}$ and $s = 100$ averaged over 100,000 runs. The clock cycles for the last three moduli are $\{19.4, 31.9, 73.9\}$ for GPV and $\{5.5, 7.5, 13.1\}$ for SAMPLEG with pre-computation.

in addition to an arbitrary 60-bit modulus. Most practical schemes require no more than a 30-bit modulus [9] for lattice dimension (n) up to 1024. More advanced schemes however, like ABE-encryption [14,19], predicate encryption [34], and obfuscation [20,26], require a super-polynomial modulus often 90 or more bits (assuming the circuits in the ABE and predicate schemes are of log-depth).

For the generic, randomized nearest plane sampler, we pre-computed and stored the Gram-Schmidt orthogonalization of the basis $\mathbf{B}_q$ and we only counted the clock cycles to run the algorithm thereafter. We had two versions of SAMPLEG: the first was the algorithm as-is, and the second would store pre-computed perturbations from PERTURB(σ), one for each G-lattice sample. This version of SAMPLEG with pre-computation saved about a factor of two in clock cycles.

4   Perturbation Sampling in Cyclotomic Rings

The lattice preimage sampling algorithm of [51] requires the generation of $n(2 + \log q)$-dimensional gaussian perturbation vectors $\mathbf{p}$ with covariance $\Sigma_p = s^2 \cdot \mathbf{I} - \alpha^2 \mathbf{T} \cdot \mathbf{T}^T$ where $\mathbf{T} \in \mathbb{Z}^{(2+\log q)n \times n \log q}$ is a matrix with small entries serving as a lattice trapdoor, α is a small constant factor and s is an upper bound on the spectral norm of $\alpha\mathbf{T}$. In [51] this is accomplished using the Cholesky factorization of Σ_p, which takes $O(n \log q)^3$ precomputation and $O(n \log q)^2$ storage and running time.

The trapdoor matrix $\mathbf{T}$ of [51] has some additional structure: $\mathbf{T}^T = [\bar{\mathbf{T}}^T, \mathbf{I}]$ for some $\bar{\mathbf{T}} \in \mathbb{Z}^{2n \times n \log q}$. Moreover, when working with algebraic lattices, $\bar{\mathbf{T}} = \phi_n(\tilde{\mathbf{T}})$ is the image (under a ring embedding $\phi_n \colon R_n \to \mathbb{Z}^{n \times n}$) of some matrix $\tilde{\mathbf{T}} \in R_n^{2 \times \log q}$ with entries in a ring R_n of rank n. (Most commonly, $R_n = \mathcal{O}_{2n} = \mathbb{Z}[x]/(x^n + 1)$ is the ring of integers of the $(2n)$th cyclotomic field $\mathcal{K}_{2n}$ for $n = 2^k$ a power of two.) In [9] it is observed that, using the sparsity of Σ_p, the preprocessing storage and on-line computation cost of noise perturbation reduce to $O(n^2 \log q)$.[4] This is a factor $\log q$ improvement over a generic implementation, but it is still quadratic in the main security parameter n. This can be a significant improvement in practice, but the overall cost of the algorithm remains substantial. When using generic trapdoors $\bar{\mathbf{T}} \in \mathbb{Z}^{2n \times n \log q}$, there is little hope to improve the running time below $O(n^2 \log q)$, because just reading the matrix $\bar{\mathbf{T}}$ takes this much time. However, when using algebraic lattices, the trapdoor $\bar{\mathbf{T}} = \phi_n(\tilde{\mathbf{T}})$ admits a compact representation $\tilde{\mathbf{T}}$ consisting of only $2n \log q$ integers, so one may hope to reduce the running time to linear or quasi-linear in n.

In this section we give an alternative algorithm to generate integer perturbation vectors $\mathbf{p}$ with covariance Σ_p when $\bar{\mathbf{T}} = \phi_n(\tilde{\mathbf{T}})$. Our algorithm takes full advantage of the ring structure of R_n, compactly representing Σ_p and all other matrices generated during the execution of the algorithm as the image of matrices with entries in the ring R_n. In particular, similarly to [27,28], our algorithm has time and space complexity quasi-linear in n, but does not require any preprocessing/storage. The algorithm can be expressed in a modular way as the combination of three steps:

1. First, the problem of sampling a $O(n \log q)$-dimensional integer vectors $\mathbf{p}$ with covariance Σ_p is reduced to the problem of sampling a $2n$-dimensional integer vector with covariance expressed by a 2×2 matrix over R_n.
2. Next, the problem of sampling an integer vector with covariance in $R_n^{2 \times 2}$ is reduced to sampling two n-dimensional integer vectors, each with a covariance expressed by a single ring element in R_n.
3. Finally, if $n > 1$, the sampling problem with covariance in R_n is reduced to sampling an n-dimensional perturbation with covariance expressed by a 2×2 matrix over the smaller ring $R_{n/2}$.

Iterating the last two steps $\log n$ times reduces the original problem to sampling in $R_1 = \mathbb{Z}$. Details about each step are given in the next subsections. We remark that the algorithm is described as a recursive procedure only for simplicity of presentation and analysis, and it can be implemented just as easily using a simple nested loop, similarly to many FFT-like algorithms.

4.1   Discrete Perturbation Algorithm for Power of Two Cyclotomics

In this subsection we present the perturbation algorithm algorithm which produces $n(2 + \log q)$-dimensional perturbations from a discrete gaussian on $\mathbb{Z}^{n(2+\log q)}$ in time $\tilde{O}(n \log q)$.

---

[4] Sparsity also reduces the preprocessing running time to $O(\log q \cdot n^2 + n^3) = O(n^3)$, but still cubic in n.

The entry point of the algorithm is the SAMPLEPz procedure, which takes as input two integer parameters n, q, matrices $\tilde{\mathbf{T}} \in R_n^{2 \times \log q}$, $\Sigma_2 \in R_n^{2 \times 2}$, and three positive real numbers s^2, α^2, $z = (\alpha^{-2} - s^{-2})^{-1}$, and is expected to produce an $n(2 + \log q)$-dimensional vector $\mathbf{p}$ with (non-spherical) discrete gaussian distribution $D_{\mathbb{Z}^{n(2+\log q)}, \sqrt{\Sigma_p}}$ of covariance

$$\Sigma_p = s^2 \cdot \mathbf{I} - \alpha^2 \begin{bmatrix} \phi_n(\tilde{\mathbf{T}}) \\ \mathbf{I} \end{bmatrix} \cdot \begin{bmatrix} \phi_n(\tilde{\mathbf{T}})^T & \mathbf{I} \end{bmatrix}$$

$$= \begin{bmatrix} \Sigma_2 & -\alpha^2 \phi_n(\tilde{\mathbf{T}}) \\ -\alpha^2 \phi_n(\tilde{\mathbf{T}})^T & (s^2 - \alpha^2)\mathbf{I} \end{bmatrix}.$$

The algorithm calls two subroutines:

- SAMPLEZ$(s^2 - \alpha^2)$ which samples a one-dimensional discrete gaussian variable of variance $s^2 - \alpha^2$ centered at 0, and can be implemented using any standard technique, and
- SAMPLE2Z(a, b, d), which, on input three ring elements a, b, d compactly describing a positive definite matrix

$$\Sigma_2 = \begin{bmatrix} \phi_n(a) & \phi_n(b) \\ \phi_n(b)^T & \phi_n(d) \end{bmatrix},$$

is expected to sample a $(2n)$-dimensional vector $p \leftarrow D_{\mathbb{Z}^{2n}, \sqrt{\Sigma_2}}$.

In turn, SAMPLE2Z (also described in Fig. 4) makes use of a procedure SAMPLEFz(f) which on input a ring element f with positive definite $\phi_n(f)$, returns a sample $p \leftarrow D_{\mathbb{Z}^n, \sqrt{\phi_n(f)}}$.

Efficiency. Multiplications are done in the field $\mathcal{K}_i$, for an element's dimension $i \in \{1, 2, \ldots, 2n\}$, in time $\Theta(i \log i)$ by using the Chinese remainder transform (*CRT*) [49].

By treating scalar arithmetic as constant time, SAMPLEPz has a time complexity of $\Theta(n \log n \log q)$ because the transformation by $\tilde{\mathbf{T}}$ is $\Theta(n \log n \log q)$ and SAMPLEFz has complexity $\Theta(n \log^2 n)$ (represented by the recurrence $R(n) = 2R(n/2) + 2\log n/2 + 4.5n$). The algorithm requires $2n \log q$ scalar storage for the trapdoor $\tilde{\mathbf{T}}$.

Note, SAMPLEFz is even more efficient, $\Theta(n \log n)$, if one were to store the polynomials in $\mathcal{K}_i$ in the canonical embedding (Fourier domain). One would change SAMPLEPz to give SAMPLE2Z the Fourier/canonical representations of a, b, d, c_0, c_1 and perform an inverse CRT/FFT on $\mathbf{p} = (\mathbf{p}_0, \mathbf{p}_1)$. This allows us to use the FFT's butterfly transformation to convert to the Fourier representation of $f(x) = f_0(x^2) + x f_1(x^2) \in \mathcal{K}_{2n}$ to the Fourier representation of $f_0, f_1 \in \mathcal{K}_n$ and multiplication/inversion is now linear time (we would only invert the non-zero entries in the Fourier domain since this corresponds to pulling back to the field, inverting, then pushing forward to the cyclic ring via the embedding given by the Chinese remainder theorem) [28, Lemma 1]. (Moving from the canonical

SAMPLE2Z(a, b, d, c)
   let $c(x) = c_0(x^2) + x \cdot c_1(x^2)$
   $q_1 \leftarrow$ SAMPLEFZ(d, c_1)
   $c_0 := c_0 + bd^{-1}(q_1 - c_1)$
   $q_0 \leftarrow$ SAMPLEFZ$(a - bd^{-1}b^*, c_0)$
   return (q_0, q_1)

SAMPLEPZ$(n, q, s, \alpha, \tilde{\mathbf{T}}, \Sigma_2, z)$
   for $i = 0, \ldots, (n \log q - 1)$:
     $q_i \leftarrow$ SAMPLEZ$(s^2 - \alpha^2)$
   $(c_0, c_1) := -\frac{\alpha^2}{s^2 - \alpha^2} \tilde{\mathbf{T}} \mathbf{q}$
   $c'(x) := c_0(x^2) + x \cdot c_1(x^2))$
   $\mathbf{p} \leftarrow$ SAMPLE2Z(a, b, d, c')
   return $(\mathbf{p}, \mathbf{q})$

SAMPLEFZ(f, c)
   if $\dim(f) = 1$ return SAMPLEZ(f, c)
   else let $f(x) = f_0(x^2) + x \cdot f_1(x^2)$
     $(q_0, q_1) \leftarrow$ SAMPLE2Z(f_0, f_1, f_0, c)
     let $q(x) = q_0(x^2) + x \cdot q_1(x^2)$
     return q

Fig. 4. Sampling algorithm SAMPLEPZ for integer perturbations where $\mathbf{T} = \phi_n(\tilde{\mathbf{T}})$ is a compact trapdoor over a power of two cyclotomic ring. Note, $\tilde{\mathbf{T}}_i$ is a *row* vector over R_n for each $i \in \{0, 1\}$. The algorithm uses a subroutine SAMPLEZ(σ^2, t) which samples a discrete gaussian over $\mathbb{Z}$ with variance σ^2 centered at t. The scalar $z = (\alpha^{-2} - s^{-2})^{-1}$.

embedding to the FFT domain is linear time since we place zeros for the non-primitive roots of unity [28, Sect. A.2].) This, however, does not change the asymptotic time complexity of SAMPLEPZ since generating $\mathbf{q}$ in the canonical embedding is now $\Theta(n \log n \log q)$.

Correctness. One would use Peikert's convolution theorem, Theorem 3, in an initial attempt to prove the correctness of the algorithms in Fig. 4. However, this would only ensure the correctness of the marginal distributions of $\mathbf{p}$ in SAMPLEPZ and q_0 in SAMPLE2Z and not their respective joint distributions, $(\mathbf{p}, \mathbf{q})$ and (q_0, q_1). Even if it were enough, tracking the Σ_3 condition in Theorem 3 through the recursive calls of the algorithms above is tedious. Instead, we derive a convolution lemma without a Σ_3 condition for the joint distribution of our discrete gaussian convolutions on the simple lattice $\mathbb{Z}^n$.

First, we show the gaussian function $\rho_{\sqrt{\Sigma}}(\cdot)$ factors in a useful manner with respect to a Schur complement decomposition.

Lemma 5. *Let* $\Sigma = \begin{bmatrix} \mathbf{A} & \mathbf{B} \\ \mathbf{B}^T & \mathbf{D} \end{bmatrix} \succ \mathbf{0}$ *be a positive definite with* $\mathbf{A} \in \mathbb{R}^{n \times n}$ *and* $\mathbf{D} \in \mathbb{R}^{m \times m}$ *and* $\Sigma/\mathbf{D} = \mathbf{A} - \mathbf{BD}^{-1}\mathbf{B}^T$ *is* $\mathbf{D}$*'s Schur complement, and let* $\mathbf{x}_1 \in \mathbb{R}^n$ *and* $\mathbf{x}_2 \in \mathbb{R}^m$ *be arbitrary. Then, the gaussian function* $\rho_{\sqrt{\Sigma}}(\mathbf{x})$ *factors as* $\rho_{\sqrt{\Sigma/\mathbf{D}}}(\mathbf{x}_1 - \mathbf{BD}^{-1}\mathbf{x}_2) \cdot \rho_{\sqrt{\mathbf{D}}}(\mathbf{x}_2) = \rho_{\sqrt{\Sigma}}(\mathbf{x})$ *where* $\mathbf{x} = (\mathbf{x}_1, \mathbf{x}_2) \in \mathbb{R}^{n+m}$.

Proof (Sketch). This is seen through defining the inverse of Σ in terms of $\Sigma/\mathbf{D}$ and writing out $\rho_{\sqrt{\Sigma}}(\mathbf{x})$ in terms of $\Sigma/\mathbf{D}$. The matrix factorization

$$\Sigma = \begin{bmatrix} \mathbf{I} & \mathbf{BD}^{-1} \\ \mathbf{0} & \mathbf{I} \end{bmatrix} \begin{bmatrix} \Sigma/\mathbf{D} & \mathbf{0} \\ \mathbf{0} & \mathbf{D} \end{bmatrix} \begin{bmatrix} \mathbf{I} & \mathbf{0} \\ \mathbf{D}^{-1}\mathbf{B}^T & \mathbf{I} \end{bmatrix}$$

yields the formula for Σ^{-1} needed to show the result. $\square$

A consequence of the above lemma is that the gaussian sum $\rho_{\sqrt{\Sigma}}(\mathbb{Z}^{n+m})$ expands in terms of the gaussian functions $\rho_{\sqrt{\mathbf{D}}}(\cdot)$ and $\rho_{\sqrt{\Sigma/\mathbf{D}}}(\cdot)$,

$$\rho_{\sqrt{\Sigma}}(\mathbb{Z}^{n+m}) = \sum_{\mathbf{y}_2 \in \mathbb{Z}^m} \rho_{\sqrt{\mathbf{D}}}(\mathbf{y}_2) \cdot \rho_{\sqrt{\Sigma/\mathbf{D}}}(\mathbb{Z}^n - \mathbf{B}\mathbf{D}^{-1}\mathbf{y}_2).$$

We will use the following lemma for the correctness proof. It states that if a discrete gaussian on the integer lattice is wide enough in its slimmest direction, then the lower dimensional discrete gaussians with covariance shaped with principal submatrices of the original are wide enough on their respective $\mathbb{Z}^{n'}$s.

Lemma 6. *Let $\epsilon > 0$, $\Sigma \succ 0$ be a positive definite matrix in $\mathbb{R}^{n \times n}$, and let $I_0 \subset [n]$ be an arbitrary, non-empty subset. If $\Sigma \succeq \eta_\epsilon^2(\mathbb{Z}^n)$, then $\Sigma[I_0] \succeq \eta_\epsilon^2(\mathbb{Z}^{|I_0|})$ and $\Sigma/\bar{I}_0 \succeq \eta_\epsilon^2(\mathbb{Z}^{n-|I_0|})$ for any principal submatrix - Schur complement pair, $(\Sigma[I_0], \Sigma/\bar{I}_0)$, of Σ.*

Proof. Note, a consequence of $\Sigma \succeq \eta_\epsilon^2(\mathbb{Z}^n)$ is that Σ's minimum eigenvalue, $\lambda_{min}(\Sigma)$, is greater than $\eta_\epsilon^2(\mathbb{Z}^n)$. Let $\mathbf{M} := \Sigma[I_0] \in \mathbb{R}^{n_0 \times n_0}$ for $n_0 = |I_0|$. $\mathbf{M}$ is diagonalizable so let $\mathbf{M} = \mathbf{Q}^T \mathbf{\Lambda} \mathbf{Q}$ be its diagonalization. Notice, we have the following inequality from the interlacing theorems which imply $\lambda_{min}(\mathbf{M}) \geq \lambda_{min}(\Sigma)$,

$$\mathbf{x}^T \mathbf{M} \mathbf{x} = \mathbf{x}^T \mathbf{Q}^T \mathbf{\Lambda} \mathbf{Q} \mathbf{x} = \mathbf{y}^T \mathbf{\Lambda} \mathbf{y} = \sum_{i \in [n_0]} \lambda_i y_i^2 \geq \lambda_{min}(\Sigma) \|\mathbf{y}\|^2 = \lambda_{min}(\Sigma) \|\mathbf{x}\|^2.$$

Next, we can bound the quantity $\rho_{\sqrt{\mathbf{M}^{-1}}}((\mathbb{Z}^{n_0})^*) = \rho_{\sqrt{\mathbf{M}^{-1}}}(\mathbb{Z}^{n_0})$ by $1 + \epsilon$:

$$\rho_{\sqrt{\mathbf{M}^{-1}}}(\mathbb{Z}^{n_0}) = \sum_{\mathbf{x} \in \mathbb{Z}^{n_0}} e^{-\pi \mathbf{x}^T \mathbf{M} \mathbf{x}} \leq \sum_{\mathbf{x} \in \mathbb{Z}^{n_0}} e^{-\pi \lambda_{min}(\Sigma) \|\mathbf{x}\|^2} \tag{2}$$

$$\leq \sum_{\mathbf{x} \in \mathbb{Z}^n} e^{-\pi \lambda_{min}(\Sigma) \|\mathbf{x}\|^2} \leq 1 + \epsilon. \tag{3}$$

The jump from $\mathbb{Z}^{n_0}$ to $\mathbb{Z}^n$ comes from the relation $\mathbb{Z}^{n_0} \subset \mathbb{Z}^n$. The proof for the Schur complement is identical. $\qquad\square$

Next, we state and prove our main convolution lemma.

Lemma 7. *For any real $0 < \epsilon \leq 1/2$, positive integers n, m, vector $\mathbf{c} = (\mathbf{c}_1, \mathbf{c}_2) \in \mathbb{R}^{n+m}$, and positive definite matrix $\Sigma = \begin{bmatrix} \mathbf{A} & \mathbf{B} \\ \mathbf{B}^T & \mathbf{D} \end{bmatrix} \succeq \eta_\epsilon^2(\mathbb{Z}^{n+m})$, $\mathbf{A} \in \mathbb{Z}^{n \times n}$, $\mathbf{B} \in \mathbb{Z}^{n \times m}$, and $\mathbf{D} \in \mathbb{Z}^{m \times m}$ (where $\Sigma/\mathbf{D} = \mathbf{A} - \mathbf{B}\mathbf{D}^{-1}\mathbf{B}^T$ is the Schur complement of $\mathbf{D}$) the random process*

- *$\mathbf{x}_2 \leftarrow D_{\mathbb{Z}^m, \sqrt{\mathbf{D}}, \mathbf{c}_2}$.*
- *$\mathbf{x}_1 \leftarrow D_{\mathbb{Z}^n, \sqrt{\Sigma/\mathbf{D}}, \mathbf{c}_1 + \mathbf{B}\mathbf{D}^{-1}(\mathbf{x}_2 - \mathbf{c}_2)}$.*

produces a vector $\mathbf{x} = (\mathbf{x}_1, \mathbf{x}_2) \in \mathbb{Z}^{n+m}$ *such that the Rényi divergence of order infinity of* $D_{\mathbb{Z}^{n+m}, \sqrt{\Sigma}, \mathbf{c}}$ *from* $\mathbf{x}$ *is less than or equal to* $1 + 4\epsilon$.

Proof. First, we write out the probability and use Lemma 5 to simplify the numerator. Let $\mathbf{x}' = (\mathbf{x}'_1, \mathbf{x}'_2)$ below.

$$\Pr[\mathbf{x}_1 = \mathbf{x}'_1, \mathbf{x}_2 = \mathbf{x}'_2] = \frac{\rho_{\sqrt{\Sigma/\mathbf{D}}}(\mathbf{x}'_1 - \mathbf{c}_1 - \mathbf{BD}^{-1}(\mathbf{x}'_2 - \mathbf{c}_2)) \cdot \rho_{\sqrt{\mathbf{D}}}(\mathbf{x}'_2 - \mathbf{c}_2)}{\rho_{\sqrt{\Sigma/\mathbf{D}}}(\mathbb{Z}^n - \mathbf{c}_1 - \mathbf{BD}^{-1}(\mathbf{x}'_2 - \mathbf{c}_2)) \cdot \rho_{\sqrt{\mathbf{D}}}(\mathbb{Z}^m - \mathbf{c}_2)}$$

$$= \frac{\rho_{\sqrt{\Sigma}}(\mathbf{x}' - \mathbf{c})}{\rho_{\sqrt{\Sigma/\mathbf{D}}}(\mathbb{Z}^n - \mathbf{c}_1 - \mathbf{BD}^{-1}(\mathbf{x}'_2 - \mathbf{c}_2)) \cdot \rho_{\sqrt{\mathbf{D}}}(\mathbb{Z}^m - \mathbf{c}_2)}$$

Regarding the denominator, we use Lemma 6 to see that $\Sigma/\mathbf{D} \succeq \eta_\epsilon^2(\mathbb{Z}^n)$ since $\Sigma \succeq \eta_\epsilon^2(\mathbb{Z}^{n+m})$. Now, we can use Lemma 3 for the first gaussian sum (dependent on $\mathbf{x}'_2$) in the denominator to see,

$$\Pr[\mathbf{x}_1 = \mathbf{x}'_1, \mathbf{x}_2 = \mathbf{x}'_2] \in \alpha \cdot D_{\mathbb{Z}^{n+m}, \sqrt{\Sigma}, \mathbf{c}}(\mathbf{x}') \cdot \left[\left(\frac{1-\epsilon}{1+\epsilon} \right), 1 \right]^{-1}$$

where $\alpha = \frac{\rho_{\sqrt{\Sigma}}(\mathbb{Z}^{n+m} - \mathbf{c})}{\rho_{\sqrt{\Sigma/\mathbf{D}}}(\mathbb{Z}^n) \cdot \rho_{\sqrt{\mathbf{D}}}(\mathbb{Z}^m - \mathbf{c}_2)}$.

Next, we show $\alpha \approx 1$. Using Lemma 5 we expand

$$\rho_{\sqrt{\Sigma}}(\mathbb{Z}^{n+m} - \mathbf{c}) = \sum_{\mathbf{y}_2 \in \mathbb{Z}^m} \rho_{\sqrt{\mathbf{D}}}(\mathbf{y}_2 - \mathbf{c}_2) \cdot \rho_{\sqrt{\Sigma/\mathbf{D}}}(\mathbb{Z}^n - \mathbf{c}_1 - \mathbf{BD}^{-1}(\mathbf{y}_2 - \mathbf{c}_2)).$$

The sum $\rho_{\sqrt{\Sigma/\mathbf{D}}}(\mathbb{Z}^n - \mathbf{c}_1 - \mathbf{BD}^{-1}(\mathbf{y}_2 - \mathbf{c}_2))$ is approximately $\rho_{\sqrt{\Sigma/\mathbf{D}}}(\mathbb{Z}^n)$ because $\Sigma/\mathbf{D} \succeq \eta_\epsilon^2(\mathbb{Z}^n)$ as a consequence of Lemma 6 and $\Sigma \succeq \eta_\epsilon^2(\mathbb{Z}^{n+m})$. In other words,

$$\rho_{\sqrt{\Sigma/\mathbf{D}}}(\mathbb{Z}^n - \mathbf{c}_1 - \mathbf{BD}^{-1}(\mathbf{y}_2 - \mathbf{c}_2)) \in \left[\frac{1-\epsilon}{1+\epsilon}, 1 \right] \cdot \rho_{\sqrt{\Sigma/\mathbf{D}}}(\mathbb{Z}^n)$$

and $\alpha \in \left[\left(\frac{1-\epsilon}{1+\epsilon} \right), 1 \right]$.

Finally, we have the approximation

$$\Pr[\mathbf{x}_1 = \mathbf{x}'_1, \mathbf{x}_2 = \mathbf{x}'_2] \in \left[\left(\frac{1-\epsilon}{1+\epsilon} \right), \left(\frac{1+\epsilon}{1-\epsilon} \right) \right] \cdot D_{\mathbb{Z}^{n+m}, \sqrt{\Sigma}, \mathbf{c}}(\mathbf{x}').$$

Given the restriction on $\epsilon \in (0, 1/2]$, we have the relation we desire

$$\Pr[\mathbf{x}_1 = \mathbf{x}'_1, \mathbf{x}_2 = \mathbf{x}'_2] \in [1 - 4\epsilon, 1 + 4\epsilon] \cdot D_{\mathbb{Z}^{n+m}, \sqrt{\Sigma}, \mathbf{c}}(\mathbf{x}'). \qquad \square$$

Next, we bound the Rényi divergence of order infinity between the output of SAMPLEPZ and the desired distribution. We need to ensure each discrete gaussian convolution in the algorithm is non-degenerate. We do not analyze the statistical loss from the floating point computations. As shown in Lemma 7, we need $\Sigma/\mathbf{D} \succeq \eta_\epsilon^2(\mathbb{Z}^{n_0})$ and $\mathbf{D} \succeq \eta_\epsilon^2(\mathbb{Z}^{n_1})$ at each of the n discrete gaussian convolutions. This is met through a simple condition on Σ_p as hinted to in Lemma 6.

Theorem 5. *Let $0 < \epsilon \le 1/2$. If $\Sigma_p \succeq \eta_\epsilon^2(\mathbb{Z}^{n(2+\log q)})$, then SAMPLEPZ returns a perturbation with a Rényi divergence of order infinity $R_\infty(D_{\mathbb{Z}^{n(2+\log q)}, \sqrt{\Sigma_p}} \|$ SAMPLEPZ$) \le 1 + 12n\hat{\epsilon}$.*

Proof. Since each covariance given to SAMPLEFZ is a Schur complement or a principal submatrix of a Schur complement of Σ_p, Lemma 6 and the interlacing theorems (Theorems 1 and 2) imply the conditions for Lemma 7 are met. As there are $n-1$ convolutions (inner nodes of a full binary tree of depth $\log n$), a quick induction argument shows the probability distribution of the output of SAMPLEPZ is in the interval $[(1-4\epsilon)^{3(n-1)}, (1+4\epsilon)^{3(n-1)}] \cdot D_{\mathbb{Z}^{n(2+\log q)}, \sqrt{\Sigma_p}}(\mathbf{x})$. Then, we have $R_\infty(D_{\mathbb{Z}^{n(2+\log q)}, \sqrt{\Sigma_p}} \| \text{SAMPLEPZ}) \le (1+4\epsilon)^{3(n-1)} \approx 1 + 12n\hat{\epsilon}$. $\square$

For common parameters $\epsilon = 2^{-128}$ and $n = 1024$, we have $1 - (1+4\epsilon)^{3(n-1)} \approx 2^{-114}$.

In summary, this shows the FFT-like recurrence in perturbation sampling the integer lattice with an algebraic covariance in power of two cyclotomic rings through repeated convolutions. The relative simplicity of the power of two case relies on the fact that matrix transpose corresponds to the conjugation field automorphism. Hermitian transpose corresponds to the conjugation automorphism in the general cyclotomic case. Therefore, we would use the canonical embedding for efficient perturbation sampling in general cyclotomic rings.

Acknowledgment. We thank Léo Ducas, Yuriy Polyakov, Kurt Rohloff, and Michael Walter for their helpful discussions as well as the anonymous reviewers for their helpful feedback and suggestions.

A   Missing Proofs

ML Analysis. Here we give the proof of Proposition 3.1. We restate the proposition for convenience.

Proposition 2. *Fix an $\epsilon > 0$ and let $s \ge (b+1)\cdot\eta_\epsilon(\mathbb{Z})$. For any one-dimensional sampler SAMPLEZ$_t(\bar{\sigma}, \bar{c}, s)$ that takes as inputs approximated centers $\bar{c} \in [0,1)$ and variances $\bar{\sigma} \in [s/(b+1), s \cdot b/(b+1)]$ represented as floating point numbers with mantissa length m, $\Delta_{\mathrm{ML}}(\text{SAMPLEG}^{D_{\mathbb{Z},\sigma,c}^t}, \text{SAMPLEG}^{\text{SAMPLEZ}_t(\bar{\sigma},\bar{c})}) \le 2k[O(t^2 2^{-m}) + \max_{\bar{\sigma},\bar{c}} \Delta_{\mathrm{ML}}(\text{SAMPLEZ}_t(\bar{\sigma}, \bar{c}, s), D_{\mathbb{Z},\bar{\sigma},\bar{c}}^t)]$.*

Before we begin the proof, we note that $d_{k-1} = q/b^k \in [1/b, 1]$ since $k = \lceil \log_b q \rceil$. This implies that every variance fed to SAMPLEZ is in the range $[s/(b+1), s \cdot b/(b+1)] \subseteq [s/(b+1), s]$. We restrict all truncated one-dimensional discrete gaussians to $\mathbb{Z} \cap [c - t \cdot s, c + t \cdot s]$ since it is unclear when $\mathbb{Z} \cap [c - t \cdot \sigma, c + t \cdot \sigma] = \mathbb{Z} \cap [c - t \cdot \bar{\sigma}, c + t \cdot \bar{\sigma}]$ when using floating point variances $\bar{\sigma}$. The ML distance is undefined when these two sets are not equal.

Proof. First, we use the triangle inequality on ML distance in order to pair together terms for an easier analysis.

$$\Delta_{\mathrm{ML}}(\mathrm{SAMPLEG}^{D^t_{\mathbb{Z},\sigma,c}}, \mathrm{SAMPLEG}^{\mathrm{SAMPLEZ}_t(\bar{\sigma},\bar{c},s)}) \leq$$

$$\Delta_{\mathrm{ML}}(\mathrm{SAMPLEG}^{D^t_{\mathbb{Z},\sigma,c}}, \mathrm{SAMPLEG}^{D^t_{\mathbb{Z},\bar{\sigma},c}}) + \Delta_{\mathrm{ML}}(\mathrm{SAMPLEG}^{D^t_{\mathbb{Z},\bar{\sigma},c}}, \mathrm{SAMPLEG}^{D^t_{\mathbb{Z},\bar{\sigma},\bar{c}}}) +$$

$$\Delta_{\mathrm{ML}}(\mathrm{SAMPLEG}^{D^t_{\mathbb{Z},\bar{\sigma},\bar{c}}}, \mathrm{SAMPLEG}^{\mathrm{SAMPLEZ}_t(\bar{\sigma},\bar{c},s)}).$$

Next, we use the data processing inequality on ML distance where we treat SAMPLEG as a function of $2k$ correlated samples from a one-dimensional discrete gaussian sampler. From [Lemma 3.2, MW17], we get the following inequality:

$$\Delta_{\mathrm{ML}}(\mathrm{SAMPLEG}^{D^t_{\mathbb{Z},\sigma,c}}, \mathrm{SAMPLEG}^{\mathrm{SAMPLEZ}_t(\bar{\sigma},\bar{c},s)}) \leq$$

$$2k \cdot \max_{\sigma_i, c_i} [\Delta_{\mathrm{ML}}(D^t_{\mathbb{Z},\sigma_1,c_1}, D^t_{\mathbb{Z},\bar{\sigma}_1,c_1}) + \Delta_{\mathrm{ML}}(D^t_{\mathbb{Z},\bar{\sigma}_2,c_2}, D^t_{\mathbb{Z},\bar{\sigma}_2,\bar{c}_2}) +$$

$$\Delta_{\mathrm{ML}}(D^t_{\mathbb{Z},\bar{\sigma}_3,\bar{c}_3}, \mathrm{SAMPLEZ}_t(\bar{\sigma}_3,\bar{c}_3,s))].$$

The maximum is taken over all $c_i \in [0,1)$ and $\sigma_i \in [s/(b+1), s \cdot b/(b+1)]$. Let $\mathbb{Z}^t = \mathbb{Z} \cap [c - t \cdot s, c + t \cdot s]$. We bound $\max_{\sigma_1, c_1} \Delta_{\mathrm{ML}}(D^t_{\mathbb{Z},\sigma_1,c_1}, D^t_{\mathbb{Z},\bar{\sigma}_1,c_1})$ as follows:

$$\max_{\sigma_1, c_1} \Delta_{\mathrm{ML}}(D^t_{\mathbb{Z},\sigma_1,c_1}, D^t_{\mathbb{Z},\bar{\sigma}_1,c_1}) = \max_{\sigma_1, c_1, x \in \mathbb{Z}^t} |\ln D^t_{\mathbb{Z},\sigma_1,c_1}(x) - \ln D^t_{\mathbb{Z},\bar{\sigma}_1,c_1}(x)|$$

$$= \max_{\sigma_1, c_1, x \in \mathbb{Z}^t} \left| \pi(x - c)^2 \left[\frac{1}{\sigma_1^2} - \frac{1}{\bar{\sigma}_1^2} \right] + \ln \frac{\rho_{\bar{\sigma}_1, c_1}(\mathbb{Z})}{\rho_{\sigma_1, c_1}(\mathbb{Z})} \right|.$$

Since $\sigma_1, \bar{\sigma}_1 \geq \eta_\epsilon(\mathbb{Z})$, we can approximate $\rho_{\sigma_1,c}(\mathbb{Z}) \in [(1 - \epsilon)^2, (1 + \epsilon)^2] \cdot \sigma$ and $\rho_{\bar{\sigma}_1,c}(\mathbb{Z}) \in [(1 - \epsilon)^2, (1 + \epsilon)^2] \cdot \bar{\sigma}$. Using the bound on the relative error of $\bar{\sigma}_1$ ($\bar{\sigma}_1 \in [1 - 2^{-m}, 1 + 2^{-m}] \cdot \sigma_1$), we can bound the expression with a simplified form below.

$$\max_{\sigma_1, c_1} \Delta_{\mathrm{ML}}(D^t_{\mathbb{Z},\sigma_1,c_1}, D^t_{\mathbb{Z},\bar{\sigma}_1,c_1}) \leq$$

$$\max_{\sigma_1} \left| \pi \frac{t^2 s^2}{\sigma_1^2} \cdot \frac{\bar{\sigma}_1^2 - \sigma_1^2}{\sigma_1^2} + 2\hat{\epsilon} + 2^{-\hat{}m} \right| \leq$$

$$\pi t^2 (b+1)^2 (2^{-m+1} + 2^{-2m}) + \hat{\epsilon} + 2^{-\hat{}m}.$$

The proof for $\Delta_{\mathrm{ML}}(D^t_{\mathbb{Z},\bar{\sigma}_2,c_2}, D^t_{\mathbb{Z},\bar{\sigma}_2,\bar{c}_2})$ is nearly identical except we get a term linear in t, yielding a bound of $O(t \cdot 2^{-m})$. $\square$

B   QR Factorization for the Basis $\mathbf{B}_q$

Here we show that despite $\mathbf{B}_q$ having a sparse $\mathbf{R}$ matrix in its QR-factorization, this does not lead to an alternative $\Theta(\log q)$-time sampling algorithm for the applications we are concerned with. The QR-factorization of a non-singular matrix $\mathbf{S}$ is $\mathbf{S} = \mathbf{QR}$ where $\mathbf{Q}$ is orthogonal and $\mathbf{R}$ is upper-triangular.

The motivation for such an algorithm comes from generic lattice algorithms, like BKZ lattice reduction, where we view the vector space holding our lattice in the basis given by the upper-triangular $\mathbf{R}$ since $\mathbf{Q}$ is orthogonal. The sparsity of $\mathbf{R}$ yields clear computational advantages.

In the G-lattice setting, the basis $\mathbf{B}_q = \mathbf{QR}$ *always* has a sparse $\mathbf{R}$ matrix (though $\mathbf{Q}$ is not sparse). This leads to a linear time algorithm to sample $D_{\mathcal{L}(\mathbf{R}),s}$ by using the canonical randomized nearest plane algorithm and a linear time algorithm for applications *if* we can view the ambient vector space in terms of $\mathbf{R}$ as a basis. Unfortunately, we cannot do this in the G-lattice setting.

Recall the general G-lattice paradigm: we have a secret trapdoor matrix $\mathbf{T}$ with small integer entries, a public psuedo-random matrix $\mathbf{A} = [\hat{\mathbf{A}} | \mathbf{G} - \hat{\mathbf{A}} \cdot \mathbf{T}]$, and we want to return a short vector in $\Lambda_{\mathbf{u}}^{\perp}(\mathbf{A}) = \{\mathbf{x} \in \mathbb{Z}^m : \mathbf{A} \cdot \mathbf{x} = \mathbf{u} \mod q\}$. The way we sample $\Lambda_{\mathbf{u}}^{\perp}(\mathbf{A})$ is as follows:

1. sample the perturbation $\mathbf{p} \sim D_{\mathbb{Z}^{n(2+\log q)}, \sqrt{s^2 \mathbf{I} - \mathbf{MM}^T}}$ where $\mathbf{M} = \begin{bmatrix} \mathbf{T} \\ \mathbf{I} \end{bmatrix}$
2. set the new coset $\mathbf{v} := \mathbf{u} - \mathbf{Ap} \mod q$
3. sample the G-lattice $\mathbf{y} \sim D_{\Lambda^{\perp}(\mathbf{G})_{\mathbf{v}}, s} = D_{\mathcal{L}(\mathbf{B}_q)+\mathbf{v}, s}$ where $\mathbf{G} = \mathbf{I}_n \otimes \mathbf{g}^T$
4. return $\mathbf{p} + \begin{bmatrix} \mathbf{T} \\ \mathbf{I} \end{bmatrix} \mathbf{y}$.

Next, we only consider the zero-coset of $\Lambda^{\perp}(\mathbf{G})$ for simplicity. Usually $\mathbf{y} = (\mathbf{I}_n \otimes \mathbf{B}_q)\mathbf{z}$ for $\mathbf{z} \in \mathbb{Z}^{n \log q}$. But if we were to use the sparsity of $\mathbf{R}$, then

$$\mathbf{y} = (\mathbf{I}_n \otimes \mathbf{R})\mathbf{z} = (\mathbf{I}_n \otimes \mathbf{Q}^T \mathbf{B}_q)\mathbf{z}.$$

Therefore, we would have to apply $\mathbf{Q}' := \mathbf{I}_n \otimes \mathbf{Q}$ as a linear transformation to $\mathbf{y}$ ($\Theta(n \log^2 q)$ time) yielding a quadratic increase (in $\log q$) in the last step as well as a quadratic increase in storage.

References

1. Agrawal, S.: Stronger security for reusable garbled circuits, general definitions and attacks. In: Katz, J., Shacham, H. (eds.) CRYPTO 2017. LNCS, vol. 10401, pp. 3–35. Springer, Cham (2017). https://doi.org/10.1007/978-3-319-63688-7_1
2. Agrawal, S., Boneh, D., Boyen, X.: Efficient lattice (H)IBE in the standard model. In: Gilbert, H. (ed.) EUROCRYPT 2010. LNCS, vol. 6110, pp. 553–572. Springer, Heidelberg (2010). https://doi.org/10.1007/978-3-642-13190-5_28
3. Agrawal, S., Boneh, D., Boyen, X.: Lattice basis delegation in fixed dimension and shorter-ciphertext hierarchical IBE. In: Rabin [59], pp. 98–115
4. Agrawal, S., Boyen, X., Vaikuntanathan, V., Voulgaris, P., Wee, H.: Functional encryption for threshold functions (or fuzzy IBE) from lattices. In: Fischlin et al. [29], pp. 280–297
5. Ajtai, M.: Generating hard instances of the short basis problem. In: Wiedermann, J., van Emde Boas, P., Nielsen, M. (eds.) ICALP 1999. LNCS, vol. 1644, pp. 1–9. Springer, Heidelberg (1999). https://doi.org/10.1007/3-540-48523-6_1

6. Alperin-Sheriff, J., Peikert, C.: Circular and KDM security for identity-based encryption. In: Fischlin et al. [29], pp. 334–352

7. Alwen, J., Peikert, C.: Generating shorter bases for hard random lattices. Theor. Comput. Syst. **48**(3), 535–553 (2011)

8. Bai, S., Langlois, A., Lepoint, T., Stehlé, D., Steinfeld, R.: Improved security proofs in lattice-based cryptography: using the Rényi divergence rather than the statistical distance. In: Iwata, T., Cheon, J.H. (eds.) ASIACRYPT 2015. LNCS, vol. 9452, pp. 3–24. Springer, Heidelberg (2015). https://doi.org/10.1007/978-3-662-48797-6_1

9. El Bansarkhani, R.E., Buchmann, J.A.: Improvement and efficient implementation of a lattice-based signature scheme. In: Lange, T., Lauter, K., Lisoněk, P. (eds.) SAC 2013. LNCS, vol. 8282, pp. 48–67. Springer, Heidelberg (2014). https://doi.org/10.1007/978-3-662-43414-7_3

10. Bellare, M., Kiltz, E., Peikert, C., Waters, B.: Identity-based (lossy) trapdoor functions and applications. In: Pointcheval and Johansson [58], pp. 228–245

11. Bendlin, R., Krehbiel, S., Peikert, C.: How to share a lattice trapdoor: threshold protocols for signatures and (H)IBE. In: Jacobson, M., Locasto, M., Mohassel, P., Safavi-Naini, R. (eds.) ACNS 2013. LNCS, vol. 7954, pp. 218–236. Springer, Heidelberg (2013). https://doi.org/10.1007/978-3-642-38980-1_14

12. Boneh, D., Freeman, D.M.: Homomorphic signatures for polynomial functions. In: Paterson, K.G. (ed.) EUROCRYPT 2011. LNCS, vol. 6632, pp. 149–168. Springer, Heidelberg (2011). https://doi.org/10.1007/978-3-642-20465-4_10

13. Boneh, D., Freeman, D.M.: Linearly homomorphic signatures over binary fields and new tools for lattice-based signatures. In: Catalano, D., Fazio, N., Gennaro, R., Nicolosi, A. (eds.) PKC 2011. LNCS, vol. 6571, pp. 1–16. Springer, Heidelberg (2011). https://doi.org/10.1007/978-3-642-19379-8_1

14. Boneh, D., Gentry, C., Gorbunov, S., Halevi, S., Nikolaenko, V., Segev, G., Vaikuntanathan, V., Vinayagamurthy, D.: Fully key-homomorphic encryption, arithmetic circuit ABE and compact garbled circuits. In: Nguyen, P.Q., Oswald, E. (eds.) EUROCRYPT 2014. LNCS, vol. 8441, pp. 533–556. Springer, Heidelberg (2014). https://doi.org/10.1007/978-3-642-55220-5_30

15. Boneh, D., Kim, S., Nikolaenko, V.: Lattice-based DAPS and generalizations: self-enforcement in signature schemes. In: Gollmann, D., Miyaji, A., Kikuchi, H. (eds.) ACNS 2017. LNCS, vol. 10355, pp. 457–477. Springer, Cham (2017). https://doi.org/10.1007/978-3-319-61204-1_23

16. Boyen, X.: Attribute-based functional encryption on lattices. In: Sahai, A. (ed.) TCC 2013. LNCS, vol. 7785, pp. 122–142. Springer, Heidelberg (2013). https://doi.org/10.1007/978-3-642-36594-2_8

17. Boyen, X., Li, Q.: Attribute-based encryption for finite automata from LWE. In: Au, M.-H., Miyaji, A. (eds.) ProvSec 2015. LNCS, vol. 9451, pp. 247–267. Springer, Cham (2015). https://doi.org/10.1007/978-3-319-26059-4_14

18. Boyen, X., Li, Q.: Towards tightly secure lattice short signature and id-based encryption. In: Cheon, J.H., Takagi, T. (eds.) ASIACRYPT 2016. LNCS, vol. 10032, pp. 404–434. Springer, Heidelberg (2016). https://doi.org/10.1007/978-3-662-53890-6_14

19. Brakerski, Z., Vaikuntanathan, V.: Circuit-ABE from LWE: unbounded attributes and semi-adaptive security. In: Robshaw, M., Katz, J. (eds.) CRYPTO 2016. LNCS, vol. 9816, pp. 363–384. Springer, Heidelberg (2016). https://doi.org/10.1007/978-3-662-53015-3_13

20. Brakerski, Z., Vaikuntanathan, V., Wee, H., Wichs, D.: Obfuscating conjunctions under entropic ring LWE. In: Sudan, M. (ed.) Proceedings of the 2016 ACM Conference on Innovations in Theoretical Computer Science, pp. 147–156. ACM (2016)
21. Camenisch, J., Neven, G., Rückert, M.: Fully anonymous attribute tokens from lattices. In: Visconti, I., De Prisco, R. (eds.) SCN 2012. LNCS, vol. 7485, pp. 57–75. Springer, Heidelberg (2012). https://doi.org/10.1007/978-3-642-32928-9_4
22. Canetti, R., Chen, Y.: Constraint-hiding constrained PRFs for NC^1 from LWE. In: Coron, J.-S., Nielsen, J.B. (eds.) EUROCRYPT 2017. LNCS, vol. 10210, pp. 446–476. Springer, Cham (2017). https://doi.org/10.1007/978-3-319-56620-7_16
23. Cash, D., Hofheinz, D., Kiltz, E., Peikert, C.: Bonsai trees, or how to delegate a lattice basis. J. Cryptol. **25**(4), 601–639 (2012)
24. Clear, M., McGoldrick, C.: Multi-identity and multi-key leveled FHE from learning with errors. In: Gennaro and Robshaw [30], pp. 630–656
25. Dai, W., Doröz, Y., Polyakov, Y., Rohloff, K., Sajjadpour, H., Savas, E., Sunar, B.: Implementation and evaluation of a lattice-based key-policy ABE scheme. IACR Cryptology ePrint Archive, 2017:601 (2017)
26. Davidson, A.: Obfuscation of bloom filter queries from ring-LWE. IACR Cryptology ePrint Archive, 2017:448 (2017)
27. Ducas, L., Nguyen, P.Q.: Faster gaussian lattice sampling using lazy floating-point arithmetic. In: Wang, X., Sako, K. (eds.) ASIACRYPT 2012. LNCS, vol. 7658, pp. 415–432. Springer, Heidelberg (2012). https://doi.org/10.1007/978-3-642-34961-4_26
28. Ducas, L., Prest, T.: Fast fourier orthogonalization. In: Abramov, S.A., Zima, E.V., Gao, X. (eds.) Proceedings of the ACM on International Symposium on Symbolic and Algebraic Computation, ISSAC 2016, pp. 191–198. ACM (2016)
29. Fischlin, M., Buchmann, J.A., Manulis, M. (eds.): Public Key Cryptography - PKC 2012. LNCS, vol. 7293. Springer, Heidelberg (2012)
30. Gennaro, R., Robshaw, M. (eds.): CRYPTO 2015. LNCS, vol. 9216. Springer, Heidelberg (2015)
31. Gentry, C., Peikert, C., Vaikuntanathan, V.: Trapdoors for hard lattices and new cryptographic constructions. In: Dwork, C. (ed.) Proceedings of the 40th Annual ACM Symposium on Theory of Computing, pp. 197–206. ACM (2008)
32. Gentry, C., Sahai, A., Waters, B.: Homomorphic encryption from learning with errors: conceptually-simpler, asymptotically-faster, attribute-based. In: Canetti, R., Garay, J.A. (eds.) CRYPTO 2013. LNCS, vol. 8042, pp. 75–92. Springer, Heidelberg (2013). https://doi.org/10.1007/978-3-642-40041-4_5
33. Gorbunov, S., Vaikuntanathan, V., Wee, H.: Attribute-based encryption for circuits. J. ACM **62**(6), 45:1–45:33 (2015). Prelim. version in STOC 2013
34. Gorbunov, S., Vaikuntanathan, V., Wee, H.: Predicate encryption for circuits from LWE. In: Gennaro and Robshaw [30], pp. 503–523
35. Gorbunov, S., Vaikuntanathan, V., Wichs, D.: Leveled fully homomorphic signatures from standard lattices. In: Servedio, R.A., Rubinfeld, R. (eds.) Proceedings ACM on Symposium on Theory of Computing, STOC 2015, pp. 469–477. ACM (2015)
36. Gordon, S.D., Katz, J., Vaikuntanathan, V.: À group signature scheme from lattice assumptions. In: Abe, M. (ed.) ASIACRYPT 2010. LNCS, vol. 6477, pp. 395–412. Springer, Heidelberg (2010). https://doi.org/10.1007/978-3-642-17373-8_23
37. Gür, K.D., Polyakov, Y., Rohloff, K., Ryan, G.W., Savaş, E., Sajjadpour, H.: Efficient implementation of gaussian sampling for trapdoor lattices and its applications. Pers. Commun. (2017, in preparation)

38. Gur, K.D., Polyakov, Y., Rohloff, K., Ryan, G.W., Savas, E.: Implementation and evaluation of improved gaussian sampling for lattice trapdoors. IACR Cryptology ePrint Archive, 2017:285 (2017)

39. Halevi, S., Halevi, T., Shoup, V., Stephens-Davidowitz, N.: Implementing BP-obfuscation using graph-induced encoding. In: Proceedings of the 2017 ACM SIGSAC Conference on Computer and Communications Security, CCS 2017, Dallas, TX, USA, 30 October–03 November 2017, pp. 783–798 (2017)

40. Howe, J., Pöppelmann, T., O'Neill, M., O'Sullivan, E., Güneysu, T.: Practical lattice-based digital signature schemes. ACM Trans. Embed. Comput. Syst. 14, 41 (2015)

41. Karney, C.F.F.: Sampling exactly from the normal distribution. ACM Trans. Math. Softw. 42(1), 3:1–3:14 (2016)

42. Kim, S., Wu, D.J.: Watermarking cryptographic functionalities from standard lattice assumptions. In: Katz, J., Shacham, H. (eds.) CRYPTO 2017. LNCS, vol. 10401, pp. 503–536. Springer, Cham (2017). https://doi.org/10.1007/978-3-319-63688-7_17

43. Klein, P.N.: Finding the closest lattice vector when it's unusually close. In: Shmoys, D.B. (ed.) Proceedings of ACM-SIAM Symposium on Discrete Algorithms, pp. 937–941. ACM/SIAM (2000)

44. Laguillaumie, F., Langlois, A., Libert, B., Stehlé, D.: Lattice-based group signatures with logarithmic signature size. In: Sako, K., Sarkar, P. (eds.) ASIACRYPT 2013. LNCS, vol. 8270, pp. 41–61. Springer, Heidelberg (2013). https://doi.org/10.1007/978-3-642-42045-0_3

45. Langlois, A., Ling, S., Nguyen, K., Wang, H.: Lattice-based group signature scheme with verifier-local revocation. In: Krawczyk, H. (ed.) PKC 2014. LNCS, vol. 8383, pp. 345–361. Springer, Heidelberg (2014). https://doi.org/10.1007/978-3-642-54631-0_20

46. Ling, S., Nguyen, K., Stehlé, D., Wang, H.: Improved zero-knowledge proofs of knowledge for the ISIS problem, and applications. In: Kurosawa, K., Hanaoka, G. (eds.) PKC 2013. LNCS, vol. 7778, pp. 107–124. Springer, Heidelberg (2013). https://doi.org/10.1007/978-3-642-36362-7_8

47. Lyubashevsky, V., Micciancio, D., Peikert, C., Rosen, A.: SWIFFT: a modest proposal for FFT hashing. In: Nyberg, K. (ed.) FSE 2008. LNCS, vol. 5086, pp. 54–72. Springer, Heidelberg (2008). https://doi.org/10.1007/978-3-540-71039-4_4

48. Lyubashevsky, V., Peikert, C., Regev, O.: On ideal lattices and learning with errors over rings. J. ACM 60(6), 43 (2013)

49. Lyubashevsky, V., Peikert, C., Regev, O.: A toolkit for ring-LWE cryptography. In: Johansson, T., Nguyen, P.Q. (eds.) EUROCRYPT 2013. LNCS, vol. 7881, pp. 35–54. Springer, Heidelberg (2013). https://doi.org/10.1007/978-3-642-38348-9_3

50. Micciancio, D.: Generalized compact knapsacks, cyclic lattices, and efficient one-way functions. Comput. Complex. 16(4), 365–411 (2007)

51. Micciancio, D., Peikert, C.: Trapdoors for lattices: Simpler, tighter, faster, smaller. In: Pointcheval and Johansson [58], pp. 700–718

52. Micciancio, D., Regev, O.: Worst-case to average-case reductions based on gaussian measures. SIAM J. Comput. 37(1), 267–302 (2007)

53. Micciancio, D., Walter, M.: Gaussian sampling over the integers: efficient, generic, constant-time. In: Katz, J., Shacham, H. (eds.) CRYPTO 2017. LNCS, vol. 10402, pp. 455–485. Springer, Cham (2017). https://doi.org/10.1007/978-3-319-63715-0_16

54. Nguyen, P.Q., Zhang, J., Zhang, Z.: Simpler efficient group signatures from lattices. In: Katz, J. (ed.) PKC 2015. LNCS, vol. 9020, pp. 401–426. Springer, Heidelberg (2015). https://doi.org/10.1007/978-3-662-46447-2_18

55. Peikert, C.: An efficient and parallel gaussian sampler for lattices. In: Rabin [59], pp. 80–97

56. Peikert, C.: A decade of lattice cryptography. Found. Trends Theor. Comput. Sci. **10**(4), 283–424 (2016)

57. Peikert, C., Vaikuntanathan, V., Waters, B.: A framework for efficient and composable oblivious transfer. In: Wagner, D. (ed.) CRYPTO 2008. LNCS, vol. 5157, pp. 554–571. Springer, Heidelberg (2008). https://doi.org/10.1007/978-3-540-85174-5_31

58. Pointcheval, D., Johansson, T. (eds.): Advances in Cryptology - EUROCRYPT 2012, vol. 7237. Springer, Heidelberg (2012). Proceedings of 31st Annual International Conference on the Theory and Applications of Cryptographic Techniques, Cambridge, UK, 15–19 April 2012

59. Rabin, T. (ed.): Advances in Cryptology - CRYPTO 2010. Lecture Notes in Computer Science, vol. 6223. Springer, Heidelberg (2010). 30th Annual Cryptology Conference, Santa Barbara, CA, USA, 15–19 August 2010, Proceedings

60. Wee, H.: Public key encryption against related key attacks. In: Fischlin et al. [29], pp. 262–279

61. Zhang, F.: The Schur Complement and Its Applications, vol. 4. Springer Science, Heidelberg (2006)

Short, Invertible Elements in Partially Splitting Cyclotomic Rings and Applications to Lattice-Based Zero-Knowledge Proofs

Vadim Lyubashevsky[1]([✉]) and Gregor Seiler[1,2]

[1] IBM Research - Zurich, Rüschlikon, Switzerland
vadim.lyubash@gmail.com
[2] ETH Zurich, Zurich, Switzerland

Abstract. When constructing practical zero-knowledge proofs based on the hardness of the Ring-LWE or the Ring-SIS problems over polynomial rings $\mathbb{Z}_p[X]/(X^n+1)$, it is often necessary that the challenges come from a set $\mathcal{C}$ that satisfies three properties: the set should be large (around 2^{256}), the elements in it should have small norms, and all the non-zero elements in the difference set $\mathcal{C} - \mathcal{C}$ should be invertible. The first two properties are straightforward to satisfy, while the third one requires us to make efficiency compromises. We can either work over rings where the polynomial $X^n + 1$ only splits into two irreducible factors modulo p, which makes the speed of the multiplication operation in the ring sub-optimal; or we can limit our challenge set to polynomials of smaller degree, which requires them to have (much) larger norms.

In this work we show that one can use the optimal challenge sets $\mathcal{C}$ and still have the polynomial $X^n + 1$ split into more than two factors. This comes as a direct application of our more general result that states that all non-zero polynomials with "small" coefficients in the cyclotomic ring $\mathbb{Z}_p[X]/(\Phi_m(X))$ are invertible (where "small" depends on the size of p and how many irreducible factors the mth cyclotomic polynomial $\Phi_m(X)$ splits into). We furthermore establish sufficient conditions for p under which $\Phi_m(X)$ will split in such fashion.

For the purposes of implementation, if the polynomial $X^n + 1$ splits into k factors, we can run FFT for $\log k$ levels until switching to Karatsuba multiplication. Experimentally, we show that increasing the number of levels from one to three or four results in a speedup by a factor of $\approx 2 - 3$. We point out that this improvement comes completely for free simply by choosing a modulus p that has certain algebraic properties. In addition to the speed improvement, having the polynomial split into many factors has other applications – e.g. when one embeds information into the Chinese Remainder representation of the ring elements, the more the polynomial splits, the more information one can embed into an element.

© International Association for Cryptologic Research 2018
J. B. Nielsen and V. Rijmen (Eds.): EUROCRYPT 2018, LNCS 10820, pp. 204–224, 2018.
https://doi.org/10.1007/978-3-319-78381-9_8

1   Introduction

Cryptography based on the presumed hardness of the Ring/Module-SIS and Ring/Module-LWE problems [Mic07, PR06, LM06, LPR10, LS15] is seen as a very likely replacement of traditional cryptography after the eventual coming of quantum computing. There already exist very efficient basic public key primitives, such as encryption schemes and digital signatures, based on the hardness of these problems. For added efficiency, most practical lattice-based constructions work over polynomial rings $\mathbb{Z}_p[X]/(f(X))$ where $f(X)$ is the cyclotomic polynomial $f(X) = X^n + 1$ and p is chosen in such a way that the $X^n + 1$ splits into n linear factors modulo p. With such a choice of parameters, multiplication in the ring can be performed very efficiently via the Number Theoretic Transform, which is an analogue of the Fast Fourier Transform that works over a finite field. Some examples of practical implementations that utilize NTT implementations of digital signatures and public key encryption based on the Ring-LWE problem can be found in [GLP12, PG13, ADPS16, BDK+17, DLL+17].

Constructions of more advanced lattice-based primitives sometimes require that the underlying ring has additional properties. In particular, *practical* protocols that utilize zero-knowledge proofs often require that elements with small coefficients are invertible (e.g. [BKLP15, BDOP16, LN17, DLNS17]). This restriction, which precludes using rings where $X^n + 1$ splits completely modulo p, stems from the structure of *approximate* zero-knowledge proofs, and we sketch this intuition below.

1.1   Approximate Zero-Knowledge Proofs

Abstractly, in a zero-knowledge proof the prover wants to prove the knowledge of s that satisfies the relation $f(s) = t$, where f and t are public. In the lattice setting, the function

$$f(s) := As \tag{1}$$

where A is a random matrix over some ring (the ring is commonly $\mathbb{Z}_p$ or $\mathbb{Z}_p[X]/(X^n + 1)$) and s is a vector over that same ring, where the coefficients of all (or almost all) the elements comprising s are bounded by some small value $\ll p$.

The function f in (1) satisfies the property that $f(s_1) + f(s_2) = f(s_1 + s_2)$ and for any c in the ring and any vector s over the ring we have $f(sc) = c \cdot f(s)$. The zero-knowledge proof for attempting to prove the knowledge of s proceeds as follows:

The Prover first chooses a "masking parameter" y and sends $w := f(y)$ to the Verifier. The Verifier picks a random challenge c from a subset of the ring and sends it to the prover (in a non-interactive proof, the Prover himself would generate $c := H(t, w)$, where H is a cryptographic hash function). The Prover then computes $z := sc + y$ and sends it to the Verifier.[1]

---

[1] In lattice-based schemes, it is important to keep the coefficients of z small, and so y must be chosen to have small coefficients as well. This can lead to the distribution of z being dependent on sc, which leaks some information about s. This problem is solved in [Lyu09, Lyu12] via various rejection-sampling procedures. How this is done is not important to this paper, and so we ignore this step.

The Verifier checks that $f(z) = ct + w$ and, crucially, it also checks to make sure that the coefficients of z are small. If these checks pass, then the Verifier accepts the proof. To show that the protocol is a proof of knowledge, one can rewind the Prover to just after his first move and send a different challenge c', and get a response z' such that $f(z') = c't + w$. Combined with the first response, we extract the equation

$$f(\bar{s}) = \bar{c}t \qquad (2)$$

where $\bar{s} = z - z'$ and $\bar{c} = c - c'$.

Notice that while the prover started with the knowledge of an s with small coefficients such that $f(s) = t$, he only ends up proving the knowledge of an $\bar{s}$ with larger coefficients such that $f(\bar{s}) = \bar{c}t$. If $\bar{c}$ also has small coefficients, then this type of proof is good enough in many (but not all) situations.

Applications of Approximate Zero-Knowledge Proofs. As a simple example of the utility of approximate zero-knowledge proofs, we consider commitment schemes where a commitment to a message m involves choosing some randomness r, and outputting $f(s) = t$, where s is defined as $\begin{bmatrix} r \\ m \end{bmatrix}$ where r and m have small coefficients.[2] Using the zero-knowledge proof from Sect. 1.1, one can prove the knowledge of an $\bar{s}$ and $\bar{c}$ such that $f(\bar{s}) = \bar{c}t$. If $\bar{c}$ is invertible in the ring, then we can argue that this implies that if t is later opened to any valid commitment s' where $f(s') = t$, then it must be $s' = \bar{s}/\bar{c}$.

The sketch of the argument is as follows: If we extract $\bar{s}, \bar{c}$ and the commitment is opened with s' such that $f(s') = t$, then multiplying both sides by $\bar{c}$ results in $f(\bar{c}s') = \bar{c}t$. Combining this with what was extracted from the zero-knowledge proof, we obtain that

$$f(\bar{c}s') = f(\bar{s}). \qquad (3)$$

If $s' \neq \bar{s}/\bar{c}$, then $\bar{c}s' \neq \bar{s}$ and we found a collision (with small coefficients) for the function f. Such a collision implies a solution to the (Ring-)SIS problem, or, depending on the parameters, may simply not exist (and the scheme can thus be based on (Ring-)LWE).

There are more intricate examples involving commitment schemes (see e.g. [BKLP15, BDOP16]) as well as other applications of such zero knowledge proofs, (e.g. to verifiable encryption [LN17] and voting protocols [DLNS17]) which require that the $\bar{c}$ be invertible.

The Challenge Set and its Effect on the Proof. The challenge c is drawn uniformly from some domain $\mathcal{C}$ which is a subset of $\mathbb{Z}_p[X]/(X^n + 1)$. In order to have small soundness error, we would like $\mathcal{C}$ to be large. When building non-interactive schemes that should remain secure against quantum computers, one

---

[2] It was shown in [BKLP15, BDOP16] that one actually does not need the message m to have small coefficients, but for simplicity we assume here that it still has them.

should have $|\mathcal{C}|$ be around 2^{256}. On the other hand, we also would like c to have a small norm. The reason for the latter is that the honest prover computes $z := sc + y$ and so the $\bar{s}$ that is extracted from the Prover in (2) is equal to $z - z'$, and must also therefore depend on $\|sc\|$. Thus, the larger the norms of c, c' are, the larger the extracted solution $\bar{s}$ will be, and the easier the corresponding (Ring-)SIS problem will be.

As a running example, suppose that we're working over the polynomial ring $\mathbb{Z}_p[X]/(X^{256} + 1)$. If invertibility were not an issue, then a simple and nearly optimal way (this way of choosing the challenge set dates back to at least the original paper that proposed a Fiat-Shamir protocol over polynomial rings [Lyu09]) to choose $\mathcal{C}$ of size 2^{256} would be to define

$$\mathcal{C} = \{c \in R_p^{256} : \|c\|_\infty = 1, \|c\|_1 = 60\}. \tag{4}$$

In other words, the challenges are ring elements consisting of exactly 60 non-zero coefficients which are ± 1.[3] The l_2 norm of such elements is $\sqrt{60}$.

If we take invertibility into consideration, then we need the difference set $\mathcal{C} - \mathcal{C}$ (excluding 0) to consist only of invertible polynomials. There are some folklore ways of creating sets all of whose non-zero differences are invertible (c.f. [SSTX09,BKLP15]). If the polynomial $X^{256} + 1$ splits into k irreducible polynomials modulo p, then all of these polynomials must have degree $256/k$. It is then easy to see, via the Chinese Remainder Theorem that every non-zero polynomial of degree less than $256/k$ is invertible in the ring $\mathbb{Z}_p[X]/(X^{256} + 1)$. We can therefore define the set

$$\mathcal{C}' = \{c \in R_p^{256} : \deg(c) < 256/k, \|c\|_\infty \le \gamma\},$$

where $\gamma \approx 2^{k-1}$ in order for the size of the set to be greater than 2^{256}. The l_2 norm of elements in this set is $\sqrt{256/k} \cdot \gamma$. If we, for example, take $k = 8$, then this norm becomes $\sqrt{32} \cdot 2^7 \approx 724$, which is around 90 times larger than the norms of the challenges in the set defined in (4). It is therefore certainly not advantageous to increase the norm of the challenge by this much only to decrease the running time of the computation. In particular, the security of the scheme will decrease and one will need to increase the ring dimension to compensate, which will in turn negate any savings in running time. For example, the extracted solution to the SIS instance in (3) is $\bar{c}s' - \bar{s}$, and its size heavily depends on the size of the coefficients in $\bar{c}$. A much more desirable solution would be to have the polynomial $X^n + 1$ split, but still be able to use the challenge set from (4).

1.2 Our Contribution

Our main result is a general theorem (Theorem 1.1) about the invertibility of polynomials with small coefficients in polynomial rings $\mathbb{Z}_p[X]/(\Phi_m(X))$, where $\Phi_m(X)$ is the mth cyclotomic polynomial. The theorem states that if a non-zero polynomial has small coefficients (where "small" is related to the prime p and

---

[3] The size of this set is $\binom{256}{60} \cdot 2^{60} > 2^{256}$.

the number of irreducible factors of $\Phi_m(X)$ modulo p), then it's invertible in the ring $\mathbb{Z}_p[X]/(\Phi_m(X))$. For the particular case of $\Phi_m(X) = X^n + 1$, we show that the polynomial $X^n + 1$ can split into several (in practice up to 8 or 16) irreducible factors and we can still use the optimal challenge sets, like ones of the form from (4). This generalizes and extends a result in [LN17] which showed that one can use the optimal set when $X^n + 1$ splits into two factors. We also show, in Sect. 3.3, some methods for creating challenge sets that are slightly sub-optimal, but allow for the polynomial to split further.

The statement of Theorem 1.1 uses notation from Definition 2.1, while the particular case of $X^n + 1$ in Corollary 1.2 is self-contained. We therefore recommend the reader to first skim the Corollary statement. The proofs of the Theorem and the Corollary are given at the end of Sect. 3.2. For completeness, we also state sufficient conditions for invertibility based on the ℓ_2-norm of the polynomial. This is an intermediate result that we need on the way to obtaining our main result about the invertibility of polynomials with small coefficients (i.e. based on the ℓ_∞ norm of the polynomial), but it could be of independent interest.

Theorem 1.1. *Let $m = \prod p_i^{e_i}$ for $e_i \geq 1$ and $z = \prod p_i^{f_i}$ for any $1 \leq f_i \leq e_i$. If p is a prime such that $p \equiv 1 \pmod{z}$ and $\mathrm{ord}_m(p) = m/z$, then the polynomial $\Phi_m(X)$ factors as*

$$\Phi_m(X) \equiv \prod_{j=1}^{\phi(z)} (X^{m/z} - r_j) \pmod{p}$$

for distinct $r_j \in \mathbb{Z}_p^*$ where $X^{m/z} - r_j$ are irreducible in the ring $\mathbb{Z}_p[X]$. Furthermore, any $\mathbf{y}$ in $\mathbb{Z}_p[X]/(\Phi_m(X))$ that satisfies either

$$0 < \|\mathbf{y}\|_\infty < \frac{1}{s_1(z)} \cdot p^{1/\phi(z)}$$

or

$$0 < \|\mathbf{y}\| < \frac{\sqrt{\phi(m)}}{s_1(m)} \cdot p^{1/\phi(z)}$$

has an inverse in $\mathbb{Z}_p[X]/(\Phi_m(X))$.

The above theorem gives sufficient conditions for p so that all polynomials with small coefficients in $\mathbb{Z}_p[X]/(\Phi_m(X))$ are invertible, but it does not state anything about whether there exist such p. In Theorem 2.5, we show that if we additionally put the condition on m and z that $8|m \Rightarrow 4|z$, then there are indeed infinitely many primes p that satisfy these conditions. In practical lattice constructions involving zero-knowledge proofs, we would normally use a modulus of size at least 2^{20}, and we experimentally confirmed (for various cyclotomic polynomials) that one can indeed find many such primes that are of that size.

Specializing the above to the ring $\mathbb{Z}_p[X]/(X^n + 1)$, we obtain the following corollary:

Corollary 1.2. *Let $n \geq k > 1$ be powers of 2 and $p = 2k + 1 \pmod{4k}$ be a prime. Then the polynomial $X^n + 1$ factors as*

$$X^n + 1 \equiv \prod_{j=1}^{k} (X^{n/k} - r_j) \pmod{p}$$

for distinct $r_j \in \mathbb{Z}_p^*$ where $X^{n/k} - r_j$ are irreducible in the ring $\mathbb{Z}_p[X]$. Furthermore, any $\mathbf{y}$ in $\mathbb{Z}_p[X]/(X^n + 1)$ that satisfies either

$$0 < \|\mathbf{y}\|_\infty < \frac{1}{\sqrt{k}} \cdot p^{1/k}$$

or

$$0 < \|\mathbf{y}\| < p^{1/k}$$

has an inverse in $\mathbb{Z}_p[X]/(X^n + 1)$.

As an application of this result, suppose that we choose $k = 8$ and a prime p congruent to 17 (mod 32) such that $p > 2^{20}$. Furthermore, suppose that we perform our zero-knowledge proofs over the ring $\mathbb{Z}_p[X]/(X^n + 1)$ (where n is a power of 2 greater than 8), and prove the knowledge of $\bar{s}, \bar{c}$ such that $f(\bar{s}) = \bar{c}t$ where $\|\bar{c}\|_\infty \leq 2$ (i.e. the challenges c are taken such that $\|c\|_\infty = 1$). Then the above theorem states that $X^n + 1$ factors into 8 polynomials and $\bar{c}$ will be invertible in the ring since $\frac{1}{\sqrt{8}} \cdot p^{1/8} > 2$.

Having $p > 2^{20}$ is quite normal for the regime of zero-knowledge proofs, and therefore having the polynomial $X^n + 1$ split into 8 factors should be possible in virtually every application. If we would like it to split further into 16 or 32 factors, then we would need $p > 2^{48}$ or, respectively, $p > 2^{112}$. In Sect. 3.3 we describe how our techniques used to derive Theorem 1.1 can also be used in a somewhat "ad-hoc" fashion to create different challenge sets $\mathcal{C}$ that are nearly-optimal (in terms of the maximal norm), but allow $X^n + 1$ to split with somewhat smaller moduli than implied by Theorem 1.1.

In Sect. 4, we describe how one would combine the partially-splitting FFT algorithm with a Karatsuba multiplication algorithm to efficiently multiply in a partially-splitting ring. For primes of size between 2^{20} and 2^{29}, one obtains a speed-up of about a factor of 2 by working over rings where $X^n + 1$ splits into 8 versus just 2 factors.

In addition to the speed improvement, there are applications whose usability can be improved by the fact that we work over rings $\mathbb{Z}_p[X]/(X^n + 1)$ where $X^n + 1$ splits into more factors. For example, [BKLP15] constructed a commitment scheme and zero-knowledge proofs of knowledge that allows to prove the fact that $\mathbf{c} = \mathbf{ab}$ when $\text{Commit}(\mathbf{a})$, $\text{Commit}(\mathbf{b})$, $\text{Commit}(\mathbf{c})$ are public (the same holds for addition). An application of this result is the verifiability of circuits. For this application, one only needs commitments of 0's and 1's, thus if we work over a ring where $X^n + 1$ splits into k irreducible factors, one can embed k bits into each Chinese Remainder coefficient of $\mathbf{a}$ and $\mathbf{b}$, and therefore proving that $\mathbf{c} = \mathbf{ab}$ implies that all k multiplications of the bits were performed correctly. Thus the larger k is, the more multiplications one can prove in parallel. Unfortunately k cannot be set too large without ruining the necessary property that the difference of any two distinct challenges is invertible or increasing the ℓ_2-norm of the challenges as described in Sect. 1.1. Our result therefore allows to prove products of 8 (or 16) commitments in parallel without having to increase the parameters of the scheme to accommodate the larger challenges.

2   Cyclotomics and Lattices

2.1   Cyclotomic Polynomials

Definition 2.1. *For any integer $m > 1$, we write*

$$\phi(m) = m \cdot \prod_{p \, is \, prime \, \wedge \, p \, | \, m} \frac{p-1}{p}$$

$$\delta(m) = \prod_{p \, is \, prime \, \wedge \, p \, | \, m} p$$

$$\tau(m) = \begin{cases} m, & if \, m \, is \, odd \\ m/2, & if \, m \, is \, even \end{cases}$$

$$s_1(m) = largest \, singular \, value \, of \, the \, matrix \, in \, (7)$$

$$ord_m(n) = \min\{k \ : \ k > 0 \, and \, n^k \, \bmod \, m = 1\}$$

The function $\phi(m)$ is the Euler phi function, $\delta(m)$ is sometimes referred to as the *radical* of m, and $\tau(m)$ is a function that sometimes comes into play when working with the geometry of cyclotomic rings. The function $ord_m(n)$ is the order of an element n in the multiplicative group $\mathbb{Z}_m^*$. In the special case of $m = 2^k$, we have $\phi(m) = \tau(m) = 2^{k-1}$ and $\delta(m) = 2$.

The mth cyclotomic polynomial, written as $\Phi_m(X)$, is formally defined to be

$$\Phi_m(X) = \prod_{i=1}^{\phi(m)} (X - \omega_i),$$

where ω_i are the mth complex primitive roots of unity (of which there are $\phi(m)$ many). Of particular interest in practical lattice cryptography is the cyclotomic polynomial $\Phi_{2^k}(X) = X^{2^{k-1}} + 1$.

If p is some prime and $r_1, \ldots, r_{\phi(m)}$ are elements in $\mathbb{Z}_p^*$ such that $ord_p(r_j) = \phi(m)$, then one can write

$$\Phi_m(X) \equiv \prod_{j=1}^{\phi(m)} (X - r_j) \pmod{p}.$$

For any $m > 1$, it is known that we can express the cyclotomic polynomial $\Phi_m(X)$ as

$$\Phi_m(X) = \Phi_{\delta(m)}\left(X^{m/\delta(m)}\right), \tag{5}$$

and the below Lemma is a generalization of this statement.

Lemma 2.2. *Let* $m = \prod p_i^{e_i}$ *for* $e_i \geq 1$ *and* $z = \prod p_i^{f_i}$ *for any* $1 \leq f_i \leq e_i$. *Then*

$$\Phi_m(X) = \Phi_z(X^{m/z}).$$

Proof. By (5), and the fact that $\delta(m) = \delta(z)$, we can rewrite $\Phi_m(X)$ as

$$\Phi_m(X) = \Phi_{\delta(m)}(X^{m/\delta(m)}) = \Phi_{\delta(m)}(X^{z/\delta(m)})(X^{m/z})$$
$$= \Phi_{\delta(z)}(X^{z/\delta(z)})(X^{m/z}) = \Phi_z(X^{m/z}). \tag{6}$$

$\square$

2.2   The Splitting of Cyclotomic Polynomials

In Theorem 2.3, we give the conditions on the prime p such that the polynomial $\Phi_m(X)$ splits into irreducible factors $X^{m/k} - r$ modulo p. In Theorem 2.5, we then show that when m and k satisfy an additional relation, there are infinitely many p that satisfy the necessary conditions of Theorem 2.3.

Theorem 2.3. *Let* $m = \prod p_i^{e_i}$ *for* $e_i \geq 1$ *and* $z = \prod p_i^{f_i}$ *for any* $1 \leq f_i \leq e_i$. *If* p *is a prime such that* $p \equiv 1 \pmod{z}$ *and* $\mathrm{ord}_m(p) = m/z$, *then the polynomial* $\Phi_m(X)$ *factors as*

$$\Phi_m(X) \equiv \prod_{j=1}^{\phi(z)} (X^{m/z} - r_j) \pmod{p}$$

for distinct $r_j \in \mathbb{Z}_p^*$ *where* $X^{m/z} - r_j$ *are irreducible in* $\mathbb{Z}_p[X]$.

Proof. Since p is a prime and $p \equiv 1 \pmod{z}$, there exists an element r such that $\mathrm{ord}_p(r) = z$. Furthermore, for all the $\phi(z)$ integers $1 < i < z$ such that $\gcd(i, z) = 1$, we also have $\mathrm{ord}_p(r^i) = z$. We therefore have, by definition of Φ, that

$$\Phi_z(X) \equiv \prod_{j=1}^{\phi(z)} (X - r_j) \pmod{p}.$$

Applying Lemma 2.2, we obtain that

$$\Phi_m(X) \equiv \prod_{j=1}^{\phi(z)} (X^{m/z} - r_j) \pmod{p}.$$

We now need to prove that the terms $X^{m/z} - r_j$ are irreducible modulo p. Suppose they are not and $X^{m/z} - r_j$ has an irreducible divisor f of degree $d < \frac{m}{z}$. Then f defines an extension field of $\mathbb{Z}_p$ of degree d, i.e. a finite field with p^d elements that all satisfy $X^{p^d} = X$. Hence f divides $X^{p^d} - X$. Now, from $\mathrm{ord}_m(p) = \frac{m}{z} > d$ it follows that we can write $p^d = am + b$ where $b \neq 1$. Thus

$$X^{p^d} - X = X^{am+b} - X = X(X^{am+(b-1)} - 1).$$

If we now consider an extension field of $\mathbb{Z}_p$ in which f splits, the roots of f are also roots of $X^{am+(b-1)} - 1$ and therefore have order dividing $am + (b-1)$. This is a contradiction. As a divisor of $X^{m/z} - r_j$ (and therefore of Φ_m), f has only roots of order m.   □

In the proof of Theorem 2.5 we need a small result about the multiplicative order of odd integers modulo powers of 2. Since we also need this later in the proof of Corollary 1.2, we state this result in the next lemma.

Lemma 2.4. *Let* $a \equiv 1 + 2^f \pmod{2^{f+1}}$ *for* $f \geq 2$. *Then the order of* a *in the group of units modulo* 2^e *for* $e \geq f$ *is equal to* 2^{e-f}, *i.e.* $\mathrm{ord}_{2^e}(a) = 2^{e-f}$.

Proof. We can write $a = 1 + 2^f k_1$ with some odd $k_1 \in \mathbb{Z}$. Then notice $a^2 = 1 + 2^{f+1}k_1 + 2^{2f}k_1^2 = 1 + 2^{f+1}(k_1 + 2^{f-1}k_1^2) = 1 + 2^{f+1}k_2$ with odd $k_2 = k_1 + 2^{f-1}k_1^2$. It follows iteratively that $a^{2^{e-f}} = 1 + 2^e k_{2^{e-f}} \equiv 1 \pmod{2^e}$, which implies the order of a modulo 2^e divides 2^{e-f}, but $a^{2^{e-f-1}} = 1 + 2^{e-1}k_{2^{e-f-1}} \not\equiv 1 \pmod{2^e}$ since $k_{2^{e-f-1}}$ is odd. So, the multiplicative order of a modulo 2^e must be 2^{e-f}.   □

Theorem 2.5. *Let* $m = \prod p_i^{e_i}$ *for* $e_i \geq 1$ *and* $z = \prod p_i^{f_i}$ *for any* $1 \leq f_i \leq e_i$. *Furthermore, assume that if* m *is divisible by 8, then* z *is divisible by 4. Then there are infinitely many primes* p *such that* $p \equiv 1 \pmod{z}$ *and* $\mathrm{ord}_m(p) = m/z$.

Proof. First we show that an integer not necessarily prime exists that fulfills the two conditions. By the Chinese remainder theorem it suffices to find integers a_i such that $a_i \bmod p_i^{f_i} = 1$ and $\mathrm{ord}_{p_i^{e_i}}(a_i) = p_i^{e_i - f_i}$. First consider the odd primes $p_i \neq 2$. It is easy to show that if g is a generator modulo p_i then either g or $g + p_i$, say g', is a generator modulo every power of p_i (c.f. [Coh00, Lemma 1.4.5]). Define $a_i = (g')^{(p_i-1)p_i^{f_i-1}}$. Then, since g' has order $(p_i - 1)p_i^{f_i-1}$ modulo $p_i^{f_i}$ and order $(p_i - 1)p_i^{e_i-1} \bmod p_i^{e_i}$, it follows that $a_i \bmod p_i^{f_i} = 1$ and

$$\mathrm{ord}_{p_i^{e_i}}(a_i) = \frac{(p_i - 1)p_i^{e_i-1}}{(p_i - 1)p_i^{f_i-1}} = p_i^{e_i - f_i}$$

as we wanted. Next, consider $p = 2$ and the case where m is divisible by 8; that is, $e_1 \geq 3$. This implies $f_1 \geq 2$. From Lemma 2.4 we see that 5 is a generator of a cyclic subgroup of $\mathbb{Z}_{2^e}^\times$ of index 2 for every $e \geq 3$, i.e. $\mathrm{ord}_{2^e}(5) = 2^{e-2}$. Therefore, $5^{2^{f_1-2}} \bmod 2^{f_1} = 1$ and

$$\mathrm{ord}_{2^{e_1}}(5^{2^{f_1-2}}) = \frac{2^{e_1-2}}{2^{f_1-2}} = 2^{e_1 - f_1}.$$

Hence $a_1 = 5^{2^{f_1-2}}$ is a valid choice in this case. If $e_1 = 2$, note that 3 is a generator modulo 4 and $a_1 = 3^{2^{f_1-1}}$ is readily seen to work. When $e_1 = f_1 = 1$, take $a_1 = 1$. So, there exists an integer a that fulfills our two conditions and in fact every integer congruent to $a \mod m$ does. By Dirichlet's theorem on arithmetic progressions, there are infinitely many primes among the $a + lm$ $(l \in \mathbb{Z})$. $\square$

As an experimental example consider $m = 2^2 3^3 7 = 756$ and $z = 2 \cdot 3 \cdot 7 = 42$. Then Φ_m splits into 12 polynomials modulo primes of the form in Theorem 2.5. There are 2058 primes of this form between 2^{20} and 2^{21}.

2.3   The Vandermonde Matrix

To each cyclotomic polynomial $\Phi_m(X)$ with roots of unity $\omega_1, \ldots, \omega_{\phi(m)}$, we associate the Vandermonde matrix

$$
\mathbf{V}_m =
\begin{bmatrix}
1 & \omega_1 & \omega_1^2 & \cdots & \omega_1^{\phi(m)-1} \\
1 & \omega_2 & \omega_2^2 & \cdots & \omega_2^{\phi(m)-1} \\
& & \cdots & & \\
1 & \omega_{\phi(m)} & \omega_{\phi(m)}^2 & \cdots & \omega_{\phi(m)}^{\phi(m)-1}
\end{bmatrix}
\in \mathbb{C}^{\phi(m) \times \phi(m)}.
\tag{7}
$$

The important property for us in this paper is the largest singular value of $\mathbf{V}_m$, which we write as

$$
s_1(m) = \max_{\mathbf{u} \in \mathbb{C}^{\phi(m)}} \frac{\|\mathbf{V}_m \mathbf{u}\|}{\|\mathbf{u}\|}.
\tag{8}
$$

It was shown in [LPR13, Lemma 4.3] that when $m = p^k$ for any prime p and positive integer k, then

$$
s_1(m) = \sqrt{\tau(m)}.
\tag{9}
$$

We do not know of a theorem analogous to (9) that holds for all m, and so we numerically computed $s_1(m)$ for all $m < 3000$ and observed that $s_1(m) \leq \sqrt{\tau(m)}$ was always satisfied. Furthermore, for most m, we still had the equality $s_1(m) = \sqrt{\tau(m)}$. The only exceptions where $s_1(m) < \sqrt{\tau(m)}$ were integers that have at least 3 distinct odd prime factors. As an example, Table 1 contains a list of all such values up to 600 for which $s_1(m) < \sqrt{\tau(m)}$. We point out that while it appears that having three prime factors is a necessary condition for m to appear in the table, it is not sufficient. For example, $255 = 3 \cdot 5 \cdot 17$, but still $s_1(255) = \sqrt{\tau(255)} = \sqrt{255}$.

For all practical sizes of m used in cryptography, the value $s_1(m)$ is fairly easy to compute numerically using basic linear algebra software (e.g. MATLAB, Scilab, etc.), and we will state all our results in terms of $s_1(m)$. Nevertheless, being able to relate $s_1(m)$ to $\tau(m)$ certainly simplifies the calculation. Based on our numerical observations, we formulate the following conjecture:

Conjecture 2.6. For all positive integers m, $s_1(m) \leq \sqrt{\tau(m)}$.

Table 1. Values of m less than 600 for which $s_1(m) \neq \sqrt{\tau(m)}$.

m	$s_1(m)$	$\sqrt{\tau(m)}/s_1(m)$
$105 = 3 \cdot 5 \cdot 7$	9.952	1.0296172
$165 = 3 \cdot 5 \cdot 11$	12.785	1.0046612
$195 = 3 \cdot 5 \cdot 13$	13.936	1.0019718
$210 = 2 \cdot 3 \cdot 5 \cdot 7$	9.952	1.0296172
$315 = 3^2 \cdot 5 \cdot 7$	17.237	1.0296172
$330 = 2 \cdot 3 \cdot 5 \cdot 11$	12.785	1.0046612
$390 = 2 \cdot 3 \cdot 5 \cdot 13$	13.936	1.0019718
$420 = 2^2 \cdot 3 \cdot 5 \cdot 7$	14.074	1.0296172
$495 = 3^2 \cdot 5 \cdot 11$	22.145	1.0046612
$525 = 3 \cdot 5^2 \cdot 7$	22.253	1.0296172
$585 = 3^2 \cdot 5 \cdot 13$	24.139	1.0019718

2.4   Cyclotomic Rings and Ideal Lattices

Throughout the paper, we will write R_m to be the *cyclotomic ring* $\mathbb{Z}[X]/(\varPhi_m(X))$ and $R_{m,p}$ to be the ring $\mathbb{Z}_p[X]/(\varPhi_m(X))$, with the usual polynomial addition and multiplication operations. We will denote by normal letters elements in $\mathbb{Z}$ and by bold letters elements in R_m. For an odd p, an element $\mathbf{w} \in R_{m,p}$ can always be written as $\sum_{i=0}^{\phi(m)-1} w_i X^i$ where $|w_i| \leq (p-1)/2$. Using this representation, for $\mathbf{w} \in R_{m,p}$ (and in R_m), we will define the lengths of elements as

$$\|\mathbf{w}\|_\infty = \max_i |w_i| \text{ and } \|\mathbf{w}\| = \sqrt{\sum_i |w_i|^2}.$$

Just as for vectors over $\mathbb{Z}$, the norms satisfy the inequality $\|\mathbf{w}\| \leq \sqrt{\phi(m)} \cdot \|\mathbf{w}\|_\infty$.

Another useful definition of length is with respect to the *embedding norm* of an element in R_m. If $\omega_1, \ldots, \omega_{\phi(m)}$ are the complex roots of $\varPhi_m(X)$, then the embedding norm of $\mathbf{w} \in R_m$ is

$$\|\mathbf{w}\|_e = \sqrt{\sum_i \mathbf{w}(\omega_i)^2}.$$

If we view of $\mathbf{w} = \begin{bmatrix} w_0 \\ w_1 \\ \cdots \\ w_{\phi(m)-1} \end{bmatrix}$ as a vector over $\mathbb{Z}^{\phi(m)}$, then the above definition is equivalent to

$$\|\mathbf{w}\|_e = \sqrt{\sum_i \mathbf{w}(\omega_i)^2} = \|\mathbf{V}_m \mathbf{w}\|$$

due to the fact that the ith position of $\mathbf{V}_m\mathbf{w}$ is $\mathbf{w}(\omega_i)$. This gives a useful relationship between the $\|\cdot\|_e$ and $\|\cdot\|$ norms as

$$\|\mathbf{w}\|_e \leq s_1(m) \cdot \|\mathbf{w}\|. \tag{10}$$

An integer lattice of dimension n is an additive sub-group of $\mathbb{Z}^n$. For the purposes of this paper, all lattices will be full-rank. The determinant of a full-rank integer lattice Λ of dimension n is the size of the quotient group $|\mathbb{Z}^n/\Lambda|$. We write $\lambda_1(\Lambda)$ to denote the Euclidean length of the shortest non-zero vector in Λ.

If $\mathcal{I}$ is an ideal in the polynomial ring R_m, then it is also an additive sub-group of $\mathbb{Z}^{\phi(m)}$, and therefore a $\phi(m)$-dimensional lattice (it can be shown that such lattices are always full-rank). Such lattices are therefore sometimes referred to as *ideal lattices*. For any ideal lattice Λ of the ring R_m, there exists a lower bound on the embedding norm of its vectors (c.f. [PR07, Lemma 6.2])

$$\forall \mathbf{w} \in \Lambda, \|\mathbf{w}\|_e \geq \sqrt{\phi(m)} \cdot \det(\Lambda)^{1/\phi(m)}.$$

Combining the above with (10) yields the following lemma:

Lemma 2.7. *If Λ is an ideal lattice in R_m, then*

$$\lambda_1(\Lambda) \geq \frac{\sqrt{\phi(m)}}{s_1(m)} \cdot \det(\Lambda)^{1/\phi(m)}.$$

3   Invertible Elements in Cyclotomic Rings

The main goal of this section is to prove Theorem 1.1. To this end, we first prove Lemma 3.1, which proves the Theorem for the ℓ_2 norm. Unfortunately directly applying this Lemma to prove the ℓ_∞ part of the Theorem 1.1 by using the relationship between the ℓ_2 and ℓ_∞ norms is sub-optimal. In Sect. 3.2 we instead show that by writing elements of partially-splitting rings $R_{m,p}$ as sums of polynomials over smaller, fully-splitting rings, one can obtain a tighter bound. We prove in Lemma 3.2 that if any of the parts of $\mathbf{y} \in R_{m,p}$ is invertible in the smaller fully-splitting ring, then the polynomial $\mathbf{y}$ is invertible in $R_{m,p}$. The full proof of Theorem 1.1 will follow from this Lemma, the special case of Lemma 3.1 applicable to fully-splitting rings, and Theorem 2.3.

3.1   Invertibility and the ℓ_2 Norm

Our main result only needs a special case of the below Lemma corresponding to when $\Phi_m(X)$ fully splits, but we prove a more general statement since it doesn't bring with it any additional complications.

Lemma 3.1. *Let $m = \prod p_i^{e_i}$ for $e_i \geq 1$ and $z = \prod p_i^{f_i}$ for any $1 \leq f_i \leq e_i$ such that*

$$\Phi_m(X) \equiv \prod_{i=1}^{\phi(z)} (X^{m/z} - r_i) \pmod{p}$$

for some distinct $r_i \in \mathbb{Z}_p^*$ where $X^{m/z} - r_i$ are irreducible in $\mathbb{Z}_p[X]$, and let
$\mathbf{y}$ be any element in the ring $R_{m,p}$. If $0 < \|\mathbf{y}\| < \frac{\sqrt{\phi(m)}}{s_1(m)} \cdot p^{1/\phi(z)}$, then $\mathbf{y}$ is
invertible in $R_{m,p}$.

Proof. Suppose that $\mathbf{y}$ is not invertible in $R_{m,p}$. By the Chinese Remainder Theorem, this implies that for (at least) one i, $\mathbf{y} \mod (X^{m/z} - r_i, p) = 0$. For an i for which $\mathbf{y} \mod (X^{m/z} - r_i, p) = 0$ (if there is more than one such i, pick one of them arbitrarily) define the set

$$\Lambda = \left\{ \mathbf{z} \in R_m : \mathbf{z} \mod \left(X^{m/z} - r_i, p \right) = 0 \right\}.$$

Notice that Λ is an additive group. Also, because $X^{m/z} - r_i$ is a factor of $\Phi_m(X)$ modulo p, for any polynomial $\mathbf{z} \in \Lambda$, the polynomial $\mathbf{z} \cdot X \in R_m$ is also in Λ. This implies that Λ is an ideal of R_m, and so an ideal lattice in the ring R_m. By looking at the Chinese Remainder representation modulo p of all the elements in Λ (they have 0 in the coefficient corresponding to modulo $X^{m/z} - r_i$, and are arbitrary in all other coefficients), one can see that $|\mathbb{Z}^{\phi(m)}/\Lambda| = p^{m/z} = p^{\phi(m)/\phi(z)}$, which is the determinant of Λ. By Lemma 2.7, we then know that $\lambda_1(\Lambda) \geq \frac{\sqrt{\phi(m)}}{s_1(m)} \cdot p^{1/\phi(z)}$.

Since $\mathbf{y} \mod (X^{m/z} - r_i, p) = 0$ and $0 < \|\mathbf{y}\|$, we know that $\mathbf{y}$ is a non-zero vector in Λ. But we also have by our hypothesis that $\|\mathbf{y}\| < \frac{\sqrt{\phi(m)}}{s_1(m)} \cdot p^{1/\phi(z)} \leq \lambda_1(\Lambda)$, which is impossible.

□

One can see that a direct application of Lemma 3.1 gives a weaker bound than what we are claiming in Theorem 1.1 – we can only conclude that all vectors $\mathbf{y}$ such that

$$\|\mathbf{y}\|_\infty \leq \frac{1}{s_1(m)} \cdot p^{1/\phi(z)}$$

are invertible. Since $z \ll m$, having $s_1(m)$ vs. $s_1(z)$ in the denominator makes a very noticeable difference in the tightness of the result (for example, if m, z are powers of 2, then $s_1(m) = \sqrt{m/2}$ and $s_1(z) = \sqrt{z/2}$). In Sect. 3.2, we instead break up $\mathbf{y}$ into a sum of elements in smaller rings $R_{z,p}$ and prove that only some of these parts, need to be invertible in $R_{z,p}$ in order for the entire element $\mathbf{y}$ to be invertible in $R_{m,p}$.

We point out that Lemma 3.1 was already implicit in [SS11, Lemma 8] for $\Phi_m(X) = X^n + 1$. To obtain a bound in the ℓ_∞ norm, the authors of that work then applied the norm inequality between the ℓ_2 and ℓ_∞ norms to obtain the bound that we described above. Using the more refined approach in the current paper, however, that bound can be tightened and would immediately produce an improvement in the main result of [SS11] which derives the statistical closeness of a particular distribution to uniform. Such applications are therefore another area in which our main result can prove useful.

3.2   Partially-Splitting Rings

In this section, we will be working with rings $R_{m,p}$ where p is chosen such that the polynomial $\Phi_m(X)$ factors into k irreducible polynomials of the form $X^{\phi(m)/k} - r_i$. Theorem 2.3 states the sufficient conditions on m, k, p in order to obtain such a factorization. Throughout this section, we will use the following notation: suppose that

$$\mathbf{y} = \sum_{j=0}^{\phi(m)-1} y_j X^j$$

is an element of the ring $R_{m,p}$, where the value p is chosen as above. Then for all integers $0 \leq i < \phi(m)/k - 1$, we define the polynomials $\mathbf{y}'_i$ as

$$\mathbf{y}'_i = \sum_{j=0}^{k-1} y_{j\phi(m)/k+i} X^j. \tag{11}$$

For example, if $\phi(m) = 8$ and $k = 4$, then for $\mathbf{y} = \sum_{i=0}^{7} y_i X^i$, we have $\mathbf{y}'_0 = y_0 + y_2 X + y_4 X^2 + y_6 X^3$ and $\mathbf{y}'_1 = y_1 + y_3 X + y_5 X^2 + y_7 X^3$.

The intuition behind the definition in (11) is that one can write $\mathbf{y}$ in terms of the $\mathbf{y}'_i$ as

$$\mathbf{y} = \sum_{i=0}^{\phi(m)/k-1} \mathbf{y}'_i(X^{\phi(m)/k}) \cdot X^i.$$

Then to calculate $\mathbf{y} \bmod (X^{\phi(m)/k} - r_j)$ where $(X^{\phi(m)/k} - r_j)$ is one of the irreducible factors of $\Phi_m(X)$ modulo p, we have

$$\mathbf{y} \bmod (X^{\phi(m)/k} - r_j) = \sum_{i=0}^{\phi(m)/k-1} \mathbf{y}'_i(r_j) \cdot X^i \tag{12}$$

simply because we plug in r_j for every $X^{\phi(m)/k}$.

Lemma 3.2. *Let $m = \prod p_i^{e_i}$ for $e_i \geq 1$ and $z = \prod p_i^{f_i}$ for any $1 \leq f_i \leq e_i$, and suppose that we can write*

$$\Phi_m(X) \equiv \prod_{j=1}^{\phi(z)} (X^{m/z} - r_j) \pmod{p} \tag{13}$$

for distinct $r_j \in \mathbb{Z}_p^*$ where $(X^{m/z} - r_j)$ are irreducible in $\mathbb{Z}_p[X]$. Let $\mathbf{y}$ be a polynomial in $R_{m,p}$ and define the associated $\mathbf{y}'_i$ as in (11), where $k = \phi(z)$. If some $\mathbf{y}'_i$ is invertible in $R_{z,p}$, then $\mathbf{y}$ is invertible in $R_{m,p}$.

Proof. By the Chinese Remainder Theorem, the polynomial $\mathbf{y}$ is invertible in $R_{m,p}$ if and only if $\mathbf{y} \bmod (X^{m/z} - r_j) \neq 0$ for all $r_1, \ldots, r_k$. When we use $k = \phi(z)$, (12) can be rewritten as

$$\mathbf{y} \bmod (X^{m/z} - r_j) = \sum_{i=0}^{m/z-1} \mathbf{y}_i'(r_j) \cdot X^i.$$

To show that $\mathbf{y}$ is invertible, it is therefore sufficient to show that

$$\exists i \text{ s.t } \forall j, \ \mathbf{y}_i'(r_j) \bmod p \neq 0.$$

Let i be such that $\mathbf{y}_i'$ is invertible in the ring $R_{z,p}$. From (13) and Lemma 2.2 we have that

$$\Phi_z(X) \equiv \prod_{j=1}^{\phi(z)} (X - r_j) \pmod{p},$$

and so the ring $R_{z,p}$ is fully-splitting. Since $\mathbf{y}_i'$ is invertible in $R_{z,p}$, the Chinese Remainder Theorem implies that for all $1 \leq j \leq \phi(z)$, $\mathbf{y}_i'(r_j) \bmod p \neq 0$, and therefore $\mathbf{y}$ is invertible in $R_{m,p}$.   □

Theorem 1.1 now follows from the combination of Theorem 2.3, and Lemmas 3.1 and 3.2.

Proof (Theorem 1.1). For the conditions on m, z, and p, it follows from Theorem 2.3 that the polynomial $\Phi_m(X)$ can be factored into irreducible factors modulo p as $\prod_{j=1}^{\phi(z)} (X^{m/z} - r_j)$. Lemma 2.2 then states that $\Phi_z(X) \equiv \prod_{j=1}^{\phi(z)} (X - r_j)$ (mod p).

For any $\mathbf{y} \in R_{m,p}$, let the $\mathbf{y}_i'$ be defined as in (11) where $k = \phi(z)$. If $0 < \|\mathbf{y}\|_\infty < \frac{1}{s_1(z)} \cdot p^{1/\phi(z)}$, then because each $\mathbf{y}_i'$ consists of $\phi(z)$ coefficients, we have that for all i, $\|\mathbf{y}_i'\| < \frac{\sqrt{\phi(z)}}{s_1(z)} \cdot p^{1/\phi(z)}$. Since $\mathbf{y} \neq 0$, it must be that for some i, $\mathbf{y}_i' \neq 0$.

Lemma 3.1 therefore implies that the non-zero $\mathbf{y}_i'$ is invertible in $R_{z,p}$. In turn, Lemma 3.2 implies that $\mathbf{y}$ is invertible in $R_{m,p}$.   □

Proof. (Of Corollary 1.2) If $n \geq k > 1$ are powers of 2, then we set $m = 2n$ and $z = 2k$ in Theorem 1.1. Then $\Phi_m(X) = X^n + 1$ and the condition that $p \equiv 2k + 1 \pmod{4k}$, i.e. $p \equiv z + 1 \pmod{2z}$, implies $p \equiv 1 \pmod{z}$. Now we need to show that $\text{ord}_m(p) = m/z$, but this follows immediately from Lemma 2.4 by setting $m = 2^e$ and $z = 2^f$ and noting that $f \geq 2$. Finally, from (9) we have $s_1(z) = \sqrt{\tau(z)} = \sqrt{\frac{z}{2}} = \sqrt{k}$ and $s_1(m) = \sqrt{n}$. Therefore the upper bounds for the $\| \cdot \|_\infty$ and $\| \cdot \|$ inequalities read $\frac{1}{\sqrt{k}} p^{1/k} = \frac{1}{s_1(z)} p^{1/k}$ and $p^{1/k} = \frac{\sqrt{n}}{s_1(m)} p^{1/k}$, respectively, as in Theorem 1.1.   □

3.3   Example of "Ad-Hoc" Applications of Lemma 3.2

Using Lemma 3.2, as we did in the proof of Theorem 1.1 above, gives a clean statement as to a sufficient condition under which polynomials are invertible in a partially-splitting ring. One thing to note is that putting a bound on the ℓ_∞ norm does not take into account the other properties that our challenge space may have. For example, our challenge space in (4) is also sparse, in addition to having the ℓ_∞ norm bounded by 1. Yet we do not know how to use this sparseness to show that one can let $\Phi_m(X)$ split further while still maintaining the invertibility of the set $C - C$.

In some cases, however, there are ways to construct challenge sets that are more in line with Lemma 3.2 and will allow further splitting. We do not see a simple way in which to systematize these ideas, and so one would have to work out the details on a case-by-case basis. Below, we give such an example for the case in which we are working over the ring $\mathbb{Z}_p[X]/(X^{256}+1)$ and would like to have the polynomial $X^{256}+1$ split into 16 irreducible factors. If we would like to have X^n+1 split into 16 factors modulo p and the set $C - C$ to have elements whose infinity norm is bounded by 2, then applying Theorem 1.1 directly implies that we need to have $2 < \frac{1}{\sqrt{16}} \cdot p^{1/16}$, which implies $p > 2^{48}$.

We will now show how one can lower the requirement on p in order to achieve a split into 16 factors by altering the challenge set C in (4).

For a polynomial $\mathbf{y} \in \mathbb{Z}_p[X]/(X^{256}+1)$, define the $\mathbf{y}_i'$ as in (11). Define $\mathcal{D}$ as

$$\mathcal{D} = \{\mathbf{y} \in \mathbb{Z}_p[X]/(X^{256}+1) : \|\mathbf{y}_i\|_\infty = 1 \text{ and } \forall 1 \leq i \leq 16, \|\mathbf{y}_i'\| = 2\} \quad (14)$$

In other words, $\mathcal{D}$ is the set of polynomials $\mathbf{y}$, such that every $\mathbf{y}_i'$ has exactly 4 non-zero elements that are ± 1. The size of $\mathcal{D}$ is $\left(\binom{16}{4} \cdot 2^4\right)^{16} \approx 2^{237}$, which should be enough for practical quantum security. The ℓ_2 norm of every element in $\mathcal{D}$ is exactly $\sqrt{64} = 8$. For a fair comparison, we should redefine the set C so that it also has size 2^{237}. The only change that one must make to the definition in (4) is to lower the ℓ_1 norm to 53 from 60. Thus all elements in C have ℓ_2 norm $\sqrt{53}$. The elements in set $\mathcal{D}$ therefore have norm that is larger by a factor of about 1.1. It then depends on the application as to whether having X^n+1 split into 16 rather than 8 factors is worth this modest increase. We will now prove that for primes $p > 2^{30.5}$ of a certain form, $X^{256}+1$ will split into 16 irreducible factors modulo p and all the non-zero elements in $\mathcal{D} - \mathcal{D}$ will be invertible. Therefore if our application calls for a modulus that is larger than $2^{30.5}$ but smaller than 2^{48}, we can use the challenge set $\mathcal{D}$ and the below lemma.

Lemma 3.3. *Suppose that $p > 2^{16 \log_2 \sqrt{14}} \approx 2^{30.5}$ is a prime congruent to 33 (mod 64). Then the polynomial $X^{256}+1$ splits into 16 irreducible polynomials of the form $X^{16} + r_j$ modulo p, and any non-zero polynomial $\mathbf{y} \in \mathcal{D} - \mathcal{D}$ (as defined in (14)) is invertible in the ring $\mathbb{Z}_p[X]/(X^{256}+1)$.*

Proof. The fact that $X^{256}+1$ splits into 16 irreducible factors follows directly from Theorem 2.3. Notice that for any $\mathbf{y} \in \mathcal{D} - \mathcal{D}$, the maximum ℓ_2 norm of $\mathbf{y}_i'$ is bounded by 4. Furthermore, the degree of each $\mathbf{y}_i'$ is $256/16 = 16$. Thus an immediate consequence of Lemmas 3.2 and 3.1 is that if $p > 2^{32}$, then any non-zero element in $\mathcal{D} - \mathcal{D}$ is invertible. To slightly improve the lower bound, we can observe that the $\mathbf{y}_i'$ of norm 4 are polynomials in $\mathbb{Z}_p[X]/(X^{16}+1)$ with exactly four 2's in them. But such elements can be written as a product of 2

Table 2. CPU cycles of our FFT-accelerated multiplication algorithm for $\mathbb{Z}_p[X]/(X^{256} + 1)$ using Karatsuba multiplication for the base case. Both the FFT and Karatsuba are plain C implementations.

Number of FFT levels	Primes			
	$2^{20} - 2^{14} + 1$	$2^{23} - 2^{13} + 1$	$2^{25} - 2^{12} + 1$	$2^{27} - 2^{11} + 1$
0	123677	123717	134506	144913
1	83820	83778	91775	97641
2	55378	55700	63148	65778
3	38111	38061	43116	43282
4	27374	27626	31782	30836
5	21968	21955	26406	24937
6	17076	17007	21518	19811
7	15149	15144	20483	18026
8	16875	16893	22329	20299

and a polynomial with 4 ±1's in it. So if both of those are invertible, so is the product. The maximum norm of these polynomials is 2 and so they are not the elements that set the lower bound. The next largest element in $\mathcal{D} - \mathcal{D}$ is one that has three 2's and two ±1's. The norm of such elements is $\sqrt{14}$. Thus for all $p > 2^{16 \cdot \log_2(\sqrt{14})} \approx 2^{30.5}$, the $\mathbf{y}'_i$ will be invertible in $\mathbb{Z}_p[X]/(X^{16} + 1)$, and thus every non-zero element in $\mathcal{D} - \mathcal{D}$ will be invertible in $\mathbb{Z}_p[X]/(X^{256} + 1)$.   □

Table 3. CPU cycles of our FFT-accelerated multiplication algorithm for $\mathbb{Z}_p[X]/(X^{256} + 1)$ using FLINT for base case multiplication. The FFT implementation is a highly optimized AVX2-based implementation.

Number of FFT levels	Primes			
	$2^{20} - 2^{14} + 1$	$2^{23} - 2^{13} + 1$	$2^{25} - 2^{12} + 1$	$2^{27} - 2^{11} + 1$
0	28245	31574	33642	35397
1	27168	29343	31419	32613
2	20989	23158	24915	25677
3	20521	22038	23582	23757
4	22543	23695	25016	24628
5	24473	24715	25337	30366
6	13578	13572	14307	13543
7	13981	14020	14522	13986
8	3873	3844	3847	3857

4   Polynomial Multiplication Implementation

We now describe in more detail the computational advantage of having the modulus Φ_m split into as many factors as possible and present our experimental results. We focus on the case where m is a power of two and write $n = \phi(m) = m/2$. In this case one can use the standard radix-2 FFT-trick to speed up the multiplication. Note that for other m, one can also exploit the splitting in a divide-and-conquer fashion similar to the radix-2 FFT.

Suppose that $\mathbb{Z}_p$ contains a fourth root of unity r so that we can write

$$X^n + 1 = (X^{n/2} + r)(X^{n/2} - r).$$

Then, in algebraic language, the FFT (or NTT) is based on the Chinese remainder theorem, which says that $R_{m,p} = \mathbb{Z}_p[X]/(X^n+1)$ is isomorphic to the direct product of $\mathbb{Z}_p[X]/(X^{n/2}+r)$ and $\mathbb{Z}_p[X]/(X^{n/2}-r)$. To multiply two polynomials in $R_{m,p}$ one can first reduce them modulo the two factors of the modulus, then multiply the resulting polynomials in the smaller rings, and finally invert the Chinese remainder map in order to obtain the product of the original polynomials. This is called the (radix-2) FFT-trick (see [Ber01] for a very good survey). Note that reducing a polynomial of degree less than n modulo the two sparse polynomials $X^{n/2} \pm r$ is very easy and takes only $\frac{n}{2}$ multiplications, $\frac{n}{2}$ additions and $\frac{n}{2}$ subtractions. If $\mathbb{Z}_p$ contains higher roots so that $X^n + 1$ splits further, then one can apply the FFT-trick recursively to the smaller rings. What is usually referred to as the number theoretic transform (NTT) is the case where $\mathbb{Z}_p$ contains a $2n$-th root of unity so that X^n+1 splits completely into linear factors. This reduces multiplication in $R_{m,p}$ to just multiplication in $\mathbb{Z}_p$.

As we are interested in the case where the modulus does not split completely, we need to be able to multiply in rings of the form $\mathbb{Z}_p[X]/(X^{n/k}-r_j)$ with $k < n$. As is common in cryptographic applications (see, for example [BCLvV17]), we will use the Karatsuba multiplication algorithm to perform this operation. For both the FFT and the Karatsuba multiplication, we have written a relatively straight-forward C implementation.

In Table 2 we give the measurements of our experiments. We have performed multiplications in $R_{512,p} = \mathbb{Z}_p[X]/(X^{256}+1)$ for four completely splitting primes between 2^{20} and 2^{30}. For each prime we have used between 0 and 8 levels of FFT before switching to Karatsuba multiplication. 0 levels of FFT means that no FFT stage was used at all and the input polynomials were directly multiplied via Karatsuba multiplication. In the other extreme of 8 levels of FFT, no Karatsuba multiplication was used and the corresponding measurements reflect the speed of our full number theoretic transform down to linear factors with pointwise multiplication as the base case. As one more example, when performing 3 levels of FFT, we were multiplying 8 polynomials each of degree less then 32 via Karatsuba multiplication. The listed numbers are numbers of CPU cycles needed for the whole multiplication. They are the medians of 10000 multiplications each. The tests where performed on a laptop equipped with an Intel Skylake i7 CPU running at 3.4 GHz. The cycle counter in this CPU ticks at a constant rate of

2.6 GHz. As one can see, being able to use a prime p so that $X^n + 1$ splits into more than two factors is clearly advantageous. For instance, by allowing $X^n + 1$ to split into 8 factors compared to just 2, we achieve a speedup of about a factor of two.

We have also experimented with highly-optimized polynomial multiplication algorithms provided by a popular computer algebra library FLINT [HJP13] and PARI [The16]. FLINT employs various forms of Kronecker substitution for the task of polynomial multiplication. For these experiments we used a fast vectorized FFT implementation written in assembler language with AVX2 instructions. For completeness, Table 3 gives the measurements for the tests with FLINT. Unfortunately, each call of the FLINT multiplication function produces additional overhead costs such as deciding on one of several algorithms and computing complex roots for the FFT used in Kronecker substitution. These additional costs are highly significant for our small polynomials. So for every additional stage of our FFT, one needs to multiply twice as many polynomials with FLINT, and hence FLINT spends twice as much time on these auxiliary tasks that one would not have in an actual cryptographic implementation specialized to a particular prime and modulus. This is especially inefficient when the number of FFT levels is large. There nearly all of the time is spend on these tasks as one can see in Table 3 by comparing the cycle counts of 7 and 8 stages of FFT. Note that for 7 stages of FFT, FLINT is used for the trivial task of multiplying polynomials of degree one.

While we were not able to do a meaningful analysis for the combination of our highly-optimized FFT with FLINT, one can see that at level 0 (where the amount of overhead it does is the lowest), FLINT outperforms our un-optimized Karatsuba multiplication by a factor between 4 and 5, while looking at Level 8 shows that our AVX-optimized FFT outperforms the non-optimized version by approximately the same margin. It is then reasonable to assume that one can improve non-FFT multiplication by approximately the same factor as we improved the FFT multiplication, and therefore the improvement going from level 1 and 3 would still be approximately a factor 2 in a routine where both Karatsuba and FFT multiplication were highly optimized.

Acknowledgements. We thank Rafaël del Pino for pointing out an improvement to Lemma 3.3. We also thank the anonymous reviewers for their advice on improving the paper. This work is supported by the SNSF ERC Transfer Grant CRETP2-166734 – FELICITY and the H2020 Project Safecrypto.

References

[ADPS16] Alkim, E., Ducas, L., Pöppelmann, T., Schwabe, P.: Post-quantum key exchange - a new hope. In: USENIX, pp. 327–343 (2016)

[BCLvV17] Bernstein, D.J., Chuengsatiansup, C., Lange, T., van Vredendaal, C.: NTRU prime: reducing attack surface at low cost. In: Adams, C., Camenisch, J. (eds.) SAC 2017. LNCS, vol. 10719, pp. 235–260. Springer, Cham (2018). https://doi.org/10.1007/978-3-319-72565-9_12

[BDK+17] Bos, J.W., Ducas, L., Kiltz, E., Lepoint, T., Lyubashevsky, V., Schanck, J.M., Schwabe, P., Stehlé, D.: CRYSTALS - Kyber: a CCA-secure module-lattice-based KEM. IACR Cryptology ePrint Archive, 2017:634 (2017). To appear in Euro S&P 2018

[BDOP16] Baum, C., Damgård, I., Oechsner, S., Peikert, C.: Efficient commitments and zero-knowledge protocols from ring-sis with applications to lattice-based threshold cryptosystems. IACR Cryptology ePrint Archive, 2016:997 (2016)

[Ber01] Bernstein, D.J.: Multidigit Multiplication for Mathematicians (2001)

[BKLP15] Benhamouda, F., Krenn, S., Lyubashevsky, V., Pietrzak, K.: Efficient zero-knowledge proofs for commitments from learning with errors over rings. In: Pernul, G., Ryan, P.Y.A., Weippl, E. (eds.) ESORICS 2015. LNCS, vol. 9326, pp. 305–325. Springer, Cham (2015). https://doi.org/10.1007/978-3-319-24174-6_16

[Coh00] Cohen, H.: A Course in Computational Algebraic Number Theory. Graduate Texts in Mathematics. Springer, Heidelberg (2000)

[DLL+17] Ducas, L., Lepoint, T., Lyubashevsky, V., Schwabe, P., Seiler, G., Stehlé, D.: CRYSTALS - Dilithium: digital signatures from module lattices. IACR Cryptology ePrint Archive, 2017:633 (2017). To appear in TCHES 2018

[DLNS17] Del Pino, R., Lyubashevsky, V., Neven, G., Seiler, G.: Practical quantum-safe voting from lattices. In: CCS (2017)

[GLP12] Güneysu, T., Lyubashevsky, V., Pöppelmann, T.: Practical lattice-based cryptography: a signature scheme for embedded systems. In: Prouff, E., Schaumont, P. (eds.) CHES 2012. LNCS, vol. 7428, pp. 530–547. Springer, Heidelberg (2012). https://doi.org/10.1007/978-3-642-33027-8_31

[HJP13] Hart, W., Johansson, F., Pancratz, S.: FLINT: Fast Library for Number Theory, Version 2.4.0 (2013). http://flintlib.org

[LM06] Lyubashevsky, V., Micciancio, D.: Generalized compact knapsacks are collision resistant. ICALP **2**, 144–155 (2006)

[LN17] Lyubashevsky, V., Neven, G.: One-shot verifiable encryption from lattices. In: Coron, J.-S., Nielsen, J.B. (eds.) EUROCRYPT 2017. LNCS, vol. 10210, pp. 293–323. Springer, Cham (2017). https://doi.org/10.1007/978-3-319-56620-7_11

[LPR10] Lyubashevsky, V., Peikert, C., Regev, O.: On ideal lattices and learning with errors over rings. In: Gilbert, H. (ed.) EUROCRYPT 2010. LNCS, vol. 6110, pp. 1–23. Springer, Heidelberg (2010). https://doi.org/10.1007/978-3-642-13190-5_1

[LPR13] Lyubashevsky, V., Peikert, C., Regev, O.: A toolkit for ring-LWE cryptography. In: Johansson, T., Nguyen, P.Q. (eds.) EUROCRYPT 2013. LNCS, vol. 7881, pp. 35–54. Springer, Heidelberg (2013). https://doi.org/10.1007/978-3-642-38348-9_3

[LS15] Langlois, A., Stehlé, D.: Worst-case to average-case reductions for module lattices. Des. Codes Crypt. **75**(3), 565–599 (2015)

[Lyu09] Lyubashevsky, V.: Fiat-shamir with aborts: applications to lattice and factoring-based signatures. In: Matsui, M. (ed.) ASIACRYPT 2009. LNCS, vol. 5912, pp. 598–616. Springer, Heidelberg (2009). https://doi.org/10.1007/978-3-642-10366-7_35

[Lyu12] Lyubashevsky, V.: Lattice signatures without trapdoors. In: Pointcheval, D., Johansson, T. (eds.) EUROCRYPT 2012. LNCS, vol. 7237, pp. 738–755. Springer, Heidelberg (2012). https://doi.org/10.1007/978-3-642-29011-4_43

[Mic07] Micciancio, D.: Generalized compact knapsacks, cyclic lattices, and efficient one-way functions. Comput. Complex. **16**(4), 365–411 (2007)

[PG13] Pöppelmann, T., Güneysu, T.: Towards practical lattice-based public-key encryption on reconfigurable hardware. In: Lange, T., Lauter, K., Lisoněk, P. (eds.) SAC 2013. LNCS, vol. 8282, pp. 68–85. Springer, Heidelberg (2014). https://doi.org/10.1007/978-3-662-43414-7_4

[PR06] Peikert, C., Rosen, A.: Efficient collision-resistant hashing from worst-case assumptions on cyclic lattices. In: Halevi, S., Rabin, T. (eds.) TCC 2006. LNCS, vol. 3876, pp. 145–166. Springer, Heidelberg (2006). https://doi.org/10.1007/11681878_8

[PR07] Peikert, C., Rosen, A.: Lattices that admit logarithmic worst-case to average-case connection factors. In: STOC, pp. 478–487 (2007)

[SS11] Stehlé, D., Steinfeld, R.: Making NTRU as secure as worst-case problems over ideal lattices. In: Paterson, K.G. (ed.) EUROCRYPT 2011. LNCS, vol. 6632, pp. 27–47. Springer, Heidelberg (2011). https://doi.org/10.1007/978-3-642-20465-4_4

[SSTX09] Stehlé, D., Steinfeld, R., Tanaka, K., Xagawa, K.: Efficient public key encryption based on ideal lattices. In: Matsui, M. (ed.) ASIACRYPT 2009. LNCS, vol. 5912, pp. 617–635. Springer, Heidelberg (2009). https://doi.org/10.1007/978-3-642-10366-7_36

[The16] The PARI Group, Univ. Bordeaux. PARI/GP version 2.9.0 (2016)

Random Oracle Model

Random Oracles and Non-uniformity

Sandro Coretti[1]([✉]), Yevgeniy Dodis[1], Siyao Guo[2], and John Steinberger[3]

[1] New York University, New York, USA
{corettis,dodis}@nyu.edu
[2] Northeastern University, Boston, USA
s.guo@neu.edu
[3] Geneva, Switzerland

Abstract. We revisit security proofs for various cryptographic primitives in the *auxiliary-input random-oracle model* (AI-ROM), in which an attacker $\mathcal{A}$ can compute arbitrary S bits of leakage about the random oracle $\mathcal{O}$ before attacking the system and then use additional T oracle queries to $\mathcal{O}$ during the attack. This model has natural applications in settings where traditional random-oracle proofs are not useful: (a) security against non-uniform attackers; (b) security against preprocessing. We obtain a number of new results about the AI-ROM:

– Unruh (CRYPTO'07) introduced the *pre-sampling technique*, which generically reduces security proofs in the AI-ROM to a much simpler *P-bit-fixing random-oracle model* (BF-ROM), where the attacker can arbitrarily fix the values of $\mathcal{O}$ on some P coordinates, but then the remaining coordinates are chosen at random. Unruh's security loss for this transformation is $\sqrt{ST/P}$. We improve this loss to the *optimal* value $O(ST/P)$, obtaining nearly tight bounds for a variety of indistinguishability applications in the AI-ROM.
– While the basic pre-sampling technique cannot give tight bounds for unpredictability applications, we introduce a novel "multiplicative version" of pre-sampling, which allows to dramatically reduce the size of P of the pre-sampled set to $P = O(ST)$ and yields nearly tight security bounds for a variety of unpredictability applications in the AI-ROM. Qualitatively, it validates Unruh's "polynomial pre-sampling conjecture"—disproved in general by Dodis *et al.* (EUROCRYPT'17)—for the special case of unpredictability applications.
– Using our techniques, we reprove nearly all AI-ROM bounds obtained by Dodis *et al.* (using a much more laborious compression technique), but we also apply it to many settings where the compression technique is either inapplicable (e.g., computational reductions) or appears intractable (e.g., Merkle-Damgård hashing).
– We show that for any *salted* Merkle-Damgård hash function with m-bit output there exists a collision-finding circuit of size $\Theta(2^{m/3})$ (taking salt as the input), which is significantly below the $2^{m/2}$ birthday security conjectured against uniform attackers.
– We build two compilers to generically extend the security of applications proven in the traditional ROM to the AI-ROM. One compiler simply prepends a public salt to the random oracle, showing that *salting generically provably defeats preprocessing.*

© International Association for Cryptologic Research 2018
J. B. Nielsen and V. Rijmen (Eds.): EUROCRYPT 2018, LNCS 10820, pp. 227–258, 2018.
https://doi.org/10.1007/978-3-319-78381-9_9

Overall, our results make it much easier to get concrete security bounds in the AI-ROM. These bounds in turn give concrete conjectures about the security of these applications (in the standard model) against *non-uniform* attackers.

1 Introduction

We start by addressing the two main themes of this work—non-uniformity and random oracles—in isolation, before connecting them to explain the main motivation for this work.

Non-uniformity. Modern cryptography (in the "standard model") usually models the attacker $\mathcal{A}$ as non-uniform, meaning that it is allowed to obtain some arbitrary (but bounded) "advice" before attacking the system. The main rationale to this modeling comes from the realization that a determined attacker will know the security parameter n of the system in advance and might be able to invest a significant amount of preprocessing to do something "special" for this fixed value of n, especially if n is not too large (for reasons of efficiency), or the attacker needs to break a lot of instances online (therefore amortizing the one-time offline cost). Perhaps the best known example of such attacks comes from *rainbow tables* ([31,46]; see also [38, Sect. 5.4.3]) for inverting arbitrary functions; the idea is to use one-time preprocessing to initialize a clever data structure in order to dramatically speed up brute-force inversion attacks. Thus, restricting to uniform attackers might not accurately model realistic preprocessing attacks one would like to protect against. However, there are other, more technical, reasons why this choice is convenient:

- Adleman [2] showed that non-uniform polynomial-time attackers can be assumed to be deterministic (formally, $BPP/poly = P/poly$), which is handy for some proofs.
- While many natural reductions in cryptography are uniform, there are several important cases where the only known (or even possible!) reduction is non-uniform. Perhaps the best known example are zero-knowledge proofs [27,28], which are *not* closed under *sequential composition* unless one allows non-uniform attackers (and simulators; intuitively, in order to use the simulator for the second zero-knowledge proof, one must use the output of the first proof's simulator as an auxiliary input to the verifier).[1] Of course, being a special case of general protocol composition, this means that any work—either using zero-knowledge as a subroutine or generally dealing with protocol composition—must use security against *non-uniform* attackers in order for the composition to work.

---

[1] There are some workarounds (see [26]) that permit one to define zero-knowledge under uniform attackers, but they are much harder to work with than assuming non-uniformity, and, as a result, were not adopted by the community.

– The non-uniform model of computation has many applications in complexity theory, such as the famous "hardness-vs-randomness" connection (see [33–36, 45]), which roughly states that *non-uniform* hardness implies non-trivial de-randomization. Thus, by defining cryptographic attackers as non-uniform machines, any lower bounds for such cryptographic applications might yield exciting de-randomization results.

Of course, despite the pragmatic, definitional, and conceptual advantages of non-uniformity, one must ensure that one does not make the attacker "too powerful," so that it can (unrealistically) solve problems which one might use in cryptographic applications. Fortunately, although non-uniform attackers can solve undecidable problems (by encoding the input in unary and outputting solutions in the non-uniform advice), the common belief is that non-uniformity cannot solve interesting "hard problems" in polynomial time. As one indirect piece of evidence, the Karp-Lipton theorem [37] shows that if NP has polynomial-size circuits, then the polynomial hierarchy collapses. And, of course, the entire field of cryptography is successfully based on the assumption that many hard problems cannot be solved even on average by polynomially sized circuits, and this belief has not been seriously challenged so far.

Hence, by and large it is believed by the theoretical community that *non-uniformity is the right cryptographic modeling of attackers*, despite being overly conservative and including potentially unrealistic attackers.

The Random-Oracle Model. Hash functions are ubiquitous in cryptography. They are widely used to build one-way functions (OWFs), collision-resistant hash functions (CRHFs), pseudorandom functions/generators (PRFs/PRGs), message authentication codes (MACs), etc. Moreover, they are often used together with other computational assumptions to show security of higher-level applications. Popular examples include Fiat-Shamir heuristics [1, 23] for signature schemes (e.g., Schnorr signatures [49]), full-domain-hash signatures [8], or trapdoor functions (TDFs) [8] and OAEP [9] encryption, among many others.

For each such application Q, one can wonder how to assess its security ε when instantiated with a concrete hash function H, such as SHA-3. Given our inability to prove unconditional lower bounds, the traditional approach is the following: Instead of proving an upper bound on ε for some specific H, one analyzes the security of Q assuming H is a *truly random (aka "ideal") function* $\mathcal{O}$. Since most Q are only secure against computationally bounded attackers, one gives the attacker $\mathcal{A}$ oracle access to $\mathcal{O}$ and limits the number of oracle queries that $\mathcal{A}$ can make by some parameter T. This now becomes the traditional *random-oracle model (ROM)*, popularized by the seminal paper of Bellare and Rogaway [8].

The appeal of the ROM stems from two aspects. First, it leads to very clean and intuitive security proofs for many primitives that resisted standard-model analysis under natural security assumptions (see some concrete examples below). Second, this resulting ROM analysis is *independent of the tedious specifics of H*, is done only *once* for a given hash-based application, and also provides (for non-pathological Q's) the *best possible* security one might hope to achieve with any concrete function H. In particular, we hope that a specific hash function H we use is sufficiently "well-designed" that it (essentially) matches this idealized bound. If it does, then our bound on ε was accurate anyway; and, if it does not, this usually serves as strong evidence that we should not use this particular H, rather than the indication that the idealized analysis was the wrong way to guess the exact security of Q. Ironically, in theory we know that the optimistic methodology above is false [5,11,12,29,44], and some applications secure in the ROM will be insecure for any instantiation of H, let alone maintain the idealized bound on ε. Fortunately, all counterexamples of this kind are rather artificial, and do not shed much light on the security of concrete schemes used in practice, such as the use of hash functions as OWFs, CRHFs, PRFs, PRGs, MACs, and also as parts of natural signature and encryption schemes used in practice [8,9,23,49]. In other words, despite purely theoretical concerns, the following *random-oracle methodology* appears to be a good way for practitioners to assess the best possible security level of a given (natural) application Q.

> **Random-oracle methodology.** *For "natural" applications of hash functions, the concrete security proven in the random-oracle model is the right bound even in the standard model, assuming the "best possible" concrete hash function H is chosen.*

Random Oracles and Non-uniformity. The main motivation for this work is to examine the soundness of the above methodology, while also being consistent with the fact that attackers should be modeled as *non-uniform*. We stress that we are not addressing the conceptual question of whether non-uniform security is the "right" way to model attackers in cryptography, as this is the subject of a rather heated on-going debate between theoreticians and practitioners; see [10,48] for some discussion on the subject. Instead, assuming we want to model attackers as non-uniform (for the reasons stated above and to be consistent with the theoretical literature), and assuming we want to have a way of correctly assessing the *concrete*, non-asymptotic security for important uses of hash functions in applications, we ask: is the random oracle methodology a sound way to achieve this goal? Unfortunately, with the traditional modeling of the random oracle, the answer is a resounding "NO," even for the *most basic* usages of hash functions, as can be seen from the following examples.

(i) In the standard model, no single function H can be collision-resistant, as a non-uniform attacker can trivially hardwire a collision. In contrast, a single (non-salted) random oracle $\mathcal{O}$ is trivially collision-resistant in the ROM, with excellent exact security $O(T^2/M)$, where M is the range of $\mathcal{O}$. This is why in the standard model one considers a *family* of collision-resistant hash functions whose public key, which we call *salt*, is chosen *after* $\mathcal{A}$ gets its non-uniform advice. Interestingly, one of the results in this paper will show that the large gap (finding collisions in time $M^{1/2}$ vs. $M^{1/3}$) between uniform and non-uniform security exists for the popular Merkle-Damgård construction *even if salting is allowed*.

(ii) In the standard model, no PRG candidate $H(x)$ can have security better than $2^{-n/2}$ even against linear-time (in n) attackers [3,10,20], where n is the seed-length of x. In contrast, an expanding random oracle $\mathcal{O}(x)$ can be trivially shown to be $(T/2^n)$-secure PRG in the traditional ROM, easily surpassing the $2^{-n/2}$ barrier in the standard model (even for huge T up to $2^{n/2}$, let alone polynomial T).

(iii) The seminal paper of Hellman [31], translated to the language of non-uniform attackers, shows that a random function $H : [N] \rightarrow [N]$ can be inverted with constant probability using a non-uniform attacker of size $O\left(N^{2/3}\right)$, while Fiat and Naor [22] extended this attack to show that every (even non-random) function H can be inverted with constant probability by circuits of size at most $N^{3/4}$. In contrast, if one models H as a random oracle $\mathcal{O}$, one can trivially show that $\mathcal{O}$ is a OWF with security $O\left(T/N\right)$ in the traditional ROM. For example, setting $T = N^{2/3}$ (or even $T = N^{3/4}$), one would still get negligible security $N^{-1/3}$ (or $N^{-1/4}$), contradicting the concrete non-uniform attacks mentioned above.

To put it differently, once non-uniformity is allowed in the standard model, the separations between the random-oracle model and the standard model are *no longer* contrived and artificial but rather lead to *impossibly good* exact security of *widely deployed* applications.

Auxiliary-Input ROM. The above concern regarding the random-oracle methodology is not new and was extensively studied by Unruh [51] and Dodis *et al.* [18]. Fortunately, these works offered a simple solution, by extending the traditional ROM to also allow for oracle-dependent *auxiliary input*. The resulting model, called the *auxiliary-input random-oracle model (AI-ROM)*, is parameterized by two parameters S ("space") and T ("time") and works as follows: First, as in the traditional random-oracle model, a function $\mathcal{O}$ is chosen uniformly from the space of functions with some domain and range. Second, the attacker $\mathcal{A}$ in the AI-ROM consists of two entities $\mathcal{A}_1$ and $\mathcal{A}_2$. The first-stage attacker $\mathcal{A}_1$ is computationally unbounded, gets full access to the random oracle $\mathcal{O}$, and computes some "non-uniform" advice z of size S. This advice is then passed to the second-stage attacker $\mathcal{A}_2$, who may make up to T queries to oracle $\mathcal{O}$ (and, unlike $\mathcal{A}_1$, might have additional application-specific restrictions, like bounded running time, etc.). This naturally maps to the preprocessing model discussed earlier and can also be used to analyze security against non-uniform circuits of size C by setting $S = T = C$.[2] Indeed, none of the concerns expressed in examples (i)–(iii) remain valid in AI-ROM: (i) $\mathcal{O}$ itself is no longer collision-resistant since $\mathcal{A}_1$ can precompute a collision; (ii)–(iii) the generic non-uniform PRG or OWF attacks mentioned earlier can also be performed on $\mathcal{O}$ itself (by letting $\mathcal{A}_1$ treat $\mathcal{O}$ as any other function H and computing the corresponding advice for $\mathcal{A}_2$). In sum, the AI-ROM model allows us to restate the modified variant of the random oracle methodology as follows:

---

[2] But separating S and T can also model non-uniform RAM computation with memory S and query complexity T.

AI-Random-Oracle Methodology. *For "natural" applications of hash functions, the concrete security proven in the* <u>AI-ROM</u> *is the right bound even in the standard model against* <u>non-uniform</u> *attackers, assuming the "best possible" concrete hash function* H *is chosen.*

Dealing with Auxiliary Information. The AI-ROM yields a clean and elegant way towards obtaining meaningful non-uniform bounds for natural applications. Unfortunately, obtaining such bounds is considerably more difficult than in the traditional ROM. In retrospect, such difficulties are expected, since we already saw several examples showing that non-uniform attackers are *very powerful* when exact security matters, which means that the security bounds obtained in the AI-ROM might often be noticeably weaker than in the traditional ROM. From a technical point, the key difficulty is this: *conditioned on the leaked value z, which can depend on the entire function table of $\mathcal{O}$ in some non-trivial manner, many of the individual values $\mathcal{O}(x)$ are no longer random to the attacker.* And this ruins many of the key techniques utilized in the traditional ROM, such as: (1) *lazy sampling*, which allows the reduction to sample the not-yet-queried values of $\mathcal{O}$ at random, as needed, without worrying that such lazy sampling will be inconsistent with the past; (2) *programmability*, which allows the reduction to dynamically define some value of $\mathcal{O}$ in a special (still random) way, as this might be inconsistent with the leakage value z it has to produce *before* knowing how and where to program $\mathcal{O}$; (3) *distinguishing-to-extraction argument*, which states that the attacker cannot distinguish the value of $\mathcal{O}$ from random without explicitly querying it (which again is false given auxiliary input). For these reasons, new techniques are required for dealing with the AI-ROM. Fortunately, two such techniques are known:

– *Pre-sampling technique.* This beautiful technique was introduced in the original, pioneering work of Unruh [51]. From our perspective, we will present Unruh's pre-sampling technique in a syntactically different (but technically equivalent) way which will be more convenient for our presentation. Specifically, Unruh implicitly introduced an intermediate oracle model, which we term the *bit-fixing random-oracle model (BF-ROM)*,[3] which can be arbitrarily fixed on some P coordinates, but then the remaining coordinates are chosen at random and independently of the fixed coordinates. Moreover, the non-uniform S-bit advice of the attacker can only depend on the P fixed points, but not on the remaining truly random points. Intuitively, dealing with the BF-ROM—at least when P is small—appears to be much easier than with the AI-ROM, as many of the traditional ROM proof techniques can be adapted provided that one avoids the "pre-sampled" set. Quite remarkably, for any value P, Unruh showed that any (S, T)-attack in the AI-ROM will have similar advantage in (appropriately chosen) P-BF-ROM, up to an additive loss of $\delta(S, T, P)$, which Unruh upper bounded by $\sqrt{ST/P}$. This yields a general recipe for dealing with the AI-ROM: (a) prove security $\varepsilon(S, T, P)$ of the

---

[3] This naming in inspired by the bit-fixing source [13] from complexity theory.

given application in the P-BF-ROM;[4] (b) optimize for the right value of P by balancing $\varepsilon(S,T,P)$ and $\delta(S,T,P)$ (while also respecting the time and other constraints of the attacker).

- *Compression technique.* Unfortunately, Dodis *et al.* [18] showed that the concrete security loss $\delta(S,T,P) = \sqrt{ST/P}$ proven by Unruh is not strong enough to get tight bounds for any of the basic applications of hash functions, such as building OWFs, PRGs, PRFs, (salted) CRHFs, and MACs. To remedy the situation, Dodis *et al.* [18] showed a different, less general technique for dealing with the AI-ROM, by adapting the *compression paradigm*, introduced by Gennaro and Trevisan [24,25] in the context of black-box separations, to the AI-ROM. The main idea is to argue that if some AI-ROM attacker succeeds with high probability in breaking a given scheme, then that attacker can be used to reversibly encode (i.e., compress) a random oracle beyond what is possible from an information-theoretic point of view. Since we are considering attackers who perform preprocessing, our encoding must include the S-bit auxiliary information produced by the attacker. Thus, the main technical challenge in applying this technique is to ensure that the constructed encoding compress by (significantly) more than S bits. Dodis *et al.* [18] proceeded by successfully applying this idea to show nearly tight (and always better than what was possible by pre-sampling) bounds for a variety of natural applications, including OWFs, PRGs, PRFs, (salted) CRHFs, and MACs.

Pre-sampling or Compression? The pre-sampling and compression techniques each have their pros and cons, as discussed below.

On a positive, pre-sampling is very general and seems to apply to most applications, as analyzing the security of schemes in BF-ROM is not much harder than in the traditional ROM. Moreover, as shown by Unruh, the pre-sampling technique appears at least "partially friendly" to computational applications of random oracles (specifically, Unruh applied it to OAEP encryption [9]). Indeed, if the size P of the pre-sampled set is not too large, then it can be hardwired as part of non-uniform advice to the (efficient) reduction to the computational assumption. In fact, in the asymptotic domain Unruh even showed that the resulting security remains "negligible in security parameter λ," despite not being smaller than any concrete negligible function (like the inverse Ackermann function).[5]

On a negative, the *concrete* security bounds which are currently obtainable using this technique are vastly suboptimal, largely due to the big security loss $\sqrt{ST/P}$ incurred by using Unruh's bound [51]. Moreover, for com-

---

[4] Observe that the parameter S is still meaningful here. $\mathcal{A}_1$ fixes $\mathcal{O}$ at P points but only passes S bits of advice to $\mathcal{A}_2$. While none of information-theoretic proofs in this paper really use this, for computational reductions S "passes through" for the final non-uniform attacker against the computational assumption, and it is necessary to have $S \ll P$ in this case.

[5] Any AI-ROM attacker of size $t = t(\lambda)$ getting inverse polynomial advantage $\delta = 1/p(\lambda)$ for infinitely many λ's has advantage $\delta - \sqrt{ST/P}$ in the BF-ROM, which can be made to be $\delta/2$ by suitably choosing $P \approx O(t^2/\delta^2)$, which is polynomial and therefore suited for a reduction to a computational hardness assumption. .

putational applications, the value of P cannot be made larger than the size of attacker for the corresponding computational assumption. Hence, for fixed ("non-asymptotic"; see Footnote 5) polynomial-size attackers, the loss $\sqrt{ST/P}$ cannot be made negligible. Motivated by this, Unruh conjectured that the security loss of pre-sampling can be improved by a tighter proof. Dodis *et al.* [18] showed that the best possible security loss is at most ST/P. For computational applications, this asymptotically disproves Unruh's conjecture, as ST/P is still non-negligible for polynomial values of P (although we will explain shortly that the situation is actually more nuanced).

Moving to the compression technique, we already mentioned that it led Dodis *et al.* [18] to establishing nearly tight AI-ROM bounds for several information-theoretic applications of random oracles. Unfortunately, each proof was noticeably more involved than the original ROM proof, or than the proof in the BF-ROM one would do if applying the more intuitive pre-sampling technique. Moreover, each primitive required a completely different set of algorithmic insights to get the required level of compression. And it is not entirely clear how far this can go. For example, we do not see any way to apply the compression paradigm to relatively basic applications of hash functions beyond using the hash function *by itself* as a given primitive; e.g., to show AI-ROM security of the classical Merkle-Damgård paradigm [16,42] (whose *tight* AI-ROM security we will later establish in this work). Moreover, unlike pre-sampling, the compression paradigm cannot be applied at all to computational applications, as the compressor and the decompressor are computationally unbounded.

1.1   Our Results

We obtain a number of results about dealing with the AI-ROM, which, at a high-level, take the best features from pre-sampling (simplicity, generality) and compression (tightness).

Improving Unruh. Recall, Unruh [51] showed that one can move from the AI-ROM to the P-BF-ROM at the additive cost $\delta(S,T,P) \leq \sqrt{ST/P}$, and Dodis *et al.* [18] showed that $\delta(S,T,P) = \Omega(ST/P)$ in general. We show that the true additive error bound is indeed $\delta(S,T,P) = \Theta(ST/P)$, therefore improving Unruh's bound by a quadratic factor; see Theorem 1. Namely, the effect of S bits of auxiliary information $z = z(\mathcal{O})$ against an attacker making T adaptive random-oracle queries can be simulated to within an additive error $O(ST/P)$ by fixing the value of the random oracle on P points (which depend on the function z), and picking the other points at random and independently of the auxiliary information.

While the quadratic improvement might appear "asymptotically small," we show that it already matches the near-tight bound for all indistinguishability applications (specifically, PRGs and PRFs) proved by [18] using much more laborious compression arguments. For example, to match the $\varepsilon = O(\sqrt{ST/N} + T/N)$ bound for PRGs with seed domain N, we show using a simple argument that the random oracle is $\varepsilon' = O(P/N + T/N)$-secure in the P-BF-ROM, where the first term corresponds to the seed being chosen from the pre-sampled set, and the second term corresponds to the probability of querying the oracle on the seed in the attack stage. Setting $P = O(\sqrt{STN})$ to balance the P/N and ST/P terms, we immediately get our final bound, which matches that of [18]. For illustrative purposes, we also apply our improved bound to argue the AI-ROM security of a couple of indistinguishability applications not considered by [18]. First, we show an improved—compared to its use as a (standard) PRF—bound for the random oracle as a *weak* PRF, which is enough for chosen-plaintext secure symmetric-key encryption. Our proof is a very simple adaptation of the PRF proof in the BF-ROM, while we believe the corresponding compression proof, if possible at all, would involve noticeable changes to the PRF proof of [18] (due to the need for better compression to get the improved bound). Second, we also apply it to a typical example of a computational application, namely, the (KEM-variant of the) TDF-based public-key encryption scheme $\mathsf{Enc}_f(m; x) = (f(x), \mathcal{O}(x) \oplus m)$ from the original Bellare-Rogaway paper [8], where f is a trapdoor permutation (part of the public key, while the inverse is the secret key) and x is the randomness used for encryption. Recall that the compression technique cannot be applied to such applications.

To sum up, we conjecture that the improved security bound ST/P should be sufficient to get good bounds for most natural indistinguishability applications; these bounds are either tight, or at least they match those attainable via compression arguments (while being much simpler and more general).

Improved Pre-sampling for Unpredictability Applications.   Even with our improved bound of ST/P for pre-sampling, we will not match the nearly tight compression bounds obtained by Dodis *et al.* [18] for OWFs and MACs. In particular, finding the optimal value of P will result in "square root terms" which are not matched by any existing attacks. As our key insight, we notice that this is not due to the limitations of pre-sampling (i.e., going through the BF-ROM), but rather to the fact that achieving an *additive* error is unnecessarily restrictive for *unpredictability* applications. Instead, we show that if one is happy with a multiplicative factor of 2 in the probability of breaking the system, then one can achieve this *generically* by setting the pre-sampling set size $P \approx ST$; see Theorem 2.

This has a number of implications. First, with this multiplicative pre-sampling technique, we can easily match the compression bounds for the OWF and MAC unpredictability applications considered by Dodis *et al.* [18], but with much simpler proofs. Second, we also apply it to a natural information-theoretic application where we believe the compression technique will fail to get a good bound; namely, building a (salted) CHRF family via the Merkle-Damgård paradigm, where the salt is the initialization vector for the construction (see Theorem 3). The salient feature of this example is that the random oracle is applied in iteration, which poses little difficulties to adapting the standard-ROM proof to the BF-ROM, but seems to completely blow up the complexity of the compression arguments, as there are too many possibilities for the attacker to cause a collision for different salts when the number of blocks is greater than 1.[6]

---

[6] The same difficulty of compression should also apply to indistinguishability applications of Merkle-Damgård, such as building PRFs [6].

The resulting AI-ROM bound $O(ST^2/M)$ becomes vacuous for circuits of size roughly $M^{1/3}$, where M is the range of the compression function. This bound is well below the conjectured $M^{1/2}$ birthday security of CRHFs based on Merkle-Damgård against uniform attackers. Quite unexpectedly, we show that $M^{1/3}$ security we prove *is tight*: there exists a (non-uniform) collision-finding attack implementable by a circuit of size $O\left(M^{1/3}\right)$ (see Theorem 4)! This example illustrates once again the surprising power of non-uniformity.

Implications to Computational Reductions. Recall that, unlike compression techniques, pre-sampling can be applied to computational reductions, by "hard-wiring" the pre-sampling set of size P into the attacker breaking the computational assumption. However, this means that P cannot be made larger than the maximum allowed running time t of such an attacker. Since standard pre-sampling incurs additive cost $\Omega(ST/P)$, one cannot achieve final security better that ST/t, irrespective of the value of ε in the (t, ε)-security of the corresponding computational assumption. For example, when t is polynomial (in the security parameter) and $\varepsilon \ll 1/t$ is exponentially small, we only get inverse polynomial security (at most ST/t) when applying standard pre-sampling. In contrast, the multiplicative variant of pre-sampling sets the list size to be roughly $P \approx ST$, which is polynomial for polynomial S and T and can be made smaller than the complexity t of the standard model attacker for the computational assumption we use. Thus, when t is polynomial and ε is exponentially small, we will get negligible security using multiplicative pre-sampling. For a concrete illustrative example, see the bound in Theorem 5 when we apply our improved pre-sampling to the natural *computational unpredictability* application of Schnorr signatures [49].[7] To put it differently, while the work of Dodis *et al.* [18] showed that Unruh's "pre-sampling conjecture" is false in general—meaning that negligible security is not possible with a polynomial list size P—we show that it is qualitatively true for *unpredictability applications*, where the list size can be made polynomial (roughly ST).

Moreover, we show that in certain computational *indistinguishability* applications, we can still apply our improved pre-sampling technique *inside the reduction*, and get final security higher than the ST/t barrier mentioned above. We illustrate this phenomenon in our analysis of TDF encryption (cf. Theorem 6) by separating the probability of the attacker's success into 2 disjoint events: (1) the attacker, given ciphertext $f(x)$, managed to query the random oracle on the TDP preimage x; (2) the attacker succeeds in distinguishing the value $\mathcal{O}(x)$ from random without querying $\mathcal{O}(x)$. Now, for the event (1), we can reduce to the TDP security with polynomial list size using our improved *multiplicative* pre-sampling (since is an *unpredictability* event), while for the event (2), we can prove *information-theoretic* security using standard additive pre-sampling, without the limitation of having to upper bound P by the running time of the TDP attacker. It is an interesting open question to classify precisely the type

---

[7] Interestingly, general Fiat-Shamir transform is *not* secure in AI-ROM, and thus our proof used the specifics of Schnorr's signatures.

of indistinguishability applications where such "hybrid" reduction technique can be applied.

Going to the Traditional ROM. So far, the general paradigm we used is to reduce the hard-to-analyze security of *any* scheme in the AI-ROM to the much simpler and proof-friendly security of the *same* scheme in the BF-ROM. However, an even simpler approach, if possible, would be to reduce the security in the AI-ROM all the way to the traditional ROM. Of course, we know that this is impossible without any modifications to the scheme, as we have plenty of examples where the AI-ROM security of the scheme is much weaker than its ROM security (or even disappears completely). Still, when a simple modification is possible without much inconvenience to the users, reducing to the ROM has a number of obvious advantages over the BF-ROM:

- While much simpler than in the AI-ROM, one must still prove a security bound in BF-ROM. It would be much easier if one could just utilize an already proven result in ROM and seamlessly "move it" to the AI-ROM at a small cost.
- Some natural schemes secure in the traditional ROM are *insecure* in the BF-ROM (and also in the AI-ROM) without any modifications. Simple example include the general Fiat-Shamir heuristic [1,23] or the FDH signature scheme [8] (see the full version of this paper [15]). Thus, to extend such schemes to the AI-ROM, we must modify them anyway, so we might as well try to generically ensure that ROM security is already enough.

As our next set of results, we show two simple compilers which build a hash function $\mathcal{O}'$ to be used in AI-ROM application out of hash function $\mathcal{O}$ used in the traditional ROM application. Both results are in the common-random-string model. This means that they utilize a public random string (which we call salt and denote a) chosen *after* the auxiliary information about $\mathcal{O}$ is computed by the attacker. The honest parties are then assumed to have reliable access to this a value. We note that in basic applications, such as encryption and authentication, the salt can simply be chosen at key generation and be made part of the public key/parameters, so this comes at a small price indeed.

The first transformation analyzed in Sect. 6.1 is simply *salting*; namely $\mathcal{O}'_a(x) = \mathcal{O}(a,x)$, where a is a public random string chosen from the domain of size K. This technique is widely used in practice (going back to password hashing [43]), and was analyzed by Dodis *et al.* [18] in the context of AI-ROM, by applying the compression argument to show that *salting provably defeats pre-processing* for the few natural applications they consider (OWFs, PRGs, PRFs, and MACs). What our work shows is that salting provably defeats pre-processing *generically*, as opposed to a few concrete applications analyzed by [18].[8] Namely,

---

[8] Of course, by performing a direct analysis of the *salted* scheme (e.g., using Theorems 1 or 2), we might get better exact security bounds than by using our general result; namely, shorter salt would be enough to get the claimed amount of security. Still, for settings where obtaining the smallest possible salt value is not critical, the simplicity and generality of our compilers offer a convenient and seamless way to argue security in AI-ROM without doing a direct analysis.

by making the salt domain K large enough, one gets almost the same security in AI-ROM than in the traditional ROM. To put differently, when salting is possible, one gets the *best of both worlds: security against non-uniform attacks, but with exact security matching that in the traditional ROM.*

The basic salting technique sacrificed a relatively large factor of K from the domain of the random oracle $\mathcal{O}$ in order to build $\mathcal{O}'$ (for K large enough to bring the "salting error" down). When the domain of $\mathcal{O}$ is an expensive resource, in Sect. 6.2 we also design a more domain-efficient compiler, which only sacrifices a small factor $k \geq 2$ in the domain of $\mathcal{O}$, at the cost that each evaluation of $\mathcal{O}'$ takes $k \geq 2$ evaluations of $\mathcal{O}$ (and the "salting error" decays exponentially in k). This transformation is based on the adaptation of the technique of Maurer [41], originally used in the context of key-agreement with randomizers. While the basic transformation needs $O(k \log N)$ bits of public salt, we also show than one can reduce the number of random bits to $O(k + \log N)$. And since we do not envision k to be larger than $O(\log N)$ for any practical need, the total length of the salt is always $O(\log N)$.

Our Main Lemma. The key technical contribution of our work is Lemma 1, proved in Sect. 2.1, which roughly shows that a random oracle with auxiliary input is "close" to the convex combination of "P-bit-fixing sources" (see Definition 1). Moreover, we give both additive and multiplicative versions of this "closeness," so that we can later use different parameters to derive our Theorem 1 (for indistinguishability applications in the AI-ROM) and Theorem 2 (for unpredictability applications in the AI-ROM) in Sect. 2.2.

1.2 Other Related Work

Most of the related work was already mentioned earlier. The realization that multiplicative error is enough for unpredictability applications, and this can lead to non-trivial savings, is related to the work of Dodis *et al.* [19] in the context of improved entropy loss of key derivation schemes. Tessaro [50] generalized Unruh's presampling techniques to the random-*permutation* model, albeit without improving the tightness of the bound.

De *et al.* [17] study the effect of salting for inverting a *permutation* $\mathcal{O}$ as well as for a specific pseudorandom generator based on one-way permutations. Chung *et al.* [14] study the effects of salting in the design of collision-resistant hash functions, and used Unruh's pre-sampling technique to argue that salting defeats preprocessing in this important case. Using salting to obtain non-uniform security was also advocated by Mahmoody and Mohammed [40], who used this technique for obtaining non-uniform black-box separation results.

Finally, the extensive body of work on the *bounded storage model* [4,21,41,52] is related to the special case of AI-ROM, where all T queries in the second stage are done by the challenger to derive the key (so that one tries to minimize T to ensure local computability), but the actual attacker is not allowed any such queries after S-bit preprocessing.

2   Dealing with Auxiliary Information

Since an attacker with oracle-dependent auxiliary input may obtain the output of arbitrary functions evaluated on a random oracle's function table, it is not obvious how the security of schemes in the *auxiliary-input random-oracle model (AI-ROM)* can be analyzed. To remedy this situation, Unruh [51] introduced the *bit-fixing random-oracle model (BF-ROM)*, in which the oracle is fixed on a subset of the coordinates and uniformly random and independent on the remaining ones, and showed that such an oracle is indistinguishable from an AI-RO.

In Sect. 2.1, we improve the security bounds proved by Unruh [51] in the following two ways: First, we show that a BF-RO is indistinguishable from an AI-RO up to an additive term of roughly ST/P, where P is the size of the fixed portion of the BF-RO; this improves greatly over Unruh's bound, which was in the order of $\sqrt{ST/P}$. Second, we prove that the probability that any distinguisher outputs 1 in the AI-ROM is at most twice the probability that said distinguisher outputs 1 in the BF-ROM—already when P is roughly equal to ST.

Section 2.2 contains the formalizations of the AI and BF-ROMs, attackers with oracle-dependent advice, and the notion of application. As a consequence of the connections between the two models, the security of *any* application in the BF-ROM translates to the AI-ROM at the cost of the ST/P term, and, additionally, the security of *unpredictability* applications translates at the mere cost of a multiplicative factor of 2 (as long as $P \geq ST$). The corresponding theorems and their proofs can also be found in Sect. 2.2.

2.1   Replacing Auxiliary Information by Bit-Fixing

In this section, we show that any random oracle about which an attacker may have a certain amount of auxiliary information can be replaced by a suitably chosen convex combination of bit-fixing sources. This substitution comes at the price of either an additive term to the distinguishing advantage or a multiplicative one to the probability that a distinguisher outputs 1. To that end, consider the following definition:

Definition 1. *An (N, M)-source is a random variable X with range $[M]^N$. A source is called*

- $(1 - \delta)$-*dense if for every subset $I \subseteq [N]$,*

$$H_\infty(X_I) \geq (1 - \delta) \cdot |I| \cdot \log M = (1 - \delta) \cdot \log M^{|I|}.$$

- $(P, 1 - \delta)$-*dense if it is fixed on at most P coordinates and is $(1 - \delta)$-dense on the rest,*
- P-*bit-fixing if it is fixed on at most P coordinates and uniform on the rest.*

That is, the min-entropy of every subset of the function table of a δ-dense source is at most a fraction of δ less than what it would be for a uniformly random one.

Lemma 1. *Let X be distributed uniformly over $[M]^N$ and $Z := f(X)$, where $f : [M]^N \to \{0,1\}^S$ is an arbitrary function. For any $\gamma > 0$ and $P \in \mathbb{N}$, there exists a family $\{Y_z\}_{z \in \{0,1\}^S}$ of convex combinations Y_z of P-bit-fixing (N, M)-sources such that for any distinguisher D taking an S-bit input and querying at most $T < P$ coordinates of its oracle,*

$$\left| P\left[\mathcal{D}^X(f(X)) = 1 \right] - P\left[\mathcal{D}^{Y_{f(X)}}(f(X)) = 1 \right] \right| \leq \frac{(S + \log 1/\gamma) \cdot T}{P} + \gamma$$

and

$$P\left[\mathcal{D}^X(f(X)) = 1 \right] \leq 2^{(S + \log 1/\gamma)T/P} \cdot P\left[\mathcal{D}^{Y_{f(X)}}(f(X)) = 1 \right] + \gamma.$$

Lemma 1 is proved using a technique (cf. the first claim in the proof) put forth by Göös *et al.* [30] in the area of communication complexity. The technique was also adopted in a paper by Kothari *et al.* [39], who gave a simplified argument for decomposing high-entropy sources into bit-fixing sources with constant density (cf. Definition 1). For self-containment, the full version of this paper [15] contains a proof of this decomposition technique. Furthermore, the proof uses the well-known H-coefficient technique by Patarin [47], while following a recent re-formulation of it due to Hoang and Tessaro [32].

Proof. Fix an arbitrary $z \in \{0,1\}^S$ and let X_z be the distribution of X conditioned on $f(X) = z$. Let $S_z = N \log M - H_\infty(X_z)$ be the min-entropy deficiency of X_z. Let $\gamma > 0$ be arbitrary.

Claim 1. For every $\delta > 0$, X_z is γ-close to a convex combination of finitely many $(P', 1 - \delta)$-dense sources for

$$P' = \frac{S_z + \log 1/\gamma}{\delta \cdot \log M}.$$

The proof of the above claim can be found in the full version of this paper [15].

Let X'_z be the convex combination of $(P', 1 - \delta)$-dense sources that is γ-close to X_z for a $\delta = \delta_z$ to be determined later. For every $(P', 1 - \delta)$ source X' in said convex combination, let Y' be the corresponding P'-bit-fixing source Y', i.e., X' and Y' are fixed on the same coordinates to the same values. The following claim bounds the distinguishing advantage between X' and Y' for any T-query distinguisher.

Claim 2. For any $(P', 1 - \delta)$-dense source X' and its corresponding P'-bit-fixing source Y', it holds that for any (adaptive) distinguisher $\mathcal{D}$ that queries at most T coordinates of its oracle,

$$\left| P\left[\mathcal{D}^{X'} = 1 \right] - P\left[\mathcal{D}^{Y'} = 1 \right] \right| \leq T\delta \cdot \log M,$$

and

$$P\left[\mathcal{D}^{X'} = 1 \right] \leq M^{T\delta} \cdot P\left[\mathcal{D}^{Y'} = 1 \right].$$

Proof. Assume without loss of generality that $\mathcal{D}$ is deterministic and does not query any of the fixed positions. Let $T_{X'}$ and $T_{Y'}$ be the random variables corresponding to the transcripts containing the query/answer pairs resulting from $\mathcal{D}$'s interaction with X' and Y', respectively. For a fixed transcript τ, denote by $p_{X'}(\tau)$ and $p_{Y'}(\tau)$ the probabilities that X' and Y', respectively, produce the answers in τ if the queries in τ are asked. Observe that these probabilities depend only on X' resp. Y' and are independent of $\mathcal{D}$.

Observe that for every transcript τ,

$$p_{X'}(\tau) \leq M^{-(1-\delta)T} \quad \text{and} \quad p_{Y'}(\tau) = M^{-T} \tag{1}$$

as X' is $(1-\delta)$-dense and Y' is uniformly distributed.

Since $\mathcal{D}$ is deterministic, $P[T_{X'} = \tau] \in \{0, p_{X'}(\tau)\}$, and similarly, $P[T_{Y'} = \tau] \in \{0, p_{Y'}(\tau)\}$. Denote by $\mathcal{T}_X$ the set of all transcripts τ for which $P[T_{X'} = \tau] > 0$. For such τ, $P[T_{X'} = \tau] = p_{X'}(\tau)$ and also $P[T_{Y'} = \tau] = p_{Y'}(\tau)$. Towards proving the first part of the lemma, observe that

$$
\left| P\left[\mathcal{D}^{X'} = 1\right] - P\left[\mathcal{D}^{Y'} = 1\right] \right| \leq \mathrm{SD}(T_{X'}, T_{Y'})
$$

$$
= \sum_{\tau} \max\left\{0, P[T_{X'} = \tau] - P[T_{Y'} = \tau]\right\}
$$

$$
= \sum_{\tau \in \mathcal{T}_X} \max\left\{0, p_{X'}(\tau) - p_{Y'}(\tau)\right\}
$$

$$
= \sum_{\tau \in \mathcal{T}_X} p_{X'}(\tau) \cdot \max\left\{0, 1 - \frac{p_{Y'}(\tau)}{p_{X'}(\tau)}\right\}
$$

$$
\leq 1 - M^{-T\delta} \leq T\delta \cdot \log M,
$$

where the first sum is over all possible transcripts and where the last inequality uses $2^{-x} \geq 1 - x$ for $x \geq 0$.

As for the second part of the lemma, observe that due to (1) and the support of $T_{X'}$ being a subset of $T_{Y'}$,

$$P[T_{X'} = \tau] \leq M^{T\delta} \cdot P[T_{Y'} = \tau]$$

for any transcript τ. Let $\mathcal{T}_\mathcal{D}$ be the set of transcripts where $\mathcal{D}$ outputs 1. Then,

$$P[\mathcal{D}^{X'} = 1] = \sum_{\tau \in \mathcal{T}_\mathcal{D}} P[T_{X'} = \tau] \leq M^{T\delta} \cdot \sum_{\tau \in \mathcal{T}_\mathcal{D}} P[T_{Y'} = \tau] = M^{T\delta} \cdot P[\mathcal{D}^{Y'} = 1].$$

$\square$

Let Y'_z be obtained by replacing every X' by the corresponding Y' in X'_z. Setting $\delta_z = (S_z + \log 1/\gamma)/(P \log M)$, Claims 1 and 2 imply

$$\left| P\left[\mathcal{D}^{X_z}(z) = 1\right] - P\left[\mathcal{D}^{Y'_z}(z) = 1\right] \right| \leq \frac{(S_z + \log 1/\gamma) \cdot T}{P} + \gamma, \tag{2}$$

as well as

$$P\big[\mathcal{D}^{X_z}(z) = 1\big] \;\leq\; 2^{(S_z + \log 1/\gamma)T/P} \cdot P\big[\mathcal{D}^{Y_z'}(z) = 1\big] + \gamma\,. \tag{3}$$

Moreover, note that for the above choice of δ_z, $P' = P$, i.e., the sources Y' are fixed on at most P coordinates, as desired.

Claim 3. $\mathbf{E}_z[S_z] \leq S$ and $\mathbf{E}_z[2^{S_z T/P}] \leq 2^{ST/P}$.

The proof of the above claim can be found in the full version of this paper [15]. The lemma now follows (using $Y_z := Y_z'$) by taking expectations over z of (2) and (3) and applying the above claim.   □

2.2 From the BF-ROM to the AI-ROM

Capturing the Models. Before Lemma 1 from the preceding section can be used to show how security proofs in the BF-ROM can be transferred to the AI-ROM, it is necessary to formally define the two models as well as attackers with oracle-dependent advice and the notion of an application. The high-level idea is to consider two-stage attackers $\mathcal{A} = (\mathcal{A}_1, \mathcal{A}_2)$ and (single-stage) challengers C with access to an oracle $\mathcal{O}$. Oracles have two interfaces pre and main, where pre is accessible only to $\mathcal{A}_1$, which may pass auxiliary information to $\mathcal{A}_2$, and both $\mathcal{A}_2$ and C may access main.

Oracles. An oracle $\mathcal{O}$ has two interfaces $\mathcal{O}$.pre and $\mathcal{O}$.main, where $\mathcal{O}$.pre is accessible only once before any calls to $\mathcal{O}$.main are made. Oracles used in this work are:

- *Random oracle* RO(N, M): Samples a random function table $F \leftarrow \mathcal{F}_{N,M}$, where $\mathcal{F}_{N,M}$ is the set of all functions from $[N]$ to $[M]$; offers no functionality at $\mathcal{O}$.pre; answers queries $x \in [N]$ at $\mathcal{O}$.main by the corresponding value $F[x] \in [M]$.
- *Auxiliary-input random oracle* AI-RO(N, M): Samples a random function table $F \leftarrow \mathcal{F}_{N,M}$; outputs F at $\mathcal{O}$.pre; answers queries $x \in [N]$ at $\mathcal{O}$.main by the corresponding value $F[x] \in [M]$.
- *Bit-Fixing random oracle* BF-RO(P, N, M): Samples a random function table $F \leftarrow \mathcal{F}_{N,M}$; takes a list at $\mathcal{O}$.pre of at most P query/answer pairs that override F in the corresponding positions; answers queries $x \in [N]$ at $\mathcal{O}$.main by the corresponding value $F[x] \in [M]$.
- *Standard model:* Neither interface offers any functionality.

The parameters N, M are occasionally omitted in contexts where they are of no relevance. Similarly, whenever evident from the context, explicitly specifying which interface is queried is omitted.

Attackers with Oracle-Dependent Advice. Attackers $\mathcal{A} = (\mathcal{A}_1, \mathcal{A}_2)$ consist of a preprocessing procedure $\mathcal{A}_1$ and a main algorithm $\mathcal{A}_2$, which carries out the actual attack using the output of $\mathcal{A}_1$. Correspondingly, in the presence of an oracle $\mathcal{O}$, $\mathcal{A}_1$ interacts with $\mathcal{O}$.pre and $\mathcal{A}_2$ with $\mathcal{O}$.main.

Definition 2. *An (S, T)-attacker $\mathcal{A} = (\mathcal{A}_1, \mathcal{A}_2)$ in the $\mathcal{O}$-model consists of two procedures*

- $\mathcal{A}_1$, *which is computationally unbounded, interacts with $\mathcal{O}$.pre, and outputs an S-bit string, and*

- $\mathcal{A}_2$, *which takes an S-bit auxiliary input and makes at most T queries to $\mathcal{O}$.main.*

In certain contexts, additional restrictions may be imposed on $\mathcal{A}_2$, captured by some parameters p. $\mathcal{A}$ is referred to as (S, T, p)-attacker in such cases. Examples of such parameters include time and space requirements of $\mathcal{A}_2$ or a limit on the number of queries of a particular type that $\mathcal{A}_2$ makes to a challenger it interacts with. Observe that the parameter S is meaningful also in the standard model, where it measures the length of standard non-uniform advice to the attacker. The parameter T, however, is not relevant as there is no random oracle to query in the attack stage. Consequently, standard-model attackers with resources p are referred to as $(S, *, p)$-attackers.

Applications. Let $\mathcal{O}$ be an arbitrary oracle. An application G in the $\mathcal{O}$-model is defined by specifying a challenger C, which is an oracle algorithm that has access to $\mathcal{O}$.main, interacts with the main stage $\mathcal{A}_2$ of an attacker $\mathcal{A} = (\mathcal{A}_1, \mathcal{A}_2)$, and outputs a bit at the end of the interaction. The *success* of $\mathcal{A}$ on G in the $\mathcal{O}$-model is defined as

$$\mathrm{Succ}_{G,\mathcal{O}}(\mathcal{A}) := \mathsf{P}\big[\mathcal{A}_2^{\mathcal{O}.\mathsf{main}}(\mathcal{A}_1^{\mathcal{O}.\mathsf{pre}}) \leftrightarrow \mathsf{C}^{\mathcal{O}.\mathsf{main}} = 1\big],$$

where $\mathcal{A}_2^{\mathcal{O}.\mathsf{main}}(\mathcal{A}_1^{\mathcal{O}.\mathsf{pre}}) \leftrightarrow \mathsf{C}^{\mathcal{O}.\mathsf{main}}$ denotes the bit output by C after its interaction with the attacker. This work considers two types of applications, captured by the next definition.

Definition 3. *For an* indistinguishability *application G in the $\mathcal{O}$-model, the advantage of an attacker $\mathcal{A}$ is defined as*

$$\mathrm{Adv}_{G,\mathcal{O}}(\mathcal{A}) := 2\left|\mathrm{Succ}_{G,\mathcal{O}}(\mathcal{A}) - \frac{1}{2}\right|.$$

For an unpredictability *application G, the advantage is defined as*

$$\mathrm{Adv}_{G,\mathcal{O}}(\mathcal{A}) := \mathrm{Succ}_{G,\mathcal{O}}(\mathcal{A}).$$

An application G is said to be $((S, T, p), \varepsilon)$-secure in the $\mathcal{O}$-model if for every (S, T, p)-attacker $\mathcal{A}$,

$$\mathrm{Adv}_{G,\mathcal{O}}(\mathcal{A}) \leq \varepsilon.$$

Combined Query Complexity. In order to enlist Lemma 1 for proving Theorems 1 and 2 below, the interaction of some attacker $\mathcal{A} = (\mathcal{A}_1, \mathcal{A}_2)$ with a challenger C in the $\mathcal{O}$-model must be "merged" into a single entity $\mathcal{D} = (\mathcal{D}_1, \mathcal{D}_2)$ that interacts with oracle $\mathcal{O}$. That is, $\mathcal{D}_1^{(\cdot)} := \mathcal{A}_1^{(\cdot)}$ and $\mathcal{D}_2^{(\cdot)}(z) := \mathcal{A}_2^{(\cdot)}(z) \leftrightarrow \mathsf{C}^{(\cdot)}$ for $z \in \{0,1\}^S$. $\mathcal{D}$ is called the *combination of $\mathcal{A}$ and C*, and the number of queries it makes to its oracle is referred to as *the combined query complexity of $\mathcal{A}$ and C*. For all applications in this work there exists an upper bound $T_G^{\mathsf{comb}} = T_G^{\mathsf{comb}}(S, T, p)$ on the combined query complexity of any attacker and the challenger.

Additive Error for Arbitrary Applications. Using the first part of Lemma 1, one proves the following theorem, which states that the security of any application translates from the BF-ROM to the AI-ROM at the cost of an additive term of roughly ST/P, where P is the maximum number of coordinates an attacker $\mathcal{A}_1$ is allowed to fix in the BF-ROM.

Theorem 1. *For any $P \in \mathbb{N}$ and every $\gamma > 0$, if an application G is $((S, T, p), \varepsilon')$-secure in the* BF-RO$(P)$*-model, then it is $((S, T, p), \varepsilon)$-secure in the* AI-RO*-model, for*

$$\varepsilon \leq \varepsilon' + \frac{(S + \log \gamma^{-1}) \cdot T_G^{\mathsf{comb}}}{P} + \gamma,$$

where T_G^{comb} is the combined query complexity corresponding to G.

Proof. Fix P as well as γ. Set BF-RO $:=$ BF-RO(P) and let G be an arbitrary application and C the corresponding challenger. Moreover, fix an (S, T)-attacker $\mathcal{A} = (\mathcal{A}_1, \mathcal{A}_2)$, and let $\{Y_z\}_{z \in \{0,1\}^S}$ be the family of distributions guaranteed to exist by Lemma 1, where the function f is defined by $\mathcal{A}_1$. Consider the following (S, T)-attacker $\mathcal{A}' = (\mathcal{A}_1', \mathcal{A}_2')$ (expecting to interact with BF-RO):

- $\mathcal{A}_1'$ internally simulates $\mathcal{A}_1$ to compute $z \leftarrow \mathcal{A}_1^{\mathsf{AI\text{-}RO.pre}}$. Then, it samples one of the P-bit-fixing sources Y' making up Y_z and presets BF-RO to match Y' on the at most P points where Y' is fixed. The output of $\mathcal{A}_1'$ is z.

- $\mathcal{A}_2'$ works exactly as $\mathcal{A}_2$.

Let $\mathcal{D}$ be the combination of $\mathcal{A}_2 = \mathcal{A}_2'$ and C. Hence, $\mathcal{D}$ is a distinguisher taking an S-bit input and making at most T_G^{comb} queries to its oracle. Therefore, by the first part of Lemma 1,

$$\mathrm{Succ}_{G, \mathsf{AI\text{-}RO}}(\mathcal{A}) \leq \mathrm{Succ}_{G, \mathsf{BF\text{-}RO}}(\mathcal{A}') + \frac{(S + \log \gamma^{-1}) \cdot T_G^{\mathsf{comb}}}{P} + \gamma.$$

Since there is only an additive term between the two success probabilities, the above inequality implies

$$\mathrm{Adv}_{G, \mathsf{AI\text{-}RO}}(\mathcal{A}) \leq \mathrm{Adv}_{G, \mathsf{BF\text{-}RO}}(\mathcal{A}') + \frac{(S + \log \gamma^{-1}) \cdot T_G^{\mathsf{comb}}}{P} + \gamma$$

for both indistinguishability and unpredictability applications. $\quad\square$

Multiplicative Error for Unpredictability Applications. Using the second part of Lemma 1, one proves the following theorem, which states that the security of any *unpredictability* application translates from the BF-ROM to the AI-ROM at the cost of a multiplicative factor of 2, provided that $\mathcal{A}_1$ is allowed to fix roughly ST coordinates in the BF-ROM.

Theorem 2. *For any $P \in \mathbb{N}$ and every $\gamma > 0$, if an* unpredictability *application G is $((S, T, p), \varepsilon')$-secure in the* BF-RO$(P, N, M)$-model for

$$P \geq (S + \log \gamma^{-1}) \cdot T_G^{\text{comb}},$$

then it is $((S, T, p), \varepsilon)$-secure in the AI-RO(N, M)-model for

$$\varepsilon \leq 2\varepsilon' + \gamma,$$

where T_G^{comb} is the combined query complexity corresponding to G.

Proof. Using the same attacker $\mathcal{A}'$ as in the proof of Theorem 1 and applying the second part of Lemma 1, one obtains, for any $P \geq (S + \log \gamma^{-1}) \cdot T_G^{\text{comb}}$,

$$\text{Succ}_{G,\text{AI-RO}}(\mathcal{A}) \leq 2^{(S + \log 1/\gamma)T_G^{\text{comb}}/P} \cdot \text{Succ}_{G,\text{BF-RO}}(\mathcal{A}') + \gamma$$
$$\leq 2 \cdot \text{Succ}_{G,\text{BF-RO}}(\mathcal{A}') + \gamma,$$

which translates into

$$\text{Adv}_{G,\text{AI-RO}}(\mathcal{A}) \leq 2 \cdot \text{Adv}_{G,\text{BF-RO}}(\mathcal{A}') + \gamma$$

for unpredictability applications. $\qquad\square$

The Security of Applications in the AI-ROM. The connections between the *auxiliary-input random-oracle model (AI-ROM)* and the *bit-fixing random-oracle model (BF-ROM)* established above suggest the following approach to proving the security of particular applications in the AI-ROM: first, deriving a security bound in the easy-to-analyze BF-ROM, and then, depending on whether one deals with an indistinguishability or an unpredictability application, generically inferring the security of the schemes in the AI-ROM, using Theorems 1 or 2.

The three subsequent sections deal with various applications in the AI-ROM: Sect. 3 is devoted to security analyses of basic primitives, where "basic" means that the oracle is directly used as the primitive; Sect. 4 deals with the collision resistance of hash functions built from a random compression function via the Merkle-Damgård construction (MDHFs); and, finally, Sect. 5 analyzes several cryptographic schemes with computational security.

3   Basic Applications in the AI-ROM

This section treats the AI-ROM security of one-way functions (OWFs), pseudorandom generators (PRGs), normal and weak pseudorandom functions

(PRFs and wPRFs), and message-authentication codes (MACs). More specifically, the applications considered are:

- *One-way functions:* For an oracle $\mathcal{O} : [N] \to [M]$, given $y = \mathcal{O}(x)$ for a uniformly random $x \in [N]$, find a preimage x' with $\mathcal{O}(x') = y$.
- *Pseudo-random generators:* For an oracle $\mathcal{O} : [N] \to [M]$ with $M > N$, distinguish $y = \mathcal{O}(x)$ for a uniformly random $x \in [N]$ from a uniformly random element of $[M]$.
- *Pseudo-random functions:* For an oracle $\mathcal{O} : [N] \times [L] \to [M]$, distinguish oracle access to $\mathcal{O}(s, \cdot)$ for a uniformly random $s \in [N]$ from oracle access to a uniformly random function $F : [L] \to [M]$.
- *Weak pseudo-random functions:* Identical to PRFs, but the inputs to the oracle are chosen uniformly at random and independently.
- *Message-authentication codes:* For an oracle $\mathcal{O} : [N] \times [L] \to [M]$, given access to an oracle $\mathcal{O}(s, \cdot)$ for a uniformly random $s \in [N]$, find a pair (x, y) such that $\mathcal{O}(s, x) = y$ for an x on which $\mathcal{O}(s, \cdot)$ was not queried.

	AI-ROM security	Bound in [18]	Lower bound
OWFs	$\frac{ST}{N} + \frac{T}{N}$	Same	$\min\left\{\frac{ST}{N}, \left(\frac{S^2T}{N^2}\right)^{1/3}\right\} + \frac{T}{N}$
PRGs	$\left(\frac{ST}{N}\right)^{1/2} + \frac{T}{N}$	Same	$\left(\frac{S}{N}\right)^{1/2} + \frac{T}{N}$
PRFs	$\left(\frac{S(T+q_{\text{prf}})}{N}\right)^{1/2} + \frac{T}{N}$	Same	$\left(\frac{S}{N}\right)^{1/2} + \frac{T}{N}$
wPRFs	$\left(\frac{S(T+q_{\text{prf}})q_{\text{prf}}}{LN}\right)^{1/2} + \frac{T}{N}$	Not analyzed	Not known
MACs	$\frac{S(T+q_{\text{sig}})}{N} + \frac{T}{N} + \frac{1}{M}$	$\frac{S(T+q_{\text{sig}})}{N} + \frac{T}{N} + \frac{T}{M}$	$\min\left\{\frac{ST}{N}, \left(\frac{S^2T}{N^2}\right)^{1/3}\right\} + \frac{T}{N}$

Table 1. Asymptotic upper and lower bounds on the security of basic primitives against (S, T)-attackers in the AI-ROM, where q_{prf} and q_{sig} denote PRF and signing queries, respectively, and where (for simplicity) $N = M$ for OWFs. Observe that attacks against OWFs also work against PRGs and PRFs.

The asymptotic bounds for the applications in question are summarized in Table 1. For OWFs, PRGs, PRFs, and MACs, the resulting bounds match the corresponding bounds derived by Dodis *et al.* [18], who used (considerably) more involved compression arguments; weak PRFs have not previously been analyzed.

The precise statements and the corresponding proofs can be found in the full version of this paper [15]; the proofs all follow the paradigm outlined in Sect. 2.2 of first assessing the security of a particular application in the BF-ROM and then generically inferring the final bound in the AI-ROM using Theorems 1 or 2.

4   Collision Resistance in the AI-ROM

A prominent application missing from Sect. 3 is that of *collision resistance*, i.e., for an oracle $\mathcal{O} : [N] \times [L] \to M$, given a uniformly random salt value $a \in [N]$, finding two distinct $x, x' \in [L]$ such that $\mathcal{O}(a, x) = \mathcal{O}(a, x')$. The reason for this omission is that in the BF-ROM, the best possible bound is easily seen to be in the order of $P/N + T^2/M$. Even applying Theorem 2 for unpredictability applications with $P \approx ST$ results in a final AI-ROM bound of roughly $ST/N + T^2/M$, which is inferior to the optimal bound of $S/N + T^2/M$ proved by Dodis et al. [18] using compression.

However, hash functions used in practice, most notably SHA-2, are based on the Merkle-Damgård mode of operation for a compression function $\mathcal{O} : [M] \times [L] \to [M]$, modeled as a random oracle here. Specifically, a B-block message $y = (y_1, \ldots, y_B)$ with $y_j \in [L]$ is hashed to $\mathcal{O}^B(y)$, where

$$\mathcal{O}^1(y_1) = \mathcal{O}(a, y_1) \quad \text{and} \quad \mathcal{O}^j(y_1, \ldots, y_j) = \mathcal{O}(\mathcal{O}^{j-1}(y_1, \ldots, y_{j-1}), y_j) \quad \text{for} \quad j > 1.$$

While—as pointed out above—Dodis et al. [18] provide a tight bound for the one-block case, it is not obvious at all how their compression-based proof can be extended to deal with even two-block messages. Fortunately, no such difficulties appear when we apply our technique of going through the BF-ROM model, allowing us to derive a bound in Theorem 3 below.

Formally, the collision resistance of Merkle-Damgård hash functions (MDHFs) in the $\mathcal{O}(ML, M)$-model is captured by the application $G^{\mathsf{MDHF}, M, L}$, which is defined via the following challenger $\mathsf{C}^{\mathsf{MDHF}, M, L}$: It initially chooses a public initialization vector (IV) $a \in [M]$ uniformly at random and sends it to the attacker. The attacker wins if he submits $y = (y_1, \ldots, y_B)$ and $y' = (y_1', \ldots, y_{B'}')$ such that $y \neq y'$ and $\mathcal{O}^B(y) = \mathcal{O}^{B'}(y')$.

For attackers $\mathcal{A} = (\mathcal{A}_1, \mathcal{A}_2)$ in the following theorem, we make the simplifying assumption that $T > \max(B, B')$. We prove the following bound on the security of MDHFs in the AI-ROM:

Theorem 3. *Application $G^{\mathsf{MDHF}, M, L}$ is $((S, T, B), \varepsilon)$-secure in the AI-RO(ML, M)-model, where*

$$\varepsilon = \tilde{O}\left(\frac{ST^2}{M} + \frac{T^2}{M}\right).$$

The proof of Theorem 3 is provided in the full version of this paper [15].

Observe that if S and T are taken to be the circuit size, the bound in Theorem 3 becomes vacuous for circuits of size $M^{1/3}$, i.e., it provides security only *well below* the birthday bound and may therefore seem extremely loose. Quite surprisingly, however, it is tight:

Theorem 4. *There exists an (S, T)-attacker $\mathcal{A} = (\mathcal{A}_1, \mathcal{A}_2)$ against application $G := G^{\mathsf{MDHF}, M, L}$ in the $\mathcal{O} := \mathsf{AI\text{-}RO}(ML, M)$-model with advantage at least*

$$\mathrm{Adv}_{G, \mathcal{O}}(\mathcal{A}) = \tilde{\Omega}\left(\frac{ST^2}{M} + \frac{1}{M}\right),$$

assuming $ST^2 \leq M/2$ and $L \geq M$.

The attack is loosely based on rainbow tables [31] and captured by the following (S,T)-attacker $\mathcal{A} = (\mathcal{A}_1, \mathcal{A}_2)$:

- $\mathcal{A}_1$: Obtain the function table $F : [M] \times [L] \to [M]$ from $\mathcal{O}$. For $i = 1, \ldots, m :=$ $S/(3\lceil \log L \rceil)$, proceed as follows:
  1. Choose $a_{i,0} \in [M]$ uniformly at random.
  2. Compute $a_{i,\ell-1} \leftarrow F^{(\ell-1)}(a_{i,0}, 0)$, where $\ell := \lfloor T/2 \rfloor$.[9]
  3. Find values $x_i \neq x_i'$ such that $a_{i,\ell} := F(a_{i,\ell-1}, x_i) = F(a_{i,\ell-1}, x_i')$; abort if no such values exist.

  Output the triples $(a_{i,\ell-1}, x_i, x_i')$ for $i = 1, \ldots, m$.
- $\mathcal{A}_2$: Obtain the public initialization vector a from $\mathsf{C}^{\mathsf{MDHF},M,L}$ and the m triples output by $\mathcal{A}_1$. Proceed as follows:
  1. If $a = a_{i,\ell-1}$ for some i, return (x_i, x_i').
  2. Otherwise, set $\tilde{a} \leftarrow a$ and for $j = 1, \ldots, T$, proceed as follows:
     (a) Query $\tilde{a} \leftarrow \mathcal{O}(\tilde{a}, 0)$.
     (b) If $\tilde{a} = a_{i,\ell-1}$ for some i, return $(0^j \| x_i, 0^j \| x_i')$; otherwise return $(0, 1)$.

The analysis of the attack can be found in the full version of this paper [15]. It should be noted that in practice hash functions use a fixed IV a, and, therefore—in contrast to, e.g., function inversion, where usually the cost of a single preprocessing stage can be amortized over many inversion challenges—the rather sizeable amount of preprocessing required by the attack to just find a collision may not be justified. However, in some cases, the hash function used in a particular application (relying on collision-resistance) is salted by prepending a random salt value to the input. Such salting essentially corresponds to the random-IV setting considered here, and, therefore, the attack becomes relevant again as one might be able to break many instances of the application using a single preprocessing phase.

5 Computationally Secure Applications in the AI-ROM

This section illustrates the bit-fixing methodology on two typical computationally secure applications: (1) Schnorr signatures [49], where Theorem 2 can be applied since forging signatures is an unpredictability application, and (2) trapdoor-function (TDF) key-encapsulation (KEM) [8], where an approach slightly more involved than merely analyzing security in the BF-ROM and applying Theorem 1 is required in order to get a tighter security reduction; see below.

(Please refer to Sect. A of the appendix for the definitions of digital signatures, KEMs, TDFs, and other standard concepts used in this section.)

Fiat-Shamir with Schnorr. Let G be a cyclic group of prime order $|G| = N$. The Schnorr signature scheme $\Sigma = (\mathsf{Gen}, \mathsf{Sig}, \mathsf{Vfy})$ in the $\mathcal{O}(N^2, N)$-model works as follows:

- *Key generation:* Choose $x \in \mathbb{Z}_N$ uniformly at random, compute $y \leftarrow g^x$, and output $\mathsf{sk} := x$ and $\mathsf{vk} := y$.

---

[9] $F^{(k)}$ stands for the k-fold application of F, and, for the sake of concreteness, let $[L] = \{0, \ldots, L-1\}$.

- *Signing:* To sign a message $m \in [N]$ with key $\mathsf{sk} = x$, pick $r \in \mathbb{Z}_N$ uniformly at random, compute $a \leftarrow g^r$, query $c \leftarrow \mathcal{O}(a, m)$, set $z \leftarrow r + cx$, and output $\sigma := (a, z)$.
- *Verification:* To verify a signature $\sigma = (a, z)$ for a message m with key $\mathsf{vk} = y$, query $c \leftarrow \mathcal{O}(a, m)$, and check whether $g^z \overset{?}{=} a y^c$. If the check succeeds and $c \neq 0$, accept the signature, and reject it otherwise.

For attackers $\mathcal{A} = (\mathcal{A}_1, \mathcal{A}_2)$ in Theorem 5, which assesses the security of Fiat-Shamir with Schnorr in the AI-ROM, we make the running time t and space complexity s of $\mathcal{A}_2$ explicit. Moreover, if $\mathcal{A}$ is an attacker against $G^{\mathsf{DS},\Sigma}$, there is an additional parameter q_{sig} that restricts $\mathcal{A}_2$ to making at most q_{sig} signing queries. The proof of Theorem 5 is provided in the full version of this paper [15].

Theorem 5. *Assume* $G^{\mathsf{DL},G}$ *for a prime* $|G| = N$ *is* $((S', *, t', s'), \varepsilon')$*-secure, and let* $\Sigma = (\mathsf{Gen}, \mathsf{Sig}, \mathsf{Vfy})$ *be the Schnorr scheme. Then, for any* $T, q_{\mathsf{sig}} \in \mathbb{N}$, $G^{\mathsf{DS},\Sigma}$ *is* $((S, T, t, s, q_{\mathsf{sig}}), \varepsilon)$*-secure in the* AI-RO$(N^2, N)$*-model for*

$$\varepsilon = \tilde{O}\left(\sqrt{T\varepsilon'} + \frac{S q_{\mathsf{sig}}(q_{\mathsf{sig}} + T)}{N}\right),$$

any $S \leq S'/\tilde{O}(T + q_{\mathsf{sig}})$, $t \leq t' - \tilde{O}(S(T + q_{\mathsf{sig}}))$, *and* $s \leq s' - \tilde{O}(S(T + q_{\mathsf{sig}}))$.

For comparison, note that the security of Schnorr signatures in the standard ROM is $O\left(\sqrt{T\varepsilon'} + \frac{q_{\mathsf{sig}}(q_{\mathsf{sig}}+T)}{N}\right)$, i.e., in the AI-ROM the second term worsens by a factor of S.

TDF Key Encapsulation. Let F be a trapdoor family (TDF) generator. TDF encryption is a key-encapsulation mechanism $\Pi = (\mathsf{Gen}, \mathsf{Enc}, \mathsf{Dec})$ that works as follows:

- *Key generation:* Run the TDF generator to obtain $(f, f^{-1}) \leftarrow F$, where $f, f^{-1} : [N] \rightarrow [N]$. Set the public key $\mathsf{pk} := f$ and the secret key $\mathsf{sk} := f^{-1}$.
- *Encapsulation:* To encapsulate a key with public key $\mathsf{pk} = f$, choose $x \in [N]$, query $k \leftarrow \mathcal{O}(x)$, compute $y \leftarrow f(x)$, and output $(c, k) \leftarrow (y, k)$.
- *Decapsulation:* To decapsulate a ciphertext $c = y$ with secret key $\mathsf{sk} = f^{-1}$, output $k \leftarrow \mathcal{O}(f^{-1}(y))$.

Theorem 6 deals with the security of TDF key encapsulation in the AI-ROM. Once again, for attackers $\mathcal{A} = (\mathcal{A}_1, \mathcal{A}_2)$, the running time t and space complexity s of $\mathcal{A}_2$ is made explicit. The proof of Theorem 6 is provided in the full version of this paper [15].

Theorem 6. *Let* Π *be TDF encapsulation. If* $G^{\mathsf{TDF},F}$ *is* $((S', *, t', s'), \varepsilon')$*-secure, then, for any* $T \in \mathbb{N}$, $G^{\mathsf{KEM\text{-}CPA},\Pi}$ *is* $((S, T, t, s), \varepsilon)$*-secure in the* AI-RO$(N, N)$*-model, where*

$$\varepsilon = \tilde{O}\left(\varepsilon' + \sqrt{\frac{ST}{N}}\right)$$

and $S = S' - \tilde{O}(ST)$, $t = t' - \tilde{O}(t_{\mathsf{tdf}} \cdot T)$, and $s = s' - \tilde{O}(ST)$, where t_{tdf} is the time required to evaluate the TDF.

Moreover, $G^{\mathsf{KEM\text{-}CCA},\Pi}$ is $((S, T, t, s), \varepsilon)$-secure with the same parameters, except that $t = t' - \tilde{O}(t_{\mathsf{tdf}} \cdot ST)$.

Observe that the above security bound corresponds simply to the sum of the security of the TDF and the security of $\mathcal{O}$ as a PRG (cf. Sect. 3); in the standard random-oracle model, the security of TDF encryption is simply upper bounded by $O(\varepsilon')$ (cf. Sect. A.2).

An important point about the proof of Theorem 6 is that it does not follow the usual paradigm of deriving the security of TDF encryption in the BF-ROM and thereafter applying Theorem 1 (for CPA/CCA security is an indistinguishability application). Doing so—as Unruh does for RSA-OAEP [51] (but in an "asymptotic sense," as explained in Footnote 5)—would immediately incur an additive error of $ST/P \leq ST/t'$, since the size of the list P is upper bounded by the TDF attacker size t'. So the naive application Theorem 1 would result in poor exact security.

Instead, our tighter proof of Theorem 6 considers two hybrid experiments (one of which is the original CPA/CCA security game in the AI-ROM). The power of the BF-ROM is used twice—with different list sizes: (1) to argue the indistinguishability of the two experiments and (2) to upper bound the advantage of the attacker in the second hybrid. Crucially, a reduction to TDF security is only required for (1), which has an unpredictability flavor and can therefore get by with a list size of roughly $P \approx ST$; observe that this is polynomial for efficient (S, T)-attackers. The list size for (2) is obtained via the usual balancing between ST/P and the security bound in the BF-ROM.[10]

6   Salting Defeats Auxiliary Information

There exist schemes that are secure in the standard ROM but not so in the AI-ROM. A simple example is if the random oracle itself is directly used as a collision-resistant hash function $\mathcal{O} : [N] \to [M]$ for some N and M: in the ROM, $\mathcal{O}$ is easily seen to be collision-resistant, while in the AI-ROM, the first phase $\mathcal{A}_1$ of an attacker $\mathcal{A} = (\mathcal{A}_1, \mathcal{A}_2)$ (cf. Sect. 2.2) can simply leak a collision to $\mathcal{A}_2$, which then outputs it, thereby breaking the collision-resistance property.

The full version of this paper [15] briefly highlights two schemes with computational security where the above phenomenon can be observed as well. The first one is a generic transformation of an identification scheme into a signature scheme using the so-called *Fiat-Shamir transform*, and the second one is the well-known *full-domain hash*.[11]

---

[10] A similar approach also works to improve the security bounds of [51] for RSA-OAEP in the AI-ROM.

[11] By virtue of Theorem 2, the existence of attacks in the AI-ROM against the above schemes obviously implies that these schemes cannot be secure in the BF-ROM either. It is also relatively straight-forward to devise direct attacks in the BF-ROM.

To remedy the situation with schemes such as those mentioned above, in this section we prove that the security of any *standard* ROM scheme can be carried over to the BF-ROM by sacrificing part of the domain of the BF-RO for *salting*. First, in Sect. 6.1, we analyze the standard way of salting a random oracle by prefixing a randomly chosen (public) value to every oracle query. Second, in Sect. 6.2, we also show how to adapt a technique by Maurer [41], originally used in the context of key-agreement with randomizers, to obtain a more domain-efficient salting technique, albeit with a longer salt value; the salt length can be reduced by standard derandomization techniques based on random walks on expander graphs.

6.1   Standard Salting

The standard way of salting a scheme is to simply prepend a public salt value to every oracle query: Consider an arbitrary application G with the corresponding challenger C. Let $\mathsf{C}_{\mathsf{salt}}$ be the challenger that is identical to C except that it initially chooses a uniformly random value $a \in [K]$, outputs a to $\mathcal{A}_2$, and prepends a to every oracle query. Denote the corresponding application by G_{salt}. Observe that the salt value a is chosen after the first stage $\mathcal{A}_1$ of the attack, and, hence, as long as the first stage $\mathcal{A}_1$ of the attacker in the BF-ROM does not prefix a position starting with a, it is as if the scheme were executed in the standard ROM. Moreover, note that the time and space complexities s and t, respectively, of $\mathcal{A}_2$ increase roughly by P due to the security reduction used in the proof.

Theorem 7. *For any $P \in \mathbb{N}$, if an application G is $((S',T',t',s'),\varepsilon')$-secure in the $\mathsf{RO}(N,M)$-model, then G_{salt} is $((S,T,t,s),\varepsilon)$-secure in the $\mathsf{BF\text{-}RO}(P,NK,M)$-model for*

$$\varepsilon = \varepsilon' + \frac{P}{K},$$

$S = S' - \tilde{O}(P)$, $T = T'$, $t = t' - \tilde{O}(P)$, *and* $s = s' - \tilde{O}(P)$.

The proof of Theorem 7 is provided in the full version of this paper [15]. Combining Theorem 7 with Theorems 1 and 2 from Sect. 2.2 yields the following corollaries:

Corollary 1. *For any $P \in \mathbb{N}$ and every $\gamma > 0$, if an arbitrary application G is $((S',T',t',s'),\varepsilon')$-secure in the $\mathsf{RO}(N,M)$-model, then G_{salt} is $((S,T,t,s),\varepsilon)$-secure in the $\mathsf{AI\text{-}RO}(NK,M)$-model for*

$$\varepsilon = \varepsilon' + \frac{P}{K} + \frac{(S + \log \gamma^{-1}) \cdot T^{\mathsf{comb}}_{G_{\mathsf{salt}}}}{P} + \gamma$$

and any $S = S' - \tilde{O}(P)$, $T = T'$, $t = t' - \tilde{O}(P)$, *and* $s = s' - \tilde{O}(P)$, *where* $T^{\mathsf{comb}}_{G_{\mathsf{salt}}}$ *is the combined query complexity corresponding to* G_{salt}.

Corollary 2. *For every $\gamma > 0$, if an unpredictability application G is $((S', T', t', s'), \varepsilon')$-secure in the $\mathsf{RO}(N, M)$-model, then G_{salt} is $((S, T, t, s), \varepsilon)$-secure in the $\mathsf{AI\text{-}RO}(NK, M)$-model for*

$$\varepsilon = 2\varepsilon' + \frac{2(S + \log\gamma^{-1}) \cdot T_{G_{\mathsf{salt}}}^{\mathsf{comb}}}{K} + \gamma$$

and any $S = S'/\tilde{O}(T_{G_{\mathsf{salt}}}^{\mathsf{comb}})$, $T = T'$, $t' = t - \tilde{O}(P)$, and $s' = s - \tilde{O}(P)$, where $P = (S + \log\gamma^{-1})T_{G_{\mathsf{salt}}}^{\mathsf{comb}}$ and where $T_{G_{\mathsf{salt}}}^{\mathsf{comb}}$ is the combined query complexity corresponding to G_{salt}.

Applications. In the full version of this paper [15], we briefly discuss how salting affects the security of the applications presented in Sects. 3 to 5. We also provide examples to illustrate that directly analyzing a salted scheme in the BF-ROM can lead to much better bounds than combining a standard-ROM security bound with one of the above corollaries.

6.2 Improved Salting

One way to think of salting is to view the function table of $\mathsf{BF\text{-}RO}(KN, M)$ as a $(K \times N)$-matrix and let the challenger in the salted application randomly pick and announce the row to be used for oracle queries. However, K has to be around the same size as N to obtain meaningful bounds. In this section, based on a technique by Maurer [41], we provide a more domain-efficient means of salting, where the security will decay exponentially (as opposed to inverse linearly) with the domain expansion factor K, at the cost that each evaluation of the derived random oracle will cost K evaluations (as opposed to 1 evaluation) of the original random oracle.

Consider an arbitrary application G with corresponding challenger C. Let $\mathsf{C}_{\mathsf{salt}'}$ be the challenger works as follows: It initially chooses a uniformly random value $a = (a_1, \ldots, a_K) \in [N]^K$ and outputs a to $\mathcal{A}_2$. Then, it internally runs C, forwards all messages between the attacker and C, but answers the queries $x \in [N]$ that C makes to the oracle by

$$\sum_{i=1}^{K} \mathsf{BF\text{-}RO.main}(i, x + a_i),$$

where addition is in $\mathbb{Z}_N$ and $\mathbb{Z}_M$, respectively. In other words, the function table of $\mathsf{BF\text{-}RO}$ is arranged as a $K \times N$ matrix, the ith row is shifted by a_i, and queries x are answered by computing the sum modulo M of all the values in the xth column of the shifted matrix, denoted F_a. Denote the corresponding application by $G_{\mathsf{salt}'}$. The proof of the following theorem is provided in the full version of this paper [15]. Moreover, we present a means of reducing the size of the public salt value.

Theorem 8. *For any* $P \in \mathbb{N}$, *if an application* G *is* $((S', T', t', s'), \varepsilon')$-*secure in the* $\mathsf{RO}(N, M)$-*model, then* $G_{\mathsf{salt}'}$ *is* $((S, T, t, s), \varepsilon)$-*secure in the* $\mathsf{BF\text{-}RO}(P, NK, M)$-*model for*

$$\varepsilon' = \varepsilon + N \cdot \left(\frac{P}{KN}\right)^K,$$

$S = S' - \tilde{O}(P)$, $T = T'$, $t = t' - \tilde{O}(P)$, *and* $s = s' - \tilde{O}(P)$.

In particular, assuming $P \le KN/2$, setting $K = O(\log N)$ will result in additive error $N(P/NK)^K = o(\frac{1}{N})$ and domain size $O(N \log N)$. But if $P \le N^{1-\Omega(1)}$, setting $K = O(1)$ will result in the same additive error $o(\frac{1}{N})$ in the original domain of near-optimal size $O(N)$. Hence, for most practical purposes, the efficiency slowdown K (in both the domain size and the complexity of oracle evaluation) is at most $O(\log N)$ and possibly constant.

Combining the above results with those in Sect. 2.2 yields the following corollaries:

Corollary 3. *For any* $P \in \mathbb{N}$ *and every* $\gamma > 0$, *if an application* G *is* $((S', T', t', s'), \varepsilon')$-*secure in the* $\mathsf{RO}(N, M)$-*model, then* $G_{\mathsf{salt}'}$ *is* $((S, T, t, s), \varepsilon)$-*secure in the* $\mathsf{AI\text{-}RO}(NK, M)$-*model for*

$$\varepsilon = \varepsilon' + N \cdot \left(\frac{P}{KN}\right)^K + \frac{(S + \log \gamma^{-1}) \cdot T_{G_{\mathsf{salt}'}}^{\mathsf{comb}}}{P} + \gamma$$

and any $S = S' - \tilde{O}(P)$, $T = T'$, $t = t' - \tilde{O}(P)$, *and* $s = s' - \tilde{O}(P)$, *where* $T_{G_{\mathsf{salt}'}}^{\mathsf{comb}}$ *is the combined query complexity corresponding to* $G_{\mathsf{salt}'}$.

Corollary 4. *For every* $\gamma > 0$, *if an application* G *is* $((S', T', t', s'), \varepsilon')$-*secure in the* $\mathsf{RO}(N, M)$-*model, then* $G_{\mathsf{salt}'}$ *is* $((S, T, t, s), \varepsilon)$-*secure in the* $\mathsf{AI\text{-}RO}(NK, M)$-*model for*

$$\varepsilon = 2\varepsilon + 2N \cdot \left(\frac{(S + \log \gamma^{-1}) T_{G_{\mathsf{salt}'}}^{\mathsf{comb}}}{KN}\right)^K$$

and any $S = S'/\tilde{O}(T_{G_{\mathsf{salt}'}}^{\mathsf{comb}})$, $T = T'$, $t' = t - \tilde{O}(P)$, *and* $s' = s - \tilde{O}(P)$, *where* $P = (S + \log \gamma^{-1}) T_{G_{\mathsf{salt}'}}^{\mathsf{comb}}$ *and where* $T_{G_{\mathsf{salt}'}}^{\mathsf{comb}}$ *is the combined query complexity corresponding to* $G_{\mathsf{salt}'}$.

Acknowledgments. The authors thank Mika Göös for pointing out the decomposition lemma for high-entropy sources in [30], Andrej Bogdanov for discussions about derandomization using random walks, and Daniel Wichs for suggestions on proving the security of computationally secure schemes in the AI-ROM. Sandro Coretti is supported by NSF grants 1314568 and 1319051. Yevgeniy Dodis is partially supported by gifts from VMware Labs and Google, and NSF grants 1619158, 1319051, 1314568. Siyao Guo is supported by NSF grants CNS1314722 and CNS-1413964; this work was done partially while the author was visiting the Simons Institute for the Theory of Computing at UC Berkeley.

A Standard-ROM Definitions and Security

A.1 Fiat-Shamir with Schnorr

Digital Signature Schemes. A digital signature scheme is a triple of algorithms $\Sigma = (\mathsf{Gen}, \mathsf{Sig}, \mathsf{Vfy})$, where Gen generates a signing key sk and a verification key vk, Sig takes a signing key sk and a message m and outputs a signature σ, and Vfy takes a verification key vk, a message m, and a signature σ and outputs a single bit, indicating whether σ is valid. In the $\mathcal{O}$-oracle model, all three algorithms may make calls to $\mathcal{O}.\mathsf{main}$.

The application of digital signatures $G^{\mathsf{DS},\Sigma}$ is defined via the following challenger $\mathsf{C}^{\mathsf{DS},\Sigma}$, which captures the (standard) EUF-CMA security of a digital signature scheme: Initially, $\mathsf{C}^{\mathsf{DS},\Sigma}$ generates a key pair $(\mathsf{sk},\mathsf{vk}) \leftarrow \mathsf{Gen}$ and passes vk to the attacker. Then, the attacker may repeatedly submit signature queries m to the challenger, who answers them by the corresponding signature $\sigma \leftarrow \mathsf{Sig}_\sigma(m)$. In the end, the challenger outputs 1 if and only if the attacker submits a pair (m^*, σ^*) with $\mathsf{Vfy}_{\mathsf{vk}}(m^*, \sigma^*) = 1$ and such that no signature query was asked for m^*.

The Discrete-Logarithm Problem. The discrete-logarithm problem in a group $G = \langle g \rangle$ can be phrased as an application $G^{\mathsf{DL},G}$, defined via the challenger $\mathsf{C}^{\mathsf{DL},G}$ that picks a uniformly random $x \in \mathbb{Z}_{|G|}$, passes $y := g^x$ to the attacker, and outputs 1 if and only if the attacker finds x. Observe that $G^{\mathsf{DL},G}$ is a *standard-model* application.

Schnorr Signatures in the Standard ROM. In the standard ROM, using the forking lemma as stated by Bellare and Neven [7], one can show the following security bound for Schnorr signatures.

Theorem 9. *Assume $G^{\mathsf{DL},G}$ for $|G| = N$ is $((S, *, t', s'), \varepsilon')$-secure, and let $\Sigma = (\mathsf{Gen}, \mathsf{Sig}, \mathsf{Vfy})$ be the Schnorr scheme. Then, $G^{\mathsf{DS},\Sigma}$ is $((S, T, t, s, q_{\mathsf{sig}}), \varepsilon)$-secure in the $\mathsf{RO}(N^2, N)$-model for*

$$\varepsilon = O\left(\sqrt{T\varepsilon'} + \frac{q_{\mathsf{sig}}(q_{\mathsf{sig}} + T)}{N}\right),$$

where $t = \Omega(t')$ and $s = \Omega(s')$.

A.2 TDF Encryption

Key-Encapsulation Mechanisms. A key-encapsulation mechanism (KEM) is a triple of algorithms $\Pi = (K, E, D)$, where K generates a public key pk and a secret key sk, E takes a public key pk and outputs a ciphertext c and a key k, and D takes a secret key sk and a ciphertext c and outputs a key k. In the $\mathcal{O}$-oracle model, all three algorithms may make calls to $\mathcal{O}.\mathsf{main}$.

The application corresponding to CPA security for KEMs $G^{\mathsf{KEM\text{-}CPA},\Pi}$ is defined via the following challenger $\mathsf{C}^{\mathsf{KEM\text{-}CPA},\Pi}$, which captures the (standard)

CCA security of a KEM scheme: Initially, $C^{\mathsf{KEM\text{-}CPA},\Pi}$ generates a key pair $(\mathsf{pk}, \mathsf{sk}) \leftarrow K$ and passes pk to the attacker. Then, the challenger chooses a random bit b as well as a random key k_1, computes $(c, k_0) \leftarrow E_{\mathsf{pk}}$, and returns the challenge (c, k_b). In the end, the challenger outputs 1 if and only if the attacker submits a bit b' with $b' = b$.

To capture CCA security, one considered the application $C^{\mathsf{KEM\text{-}CCA},\Pi}$ defined by the challenger $C^{\mathsf{KEM\text{-}CCA},\Pi}$ that proceeds as $C^{\mathsf{KEM\text{-}CPA},\Pi}$, except that the attacker gets to ask decryption queries c', which the challenger answers with $k' \leftarrow D_{\mathsf{sk}}(c')$, provided $c' \neq c$.

Trapdoor Functions. The inversion problem for a trapdoor function generator F can be phrased as an application $G^{\mathsf{TDF},F}$, defined via the challenger $C^{\mathsf{TDF},F}$ that generates $(f, f^{-1}) \leftarrow F$, picks a random x, passes $y := f(x)$ to the attacker, and outputs 1 if and only if the attacker finds x. Observe that $G^{\mathsf{TDF},F}$ is a *standard-model* application.

The security of TDF key encapsulation in the standard ROM. In the standard ROM, one can show the following security bound for TDF encryption.

Theorem 10. *Let Π be TDF key encapsulation. If $G^{\mathsf{TDF},F}$ is $((S', *, t', s'), \varepsilon')$-secure, then $G^{\mathsf{KEM\text{-}CPA},\Pi}$ is $((S, T, t, s), \varepsilon)$-secure in the $\mathsf{RO}(N, N)$, where*

$$\varepsilon = O(\varepsilon')$$

and $S = S'$, $t = \Omega(t')$, and $s = \Omega(s')$.

References

1. Abdalla, M., An, J.H., Bellare, M., Namprempre, C.: From identification to signatures via the Fiat-Shamir transform: minimizing assumptions for security and forward-security. In: Knudsen, L.R. (ed.) EUROCRYPT 2002. LNCS, vol. 2332, pp. 418–433. Springer, Heidelberg (2002). https://doi.org/10.1007/3-540-46035-7_28

2. Adleman, L.M.: Two theorems on random polynomial time. In: 19th Annual Symposium on Foundations of Computer Science, Ann Arbor, Michigan, USA, 16–18 October 1978, pp. 75–83 (1978)

3. Alon, N., Goldreich, O., Håstad, J., Peralta, R.: Simple construction of almost k-wise independent random variables. Random Struct. Algorithms **3**(3), 289–304 (1992)

4. Aumann, Y., Ding, Y.Z., Rabin, M.O.: Everlasting security in the bounded storage model. IEEE Trans. Inf. Theory **48**(6), 1668–1680 (2002)

5. Bellare, M., Boldyreva, A., Palacio, A.: An uninstantiable random-oracle-model scheme for a hybrid-encryption problem. In: Cachin, C., Camenisch, J.L. (eds.) EUROCRYPT 2004. LNCS, vol. 3027, pp. 171–188. Springer, Heidelberg (2004). https://doi.org/10.1007/978-3-540-24676-3_11

6. Bellare, M., Canetti, R., Krawczyk, H.: Pseudorandom functions revisited: the cascade construction and its concrete security. In: 37th Annual Symposium on Foundations of Computer Science, FOCS 1996, Burlington, Vermont, USA, 14–16 October 1996, pp. 514–523 (1996)

7. Bellare, M., Neven, G.: Multi-signatures in the plain public-key model and a general forking lemma. In: Proceedings of the 13th ACM Conference on Computer and Communications Security, CCS 2006, Alexandria, VA, USA, 30 October–3 November 2006, pp. 390–399 (2006)

8. Bellare, M., Rogaway, P.: Random oracles are practical: a paradigm for designing efficient protocols. In: Proceedings of the 1st ACM Conference on Computer and Communications Security, CCS 1993, Fairfax, Virginia, USA, 3–5 November 1993, pp. 62–73 (1993)

9. Bellare, M., Rogaway, P.: Optimal asymmetric encryption. In: De Santis, A. (ed.) EUROCRYPT 1994. LNCS, vol. 950, pp. 92–111. Springer, Heidelberg (1995). https://doi.org/10.1007/BFb0053428

10. Bernstein, D.J., Lange, T.: Non-uniform cracks in the concrete: the power of free precomputation. In: Sako, K., Sarkar, P. (eds.) ASIACRYPT 2013. LNCS, vol. 8270, pp. 321–340. Springer, Heidelberg (2013). https://doi.org/10.1007/978-3-642-42045-0_17

11. Canetti, R., Goldreich, O., Halevi, S.: On the random-oracle methodology as applied to length-restricted signature schemes. In: Naor, M. (ed.) TCC 2004. LNCS, vol. 2951, pp. 40–57. Springer, Heidelberg (2004). https://doi.org/10.1007/978-3-540-24638-1_3

12. Canetti, R., Goldreich, O., Halevi, S.: The random oracle methodology, revisited. J. ACM 51(4), 557–594 (2004)

13. Chor, B., Goldreich, O., Håstad, J., Friedman, J., Rudich, S., Smolensky, R.: The bit extraction problem of t-resilient functions (preliminary version). In: 26th Annual Symposium on Foundations of Computer Science, Portland, Oregon, USA, 21–23 October 1985, pp. 396–407 (1985)

14. Chung, K.-M., Lin, H., Mahmoody, M., Pass, R.: On the power of nonuniformity in proofs of security. In: Innovations in Theoretical Computer Science, ITCS 2013, Berkeley, CA, USA, 9–12 January 2013, pp. 389–400 (2013)

15. Coretti, S., Dodis, Y., Guo, S., Steinberger, J.: Random oracles and non-uniformity (full version of this paper). Cryptology ePrint Archive, Report 2017/937 (2017). https://eprint.iacr.org/2017/937

16. Damgård, I.B.: A design principle for hash functions. In: Brassard, G. (ed.) CRYPTO 1989. LNCS, vol. 435, pp. 416–427. Springer, New York (1990). https://doi.org/10.1007/0-387-34805-0_39

17. De, A., Trevisan, L., Tulsiani, M.: Time space tradeoffs for attacks against one-way functions and PRGs. In: Rabin, T. (ed.) CRYPTO 2010. LNCS, vol. 6223, pp. 649–665. Springer, Heidelberg (2010). https://doi.org/10.1007/978-3-642-14623-7_35

18. Dodis, Y., Guo, S., Katz, J.: Fixing cracks in the concrete: random oracles with auxiliary input, revisited. In: Coron, J.-S., Nielsen, J.B. (eds.) EUROCRYPT 2017. LNCS, vol. 10211, pp. 473–495. Springer, Cham (2017). https://doi.org/10.1007/978-3-319-56614-6_16

19. Dodis, Y., Pietrzak, K., Wichs, D.: Key derivation without entropy waste. In: Nguyen, P.Q., Oswald, E. (eds.) EUROCRYPT 2014. LNCS, vol. 8441, pp. 93–110. Springer, Heidelberg (2014). https://doi.org/10.1007/978-3-642-55220-5_6

20. Dodis, Y., Steinberger, J.: Message authentication codes from unpredictable block ciphers. In: Halevi, S. (ed.) CRYPTO 2009. LNCS, vol. 5677, pp. 267–285. Springer, Heidelberg (2009). https://doi.org/10.1007/978-3-642-03356-8_16

21. Dziembowski, S., Maurer, U.M.: Tight security proofs for the bounded-storage model. In: Proceedings on 34th Annual ACM Symposium on Theory of Computing, Montréal, Québec, Canada, 19–21 May 2002, pp. 341–350 (2002)

22. Fiat, A., Naor, M.: Rigorous time/space trade-offs for inverting functions. SIAM J. Comput. **29**(3), 790–803 (1999)
23. Fiat, A., Shamir, A.: How to prove yourself: practical solutions to identification and signature problems. In: Odlyzko, A.M. (ed.) CRYPTO 1986. LNCS, vol. 263, pp. 186–194. Springer, Heidelberg (1987). https://doi.org/10.1007/3-540-47721-7_12
24. Gennaro, R., Gertner, Y., Katz, J., Trevisan, L.: Bounds on the efficiency of generic cryptographic constructions. SIAM J. Comput. **35**(1), 217–246 (2005)
25. Gennaro, R., Trevisan, L.: Lower bounds on the efficiency of generic cryptographic constructions. In: 41st Annual Symposium on Foundations of Computer Science, FOCS 2000, Redondo Beach, California, USA, 12–14 November 2000, pp. 305–313 (2000)
26. Goldreich, O.: A uniform-complexity treatment of encryption and zero-knowledge. J. Cryptol. **6**(1), 21–53 (1993)
27. Goldreich, O., Krawczyk, H.: On the composition of zero-knowledge proof systems. SIAM J. Comput. **25**(1), 169–192 (1996)
28. Goldreich, O., Oren, Y.: Definitions and properties of zero-knowledge proof systems. J. Cryptol. **7**(1), 1–32 (1994)
29. Goldwasser, S., Kalai, Y.T.: On the (in)security of the Fiat-Shamir paradigm. In: Proceedings of 44th Symposium on Foundations of Computer Science (FOCS 2003), Cambridge, MA, USA, 11–14 October 2003, pp. 102–113 (2003)
30. Göös, M., Lovett, S., Meka, R., Watson, T., Zuckerman, D.: Rectangles are non-negative juntas. SIAM J. Comput. **45**(5), 1835–1869 (2016)
31. Hellman, M.E.: A cryptanalytic time-memory trade-off. IEEE Trans. Inf. Theory **26**(4), 401–406 (1980)
32. Hoang, V.T., Tessaro, S.: Key-alternating ciphers and key-length extension: exact bounds and multi-user security. In: Robshaw, M., Katz, J. (eds.) CRYPTO 2016. LNCS, vol. 9814, pp. 3–32. Springer, Heidelberg (2016). https://doi.org/10.1007/978-3-662-53018-4_1
33. Impagliazzo, R.: Hardness as randomness: a survey of universal derandomization. CoRR, cs.CC/0304040 (2003)
34. Impagliazzo, R., Nisan, N., Wigderson, A.: Pseudorandomness for network algorithms. In: Proceedings of the Twenty-Sixth Annual ACM Symposium on Theory of Computing, Montréal, Québec, Canada, 23–25 May 1994, pp. 356–364 (1994)
35. Impagliazzo, R., Wigderson, A.: P = BPP if E requires exponential circuits: derandomizing the XOR lemma. In: Proceedings of the Twenty-Ninth Annual ACM Symposium on the Theory of Computing, El Paso, Texas, USA, 4–6 May 1997, pp. 220–229 (1997)
36. Kabanets, V.: Derandomization: a brief overview. Bull. EATCS **76**, 88–103 (2002)
37. Karp, R.M., Lipton, R.J.: Some connections between nonuniform and uniform complexity classes. In: Proceedings of the 12th Annual ACM Symposium on Theory of Computing, Los Angeles, California, USA, 28–30 April 1980, pp. 302–309 (1980)
38. Katz, J., Lindell, Y.: Introduction to Modern Cryptography. Chapman and Hall/CRC Press, Boca Raton (2007)
39. Kothari, P., Meka, R., Raghavendra, P.: Approximating rectangles by juntas and weakly-exponential lower bounds for LP relaxations of CSPs. In: STOC (2017)
40. Mahmoody, M., Mohammed, A.: On the power of hierarchical identity-based encryption. In: Fischlin, M., Coron, J.-S. (eds.) EUROCRYPT 2016. LNCS, vol. 9666, pp. 243–272. Springer, Heidelberg (2016). https://doi.org/10.1007/978-3-662-49896-5_9
41. Maurer, U.M.: Conditionally-perfect secrecy and a provably-secure randomized cipher. J. Cryptol. **5**(1), 53–66 (1992)

42. Merkle, R.C.: A certified digital signature. In: Brassard, G. (ed.) CRYPTO 1989. LNCS, vol. 435, pp. 218–238. Springer, New York (1990). https://doi.org/10.1007/0-387-34805-0_21

43. Morris, R., Thompson, K.: Password security - a case history. Commun. ACM **22**(11), 594–597 (1979)

44. Nielsen, J.B.: Separating random oracle proofs from complexity theoretic proofs: the non-committing encryption case. In: Yung, M. (ed.) CRYPTO 2002. LNCS, vol. 2442, pp. 111–126. Springer, Heidelberg (2002). https://doi.org/10.1007/3-540-45708-9_8

45. Nisan, N., Wigderson, A.: Hardness vs. randomness (extended abstract). In: 29th Annual Symposium on Foundations of Computer Science, White Plains, New York, USA, 24–26 October 1988, pp. 2–11 (1988)

46. Oechslin, P.: Making a faster cryptanalytic time-memory trade-off. In: Boneh, D. (ed.) CRYPTO 2003. LNCS, vol. 2729, pp. 617–630. Springer, Heidelberg (2003). https://doi.org/10.1007/978-3-540-45146-4_36

47. Patarin, J.: The "Coefficients H" technique. In: Avanzi, R.M., Keliher, L., Sica, F. (eds.) SAC 2008. LNCS, vol. 5381, pp. 328–345. Springer, Heidelberg (2009). https://doi.org/10.1007/978-3-642-04159-4_21

48. Rogaway, P.: Formalizing human ignorance. In: Nguyen, P.Q. (ed.) VIETCRYPT 2006. LNCS, vol. 4341, pp. 211–228. Springer, Heidelberg (2006). https://doi.org/10.1007/11958239_14

49. Schnorr, C.P.: Efficient identification and signatures for smart cards. In: Quisquater, J.-J., Vandewalle, J. (eds.) EUROCRYPT 1989. LNCS, vol. 434, pp. 688–689. Springer, Heidelberg (1990). https://doi.org/10.1007/3-540-46885-4_68

50. Tessaro, S.: Security amplification for the cascade of arbitrarily weak PRPs: tight bounds via the interactive hardcore lemma. In: Ishai, Y. (ed.) TCC 2011. LNCS, vol. 6597, pp. 37–54. Springer, Heidelberg (2011). https://doi.org/10.1007/978-3-642-19571-6_3

51. Unruh, D.: Random oracles and auxiliary input. In: Menezes, A. (ed.) CRYPTO 2007. LNCS, vol. 4622, pp. 205–223. Springer, Heidelberg (2007). https://doi.org/10.1007/978-3-540-74143-5_12

52. Vadhan, S.P.: Constructing locally computable extractors and cryptosystems in the bounded-storage model. J. Cryptol. **17**(1), 43–77 (2004)

Another Step Towards Realizing Random Oracles: Non-malleable Point Obfuscation

Ilan Komargodski[1](✉) and Eylon Yogev[2](✉)

[1] Cornell Tech, New York, NY 10044, USA
komargodski@cornell.edu
[2] Weizmann Institute of Science, Rehovot 76100, Israel
eylon.yogev@weizmann.ac.il

Abstract. The random oracle paradigm allows us to analyze the security of protocols and construction in an idealized model, where all parties have access to a truly random function. This is one of the most successful and well-studied models in cryptography. However, being such a strong idealized model, it is known to be susceptible to various weaknesses when implemented naively in "real-life", as shown by Canetti, Goldreich and Halevi (J. ACM 2004).

As a counter-measure, one could try to identify and implement only one or few of the properties a random oracle possesses that are needed for a specific setting. Such a systematic study was initiated by Canetti (CRYPTO 1997), who showed how to implement the property that the output of the function does not reveal anything regarding the input by constructing a point function obfucator. This property turned out to suffice in many follow-up works and applications.

In this work, we tackle another natural property of random oracles and implement it in the standard model. The property we focus on is *non-malleability*, where it is guaranteed that the output on an input cannot be used to generate the output on any related point. We construct a point-obfuscator that is both point-hiding (à la Canetti) *and* is non-malleable. The cost of our construction is a single exponentiation on top of Canetti's construction and could be used for any application where point obfuscators are used and obtain improved security guarantees. The security of our construction relies on variants of the DDH and power-DDH assumptions.

On the technical side, we introduce a new technique for proving security of a construction based on a DDH-like assumption. We call this technique "double-exponentiation" and believe it will be useful in the future.

I. Komargodski—Supported in part by a Packard Foundation Fellowship and AFOSR grant FA9550-15-1-0262. Initial parts of this work were done at the Weizmann Institute of Science, supported in part by a grant from the Israel Science Foundation (no. 950/16) and by a Levzion Fellowship.

E. Yogev—Supported in part by a grant from the Israel Science Foundation (no. 950/16).

© International Association for Cryptologic Research 2018
J. B. Nielsen and V. Rijmen (Eds.): EUROCRYPT 2018, LNCS 10820, pp. 259–279, 2018.
https://doi.org/10.1007/978-3-319-78381-9_10

1   Introduction

The Random Oracle model [6] is one of the most well studied models in the cryptographic literature. In this model, everyone has access to a single random function. It is usually possible to show clean and simple constructions that are information-theoretically secure in this idealized model. Also, in many cases it allows to prove unconditional lower bounds.

One major question is when (and under what assumptions) can we replace the Random Oracle with a "real life" object. It is known that such a transformation is impossible in the general case, but the counter examples are usually quite contrived [13,17,26]. This leaves the possibility that for specific applications of a Random Oracle such a transformation could possibly exist. One of the obstacles in answering the aforementioned question is that it seems hard to formalize and list all the properties such a generic transformation should preserve. In practice, this difficulty is circumvented by replacing the Random Oracle with an ad-hoc "cryptogrpahic hash function" (e.g., MD5, SHA-1, SHA-256) which results with protocols and constructions that have no provable security guarantees, and often tend to be broken [39,41,42].

Motivated by the above, Canetti [15] initiated the systematic study of identifying useful properties of a Random Oracle and then realizing them in the standard model. In his work, he focused on one property called "point obfuscation" (or "oracle hashing"). This property ensures that when the Random Oracle is applied on an input, the output value is completely uncorrelated to the input, and at the same time, it is possible to verify whether a given output was generated from a given input. Canetti formally defined this notion and gave a construction of such a primitive in the standard model based on a variant of the decisional Diffie-Hellman assumption (DDH). Since then, other instantiations of this primitive were suggested. Wee [43] gave a construction whose security is based on a strong notion of one-way permutations, Goldwasser et al. [27] gave a construction based on the Learning With Errors assumption, and more recently Bellare and Stepanovs [7] proposed a framework for constructing point obfuscators. The latter result gives a generic construction of point obfuscators based on either (1) indistinguishability obfuscation [3,24] and any one-way function, (2) deterministic public-key encryption [4], or (3) UCEs [5].

While hiding the point is a natural and useful goal, there are many setting where this is not enough to replace a Random Oracle. One other natural property we wish to realize in "real life" is that of *non-malleability*: given the value of a Random Oracle on a random point x, it is infeasible to get the value of the Random Oracle at any "related" point (e.g., the point $x + 1$). The work of Canetti and Varia [19] identified this property and the goal of realizing it. Their work provided definitions (of non-malleable obfuscation for general circuits, and not only for point functions) and constructions of non-malleable (multi) point obfuscators in the random oracle model.

In this work, we focus on construction of non-malleable point obfuscators in the plain model. Observe that many of the known constructions of point obfuscators *are* malleable. For example, let us recall the construction of Canetti [15] which involves a group $\mathcal{G}$ with a generator $g \in \mathcal{G}$. For an input point x and randomness r (interpreted as a random group element) the obfuscation is:

$$O(x; r) = (r, r^x).$$

Indeed, the obfuscation of $x + 1$ can be computed by multiplying r^x by r and outputting the pair (r, r^{x+1}). In other words, the obfuscation of a point is *malleable*. The point obfuscators of Wee [43] and of Goldwasser et al. [27] admit similar attacks (i.e., they are malleable).[1]

Thus, we ask whether we can remedy this situation and provide a construction of a secure point obfuscator in the plain model that is provably non-malleable under simple and concrete assumptions. We view this as a necessary and vital step towards understanding the possibility for realizing a Random Oracle in "real life".

1.1 Our Results

We provide a construction of a secure point obfuscator that is non-malleable for a wide class of mauling functions. Our notion of non-malleability is parametrized by a distribution $\mathcal{X}$ over the input domain X and by a class of possible mauling attacks $\mathcal{F} = \{f \colon X \to X\}$. Roughly speaking, our notion guarantees that for every function $f \in \mathcal{F}$, any polynomial-time adversary, when given the obfuscation of a point $x \leftarrow \mathcal{X}$, cannot generate the obfuscation of the point $f(x)$.[2]

We give a construction of a (public-coin[3]) point obfuscator that is non-malleable for any well-spread distribution $\mathcal{X}$ (i.e., a distribution that has superlogarithmic min-entropy) and the class of mauling functions $\mathcal{F}$ which can be described by univariate polynomials of bounded polynomial degree (in the security parameter). Our construction involves a group $\mathcal{G}$ with a generator $g \in \mathcal{G}$. For an input point x and randomness r (interpreted as a random group element) the obfuscation is:

$$O(x; r) = (r, r^{g^{h(x)}}),$$

where $h(x) = x^4 + x^3 + x^2 + x$. We prove security and non-malleability of the above point obfuscator under variants of the DDH and power-DDH assumptions (see Sect. 2.2). We also present two ways to support more general mauling functions $\mathcal{F}$ by strengthening the underlying security assumption (yet the

---

[1] The work of [7] is an exception since it gives constructions based on generic primitives so we need non-malleability of the underlying building block. The required notion of non-malleability is usually very strong. Consider, for example, their construction from DPKE, where the point function obfuscation includes a ciphertext and a public-key (of some encryption scheme). To get non-malleability for the point obfuscator we need non-malleability for the DPKE for an adversary that can maul not only the ciphertext but also the public-key.

[2] We also require that the obfuscation that the adversary outputs is verifiable, that is, it looks like an obfuscation of the value $f(x)$ (i.e., it comes from the "same family" of circuits). This prevents trivial attacks that treat the input circuit as a black-box.

[3] An obfuscator is public-coin if the random bits used for the obfuscation are given as part of the output of the obfuscator.

construction remains the same). First, we show how to support a larger class of mauling function by assuming (sub-)exponential security of the underlying assumption. Second, we show that our construction is secure against any mauling function f for which one cannot distinguish the triple $(g, g^x, g^{h(f(x))})$ from a triple (g, g^{r_1}, g^{r_2}), where r_1, r_2 are random exponents. We do not have a simple characterization of the functions f for which this assumption holds.

In terms of efficiency, our construction is quite efficient: it involves only two group exponentiation (Canetti's construction requires a single exponentiation), does not rely on any setup assumptions, and does not rely on expensive machinery such as zero-knowledge proofs, which are usually employed to achieve non-malleability. Moreover, it satisfies the same *privacy* guarantees as of Canneti's obfuscator. As such, our point obfuscator can be used in any application where point obfuscators are used. These include encryption schemes [15], storing passwords [19,40], reusable fuzzy extractors [16], round-efficient zero-knowledge proofs and arguments [10], and more.

Applications to Non-interactive Non-malleable Commitments. It is possible to view our obfuscator as a non-interactive non-malleable commitment that is secure when committing to strings that come from a distribution with super-logarithmic entropy. To commit to a string x, compute the obfuscation of x and that would be the commitment. The opening is x itself (and thus for security it has to have entropy). The resulting commitment scheme is computationally hiding by the security of the point obfuscator, and also non-malleable against a large class of mauling functions.

Previously, constructions of non-interactive non-malleable commitments (in the plain model, without any setup assumptions) required an ad-hoc and non-standard primitive called "adaptive injective one-way functions" that has built-in some form of non-malleability [34]. More recent works provide constructions that are secure against uniform adversaries [33] or ensure limited forms of non-malleability ("with respect to opening") [31]. These constructions, however, allow to commit on worst-case inputs and handle arbitrary mauling functions.

1.2   Related Work

Non-malleable Cryptography. Non-malleability was introduced as a measure to augment and strengthen cryptographic primitives (such as encryption schemes or commitment schemes) in such a way that it does not only guarantee privacy, but also that it is hard to manipulate a given ciphertext (or commitment) of one value into a ciphertext of another.

Non malleability was first defined in the seminal work of Dolev, Dwork, and Naor [22] where they presented a non-malleable public-key encryption scheme, a non-malleable string commitment scheme, a non-malleable zero-knowledge protocol. Since then, there has been a long line of works on non-malleability. See [12, 29–33, 33, 35–37] to name just a few.

A particular type of non-malleable protocols (or primitives) that may a-priori be related to non-malleable point obfuscators are *non-interactive* commitments and encryption schemes. These were the focus of multiple works (see, for example, [21–23,38] and some of the references given above). However, these notions do not imply point obfuscators as they do not support public verification on a given input (without revealing the randomness which completely breaks security).

In the context of obfuscation, the only work we are aware of is that of Canetti and Varia [19] who gave several incomparable definitions for non-malleable obfuscation. They also gave a construction of a (multi-bit) non-malleable point obfuscator (under each definition), however, their construction is in the Random Oracle model.

Obfuscation with High Min-Entropy. Canetti, Micciancio and Reingold [18] gave a construction of a point obfuscator that satisfies a relaxed notion of security where the input is guaranteed to come from a source with high min-entropy. Their underlying assumption is any collision resistant hash function. There is a significant (qualitative) difference between this notion and the original notion of Canetti [15] that we consider in this work. We refer to Wee [43, Section 1.3] for an elaborate discussion.

Boldyreva et al. [11] showed how to make the point obfuscator of [18] non-malleable using non-interactive zero-knowledge proofs (assuming a common reference string). Following the work of Boldyreva et al., Baecher et al. [2] presented a game-based definition of non-malleability which is very similar to ours (see also [20]). However, they did not provide new constructions in the plain model.

1.3   Our Techniques

Our starting point is Canetti's point function construction [15], who presented a construction under a variant of the DDH assumption (and no random oracles). Recall that the DDH assumption involves a group ensemble $\mathcal{G} = \{\mathbb{G}_\lambda\}_{\lambda \in \mathbb{N}}$ with a generator g and it asserts that (g^x, g^y, g^{xy}) is computationally indistinguishable from a sequence of random group elements, where x and y are chosen uniformly at random. Canetti's variant is that the foregoing indistinguishability holds even if x has high enough min-entropy (yet y is completely random). For an input point x and using randomness r, viewed as a random group element of $\mathbb{G}_\lambda$, Canetti's construction is:

$$O(x; r) = r, r^x.$$

As we mentioned, it is easy to modify r^x to get r^{x+1}, giving an obfuscation of the point $x + 1$. Let us first focus on the goal of modifying the construction such that it is non-malleable against this function: $f(x) = x + 1$. Towards this end, we change the construction to be:

$$O(x; r) = r, r^{x^2}.$$

The claim is that under a suitable variant of the power-DDH assumptions this is a non-malleable point obfuscator against the function f. Roughly speaking, we assume that $(g^x, g^{x^2}, g^{x^3}, \ldots)$ is indistinguishable from a sequence of random group elements, where x comes from a distribution with high enough min-entropy. Assume first that the adversary outputs a point obfuscation of $x + 1$ under the *same randomness* r as she received. That is, on input r, w, the output is r, w' for an element $w' \in \mathbb{G}$. Later, we show how to handle adversaries that output an obfuscation of $x + 1$ under new randomness.

The point obfuscation of $x + 1$ under this construction (with the same randomness r) is (r, w), where $w = r^{x^2 + 2x + 1}$. Suppose that there is an adversary $\mathcal{A}$ that given r, r^{x^2} can compute w, then we show how to break the security of our assumption. We are given a challenge (g, g^{z_1}, g^{z_2}), where either $z_i = x^i$ or $z_i = r_i$ and each r_i is chosen at random. Then, we can run the adversary on the input g^s, g^{sz_2}, for a random s to get w. We compute $w' = g^{s(z_2 + 2z_1 + 1)}$ and compare it to w. If $w = w'$ we output 1, and otherwise we output a random bit. In the case that $z_i = x^i$, the adversary gets g^s, g^{sx^2} which is exactly the distribution of a point obfuscation of x and thus will output $w = g^{s(x^2 + 2x + 1)} = w'$ with some non-negligible probability. Otherwise, the adversary gets g^{sr_2} for a random r_2 and the probability that she outputs $w' = g^{s(r_2 + 2r_1 + 1)}$ is negligible as she has no information regarding r_2 (this is true even for an unbounded adversaries). Overall, we have a non-negligible advantage in distinguishing the two cases.

While the above construction is non-malleable against the function $f(x) = x + 1$, it is malleable for the function $f(x) = 2x$. Indeed, given r^{x^2} one can simply compute $(r^{x^2})^4 = r^{4x^2} = r^{(2x)^2}$ which is a valid obfuscation of the point $2x$. Our second observation is that we can modify the construction to resist this attack by defining:

$$\mathcal{O}(x; r) = r, r^{x^2 + x}.$$

The proof of non-malleability is similar to the proof above; we run the adversary $\mathcal{A}$ on $g^s, g^{s(z_2 + z_1)}$ to get w, and compute $w' = g^{s(4z_2 + 2z_1)}$. If $z_i = x^i$, then the adversary sees exactly the distribution of a point obfuscation of x and thus will output $w = g^{s(4z_2 + 2z_1)} = w'$ with some non-negligible probability. Otherwise, the adversary gets $g^{s(r_2 + r_1)}$ for random r_i's. We bound the probability that $\mathcal{A}$ outputs $w' = g^{r(4r_2 + 2r_1)}$. This is again an information theoretic argument where we assume that the adversary gets $r_2 + r_1$ and needs to compute $4r_2 + 2r_1$. The argument follows since the adversary has only information regarding the sum $r_2 + r_1$ which leaves the random variable corresponding to $4r_2 + 2r_1$ with high min-entropy (given the adversary's view), and thus the probability of outputting $w = w'$ is negligible.

One important thing to notice is that the proof relied on the fact that the adversary only had the sum $r_1 + r_2$ which is a linear combination of (r_1, r_2) with the coefficients $(1, 1)$ but the final goal was to output a different combination with the coefficients $(4, 2)$, which are linearly independent of $(1, 1)$. That is, the key observation is that for $h(x) = x^2 + x$ the polynomial $h(f(x))$ for $f(x) = 2x$ has (non-free) coefficients which are not all the same. Generalizing this argument, we can show that the construction is non-malleable against any linear function $f(x) = ax + b$ for any constants a, b such that the function $h(f(x))$ written as a polynomial over x has at least 2 different (non-free) coefficients. For non-linear functions, a similar proof works but the running time of the security reduction (that is, the loss in the security of our scheme) will be proportional to the degree of $f(x)$.

Given the above observation, we can easily check if our construction is non-malleable for a function f by computing the polynomial $h(f(x))$. It turns our that the above construction is actually *malleable* for a simple function such as $f(x) = 3x + 1$. Indeed, $h(f(x)) = (3x + 1)^2 + (3x + 1) = 9x^2 + 9x + 2$ has the same two non-free coefficients. In order to eliminate more functions f, we need to add more constraints to the set of equations which translates to taking a higher degree of polynomial $h(x)$. That is, we define $h(x) = x^3 + x^2 + x$, and construct the obfuscator:

$$\mathcal{O}(x; r) = r, r^{x^3 + x^2 + x}.$$

For a function f to be malleable under this construction, it must hold that the polynomial $h(f(x))$ has all *three* non-free coefficients equal. However, there is still single function that satisfies this condition (the function is $f(x) = -x - 2 \cdot 3^{-1}$, where 3^{-1} is the inverse of 3 in the relevant group). As a final step, we modify the construction to be of one degree higher and this does eliminate all possible functions f. Thus, we define the construction:

$$\mathcal{O}(x; r) = r, r^{x^4 + x^3 + x^2 + x}.$$

The Double Exponentiation. In our exposition above, we assumed that the adversary "uses the same randomness she received". That is, on input r, w she mauls the point and outputs r, w'. Suppose now that the adversary is allowed to output r', w', where r' might be arbitrary. Recall that the issue is that we cannot simulate the power of w' from the challenge under the randomness r' to check consistency (since we do not know the discrete log of r'). Let us elaborate on this in the simple case where the obfuscation is r, r^x (and not the degree 4 polynomial in the exponent; this is just for simplicity). When the malleability adversary gets r, r^x and returns r, w', it is easy to check that $w' = r^{f(x)}$ by recomputing this value since we know the discrete log of r. However, when it return r', w', it is hard to recompute $r'^{f(x)}$ since we do not know the discrete log of r (and only get the value x in the exponent from the challenge).

In other words, we need to be able (in the security proof) to compute the obfuscation of some input that depends on the exponents from the challenge under randomness that comes from the adversary's mauled obfuscation. If we knew either the discrete log of the challenge or the discrete log of the randomness used by the adversary we would be done.

In the description above we actually used this property. Since we assumed that the adversary outputs the same randomness r (that we chose and know the discrete log of), we could use $r = g^s$ to compute the obfuscation of the challenge we received. However, if the adversary outputs randomness r', then not only we no longer know the discrete log of r' (and this is hard to compute), but we also do not have the discrete log of the challenge.

Thus, we need to modify our construction such that we can compute the obfuscation of x given only g^x and while given the public coins r explicitly (without given their discrete log). Towards this end, we introduce a new technique that we call "double exponentiation". Consider any mapping of the group elements $\mathbb{G}_\lambda \to \mathbb{Z}_q^*$ where q is the order of $\mathbb{G}_\lambda$ (e.g., their binary representation as strings). Then, we define the final version of our construction:

$$\mathcal{O}(x; r) = r, r^{g^{x^4 + x^3 + x^2 + x}}.$$

One can observe that it is possible to compute the obfuscation of x given only $g^{x^4 + x^3 + x^2 + x}$ and given r by a single exponentiation. In addition, the construction is still efficient, consists of just two group elements, and involves only two exponentiations.

A final remark about security. Proving that the resulting construction is still a point obfuscator is not immediate a-priori. Our proof works by a reduction to the security of Canetti's construction via an intermediate notion of security called virtual gray-box obfuscation [8]. We refer to Sect. 4 for more details.

2   Preliminaries

For a distribution X we denote by $x \leftarrow X$ the process of sampling a value x from the distribution X. Similarly, for a set $\mathcal{X}$ we denote by $x \leftarrow \mathcal{X}$ the process of sampling a value x from the uniform distribution over $\mathcal{X}$. For a randomized function f and an input $x \in \mathcal{X}$, we denote by $y \leftarrow f(x)$ the process of sampling a value y from the distribution $f(x)$. For an integer $n \in \mathbb{N}$ we denote by $[n]$ the set $\{1, \dots, n\}$.

Throughout the paper, we denote by λ the security parameter. A function $\mathsf{negl} \colon \mathbb{N} \to \mathbb{R}^+$ is *negligible* if for every constant $c > 0$ there exists an integer N_c such that $\mathsf{negl}(\lambda) < \lambda^{-c}$ for all $\lambda > N_c$. Two sequences of random variables $X = \{X_\lambda\}_{\lambda \in \mathbb{N}}$ and $Y = \{Y_\lambda\}_{\lambda \in \mathbb{N}}$ are *computationally indistinguishable* if for any probabilistic polynomial-time algorithm $\mathcal{A}$ there exists a negligible function $\mathsf{negl}(\cdot)$ such that $\left| \Pr[\mathcal{A}(1^\lambda, X_\lambda) = 1] - \Pr[\mathcal{A}(1^\lambda, Y_\lambda) = 1] \right| \leq \mathsf{negl}(\lambda)$ for all sufficiently large $\lambda \in \mathbb{N}$.

2.1   Point Obfuscation

For an input $x \in \{0,1\}^n$, the point function $I_x \colon \{0,1\}^n \to \{0,1\}$ outputs 1 on input x and 0 everywhere else. A point obfuscator is a compiler that gets a point x as input and outputs a circuit that has the same functionality as I_x but where x is (supposedly) computationally hidden. Let us recall the definition of security of Canetti [15] (called there *oracle simulation*).

Definition 1 (Functional Equivalence). *We say that two circuits C and C' are* functionally equivalent *and denote it by $C \equiv C'$ if they compute the same function (i.e., $\forall x : C(x) = C'(x)$).*

Definition 2 (Point Obfuscation). *A point obfuscator $\mathcal{O}$ for a domain $X = \{X_\lambda\}_{\lambda \in \mathbb{N}}$ of inputs is a probabilistic polynomial-time algorithm that gets as input a point $x \in X_\lambda$, and outputs a circuit C such that*

1. **Completeness:** *For all $\lambda \in \mathbb{N}$ and all $x \in X_\lambda$, it holds that*

$$\Pr[\mathcal{O}(x) \equiv I_x] = 1,$$

where the probabilities are over the internal randomness of $\mathcal{O}$.

2. **Soundness:** *For every probabilistic polynomial-time algorithm $\mathcal{A}$, and any polynomial function $p(\cdot)$, there exists a probabilistic polynomial-time simulator $\mathcal{S}$, such that for every $x \in X_\lambda$, any predicate $P \colon X_\lambda \to \{0,1\}$, and all large enough $\lambda \in \mathbb{N}$,*

$$\left| \Pr[\mathcal{A}(\mathcal{O}(x)) = P(x)] - \Pr[\mathcal{S}^{I_x}(1^\lambda) = P(x)] \right| \le \frac{1}{p(\lambda)},$$

where the probabilities are over the internal randomness of $\mathcal{A}$ and $\mathcal{O}$, and $\mathcal{S}$, respectively.

The obfuscation is called public coin *if it publishes its internal coin tosses as part of its output.*

Indistinguishability-Based Security. Another way to formalize the security of a point obfuscator is via an indistinguishability-based security definition (rather than simulation-based). Canetti [15] suggested such a definition (termed *distributional indistinguishability* there): the input comes from a distribution $\mathcal{X}_\lambda$ over the input space X_λ and the guarantee is that for any adversary $\mathcal{A}$ that outputs a single bit, the following two distributions are computationally indistinguishable:

$$(x, \mathcal{A}(\mathcal{O}(x; r))) \approx_c (x, \mathcal{A}(\mathcal{O}(y; r))), \tag{1}$$

where r is the randomness (chosen uniformly) for the point obfuscator and x and y are chosen independently from $\mathcal{X}_\lambda$.

One of Canetti's results [15, Theorem 4] was that the indisinguishability-based definition is *equivalent* to the simulation-based definition given in Eq. 1 if the indisinguishability-based security holds with respect to all distributions that have super-logarithmic min-entropy (over the message space). Such a distribution is called a *well-spread distribution*:

Definition 3 (Well-Spread Distribution). *An ensemble of distributions $\mathcal{X} = \{\mathcal{X}_\lambda\}_{\lambda \in \mathbb{N}}$, where $\mathcal{X}_\lambda$ is over $\{0,1\}^\lambda$, is* well-spread *if*

1. *it is efficiently and uniformly samplable – there is a probabilistic polynomial-time algorithm that given 1^λ as input, outputs a sample according to $\mathcal{X}_\lambda$.*
2. *for all large enough $\lambda \in \mathbb{N}$, it has super-logarithmic min-entropy. Namely,*

$$\mathsf{H}_\infty(\mathcal{X}_\lambda) = - \min_{x \in \{0,1\}^\lambda} \log_2 \Pr[X = x] \ge \omega(\log \lambda).$$

Canetti's Construction. In [15], Canetti provided a construction that satisfies Definition 2. In his construction, the domain of inputs X_λ is $\mathbb{Z}_p$ for prime $p \approx 2^\lambda$. Let $\mathcal{G} = \{\mathbb{G}_\lambda\}_{\lambda \in \mathbb{N}}$ be a group ensemble with uniform and efficient representation and operations, where each $\mathbb{G}_\lambda$ is a group of prime order $p \in (2^\lambda, 2^{\lambda+1})$. The *public coin* point obfuscator $\mathcal{O}$ for points in the domain $\mathbb{Z}_p$ is defined as follows: $\mathcal{O}(I_x)$ samples a random generator $r \leftarrow \mathbb{G}_\lambda^*$ and outputs the pair (r, r^x). Evaluation of the obfuscation at point z is done by checking whether $r^x = r^z$.

Canetti proved that this construction satisfies Eq. 1 for any well-spread distribution under the strong variant of the DDH assumption, that we review below (see Assumption 3). Thereby, the result is that under the same assumption his construction satisfies Definition 2, as well.

2.2 Hardness Assumptions

The DDH and Power-DDH Assumptions. The DDH assumption says that in a suitable group, the triple of elements (g^x, g^y, g^{xy}) is pseudorandom for random x and y. The power-DDH assumption says that the power sequence $(g, g^x, g^{x^2}, \ldots, g^{x^t})$ is pseudorandom, for a random x and a polynomially bounded t. While the power-DDH assumption is less common in the literature, there are many works that explicitly rely on it (see, for example, [1,14,25,28]). To the best of our knowledge, the power-DDH assumption is *incomparable* to the DDH assumption.

Throughout this section, let $\mathcal{G} = \{\mathbb{G}_\lambda\}_{\lambda \in \mathbb{N}}$ be a group ensemble with uniform and efficient representation and operations, where each $\mathbb{G}_\lambda$ is a group of prime order $p \in (2^{\lambda-1}, 2^\lambda)$.

Assumption 1 (DDH). *The DDH assumption asserts that for the group $\mathbb{G}_\lambda$ with associated generator g, the ensembles (g^x, g^y, g^{xy}) and (g^x, g^y, g^z) are computationally indistinguishable, where $x, y, z \leftarrow \mathbb{Z}_p^*$.*

Assumption 2 (Power-DDH). *The power-DDH assumption asserts that for the group $\mathbb{G}_\lambda$ with associated generator g, for every polynomially bounded function $t(\cdot)$, the ensembles $(g, g^x, g^{x^2} \ldots, g^{x^t})$ and $(g, g^{r_1}, g^{r_2} \ldots, g^{r_t})$ are computationally indistinguishable, where $x, r_1, \ldots, r_t \leftarrow \mathbb{Z}_p^*$.*

We need an even stronger variant of both assumptions. The strong variant that we need, first proposed by Canetti [15], roughly, says that DDH is hard not only if x, y and z are chosen uniformly at random, but even if x is chosen from a distribution with enough min-entropy (i.e., a well-spread distribution; see Definition 3). Analogously, we define a strong variant of the power-DDH assumption where x is chosen from such a distribution rather than from the uniform one.

Assumption 3 (Strong DDH and power-DDH). *The* strong *variant of the DDH and power-DDH assumptions is when the two distributions are computationally indistinguishable even if x is chosen uniformly from a well-spread distribution X_λ (rather than from $\mathbb{Z}_p^*$).*

3   Non-malleable Point Obfuscation

We define non-malleability of point function obfuscators. Such obfuscators not only hide the obfuscated point, but they also (informally) ensure that an obfuscation of a point x cannot be transformed into an obfuscation of a related (yet different) point.

There are several ways to formalize this notion of security. We focus on a notion of security where the objective of the adversary, given an obfuscation of x, is to come up with a circuit (of prescribed structure) that is a point function on a related point (a similar definition is given in [2]). We discuss the relation to the notions of Canetti and Varia [19] below.

Definition 4 (Verifier). *A PPT algorithm $\mathcal{V}$ for a point obfuscator $\mathcal{O}$ for the ensemble of domains $\{X_\lambda\}_{\lambda \in \mathbb{N}}$ domain is called a* verifier *if for all $\lambda \in \mathbb{N}$ and all $x \in X_\lambda$, it holds that $\Pr[\mathcal{V}(\mathcal{O}(x)) = 1] = 1$, where the probability is taken over the randomness of $\mathcal{V}$ and $\mathcal{O}$.*

Notice that there is no guarantee as to what $\mathcal{V}$ is suppose to output when its input is not a valid obfuscation. In particular, a verifier that always outputs 1 is a legal verifier. In many cases, including the obfuscator of Canetti [15] and our own, one can define a meaningful verifier.

Definition 5 (Non-malleable Point Function). *Let $\mathcal{O}$ be a point obfuscator for an ensemble of domains $\{X_\lambda\}_{\lambda \in \mathbb{N}}$ with an associated verifier $\mathcal{V}$. Let $\{\mathcal{F}_\lambda\}_{\lambda \in \mathbb{N}} = \{f \colon X_\lambda \to X_\lambda\}_{\lambda \in \mathbb{N}}$ be an ensemble of families of functions, and let $\{\mathcal{X}_\lambda\}_{\lambda \in \mathbb{N}}$ be an ensemble of distributions over X.*

The point obfuscator $\mathcal{O}$ is a non-malleable obfuscator *for $\mathcal{F}$ and $\mathcal{X}$ if for any polynomial-time adversary $\mathcal{A}$, there exists a negligible function $\mathsf{negl}(\cdot)$, such that for any $\lambda \in \mathbb{N}$ it holds that:*

$$\Pr\left[\mathcal{V}(C) = 1,\ f \in \mathcal{F}_\lambda,\ \text{and}\ I_{f(x)} \equiv C \ \middle|\ \begin{array}{l} x \leftarrow \mathcal{X}_\lambda \\ (C, f) \leftarrow \mathcal{A}(\mathcal{O}(x)) \end{array}\right] \leq \mathsf{negl}(\lambda).$$

That is, the adversary $\mathcal{A}$, given an obfuscation of a point x sampled from $\mathcal{X}_\lambda$, cannot output a function $f \in \mathcal{F}_\lambda$ and a valid-looking obfuscation of the point $f(x)$, except with negligible probability.

The verifier $\mathcal{V}$. We require that an attacker outputs an obfuscation with a prescribed structure so that it passes the verifier $\mathcal{V}$. Without such a requirement, there is a trivial attack for the adversary: use the given circuit $\widehat{C}_w$ to create a new circuit that gets x, computes $f^{-1}(x)$ and then applies $\widehat{C}_w$ on this value. The result is a circuit that accepts the point $f(w)$.

In general, it might be hard to come up with a verifier $\mathcal{V}$ that tests whether a given circuit is legal, but here we are interested in the case where this can be done efficiently. In our case, it will be very easy to define $\mathcal{V}$ since a "valid-looking" obfuscation will consist of all pairs of group elements (in some given group).

Adaptivity of f. We stress that our definition is adaptive with respect to the family $\mathcal{F}_\lambda$. That is, the adversary first gets to see the obfuscation $\mathcal{O}(x)$ of the point x and then may chose the function it wishes to maul to. This definition is stronger than a static version in which the function f is fixed and known in advance (before the adversary sees the challenge).

3.1   Relation to Canetti-Varia

The work of Canetti and Varia [19] presented a systematic study of non-malleable obfuscation both specifically for point functions and also for general functionalities. They gave two definitions for non-malleability, called *functional non-malleability* and *verifiable non-malleability*.

The verifiable non-malleability definition is more related to ours since there they also require that there is a verifier $\mathcal{V}$ that gets an alleged obfucated circuit and checks whether it is a legitimate output of the obfuscator. Recall that the obfuscator of Canetti (as well as ours) has this property: An obfuscation can be verified by simply checking whether the obfuscation consists of two group elements in the desired group.

The verifiable non-malleability notion of Canetti and Varia asserts that, roughly, whatever mauling attack one can apply on an obfuscation, there exists a simulator that has only oracle access to the input circuit and outputs a "similarly mauled" obfuscation. To prevent trivial attacks (that treat the input circuit as a black-box), they allow the simulator to output a circuit that has oracle gates to its own oracle (namely, to the input circuit). The verifiability ensures that the output of the adversary (and the simulator) have a "legal" structure. The precise definition is subtle and it captures a wide range of mauling attacks in a meaningful way. We refer to [19] for their elaborate discussions on the matter. We provide their formal definition, restricted to point functions next.

Definition 6 (Verifiable Non-malleable Point Obfuscation [19]). *Let $\mathcal{O}$ be a point obfuscator for a domain $X = \{X_\lambda\}_{\lambda \in \mathbb{N}}$ with an associated verifier $\mathcal{V}$. For every PPT adversary $\mathcal{A}$ and every polynomial $p(\cdot)$, there exists a PPT simulator $\mathcal{S}$ such that for all sufficiently large $\lambda \in \mathbb{N}$, for any input $x \in X_\lambda$ and any polynomial-time computable relation $E \colon X_\lambda \times X_\lambda \to \{0,1\}$ (that may depend on x), it holds that*

$$\Pr\left[C \neq \mathcal{O}(x) \wedge \mathcal{V}(C) = 1 \wedge (\exists y \in X_\lambda \colon I_y \equiv C \wedge E(x,y) = 1) \mid C \leftarrow \mathcal{A}(\mathcal{O}(x))\right] -$$

$$\Pr\left[\mathcal{V}(C) = 1 \wedge (\exists y \in X_\lambda \colon I_y \equiv C^{I_x} \wedge E(x,y) = 1) \mid C \leftarrow \mathcal{S}^{I_x}(1^\lambda)\right] \leq \frac{1}{p(\lambda)}.$$

We observe that our definition is related to the above definition albeit with the following modifications. First, the input for our obfuscator is sampled from a well-spread distribution, rather than being worst-case. Second, the non-malleablility in our definition is parametrized with a family of functions, whereas the above definition requires non-malleability for all possible relations. The modified definition is given next.

Definition 7 (). *Let $\mathcal{O}$ be a point obfuscator for a domain $X = \{X_\lambda\}_{\lambda \in \mathbb{N}}$ with an associated verifier $\mathcal{V}$. Let $\{\mathcal{F}_\lambda\}_{\lambda \in \mathbb{N}} = \{f : X_\lambda \to X_\lambda\}_{\lambda \in \mathbb{N}}$ be an ensemble of families of functions, and let $\{\mathcal{X}_\lambda\}_{\lambda \in \mathbb{N}}$ be an ensemble of distributions over X. For every PPT adversary $\mathcal{A}$ and every polynomial $p(\cdot)$, there exists a PPT simulator $\mathcal{S}$ such that for all sufficiently large $\lambda \in \mathbb{N}$, for any function $f \in \mathcal{F}_\lambda$, it holds that*

$$\Pr_{x \leftarrow \mathcal{X}_\lambda} \left[C \neq \mathcal{O}(x), \mathcal{V}(C) = 1 \text{ and } I_{f(x)} \equiv C \mid C \leftarrow \mathcal{A}(\mathcal{O}(x)) \right] -$$

$$\Pr_{x \leftarrow \mathcal{X}_\lambda} \left[\mathcal{V}(C) = 1 \text{ and } I_{f(x)} \equiv C^{I_x} \mid C \leftarrow \mathcal{S}^{I_x}(1^\lambda) \right] \leq \frac{1}{p(\lambda)}.$$

Definition 7 is a special case of Definition 6 since it has restrictions on the input to the obfuscator and the set of relations it supports. In the next claim, we show that our notion of non-malleability from Definition 5 implies the notion from Definition 7.

Claim. A point obfuscator satisfying Definition 5 with respect to an ensemble of families of functions $\mathcal{F}$ and an ensemble of distributions $\mathcal{X}$ also satisfies Definition 7 with respect to $\mathcal{F}$ and $\mathcal{X}$.

Proof. Let $\mathcal{O}$ be an obfuscator that satisfies Definition 5 with respect to the function in $\mathcal{F}$ and the distribution $\mathcal{X}$. Thus, for any $f \in \mathcal{F}$, there is no PPT adversary that can generate a valid-looking circuit C such that $I_{f(x)} \equiv C$ for $x \leftarrow \mathcal{X}$, except with negligible probability. Namely,

$$\Pr_{x \leftarrow \mathcal{X}} \left[C \neq \mathcal{O}(x), \mathcal{V}(C) = 1 \text{ and } I_{f(x)} \equiv C \mid C \leftarrow \mathcal{A}(\mathcal{O}(x)) \right] \leq \mathsf{negl}(\lambda).$$

Hence, a simulator that does nothing (say, outputs $\perp$) will satisfy security requirement of Definition 7.

A Discussion. Our definition is thus, morally, equivalent to the strong definition of [19], albeit with the assumption that the input comes from a well-spread distribution and the mauling is restricted to functions rather than relations. Getting a construction in the plain model that resolves these two issues is left as an open problem.

Lastly, observe that in the above proof, the simulator is in fact independent of the adversary $\mathcal{A}$ and independent of the distinguishability gap (the polynomial $p(\cdot)$). Thus, we actually get one simulator for all adversaries and the computational distance between the output of the adversary and the output of the simulator is negligible.

4   Our Obfuscator

Let $\lambda \in \mathbb{N}$ be the security parameter and let $X_\lambda = \mathbb{Z}_{2^\lambda}$ be the domain. Let $\mathcal{F}_{\mathsf{poly}} = \{f : X_\lambda \to X_\lambda\}_{\lambda \in \mathbb{N}}$ be the ensemble of classes of all functions that can be computed by polynomials of degree $\mathsf{poly}(\lambda)$, except the constant functions and the identity function.

Let $\mathcal{G} = \{\mathbb{G}_\lambda\}_{\lambda \in \mathbb{N}}$ be a group ensemble with uniform and efficient representation and operations, where each $\mathbb{G}_\lambda$ is a group of prime order $q \in (2^{\lambda-1}, 2^\lambda)$. We assume that for every $\lambda \in \mathbb{N}$ there is a canonical and efficient mapping between the elements of $\mathbb{G}_\lambda$ and the domain X_λ. Let g be the generator of the group $\mathbb{G}_{5\lambda}$. Our obfuscator gets as input an element $x \in X_\lambda$ and randomness $r \in \mathbb{G}_{5\lambda}$ and computes:

$$\mathcal{O}(x; r) = \left(r, r^{g^{x^4 + x^3 + x^2 + x}}\right).$$

The verifier $\mathcal{V}$ for a valid-looking obfuscation is the natural one: it checks whether the obfuscation consists of merely two group elements in $\mathbb{G}_{5\lambda}$. In the next two theorems we show that our obfuscator is both secure and non-malleable. The first part is based on the strong DDH assumption (Assumptions 1 and 3) and the second is based on (Assumptions 2 and 3). Thus, overall, our obfuscator is both secure and non-malleable under the assumption that there is a group where the strong DDH and strong power-DDH assumptions hold.

Theorem 4. *Under the strong DDH assumption (Assumptions 1 and 3), the obfuscator $\mathcal{O}$ above is a point obfuscator according to Definition 2.*

Theorem 5. *Let X_λ be any well-spread distribution over X_λ. Under the strong power-DDH assumption (Assumptions 2 and 3), the obfuscator $\mathcal{O}$ above is non-malleable according to Definition 5 for the family of functions $\mathcal{F}_{\mathsf{poly}}$ and the distribution X_λ.*

The proofs of these theorems appear in the following two subsections.

4.1   Proof of Theorem 4

For completeness, we first notice that for any $x \in X_\lambda$ it holds that $x^4 + x^3 + x^2 + x \leq 2^{5\lambda}$ and thus for any distinct $x, y \in X_\lambda$ it holds that $y^4 + y^3 + y^2 + y \neq x^4 + x^3 + x^2 + x$. Therefore, we get that for every $x \in X_\lambda$ it holds that $\mathcal{O}(x) \equiv I_x$, as required.

To prove soundness, we reduce to the security of our construction to the security of the r, r^x construction of Canetti [15]. We prove the following claim general claim regarding point function obfuscators.

Claim. Let $f \colon X_\lambda \to X'_\lambda$ be an injective polynomial-time computable function, and let $\mathcal{O}$ be a secure point obfuscator. Then, $\mathcal{O}'(x) = \mathcal{O}(f(x))$ is also a secure point obfuscator.

Proof. We prove that for any probabilistic polynomial-time algorithm $\mathcal{A}$, there is a probabilistic polynomial-time simulator $\mathcal{S}$ and a negligible function $\mathsf{negl}(\cdot)$, such that for all $x \in X_\lambda$ and all $\lambda \in \mathbb{N}$,

$$\left| \Pr_{\mathcal{A}, \mathcal{O}}[\mathcal{A}(\mathcal{O}'(x)) = 1] - \Pr_{\mathcal{S}}[\mathcal{S}^{I_x}(1^\lambda)) = 1] \right| \leq \mathsf{negl}(\lambda),$$

where the probabilities are over the internal randomness of $\mathcal{A}, \mathcal{O}$ and $\mathcal{S}$.

Let $\mathcal{A}$ be such an adversary and let $\mathcal{S}$ be the corresponding simulator whose existence is guaranteed by the fact that $\mathcal{O}$ is a secure point obfuscator. It holds that for every $x \in X_\lambda$:

$$\left| \Pr_{\mathcal{A},\mathcal{O}}[\mathcal{A}(\mathcal{O}(x)) = 1] - \Pr_{\mathcal{S}}[\mathcal{S}^{I_x}(1^\lambda)) = 1] \right| \le \mathsf{negl}(\lambda),$$

As a first step, we construct a simulator $\mathcal{S}'$ that is inefficient yet makes only a polynomial-number of queries to its oracle (we will get rid of this assumption later using a known transformation). We define a simulator $\mathcal{S}'$ (with oracle access to I_x) that works by simulating $\mathcal{S}$ as follows. When $\mathcal{S}$ performs a query y to its oracles, then $\mathcal{S}'$ finds x' such that $f(x') = y$. If no such x' exists, then $\mathcal{S}'$ replies with 0. Otherwise, if $\mathcal{S}'$ found such an x', then it performs the query to its oracle with x' and answers with the reply of the oracle. Since f is injective, we have that $f(x) = y$ if and only if $x' = x$. Thus, it holds that

$$\Pr_{\mathcal{S}}[\mathcal{S}^{I_{f(x)}}(1^\lambda)) = 1] = \Pr_{\mathcal{S}'}[\mathcal{S}'^{I_x}(1^\lambda)) = 1].$$

Thus, we get that

$$\left| \Pr_{\mathcal{A},\mathcal{O}}[\mathcal{A}(\mathcal{O}'(x)) = 1] - \Pr_{\mathcal{S}'}[\mathcal{S}'^{I_x}(1^\lambda)) = 1] \right| \le \mathsf{negl}(\lambda).$$

We are left to take care of the fact that the simulator is inefficient. For this we use a result of Bitansky and Canetti [8] who showed that this can be solved generically. Let us elaborate.

Bitansky and Canetti called obfuscators in which the simulation is inefficient yet the number of queries is bounded by a polynomial as *gray-box obfuscation*. This is in contrast to virtual-black box obfuscation where the simulator is required to be both efficient in its running time and the number of queries and indistinguishability obfuscation [3,24], which can be phrased as a simulation-based definition where the simulator is unbounded in both running time and number of queries (see [8, Proposition 3.1]). One of the main results of Bitansky and Canetti was that for point functions, the virtual-black box and virtual-gray box notions are equivalent: a simulator that runs in unbounded time yet makes a polynomial number of queries can be turned into one that runs in polynomial-time and makes a polynomial number of queries.[4]

Using their result for our construction we obtain a simulator that works in polynomial-time and makes a polynomial number of queries to its oracle. This completes the claim.

We finish the proof by applying the claim with $f(x) = g^{x^4 + x^3 + x^2 + x}$, noticing that this function is injective and efficiently computable.

---

[4] See [9] for more general families of functions where a similar equivalence holds.

4.2   Proof of Theorem 5

Assume that there exists an adversary $\mathcal{A}$, and a distribution $\mathcal{X}_\lambda$ such that given an obfuscation of a point $x \leftarrow \mathcal{X}_\lambda$, the adversary $\mathcal{A}$ outputs a function $f \in \mathcal{F}_{\mathsf{poly}}$ and a valid-looking obfuscation (i.e., an obfuscation that passes the verification of $\mathcal{V}$) of $f(x)$ with probability at least $\varepsilon > 0$. Denote by $t = t(\lambda)$ the degree of f (written as a polynomial over X_λ). We show how to construct an adversary $\mathcal{A}'$ that breaks the strong power-DDH assumption for the power sequence of length $T = 4t$.

Suppose we are given $(g^{z_0}, g^{z_1}, \ldots, g^{z_T})$, where $z_0 = 1$ and either $\forall i \in [T] :$ $z_i = x^i$ for a random $x \leftarrow X_\lambda$ or $\forall i \in [T] : z_i = r_i$ for random $r_1, \ldots, r_t \leftarrow X_\lambda$. Our goal is to show that $\mathcal{A}'$ can distinguish between the two cases. The algorithm $\mathcal{A}'$, on input $(g^{z_0}, \ldots, g^{z_T})$, first samples a random generator $r \leftarrow \mathbb{G}$ and computes $g^{z_1+z_2+z_3+z_4}$. Then, it runs $\mathcal{A}$ on the input pair $(r, r^{g^{z_1+z_2+z_3+z_4}})$ to get a function f and an output pair $(r_\mathcal{A}, w_\mathcal{A})$. We assume that we are given the coefficients of the polynomial that represents the function f, as otherwise we can learns these coefficients by interpolation of random evaluations of f (according to the distribution of the inputs $\mathcal{X}_\lambda$).

Let $h(x) = x^4 + x^3 + x^2 + x$ and let us write the polynomial $h(f(x))$ as a polynomial of degree at most $4t$ with coefficients b_i:

$$h(f(x)) = (f(x))^4 + (f(x))^3 + (f(x))^2 + f(x) = \sum_{i=0}^{4t} b_i x^i.$$

Using these values, it computes $u = g^{\sum_{i=0}^{T} b_i z_i}$ and $w_{\mathsf{real}} = r^u_\mathcal{A}$. Finally, the adversary $\mathcal{A}'$ outputs 1 if and only if $w_{\mathsf{real}} = w_\mathcal{A}$. The precise description of $\mathcal{A}'$ is given in Fig. 1.

The algorithm $\mathcal{A}'(g^{z_0}, g^{z_1}, \ldots, g^{z_T})$:

1. Choose a random generator $r \leftarrow \mathbb{G}$ and compute $g^{z_1+z_2+z_3+z_4}$.
2. $(f, r_\mathcal{A}, w_\mathcal{A}) \leftarrow \mathcal{A}(r, r^{g^{z_1+z_2+z_3+z_4}})$.
3. Compute the coefficients b_i for $i \in [T]$ of $h(f(x))$.
4. Compute $w_{\mathsf{real}} = r_\mathcal{A}^{g^{\sum_{i=0}^{T} b_i z_i}}$.
5. If $w_{\mathsf{real}} = w_\mathcal{A}$, then output 1. Otherwise, output 0.

Fig. 1. The adversary $\mathcal{A}'$ that breaks the power-DDH assumption.

We argue that $\mathcal{A}'$ successfully breaks the power-DDH assumption.

The Real Case. Observe that if $z_i = x^i$ for each $i \in [T]$, then the distribution that $\mathcal{A}$ sees is exactly the distribution $(r, r^{g^{x^4+x^3+x^2+x}})$ and thus with probability at least ε, the adversary $\mathcal{A}$ will maul the point obfuscation of x to a point obfuscation of $f(x)$. That is,

$$w_{\mathcal{A}} = r^{g^{h(f(x))}}_{\mathcal{A}} = r^{g^{\sum_{i=0}^{T} b_i x^i}}_{\mathcal{A}} = r^{g^{\sum_{i=0}^{T} b_i z_i}}_{\mathcal{A}} = w_{\mathsf{real}}.$$

Thus, $\mathcal{A}'$ will output 1 with probability at least ε.

The Random Case. Suppose that $z_i = r_i$ is random for each $i \in [T]$. We show that the probability that $w_{\mathsf{real}} = w_{\mathcal{A}}$ is negligible (in λ). This is an information theoretic claim that holds against unbounded adversaries. The adversary $\mathcal{A}$ holds r and $r^{g^{r_1 + r_2 + r_3 + r_4}}$ and let us even assume that she knows $s = r_1 + r_2 + r_3 + r_4$. In order for $\mathcal{A}'$ to succeed, she needs to be able to compute $s' = \sum_{i=0}^{T} b_i r_i$ (recall that $\mathcal{A}'$ is unbounded). We show that the min-entropy of this value s' given all the information of the adversary is high and therefore it cannot guess it with noticeable probability. Denote by $\mathsf{view}(\mathcal{A})$ a random variables that correspond to the view of $\mathcal{A}$ and denote by S' a random variable that corresponds to the value of s'.

We first show that if the degree of f (denoted above by t) is at least 2, then the min-entropy of S' is at least λ. This means that $\mathcal{A}'$ will be able to guess it with only negligible probability.

Claim. If $t \geq 2$, then $\mathsf{H}_\infty(S' \mid \mathsf{view}(\mathcal{A})) \geq \lambda$.

Proof. If the degree of f is at least 2, then the degree of $h(f(\cdot))$ is at least 5 and thus there exist $i > 4$ such that $b_i \neq 0$. In this case, since r_i is uniform in $\mathcal{X}_\lambda$, then the random variable s' has min-entropy λ given the view of $\mathcal{A}$.

The case where f is a linear function (i.e., a degree 1 polynomial) is slightly harder to handle and here we use properties of the exact choice of our degree 4 polynomial. Let f be written as $f(x) = ax + b$ for some fixed $a, b \in X_\lambda$. We expand the polynomial $h(f(x))$ and rewrite it by grouping terms:

$$\begin{aligned}
h(f(x)) &= (ax + b)^4 + (ax + b)^3 + (ax + b)^2 + (ax + b) \\
&= a^4 x^4 + (4a^3 b + a^3)x^3 + (6a^2 b^2 + 3a^2 b + a^2)x^2 \\
&\quad + (4ab^3 + 3ab^2 + 2ab + a)x + b^4 + b^3 + b^2 + b.
\end{aligned}$$

We show that the coefficients of $h(f(\cdot))$ cannot be all identical.

Claim. The coefficients of h are not all identical.

Proof. If they were identical, then

$$a^4 = 4a^3 b + a^3 = 6a^2 b^2 + 3a^2 b + a^2 = 4ab^3 + 3ab^2 + 2ab + a.$$

Solving this set of equations gives that the only solutions are $a = 0, b = *$ (i.e., b is arbitrary) and $a = 1, b = 0$ (i.e., the identity function). However, these are illegal according to our definition of $\mathcal{F}_{\mathsf{poly}}$: this class contains neither constant functions nor the identity function.

Using the fact that the coefficients are not all identical, we claim that the min-entropy of S' is at least λ even given the view of $\mathcal{A}$. Thus, again, the probability of guessing correctly the value is negligible.

Claim. Let $R_1, R_2, R_3, R_4 \leftarrow X_\lambda$ be random variable whose distribution is uniform from $\mathcal{X}_\lambda$, and let their sum be $S = R_1 + R_2 + R_3 + R_4 \in X_{6\lambda}$. Let $b_1, b_2, b_3, b_4 \in X_\lambda$ be arbitrary constants such that at least two of them are different. Let $S' = b_1 R_1 + b_2 R_2 + b_3 R_3 + b_4 R_4$. Then, $\mathsf{H}_\infty(S' \mid S) \geq \lambda$.

Proof. We lower bound the min entropy by computing $\Pr[S' = s' \mid S = s]$ for each $s, s' \in X_\lambda$. This probability is exactly the fraction of possible r_1, r_2, r_3, r_4 such that $r_1 + r_2 + r_3 + r_4 = s$ and $b_1 r_1 + b_2 r_2 + b_3 r_3 + b_4 r_4 = s'$. Writing this in matrix form we have

$$\underbrace{\begin{bmatrix} 1 & 1 & 1 & 1 \\ b_1 & b_2 & b_3 & b_4 \end{bmatrix}}_{A} \cdot \begin{bmatrix} r_1 \\ r_2 \\ r_3 \\ r_4 \end{bmatrix} = \begin{bmatrix} s \\ s' \end{bmatrix}.$$

Denote by Q the size of the support of $\mathcal{X}_\lambda$ and notice that $Q \geq 2^\lambda$. Since $\mathcal{X}_\lambda$ is well-spread, its min-entropy is super logarithmic in λ and thus the support size is super polynomial in λ. Since not all the b_i's are equal, we have that A's rank is 2, and thus the solution dimension is 2 for each $s' \in X_\lambda$ and the number of possible solutions is Q^2 out of the total Q^4 possibilities. Altogether, we get that for every $s' \in X_\lambda$, it holds that $\Pr[S' = s' \mid S = s] = Q^2/Q^4 \leq 1/Q < 1/2^\lambda$. Thus, the min-entropy is at least λ.

Combining the above, we get that overall, the probability of distinguishing is:

$$\left| \Pr[\mathcal{A}'(g^{x^1}, \ldots, g^{x^T}) = 1] - \Pr[\mathcal{A}'(g^{r_1}, \ldots, g^{r_T}) = 1] \right| \geq \varepsilon - \mathsf{negl}(\lambda)$$

which contradicts the security of the power-DDH assumption.

4.3   Supporting More Functions

In our construction above, we have shown how to get a point function obfuscator that is non-malleable against any function that can be written as a univariate polynomial of a polynomial degree. The reason that there is a bound on the degree of the polynomial is that the security reduction runs in time that is proportional to the degree. In particular, to be resilient against a function f of degree t we had to construct $g^{h(f(x))}$ in the reduction given the sequence $\{g^{x^i}\}_{i=0}^{4t}$ (recall that $h(x) = x^4 + x^3 + x^2 + x$).

Exponential Security. Suppose that the min-entropy of the inputs is k. Thus, the support-size of the distribution is at most 2^k and hence any function can be written as a polynomial of degree at most 2^k. That is, we can assume without loss of generality that the mauling function is described by a degree $t \leq 2^k$ polynomial. Thus, if we assume an exponential version of the strong power-DDH assumption, where the adversary's running time and advantage are bounded by $2^{O(k)}$ and $2^{-\Omega(k)}$, respectively, we can support functions of exponential degree (in k).

Uber Assumption. Instead of building the polynomial $h(f(x))$ in the proof monomial by monomial in order to break the power-DDH assumption, we can, alternatively, modify our assumption to get a more direct security proof without the large security loss. Concretely, instead of having the reduction computing $g^{h(f(x))}$ given $\{g^{z_i}\}_{i=0}^{4t}$, where t is the degree f, we assume an "uber" power-DDH assumption that is parametrized by a class of functions $\mathcal{F} = \{f \colon \mathbb{Z}_p \to \mathbb{Z}_p\}$ (and thus can thought of as a collection of assumptions, one per $f \in \mathcal{F}$). The assumption says that for any $f \in \mathcal{F}$, the following two distributions are computationally-indistinguishable:

$$(g, g^x, g^{h(f(x))}) \approx_c (g, g^x, g^s),$$

where $x \leftarrow \mathcal{X}$ and $s \leftarrow \mathbb{Z}_p^*$ is chosen at random. Having such an assumption for a class of mauling functions $\mathcal{F}$, implies that our construction is non-malleable for the same class $\mathcal{F}$.

Acknowledgments. We thank the anonymous reviewers of EUROCRYPT 2018 for their elaborate and useful comments. We are grateful to Ran Canetti for multiple useful suggestions and feedback about this work. Thanks to Nir Bitansky, Abhishek Jain, and Omer Paneth for multiple discussions. Lastly, we thank Moni Naor for his encouragement, support and advice.

References

1. Applebaum, B., Harnik, D., Ishai, Y.: Semantic security under related-key attacks and applications. In: Innovations in Computer Science - ICS. pp. 45–60 (2011)
2. Baecher, P., Fischlin, M., Schröder, D.: Expedient non-malleability notions for hash functions. In: Kiayias, A. (ed.) CT-RSA 2011. LNCS, vol. 6558, pp. 268–283. Springer, Heidelberg (2011). https://doi.org/10.1007/978-3-642-19074-2_18
3. Barak, B., Goldreich, O., Impagliazzo, R., Rudich, S., Sahai, A., Vadhan, S.P., Yang, K.: On the (im)possibility of obfuscating programs. Journal of the ACM **59**(2), 6 (2012). Preliminary version appeared in CRYPTO 2001
4. Bellare, M., Boldyreva, A., O'Neill, A.: Deterministic and efficiently searchable encryption. In: Menezes, A. (ed.) CRYPTO 2007. LNCS, vol. 4622, pp. 535–552. Springer, Heidelberg (2007). https://doi.org/10.1007/978-3-540-74143-5_30
5. Bellare, M., Hoang, V.T., Keelveedhi, S.: Instantiating random oracles via UCEs. In: Canetti, R., Garay, J.A. (eds.) CRYPTO 2013. LNCS, vol. 8043, pp. 398–415. Springer, Heidelberg (2013). https://doi.org/10.1007/978-3-642-40084-1_23
6. Bellare, M., Rogaway, P.: Random oracles are practical: a paradigm for designing efficient protocols. In: Proceedings of the 1st ACM Conference on Computer and Communications Security, CCS, pp. 62–73 (1993)
7. Bellare, M., Stepanovs, I.: Point-function obfuscation: a framework and generic constructions. In: Kushilevitz, E., Malkin, T. (eds.) TCC 2016. LNCS, vol. 9563, pp. 565–594. Springer, Heidelberg (2016). https://doi.org/10.1007/978-3-662-49099-0_21
8. Bitansky, N., Canetti, R.: On strong simulation and composable point obfuscation. J. Cryptol. **27**(2), 317–357 (2014)

9. Bitansky, N., Canetti, R., Kalai, Y.T., Paneth, O.: On virtual grey box obfuscation for general circuits. In: Garay, J.A., Gennaro, R. (eds.) CRYPTO 2014. LNCS, vol. 8617, pp. 108–125. Springer, Heidelberg (2014). https://doi.org/10.1007/978-3-662-44381-1_7

10. Bitansky, N., Paneth, O.: Point obfuscation and 3-round zero-knowledge. In: Cramer, R. (ed.) TCC 2012. LNCS, vol. 7194, pp. 190–208. Springer, Heidelberg (2012). https://doi.org/10.1007/978-3-642-28914-9_11

11. Boldyreva, A., Cash, D., Fischlin, M., Warinschi, B.: Foundations of non-malleable hash and one-way functions. In: Matsui, M. (ed.) ASIACRYPT 2009. LNCS, vol. 5912, pp. 524–541. Springer, Heidelberg (2009). https://doi.org/10.1007/978-3-642-10366-7_31

12. Brenner, H., Goyal, V., Richelson, S., Rosen, A., Vald, M.: Fast non-malleable commitments. In: Proceedings of the 22nd ACM SIGSAC Conference on Computer and Communications Security, CCS, pp. 1048–1057. ACM (2015)

13. Brzuska, C., Farshim, P., Mittelbach, A.: Random-oracle uninstantiability from indistinguishability obfuscation. In: Dodis, Y., Nielsen, J.B. (eds.) TCC 2015. LNCS, vol. 9015, pp. 428–455. Springer, Heidelberg (2015). https://doi.org/10.1007/978-3-662-46497-7_17

14. Camenisch, J., Neven, G., Shelat, A.: Simulatable adaptive oblivious transfer. In: Naor, M. (ed.) EUROCRYPT 2007. LNCS, vol. 4515, pp. 573–590. Springer, Heidelberg (2007). https://doi.org/10.1007/978-3-540-72540-4_33

15. Canetti, R.: Towards realizing random oracles: hash functions that hide all partial information. In: Kaliski, B.S. (ed.) CRYPTO 1997. LNCS, vol. 1294, pp. 455–469. Springer, Heidelberg (1997). https://doi.org/10.1007/BFb0052255

16. Canetti, R., Fuller, B., Paneth, O., Reyzin, L., Smith, A.: Reusable fuzzy extractors for low-entropy distributions. In: Fischlin, M., Coron, J.-S. (eds.) EUROCRYPT 2016. LNCS, vol. 9665, pp. 117–146. Springer, Heidelberg (2016). https://doi.org/10.1007/978-3-662-49890-3_5

17. Canetti, R., Goldreich, O., Halevi, S.: The random oracle methodology, revisited. J. ACM **51**(4), 557–594 (2004)

18. Canetti, R., Micciancio, D., Reingold, O.: Perfectly one-way probabilistic hash functions (preliminary version). In: Proceedings of the Thirtieth Annual ACM Symposium on the Theory of Computing, STOC, pp. 131–140. ACM (1998)

19. Canetti, R., Varia, M.: Non-malleable obfuscation. In: Reingold, O. (ed.) TCC 2009. LNCS, vol. 5444, pp. 73–90. Springer, Heidelberg (2009). https://doi.org/10.1007/978-3-642-00457-5_6

20. Chen, Y., Qin, B., Zhang, J., Deng, Y., Chow, S.S.M.: Non-malleable functions and their applications. In: Cheng, C.-M., Chung, K.-M., Persiano, G., Yang, B.-Y. (eds.) PKC 2016. LNCS, vol. 9615, pp. 386–416. Springer, Heidelberg (2016). https://doi.org/10.1007/978-3-662-49387-8_15

21. Crescenzo, G.D., Katz, J., Ostrovsky, R., Smith, A.D.: Efficient and non-interactive non-malleable commitment. In: Advances in Cryptology - EUROCRYPT, pp. 40–59 (2001)

22. Dolev, D., Dwork, C., Naor, M.: Nonmalleable cryptography. SIAM Rev. **45**(4), 727–784 (2003)

23. Fischlin, M., Fischlin, R.: Efficient non-malleable commitment schemes. J. Cryptol. **24**(1), 203–244 (2011)

24. Garg, S., Gentry, C., Halevi, S., Raykova, M., Sahai, A., Waters, B.: Candidate indistinguishability obfuscation and functional encryption for all circuits. In: 54th Annual IEEE Symposium on Foundations of Computer Science, FOCS, pp. 40–49 (2013)

25. Gentry, C.: Practical identity-based encryption without random oracles. In: Vaudenay, S. (ed.) EUROCRYPT 2006. LNCS, vol. 4004, pp. 445–464. Springer, Heidelberg (2006). https://doi.org/10.1007/11761679_27

26. Goldwasser, S., Kalai, Y.T.: On the (in)security of the Fiat-Shamir paradigm. In: 44th Annual IEEE Symposium on Foundations of Computer Science FOCS, pp. 102–113 (2003)

27. Goldwasser, S., Kalai, Y.T., Peikert, C., Vaikuntanathan, V.: Robustness of the learning with errors assumption. In: Innovations in Computer Science - ICS 2010, pp. 230–240. Tsinghua University Press (2010)

28. Golle, P., Jarecki, S., Mironov, I.: Cryptographic primitives enforcing communication and storage complexity. In: Blaze, M. (ed.) FC 2002. LNCS, vol. 2357, pp. 120–135. Springer, Heidelberg (2003). https://doi.org/10.1007/3-540-36504-4_9

29. Goyal, V., Khurana, D., Sahai, A.: Breaking the three round barrier for non-malleable commitments. In: IEEE 57th Annual Symposium on Foundations of Computer Science, FOCS, pp. 21–30. IEEE Computer Society (2016)

30. Goyal, V., Pandey, O., Richelson, S.: Textbook non-malleable commitments. In: Proceedings of the 48th Annual ACM SIGACT Symposium on Theory of Computing, STOC, pp. 1128–1141. ACM (2016)

31. Khurana, D., Sahai, A.: How to achieve non-malleability in one or two rounds. In: 58th IEEE Annual Symposium on Foundations of Computer Science, FOCS, pp. 564–575. IEEE Computer Society (2017)

32. Lin, H., Pass, R.: Constant-round nonmalleable commitments from any one-way function. J. ACM 62(1), 5:1–5:30 (2015)

33. Lin, H., Pass, R., Soni, P.: Two-round concurrent non-malleable commitment from time-lock puzzles. IACR Cryptology ePrint Archive 2017, 273 (2017)

34. Pandey, O., Pass, R., Vaikuntanathan, V.: Adaptive one-way functions and applications. In: Wagner, D. (ed.) CRYPTO 2008. LNCS, vol. 5157, pp. 57–74. Springer, Heidelberg (2008). https://doi.org/10.1007/978-3-540-85174-5_4

35. Pass, R.: Unprovable security of perfect NIZK and non-interactive non-malleable commitments. Comput. Complex. 25(3), 607–666 (2016)

36. Pass, R., Rosen, A.: Concurrent nonmalleable commitments. SIAM J. Comput. 37(6), 1891–1925 (2008)

37. Pass, R., Shelat, A., Vaikuntanathan, V.: Construction of a non-malleable encryption scheme from any semantically secure one. In: Advances in Cryptology - CRYPTO, pp. 271–289 (2006)

38. Sahai, A.: Non-malleable non-interactive zero knowledge and adaptive chosen-ciphertext security. In: 40th Annual Symposium on Foundations of Computer Science, FOCS, pp. 543–553 (1999)

39. Stevens, M., Bursztein, E., Karpman, P., Albertini, A., Markov, Y.: The first collision for full SHA-1. IACR Cryptol. ePrint Archive 2017, 190 (2017)

40. Wagner, D.A., Goldberg, I.: Proofs of security for the Unix password hashing algorithm. In: Advances in Cryptology - ASIACRYPT, pp. 560–572 (2000)

41. Wang, X., Yin, Y.L., Yu, H.: Finding collisions in the full SHA-1. In: Advances in Cryptology - CRYPTO, pp. 17–36 (2005)

42. Wang, X., Yu, H.: How to break MD5 and other hash functions. In: Advances in Cryptology - EUROCRYPT, pp. 19–35 (2005)

43. Wee, H.: On obfuscating point functions. In: STOC, pp. 523–532. ACM (2005)

The Wonderful World of Global Random Oracles

Jan Camenisch[1]($\boxtimes$) , Manu Drijvers[1,2], Tommaso Gagliardoni[1],
Anja Lehmann[1], and Gregory Neven[1]

[1] IBM Research – Zurich, Säumerstrasse 4, 8803 Rüschlikon, Switzerland
{jca,mdr,tog,anj,nev}@zurich.ibm.com
[2] Department of Computer Science, ETH Zurich, 8092 Zurich, Switzerland

Abstract. The random-oracle model by Bellare and Rogaway (CCS'93) is an indispensable tool for the security analysis of practical cryptographic protocols. However, the traditional random-oracle model fails to guarantee security when a protocol is composed with arbitrary protocols that use the *same* random oracle. Canetti, Jain, and Scafuro (CCS'14) put forth a *global* but non-programmable random oracle in the Generalized UC framework and showed that some basic cryptographic primitives with composable security can be efficiently realized in their model. Because their random-oracle functionality is non-programmable, there are many practical protocols that have no hope of being proved secure using it. In this paper, we study alternative definitions of a global random oracle and, perhaps surprisingly, show that these allow one to prove GUC-secure existing, very practical realizations of a number of essential cryptographic primitives including public-key encryption, non-committing encryption, commitments, Schnorr signatures, and hash-and-invert signatures. Some of our results hold generically for any suitable scheme proven secure in the traditional ROM, some hold for specific constructions only. Our results include many highly practical protocols, for example, the folklore commitment scheme $\mathcal{H}(m\|r)$ (where m is a message and r is the random opening information) which is far more efficient than the construction of Canetti et al.

1 Introduction

The random-oracle model (ROM) [3] is an overwhelmingly popular tool in cryptographic protocol design and analysis. Part of its success is due to its intuitive idealization of cryptographic hash functions, which it models through calls to an external oracle that implements a random function. Another important factor is its capability to provide security proofs for highly practical constructions of important cryptographic building blocks such as digital signatures, public-key encryption, and key exchange. In spite of its known inability to provide provable guarantees when instantiated with a real-world hash function [14], the ROM is still widely seen as convincing evidence that a protocol will resist attacks in practice.

© International Association for Cryptologic Research 2018
J. B. Nielsen and V. Rijmen (Eds.): EUROCRYPT 2018, LNCS 10820, pp. 280–312, 2018.
https://doi.org/10.1007/978-3-319-78381-9_11

Most proofs in the ROM, however, are for property-based security notions, where the adversary is challenged in a game where he faces a single, isolated instance of the protocol. Security can therefore no longer be guaranteed when a protocol is composed. Addressing this requires composable security notions such as Canetti's Universal Composability (UC) framework [10], which have the advantage of guaranteeing security even if protocols are arbitrarily composed.

UC modeling. In the UC framework, a random oracle is usually modeled as an ideal functionality that a protocol uses as a subroutine in a so-called *hybrid model*, similarly to other setup constructs such as a common reference string (CRS). For example, the random-oracle functionality $\mathcal{F}_{RO}$ [21] simply assigns a random output value h to each input m and returns h. In the security proof, the simulator executes the code of the subfunctionality, which enables it to observe the queries of all involved parties and to program any random-looking values as outputs. Setup assumptions play an important role for protocols in the UC model, as many important cryptographic primitives such as commitments simply cannot be achieved [13]; other tasks can, but have more efficient instantiations with a trusted setup.

An important caveat is that this way of modeling assumes that each instance of each protocol uses its own separate and independent instance of the subfunctionality. For a CRS this is somewhat awkward, because it raises the question of how the parties should agree on a common CRS, but it is even more problematic for random oracles if all, supposedly independent, instances of $\mathcal{F}_{RO}$ are replaced in practice with the *same* hash function. This can be addressed using the Generalized UC (GUC) framework [12] that allows one to model different protocol instances sharing access to global functionalities. Thus one can make the setup functionality globally available to all parties, meaning, including those outside of the protocol execution as well as the external environment.

Global UC random oracle. Canetti et al. [15] indeed applied the GUC framework to model globally accessible random oracles. In doing so, they discard the globally accessible variant of $\mathcal{F}_{RO}$ described above as of little help for proving security of protocols because it is too "strict", allowing the simulator neither to observe the environment's random-oracle queries, nor to program its answers. They argue that any shared functionality that provides only public information is useless as it does not give the simulator any advantage over the real adversary. Instead, they formulate a global random-oracle functionality that grants the ideal-world simulator access to the list of queries that the environment makes outside of the session. They then show that this shared functionality can be used to design a reasonably efficient GUC-secure commitment scheme, as well as zero-knowledge proofs and two-party computation. However, their global random-oracle functionality rules out security proofs for a number of practical protocols, especially those that require one to program the random oracle.

Our Contributions. In this paper, we investigate different alternative formulations of globally accessible random-oracle functionalities and protocols that can be proven secure with respect to these functionalities. For instance, we show that the simple variant discarded by Canetti et al. surprisingly suffices to prove the GUC-security of a number of truly practical constructions for useful cryptographic primitives such as digital signatures and public-key encryption. We achieve these results by carefully analyzing the minimal capabilities that the *simulator* needs in order to simulate the real-world (hybrid) protocol, while fully exploiting the additional capabilities that one has in proving the indistinguishability between the real and the ideal worlds. In the following, we briefly describe the different random-oracle functionalities we consider and which we prove GUC-secure using them.

Strict global random oracle. First, we revisit the strict global random-oracle functionality $\mathcal{G}_{\mathsf{sRO}}$ described above and show that, in spite of the arguments of Canetti et al. [15], it actually suffices to prove the GUC-security of many practical constructions. In particular, we show that any digital signature scheme that is existentially unforgeable under chosen-message attack in the traditional ROM also GUC-realizes the signature functionality with $\mathcal{G}_{\mathsf{sRO}}$, and that any public-key encryption (PKE) scheme that is indistinguishable under adaptive chosen-ciphertext attack in the traditional ROM GUC-realizes the PKE functionality under $\mathcal{G}_{\mathsf{sRO}}$ with static corruptions.

This result may be somewhat surprising as it includes many schemes that, in their property-based security proofs, rely on invasive proof techniques such as rewinding, observing, and programming the random oracle, all of which are tools that the GUC simulator is not allowed to use. We demonstrate, however, that none of these techniques are needed during the simulation of the protocol, but rather only show up when proving indistinguishability of the real and the ideal worlds, where they are allowed. A similar technique was used It also does not contradict the impossibility proof of commitments based on global setup functionalities that simply provide public information [12,13] because, in the GUC framework, signatures and PKE do not imply commitments.

Programmable global random oracles. Next, we present a global random-oracle functionality $\mathcal{G}_{\mathsf{pRO}}$ that allows the simulator as well as the real-world adversary to program arbitrary points in the random oracle, as long as they are not yet defined. We show that it suffices to prove the GUC-security of Camenisch et al.'s non-committing encryption scheme [8], i.e., PKE scheme secure against adaptive corruptions. Here, the GUC simulator needs to produce dummy ciphertexts that can later be made to decrypt to a particular message when the sender or the receiver of the ciphertext is corrupted. The crucial observation is that, to embed a message in a dummy ciphertext, the simulator only needs to program the random oracle at *random* inputs, which have negligible chance of being already queried or programmed. Again, this result is somewhat surprising as $\mathcal{G}_{\mathsf{pRO}}$ does not give the simulator any advantage over the real adversary either.

We also define a restricted variant $\mathcal{G}_{\mathsf{rpRO}}$ that, analogously to the observable random oracle of Canetti et al. [15], offers programming subject to some restrictions, namely that protocol parties can check whether the random oracle was programmed on a particular point. If the adversary tries to cheat by programming the random oracle, then honest parties have a means of detecting this misbehavior. However, we will see that the simulator can hide its programming from the adversary, giving it a clear advantage over the real-world adversary. We use it to GUC-realize the commitment functionality through a new construction that, with only two exponentiations per party and two rounds of communication, is considerably more efficient than the one of Canetti et al. [15], which required five exponentiations and five rounds of communication.

Programmable and observable global random oracle. Finally, we describe a global random-oracle functionality $\mathcal{G}_{\mathsf{rpoRO}}$ that combines the restricted forms of programmability and observability. We then show that this functionality allows us to prove that commitments can be GUC-realized by the most natural and efficient random-oracle based scheme where a commitment $c = \mathcal{H}(m\|r)$ is the hash of the random opening information r and the message m.

Transformations between different oracles. While our different types of oracles allow us to securely realize different protocols, the variety in oracles partially defies the original goal of modeling the situation where all protocols use the *same* hash function. We therefore explore some relations among the different types by presenting efficient protocol transformations that turn any protocol that securely realizes a functionality with one type of random oracle into a protocol that securely realizes the same functionality with a different type.

Other Related Work. Dodis et al. [17] already realized that rewinding can be used in the indistinguishability proof in the GUC model, as long as it's not used in the simulation itself. In a broader sense, our work complements existing studies on the impact of programmability and observability of random oracles in security reductions. Fischlin et al. [18] and Bhattacharyya and Mukherjee [6] have proposed formalizations of non-programmable and weakly-programmable random oracles, e.g., only allowing non-adaptive programmability. Both works give a number of possibility and impossibility results, in particular that full-domain hash (FDH) signatures can only be proven secure (via black-box reductions) if the random oracle is fully programmable [18]. Non-observable random oracles and their power are studied by Ananth and Bhaskarin [1], showing that Schnorr and probabilistic RSA-FDH signatures can be proven secure. All these works focus on the use of random oracles in individual reductions, whereas our work proposes globally re-usable random-oracle functionalities within the UC framework. The strict random oracle functionality $\mathcal{G}_{\mathsf{sRO}}$ that we analyze is comparable to a non-programmable and non-observable random oracle, so our result that any unforgeable signature scheme is also GUC-secure w.r.t. $\mathcal{G}_{\mathsf{sRO}}$ may seem to contradict the above results. However, the $\mathcal{G}_{\mathsf{sRO}}$ functionality imposes these restrictions only for the GUC simulator, whereas the reduction can fully program the random oracle.

Summary. Our results clearly paint a much more positive picture for global random oracles than was given in the literature so far. We present several formulations of globally accessible random-oracle functionalities that allow to prove the composable security of some of the most efficient signature, PKE, and commitment schemes that are currently known. We even show that the most natural formulation, the strict global random oracle $\mathcal{G}_{\mathsf{sRO}}$ that was previously considered useless, suffices to prove GUC-secure a large class of efficient signature and encryption schemes. By doing so, our work brings the (composable) ROM back closer to its original intention: to provide an *intuitive* idealization of hash functions that enables to prove the security of *highly efficient* protocols. We expect that our results will give rise to many more practical cryptographic protocols that can be proven GUC-secure, among them known protocols that have been proven secure in the traditional ROM model.

2   Preliminaries

In the rest of this work, we use "iff" for "if and only if", "w.l.o.g." for "without loss of generality", and $n \in \mathbb{N}$ to denote the security parameter. A function $\varepsilon(n)$ is *negligible* if it is asymptotically smaller than $1/p(n)$ for every polynomial function p. We denote by $x \xleftarrow{\$} X$ that x is a sample from the uniform distribution over the set X. When A is a probabilistic algorithm, then $y := \mathsf{A}(x; r)$ means that y is assigned the outcome of a run of A on input x with coins r. Two distributions X and Y over a domain $\Sigma(n)$ are said to be *computationally indistinguishable*, written $X \approx Y$, if for any PPT algorithm $\mathcal{A}$, $|\mathcal{A}(X(s)) - \mathcal{A}(Y(s))|$ is negligible for all $s \in \Sigma(n)$.

2.1   The Basic and Generalized UC Frameworks

Basic UC. The universal composability (UC) framework [9,10] is a framework to define and prove the security of protocols. It follows the simulation-based security paradigm, meaning that security of a protocol is defined as the simulatability of the protocol based on an ideal functionality $\mathcal{F}$. In an imaginary ideal world, parties hand their protocol inputs to a trusted party running $\mathcal{F}$, where $\mathcal{F}$ by construction executes the task at hand in a secure manner. A protocol π is considered a secure realization of $\mathcal{F}$ if the real world, in which parties execute the real protocol, is indistinguishable from the ideal world. Namely, for every real-world adversary $\mathcal{A}$ attacking the protocol, we can design an ideal-world attacker (simulator) $\mathcal{S}$ that performs an equivalent attack in the ideal world. As the ideal world is secure by construction, this means that there are no meaningful attacks on the real-world protocol either.

One of the goals of UC is to simplify the security analysis of protocols, by guaranteeing secure composition of protocols and, consequently, allowing for modular security proofs. One can design a protocol π assuming the availability of an ideal functionality $\mathcal{F}'$, i.e., π is a $\mathcal{F}'$-hybrid protocol. If π securely realizes $\mathcal{F}$, and another protocol π' securely realizes $\mathcal{F}'$, then the composition theorem guarantees that π composed with π' (i.e., replacing π' with $\mathcal{F}'$) is a secure realization of $\mathcal{F}$.

Security is defined through an interactive Turing machine (ITM) $\mathcal{Z}$ that models the environment of the protocol and chooses protocol inputs to all participants. Let $\mathrm{EXEC}_{\pi,\mathcal{A},\mathcal{Z}}$ denote the output of $\mathcal{Z}$ in the real world, running with protocol π and adversary $\mathcal{A}$, and let $\mathrm{IDEAL}_{\mathcal{F},\mathcal{S},\mathcal{Z}}$ denote its output in the ideal world, running with functionality $\mathcal{F}$ and simulator $\mathcal{S}$. Protocol π securely realizes $\mathcal{F}$ if for every polynomial-time adversary $\mathcal{A}$, there exists a simulator $\mathcal{S}$ such that for every environment $\mathcal{Z}$, $\mathrm{EXEC}_{\pi,\mathcal{A},\mathcal{Z}} \approx \mathrm{IDEAL}_{\mathcal{F},\mathcal{S},\mathcal{Z}}$.

Generalized UC. A Basic UC protocol using random oracles is modeled as a $\mathcal{F}_{\mathrm{RO}}$-hybrid protocol. Since an instance of a Basic UC functionality can only be used by a single protocol instance, this means that every protocol instance uses its own random oracle that is completely independent of other protocol instances' random oracles. As the random-oracle model is supposed to be an idealization of real-world hash functions, this is not a very realistic model: Given that we only have a handful of standardized hash functions, it's hard to argue their independence across many protocol instances.

To address these limitations of Basic UC, Canetti et al. [12] introduced the Generalized UC (GUC) framework, which allows for shared "global" ideal functionalities (denoted by $\mathcal{G}$) that can be used by all protocol instances. Additionally, GUC gives the environment more powers in the UC experiment. Let $\mathrm{GEXEC}_{\pi,\mathcal{A},\mathcal{Z}}$ be defined as $\mathrm{EXEC}_{\pi,\mathcal{A},\mathcal{Z}}$, except that the environment $\mathcal{Z}$ is no longer constrained, meaning that it is allowed to start arbitrary protocols in addition to the challenge protocol π. Similarly, $\mathrm{GIDEAL}_{\mathcal{F},\mathcal{S},\mathcal{Z}}$ is equivalent to $\mathrm{IDEAL}_{\mathcal{F},\mathcal{S},\mathcal{Z}}$ but $\mathcal{Z}$ is now unconstrained. If π is a $\mathcal{G}$-hybrid protocol, where $\mathcal{G}$ is some shared functionality, then $\mathcal{Z}$ can start additional $\mathcal{G}$-hybrid protocols, possibly learning information about or influencing the state of $\mathcal{G}$.

Definition 1. *Protocol π GUC-emulates protocol φ if for every adversary $\mathcal{A}$ there exists an adversary $\mathcal{S}$ such that for all unconstrained environments $\mathcal{Z}$,* $\mathrm{GEXEC}_{\pi,\mathcal{A},\mathcal{Z}} \approx \mathrm{GEXEC}_{\varphi,\mathcal{S},\mathcal{Z}}$.

Definition 2. *Protocol π GUC-realizes ideal functionality $\mathcal{F}$ if for every adversary $\mathcal{A}$ there exists a simulator $\mathcal{S}$ such that for all unconstrained environments $\mathcal{Z}$,* $\mathrm{GEXEC}_{\pi,\mathcal{A},\mathcal{Z}} \approx \mathrm{GIDEAL}_{\mathcal{F},\mathcal{S},\mathcal{Z}}$.

GUC gives very strong security guarantees, as the unconstrained environment can run arbitrary protocols in parallel with the challenge protocol, where the different protocol instances might share access to global functionalities. However, exactly this flexibility makes it hard to reason about the GUC experiment. To address this, Canetti et al. also introduced Externalized UC (EUC). Typically, a protocol π uses many local hybrid functionalities $\mathcal{F}$ but only uses a single shared functionality $\mathcal{G}$. Such protocols are called $\mathcal{G}$-subroutine respecting, and EUC allows for simpler security proofs for such protocols. Rather than considering unconstrained environments, EUC considers $\mathcal{G}$-externally constrained environments. Such environments can invoke only a single instance of the challenge protocol, but can additionally query the shared functionality $\mathcal{G}$ through dummy parties that are not part of the challenge protocol. The EUC experiment is equivalent to the Basic UC experiment, except that it considers $\mathcal{G}$-externally constrained environments: A $\mathcal{G}$-subroutine respecting protocol π EUC-emulates a protocol φ if for every polynomial-time adversary $\mathcal{A}$ there is an adversary $\mathcal{S}$ such that for every $\mathcal{G}$-externally constrained environment $\text{EXEC}^{\mathcal{G}}_{\pi,\mathcal{A},\mathcal{Z}} \approx \text{EXEC}^{\mathcal{G}}_{\varphi,\mathcal{S},\mathcal{Z}}$. Figure 2(b) depicts EUC-emulation and shows that this setting is much simpler to reason about than GUC-emulation: We can reason about this static setup, rather than having to imagine arbitrary protocols running alongside the challenge protocol. Canetti et al. prove that showing EUC-emulation is useful to obtain GUC-emulation.

Theorem 1. *Let π be a $\mathcal{G}$-subroutine respecting protocol, then protocol π GUC-emulates protocol φ if and only if π $\mathcal{G}$-EUC-emulates φ.*

Conventions. When specifying ideal functionalities, we will use some conventions for ease of notation. For a non-shared functionality with session id sid, we write "On input x from party $\mathcal{P}$", where it is understood the input comes from machine $(\mathcal{P}, \text{sid})$. For shared functionalities, machines from any session may provide input, so we always specify both the party identity and the session identity of machines. In some cases an ideal functionality requires immediate input from the adversary. In such cases we write "wait for input x from the adversary", which is formally defined by Camenisch et al. [7].

2.2 Basic Building Blocks

One-Way Trapdoor Permutations. A (family of) *one-way trapdoor permutations* is a tuple OWTP := (OWTP.Gen, OWTP.Sample, OWTP.Eval, OWTP.Invert) of PPT algorithms. On input n; OWTP.Gen outputs: a *permutation domain* Σ (e.g., $\mathbb{Z}_N$ for an RSA modulus N), and efficient representations of, respectively, a permutation φ in the family (e.g., an RSA public exponent e), and of its inverse φ^{-1} (e.g., an RSA secret exponent d). Security requires that no PPT adversary can invert a point $y = \varphi(x)$ for a random challenge template (Σ, φ, y) with non-negligible probability. We will often use OWTPs to generate public and secret keys for, e.g., signature schemes or encryption schemes by, e.g., setting pk $= (\Sigma, \varphi)$ and sk $= \varphi^{-1}$. W.l.o.g. in the following we assume that the representation of Σ also includes the related security parameter n, and secret keys also include the public part. Notice that, in general, OWTP.Invert also takes φ as input, although in practice this might be unnecessary, depending on the particular OWTP in exam.

Signature Schemes. A (stateless) *signature scheme* is a tuple SIG $=$ (KGen, Sign, Verify) of polynomial time algorithms, where KGen and Sign can be probabilistic and Verify is deterministic. On input the security parameter, KGen outputs a public/secret key pair (pk, sk). Sign takes as input sk (and we write this as a short-hand notation Sign_{sk}) and a message m, and outputs a signature σ. Verify takes as

---

$\mathcal{G}_{\mathsf{sRO}}$ – functionality for the strict global random oracle.

Parameters: output size $\ell(n)$
Variables: initially empty list $\mathsf{List}_{\mathcal{H}}$

1. **Query:** on input $(\mathsf{HashQuery}, m)$ from a machine $(\mathcal{P}, \mathsf{sid})$, proceed as follows.
   - Find h such that $(m, h) \in \mathsf{List}_{\mathcal{H}}$. If no such h exists, let $h \xleftarrow{\$} \{0, 1\}^{\ell(n)}$ and store (m, h) in $\mathsf{List}_{\mathcal{H}}$.
   - Output $(\mathsf{HashConfirm}, h)$ to $(\mathcal{P}, \mathsf{sid})$.

Fig. 1. The strict global random oracle functionality $\mathcal{G}_{\mathsf{sRO}}$ that does not give any extra power to anyone (mentioned but not defined by Canetti et al. [15]).

input a public key pk (and we write this as a shorthand notation $\mathsf{Verify}_{\mathsf{pk}}$), a message m and a signature σ, and outputs a single bit denoting acceptance or rejection of the signature. The standard security notion we assume for signature schemes is *existential unforgeability under chosen message attacks (EUF-CMA)* [20], which we recall here briefly. In such game-based security notion, an adversary is allowed to perform a number of signature queries, adaptively, on messages of his choice for a secret key generated by a challenger. Then, he wins the game if he manages to output a valid signature for a fresh message for that key. We say that a signature scheme is EUF-CMA secure if no PPT adversary can win this game with more than negligible probability.

Public-Key Encryption Schemes. A *public-key encryption scheme* is a tuple of PPT algorithms $\Pi = (\mathsf{KGen}, \mathsf{Enc}, \mathsf{Dec})$. On input n, KGen outputs a public/private key pair $(\mathsf{pk}, \mathsf{sk})$. Enc takes as input a public key pk (and we write this as a shorthand notation $\mathsf{Enc}_{\mathsf{pk}}$) and a plaintext m, and outputs a ciphertext c. Dec takes as input a secret key sk (and we write this as a shorthand notation $\mathsf{Dec}_{\mathsf{sk}}$) and a ciphertext c, and outputs either $\perp_m$ or a message m. The standard security notion we assume for public-key encryption schemes is *indistinguishability under adaptive chosen message attacks (IND-CCA2)* [2], which we recall here briefly. In such game-based security notion, an adversary sends a challenge plaintext of his choice to an external challenger, who generates a key pair and either responds to the adversary with an encryption of the challenge plaintext, or with the encryption of a random plaintext (having the same leakage as the original plaintext, in case we are considering corruption models), the goal of the adversary being to distinguish which is the case. We say that a PKE scheme is IND-CCA2 secure if no PPT adversary can win this game with more than negligible advantage over guessing, even if allowed to query adaptively a decryption oracle on any ciphertext of his choice – except the challenge ciphertext.

3 Strict Random Oracle

This section focuses on the so-called *strict* global random oracle $\mathcal{G}_{\mathsf{sRO}}$ depicted in Fig. 1, which is the most natural definition of a global random oracle: on a fresh

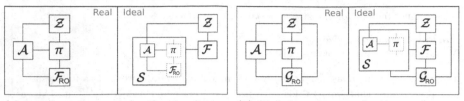

(a) Local random oracle: the simulator simulates the RO and has full control.

(b) Global random oracle: the random oracle is external to the simulator.

Fig. 2. The UC experiment with a local random oracle (a) and the EUC experiment with a global random oracle (b).

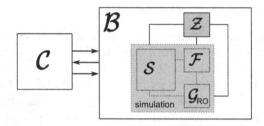

Fig. 3. Reduction $\mathcal{B}$ from a real-world adversary $\mathcal{A}$ and a black-box environment $\mathcal{Z}$, simulating all the ideal functionalities (even the global ones) and playing against an external challenger $\mathcal{C}$.

input m, a random value h is chosen, while on repeating inputs, a consistent answer is given back. This natural definition was discussed by Canetti et al. [15] but discarded as it does not suffice to realize $\mathcal{F}_{\mathsf{COM}}$. While this is true, we will argue that $\mathcal{G}_{\mathsf{sRO}}$ is still useful to realize other functionalities.

The code of $\mathcal{G}_{\mathsf{sRO}}$ is identical to that of a *local* random oracle $\mathcal{F}_{\mathrm{RO}}$ in UC. In Basic UC, this is a very strong definition, as it gives the simulator a lot of power: In the ideal world, it can simulate the random oracle $\mathcal{F}_{\mathrm{RO}}$, which gives it the ability to observe all queries and program the random oracle on the fly (cf. Fig. 2(a)). In GUC, the global random oracle $\mathcal{G}_{\mathsf{sRO}}$ is present in both worlds and the environment can access it (cf. Fig. 2(b)). In particular, the simulator is not given control of $\mathcal{G}_{\mathsf{sRO}}$ and hence cannot simulate it. Therefore, the simulator has no more power over the random oracle than explicitly offered through the interfaces of the global functionality. In the case of $\mathcal{G}_{\mathsf{sRO}}$, the simulator can neither program the random oracle, nor observe the queries made.

As the simulator obtains no relevant advantage over the real-world adversary when interacting with $\mathcal{G}_{\mathsf{sRO}}$, one might wonder how it could help in security proofs. The main observation is that the situation is different when one proves that the real and ideal world are indistinguishable. Here one needs to show that no environment can distinguish between the real and ideal world and thus, when doing so, one has full control over the global functionality. This is for instance the case when using the (distinguishing) environment in a cryptographic reduction:

as depicted in Fig. 3, the reduction algorithm $\mathcal{B}$ simulates the complete view of the environment $\mathcal{Z}$, including the global $\mathcal{G}_{sRO}$, allowing $\mathcal{B}$ to freely observe and program $\mathcal{G}_{sRO}$. As a matter of facts, $\mathcal{B}$ can also rewind the environment here – another power that the simulator $\mathcal{S}$ does not have but is useful in the security analysis of many schemes. It turns out that for some primitives, the EUC simulator does not need to program or observe the random oracle, but only needs to do so when proving that no environment can distinguish between the real and the ideal world.

This allows us to prove a surprisingly wide range of practical protocols secure with respect to $\mathcal{G}_{sRO}$. First, we prove that any signature scheme proven to be EUF-CMA in the local random-oracle model yields UC secure signatures with respect the global $\mathcal{G}_{sRO}$. Second, we show that any public-key encryption scheme proven to be IND-CCA2 secure with local random oracles yields UC secure public-key encryption (with respect to static corruptions), again with the global $\mathcal{G}_{sRO}$. These results show that highly practical schemes such as Schnorr signatures [23], RSA full-domain hash signatures [3,16], RSA-PSS signatures [5], RSA-OAEP encryption [4], and the Fujisaki-Okamoto transform [19] all remain secure when all schemes share a single hash function that is modeled as a strict global random oracle. This is remarkable, as their security proofs in the local random-oracle model involve techniques that are not available to an EUC simulator: signature schemes typically require programming of random-oracle outputs to simulate signatures, PKE schemes typically require observing the adversary's queries to simulate decryption queries, and Schnorr signatures need to rewind the adversary in a forking argument [22] to extract a witness. However, it turns out, this rewinding is only necessary in the reduction $\mathcal{B}$ showing that no distinguishing environment $\mathcal{Z}$ can exist and we can show that all these schemes can safely be used in composition with arbitrary protocols and with a natural, globally accessible random-oracle functionality $\mathcal{G}_{sRO}$.

3.1 Composable Signatures Using $\mathcal{G}_{sRO}$

Let $\mathsf{SIG} = (\mathsf{KGen}, \mathsf{Sign}, \mathsf{Verify})$ be an EUF-CMA secure signature scheme in the ROM. We show that this directly yields a secure realization of UC signatures $\mathcal{F}_{SIG}$ with respect to a strict global random oracle $\mathcal{G}_{sRO}$. We assume that SIG uses a single random oracle that maps to $\{0,1\}^{\ell(n)}$. Protocols requiring multiple random oracles or mapping into different ranges can be constructed using standard domain separation and length extension techniques.

We define π_{SIG} to be SIG phrased as a GUC protocol. Whenever an algorithm of SIG makes a call to a random oracle, π_{SIG} makes a call to $\mathcal{G}_{sRO}$.

1. On input (KeyGen, sid), signer $\mathcal{P}$ proceeds as follows.
   - Check that $sid = (\mathcal{P}, sid')$ for some sid', and no record (sid, sk) exists.
   - Run $(pk, sk) \leftarrow \mathsf{SIG.KGen}(n)$ and store (sid, sk).
   - Output $(\mathsf{KeyConf}, sid, pk)$.
2. On input (Sign, sid, m), signer $\mathcal{P}$ proceeds as follows.
   - Retrieve record (sid, sk), abort if no record exists.

$\mathcal{F}_{\mathsf{SIG}}$ – functionality for public-key signatures.
Variables: initially empty records keyrec and sigrec.

1. **Key Generation.** On input (KeyGen, sid) from a party $\mathcal{P}$.
   - If sid $\neq (\mathcal{P}, \mathsf{sid}')$ or a record (keyrec, sid, pk) exists, then abort.
   - Send (KeyGen, sid) to $\mathcal{A}$ and wait for (KeyConf, sid, pk) from $\mathcal{A}$. If a record (sigrec, sid, $*$, $*$, pk, $*$) exists, abort (*Consistency*).
   - Create record (keyrec, sid, pk).
   - Output (KeyConf, sid, pk) to $\mathcal{P}$.
2. **Signature Generation.** On input (Sign, sid, m) from $\mathcal{P}$.
   - If sid $\neq (\mathcal{P}, \mathsf{sid}')$ or no record (keyrec, sid, pk) exists, then abort.
   - Send (Sign, sid, m) to $\mathcal{A}$, and wait for (Signature, sid, σ) from $\mathcal{A}$.
   - If a record (sigrec, sid, m, σ, pk, false) exists, then abort.
   - Create record (sigrec, sid, m, σ, pk, true) (*Completeness*).
   - Output (Signature, sid, σ) to $\mathcal{P}$.
3. **Signature Verification.** On input (Verify, sid, m, σ, pk$'$) from some party $\mathcal{V}$.
   - If a record (sigrec, sid, m, σ, pk$'$, b) exists, set $f \leftarrow b$ (*Consistency*).
   - Else, if a record (keyrec, sid, pk) exists, $\mathcal{P}$ is honest, and no record (sigrec, sid, m, $*$, pk, true) exists, set $f \leftarrow 0$ (*Unforgeability*).
   - Else, send (Verify, sid, m, σ, pk$'$) to $\mathcal{A}$ and wait for (Verified, sid, b), and set $f \leftarrow b$.
   - Create a record (sigrec, sid, m, σ, pk$'$, f) and output (Verified, sid, f) to $\mathcal{V}$.

Fig. 4. The signature functionality $\mathcal{F}_{\mathsf{SIG}}$ due to Canetti [11]

   - Output (Signature, sid, σ) with $\sigma \leftarrow \mathsf{SIG.Sign}(\mathsf{sk}, m)$.
3. On input (Verify, sid, m, σ, pk$'$) a verifier $\mathcal{V}$ proceeds as follows.
   - Output (Verified, sid, f) with $f \leftarrow \mathsf{SIG.Verify}(\mathsf{pk}', \sigma, m)$.

We will prove that π_{SIG} will realize UC signatures. There are two main approaches to defining a signature functionality: using adversarially provided algorithms to generate and verify signature objects (e.g., the 2005 version of [9]), or by asking the adversary to create and verify signature objects (e.g., [11]). For a version using algorithms, the functionality will locally create and verify signature objects using the algorithm, without activating the adversary. This means that the algorithms cannot interact with external parties, and in particular communication with external functionalities such as a global random oracle is not permitted. We could modify an algorithm-based $\mathcal{F}_{\mathsf{SIG}}$ to allow the sign and verify algorithms to communicate *only with a global random oracle*, but we choose to use an $\mathcal{F}_{\mathsf{SIG}}$ that interacts with the adversary as this does not require special modifications for signatures with global random oracles.

Theorem 2. *If* SIG *is EUF-CMA in the random-oracle model, then* π_{SIG} *GUC-realizes* $\mathcal{F}_{\mathsf{SIG}}$ *(as defined in Fig. 4) in the* $\mathcal{G}_{\mathsf{sRO}}$*-hybrid model.*

Proof. By the fact that π_{SIG} is $\mathcal{G}_{\mathsf{sRO}}$-subroutine respecting and by Theorem 1, it is sufficient to show that π_{SIG} $\mathcal{G}_{\mathsf{sRO}}$-EUC-realizes $\mathcal{F}_{\mathsf{SIG}}$. We define the UC simulator $\mathcal{S}$ as follows.

1. **Key Generation.** On input (KeyGen, sid) from $\mathcal{F}_{SIG}$, where sid $= (\mathcal{P}, \text{sid}')$ and $\mathcal{P}$ is honest.
   - Simulate honest signer "$\mathcal{P}$", and give it input (KeyGen, sid).
   - When "$\mathcal{P}$" outputs (KeyConf, sid, pk) (where pk is generated according to π_{SIG}), send (KeyConf, sid, pk) to $\mathcal{F}_{SIG}$.
2. **Signature Generation.** On input (Sign, sid, m) from $\mathcal{F}_{SIG}$, where sid $= (\mathcal{P}, \text{sid}')$ and $\mathcal{P}$ is honest.
   - Run simulated honest signer "$\mathcal{P}$" with input (Sign, sid, m).
   - When "$\mathcal{P}$" outputs (Signature, sid, σ) (where σ is generated according to π_{SIG}), send (Signature, sid, σ) to $\mathcal{F}_{SIG}$.
3. **Signature Verification.** On input (Verify, sid, m, σ, pk') from $\mathcal{F}_{SIG}$, where sid $= (\mathcal{P}, \text{sid}')$.
   - Run $f \leftarrow$ SIG.Verify(pk', σ, m), and send (Verified, sid, f) to $\mathcal{F}_{SIG}$.

We must show that π_{SIG} realizes $\mathcal{F}_{SIG}$ in the Basic UC sense, but with respect to $\mathcal{G}_{sRO}$-externally constrained environments, i.e., the environment is now allowed to access $\mathcal{G}_{sRO}$ via dummy parties in sessions unequal to the challenge session. Without loss of generality, we prove this with respect to the dummy adversary.

During key generation, $\mathcal{S}$ invokes the simulated honest signer $\mathcal{P}$, so the resulting keys are exactly like in the real world. The only difference is that in the ideal world $\mathcal{F}_{SIG}$ can abort key generation in case the provided public key pk already appears in a previous sigrec record. But if this happens it means that $\mathcal{A}$ has successfully found a collision in the public key space, which must be exponentially large as the signature scheme is EUF-CMA by assumption. This means that such event can only happen with negligible probability.

For a corrupt signer, the rest of the simulation is trivially correct: the adversary generates keys and signatures locally, and if an honest party verifies a signature, the simulator simply executes the verification algorithm as a real world party would do, and $\mathcal{F}_{SIG}$ does not make further checks (the unforgeability check is only made when the signer is honest). When an honest signer signs, the simulator creates a signature using the real world signing algorithm, and when $\mathcal{F}_{SIG}$ asks the simulator to verify a signature, $\mathcal{S}$ runs the real world verification algorithm, and $\mathcal{F}_{SIG}$ keeps records of the past verification queries to ensure consistency. As the real world verification algorithm is deterministic, storing verification queries does not cause a difference. Finally, when $\mathcal{S}$ provides $\mathcal{F}_{SIG}$ with a signature, $\mathcal{F}_{SIG}$ checks that there is no stored verification query exists that states the provided signature is invalid. By completeness of the signature scheme, this check will never trigger.

The only remaining difference is that $\mathcal{F}_{SIG}$ prevents forgeries: if a verifier uses the correct public key, the signer is honest, and we verify a signature on a message that was never signed, $\mathcal{F}_{SIG}$ rejects. This would change the verification outcome of a signature that would be accepted by the real-world verification algorithm. As this event is the only difference between the real and ideal world, what remains to show is that this check changes the verification outcome only with negligible probability. We prove that if there is an environment that causes this event with non-negligible probability, then we can use it to construct a forger $\mathcal{B}$ that breaks the EUF-CMA unforgeability of SIG.

Our forger $\mathcal{B}$ plays the role of $\mathcal{F}_{\mathsf{SIG}}$, $\mathcal{S}$, and even the random oracle $\mathcal{G}_{\mathsf{sRO}}$, and has black-box access to the environment $\mathcal{Z}$. Then $\mathcal{B}$ receives a challenge public key pk and is given access to a signing oracle $\mathcal{O}^{\mathsf{Sign}(\mathsf{sk},\cdot)}$ and to a random oracle RO. It responds $\mathcal{Z}$'s $\mathcal{G}_{\mathsf{sRO}}$ queries by relaying queries and responses to and from RO. It runs the code of $\mathcal{F}_{\mathsf{SIG}}$ and $\mathcal{S}$, but uses $\mathcal{O}^{\mathsf{Sign}(\mathsf{sk},\mathsf{m})}$ instead of $\mathcal{F}_{\mathsf{SIG}}$'s signature generation interface to generate signatures. If the unforgeability check of $\mathcal{F}_{\mathsf{SIG}}$ triggers for a cryptographically valid signature σ on message m, then we know that $\mathcal{B}$ made no query $\mathcal{O}^{\mathsf{Sign}(\mathsf{sk},\mathsf{m})}$, meaning that $\mathcal{B}$ can submit (σ, m) to win the EUF-CMA game.   □

3.2 Composable Public-Key Encryption Using $\mathcal{G}_{\mathsf{sRO}}$

Let PKE = (KGen, Enc, Dec) be a CCA2 secure public-key encryption scheme in the ROM. We show that this directly yields a secure realization of GUC public-key encryption $\mathcal{F}^{\mathcal{L}}_{\mathsf{PKE}}$, as recently defined by Camenisch et al. [8] and depicted in Fig. 5, with respect to a strict global random oracle $\mathcal{G}_{\mathsf{sRO}}$ and static corruptions. As with our result for signature schemes, we require that PKE uses a single random oracle that maps to $\{0,1\}^{\ell(n)}$.

We define π_{PKE} to be PKE phrased as a GUC protocol.

1. On input (KeyGen, sid, n), party $\mathcal{P}$ proceeds as follows.
   - Check that sid = $(\mathcal{P}, \mathsf{sid}')$ for some sid', and no record (sid, sk) exists.
   - Run (pk, sk) ← PKE.KGen(n) and store (sid, sk).
   - Output (KeyConf, sid, pk).
2. On input (Encrypt, sid, pk', m), party $\mathcal{Q}$ proceeds as follows.
   - Set c ← PKE.Enc(pk', m) and output (Ciphertext, sid, c).
3. On input (Decrypt, sid, c), party $\mathcal{P}$ proceeds as follows.
   - Retrieve (sid, sk), abort if no such record exist.
   - Set m ← PKE.Dec(sk, c) and output (Plaintext, sid, m).

Theorem 3. *Protocol π_{PKE} GUC-realizes $\mathcal{F}^{\mathcal{L}}_{\mathsf{PKE}}$ with static corruptions with leakage function $\mathcal{L}$ in the $\mathcal{G}_{\mathsf{sRO}}$-hybrid model if PKE is CCA2 secure with leakage $\mathcal{L}$ in the ROM.*

Proof. By the fact that π_{PKE} is $\mathcal{G}_{\mathsf{sRO}}$-subroutine respecting and by Theorem 1, it is sufficient to show that π_{PKE} $\mathcal{G}_{\mathsf{sRO}}$-EUC-realizes $\mathcal{F}^{\mathcal{L}}_{\mathsf{PKE}}$.

We define simulator $\mathcal{S}$ as follows.

1. On input (KEYGEN, sid).
   - Parse sid as $(\mathcal{P}, \mathsf{sid}')$. Note that $\mathcal{P}$ is honest, as $\mathcal{S}$ does not make KeyGen queries on behalf of corrupt parties.
   - Invoke the simulated receiver "$\mathcal{P}$" on input (KeyGen, sid) and wait for output (KeyConf, sid, pk) from "$\mathcal{P}$".
   - Send (KeyConf, sid, pk) to $\mathcal{F}^{\mathcal{L}}_{\mathsf{PKE}}$.
2. On input (Enc-M, sid, pk', m) with $m \in \mathcal{M}$.
   - $\mathcal{S}$ picks some honest party "$\mathcal{Q}$" and gives it input (Encrypt, sid, pk', m). Wait for output (Ciphertext, sid, c) from "$\mathcal{Q}$".

$\mathcal{F}_{\mathsf{PKE}}^{\mathcal{L}}$ – functionality of public-key encryption with leakage function $\mathcal{L}$.

Parameters: message space $\mathcal{M}$
Variables: initially empty records keyrec, encrec, decrec.

1. KeyGen. On input (KeyGen, sid) from party $\mathcal{P}$:
   - If sid $\neq (\mathcal{P}, \text{sid}')$ or a record (keyrec, sid, pk) exists, then **abort**.
   - Send (KeyGen, sid) to $\mathcal{A}$ and wait for (KeyConf, sid, pk) from $\mathcal{A}$.
   - Create record (keyrec, sid, pk).
   - Output (KeyConf, sid, pk) to $\mathcal{P}$.
2. Encrypt. On input (Encrypt, sid, pk', m) from party $\mathcal{Q}$ with $m \in \mathcal{M}$:
   - Retrieve record (keyrec, sid, pk) for sid.
   - If pk' = pk and $\mathcal{P}$ is honest, then:
     - Send (Enc-L, sid, pk, $\mathcal{L}(m)$) to $\mathcal{A}$, and wait for (Ciphertext, sid, c) from $\mathcal{A}$.
     - If a record (encrec, sid, $\cdot$, c) exists, then **abort**.
     - Create record (encrec, sid, m, c).
   - Else (i.e., pk' $\neq$ pk or $\mathcal{P}$ is corrupt) then:
     - Send (Enc-M, sid, pk', m) to $\mathcal{A}$, and wait for (Ciphertext, sid, c) from $\mathcal{A}$.
   - Output (Ciphertext, sid, c) to $\mathcal{Q}$.
3. Decrypt. On input (Decrypt, sid, c) from party $\mathcal{P}$:
   - If sid $\neq (\mathcal{P}, \text{sid}')$ or no record (keyrec, sid, pk) exists, then **abort**.
   - If a record (encrec, sid, m, c) for c exists:
     - Output (Plaintext, sid, m) to $\mathcal{P}$.
   - Else (i.e., if no such record exists):
     - Send (Decrypt, sid, c) to $\mathcal{A}$ and wait for (Plaintext, sid, m) from $\mathcal{A}$.
     - Create record (encrec, sid, m, c).
     - Output (Plaintext, sid, m) to $\mathcal{P}$.

Fig. 5. The PKE functionality $\mathcal{F}_{\mathsf{PKE}}^{\mathcal{L}}$ with leakage function $\mathcal{L}$ [8,9].

   - Send (Ciphertext, sid, c) to $\mathcal{F}_{\mathsf{PKE}}^{\mathcal{L}}$.
3. On input (Enc-L, sid, pk, l).
   - $\mathcal{S}$ does not know which message is being encrypted, so it chooses a dummy plaintext $m' \in \mathcal{M}$ with $\mathcal{L}(m') = l$.
   - Pick some honest party "$\mathcal{Q}$" and give it input (Encrypt, sid, pk, m'), Wait for output (Ciphertext, sid, c) from "$\mathcal{Q}$".
   - Send (Ciphertext, sid, c) to $\mathcal{F}_{\mathsf{PKE}}^{\mathcal{L}}$.
4. On input (Decrypt, sid, c).
   - Note that $\mathcal{S}$ only receives such input when $\mathcal{P}$ is honest, and therefore $\mathcal{S}$ simulates "$\mathcal{P}$" and knows its secret key sk.
   - Give "$\mathcal{P}$" input (Decrypt, sid, c) and wait for output (Plaintext, sid, m) from "$\mathcal{P}$".
   - Send (Plaintext, sid, m) to $\mathcal{F}_{\mathsf{PKE}}^{\mathcal{L}}$.

What remains to show is that $\mathcal{S}$ is a satisfying simulator, i.e., no $\mathcal{G}_{\mathsf{sRO}}$-externally constrained environment can distinguish the real protocol π_{PKE} from $\mathcal{F}_{\mathsf{PKE}}^{\mathcal{L}}$ with $\mathcal{S}$. If the receiver $\mathcal{P}$ (i.e., such that sid $= (\mathcal{P}, \text{sid}')$) is corrupt, the simulation is trivially correct: $\mathcal{S}$ only creates ciphertexts when it knows the plaintext, so it can simply follow the real protocol. If $\mathcal{P}$ is honest, $\mathcal{S}$ does not know the message for which it is computing ciphertexts, so a dummy plaintext is encrypted. When the environment submits that ciphertext for decryption by $\mathcal{P}$, the functionality $\mathcal{F}^{\mathcal{L}}_{\mathsf{PKE}}$ will still return the correct message. Using a sequence of games, we show that if an environment exists that can notice this difference, it can break the CCA2 security of PKE.

Let Game 0 be the game where $\mathcal{S}$ and $\mathcal{F}^{\mathcal{L}}_{\mathsf{PKE}}$ act as in the ideal world, except that $\mathcal{F}^{\mathcal{L}}_{\mathsf{PKE}}$ passes the full message m in Enc-L inputs to $\mathcal{S}$, and $\mathcal{S}$ returns a real encryption of m as the ciphertext. It is clear that Game 0 is identical to the real world $\mathrm{EXEC}^{\mathcal{G}}_{\pi,\mathcal{A},\mathcal{Z}}$. Let Game i for $i = 1, \ldots, q_{\mathrm{E}}$, where q_{E} is the number of Encrypt queries made by $\mathcal{Z}$, be defined as the game where for $\mathcal{Z}$'s first i Encrypt queries, $\mathcal{F}^{\mathcal{L}}_{\mathsf{PKE}}$ passes only $\mathcal{L}(m)$ to $\mathcal{S}$ and $\mathcal{S}$ returns the encryption of a dummy message m' so that $\mathcal{L}(m') = \mathcal{L}(m)$, while for the $i + 1$-st to q_{E}-th queries, $\mathcal{F}^{\mathcal{L}}_{\mathsf{PKE}}$ passes m to $\mathcal{S}$ and $\mathcal{S}$ returns an encryption of m. It is clear that Game q_{E} is identical to the ideal world $\mathrm{IDEAL}^{\mathcal{G}}_{\mathcal{F},\mathcal{S},\mathcal{Z}}$.

By a hybrid argument, for $\mathcal{Z}$ to have non-negligible probability to distinguish between $\mathrm{EXEC}^{\mathcal{G}}_{\pi,\mathcal{A},\mathcal{Z}}$ and $\mathrm{IDEAL}^{\mathcal{G}}_{\mathcal{F},\mathcal{S},\mathcal{Z}}$, there must exist an i such that $\mathcal{Z}$ distinguishes with non-negligible probability between Game $(i-1)$ and Game i. Such an environment gives rise to the following CCA2 attacker $\mathcal{B}$ against PKE.

Algorithm $\mathcal{B}$ receives a challenge public key pk as input and is given access to decryption oracle $\mathcal{O}^{\mathsf{Dec}(\mathsf{sk},\cdot)}$ and random oracle RO. It answers $\mathcal{Z}$'s queries $\mathcal{G}_{\mathsf{sRO}}(m)$ by relaying responses from its own oracle $\mathrm{RO}(m)$ and lets $\mathcal{S}$ use pk as the public key of $\mathcal{P}$. It largely runs the code of Game $(i-1)$ for $\mathcal{S}$ and $\mathcal{F}^{\mathcal{L}}_{\mathsf{PKE}}$, but lets $\mathcal{S}$ respond to inputs $(\mathsf{Dec}, \mathsf{sid}, c)$ from $\mathcal{F}^{\mathcal{L}}_{\mathsf{PKE}}$ by calling its decryption oracle $m = \mathcal{O}^{\mathsf{Decrypt}(\mathsf{sk},c)}$. Note that $\mathcal{F}^{\mathcal{L}}_{\mathsf{PKE}}$ only hands such inputs to $\mathcal{S}$ for ciphertexts c that were *not* produced via the Encrypt interface of $\mathcal{F}^{\mathcal{L}}_{\mathsf{PKE}}$, as all other ciphertexts are handled by $\mathcal{F}^{\mathcal{L}}_{\mathsf{PKE}}$ itself.

Let m_0 denote the message that Functionality $\mathcal{F}^{\mathcal{L}}_{\mathsf{PKE}}$ hands to $\mathcal{S}$ as part of the i-th Enc-L input. Algorithm $\mathcal{B}$ now sets m_1 to be a dummy message m' such that $\mathcal{L}(m') = \mathcal{L}(m_0)$ and hands (m_0, m_1) to the challenger to obtain the challenge ciphertext c^* that is an encryption of m_b. It is clear that if $b = 0$, then the view of $\mathcal{Z}$ is identical to that in Game $(i-1)$, while if $b = 1$, it is identical to that in Game i. Moreover, $\mathcal{B}$ will never have to query its decryption oracle on the challenge ciphertext c^*, because any decryption queries for c^* are handled by $\mathcal{F}^{\mathcal{L}}_{\mathsf{PKE}}$ directly. By outputting 0 if $\mathcal{Z}$ decides it runs in Game $(i-1)$ and outputting 1 if $\mathcal{Z}$ decides it runs in Game i, $\mathcal{B}$ wins the CCA2 game with non-negligible probability. $\qquad\square$

4   Programmable Global Random Oracle

We now turn our attention to a new functionality that we call the *programmable global random oracle*, denoted by $\mathcal{G}_{\mathsf{pRO}}$. The functionality simply extends the strict random oracle $\mathcal{G}_{\mathsf{sRO}}$ by giving the adversary (real-world adversary $\mathcal{A}$ and ideal-world adversary $\mathcal{S}$) the power to program input-output pairs. Because we are in GUC or EUC, that also means that the environment gets this power.

$\mathcal{G}_{\mathsf{pRO}}$ – functionality for the programmable global random oracle.

Parameters: output size $\ell(n)$
Variables: initially empty list $\mathsf{List}_{\mathcal{H}}$

1. Query: on input $(\mathsf{HashQuery}, m)$ from a machine $(\mathcal{P}, \mathsf{sid})$, proceed as follows.
   - Find h such that $(m, h) \in \mathsf{List}_{\mathcal{H}}$. If no such h exists, let $h \xleftarrow{\$} \{0,1\}^{\ell(n)}$ and store (m, h) in $\mathsf{List}_{\mathcal{H}}$.
   - Output $(\mathsf{HashConfirm}, h)$ to $(\mathcal{P}, \mathsf{sid})$.
2. Program: on input $(\mathsf{ProgramRO}, m, h)$ from adversary $\mathcal{A}$
   - If $\exists\, h' \in \{0,1\}^{\ell(n)}$ such that $(m, h') \in \mathsf{List}_{\mathcal{H}}$ and $h \neq h'$, then abort
   - Else, add (m, h) to $\mathsf{List}_{\mathcal{H}}$ and output $(\mathsf{ProgramConfirm})$ to $\mathcal{A}$

Fig. 6. The programmable global random oracle functionality $\mathcal{G}_{\mathsf{pRO}}$.

Thus, as in the case of $\mathcal{G}_{\mathsf{sRO}}$, the simulator is thus not given any extra power compared to the environment (through the adversary), and one might well think that this model would not lead to the realization of any useful cryptographic primitives either. To the contrary, one would expect that the environment being able to program outputs would interfere with security proofs, as it destroys many properties of the random oracle such as collision or preimage resistance.

As it turns out, we can actually realize public-key encryption secure against adaptive corruptions (also known as non-committing encryption) in this model: we prove that the PKE scheme of Camenisch et al. [8] GUC-realizes $\mathcal{F}_{\mathsf{PKE}}$ against adaptive corruptions in the $\mathcal{G}_{\mathsf{pRO}}$-hybrid model. The security proof works out because the simulator equivocates dummy ciphertexts by programming the random oracle on *random* points, which are unlikely to have been queried by the environment before.

4.1 The Programmable Global Random Oracle $\mathcal{G}_{\mathsf{pRO}}$

The functionality $\mathcal{G}_{\mathsf{pRO}}$ (cf. Fig. 6) is simply obtained from $\mathcal{G}_{\mathsf{sRO}}$ by adding an interface for the adversary to program the oracle on a single point at a time. To this end, the functionality $\mathcal{G}_{\mathsf{pRO}}$ keeps an internal list of preimage-value assignments and, if programming fails (because it would overwrite a previously taken value), the functionality aborts, i.e., it replies with an error message $\perp$.

Notice that our $\mathcal{G}_{\mathsf{pRO}}$ functionality does not guarantee common random-oracle properties such as collision resistance: an adversary can simply program collisions into $\mathcal{G}_{\mathsf{pRO}}$. However, this choice is by design, because we are interested in achieving security with the *weakest* form of a *programmable* global random oracle to see what can be achieved against the strongest adversary possible.

4.2 Public-Key Encryption with Adaptive Corruptions from $\mathcal{G}_{\mathsf{pRO}}$

We show that GUC-secure non-interactive PKE with adaptive corruptions (often referred to as non-committing encryption) is achievable in the hybrid $\mathcal{G}_{\mathsf{pRO}}$ model by proving the PKE scheme by Camenisch et al. [8] secure in this model. We recall the scheme in Fig. 7 based on the following building blocks:

- a family of one-way trapdoor permutations OWTP = (OWTP.Gen, OWTP.Sample, OWTP.Eval, OWTP.Invert), where domains Σ generated by OWTP.Gen(1^n) have cardinality at least 2^n;
- a block encoding scheme (EC, DC), where EC : $\{0,1\}^* \to (\{0,1\}^{\ell(n)})^*$ is an encoding function such that the number of blocks that it outputs for a given message m depends only on the leakage $\mathcal{L}(m)$, and DC its deterministic inverse (possibly rejecting with $\bot$ if no preimage exists).

π_{PKE} – public-key encryption secure against adaptive corruptions.

Parameters: block size $\ell(n)$

1. **KeyGen.** On input (KeyGen, sid) from party $\mathcal{P}$:
   - Check that sid $= (\mathcal{P}, \mathsf{sid}')$ and no record (keyrec, sid, sk) exist.
   - Sample $(\varphi, \varphi^{-1}, \Sigma) \leftarrow$ OWTP.Gen(1^n).
   - Set pk $\leftarrow (\varphi, \Sigma)$, sk $\leftarrow (\varphi, \varphi^{-1}, \Sigma)$.
   - Create record (keyrec, sid, pk, sk).
   - Output (KeyConf, sid, pk) to $\mathcal{P}$.
2. **Encrypt.** On input (Encrypt, sid, pk$'$, m) from party $\mathcal{Q}$:
   - Parse pk$'$ as (φ, Σ), get $(m_1, \ldots, m_k) \leftarrow$ EC(m) and $x \leftarrow$ OWTP.Sample(Σ).
   - Let $c_1 \leftarrow$ OWTP.Eval(Σ, φ, x), $c_{2,i} \leftarrow m_i \oplus h_i, \forall i = 1, \ldots, k$, and $c_3 \leftarrow h$ where h and all h_i are obtained as (HashConfirm, h_i) $\leftarrow \mathcal{G}_{\mathsf{pRO}}$(HashQuery, $(x\|i)$) and (HashConfirm, h) $\leftarrow \mathcal{G}_{\mathsf{pRO}}$(HashQuery, $(x\|k\|m)$), respectively.
   - Set $c \leftarrow (c_1, c_{2,1}, \ldots, c_{2,k}, c_3)$.
   - Output (Ciphertext, sid, c) to $\mathcal{Q}$.
3. **Decrypt.** On input (Decrypt, sid, c) from party $\mathcal{P}$:
   - Check that sid $= (\mathcal{P}, \mathsf{sid}')$ and (keyrec, sid, sk) exist, if not, then **abort**.
   - Parse sk as $(\varphi, \varphi^{-1}, \Sigma)$, and c as $(c_1, c_{2,1}, \ldots, c_{2,k}, c_3)$.
   - Set $x' \leftarrow$ OWTP.Invert($\Sigma, \varphi, \varphi^{-1}, c_1$), $m_i' \leftarrow c_{2,i} \oplus h_i'$ for $i = 1, \ldots, k$, and $m' \leftarrow$ DC($m_1', \ldots, m_k'$), where all h_i' are obtained as (HashConfirm, h_i') $\leftarrow \mathcal{G}_{\mathsf{pRO}}$(HashQuery, $(x'\|i)$).
   - If $m' = \bot_m$ or $h' \neq c_3$, then output (Plaintext, sid, $\bot_m$) to $\mathcal{P}$, where h' is obtained from (HashConfirm, h') $\leftarrow \mathcal{G}_{\mathsf{pRO}}$(HashQuery, $(x'\|k\|m')$).
   - Else, output (Plaintext, sid, m) to $\mathcal{P}$.

Fig. 7. Public-key encryption scheme secure against adaptive attacks [8] based on one-way permutation OWTP and encoding function (EC, DC).

Theorem 4. *Protocol π_{PKE} in Fig. 7 GUC-realizes $\mathcal{F}_{\mathsf{PKE}}$ with adaptive corruptions and leakage function $\mathcal{L}$ in the $\mathcal{G}_{\mathsf{pRO}}$-hybrid model.*

Proof. We need to show that π_{PKE} GUC-realizes $\mathcal{F}_{\mathsf{PKE}}^{\mathcal{L}}$, i.e., that, given any environment $\mathcal{Z}$ and any real-world adversary $\mathcal{A}$, there exists a simulator $\mathcal{S}$ such that the output distribution of $\mathcal{Z}$ interacting with $\mathcal{F}_{\mathsf{PKE}}^{\mathcal{L}}$, $\mathcal{G}_{\mathsf{pRO}}$, and $\mathcal{S}$ is indistinguishable from its output distribution when interacting with π_{PKE}, $\mathcal{G}_{\mathsf{pRO}}$, and $\mathcal{A}$. Because π_{PKE} is $\mathcal{G}_{\mathsf{sRO}}$-subroutine respecting, by Theorem 1 it suffices to show that π_{PKE} $\mathcal{G}_{\mathsf{pRO}}$-EUC-realizes $\mathcal{F}_{\mathsf{PKE}}^{\mathcal{L}}$.

The simulator $\mathcal{S}$ is depicted in Fig. 8. Basically, it generates an honest key pair for the receiver and responds to Enc-M and Decrypt inputs by using the honest encryption and decryption algorithms, respectively. On Enc-L inputs, however, it creates a dummy ciphertext c composed of $c_1 = \varphi(x)$ for a freshly sampled x (but rejecting values of x that were used before) and randomly chosen $c_{2,1}, \ldots, c_{2,k}$ and c_3 for the correct number of blocks k. Only when either the secret key or the randomness used for this ciphertext must be revealed to the adversary, i.e., only when either the receiver or the party $\mathcal{Q}$ who created the ciphertext is corrupted, does the simulator program the random oracle so that the dummy ciphertext decrypts to the correct message m. If the receiver is corrupted, the simulator obtains m by having it decrypted by $\mathcal{F}_{\mathsf{PKE}}$; if the encrypting party $\mathcal{Q}$ is corrupted, then m is included in the history of inputs and outputs that is handed to $\mathcal{S}$ upon corruption. The programming is done through the Program subroutine, but the simulation aborts in case programming fails, i.e., when a point needs to be programmed that is already assigned. We will prove in the reduction that any environment causing this to happen can be used to break the one-wayness of the trapdoor permutation.

We now have to show that $\mathcal{S}$ successfully simulates a real execution of the protocol π_{PKE} to a real-world adversary $\mathcal{A}$ and environment $\mathcal{Z}$. To see this, consider the following sequence of games played with $\mathcal{A}$ and $\mathcal{Z}$ that gradually evolve from a real execution of π_{PKE} to the simulation by $\mathcal{S}$.

Let Game 0 be a game that is generated by letting an ideal functionality $\mathcal{F}_0$ and a simulator $\mathcal{S}_0$ collaborate, where $\mathcal{F}_0$ is identical to $\mathcal{F}_{\mathsf{PKE}}^{\mathcal{L}}$, except that it passes the full message m along with Enc-L inputs to $\mathcal{S}_0$. The simulator $\mathcal{S}_0$ simply performs all key generation, encryption, and decryption using the real algorithms, without any programming of the random oracle. The only difference between Game 0 and the real world is that the ideal functionality $\mathcal{F}_0$ aborts when the same ciphertext c is generated twice during an encryption query for the honest public key. Because $\mathcal{S}_0$ generates honest ciphertexts, the probability that the same ciphertext is generated twice can be bounded by the probability that two honest ciphertexts share the same first component c_1. Given that c_1 is computed as $\varphi(x)$ for a freshly sampled x from Σ, and given that x is uniformly distributed over Σ which has size at least 2^n, the probability of a collision occurring over q_E encryption queries is at most $q_E^2/2^n$.

Let Game 1 to Game q_E be games for a hybrid argument where gradually all ciphertexts by honest users are replaced with dummy ciphertexts. Let Game i be the game with a functionality $\mathcal{F}_i$ and simulator $\mathcal{S}_i$ where the first $i-1$ Enc-L inputs of $\mathcal{F}_i$ to $\mathcal{S}_i$ include only the leakage $\mathcal{L}(m)$, and the remaining such inputs include the full message. For the first $i-1$ encryptions, $\mathcal{S}_i$ creates a dummy ciphertext and programs the random oracle upon corruption of the party or the receiver as done by $\mathcal{S}$ in Fig. 8, aborting in case programming fails. For the remaining Enc-L inputs, $\mathcal{S}_i$ generates honest encryptions of the real message.

Parameters: leakage function $\mathcal{L}$, hash output length $\ell(n)$
Variables: initially empty list EncL
Subroutines: Program(m, c, r) depicted in Figure 9

1. On input (KeyGen, sid) from $\mathcal{F}_{\mathsf{PKE}}^{\mathcal{L}}$:
   - Sample $r \xleftarrow{\$} \{0, 1\}^n$ and honestly generate keys with randomness r by generating $(\Sigma, \varphi, \varphi^{-1}) \leftarrow$ OWTP.Gen$(n; r)$ and setting pk $\leftarrow (\Sigma, \varphi)$, sk $\leftarrow \varphi^{-1}$.
   - Record (pk, sk, r).
   - Send (KeyConf, sid, pk) to $\mathcal{F}_{\mathsf{PKE}}^{\mathcal{L}}$.
2. On input (Enc-L, sid, pk, λ) from $\mathcal{F}_{\mathsf{PKE}}^{\mathcal{L}}$:
   - Parse pk as (Σ, φ).
   - Sample $r \xleftarrow{\$} \{0, 1\}^n$ and generate $x \leftarrow$ OWTP.Sample$(\Sigma; r)$ until x does not appear in EncL.
   - Choose a dummy plaintext m such that $\mathcal{L}(m) = \lambda$ and let k be such that $(m_1, \ldots, m_k) \leftarrow$ EC(m).
   - Generate a dummy ciphertext c with $c_1 \leftarrow$ OWTP.Eval(Σ, φ, x) and with random $c_{2,1}, \ldots, c_{2,k}, c_3 \xleftarrow{\$} \{0, 1\}^{\ell(n)}$.
   - Record $(c, \perp_m, r, x, \mathsf{pk})$ in EncL.
   - Send (Ciphertext, sid, c) to $\mathcal{F}_{\mathsf{PKE}}^{\mathcal{L}}$.
3. On input (Enc-M, sid, pk', m) from $\mathcal{F}_{\mathsf{PKE}}^{\mathcal{L}}$:
   - Sample $r \xleftarrow{\$} \{0, 1\}^n$ and produce ciphertext c honestly from m using key pk' and randomness r.
   - Send (Ciphertext, sid, c) to $\mathcal{F}_{\mathsf{PKE}}^{\mathcal{L}}$.
4. On input (Decrypt, sid, c) from $\mathcal{F}_{\mathsf{PKE}}^{\mathcal{L}}$:
   - Decrypt c honestly using the recorded secret key sk to yield plaintext m.
   - Send (Plaintext, sid, m) to $\mathcal{F}_{\mathsf{PKE}}^{\mathcal{L}}$.
5. On corruption of party $\mathcal{Q}$, receive as input from $\mathcal{F}_{\mathsf{PKE}}^{\mathcal{L}}$ the history of $\mathcal{Q}$'s inputs and outputs, then compose $\mathcal{Q}$'s state as follows and hand it to $\mathcal{F}_{\mathsf{PKE}}^{\mathcal{L}}$:
   - For every input (Encrypt, sid, pk', m) and corresponding response (Ciphertext, sid, c) in $\mathcal{Q}$'s history:
     - If pk' $\neq$ pk, then include the randomness r that $\mathcal{S}$ used in the corresponding Enc-M query into $\mathcal{Q}$'s state.
     - If pk' $=$ pk, then
       * Find $(c, \perp_m, r, x, \mathsf{pk})$ in EncL, update it to $(c, m, r, x, \mathsf{pk})$, and include r into $\mathcal{Q}$'s state.
       * Execute Program(m, c, r).
   - If $\mathcal{Q}$ is the receiver, i.e., sid $= (\mathcal{Q}, \mathsf{sid}')$, then include the randomness r used at key generation into $\mathcal{Q}$'s state, and for all remaining $(c, \perp_m, r, x, \mathsf{pk})$ in EncL do:
     - Send (Decrypt, sid, c) to $\mathcal{F}_{\mathsf{PKE}}^{\mathcal{L}}$ in name of $\mathcal{Q}$ and wait for response (Plaintext, sid, m).
     - If $m \neq \perp_m$, then execute Program(m, c, r).
     - Update record $(c, \perp_m, r, x, \mathsf{pk})$ in EncL to $(c, m, r, x, \mathsf{pk})$

Fig. 8. The EUC simulator $\mathcal{S}$ for protocol π_{PKE}.

On input (m, c, r) do the following:

- Parse $(m_1, \ldots, m_k) := \mathsf{EC}(m)$, and $c := (c_1, c_{2,1}, \ldots, c_{2,k'}, c_3)$; let $x := \mathsf{OWTP.Sample}(\Sigma; r)$.
- For $i = 1, \ldots, k$:
  - Execute $\mathcal{G}_{\mathsf{pRO}}.\mathtt{Program}(x\|i, m_i \oplus c_{2,i})$; abort if unsuccessful.
- Execute $\mathcal{G}_{\mathsf{pRO}}.\mathtt{Program}(x\|k\|m, c_3)$; abort if unsuccessful.

Fig. 9. The oracle programming routine $\mathtt{Program}$.

One can see that Game q_E is identical to the ideal world with $\mathcal{F}_{\mathsf{PKE}}^{\mathcal{L}}$ and $\mathcal{S}$. To have a non-negligible advantage distinguishing the real from the ideal world, there must exist an $i \in \{1, \ldots, q_\mathrm{E}\}$ such that $\mathcal{Z}$ and $\mathcal{A}$ can distinguish between Game $(i-1)$ and Game i. These games are actually identical, *except* in the case that abort happens during the programming of the random oracle $\mathcal{G}_{\mathsf{pRO}}$ for the i-th ciphertext, which is a real ciphertext in Game $(i-1)$ and a dummy ciphertext in Game i. We call this the ROABORT event. We show that if there exists an environment $\mathcal{Z}$ and real-world adversary $\mathcal{A}$ that make ROABORT happen with non-negligible probability ν, then we can construct an efficient algorithm $\mathcal{B}$ (the "reduction") with black-box access to $\mathcal{Z}$ and $\mathcal{A}$ that is able to invert OWTP.

Our reduction $\mathcal{B}$ must only simulate honest parties, and in particular must provide to $\mathcal{A}$ a consistent view of their secrets (randomness used for encryption, secret keys, and decrypted plaintexts, just like $\mathcal{S}$ does) when they become corrupted. Moreover, since we are not in the idealized scenario, there is no external global random oracle functionality $\mathcal{G}_{\mathsf{pRO}}$: instead, $\mathcal{B}$ simulates $\mathcal{G}_{\mathsf{pRO}}$ for all the parties involved, and answers all their oracle calls.

Upon input the OWTP challenge (Σ, φ, y), $\mathcal{B}$ runs the code of Game $(i-1)$, but sets the public key of the receiver to $\mathsf{pk} = (\Sigma, \varphi)$. Algorithm $\mathcal{B}$ answers the first $i-1$ encryption requests with dummy ciphertexts and the $(i+1)$-st to q_E-th queries with honestly generated ciphertexts. For the i-th encryption request, however, it returns a special dummy ciphertext with $c_1 = y$.

To simulate $\mathcal{G}_{\mathsf{pRO}}$, $\mathcal{B}$ maintains an initially empty list $\mathsf{List}_{\mathcal{H}}$ to which pairs (m, h) are either added by lazy sampling for HashQuery queries, or by programming for ProgramRO queries. (Remember that the environment $\mathcal{Z}$ can program entries in $\mathcal{G}_{\mathsf{pRO}}$ as well.) For requests from $\mathcal{Z}$, $\mathcal{B}$ actually performs some additional steps that we describe further below.

It answers Decrypt requests for a ciphertext $c = (c_1, c_{2,1}, \ldots, c_{2,k}, c_3)$ by searching for a pair of the form $(x\|k\|m, c_3) \in \mathsf{List}_{\mathcal{H}}$ such that $\varphi(x) = c_1$ and $m = \mathsf{DC}(c_{2,1} \oplus h_1, \ldots, c_{2,k} \oplus h_k)$, where $h_j = \mathcal{H}(x\|j)$, meaning that h_j is assigned the value of a simulated request $(\mathsf{HashQuery}, x\|j)$ to $\mathcal{G}_{\mathsf{pRO}}$. Note that at most one such pair exists for a given ciphertext c, because if a second $(x'\|k\|m', c_3) \in \mathsf{List}_{\mathcal{H}}$ would exist, then it must hold that $\varphi(x') = c_1$. Because φ is a permutation, this means that $x = x'$. Since for each $j = 1, \ldots, k$, only one pair $(x\|j, h_j) \in \mathsf{List}_{\mathcal{H}}$ can be registered, this means that $m' = \mathsf{DC}(c_{2,1} \oplus h_1, \ldots, c_{2,k} \oplus h_k) = m$ because

DC is deterministic. If such a pair $(x\|k\|m, c_3)$ exists, it returns m, otherwise it rejects by returning $\perp_m$.

One problem with the decryption simulation above is that it does not necessarily create the same entries into $\mathsf{List}_{\mathcal{H}}$ as an honest decryption would have, and $\mathcal{Z}$ could detect this by checking whether programming for these entries succeeds. In particular, $\mathcal{Z}$ could first ask to decrypt a ciphertext $c = (\varphi(x), c_{2,1}, \ldots, c_{2,k}, c_3)$ for random $x, c_{2,1}, \ldots, c_{2,k}, c_3$ and then try to program the random oracle on any of the points $x\|j$ for $j = 1, \ldots, k$ or on $x\|k\|m$. In Game $(i-1)$ and Game i, such programming would fail because the entries were created during the decryption of c. In the simulation by $\mathcal{B}$, however, programming would succeed, because no valid pair $(x\|k\|m, c_3) \in \mathsf{List}_{\mathcal{H}}$ was found to perform decryption.

To preempt the above problem, $\mathcal{B}$ checks all incoming requests HashQuery and ProgramRO by $\mathcal{Z}$ for points of the form $x\|j$ or $x\|k\|m$ against all previous decryption queries $c = (c_1, c_{2,1}, \ldots, c_{2,k}, c_3)$. If $\varphi(x) = c_1$, then $\mathcal{B}$ immediately triggers (by mean of appropriate HashQuery calls) the creation of all random-oracle entries that would have been generated by a decryption of c by computing $m' = \mathsf{DC}(c_{2,1} \oplus \mathcal{H}(x\|1), \ldots, c_{2,k} \oplus \mathcal{H}(x\|k))$ and $c_3' = \mathcal{H}(x\|k\|m')$. Only then does $\mathcal{B}$ handle $\mathcal{Z}$'s original HashQuery or ProgramRO request.

The only remaining problem is if during this procedure $c_3' = c_3$, meaning that c was previously rejected during by $\mathcal{B}$, but it becomes a valid ciphertext by the new assignment of $\mathcal{H}(x\|k\|m) = c_3' = c_3$. This happens with negligible probability, though: a random value c_3' will only hit a fixed c_3 with probability $1/|\Sigma| \leq 1/2^n$. Since up to q_D ciphertexts may have been submitted with the same first component $c_1 = \varphi(x)$ and with different values for c_3, the probability that it hits any of them is at most $q_D/2^n$. The probability that this happens for at least one of $\mathcal{Z}$'s q_H HashQuery queries or one of its q_P ProgramRO queries during the entire execution is at most $(q_H + q_P)q_D/2^n$.

When $\mathcal{A}$ corrupts a party, $\mathcal{B}$ provides the encryption randomness that it used for all ciphertexts that such party generated. If $\mathcal{A}$ corrupts the receiver or the party that generated the i-th ciphertext, then $\mathcal{B}$ cannot provide that randomness. Remember, however, that $\mathcal{B}$ is running $\mathcal{Z}$ and $\mathcal{A}$ in the hope for the ROABORT event to occur, meaning that the programming of values for the i-th ciphertext fails because the relevant points in $\mathcal{G}_{\mathsf{pRO}}$ have been assigned already. Event ROABORT can only occur at the corruption of either the receiver or of the party that generated the i-th ciphertext, whichever comes first. Algorithm $\mathcal{B}$ therefore checks $\mathsf{List}_{\mathcal{H}}$ for points of the form $x\|j$ or $x\|k\|m$ such that $\varphi(x) = y$. If ROABORT occurred, then $\mathcal{B}$ will find such a point and output x as its preimage for y. If it did not occur, then $\mathcal{B}$ gives up. Overall, $\mathcal{B}$ will succeed whenever ROABORT occurs. Given that Game $(i-1)$ and Game i are different only when ROABORT occurs, and given that $\mathcal{Z}$ and $\mathcal{A}$ have non-negligible probability of distinguishing between Game $(i-1)$ and Game i, we conclude that $\mathcal{B}$ succeeds with non-negligible probability. $\qquad\square$

5   Restricted Programmable Global Random Oracles

The strict and the programmable global random oracles, $\mathcal{G}_{sRO}$ and $\mathcal{G}_{pRO}$, respectively, do not give the simulator any extra power compared to the real world adversary/environment. Canetti and Fischlin [13] proved that it is impossible to realize UC commitments without a setup assumption that gives the simulator an advantage over the environment. This means that, while $\mathcal{G}_{sRO}$ and $\mathcal{G}_{pRO}$ allowed for security proofs of many practical schemes, we cannot hope to realize even the seemingly simple task of UC commitments with this setup. In this section, we turn our attention to programmable global random oracles that do grant an advantage to the simulator.

5.1   Restricting Programmability to the Simulator

Canetti et al. [15] defined a global random oracle that restricts observability only adversarial queries, (hence, we call it the *restricted observable global random oracle* $\mathcal{G}_{roRO}$), and show that this is sufficient to construct UC commitments. More precisely, if sid is the identifier of the challenge session, a list of so-called *illegitimate* queries for sid can be obtained by the adversary, which are queries made on inputs of the form (sid,...) by machines that are not part of session sid. If honest parties only make legitimate queries, then clearly this restricted observability will not give the adversary any new information, as it contains only queries made by the adversary. In the ideal world, however, the simulator S can observe all queries made through corrupt machines within the challenge session sid as it is the ideal-world attacker, which means it will see all legitimate queries in sid. With the observability of illegitimate queries, that means S can observe *all* hash queries of the form (sid,...), regardless of whether they are made by honest or corrupt parties, whereas the real-world attacker does not learn anything form the observe interface.

We recall the restricted observable global random oracle $\mathcal{G}_{roRO}$ due to Canetti et al. [15] in a slightly modified form in Fig. 10. In their definition, it allows *ideal functionalities* to obtain the illegitimate queries corresponding to their own session. These functionalities then allow the adversary to obtain the illegitimate queries by forwarding the request to the global random oracle. Since the adversary can spawn any new machine, and in particular an ideal functionality, the adversary can create such an ideal functionality and use it to obtain the illegitimate queries. We chose to explicitly model this adversarial power by allowing the adversary to query for the illegitimate queries directly.

Also in Fig. 10, we define a *restricted* programmable global random oracle $\mathcal{G}_{rpRO}$ by using a similar approach to restrict programming access from the real-world adversary. The adversary can program points, but parties in session sid can *check* whether the random oracle was programmed on a particular point (sid,...). In the real world, the adversary is allowed to program, but honest parties can check whether points were programmed and can, for example, reject signatures based on a programmed hash. In the ideal world, the simulator controls the corrupt parties in sid and is therefore the only entity that can check whether

$\mathcal{G}_{\mathsf{roRO}}$, $\mathcal{G}_{\mathsf{rpRO}}$, and $\mathcal{G}_{\mathsf{rpoRO}}$ – functionalities of the global random oracle with restricted programming and/or restricted observability.

Parameters: output size function ℓ.
Variables: initially empty lists $\mathsf{List}_{\mathcal{H}}$, prog.

1. **Query.** On input $(\mathsf{HashQuery}, m)$ from a machine $(\mathcal{P}, \mathsf{sid})$ or from the adversary:
   - Look up h such that $(m, h) \in \mathsf{List}_{\mathcal{H}}$. If no such h exists:
     - draw $h \xleftarrow{\$} \{0, 1\}^{\ell(n)}$
     - set $\mathsf{List}_{\mathcal{H}} := \mathsf{List}_{\mathcal{H}} \cup \{(m, h)\}$
   - Parse m as (s, m').
   - If this query is made by the adversary, or if $s \neq \mathsf{sid}$, then add (s, m', h) to the (initially empty) list of illegitimate queries $\mathcal{Q}_s$.
   - Output $(\mathsf{HashConfirm}, h)$ to the caller.
2. **Observe.** ($\mathcal{G}_{\mathsf{roRO}}$ and $\mathcal{G}_{\mathsf{rpoRO}}$ only) On input $(\mathsf{Observe}, \mathsf{sid})$ from the adversary:
   - If $\mathcal{Q}|_{\mathsf{sid}}$ does not exist yet, then set $\mathcal{Q}_{\mathsf{sid}} = \varnothing$.
   - Output $(\mathsf{ListObserve}, \mathcal{Q}_{\mathsf{sid}})$ to the adversary.
3. **Program.** ($\mathcal{G}_{\mathsf{rpRO}}$ and $\mathcal{G}_{\mathsf{rpoRO}}$ only) On input $(\mathsf{ProgramRO}, m, h)$ with $h \in \{0, 1\}^{\ell(n)}$ from the adversary:
   - If $\exists\, h' \in \{0, 1\}^{\ell(n)}$ such that $(m, h') \in \mathsf{List}_{\mathcal{H}}$ and $h \neq h'$, ignore this input.
   - Set $\mathsf{List}_{\mathcal{H}} := \mathsf{List}_{\mathcal{H}} \cup \{(m, h)\}$ and $\mathsf{prog} := \mathsf{prog} \cup \{m\}$.
   - Output $(\mathsf{ProgramConfirm})$ to the adversary.
4. **IsProgrammed:** ($\mathcal{G}_{\mathsf{rpRO}}$ and $\mathcal{G}_{\mathsf{rpoRO}}$ only) On input $(\mathsf{IsProgrammed}, m)$ from a machine $(\mathcal{P}, \mathsf{sid})$ or from the adversary:
   - If the input was given by $(\mathcal{P}, \mathsf{sid})$, parse m as (s, m'). If $s \neq \mathsf{sid}$, ignore this input.
   - Set $b \leftarrow m \in \mathsf{prog}$ and output $(\mathsf{IsProgrammed}, b)$ to the caller.

Fig. 10. The global random-oracle functionalities $\mathcal{G}_{\mathsf{roRO}}$, $\mathcal{G}_{\mathsf{rpRO}}$, and $\mathcal{G}_{\mathsf{rpoRO}}$ with restricted observability, restricted programming, and combined restricted observability and programming, respectively. Functionality $\mathcal{G}_{\mathsf{roRO}}$ contains only the **Query** and **Observe** interfaces, $\mathcal{G}_{\mathsf{rpRO}}$ contains only the **Query**, **Program**, and **IsProgrammed** interfaces, and $\mathcal{G}_{\mathsf{rpoRO}}$ contains all interfaces.

points are programmed. Note that while it typically internally simulates the real-world adversary that may want to check whether points of the form $(\mathsf{sid}, \ldots)$ are programmed, the simulator can simply "lie" and pretend that no points are programmed. Therefore, the extra power that the simulator has over the real-world adversary is programming points without being detected.

It may seem strange to offer a new interface allowing all parties to check whether certain points are programmed, even though a real-world hash function does not have such an interface. However, we argue that if one accepts a programmable random oracle as a proper idealization of a clearly non-programmable real-world hash function, then it should be a small step to accept the instantiation of the **IsProgrammed** interface that always returns "false" to the question whether any particular entry was programmed into the hash function.

5.2 UC-Commitments from $\mathcal{G}_{\mathsf{rpRO}}$

We now show that we can create a UC-secure commitment protocol from $\mathcal{G}_{\mathsf{rpRO}}$. A UC-secure commitment scheme must allow the simulator to extract the message from adversarially created commitments, and to equivocate dummy commitments created for honest committers, i.e., first create a commitment that it can open to any message after committing. Intuitively, achieving the equivocability with a programmable random oracle is simple: we can define a commitment that uses the random-oracle output, and the adversary can later change the committed message by programming the random oracle. Achieving extractability, however, seems difficult, as we cannot extract by observing the random-oracle queries. We overcome this issue with the following approach. The receiver of a commitment chooses a nonce on which we query random oracle, interpreting the random oracle output as a public key pk. Next, the committer encrypts the message to pk and sends the ciphertext to the receiver, which forms the commitment. To open, the committer reveals the message and the randomness used to encrypt it.

This solution is extractable as the simulator that plays the role of receiver can program the random oracle such that it knows the secret key corresponding to pk, and simply decrypt the commitment to find the message. However, we must take care to still achieve equivocability. If we use standard encryption, the simulator cannot open a ciphertext to any value it learns later. The solution is to use *non-committing encryption*, which, as shown in Sect. 4, can be achieved using a programmable random oracle. We use a slightly different encryption scheme, as the security requirements here are slightly less stringent than full non-committing encryption, and care must be taken that we can interpret the result of the random oracle as a public key, which is difficult for constructions based on trapdoor one-way permutations such as RSA. This approach results in a very efficient commitment scheme: with two exponentiations per party (as opposed to five) and two rounds of communication (as opposed to five), it is considerably more efficient than the one of [15].

Let $\mathsf{COM}_{\mathcal{G}_{\mathsf{rpRO}}}$ be the following commitment protocol, parametrized by a group $\mathbb{G} = \langle g \rangle$ of prime order q. We require an algorithm Embed that maps elements of $\{0,1\}^{\ell(n)}$ into $\mathbb{G}$, such that for $h \xleftarrow{\$} \{0,1\}^{\ell(n)}$, Embed($h$) is statistically close to uniform in $\mathbb{G}$. Furthermore, we require an efficiently computable probabilistic algorithm Embed^{-1}, such that for all $x \in \mathbb{G}$, $\mathsf{Embed}(\mathsf{Embed}^{-1}(x)) = x$ and for $x \xleftarrow{\$} \mathbb{G}$, $\mathsf{Embed}^{-1}(x)$ is statistically close to uniform in $\{0,1\}^{\ell(n)}$. $\mathsf{COM}_{\mathcal{G}_{\mathsf{rpRO}}}$ assumes authenticated channels $\mathcal{F}_{\mathsf{auth}}$ as defined by Canetti [9].

1. On input (Commit, sid, x), party $\mathcal{C}$ proceeds as follows.
    - Check that sid $= (\mathcal{C}, \mathcal{R}, \mathsf{sid}')$ for some $\mathcal{C}$, sid'. Send Commit to $\mathcal{R}$ over $\mathcal{F}_{\mathsf{auth}}$ by giving $\mathcal{F}_{\mathsf{auth}}$ input (Send, $(\mathcal{C}, \mathcal{R}, \mathsf{sid}, 0)$, "Commit").
    - $\mathcal{R}$, upon receiving (Sent, $(\mathcal{C}, \mathcal{R}, \mathsf{sid}, 0)$, "Commit") from $\mathcal{F}_{\mathsf{auth}}$, takes a nonce $n \xleftarrow{\$} \{0,1\}^n$ and sends the nonce back to $\mathcal{C}$ by giving $\mathcal{F}_{\mathsf{auth}}$ input (Send, $(\mathcal{R}, \mathcal{C}, \mathsf{sid}, 0)$, n).

$\mathcal{F}_{\mathsf{COM}}$ – functionality for interactive commitments.

1. **Commit:** on input (Commit, sid, x) from a party $\mathcal{C}$ proceed as follows.
   - Check that sid $= (\mathcal{C}, \mathcal{R}, \mathsf{sid}')$.
   - Store x and generate public delayed output (Receipt, sid) to $\mathcal{R}$. Ignore subsequent Commit inputs.
2. **Open:** on input (Open, sid) from $\mathcal{C}$ proceed as follows.
   - Check that a committed value x is stored.
   - Generate public delayed output (Open, sid, x) to $\mathcal{R}$.

Fig. 11. The commitment functionality $\mathcal{F}_{\mathsf{COM}}$ by Canetti [9].

- $\mathcal{C}$, upon receiving (Sent, $(\mathcal{R}, \mathcal{C}, \mathsf{sid}, 0), n$), queries $\mathcal{G}_{\mathsf{rpRO}}$ on (sid, n) to obtain h_n. It checks whether this point was programmed by giving $\mathcal{G}_{\mathsf{roRO}}$ input (IsProgrammed, (sid, n)) and aborts if $\mathcal{G}_{\mathsf{roRO}}$ returns (IsProgrammed, 1).
- Set pk $\leftarrow$ Embed(h_n).
- Pick a random $r \xleftarrow{\$} \mathbb{G}$ and $\rho \in \mathbb{Z}_q$. Set $c_1 \leftarrow g^r$, query $\mathcal{G}_{\mathsf{rpRO}}$ on (sid, pk^r) to obtain h_r and let $c_2 \leftarrow h_r \oplus x$.
- Store (r, x) and send the commitment to $\mathcal{R}$ by giving $\mathcal{F}_{\mathsf{auth}}$ input (Send, $(\mathcal{C}, \mathcal{R}, \mathsf{sid}, 1), (c_1, c_2)$).
- $\mathcal{R}$, upon receiving (Sent, $(\mathcal{C}, \mathcal{R}, \mathsf{sid}, 1), (c_1, c_2)$) from $\mathcal{F}_{\mathsf{auth}}$ outputs (Receipt, sid).

2. On input (Open, sid), $\mathcal{C}$ proceeds as follows.
   - It sends (r, x) to $\mathcal{R}$ by giving $\mathcal{F}_{\mathsf{auth}}$ input (Send, $(\mathcal{C}, \mathcal{R}, \mathsf{sid}, 2), (r, x)$).
   - $\mathcal{R}$, upon receiving (Sent, $(\mathcal{C}, \mathcal{R}, \mathsf{sid}, 1), (r, x)$):
     • Query $\mathcal{G}_{\mathsf{rpRO}}$ on (sid, n) to obtain h_n and let pk $\leftarrow$ Embed(h_n).
     • Check that $c_1 = g^r$.
     • Query $\mathcal{G}_{\mathsf{rpRO}}$ on (sid, pk^r) to obtain h_r and check that $c_2 = h_r \oplus x$.
     • Check that none of the points was programmed by giving $\mathcal{G}_{\mathsf{roRO}}$ inputs (IsProgrammed, (sid, n)) and (IsProgrammed, pk^r) and asserting that it returns (IsProgrammed, 0) for both queries.
     • Output (Open, sid, x).

$\mathsf{COM}_{\mathcal{G}_{\mathsf{rpRO}}}$ is a secure commitment scheme under the computational Diffie-Hellman assumption, which given a group $\mathbb{G}$ generated by g of prime order q, challenges the adversary to compute $g^{\alpha\beta}$ on input (g^α, g^β), with $(\alpha, \beta) \xleftarrow{\$} \mathbb{Z}_q^2$.

Theorem 5. $\mathsf{COM}_{\mathcal{G}_{\mathsf{rpRO}}}$ *GUC-realizes* $\mathcal{F}_{\mathsf{COM}}$ *(as defined in Fig. 11) in the* $\mathcal{G}_{\mathsf{roRO}}$ *and* $\mathcal{F}_{\mathsf{auth}}$ *hybrid model under the CDH assumption in* $\mathbb{G}$.

Proof. By the fact that $\mathsf{COM}_{\mathcal{G}_{\mathsf{rpRO}}}$ is $\mathcal{G}_{\mathsf{rpRO}}$-subroutine respecting and by Theorem 1, it is sufficient to show that $\mathsf{COM}_{\mathcal{G}_{\mathsf{rpRO}}}$ $\mathcal{G}_{\mathsf{rpRO}}$-EUC-realizes $\mathcal{F}_{\mathsf{COM}}$.

We describe a simulator $\mathcal{S}$ by defining its behavior in the different corruption scenarios. In all scenarios, whenever the simulated real-world adversary makes an IsProgrammed query or instructs a corrupt party to make such a query on a point that $\mathcal{S}$ has programmed, the simulator intercepts this query and simply replies (IsProgrammed, 0), lying that the point was not programmed.

When both the sender and the receiver are honest, $\mathcal{S}$ works as follows.

1. When $\mathcal{F}_{\text{COM}}$ asks $\mathcal{S}$ for permission to output (Receipt, sid):
   - Parse sid as $(\mathcal{C}, \mathcal{R}, \text{sid}')$ and let "$\mathcal{C}$" create a dummy commitment by choosing $r \overset{\$}{\leftarrow} \mathbb{Z}_q$, letting $c_1 = g^r$, choosing $c_2 \overset{\$}{\leftarrow} \{0,1\}^{\ell(n)}$.
   - When "$\mathcal{R}$" outputs (Receipt, sid), allow $\mathcal{F}_{\text{COM}}$ to proceed.
2. When $\mathcal{F}_{\text{COM}}$ asks $\mathcal{S}$ for permission to output (Open, sid, x):
   - Program $\mathcal{G}_{\text{rpRO}}$ by giving $\mathcal{G}_{\text{roRO}}$ input (ProgramRO, (sid, pk^r), $c_2 \oplus x$), such that the commitment (c_1, c_2) commits to x.
   - Give "$\mathcal{C}$" input (Open, sid) instructing it to open its commitment to x.
   - When "$\mathcal{R}$" outputs (Open, sid, x), allow $\mathcal{F}_{\text{COM}}$ to proceed.

If the committer is corrupt but the receiver is honest, $\mathcal{S}$ works as follows.

1. When the simulated receiver "$\mathcal{R}$" notices the commitment protocol starting (i.e., receives (Sent, $(\mathcal{C}, \mathcal{R}, \text{sid}, 0)$, "Commit") from "$\mathcal{F}_{\text{auth}}$"):
   - Choose nonce n as in the protocol.
   - Before sending n, choose $\text{sk} \overset{\$}{\leftarrow} \mathbb{Z}_q$ and set $\text{pk} \leftarrow g^{\text{sk}}$.
   - Program $\mathcal{G}_{\text{rpRO}}$ by giving $\mathcal{G}_{\text{rpRO}}$ input (ProgramRO, (sid, n), $\text{Embed}^{-1}(\text{pk})$). Note that this simulation will succeed with overwhelming probability as n is freshly chosen, and note that as pk is uniform in $\mathbb{G}$, by definition of Embed^{-1} the programmed value $\text{Embed}^{-1}(\text{pk})$ is uniform in $\{0,1\}^{\ell(n)}$.
   - $\mathcal{S}$ now lets "$\mathcal{R}$" execute the remainder the protocol honestly.
   - When "$\mathcal{R}$" outputs (Receipt, sid), $\mathcal{S}$ extracts the committed value from (c_1, c_2). Query $\mathcal{G}_{\text{rpRO}}$ on (sid, c_1^{sk}) to obtain h_r and set $x \leftarrow c_2 \oplus h_r$.
   - Make a query with $\mathcal{F}_{\text{COM}}$ on $\mathcal{C}$'s behalf by sending (Commit, sid, x) on $\mathcal{C}$'s behalf to $\mathcal{F}_{\text{COM}}$.
   - When $\mathcal{F}_{\text{COM}}$ asks permission to output (Receipt, sid), allow.
2. When "$\mathcal{R}$" outputs (Open, sid, x):
   - Send (Open, sid) on $\mathcal{C}$'s behalf to $\mathcal{F}_{\text{COM}}$.
   - When $\mathcal{F}_{\text{COM}}$ asks permission to output (Open, sid, x), allow.

If the receiver is corrupt but the committer is honest, $\mathcal{S}$ works as follows.

1. When $\mathcal{F}_{\text{COM}}$ asks permission to output (Receipt, sid):
   - Parse sid as $(\mathcal{C}, \mathcal{R}, \text{sid}')$.
   - Allow $\mathcal{F}_{\text{COM}}$ to proceed.
   - When $\mathcal{F}_{\text{COM}}$ receives (Receipt, sid) from $\mathcal{F}_{\text{COM}}$ as $\mathcal{R}$ is corrupt, it simulates "$\mathcal{C}$" by choosing $r \overset{\$}{\leftarrow} \mathbb{Z}_q$, computing $c_1 = g^r$, and choosing $c_2 \overset{\$}{\leftarrow} \{0,1\}^{\ell(n)}$.
2. When $\mathcal{F}_{\text{COM}}$ asks permission to output (Open, sid, x):
   - Allow $\mathcal{F}_{\text{COM}}$ to proceed.
   - When $\mathcal{S}$ receives (Open, sid, x) from $\mathcal{F}_{\text{COM}}$ as $\mathcal{R}$ is corrupt, $\mathcal{S}$ programs $\mathcal{G}_{\text{rpRO}}$ by giving $\mathcal{G}_{\text{rpRO}}$ input (ProgramRO, (sid, pk^r), $c_2 \oplus x$), such that the commitment (c_1, c_2) commits to x.
   - $\mathcal{S}$ inputs (Open, sid) to "$\mathcal{C}$", instructing it to open its commitment to x.

What remains to show is that $\mathcal{S}$ is a satisfying simulator, i.e., no $\mathcal{G}_{\mathsf{rpRO}}$-externally constrained environment can distinguish $\mathcal{F}_{\mathsf{COM}}$ and $\mathcal{S}$ from $\mathsf{COM}_{\mathcal{G}_{\mathsf{rpRO}}}$ and $\mathcal{A}$. When simulating an honest receiver, $\mathcal{S}$ extracts the committed message correctly: Given pk and $c_1 = g^r$ for some r, there is a unique value pk^r, and the message x is uniquely determined by c_2 and pk^r. Simulator $\mathcal{S}$ also simulates an honest committer correctly. When committing, it does not know the message, but can still produce a commitment that is identically distributed as long as the environment does not query the random oracle on $(\mathsf{sid}, \mathsf{pk}^r)$. When $\mathcal{S}$ later learns the message x, it must equivocate the commitment to open to x, by programming $\mathcal{G}_{\mathsf{rpRO}}$ on $(\mathsf{sid}, \mathsf{pk}^r)$, which again succeeds unless the environment makes a random oracle query on $(\mathsf{sid}, \mathsf{pk}^r)$. If there is an environment that triggers such a $\mathcal{G}_{\mathsf{rpRO}}$ with non-negligible probability, we can construct an attacker $\mathcal{B}$ that breaks the CDH problem in $\mathbb{G}$.

Our attacker $\mathcal{B}$ plays the role of $\mathcal{F}_{\mathsf{COM}}$, $\mathcal{S}$, and $\mathcal{G}_{\mathsf{rpRO}}$, and has black-box access to the environment. $\mathcal{B}$ receives CDH problem g^α, g^β and is challenged to compute $g^{\alpha\beta}$. It simulates $\mathcal{G}_{\mathsf{rpRO}}$ to return $h_n \leftarrow \mathsf{Embed}^{-1}(g^\alpha)$ on random query (sid, n). When simulating an honest committer committing with respect to this pk, set $c_1 \leftarrow g^\beta$ and $c_2 \xleftarrow{\$} \{0,1\}^{\ell(n)}$. Note that $\mathcal{S}$ cannot successfully open this commitment, but remember that we consider an environment that with non-negligible probability makes a $\mathcal{G}_{\mathsf{rpRO}}$ query on $\mathsf{pk}^r (= g^{\alpha\beta})$ before the commitment is being opened. Next, $\mathcal{B}$ will choose a random $\mathcal{G}_{\mathsf{rpRO}}$ query on (sid, m). With nonnegligible probability, we have $m = g^{\alpha\beta}$, and $\mathcal{B}$ found the solution to the CDH challenge. $\qquad\square$

5.3   Adding Observability for Efficient Commitments

While the commitment scheme $\mathsf{COM}_{\mathcal{G}_{\mathsf{rpRO}}}$ from the restricted programmable global random oracle is efficient for a composable commitment scheme, there is still a large efficiency gap between composable commitments from global random oracles and standalone commitments or commitments from local random oracles. Indeed, $\mathsf{COM}_{\mathcal{G}_{\mathsf{rpRO}}}$ still requires multiple exponentiations and rounds of interaction, whereas the folklore commitment scheme $c = \mathcal{H}(m\|r)$ for message m and random opening information r consists of computing a single hash function.

We extend $\mathcal{G}_{\mathsf{rpRO}}$ to, on top of programmability, offer the restricted observability interface of the global random oracle due to Canetti et al. [15]. With this restricted programmable *and observable* global random oracle $\mathcal{G}_{\mathsf{rpoRO}}$ (as shown in Fig. 10), we can close this efficiency gap and prove that the folklore commitment scheme above is a secure composable commitment scheme with a global random oracle.

Let $\mathsf{COM}_{\mathcal{G}_{\mathsf{rpoRO}}}$ be the commitment scheme that simply hashes the message and opening, phrased as a GUC protocol using $\mathcal{G}_{\mathsf{rpoRO}}$ and using authenticated channels, which is formally defined as follows.

1. On input $(\mathsf{Commit}, \mathsf{sid}, x)$, party C proceeds as follows.
    - Check that $\mathsf{sid} = (C, R, \mathsf{sid}')$ for some C, sid'.
    - Pick $r \xleftarrow{\$} \{0,1\}^n$ and query $\mathcal{G}_{\mathsf{rpoRO}}$ on (sid, r, x) to obtain c.

- Send c to R over $\mathcal{F}_{\mathsf{auth}}$ by giving $\mathcal{F}_{\mathsf{auth}}$ input (Send, $(C, R, \mathsf{sid}, 0), c)$.
- R, upon receiving (Sent, $(C, R, \mathsf{sid}, 0), c)$ from $\mathcal{F}_{\mathsf{auth}}$, outputs (Receipt, sid).

2. On input (Open, sid), C proceeds as follows.
   - It sends (r, x) to R by giving $\mathcal{F}_{\mathsf{auth}}$ input (Send, $(C, R, \mathsf{sid}, 2), (r, x))$.
   - R, upon receiving (Sent, $(C, R, \mathsf{sid}, 1), (r, x))$ from $\mathcal{F}_{\mathsf{auth}}$, queries $\mathcal{G}_{\mathsf{rpoRO}}$ on (sid, r, x) and checks that the result is equal to c, and checks that (sid, r, x) is not programmed by giving $\mathcal{G}_{\mathsf{rpoRO}}$ input (IsProgrammed, (sid, r, x)) and aborting if the result is not (IsProgrammed, 0). Output (Open, sid, x).

Theorem 6. $\mathsf{COM}_{\mathcal{G}_{\mathsf{rpoRO}}}$ *GUC-realizes* $\mathcal{F}_{\mathsf{COM}}$ *(as defined in Fig. 11), in the* $\mathcal{G}_{\mathsf{rpoRO}}$ *and* $\mathcal{F}_{\mathsf{auth}}$ *hybrid model.*

Proof. By the fact that $\mathsf{COM}_{\mathcal{G}_{\mathsf{rpoRO}}}$ is $\mathcal{G}_{\mathsf{rpoRO}}$-subroutine respecting and by Theorem 1, it is sufficient to show that $\mathsf{COM}_{\mathcal{G}_{\mathsf{rpoRO}}}$ $\mathcal{G}_{\mathsf{rpoRO}}$-EUC-realizes $\mathcal{F}_{\mathsf{rpo\text{-}COM}}$.

We define a simulator $\mathcal{S}$ by describing its behavior in the different corruption scenarios. For all scenarios, $\mathcal{S}$ will internally simulate $\mathcal{A}$ and forward any messages between $\mathcal{A}$ and the environment, the corrupt parties, and $\mathcal{G}_{\mathsf{rpoRO}}$. It stores all $\mathcal{G}_{\mathsf{rpoRO}}$ queries that it makes for $\mathcal{A}$ and for corrupt parties. Only when $\mathcal{A}$ directly or through a corrupt party makes an IsProgrammed query on a point that $\mathcal{S}$ programmed, $\mathcal{S}$ will not forward this query to $\mathcal{G}_{\mathsf{rpoRO}}$ but instead return (IsProgrammed, 0). When we say that $\mathcal{S}$ queries $\mathcal{G}_{\mathsf{rpoRO}}$ on a point (s, m) where s is the challenge sid, for example when simulating an honest party, it does so through a corrupt dummy party that it spawns, such that the query is not marked as illegitimate.

When both the sender and the receiver are honest, $\mathcal{S}$ works as follows.

1. When $\mathcal{F}_{\mathsf{rpo\text{-}COM}}$ asks $\mathcal{S}$ for permission to output (Receipt, sid):
   - Parse sid as (C, R, sid') and let "C" commit to a dummy value by giving it input (Commit, sid, $\perp$), except that it takes $c \xleftarrow{\$} \{0, 1\}^{\ell(n)}$ instead of following the protocol.
   - When "R" outputs (Receipt, sid), allow $\mathcal{F}_{\mathsf{rpo\text{-}COM}}$ to proceed.
2. When $\mathcal{F}_{\mathsf{rpo\text{-}COM}}$ asks $\mathcal{S}$ for permission to output (Open, sid, x):
   - Choose a random $r \xleftarrow{\$} \{0, 1\}^n$ and program $\mathcal{G}_{\mathsf{rpoRO}}$ by giving it input (ProgramRO, $(\mathsf{sid}, r, x), c)$, such that the commitment c commits to x. Note that since r is freshly chosen at random, the probability that $\mathcal{G}_{\mathsf{rpoRO}}$ is already defined on (sid, r, x) is negligible, so the programming will succeed with overwhelming probability.
   - Give "C" input (Open, sid) instructing it to open its commitment to x.
   - When "R" outputs (Open, sid, x), allow $\mathcal{F}_{\mathsf{rpo\text{-}COM}}$ to proceed.

If the committer is corrupt but the receiver is honest, $\mathcal{S}$ works as follows.

1. When simulated receiver "R" outputs (Receipt, sid):
   - Obtain the list $\mathcal{Q}_{\mathsf{sid}}$ of all random oracle queries of form $(\mathsf{sid}, \ldots)$, by combining the queries that $\mathcal{S}$ made on behalf of the corrupt parties and the simulated honest parties, and by obtaining the illegitimate queries made outside of $\mathcal{S}$ by giving $\mathcal{G}_{\mathsf{rpoRO}}$ input (Observe, sid).

- Find a non-programmed record $((\mathsf{sid}, r, x), c) \in \mathcal{Q}_{\mathsf{sid}}$. If no such record is found, set x to a dummy value.
- Make a query with $\mathcal{F}_{\mathsf{rpo\text{-}COM}}$ on C's behalf by sending $(\mathsf{Commit}, \mathsf{sid}, x)$ on C's behalf to $\mathcal{F}_{\mathsf{rpo\text{-}COM}}$.
- When $\mathcal{F}_{\mathsf{rpo\text{-}COM}}$ asks permission to output $(\mathsf{Receipt}, \mathsf{sid})$, allow.

2. When "R" outputs $(\mathsf{Open}, \mathsf{sid}, x)$:
    - Send $(\mathsf{Open}, \mathsf{sid})$ on C's behalf to $\mathcal{F}_{\mathsf{rpo\text{-}COM}}$.
    - When $\mathcal{F}_{\mathsf{rpo\text{-}COM}}$ asks permission to output $(\mathsf{Open}, \mathsf{sid}, x)$, allow.

If the receiver is corrupt but the committer is honest, $\mathcal{S}$ works as follows.

1. When $\mathcal{F}_{\mathsf{rpo\text{-}COM}}$ asks permission to output $(\mathsf{Receipt}, \mathsf{sid})$:
    - Parse sid as (C, R, sid').
    - Allow $\mathcal{F}_{\mathsf{rpo\text{-}COM}}$ to proceed.
    - When $\mathcal{S}$ receives $(\mathsf{Receipt}, \mathsf{sid})$ from $\mathcal{F}_{\mathsf{rpo\text{-}COM}}$ as R is corrupt, it simulates "C" by choosing $c \xleftarrow{\$} \{0, 1\}^{\ell(n)}$ instead of following the protocol.

2. When $\mathcal{F}_{\mathsf{rpo\text{-}COM}}$ asks permission to output $(\mathsf{Open}, \mathsf{sid}, x)$:
    - Allow $\mathcal{F}_{\mathsf{rpo\text{-}COM}}$ to proceed.
    - When $\mathcal{S}$ receives $(\mathsf{Open}, \mathsf{sid}, x)$ from $\mathcal{F}_{\mathsf{rpo\text{-}COM}}$ as R is corrupt, choose $r \xleftarrow{\$} \{0, 1\}^n$ and program $\mathcal{G}_{\mathsf{rpoRO}}$ by giving $\mathcal{F}_{\mathsf{rpo\text{-}COM}}$ input $(\mathsf{ProgramRO}, (\mathsf{sid}, r, x), c)$, such that the commitment c commits to x. Note that since r is freshly chosen at random, the probability that $\mathcal{G}_{\mathsf{rpoRO}}$ is already defined on (sid, r, x) is negligible, so the programming will succeed with overwhelming probability.
    - $\mathcal{S}$ inputs $(\mathsf{Open}, \mathsf{sid})$ to "C", instructing it to open its commitment to x.

We must show that $\mathcal{S}$ extracts the correct value from a corrupt commitment. It obtains a list of all $\mathcal{G}_{\mathsf{rpoRO}}$ queries of the form $(\mathsf{sid}, \ldots)$ and looks for a non-programmed entry (sid, r, x) that resulted in output c. If this does not exist, then the environment can only open its commitment successfully by later finding a preimage of c, as the honest receiver will check that the point was not programmed. Finding such a preimage happens with negligible probability, so committing to a dummy value is sufficient. The probability that there are multiple satisfying entries is also negligible, as this means the environment found collisions on the random oracle.

Next, we argue that the simulated commitments are indistinguishable from honest commitments. Observe that the commitment c is distributed equally to real commitments, namely uniform in $\{0, 1\}^{\ell(n)}$. The simulator can open this value to the desired x if programming the random oracle succeeds. As it first takes a fresh nonce $r \xleftarrow{\$} \{0, 1\}^n$ and programs (sid, r, x), the probability that $\mathcal{G}_{\mathsf{rpoRO}}$ is already defined on this input is negligible. $\square$

6   Unifying the Different Global Random Oracles

At this point, we have considered several notions of global random oracles that differ in whether they offer programmability or observability, and in whether this

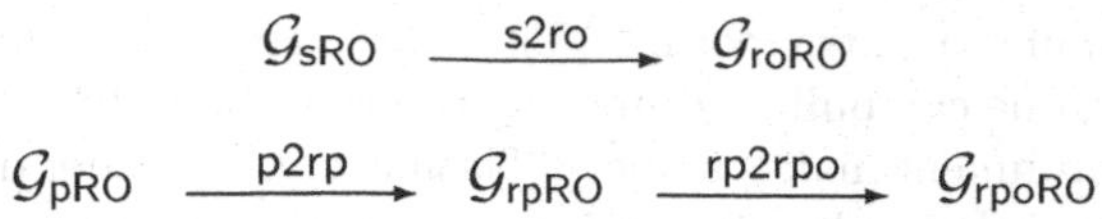

$$\mathcal{G}_{sRO} \xrightarrow{\text{s2ro}} \mathcal{G}_{roRO}$$

$$\mathcal{G}_{pRO} \xrightarrow{\text{p2rp}} \mathcal{G}_{rpRO} \xrightarrow{\text{rp2rpo}} \mathcal{G}_{rpoRO}$$

Fig. 12. Relations between different notions of global random oracles. An arrow from $\mathcal{G}$ to $\mathcal{G}'$ indicates the existence of simple transformation such that any protocol that $\mathcal{G}$-EUC-realizes a functionality $\mathcal{F}$, the transformed protocol $\mathcal{G}'$-EUC-realizes the transformed functionality $\mathcal{F}$ (cf. Theorem 7).

power is restricted to machines within the local session, or also available to other machines. Having several coexisting variants of global random oracles, each with their own set of schemes that they can prove secure, is somewhat unsatisfying. Indeed, if different schemes require different random oracles that in practice end up being replaced with the same hash function, then we're back to the problem that motivated the concept of global random oracles.

We were able to distill a number of relations and transformations among the different notions, allowing a protocol that realizes a functionality with access to one type of global random oracle to be efficiently transformed into a protocol that realizes the same functionality with respect to a different type of global random oracle. A graphical representation of our transformation is given in Fig. 12.

The transformations are very simple and hardly affect efficiency of the protocol. The s2ro transformation takes as input a $\mathcal{G}_{sRO}$-subroutine-respecting protocol π and transforms it into a $\mathcal{G}_{roRO}$-subroutine respecting protocol $\pi' = \text{s2ro}(\pi)$ by replacing each query (HashQuery, m) to $\mathcal{G}_{sRO}$ with a query (HashQuery, (sid, m)) to $\mathcal{G}_{roRO}$, where sid is the session identifier of the calling machine. Likewise, the p2rp transformation takes as input a $\mathcal{G}_{pRO}$-subroutine-respecting protocol π and transforms it into a $\mathcal{G}_{rpRO}$-subroutine respecting protocol $\pi' = \text{p2rp}(\pi)$ by replacing each query (HashQuery, m) to $\mathcal{G}_{pRO}$ with a query (HashQuery, (sid, m)) to $\mathcal{G}_{rpRO}$ and replacing each query (ProgramRO, m, h) to $\mathcal{G}_{pRO}$ with a query (ProgramRO, (sid, m), h) to $\mathcal{G}_{rpRO}$, where sid is the session identifier of the calling machine. The other transformation rp2rpo simply replaces HashQuery, ProgramRO, and IsProgrammed queries to $\mathcal{G}_{rpRO}$ with identical queries to $\mathcal{G}_{rpoRO}$.

Theorem 7. *Let π be a $\mathcal{G}_{xRO}$-subroutine-respecting protocol and let $\mathcal{G}_{yRO}$ be such that there is an edge from $\mathcal{G}_{xRO}$ to $\mathcal{G}_{yRO}$ in Fig. 12, where $x, y \in \{s, ro, p, rp, rpo\}$. Then if π $\mathcal{G}_{xRO}$-EUC-realizes a functionality $\mathcal{F}$, where $\mathcal{F}$ is an ideal functionality that does not communicate with $\mathcal{G}_{xRO}$, then $\pi' = \text{x2y}(\pi)$ is a $\mathcal{G}_{yRO}$-subroutine-respecting protocol that $\mathcal{G}_{yRO}$-EUC-realizes $\mathcal{F}$.*

Proof (sketch). We first provide some detail for the s2ro transformation. The other transformations can be proved in a similar fashion, so we only provide an intuition here.

As protocol π $\mathcal{G}_{sRO}$-EUC-realizes $\mathcal{F}$, there exists a simulator $\mathcal{S}_s$ that correctly simulates the protocol with respect to the dummy adversary. Observe that $\mathcal{G}_{roRO}$ offers the same HashQuery interface to the adversary as $\mathcal{G}_{sRO}$, and that the $\mathcal{G}_{roRO}$

only gives the simulator extra powers. Therefore, given the dummy-adversary simulator S_s for π, one can build a dummy-adversary simulator S_{ro} for s2ro(π) as follows. If the environment makes a query (HashQuery, x), either directly through the dummy adversary, or indirectly by instructing a corrupt party to make that query, S_{ro} checks whether x can be parsed as (sid, x') where sid is the challenge session. If so, then it passes a direct or indirect query (HashQuery, x') to S_s, depending whether the environment's original query was direct or indirect. If x cannot be parsed as (sid, x'), then it simply relays the query to G_{roRO}. Simulator S_{ro} relays S_s's inputs to and outputs from $\mathcal{F}$. When S_s makes a (HashQuery, x') query to G_{sRO}, S_{ro} makes a query (HashQuery, (sid, x')) to G_{roRO} and relays the response back to S_s. Finally, S_{ro} simply relays any Observe queries by the environment to G_{roRO}. Note, however, that these queries do not help the environment in observing the honest parties, as they only make legitimate queries.

To see that S_{ro} is a good simulator for s2ro(π), we show that if there exists a distinguishing dummy-adversary environment Z_{ro} for s2ro(π) and S_{ro}, then there also exists a distinguishing environment Z_s for π and S_s, which would contradict the security of π. The environment Z_s runs Z_{ro} by internally executing the code of G_{roRO} to respond to Z_{ro}'s G_{roRO} queries, except for queries (HashQuery, x) where x can be parsed as (sid, x'), for which Z_s reaches out to its own G_{sRO} functionality with a query (HashQuery, x').

The p2rp transformation is very similar to s2ro and prepends sid to random oracle queries. Moving to the *restricted* programmable RO only reduces the power of the adversary by making programming detectable to honest users through the IsProgrammed interface. The simulator, however, maintains its power to program without being detected, because it can intercept the environment's IsProgrammed queries for the challenge sid and pretend that they were not programmed. The environment cannot circumvent the simulator and query G_{rpRO} directly, because IsProgrammed queries for sid must be performed from a machine within sid.

Finally, the rp2rpo transformation increases the power of both the simulator and the adversary by adding a Observe interface. Similarly to the s2ro simulator, however, the interface cannot be used by the adversary to observe queries made by honest parties, as these queries are all legitimate.   □

Unfortunately, we were unable to come up with security-preserving transformations from non-programmable to programmable random oracles that apply to any protocol. One would expect that the capability to program random-oracle entries destroys the security of many protocols that are secure for non-programmable random oracles. Often this effect can be mitigated by letting the protocol, after performing a random-oracle query, additionally check whether the entry was programmed through the IsProgrammed interface, and rejecting or aborting if it was. While this seems to work for signature or commitment schemes where rejection is a valid output, it may not always work for arbitrary protocols with interfaces that may not be able to indicate rejection. We leave the study of more generic relations and transformations between programmable and non-programmable random oracles as interesting future work.

Acknowledgements. We thank Ran Canetti, Alessandra Scafuro, and the anonymous reviewers for their valuable comments. This work was supported by the ERC under grant PERCY (#321310) and by the EU under CHIST-ERA project USE-IT.

References

1. Ananth, P., Bhaskar, R.: Non observability in the random oracle model. In: Susilo, W., Reyhanitabar, R. (eds.) ProvSec 2013. LNCS, vol. 8209, pp. 86–103. Springer, Heidelberg (2013). https://doi.org/10.1007/978-3-642-41227-1_5
2. Bellare, M., Desai, A., Pointcheval, D., Rogaway, P.: Relations among notions of security for public-key encryption schemes. In: Krawczyk, H. (ed.) CRYPTO 1998. LNCS, vol. 1462, pp. 26–45. Springer, Heidelberg (1998). https://doi.org/10.1007/BFb0055718
3. Bellare, M., Rogaway, P.: Random oracles are practical: a paradigm for designing efficient protocols. In: ACM CCS 93
4. Bellare, M., Rogaway, P.: Optimal asymmetric encryption. In: De Santis, A. (ed.) EUROCRYPT 1994. LNCS, vol. 950, pp. 92–111. Springer, Heidelberg (1995). https://doi.org/10.1007/BFb0053428
5. Bellare, M., Rogaway, P.: The exact security of digital signatures-how to sign with RSA and Rabin. In: Maurer, U. (ed.) EUROCRYPT 1996. LNCS, vol. 1070, pp. 399–416. Springer, Heidelberg (1996). https://doi.org/10.1007/3-540-68339-9_34
6. Bhattacharyya, R., Mukherjee, P.: Non-adaptive programmability of random oracle. Theor. Comput. Sci. **592**, 97–114 (2015)
7. Camenisch, J., Enderlein, R.R., Krenn, S., Küsters, R., Rausch, D.: Universal composition with responsive environments. In: Cheon, J.H., Takagi, T. (eds.) ASIACRYPT 2016. LNCS, vol. 10032, pp. 807–840. Springer, Heidelberg (2016). https://doi.org/10.1007/978-3-662-53890-6_27
8. Camenisch, J., Lehmann, A., Neven, G., Samelin, K.: UC-secure non-interactive public-key encryption. In: IEEE CSF 2017 (2017)
9. Canetti, R.: Universally composable security: a new paradigm for cryptographic protocols. Cryptology ePrint Archive, Report 2000/067 (2000)
10. Canetti, R.: Universally composable security: a new paradigm for cryptographic protocols. In: FOCS 2001 (2001)
11. Canetti, R.: Universally composable signature, certification, and authentication. In: CSFW 2004 (2004)
12. Canetti, R., Dodis, Y., Pass, R., Walfish, S.: Universally composable security with global setup. In: Vadhan, S.P. (ed.) TCC 2007. LNCS, vol. 4392, pp. 61–85. Springer, Heidelberg (2007). https://doi.org/10.1007/978-3-540-70936-7_4
13. Canetti, R., Fischlin, M.: Universally composable commitments. In: Kilian, J. (ed.) CRYPTO 2001. LNCS, vol. 2139, pp. 19–40. Springer, Heidelberg (2001). https://doi.org/10.1007/3-540-44647-8_2
14. Canetti, R., Goldreich, O., Halevi, S.: The random oracle methodology, revisited (preliminary version). In: ACM STOC 1998 (1998)
15. Canetti, R., Jain, A., Scafuro, A.: Practical UC security with a global random oracle. In: ACM CCS 2014 (2014)
16. Coron, J.-S.: On the exact security of full domain hash. In: Bellare, M. (ed.) CRYPTO 2000. LNCS, vol. 1880, pp. 229–235. Springer, Heidelberg (2000). https://doi.org/10.1007/3-540-44598-6_14

17. Dodis, Y., Shoup, V., Walfish, S.: Efficient constructions of composable commitments and zero-knowledge proofs. In: Wagner, D. (ed.) CRYPTO 2008. LNCS, vol. 5157, pp. 515–535. Springer, Heidelberg (2008). https://doi.org/10.1007/978-3-540-85174-5_29

18. Fischlin, M., Lehmann, A., Ristenpart, T., Shrimpton, T., Stam, M., Tessaro, S.: Random oracles with(out) programmability. In: Abe, M. (ed.) ASIACRYPT 2010. LNCS, vol. 6477, pp. 303–320. Springer, Heidelberg (2010). https://doi.org/10.1007/978-3-642-17373-8_18

19. Fujisaki, E., Okamoto, T.: Secure integration of asymmetric and symmetric encryption schemes. J. Cryptol. 26(1), 80–101 (2013)

20. Goldwasser, S., Micali, S., Rivest, R.L.: A digital signature scheme secure against adaptive chosen-message attacks. SIAM J. Comput. 17(2), 281–308 (1988)

21. Nielsen, J.B.: Separating random oracle proofs from complexity theoretic proofs: the non-committing encryption case. In: Yung, M. (ed.) CRYPTO 2002. LNCS, vol. 2442, pp. 111–126. Springer, Heidelberg (2002). https://doi.org/10.1007/3-540-45708-9_8

22. Pointcheval, D., Stern, J.: Security arguments for digital signatures and blind signatures. J. Cryptol. 13(3), 361–396 (2000)

23. Schnorr, C.P.: Efficient signature generation by smart cards. J. Cryptol. 4(3), 161–174 (1991)

Fully Homomorphic Encryption

Homomorphic Lower Digits Removal and Improved FHE Bootstrapping

Hao Chen[1(✉)] and Kyoohyung Han[2]

[1] Microsoft Research, Redmond, USA
haoche@microsoft.com
[2] Seoul National University, Seoul, Korea
satanigh@snu.ac.kr

Abstract. Bootstrapping is a crucial operation in Gentry's break-through work on fully homomorphic encryption (FHE), where a homo-morphic encryption scheme evaluates its own decryption algorithm. There has been a couple of implementations of bootstrapping, among which HElib arguably marks the state-of-the-art in terms of throughput, ciphertext/message size ratio and support for large plaintext moduli.

In this work, we applied a family of "lowest digit removal" polynomi-als to design an improved homomorphic digit extraction algorithm which is a crucial part in bootstrapping for both FV and BGV schemes. When the secret key has 1-norm $h = ||s||_1$ and the plaintext modulus is $t = p^r$, we achieved bootstrapping depth $\log h + \log(\log_p(ht))$ in FV scheme. In case of the BGV scheme, we brought down the depth from $\log h + 2\log t$ to $\log h + \log t$.

We implemented bootstrapping for FV in the SEAL library. We also introduced another "slim mode", which restrict the plaintexts to batched vectors in $\mathbb{Z}_{p^r}$. The slim mode has similar throughput as the full mode, while each individual run is much faster and uses much smaller mem-ory. For example, bootstrapping takes 6.75 s for vectors over $GF(127)$ with 64 slots and 1381 s for vectors over $GF(257^{128})$ with 128 slots. We also implemented our improved digit extraction procedure for the BGV scheme in HElib.

Keywords: Homomorphic encryption · Bootstrapping
Implementation

1 Introduction

Fully Homomorphic Encryption (FHE) allows an untrusted party to evalu-ate arbitrary functions on encrypted data, without knowing the secret key.

K. Han—Supported by Institute for Information & communications Technology Pro-motion (IITP) grant funded by the Korea government (MSIT) (No. B0717-16-0098, Development of homomorphic encryption for DNA analysis and biometry authenti-cation).

© International Association for Cryptologic Research 2018
J. B. Nielsen and V. Rijmen (Eds.): EUROCRYPT 2018, LNCS 10820, pp. 315–337, 2018.
https://doi.org/10.1007/978-3-319-78381-9_12

Gentry introduced the first FHE scheme in the breakthrough work [20]. Since then, there has been a large collection of work (e.g., [6–10,13,16,19,22,31]), introducing more efficient schemes.

These schemes all follow Gentry's original blueprint, where each ciphertext is associated with a certain amount of "noise", and the noise grows as homomorphic evaluations are performed. When the noise is too large, decryption will fail to give the correct result. Therefore, if no additional measure is taken, one set of parameters can only evaluate circuits of a bounded depth. This approach is called leveled homomorphic encryption (LHE) and is used in a many works.

However, if we wish to homomorphically evaluate functions of arbitrary complexity using one single set of parameters, then we need a procedure to lower the noise in a ciphertext. This can be done via Gentry's brilliant *bootstrapping* technique. Roughly speaking, bootstrapping a ciphertext in some given scheme means running its own decryption algorithm homomorphically, using an encryption of the secret key. The result is a new ciphertext which encrypts the same message while having lower noise.

Bootstrapping is a very expensive operation. The decryption circuit of a scheme can be complex, and may not be conveniently supported by the scheme itself. Hence, in order to perform bootstrapping, one either needs to make significant optimizations to simplify the decryption circuit, or design some scheme which can handle its decryption circuit more comfortably. Among the best works on bootstrapping implementations, the work of Halevi and Shoup [25], which optimized and implemented bootstrapping over the scheme of Brakerski, Gentry and Vaikuntanathan (BGV), is arguably still the state-of-the-art in terms of throughput, ciphertext/message size ratio and flexible plaintext moduli. For example, they were able to bootstrap a vector of size 1024 over $GF(2^{16})$ within 5 min. However, when the plaintext modulus reaches 2^8, bootstrapping still takes a few hours to perform. The reason is mainly due to a digit extraction procedure, whose cost grows significantly with the plaintext modulus. The Fan-Vercauteran (FV) scheme, a scale-invariant variant of BGV, has also been implement in [1,27] and used in applications. We are not aware of any previous implementation of bootstrapping for FV.

1.1   Contributions

In this paper, we aim at improving the efficiency of bootstrapping under large prime power plaintext moduli.

- We used a family of low degree lowest-digit-removal polynomials to design an improved algorithm to remove v lowest base-p digits from integers modulo p^e. Our new algorithm has depth $v \log p + \log e$, compared to $(e-1) \log p$ in previous work.
- We then applied our algorithm to improve the digit extraction step in the bootstrapping procedure for FV and BGV schemes. Let $h = ||s||_1$ denote the 1-norm of the secret key, and assume the plaintext space is a prime

power $t = p^r$. Then for FV scheme, we achieved bootstrapping depth $\log h + \log \log_p(ht)$. In case of BGV, we have reduced the bootstrapping degree from $\log h + 2\log(t)$ to $\log h + \log t$.

- We provided a first implementation of the bootstrapping functionality for FV scheme in the SEAL library [27]. We also implemented our revised digit extraction algorithm in HElib which can directly be applied to improve HElib bootstrapping for large plaintext modulus p^r.

- We also introduced a light-weight mode of bootstrapping which we call the "slim mode" by restricting the plaintexts to a subspace. In this mode, messages are vectors where each slot only holds a value in $\mathbb{Z}_{p^r}$ instead of a degree-d extension ring. The slim mode might be more applicable in some use-cases of FHE, including machine learning over encrypted data. We implemented the slim mode of bootstrapping in SEAL and showed that in this mode, bootstrapping is about d times faster, hence we can achieve a similar throughput as in the full mode.

1.2 Application: Machine Learning over Encrypted Data

Machine learning over encrypted data is one of the signature use-cases of FHE and an active research area. Research works in this area can be divided into two categories: evaluating a pre-trained machine learning model over private testing data, or training a new model on private training data. Often times, the model evaluation requires a lower-depth circuit, and thus can be achieved using LHE. On the other hand, training a machine learning model requires a much deeper circuit, and bootstrapping becomes necessary. This may explain that there are few works in the model training direction.

In the model evaluation case (e.g. [4,5,23,24]), one encodes the data as either polynomials in R_t, or as elements of $\mathbb{Z}_t$ when batching is used. One distinguishing feature of these methods is that the scheme maintains the full precision of plaintexts as evaluations are performed, in contrast to computations over plaintext data, where floating point numbers are used and only a limited precision is maintained. This implies that the plaintext modulus t needs to be taken large enough to "hold the result".

In the training case, because of the large depth and size of the circuit, the above approach is simply infeasible: t needs to be so large that the homomorphic evaluations become too inefficient, as pointed out in [17]. Therefore, some analog of plaintext truncation needs to be performed alongside the evaluation. However, in order to perform the truncation function homomorphically, one has to express the function as a polynomial. Fortunately, our digit removal algorithm can also be used as a truncation method over $\mathbb{Z}_{p^r}$. Therefore, we think that improving bootstrapping for prime power plaintext modulus has practical importance.

There is one other work [12] which does not fall into either categories. It performs homomorphic evaluation over point numbers and outputs an approximate result. It modifies the BGV and FV schemes: instead of encoding noise and message in different parts of a ciphertexts, one puts noise in lower bits of messages, and uses modulus switching creatively as a plaintext management technique.

As a result, they could evaluate deeper circuits with smaller HE parameters. It is then an interesting question whether there exists an efficient bootstrapping algorithm for this modified scheme.

1.3   Related Works

After bootstrapping was introduced by Gentry at 2009, many methods are proposed to improve its efficiency. Existing bootstrapping implementations can be classified into three branches. The first branch [21,25] builds on top of somewhat homomorphic encryption schemes based on the RLWE problem. The second branch aims at minimizing the time to bootstrap one single bit of message after each boolean gate evaluation. Works in this direction include [3,14,15,18]. They were able to obtain very fast results: less than 0.1 s for a single bootstrapping. The last branch considers bootstrapping over integer-based homomorphic encryption schemes under the sparse subset sum problem assumption. Some works [13,16,28,31] used a squashed decryption circuit and evaluate bit-wise (or digit-wise) addition in encrypted state instead of doing a digit extraction. In [11], they show that using digit extraction for bootstrapping results in lower computational complexity while consuming a similar amount of depth as previous approaches.

Our work falls into the first branch. We aim at improving the bootstrapping procedure for the two schemes BGV and FV, with the goal of improving the throughput and after level for bootstrapping in case of large plaintext modulus. Therefore, our main point of comparison in this paper will be the work of Halevi and Shoup [25]. We note that a digit extraction procedure is used for all branches except the second one. Therefore, improving the digit extraction procedure is one of the main tasks for an efficient bootstrapping algorithm.

1.4   Roadmap

In Sect. 2, we introduce notations and necessary background on the BGV and FV schemes. In Sect. 3, after reviewing the digit extraction procedure of [25], we define the lowest digit removal polynomials, and use them to give an improved digit removal algorithm. In Sect. 4, we describe our method for bootstrapping in the FV scheme, and how our algorithm leads to an improved bootstrapping for BGV scheme when the plaintext modulus is p^r with $r > 1$. In Sect. 5, we present and discuss our performance results. Finally, in Sect. 6 we conclude with future directions. Proofs and more details regarding the SEAL implementation of bootstrapping are included in the Appendix.

2   Background

2.1   Basics of BGV and FV Schemes

First, we introduce some notations. Both BGV and FV schemes are initialized with integer parameters m, t and q. Here m is the cyclotomic field index, t is the plaintext modulus, and q is the coefficient modulus. Note that in BGV, it is required that $(t, q) = 1$.

Let $\phi_m(x)$ denote the m-th cyclotomic polynomial and let n denote its degree. We use the following common notations $R = \mathbb{Z}[x]/(\phi_m(x))$, $R_t = R/tR$, and $R_q = R/qR$. In both schemes, the message is a polynomial $m(x)$ in R_t, and the secret key s is an element of R_q. In practice, s is usually taken to be ternary (i.e., each coefficient is either -1, 0 or 1) and often sparse (i.e., the number of nonzero coefficients of s are bounded by some $h \ll n$). A ciphertext is a pair (c_0, c_1) of elements in R_q.

Decryption Formula. The decryption of both schemes starts with a dot-product with the extended secret key $(1, s)$. In BGV, we have

$$c_0 + c_1 s = m(x) + tv + \alpha q,$$

and decryption returns $m(x) = ((c_0 + c_1 s) \mod q) \mod t$. In FV, the equation is

$$c_0 + c_1 s = \Delta m(x) + v + \alpha q$$

and decryption formula is $m(x) = \lfloor \frac{(c_0 + c_1 s) \mod q}{\Delta} \rceil$.

Plaintext Space. The native plaintext space in both schemes is R_t, which consists of polynomials with degree less than n and integer coefficients between 0 and $t - 1$. Additions and multiplications of these polynomials are performed modulo both $\phi_m(x)$ and t.

A widely used plaintext-batching technique [30] turns the plaintext space into a vector over certain finite rings. Since batching is used extensively in our bootstrapping algorithm, we recall the details here. Suppose $t = p^r$ is a prime power, and assume p and m are co-prime. Then $\phi_m(x) \mod p^r$ factors into a product of k irreducible polynomials of degree d. Moreover, d is equal to the order of p in $\mathbb{Z}_m^*$, and k is equal to the size of the quotient group $\mathbb{Z}_m^*/\langle p \rangle$. For convenience, we fix a set $S = \{s_1, \ldots, s_k\}$ of integer representatives of the quotient group. Let $f(x)$ be one of the irreducible factors of $\phi_m(x) \mod p^r$, and consider the finite extension ring

$$E = \mathbb{Z}_{p^r}[x]/(f(x)).$$

Then all primitive m-th roots of unity exist in E. Fix $\zeta \in E$ to be one such root. Then we have a ring isomorphism

$$R_t \to E^k$$
$$m(x) \mapsto (m(\zeta^{s_1}), m(\zeta^{s_2}), \ldots, m(\zeta^{s_k}))$$

Using this isomorphism, we can regard the plaintexts as vectors over E, and additions/multiplications between the plaintexts are executed coefficient-wise on the components of the vectors, which are often called *slots*.

In the reset of the paper, we will move between the above two ways of viewing the plaintexts, and we will distinguish them by writing them as polynomials (no batching) and vectors (batching). For example, $\mathsf{Enc}(m(x))$ means an encryption of $m(x) \in R_t$, whereas $\mathsf{Enc}((m_1, \ldots, m_k))$ means a batch encryption of a vector $(m_1, \ldots, m_k) \in E^k$.

Modulus Switching. Modulus switching is a technique which scales a ciphertext (c_0, c_1) with modulus q to another one (c'_0, c'_1) with modulus q' that decrypts to the same message. In BGV, modulus switching is a necessary technique to reduce the noise growth. Modulus switching is not strictly necessary for FV, at least if used in the LHE mode. However, it will be of crucial use in our bootstrapping procedure. More precisely, modulus switching in BGV requires q and q' to be both co-prime to t. For simplicity, suppose $q \equiv q' \equiv 1 (\mod t)$. Then c'_i equals the closest integer polynomial to $\frac{q'}{q} c$ such that $c'_i \equiv c_i \mod t$. For FV, q and q' do not need to be co-prime to t, and modulus switching simply does scaling and rounding to integers, i.e., $c'_i = \lfloor q'/q c_i \rceil$.

We stress that modulus switching slightly increase the noise-to-modulus ratio due to rounding errors in the process. Therefore, one can not switch to arbitrarily small modulus q'. On the other hand, in bootstrapping we often like to switch to a small q'. The following lemma puts a lower bound on the size of q' for FV (the case for BGV is similar).

Lemma 1. *Suppose $c_0 + c_1 s = \Delta m + v + aq$ is a ciphertext in FV with $|v| < \Delta/4$. if $q' > 4t(1 + ||s||_1)$, and (c'_0, c'_1) is the ciphertext after switching the modulus to q', then (c'_0, c'_1) also decrypts to m.*

Proof. See appendix.

We remark that although the requirement in BGV that q and t are co-prime seems innocent, it affects the depth of the decryption circuit when t is large. Therefore, it results in an advantage for doing bootstrapping in FV over BGV. We will elaborate on this point later.

Multiply and Divide by p in Plaintext Space. In bootstrapping, we will use following functionalities: dividing by p, which takes an encryption of $pm \mod p^e$ and returns an encryption of $m \mod p^{e-1}$, and multiplying by p which is the inverse of division. In BGV scheme, multiply by p can be realized via a fast scalar multiplication $(c_0, c_1) \rightarrow ((pc_0) \mod q, (pc_1) \mod q)$. In the FV scheme, these operations are essentially free, because if $c_0 + c_1 s = \lfloor \frac{q}{p^{e-1}} \rfloor m + v + q\alpha$, then the same ciphertext satisfies $c_0 + c_1 s = \lfloor \frac{q}{p^e} \rfloor pm + v + v' + q\alpha$ for some small v'. In the rest of the paper, we will omit these operations, assuming that they are free to perform.

3   Digit Removal Algorithm

The previous method for digit extraction used certain lifting polynomials with good properties. We used a family of "lowest digit removal" polynomials which have a stronger lifting property. We then combined these lowest digit removal polynomials with the lifting polynomials to construct a new digit removal algorithm.

For convenience of exposition, we use some slightly modified notations from [25]. Fix a prime p. Let z be an integer with (balanced) base-p expansion $z = \sum_{i=0}^{e-1} z_i p^i$. For integers $i, j \geq 0$, we use $z_{i,j}$ to denote any integer with first base-p digit equal to z_i and the next j digits zero. In other words, we have $z_{i,j} \equiv z_i$ mod p^{j+1}.

3.1   Reviewing the Digit Extraction Method of Halevi and Shoup

The bootstrapping procedure in [25] consists of five main steps: modulus switching, dot product (with an encrypted secret key), linear transform, digit extraction, and another "inverse" linear transform. Among these, the digit extraction step dominates the cost in terms of both depth and work. Hence we will focus on optimizing the digit extraction. Essentially, we need the following functionality.

DigitRemove(p, e, v) : fix prime p, for two integers $v < e$ and an input u mod p^e, let $u = \sum u_i p^i$ with $|u_i| \leq p/2$ when p is odd (and $u_i = 0, 1$ when $p = 2$), returns

$$u\langle v, \dots, e-1 \rangle := \sum_{i=v}^{e-1} u_i p^i.$$

We say this functionality "removes" the v lowest significant digits in base p from an e-digits integer. To realize the above functionality over homomorphically encrypted data, the authors in [25] constructed some special polynomials $F_e(\cdot)$ with the following lifting property.

Lemma 2 (Corollary 5.5 in [25]). *For every prime p and $e \geq 1$ there exist a degree p-polynomial F_e such that for every integer z_0, z_1 with $z_0 \in [p]$ and every $1 \leq e' \leq e$ we have $F_e(z_0 + p^{e'} z_1) = z_0 \pmod{p^{e'+1}}$.*

For example, if $p = 2$, we can take $F_e(x) = x^2$. One then uses these lifting polynomials F_e to extract each digit u_i from u in a successive fashion. The digit extraction procedure is defined in Fig. 1 in [25] and can be visualized in the following diagram.

In the diagram, the top-left digit is the input. This algorithm starts with the top row. From left to right, it successively applies the lifting polynomial to obtain all the blue digits. Then the green digits on the next row can be obtained from subtracting all blue digits on the same diagonal from the input and then dividing by p. When this procedure concludes, the (i,j)-th digit of the diagram will be $u_{i,j}$. In particular, digits on the final diagonal will be $u_{i,e-1-i}$. Then we can compute

$$u\langle v, \cdots, e-1\rangle = u - \sum_{i=0}^{v-1} u_{i,e-1-i} \cdot p^i.$$

$$
\begin{array}{lllll}
u = u_{0,0} & u_{0,1} & \cdots & u_{0,r-1} & \cdots & u_{0,e-1} \\
u_{1,0} & u_{1,1} & \cdots & u_{1,r-2} & \cdots & u_{1,e-2} \\
\vdots & \vdots & & \reflectbox{$\ddots$} \\
\\
u_{e-2,0} & u_{e-2,1} \\
u_{e-1,0}
\end{array}
$$

3.2 Lowest Digit Removal Polynomials

We first stress that in the above method, it is not enough to obtain the u_i mod p. Rather, one requires $u_{i,e-1-i}$. The reason is one has to clear the higher digits to create numbers with base-p expansion $(u_i, \underbrace{0, 0, \ldots, 0}_{e-1-i})$, otherwise it will mess up the u_i' for $i' > i$. Previously, to obtain $u_{i,j}$, one needs to apply the lifting polynomial j times. Fortunately, there is a polynomial of lower degree with the same functionality, as shown in the following lemma.

Lemma 3. *Let p be a prime and $e \geq 1$. Then there exists a polynomial f of degree at most $(e-1)(p-1)+1$ such that for every integer $0 \leq x < p^e$, we have*

$$f(x) \equiv x - (x \mod p) \mod p^e,$$

where $|x \mod p| \leq (p-1)/2$ when p is odd.

Proof. We complete the proof sketch in [26] by adding in the necessary details. To begin, we introduce a function

$$F_A(x) := \sum_{j=0}^{\infty} (-1)^j \binom{A+j-1}{j}\binom{x}{A+j}.$$

This function $F_A(x)$ converges on every integer, and for $M \in \mathbb{Z}$,

$$F_A(M) = \begin{cases} 1 & \text{if } M > A \\ 0 & \text{otherwise.} \end{cases}$$

Define $\hat{f}(x)$ as

$$\hat{f}(x) = p \sum_{j=1}^{\infty} F_{j \cdot p}(x) = \sum_{m=p}^{\infty} a(m)\binom{x}{m}. \tag{1}$$

We can verify that the function $\hat{f}(x)$ satisfies the properties in the lemma (for the least residue system), but its degree is infinite. So we let

$$f(x) = \sum_{m=p}^{(e-1)(p-1)+1} a(m)\binom{x}{m}.$$

Now we will prove that the polynomial $f(x)$ has p-integral coefficients and has the same value with $\hat{f}(x)$ for $x \in \mathbb{Z}_{p^e}$.

Claim. $f(x)$ has p-integral coefficients and $a(m)\binom{x}{m}$ is multiple of p^e for all $x \in \mathbb{Z}$ when $m > (e-1)(p-1)+1$.

Proof. If we rewrite the Eq. 1,

$$\hat{f}(x) = p\sum_{j=1}^{\infty} F_{j \cdot p}(x) = p\sum_{j=1}^{\infty}\left(\sum_{i=0}^{\infty}(-1)^i\binom{jp+i-1}{i}\binom{x}{jp+i}\right).$$

By replacing the $jp+i$ to m, we arrive at the following equation:

$$a(m) = p\sum_{k=1}^{\infty}(-1)^{m-kp}\binom{m-1}{m-kp}.$$

In the equation, we can notice that the term $(-1)^{m-kp}\binom{m-1}{m-kp}$ is the coefficient of X^{m-pk} in the Taylor expansion of $(1+X)^{-kp}$. Therefore, $a(m)$ is actually the coefficient of X^m in the Taylor expansion of $\sum_{k=1}^{\infty} pX^{kp}(1+X)^{-kp}$.

$$\sum_{k=1}^{\infty} pX^{kp}(1+X)^{-kp} = p\sum_{k=1}^{\infty}(\frac{X}{X+1})^{kp} = p\frac{(1+X)^p}{(1+X)^p - X^p}$$

We can get a m-th coefficient of Taylor expansion from following equation:

$$p\frac{(1+X)^p}{(1+X)^p - X^p} = p\frac{(1+X)^p}{1+B(X)} = p(1+X)^p(1 - B(X) + B(X)^2 - \cdots).$$

Because $B(X)$ is multiple of pX, the coefficient of X^m can be obtained from a finite number of powers of $B(X)$. We can also find out the degree of $B(X)$ is $p-1$, so

$$\mathsf{Deg}(p(1+X)^p(1 - B(X) + \cdots + (-1)^{(e-2)}B(X)^{(e-2)})) = (e-1)(p-1)+1.$$

Hence these terms do not contribute to X^m. This means that $a(m)$ is m-th coefficient of

$$p(1+X)^p B(X)^{e-1}\sum_{i=0}^{\infty}(-1)^i B(X)^i$$

which is multiple of p^e (since $B(X)$ is multiple of p). ∎

By the claim above, the p-adic valuation of $a(m)$ is larger than $\frac{m}{p-1}$ and it is trivial that the p-adic valuation of $m!$ is less than $\frac{m}{p-1}$. Therefore, we proved that the coefficients of $f(x)$ are p-integral. Indeed, we proved that $a(m)\binom{x}{m}$ is multiple of p^n for any integer when $m > (e-1)(p-1)+1$. This means that $\hat{f}(x) = f(x) \bmod p^e$ for all $x \in \mathbb{Z}_{p^e}$.

As a result, the degree $(e-1)(p-1)+1$ polynomial $f(x)$ satisfies the conditions in lemma for the least residue system. For balanced residue system, we can just replace $f(x)$ by $f(x + (p-1)/2)$.   □

Note that the above polynomial $f(x)$ removes the lowest base-p digit in an integer. It is also desirable sometimes to "retain" the lowest digit, while setting all the other digits to zero. This can be easily done via $g(x) = x - f(x)$. In the rest of the paper, we will denote such polynomial that retains the lowest digit in the balanced base-p representation by $G_{e,p}(x)$ (or $G_e(x)$ if p is clear from context). In other words, if $x \in \mathbb{Z}_{p^e}$ and $x \equiv x_0 \bmod p$ with $|x_0| \leq p/2$, then $G_e(x) = x_0 \bmod p^e$.

Example 4. *When $e = 2$, we have $f(x) = -x(x-1)\cdots(x-p+1)$ and $G_2(x) = x - f(x + (p-1)/2)$.*

We recall that in the previous method, it takes degree p^{e-i-1} and $(e-i-1)$ evaluations of polynomials of degree p to obtain $u_{i,e-i}$. With our lowest digit removing polynomial, it only takes degree $(e-i-1)(p-1)+1$. As a result, by combining the lifting polynomials and lowest digit removing polynomials, we can make the digit extraction algorithm faster with lower depth.

The following diagram illustrates how our new digit removal algorithm works. First, each blue digit is obtained by evaluating a lifting polynomial to the entry on its left. Then, the red digit on each row is obtained by evaluating the remaining lowest digit polynomial to the left-most digit on its row. Green digits are obtained by subtracting all the blue digits on the same diagonal from the input, and dividing by p. Finally, in order to remove the v lowest digits, we subtract all the red digits from the input.

$$
\begin{array}{cccccc}
u_{0,0} & u_{0,1} & \cdots & u_{0,v-2} & u_{0,v-1} & u_{0,e-1} \\
u_{1,0} & u_{1,1} & \cdots & u_{1,v-2} & & u_{1,e-2} \\
\vdots & & & & & \\
u_{v-2,0} & u_{v-2,1} & & u_{v-2,e-r+1} & & \\
u_{v-1,0} & & u_{v-1,e-v} & & &
\end{array}
$$

We remark that the major difference of this procedure is that we only need to populate the top left triangle of side length v, plus the right most v-by-1 diagonal, where as the previous method needs to populate the entire triangle of side length e.

Moreover, the red digits in our method has lower depth: in the previous method, the i-th red digit is obtained by evaluating lift polynomial $(e-i-1)$ times, hence its degree is p^{e-i-1} on top of the i-th green digit. However, in our method, its degree is only $(p-1)(e-i-1)+1$ on top of the i-th green digit, which has degree at most p^i, the total degree of the algorithm is bounded by the maximum degree over all the red digits, that is

$$\max_{0 \le i < r} p^i((e-1-i)(p-1)+1).$$

Since each individual term is bounded by ep^v, the total degree of the procedure is at most ep^v. This is lower than p^{e-1} in the previous method when $v \le e-2$ and $p > e$.

3.3   Improved Algorithm for Removing Digits

We discuss one further optimization to remove v lowest digits in base p from an e-digit integer. If ℓ is an integer such that $p^\ell > (p-1)(e-1)+1$, then instead

---

Data: $x \in \mathbb{Z}_{p^e}$
Result: $x - [x]_{p^v} \bmod p^e$
// $F_i(x)$: lifting polynomial with $F_i(x + O(p^i)) = x + O(p^{i+1})$
// $G_i(x)$: lowest digit retain polynomial with $G_i(x) = [x]_p \bmod p^i$
Find largest ℓ such that $p^\ell < (p-1)(e-1)+1$;
Initialize res $= x$;
for $i \in [0, v)$ do

   // evaluate lowest digit retain polynomial
   $R_i = G_{e-i}(x')$;   // $R_i = x_i \bmod p^{e-i}$
   $R_i = R_i \cdot p^i$;   // $R_i = x_i p^i \bmod p^e$
   if $i < v-1$ then
      // evaluate lifting polynomial
      $L_{i,0} = F_1(x')$
   end
   for $j \in [0, \ell-2)$ do
      if $i + j < v - 1$ then
         $L_{i,j+1} = F_{j+2}(L_{i,j})$
      end
   end
   if $i < v-1$ then
      $x' = x$;
      for $j \in [0, i+1)$ do
         if $i - j > \ell - 2$ then
            $x' = x' - R_j$
         end
         else
            $x' = x' - L_{j,i-j}$
         end
      end
   end
   res $=$ res $- R_i$;
end
return res;

Algorithm 1. Removing v lowest digits from $x \in \mathbb{Z}_{p^e}$

of using lifting polynomials to obtain the ℓ-th digit, we can just use the result of evaluating the G_i polynomial (or, the red digit) to obtain the green digit in the next row. This saves some work and also lowers the depth of the overall procedure. This optimization is incorporated into Algorithm 1.

The depth and computation cost of Algorithm 1 is summarized in Theorem 5. The depth is simply the maximum depth of all the removed digits. To determine the computational cost to evaluate Algorithm 1 homomorphically, we need to specify the unit of measurement. Since scalar multiplication is much faster than FHE schemes than ciphertext multiplication, we choose to measure the computational cost by the number of ciphertext multiplications. The Paterson-Stockmeyer algorithm [29] evaluates a polynomial of degree d with $\sim \sqrt{2d}$ non-constant multiplications, and we use that as the base of our estimate.

Theorem 5. *Algorithm 1 is correct. Its depth is bounded above by*

$$\log(ep^v) = v \log(p) + \log(e).$$

The number of non-constant multiplications is asymptotically equal to $\sqrt{2pev}$.

Table 1 compares the asymptotic depth and number of non-constant multiplications between our method for digit removal and the method of [25]. From the table, we see that the advantage of our method grows with the difference $e - v$. In the bootstrapping scenario, we have $e - v = r$, the exponent of the plaintext modulus. Hence, our algorithm compares favorably for larger values of r.

Table 1. Complexity of DigitRemove(p, e, v)

Method	Depth	No. ciphertext multiplications
[25]	$e \log(p)$	$\frac{1}{2} e^2 \sqrt{2p}$
This work	$v \log(p) + \log(e)$	$\sqrt{2pev}$

4   Improved Bootstrapping for FV and BGV

4.1   Reviewing the Method of [25]

The bootstrapping for FV scheme follows the main steps from [25] for the BGV scheme, while we make two modifications in modulus switching and digit extraction. First, we review the procedure in [25].

Modulus Switching. One fixes some $q' < q$ and compute a new ciphertext c' which encrypts the same plaintext but has much smaller size.

$$\text{Enc}(\mathbf{m}(x)) = \text{Enc}((\mathbf{m}_0(x), \cdots, \mathbf{m}_{k-1}(x)))$$
$$\downarrow \quad \text{Modulus Switching and Dot Product}$$
$$\text{Enc}(\mathbf{m}(x) \cdot p^{e-r} + e(x))$$
$$\downarrow \quad \text{LinearTransformation}$$
$$\text{Enc}((m_0 \cdot p^{e-r} + e_0, \cdots, m_{k-1} \cdot p^{e-r} + e_{k-1})), \cdots \text{Enc}((m_{n-k} \cdot p^{e-r} + e_{n-k}, \cdots, m_{n-1} \cdot p^{e-r} + e_{n-1}))$$
$$\downarrow \quad d \text{ number of Digit Extraction}$$
$$\text{Enc}((m_0, m_1, \cdots, m_{k-1})), \text{Enc}((m_k, \cdots, m_{2k-1})), \cdots \text{Enc}((m_{n-k}, \cdots, m_{n-1}))$$
$$\downarrow \quad \text{InverseLinearTransformation}$$
$$\text{Enc}(\mathbf{m}(x)) = \text{Enc}(\mathbf{m}_0(x), \cdots, \mathbf{m}_{k-1}(x))$$

Fig. 1. Bootstrapping procedure

Dot Product with Bootstrapping Key. Here we compute homomorphically the dot product $\langle c', \mathfrak{s} \rangle$, where $\mathfrak{s}$ is an encryption of a new secret key s' under a large coefficient modulus Q and a new plaintext modulus $t' = p^e$. The result of this step is an encryption of $m + tv$ under the new parameters (s', t', Q).

Linear Transformation. Let d denote the multiplicative order of p in $\mathbb{Z}_m^*$ and $k = n/d$ be the number of slots supported in plaintext batching. Suppose the input to linear transform is an encryption of $\sum_{i=0}^{n-1} a_i x^i$, then the output of this step is d ciphertexts $C_0, \ldots, C_{d-1}$, where C_j is a batch encryption of $(a_{jk}, a_{jk+1}, \ldots, a_{jk+k-1})$.

Digit Extraction. When the above steps are done, we obtain d ciphertexts, where the first ciphertext is a batch encryption of

$$(m_0 \cdot p^{e-r} + e_0, m_1 \cdot p^{e-r} + e_1, \cdots, m_{k-1} \cdot p^{e-r} + e_{k-1}).$$

Assuming that $|e_i| \leq \frac{p^{e-r}}{2}$ for each i, we will apply Algorithm 1 to remove the lower digits e_i, resulting in d new ciphertexts encrypting Δm_i for $0 \leq i < n$ in their slots. Then we perform a free division to get d ciphertexts, encrypting m_i in their slots.

Inverse Linear Transformation. Finally, we apply another linear transformation which combines the d ciphertexts into one single ciphertext encrypting $m(x)$.

4.2   Our Modifications

FV. Suppose $t = p^r$ is a prime power, and we have a ciphertext (c_0, c_1) modulo q. Here, instead of switching to a modulus q' co-prime to p as done in BGV, we switch to $q' = p^e$, and obtain ciphertext (c'_0, c'_1) such that

$$c'_0 + c'_1 s = p^{e-r} m + v + \alpha p^e.$$

Then, one input ciphertext to the digit extraction step will be a batch encryption

$$\mathsf{Enc}((p^{e-r}m_0 + v_0, \ldots, p^{e-r}m_k + v_k))$$

under plaintext modulus p^e. Hence this step requires $\mathsf{DigitRemove}(p, e, e - r)$.

BGV. To apply our ideas to the digit extraction step in BGV bootstrapping, we simply replace the algorithm in [25] with our digit removal Algorithm 1.

4.3 Comparing Bootstrapping Complexities

The major difference in the complexities of bootstrapping between the two schemes comes from the parameter e. In case of FV, by Lemma 1, we can choose (roughly) $e = r + \log_p(||s||_1)$. On the other hand, the estimate of e for correct bootstrapping in [25] for the BGV scheme is

$$e \geq 2r + \log_p(||s||_1).$$

We can analyze the impact of this difference on the depth of digit removal, and therefore on the depth of bootstrapping. Setting $v = e - r$ in Theorem 5, the depth for the BGV case is

$$(r + \log_p(||s||_1)) \log p + \log(2r + \log_p(||s||_1)).$$

Substituting $r = \log_p(t)$ into the above formula and throwing away lower order terms, we obtain the improved depth for the digit extraction in step BGV bootstrapping as

$$\log t + \log(||s||_1) + \log(\log_p(t^2 \cdot ||s||_1)) \approx \log t + \log(||s||_1).$$

Note that the depth grows linearly with the logarithm of the plaintext modulus t. On the other hand, the depth in the FV case turns out to be

$$\log(||s||_1) + \log(\log_p(t \cdot ||s||_1)).$$

which only scales with $\log \log t$. This is smaller than BGV in the large plaintext modulus regime.

We can also compare the number of ciphertext multiplications needed for the digit extraction procedures. Replacing v with $e - r$ in the second formula in Theorem 5 and letting $e = 2r + \log_p(||s||_1)$ for BGV (resp. $e = r + \log_p(||s||_1)$ for FV), we see that the number of ciphertext multiplications for BGV is asymptotically equal to

$$\frac{\sqrt{2p}}{(\log p)^{3/2}} (2\log(t) + \log(||s||_1))^{1/2}(\log(t) + \log(||s||_1)).$$

In the FV case, the number of ciphertext multiplications is asymptotically equal to

$$\frac{\sqrt{2p}}{(\log p)^{3/2}} (\log(t) + \log(||s||_1))^{1/2} \log(||s||_1).$$

Hence when t is large, the digit extraction procedure in bootstrapping requires less work for FV than BGV.

For completeness, we also analyze the original digit extraction method in BGV bootstrapping. Recall that the previous algorithm has depth $(e-1)\log p$, and takes about $\frac{1}{2}e^2$ homomorphic evaluations of polynomials of degree p. If we use the Paterson-Stockmeyer method for polynomial evaluation, then the total amount of ciphertext multiplications is roughly $\frac{1}{2}e^2\sqrt{2p}$. Plugging in the lower bound $e \geq 2r + \log_p(||s||_1)$, we obtain an estimate of depth and work needed for the digit extraction step in the original BGV bootstrapping method in [25]. Table 2 summarizes the cost for three different methods.

Table 2. Asymptotic complexity of digit extraction step in bootstrapping. Here $h = ||s||_1$ is the 1-norm of the secret key, and $t = p^r$ is the plaintext modulus.

Method	Depth	No. ciphertext multiplications
[25] (BGV)	$2\log(t) + \log(h)$	$\frac{\sqrt{2p}}{2(\log p)^2}(2\log(t) + \log(h))^2$
This work (BGV)	$\log(t) + \log(h)$	$\frac{\sqrt{2p}}{(\log p)^{3/2}}(2\log(t) + \log(h))^{1/2}(\log(t) + \log(h))$
This work (FV)	$\log\log(t) + \log(h)$	$\frac{\sqrt{2p}}{(\log p)^{3/2}}(\log(t) + \log(h))^{1/2}\log(h)$

Fixing p and h in the last column of Table 2, we can see how the number of multiplications grows with $\log t$. The method in [25] scales by $(\log t)^2$, while our new method for BGV improves it to $(\log t)^{3/2}$. In the FV case, the number of multiplications scales by only $(\log t)^{1/2}$.

Remark 1. As another advantage of our revised BGV bootstrapping, we make a remark on security. From Table 2, we see that in order for bootstrapping to be more efficient, it is advantageous to use a secret key with smaller 1-norm. For this reason, both [25] and this work choose to use a sparse secret key, and a recent work [2] shows that sparseness can be exploited in the attacks. To resolve this, note that it is easy to keep the security level in our situation: since our method reduces the overall depth for the large plaintext modulus case, we could use a smaller modulus q, which increases the security back to a desired level.

4.4   Slim Bootstrapping Algorithm

The bootstrapping algorithm for FV and BGV is expensive also due to the d repetitions of digit extraction. For some parameters, the extension degree d can be large. However, many interesting applications requires arithmetic over $\mathbb{Z}_{p^r}$ rather than its degree-d extension ring, making it hard to utilize the full plaintext space.

Therefore we will introduce one more bootstrapping algorithm which is called "slim" bootstrapping. This bootstrapping algorithm works with the plaintext space $\mathbb{Z}_t^k$, embedded as a subspace of R_t through the batching isomorphism.

This method can be adapted using almost the same algorithm as the original bootstrapping algorithm, except that we only need to perform one digit extraction operation, hence it is roughly d times faster than the full bootstrapping algorithm. Also, we need to revise the linear transformation and inverse linear transformation slightly. We give an outline of our slim bootstrapping algorithm below (Fig. 2).

$$\text{Enc}(m_0, m_1, m_2, \cdots, m_{k-1})$$
$$\downarrow \quad \text{InverseLinearTransformation}$$
$$\text{Enc}(m_0 + m_1 x^d + \cdots + m_{k-1} x^{d(k-1)})$$
$$\downarrow \quad \text{Modulus Switching and Dot Product}$$
$$\text{Enc}(m(x) \cdot p^{e-r} + e(x))$$
$$\downarrow \quad \text{LinearTransformation}$$
$$\text{Enc}(m_0 \cdot p^{e-r} + e_0, \cdots, m_{k-1} \cdot p^{e-r} + e_{k-1})$$
$$\downarrow \quad \text{Digit Extraction}$$
$$\text{Enc}(m_0, m_1, m_2, \cdots, m_{k-1})$$

Fig. 2. Slim bootstrapping

Inverse Linear Transformation. We take as input a batch encryption of $(m_1 \ldots, m_k) \in \mathbb{Z}_{p^r}^k$. In the first step, we apply an "inverse" linear transformation to obtain an encryption of $m_1 + m_2 x^d + \ldots + m_k x^{d(k-1)}$. This can be done using k slot permutations and k plaintext multiplications.

Modulus Switching and Dot Product with Bootstrapping Key. These two steps are exactly the same as the full bootstrapping procedure. After these steps, we obtain a (low-noise) encryption of

$$(\Delta m_1 + v_1 + (\Delta m_2 + v_2)x^d + \ldots + (\Delta m_k + v_k)x^{d(k-1)}).$$

Linear Transformation. In this step, we apply another linear transformation consisting of k slot permutations and k scalar multiplications to obtain a batch encryption of $(\Delta m_1 + v_1, \ldots, \Delta m_k + v_k)$. Details of this step can be found in the appendix.

Digit Extraction. Then, we apply digit-removal algorithm to remove the noise coefficients v_i, resulting in a batch encryption of $(\Delta m_1, \ldots, \Delta m_k)$. We then execute the free division and obtain a batch encryption of $(m_1, \ldots, m_k)$. This completes the slim bootstrapping process.

5 Implementation and Performance

We implemented both the full mode and the slim mode of bootstrapping for FV in the SEAL library. We also implemented our revised digit extraction procedure in HElib. Since SEAL only supports power-of-two cyclotomic rings, and p needs to be co-prime to m in order to use batching, we can not use $p = 2$ for SEAL bootstrapping. Instead we chose $p = 127$ and $p = 257$ because they give more slots among primes of reasonable size.

The following tables in this section illustrate some results. We used sparse secrets with hamming weight 64 and 128, and we estimated security levels using Martin Albrecht's LWE estimator [2].

Table 3. Comparison of digit removal algorithms in HElib (Toshiba Portege Z30t-C laptop with 2.6 GHz CPU and 8 GB memory)

(p, e, v)	[25]		Our method	
	Timing (s)	Before/After level	Timing (s)	Before/After level
$(2, 11, 5)$	15	23/3	16	23/10
$(2, 21, 13)$	264	56/16	**239**	56/**22**
$(5, 6, 3)$	49.5	39/5	**30**	39/**13**
$(17, 4, 2)$	61.2	38/5	**35.5**	38/**14**
$(31, 3, 1)$	26.3	32/8	**12.13**	32/**18**
$(127, 3, 1)$	73.2	42/3	**38**	42/**20**

We implemented Algorithm 1 in HElib and compared with the results of the original HElib implementation for removing v digits from e digits. From Table 3, we see that for $e \geq v+2$ and large p, our digit removal procedure can outperform the current HElib implementation in both depth and work. Therefore, for these settings, we can replace the digit extraction procedure in the recryption function in HElib, and obtain a direct improvement on after level and time for recryption. When $p = 2$ and r, e are small, the current HElib implementation can be faster due to the fact that the lifting polynomial is $F_e(x) = x^2$ and squaring operation is faster than generic multiplication. Also, when $e = v + 1$, i.e., the task is to remove all digits except the highest one, our digit removal method has similar performance as the HElib counterpart.

Tables 4 and 5 present timing results for the full and slim modes of bootstrapping for FV implemented in SEAL. In both tables, the column labeled "recrypt init. time" shows the time to compute the necessary data needed in bootstrapping. The "recrypt time" column shows the time it takes to perform one bootstrapping. The before (resp. after) level shows the maximal depth of circuit that can be evaluated on a freshly encrypted ciphertext (resp. freshly bootstrapped ciphertext). Here $R(p^r, d)$ denotes a finite ring with degree d over base ring $\mathbb{Z}_{p^r}$, and $GF(p^r)$ denotes the finite field with p^r elements.

Table 4. Time table for bootstrapping for FV scheme, hw $= 128$ (Intel(R) Core(TM) i7-4770 CPU with 3.4 GHZ CPU and 32 GB memory)

Parameters					Result			
n	$\log q$	Plaintext space	Slots	Security	Fresh/After level	Recrypt time (s)	Memory usage (GB)	Recrypt init. time (s)
16384	558	$GF(127^{256})$	64	92.9	24/7	2027	8.9	193
16384	558	$GF(257^{128})$	128	92.9	22/4	1381	7.5	242
32768	806	$R(127^2, 256)$	64	126.2	32/12	21295	27.6	658
32768	806	$R(257^2, 128)$	128	126.2	23/6	11753	26.6	732

Table 5. Time table for slim bootstrapping for FV scheme, hw $= 128$ (Intel(R) Core(TM) i7-4770 CPU with 3.4 GHZ CPU and 32 GB memory)

Parameters					Result			
n	$\log q$	Plaintext space	Number of slots	Security parameter	Fresh/After level	Recrypt init. time (s)	Memory usage (GB)	Recrypt time (s)
16384	558	$\mathbb{Z}_{127}$	64	92.9	23/10	57	2.0	6.75
32768	806	$\mathbb{Z}_{127^2}$	64	126.2	25/11	59	2.0	30.2
32768	806	$\mathbb{Z}_{127^3}$	64	126.2	20/6	257	8.9	34.5
16384	558	$\mathbb{Z}_{257}$	128	92.9	22/7	59	2.0	10.8
32768	806	$\mathbb{Z}_{257}$	128	126.2	31/15	207	7.4	36.8
32768	806	$\mathbb{Z}_{257^2}$	128	126.2	23/7	196	7.4	42.1

Comparing the corresponding entries from Tables 4 and 5, we see that the slim mode of bootstrapping is either close to or more than d times faster than the full mode.

6   Future Directions

In this work, we designed bootstrapping algorithms for the FV scheme whose depth depend linearly on $\log \log t$. For the BGV scheme, we were able to improve the dependence on t from $2 \log t$ to $\log t$. One interesting direction is to explore whether we can further improve the bootstrapping depth for BGV.

We also presented a slim mode of bootstrapping, which operates on a subspace of the plaintext space equivalent to a vector over $\mathbb{Z}_{p^r}$. The slim mode has a similar throughput as the full mode while being much faster. For example, it takes less than 7 s to bootstrap a vector in $\mathbb{Z}_{127}^{64}$ with after level 10. However, the ciphertext sizes of the slim mode are the same as those of the full mode, resulting in a larger ciphertext/message expansion ratio. It would be interesting to investigate whether we could reduce the ciphertext sizes while keeping the performance results.

Acknowledgements. We wish to thank Kim Laine, Amir Jalali and Zhicong Huang for implementing significant performance optimizations to SEAL. We thank Shai Halevi for helpful discussions on bootstrapping in HElib.

A Optimizing the Linear Transform for Slim Bootstrapping

In our slim mode of bootstrapping, we used a linear transform which has the following property: the input is an encryption of $\sum m_i x^i$, and the output is a batch encryption of $(m_0, m_d, \ldots, m_{d(k-1)})$. A straightforward implementation of this functionality requires n slot permutations and n scalar multiplications. However, in the case when n is a power of 2, we can break down the linear transform into two parts, which we call coefficient selection and sparse linear transform. This reduces the number of slot permutations to $\log(d) + k$ and the number of scalar multiplications to k.

A.1 Coefficient Selection

The first part of the optimized linear transform functionality can be viewed as a coefficient selection. This process gets input $\mathsf{Enc}(m(x))$ and outputs $\mathsf{Enc}(m'(x))$ with $m'(x) = \sum_{i=0}^{n/d-1} m_{id} \cdot x^{id}$. In other words, it selects the coefficients of $m(x)$ where the exponents of x are divisible by d. The following algorithm is specified to the case when n is a power of two. Using the property that $x^n = -1$ in the ring R, we can construct an automorphism ϕ_i of R such that

$$\phi_i : X^{2^i} \to X^{n+2^i} = -X^{2^i}.$$

For example, $\phi_0(\cdot)$ negates all odd coefficients, because ϕ_0 maps X to $-X$. This means that $\frac{1}{2}(\phi_0(m(x)) + m(x))$ will remove all odd terms and double the even terms. Using this property, we construct a recursive algorithm which return $m'(x) = \sum_{i=0}^{n/d-1} m_{id} \cdot x^{id}$ for power of two d.

- For given $m(x)$, First compute $m_0(x) = m(x) + \phi_0(m(x))$.
- Recursively compute $m_i(x) = \phi_i(m_{i-1}(x)) + m_{i-1}(x)$ for $1 \leq i \leq \log_2 d$.
- Return $m'(x) = d^{-1} \cdot m_{\log_2 d} \bmod t$ for plain modulus t.

The function $\phi_i : X \to X^{\frac{n+2^i}{2^i}}$ can be evaluated homomorphically by using the same technique used in slot permutation. Another operation is just multiplying by $d^{-1} \bmod t$. Hence we can obtain $\mathsf{Enc}(m'(x))$. This process needs $\log d$ slot permutations and additions.

A.2 Sparse Linear Transform

The desired functionality of the sparse linear transform is: take as input an encryption c of $\sum m_i x^{id}$ and output a batch encryption of $(m_0, m_1 \ldots, m_{k-1})$.

We claim that this functionality can be expressed as $\sum_{i=0}^{k-1} \lambda_i \sigma_{s_i}(c)$, where λ_i are pre-computed polynomials in R_t and the s_i form a set of representatives of $\mathbb{Z}_m^*/\langle p \rangle$. This is because the input plaintext only has k nonzero coefficients $m_0, \ldots, m_{k-1}$. Hence for each i it is possible to write m_i as a linear combination of the evaluations of the input at k different roots of unity. Therefore, this step only requires k slot permutations and k plaintext multiplications. We can also adapt the babystep-giantstep method to reduce the number of slot permutations to $O(\sqrt{k})$, and we omit further details.

B Memory Usage

In our implementation of the bootstrapping procedure in SEAL, we pre-compute some data which are used in the linear transforms. The major part of the memory consumption consists of slot-permutation keys and plaintext polynomials. More precisely, each plaintext polynomial has size $n \log t$ bits, and the size of one slot-permutation key in SEAL is $(2n \log q) \cdot \lfloor \frac{\log q}{62} \rfloor$.

Here we report the number of such keys and plaintext polynomials used in our bootstrapping. In the full mode, we need $2\sqrt{n}$ slot-permutation keys, and $2\sqrt{n} + d + k$ plaintext polynomials.

On the other hand, the slim mode of bootstrapping in SEAL requires considerably less memory. Both inverse linear transform and the linear transform can be implemented via the babystep-giantstep technique, each using only $2\sqrt{k}$ slot-permutation keys and k plaintext polynomials.

C Proofs

C.1 Proof of Lemma 1

Lemma 1. *Suppose $c_0 + c_1 s = \Delta m + v + aq$ is a ciphertext in FV with $|v| < \Delta/4$. if $q' > 4t(1 + ||s||_1)$, and (c_0', c_1') is the ciphertext after switching the modulus to q', then (c_0', c_1') also decrypts to m.*

Proof. We define the invariant noise to be the term v_{inv} such that

$$\frac{t}{q}(c_0 + c_1 s) = m + v_{inv} + rt.$$

Decryption is correct as long as $||v_{inv}|| < \frac{1}{2}$. Now introducing the new modulus q', we have

$$\frac{t}{q'}\left(\frac{q'}{q}c_0 + \frac{q'}{q}c_1 s\right) = m + v_{inv} + rt.$$

Taking nearest integers of the coefficients on the left hand side, we arrive at

$$\frac{t}{q'}\left(\lfloor \frac{q'}{q}c_0 \rceil + \lfloor \frac{q'}{q}c_1 \rceil s\right) = m + v_{inv} + rt + \delta,$$

with the rounding error $||\delta|| \le t/q'(1 + ||s||_1)$. Thus the new invariant noise is

$$v_{inv'} = v_{inv} + \delta$$

We need $||\delta|| < 1/4$ for correct decryption. Hence the lower bound on q' is

$$q' > 4t(1 + ||s||_1).$$

C.2   Proof of Theorem 5

Proof. Correctness of Algorithm 1 is easy to show. In fact, the only place we deviate from the algorithm in [25] for digit extraction is that we used the digits R_i to replace $x_{i,j}$ in certain places. Since R_i has lowest digit x_i followed by $(e - i - 1)$ zeros, we can actually use it to replace $x_{i,j}$ for any $j \le e - i - 1$ and still maintain the correctness.

To analyze the depth, note that we used polynomials of degree p^i to compute $z_{i,i}$ for $0 \le i \le v - 1$. Then, to compute $z_{i,e-1-i}$, a polynomial of degree $(e - 1 - i)(p - 1) + 1$ is used. Since the final result is a sum of the terms $z_{i,e-1-i}$ for $0 \le i < v$, the degree of the entire algorithm is given by

$$\max_{0 \le i < v} p^i((e - 1 - i)(p - 1) + 1)$$

Since each individual term above is bounded by ep^v, the degree is at most ep^v. Hence the depth of the algorithm is bounded by $\log(e) + v\log(p)$.

We now estimate the amount of work of our algorithm in terms of non-constant multiplications. The work consists of two parts: evaluating lift polynomials and lowest digit removal polynomials. Let $W(n)$ denote the number of non-constant multiplications to evaluate a polynomial of degree n. Then the total work is

$$\sum_{i=1}^{v} W((e - i)(p - 1) + 1) + \ell v W(p)$$

where $\ell = \lfloor \log_p((e - 1)(p - 1) + 1) \rfloor$ is the optimization parameter used in Algorithm 1. Since we used the Paterson-Stockmeyer algorithm for polynomial evaluation, we have $W(n) \sim \sqrt{2n}$. Substituting this estimate into the above formula, we obtain

$$\sum_{i=1}^{v} \sqrt{2((e - i)(p - 1) + 1)} + \ell v\sqrt{2p}$$

$$\sim \sqrt{2p}\sum_{i=1}^{v} \sqrt{e - i} + \sqrt{2p}(1 + \log_p(e))v$$

$$\sim \sqrt{2p}v(\sqrt{e} + \log_p(e))$$

$$\sim \sqrt{2pe}v.$$

This completes the proof.

References

1. Aguilar-Melchor, C., Barrier, J., Guelton, S., Guinet, A., Killijian, M.-O., Lepoint, T.: NFLLIB: NTT-Based Fast Lattice Library. In: Sako, K. (ed.) CT-RSA 2016. LNCS, vol. 9610, pp. 341–356. Springer, Cham (2016). https://doi.org/10.1007/978-3-319-29485-8_20

2. Albrecht, M.R., Player, R., Scott, S.: On the concrete hardness of learning with errors. J. Math. Cryptol. **9**(3), 169–203 (2015)

3. Biasse, J.-F., Ruiz, L.: FHEW with efficient multibit bootstrapping. In: Lauter, K., Rodríguez-Henríquez, F. (eds.) LATINCRYPT 2015. LNCS, vol. 9230, pp. 119–135. Springer, Cham (2015). https://doi.org/10.1007/978-3-319-22174-8_7

4. Bonte, C., Bootland, C., Bos, J.W., Castryck, W., Iliashenko, I., Vercauteren, F.: Faster homomorphic function evaluation using non-integral base encoding. Cryptology ePrint Archive, Report 2017/333 (2017). http://eprint.iacr.org/2017/333

5. Bos, J.W., Castryck, W., Iliashenko, I., Vercauteren, F.: Privacy-friendly forecasting for the smart grid using homomorphic encryption and the group method of data handling. Cryptology ePrint Archive, Report 2016/1117 (2016). http://eprint.iacr.org/2016/1117

6. Bos, J.W., Lauter, K., Loftus, J., Naehrig, M.: Improved security for a ring-based fully homomorphic encryption scheme. In: Stam, M. (ed.) IMACC 2013. LNCS, vol. 8308, pp. 45–64. Springer, Heidelberg (2013). https://doi.org/10.1007/978-3-642-45239-0_4

7. Brakerski, Z.: Fully homomorphic encryption without modulus switching from classical GapSVP. In: Safavi-Naini, R., Canetti, R. (eds.) CRYPTO 2012. LNCS, vol. 7417, pp. 868–886. Springer, Heidelberg (2012). https://doi.org/10.1007/978-3-642-32009-5_50

8. Brakerski, Z., Gentry, C., Vaikuntanathan, V.: (Leveled) fully homomorphic encryption without bootstrapping. In: Proceedings of the 3rd Innovations in Theoretical Computer Science Conference, pp. 309–325. ACM (2012)

9. Brakerski, Z., Vaikuntanathan, V.: Fully homomorphic encryption from ring-LWE and security for key dependent messages. In: Rogaway, P. (ed.) CRYPTO 2011. LNCS, vol. 6841, pp. 505–524. Springer, Heidelberg (2011). https://doi.org/10.1007/978-3-642-22792-9_29

10. Brakerski, Z., Vaikuntanathan, V.: Efficient fully homomorphic encryption from (standard) LWE. SIAM J. Comput. **43**(2), 831–871 (2014)

11. Cheon, J.H., Han, K., Kim, D.: Faster bootstrapping of FHE over the integers. IACR Cryptology ePrint Archive, 2017:79 (2017)

12. Cheon, J.H., Kim, A., Kim, M., Song, Y.: Homomorphic encryption for arithmetic of approximate numbers. Cryptology ePrint Archive, Report 2016/421 (2016). http://eprint.iacr.org/2016/421

13. Cheon, J.H., Stehlé, D.: Fully homomophic encryption over the integers revisited. In: Oswald, E., Fischlin, M. (eds.) EUROCRYPT 2015. LNCS, vol. 9056, pp. 513–536. Springer, Heidelberg (2015). https://doi.org/10.1007/978-3-662-46800-5_20

14. Chillotti, I., Gama, N., Georgieva, M., Izabachène, M.: Faster fully homomorphic encryption: bootstrapping in less than 0.1 seconds. In: Cheon, J.H., Takagi, T. (eds.) ASIACRYPT 2016. LNCS, vol. 10031, pp. 3–33. Springer, Heidelberg (2016). https://doi.org/10.1007/978-3-662-53887-6_1

15. Chillotti, I., Gama, N., Georgieva, M., Izabachène, M.: Improving TFHE: faster packed homomorphic operations and efficient circuit bootstrapping (2017). https://eprint.iacr.org/2017/430

16. Coron, J.-S., Lepoint, T., Tibouchi, M.: Scale-invariant fully homomorphic encryption over the integers. In: Krawczyk, H. (ed.) PKC 2014. LNCS, vol. 8383, pp. 311–328. Springer, Heidelberg (2014). https://doi.org/10.1007/978-3-642-54631-0_18

17. Costache, A., Smart, N.P., Vivek, S., Waller, A.: Fixed point arithmetic in SHE scheme. IACR Cryptology ePrint Archive, 2016:250 (2016)

18. Ducas, L., Micciancio, D.: FHEW: bootstrapping homomorphic encryption in less than a second. In: Oswald, E., Fischlin, M. (eds.) EUROCRYPT 2015. LNCS, vol. 9056, pp. 617–640. Springer, Heidelberg (2015). https://doi.org/10.1007/978-3-662-46800-5_24

19. Fan, J., Vercauteren, F.: Somewhat practical fully homomorphic encryption. Cryptology ePrint Archive, Report 2012/144 (2012). http://eprint.iacr.org/

20. Gentry, C.: Fully homomorphic encryption using ideal lattices. STOC **9**, 169–178 (2009)

21. Gentry, C., Halevi, S., Smart, N.P.: Better bootstrapping in fully homomorphic encryption. In: Fischlin, M., Buchmann, J., Manulis, M. (eds.) PKC 2012. LNCS, vol. 7293, pp. 1–16. Springer, Heidelberg (2012). https://doi.org/10.1007/978-3-642-30057-8_1

22. Gentry, C., Sahai, A., Waters, B.: Homomorphic encryption from learning with errors: conceptually-simpler, asymptotically-faster, attribute-based. In: Canetti, R., Garay, J.A. (eds.) CRYPTO 2013. LNCS, vol. 8042, pp. 75–92. Springer, Heidelberg (2013). https://doi.org/10.1007/978-3-642-40041-4_5

23. Gilad-Bachrach, R., Dowlin, N., Laine, K., Lauter, K., Naehrig, M., Wernsing, J., Cryptonets: applying neural networks to encrypted data with high throughput and accuracy. In: International Conference on Machine Learning, pp. 201–210 (2016)

24. Graepel, T., Lauter, K., Naehrig, M.: ML confidential: machine learning on encrypted data. In: Kwon, T., Lee, M.-K., Kwon, D. (eds.) ICISC 2012. LNCS, vol. 7839, pp. 1–21. Springer, Heidelberg (2013). https://doi.org/10.1007/978-3-642-37682-5_1

25. Halevi, S., Shoup, V.: Bootstrapping for HElib. In: Oswald, E., Fischlin, M. (eds.) EUROCRYPT 2015. LNCS, vol. 9056, pp. 641–670. Springer, Heidelberg (2015). https://doi.org/10.1007/978-3-662-46800-5_25

26. Griffin, M.: Lowest degree of polynomial that removes the first digit of an integer in base p. MathOverflow, Version 08 May 2017. https://mathoverflow.net/users/61910/michaelgriffin, https://mathoverflow.net/q/269282

27. Laine, K., Player, R.: Simple encrypted arithmetic library-SEAL (v2. 0). Technical report, September 2016

28. Nuida, K., Kurosawa, K.: (Batch) fully homomorphic encryption over integers for non-binary message spaces. In: Oswald, E., Fischlin, M. (eds.) EUROCRYPT 2015. LNCS, vol. 9056, pp. 537–555. Springer, Heidelberg (2015). https://doi.org/10.1007/978-3-662-46800-5_21

29. Paterson, M.S., Stockmeyer, L.J.: On the number of nonscalar multiplications necessary to evaluate polynomials. SIAM J. Comput. **2**(1), 60–66 (1973)

30. Smart, N.P., Vercauteren, F.: Fully homomorphic SIMD operations. Des. Codes Cryptograp. **292**, 1–25 (2014)

31. van Dijk, M., Gentry, C., Halevi, S., Vaikuntanathan, V.: Fully homomorphic encryption over the integers. In: Gilbert, H. (ed.) EUROCRYPT 2010. LNCS, vol. 6110, pp. 24–43. Springer, Heidelberg (2010). https://doi.org/10.1007/978-3-642-13190-5_2

Homomorphic SIM²D Operations:
Single Instruction Much More Data

Wouter Castryck$^{(\boxtimes)}$, Ilia Iliashenko, and Frederik Vercauteren

imec-Cosic, Department of Electrical Engineering, KU Leuven, Leuven, Belgium
{wouter.castryck,ilia.iliashenko,frederik.vercauteren}@esat.kuleuven.be

Abstract. In 2014, Smart and Vercauteren introduced a packing technique for homomorphic encryption schemes by decomposing the plaintext space using the Chinese Remainder Theorem. This technique allows to encrypt multiple data values simultaneously into one ciphertext and execute Single Instruction Multiple Data operations homomorphically. In this paper we improve and generalize their results by introducing a flexible Laurent polynomial encoding technique and by using a more fine-grained CRT decomposition of the plaintext space. The Laurent polynomial encoding provides a convenient common framework for all conventional ways in which input data types can be represented, e.g. finite field elements, integers, rationals, floats and complex numbers. Our methods greatly increase the packing capacity of the plaintext space, as well as one's flexibility in optimizing the system parameters with respect to efficiency and/or security.

1 Introduction

Homomorphic encryption allows to perform arithmetic operations on encrypted data without decryption. The idea stems from [26] where the authors introduced so-called 'privacy homomorphisms' from plaintext space to ciphertext space. In 2009, Gentry [21] presented the first fully homomorphic encryption scheme (FHE) using ideal lattices. This breakthrough result was followed by several variants and improvements [6–9,20,23] all using the same blueprint. One first constructs a *somewhat* homomorphic encryption (SHE) scheme that can homomorphically evaluate arithmetic circuits of limited depth and then turns this into a fully homomorphic scheme using a bootstrapping procedure. The security of these schemes relies on the presence of a noise component in the ciphertexts. This noise grows during arithmetic operations and eventually reaches a threshold beyond which the ciphertext can no longer be decrypted correctly. The bootstrapping procedure basically reduces the inherent noise by executing

This work was supported by the European Commission under the ICT programme with contract H2020-ICT-2014-1 644209 HEAT, and through the European Research Council under the FP7/2007-2013 programme with ERC Grant Agreement 615722 MOTMELSUM. The first author thanks Ghent University for its hospitality. The authors also thank the anonymous referees for some helpful remarks.

© International Association for Cryptologic Research 2018
J. B. Nielsen and V. Rijmen (Eds.): EUROCRYPT 2018, LNCS 10820, pp. 338–359, 2018.
https://doi.org/10.1007/978-3-319-78381-9_13

the decryption circuit homomorphically. Despite considerable effort in making bootstrapping more efficient [2,11,13,19], full fledged FHE is still rather slow, so implementers typically resort to using SHE schemes for practical applications.

The efficiency of homomorphic encryption schemes can be improved significantly by a judicious choice of plaintext space and encoding techniques for the common data types such as finite field elements, integers, rationals, floats and complex numbers. Concretely, throughout this paper we assume that the plaintext space is a ring of the form

$$R_t = \mathbb{Z}_t[X]/(\overline{f}(X))$$

where $t \geq 2$ is an integer called the plaintext modulus, and $\overline{f}(X)$ is the reduction modulo t of a monic irreducible polynomial $f(X) \in \mathbb{Z}[X]$ of degree $d \geq 1$. This setting is valid for most SHE schemes whose security relies on the Ring-LWE problem.[1] The degree d together with the ciphertext modulus q and the standard deviation σ of the initial noise distribution are the main security parameters, and these are typically determined by the required security level. The noise growth is influenced by d, q, σ, but also by the plaintext modulus t. A first optimization to decrease the noise growth is therefore to use a smaller plaintext space. Several encoding techniques [3,10,12,14,18,25] have been proposed whose goal is to 'spread out' the numerical input data as evenly as possible over the whole plaintext space, allowing for a smaller value of t. A second optimization, which can be combined with the first, is to decompose the plaintext space into smaller pieces using the Chinese Remainder Theorem (CRT) and run several computations in parallel [4,27]. Smart and Vercauteren [27] described how to carry out SIMD calculations in an SHE context by viewing R_t as the CRT composition of

$$\mathbb{Z}_t[X]/(\overline{f}_1(X)) \times \mathbb{Z}_t[X]/(\overline{f}_2(X)) \times \cdots \times \mathbb{Z}_t[X]/(\overline{f}_r(X)),$$

where $\overline{f}_1(X)\overline{f}_2(X)\cdots\overline{f}_r(X)$ is a factorization of $\overline{f}(X)$ into coprime factors. In fact, they concentrate on the case $t = 2$, but the above immediate generalization is discussed in [22]. We will refer to this decomposition of R_t as a *vertical* slicing of the plaintext space.

Contributions. Our first contribution is an improvement of the above SIMD approach by utilizing a more fine-grained CRT decomposition of the plaintext space. We do this by also taking into account factorizations of the plaintext modulus t. We will refer to the CRT decomposition

$$R_t \cong \mathbb{Z}[X]/(t_1, f(X)) \times \mathbb{Z}[X]/(t_2, f(X)) \times \cdots \times \mathbb{Z}[X]/(t_s, f(X)),$$

---

[1] A recent adaptation of the FV scheme due to Chen et al. [10] uses as plaintext modulus a linear polynomial $x - a$ instead of an integer t. The resulting plaintext space $R_{x-a} = \mathbb{Z}[X]/(X^n + 1, x - a) \cong \mathbb{Z}/(a^n + 1)$ has various nice features, both in terms of noise growth and in terms of packing capacity. However, the algebraic structure of R_{x-a} becomes more restrictive for CRT decomposition, so rings of this type will not be considered in this paper.

corresponding to a factorization $t = t_1 t_2 \cdots t_s$ into coprime factors, as a *horizontal* slicing of the plaintext space. The flexibility of our method stems partly from the fact that factorisations modulo the various t_i do not imply a global factorisation modulo t. This alternative type of slicing for SIMD purposes is not new (see e.g. [4]). However, by *combining* horizontal and vertical slicing as explained in Sect. 4, the plaintext space becomes subdivided in 'bricks' as depicted in Fig. 4. In our SIMD approach, which we call SIM^2D, each data slot corresponds to a set of such bricks (called a block) rather than one vertical or horizontal slice as considered in previous works. This results in a much more flexible but, at the same time, denser packing as described in Sect. 5. In Sect. 6 we provide several tools that can help in making an optimal choice of blocks. This includes slight alterations to t and/or $f(X)$ that lead to more fine-grained decompositions.

Our second contribution is a novel encoding technique for Laurent polynomials into a plaintext space of the form $R_t = \mathbb{Z}_t[X]/(\overline{f}(X))$ that works for general $\overline{f}$ (under the mild assumption that $\overline{f}(0)$ is an invertible element of $\mathbb{Z}_t$). Previous work [15] could only deal with the very special case of 2-power cyclotomic polynomials, due to concerns of mixing of integral and fractional parts. Our encoding technique is explained in Sect. 3. Encoding elements of the Laurent polynomial ring $\mathbb{Z}[X^{\pm 1}]$ serves as a convenient common framework for all customary encoding techniques: indeed, under $X \mapsto b$ the Laurent polynomials specialize to b-ary expansions for any choice of base $b \in \mathbb{C} \setminus \{0\}$. This framework allows to encode common data types such as finite field elements, integers, rationals, floats and complex numbers. Furthermore, we show that choosing different bases b for different blocks can be useful in optimizing the data packing (see Sect. 6).

Our algorithms for encoding, packing, unpacking and decoding are easy to implement (pseudo-code is provided) and extremely flexible to use. The overall goal is to provide a set of tools which together can be used to perform SIMD in an optimal way, given the constraints on the plaintext space imposed by security, efficiency and correctness requirements.

2   Preliminaries

2.1   Basic Notation

Vectors are denoted by bold letters such as $\mathbf{a}$ and when the individual coordinates are required, we write a row vector as $(a_1, \ldots, a_k)$. For a natural number r, we denote the set $\{1, \ldots, r\}$ by $[r]$. Similarly, for any $\ell, m \in \mathbb{Z}, \ell \leq m$, the set $\{\ell, \ell+1, \ldots, m-1, m\}$ is denoted by $[\ell, m]$. The quotient ring of integers modulo a natural number t is denoted $\mathbb{Z}_t$.

2.2   Laurent Polynomials

Most common numerical types (integers, rational, real or complex numbers) are represented as (finite) power series expansions in a certain base $b \in \mathbb{C} \setminus \{0\}$, using digits that are taken from some given subset of $\mathbb{Z}$. These expansions naturally correspond to Laurent polynomials with integral coefficients, i.e. elements of the ring $\mathbb{Z}[X^{\pm 1}]$.

Most frequently, an integral base $b > 1$ with digit set $\{0, \ldots, b-1\}$ is used in practice, such as binary $b = 2$ or ternary $b = 3$. For use in SHE schemes, several variations [3,12,14,18] have been proposed. For the purposes of this paper we mention the non-integral base non-adjacent form (NIBNAF) from [3] which is a very sparse expansion with respect to a real base $b \in (1,2)$ and using the digit set $\{-1,0,1\}$. All of these expansions can be thought of as the evaluations at $X = b$ of a Laurent polynomial with integral coefficients.

Example 1. The real number 2.3 can be approximated in base $b = 2$ using digits in $\{0,1\}$ as

$$2.3 \simeq 1 \cdot 2 + 1 \cdot 2^{-2} + 1 \cdot 2^{-5} + 1 \cdot 2^{-6},$$

which is the evaluation of the Laurent polynomial

$$1 \cdot X + 1 \cdot X^{-2} + 1 \cdot X^{-5} + 1 \cdot X^{-6} \in \mathbb{Z}[X^{\pm 1}]$$

at $X = b = 2$.

Recall that in general, any Laurent polynomial $a(X) \in \mathbb{Z}[X^{\pm 1}]$ can be written as

$$a(X) = a_\ell X^\ell + \cdots + a_{m-1} X^{m-1} + a_m X^m \tag{1}$$

where $a_i \in \mathbb{Z}$ for every $i \in [\ell, m]$, $a_\ell, a_m \neq 0$ and $\ell \leq m$. For a modulus t (which will be clear from the context) we write $\bar{a}(X)$ for the Laurent polynomial in $\mathbb{Z}_t[X^{\pm 1}]$ obtained by reducing all coefficients.

Definition 1. For an integral Laurent polynomial $a(X) \in \mathbb{Z}[X^{\pm 1}]$ represented as in Eq. (1), we define the *bounding box* of $a(X)$ as the tuple (w, h) with $w = m - \ell$ and $h = \log_2(\max_i a_i - \min_i a_i + 1)$ the sizes of the exponent and the coefficient ranges of $a(X)$.

We represent the bounding box graphically with a rectangle of width w and height h (Fig. 1).

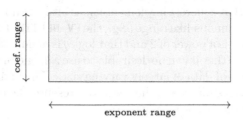

Fig. 1. The bounding box of a polynomial.

2.3 Plaintext Space

Most SHE schemes utilize quotient rings of the form

$$R = \mathbb{Z}[X]/(f(X))$$

where $f(X) \in \mathbb{Z}[X]$ is a monic irreducible polynomial of degree d. The plaintext space is typically represented as a quotient ring $R_t = \mathbb{Z}_t[X]/(\overline{f}(X))$ for an integral plaintext modulus t. Similarly, the ciphertext space is defined as $R_q = \mathbb{Z}_q[X]/(\overline{f}(X))$ where $q \gg t$. Another important parameter is the standard deviation σ of the discretized Gaussian distribution from which the SHE encryption scheme samples its noise, which is embedded into the ciphertexts.

Typically, one first sets the parameters q, d and σ, primarily as functions of the security level, in order to prevent all known attacks on the underlying lattice problems [1]. Afterwards, the plaintext modulus t is selected, subject to two constraints. Firstly, it is bounded from above, which stems from the fact that the embedded noise grows during arithmetic operations up to a critical threshold above which ciphertexts can no longer be decrypted. Since the plaintext modulus directly affects the noise growth in ciphertexts, one can find a maximal t for which the decryption remains correct while evaluating a given arithmetic circuit $\mathcal{C}$. We denote this bound by $t_{\mathcal{C}}^{\max}$. If it is impossible to satisfy this bound then one can use the Chinese Remainder Theorem to split the computation into smaller parts, as explained in Remark 3; see also [4]. Secondly, as explained in the next section, the plaintext modulus t is bounded from below by some value $t_{\mathcal{C}}^{\min}$ which depends on the input data and on the way the latter is encoded, and which ensures correct decoding.

Remark 1. The values of q, d, σ are not uniquely determined by the security level. Therefore, one can try to use the remaining freedom to target a specific value of $t_{\mathcal{C}}^{\max}$. In the remainder of the paper, we will assume that $t_{\mathcal{C}}^{\max}$ is given, and our aim is to utilize the available plaintext space in an optimal way. One motivation for targeting maximal flexibility here is that it is not clear whether preselecting a precise value of $t_{\mathcal{C}}^{\max}$ is always possible in practice (e.g., for a fixed degree and security level it turns out that the value of $t_{\mathcal{C}}^{\max}$ stabilizes as $q \to \infty$). This is further impeded by the fact that concrete implementations often do not allow q and d to be picked from some continuous-like range (e.g., the FV-NFLlib [16] and the SEAL [24] libraries require that d is a power of 2 and that $\log_2 q$ is a multiple of some integer). A second motivation is that it can be desirable to use a single SHE implementation for encrypting batches of data of largely varying sizes. The plaintext space should be chosen to fit the largest data, and the methods presented below can then be used to optimize the handling of the smaller data.

The most common choice for $f(X)$ is a cyclotomic polynomial. The nth cyclotomic polynomial $\Phi_n(X) \in \mathbb{Z}[X]$ is the minimal polynomial of a primitive nth root of unity in $\mathbb{C}$

$$\Phi_n(X) = \prod_{0 < k < n,\,(k,n)=1} (X - \zeta_n^k),$$

where $\zeta_n = e^{2\pi i/n}$. The degree of $\Phi_n(X)$ is equal to $\phi(n)$, where $\phi(n)$ is the totient function. It is always irreducible over $\mathbb{Z}$ and, additionally, $\Phi(0) = 1$ for $n \geq 3$.

Cyclotomic polynomials are often used by SHE implementers since they have very nice arithmetic properties such as fast modular reduction and simple Galois groups, which can be used to move data values in between data slots.

3   Plaintext Encoding/decoding of Laurent Polynomials

In this section we consider the problem of encoding an integral Laurent polynomial in the plaintext space and the reverse operation of decoding. We also give necessary conditions on the 'size' of the plaintext space such that a given circuit $\mathcal{C}$ can be evaluated correctly.

3.1   Encoding

Assume that the input data (integers, rationals, reals, ...) has been represented as a Laurent polynomial $a(X) \in \mathbb{Z}[X^{\pm 1}]$. Encoding such a Laurent polynomial in the plaintext space R_t has been considered in a series of recent works [3,12,14,18]. However, it was emphasized in [14] that the plaintext space should only be defined modulo a 2-power cyclotomic polynomial $f(X) = X^{2^k} + 1$ for some k. The reason for this restriction is that the authors required a small and sparse representation for X^{-1}, which in this case is given by $X^{-1} \equiv -X^{2^k - 1} \bmod f(X)$.

Here we propose a very general way of encoding Laurent polynomials which works for almost all defining polynomials f. Let $\overline{f}(X)$ denote the reduction modulo t of $f(X)$ and assume that $f(0)$ is co-prime with t, so $\overline{f}(0)$ is invertible in $\mathbb{Z}_t$. Define $\overline{g}(X)$ by writing $\overline{f}(X) = \overline{g}(X)X + \overline{f}(0)$, then it is obvious that modulo $\overline{f}(X)$ we have that $X^{-1} \equiv -\overline{g}(X)\overline{f}(0)^{-1}$.

The encoding map $\mathsf{Encd}_{\overline{f}}$ is then given by the sequence of ring homomorphisms

$$\mathbb{Z}[X^{\pm 1}] \xrightarrow{\bmod t} \mathbb{Z}_t[X^{\pm 1}] \xrightarrow{\eta_{\overline{f}}} R_t$$

with

$$\eta_{\overline{f}} : \begin{array}{l} X \mapsto X \\ X^{-1} \mapsto -\overline{g}(X)\overline{f}(0)^{-1} . \end{array}$$

Example 2. In the case of the 2-power cyclotomic polynomial $f(X) = X^{2^k} + 1$, the above map replaces negative powers X^{-j} by $-X^{d-j}$, which coincides with the approach from [14]: when expressed in terms of the basis $1, X, X^2, \ldots, X^{d-1}$ of R_t, the map $\eta_{\overline{f}}$ places the positive exponents at the low end of this range, and the negative exponents are placed at the high end.

3.2   Decoding

The crux of the construction relies on the fact that the above encoding map $\mathsf{Encd}_{\overline{f}}$ defines an isomorphism when restricted to a subset of Laurent polynomials. Indeed, if we choose a subset of $\mathbb{Z}_t[X^{\pm 1}]$ of the form

$$\mathbb{Z}_t[X^{\pm 1}]_\ell^m = \left\{ \sum_{i=\ell}^m \overline{a}_i X^i \,|\, \overline{a}_i \in \mathbb{Z}_t \right\}$$

with ℓ and m chosen such that $m - \ell + 1 = d$, then the restriction of $\eta_{\bar{f}}$ to $\mathbb{Z}_t[X^{\pm 1}]_\ell^m$ is an isomorphism between two free $\mathbb{Z}_t$-modules of rank d. The inverse of this map, denoted $\theta_{\bar{f},\ell,m}$, is easy to compute in practice, since it simply corresponds to a matrix inversion.

Thus, $\theta_{\bar{f},\ell,m}$ determines the decoding algorithm from R_t to Laurent polynomials over $\mathbb{Z}_t$. In the final step, one has to lift a Laurent polynomial from $\mathbb{Z}_t[X^{\pm 1}]$ to $\mathbb{Z}[X^{\pm 1}]$ by choosing a representative for each coefficient in a non-empty subset A of $\mathbb{Z}$ of size t. For simplicity we will always take $A = [z, z + t - 1]$ for some $z \in \mathbb{Z}$, common choices being $A = [-\lfloor (t-1)/2 \rfloor, \lceil (t-1)/2 \rceil]$ or $A = [0, t-1]$. But any set A of representatives would be possible, and in fact it can even depend on the coefficient under consideration. Together these two steps define the decoding map $\mathsf{Decd}_{\bar{f},\ell,m,A}$.

3.3   Correctness Conditions

Since homomorphic encryption aims to perform arithmetic operations on ciphertexts, one usually deals with a ciphertext being the outcome of an arithmetic circuit involving only multiplications and additions. By the homomorphic property this ciphertext corresponds to a plaintext which is the result of the same operations in the plaintext space. Given a circuit $\mathcal{C}$, the result of its evaluation on encodings of Laurent polynomials $\mathbf{a} = (a_1(X), \ldots, a_k(X))$ is denoted by $\mathcal{C}(\mathsf{Encd}_{\bar{f}}(\mathbf{a})) \in R_t$.

To guarantee correctness of circuit evaluation, one has to make sure that there exist $\ell, m \in \mathbb{Z}$ such that $m - \ell + 1 = d$ and some non-empty set $A \subsetneq \mathbb{Z}$ of size at most t such that

$$\mathsf{Decd}_{\bar{f},\ell,m,A}(\mathcal{C}(\mathsf{Encd}_{\bar{f}}(\mathbf{a}))) = \mathcal{C}(\mathbf{a}),$$

where $\mathcal{C}(\mathbf{a})$ is the result of the same circuit evaluation in $\mathbb{Z}[X^{\pm 1}]$. This implies that the bounding box (w, h) of $\mathcal{C}(\mathbf{a})$ has to satisfy $w \leq m - \ell + 1 = d$ and $h \leq \log_2 |A| = \log_2 t$. In this case, we say that *the plaintext space covers the bounding box of $\mathcal{C}(\mathbf{a})$* as shown in Fig. 2 below.

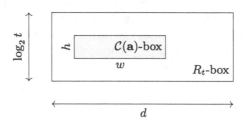

Fig. 2. The bounding box of $\mathcal{C}(\mathbf{a})$ is covered by the plaintext space R_t.

If the bounding box (w, h) of a Laurent polynomial has larger height than the plaintext space, i.e. $h > \log_2 t$, then we say that the computation *overflows modulo t*. If we end up with $w > d$ then we say that it *overflows modulo* $\overline{f}(X)$.

The parameters t and d should therefore be taken large enough to satisfy the above requirement. In practice, d is usually fixed by the security requirements of the SHE scheme. The choice for t, however, strongly depends on the arithmetic circuit $\mathcal{C}$ one is trying to evaluate. Initially, the input data of the circuit is encoded by Laurent polynomials whose bounding boxes are of height $h \le \log_2(|\{b\text{-base digits}\}|)$. During arithmetic operations the height (typically) grows to the height of the bounding box of the outcome. For a given circuit $\mathcal{C}$, this defines a lower bound for t to guarantee correct decoding, which we denote $t_{\mathcal{C}}^{\min}$. Combined with the upper bound on t from Sect. 2.3, one obtains a range for t, namely $[t_{\mathcal{C}}^{\min}, t_{\mathcal{C}}^{\max}]$.

Example 3. To illustrate encoding and decoding, we take $R_t = \mathbb{Z}_7[X]/(\overline{f}(X))$ where $\overline{f}(X) = X^9 + 4X^7 + 1$. Thus, $\overline{g}(X) = X^8 + 4X^6$ and $\mathsf{Encd}_{\overline{f}}$ maps X^{-1} to $6X^8 + 3X^6$. Let us multiply two rational numbers, $\frac{182}{243}$ and 1476. Their base-3 expansions are as follows

$$\frac{182}{243} = 2 \cdot 3^{-5} + 2 \cdot 3^{-3} + 2 \cdot 3^{-1}, \qquad 1476 = 2 \cdot 3^2 + 2 \cdot 3^6$$

or as Laurent polynomials

$$a = 2X^{-5} + 2X^{-3} + 2X^{-1}, \qquad b = 2X^2 + 2X^6.$$

Applying $\mathsf{Encd}_{\overline{f}}$ we get encodings of a and b in R_t, namely, $\overline{a} = 6X^2 + 4X^4 + 4X^6 + 5X^8$ and $\overline{b} = 2X^2 + 2X^6$. Their product is equal to

$$\overline{c} = X + 4X^3 + 5X^4 + 4X^5 + X^6 + 3X^8.$$

We take $\ell = -3$, $m = 5$ and $A = \{4, 5, \ldots, 10\}$ in order to keep the product inside the box. Now we can define $\mathsf{Decd}_{\overline{f}, -3, 5, A}$. The first step is to construct a linear operator $\theta_{\overline{f}, -3, 5}$ using the inverse of the matrix defining the restriction of $\eta_{\overline{f}}$ on $\mathbb{Z}_7[X^{\pm 1}]_{-3}^5$:

$$
\begin{bmatrix}
0 & 0 & 0 & 1 & 0 & 0 & 0 & 0 & 0 \\
0 & 0 & 0 & 0 & 1 & 0 & 0 & 0 & 0 \\
0 & 0 & 0 & 0 & 0 & 1 & 0 & 0 & 0 \\
0 & 0 & 0 & 0 & 0 & 0 & 1 & 0 & 0 \\
3 & 0 & 0 & 0 & 0 & 0 & 0 & 1 & 0 \\
0 & 3 & 0 & 0 & 0 & 0 & 0 & 0 & 1 \\
6 & 0 & 3 & 0 & 0 & 0 & 0 & 0 & 0 \\
0 & 6 & 0 & 0 & 0 & 0 & 0 & 0 & 0 \\
0 & 0 & 6 & 0 & 0 & 0 & 0 & 0 & 0
\end{bmatrix}^{-1}
=
\begin{bmatrix}
0 & 0 & 0 & 0 & 0 & 0 & 6 & 0 & 4 \\
3 & 0 & 0 & 0 & 0 & 0 & 0 & 6 & 0 \\
0 & 0 & 0 & 0 & 0 & 0 & 0 & 0 & 6 \\
1 & 0 & 0 & 0 & 0 & 0 & 0 & 0 & 0 \\
0 & 1 & 0 & 0 & 0 & 0 & 0 & 0 & 0 \\
0 & 0 & 1 & 0 & 0 & 0 & 0 & 0 & 0 \\
0 & 0 & 0 & 1 & 0 & 0 & 0 & 0 & 0 \\
0 & 0 & 0 & 0 & 1 & 0 & 3 & 0 & 2 \\
5 & 0 & 0 & 0 & 0 & 1 & 0 & 3 & 0
\end{bmatrix} \in \mathbb{Z}_7^{9 \times 9}.
$$

Then $\bar{c}$ is mapped to a Laurent polynomial $4X^{-3} + 4X^{-1} + X + 4X^3 + 4X^5 \in \mathbb{Z}_7[X^{\pm 1}]$. By looking for representatives of the coefficients in the set A we get $4X^{-3} + 4X^{-1} + 8X + 4X^3 + 4X^5 \in \mathbb{Z}[X^{\pm 1}]$ and evaluate it at $X = 3$

$$4 \cdot 3^{-3} + 4 \cdot 3^{-1} + 8 \cdot 3 + 4 \cdot 3^3 + 4 \cdot 3^5 = \frac{29848}{27},$$

which is the correct product of $\frac{182}{243}$ and 1476.

Remark 2. Note that the above condition for correct decoding *only* depends on the bounding box of the evaluation of the circuit $\mathcal{C}(\mathbf{a})$ and not on the bounding boxes of the individual inputs $a_i(X) \in \mathbb{Z}[X^{\pm 1}]$ nor on those of the intermediate values. Indeed, we always have

$$\mathcal{C}(\mathsf{Encd}_{\overline{f}}(a_1(X)), \ldots, \mathsf{Encd}_{\overline{f}}(a_k(X))) = \mathsf{Encd}_{\overline{f}}(\mathcal{C}(\mathbf{a})), \tag{2}$$

simply because $\mathsf{Encd}_{\overline{f}}$ is a ring homomorphism. This implies that the bounding boxes of the input or intermediate values should not necessarily be contained in the bounding box of the plaintext space, as long as the outcome of evaluation is.

4   Splitting the Plaintext Space

In this section we recall how the Chinese Remainder Theorem (CRT) can be used to split the plaintext space naturally along two directions: firstly, we will split horizontally for each prime power factor t_i of the plaintext modulus t and secondly, each horizontal slice will be split vertically by factoring $f(X) \bmod t_i$.

4.1   Horizontal Splitting

If t is a composite that factors into distinct prime powers $t = t_1 \ldots t_s$ then the ring R_t can be mapped via the CRT to a direct product of R_{t_i}'s resulting in the following ring isomorphism

$$\begin{aligned} \mathsf{CRT}_t : R_t &\to R_{t_1} \times \cdots \times R_{t_s} \\ \bar{a}(X) &\mapsto (\bar{a}(X) \bmod t_1, \ldots, \bar{a}(X) \bmod t_s) \end{aligned}$$

whose inverse is easy to compute. For a given index subset $I = \{i_1, \ldots, i_c\} \subseteq [s]$ the map CRT_t induces a surjective morphism

$$\mathsf{CRT}_{t_I} : R_t \to R_{t_{i_1}} \times \cdots \times R_{t_{i_c}},$$

which is well-defined via the projection map

$$\pi_{t_I} : \prod_{i \in [s]} R_{t_i} \to \prod_{i \in I} R_{t_i}$$

so that $\mathsf{CRT}_{t_I} = \pi_{t_I} \cdot \mathsf{CRT}_t$. The CRT_t can be represented as a 'horizontal' splitting of the plaintext space according to the unique factorization of t into distinct prime powers $\{t_i\}_{i \in [s]}$. Each horizontal slice in Fig. 3 corresponds to some R_{t_i}.

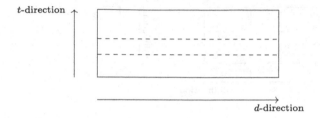

Fig. 3. CRT_t decomposition of R_t

4.2 Vertical Splitting

For each factor t_i of t we define $\overline{f}_i(X) \in \mathbb{Z}_{t_i}[X]$ to be the reduction of $f(X)$ modulo t_i. Since $f(0)$ is co-prime with t, it is also co-prime with any t_i and thus, $\overline{f}_i(0)$ is invertible.

The factorization of $\overline{f}_i(X)$ into irreducible factors modulo t_i can be computed as follows: if t_i is prime, then one can simply use factorization algorithms for polynomials over finite fields; for t_i a prime power, one first computes the factorization modulo the prime and then lifts it using Hensel's lemma to a factorization modulo t_i. The result in both cases is that we can easily obtain a factorization

$$\overline{f}_i(X) \equiv \prod_{j=1}^{r_i} \overline{f}_{ij}(X)$$

for monic irreducible polynomials $\overline{f}_{ij}(X) \in \mathbb{Z}_{t_i}[X]$. Note that the constant terms $\overline{f}_{ij}(0)$ are all invertible because their product $\overline{f}_i(0)$ is invertible. Applying the CRT in the polynomial dimension gives the following map for each t_i:

$$\mathsf{CRT}_{t_i, \overline{f}_i} : R_{t_i} \;\; \to R_{t_i, 1} \times \cdots \times R_{t_i, r_i}$$
$$\overline{a}(X) \mapsto (\overline{a}(X) \bmod \overline{f}_{i1}(X), \ldots, \overline{a}(X) \bmod \overline{f}_{i r_i}(X)).$$

Here the $R_{t_i, j}$ denotes the ring $\mathbb{Z}_{t_i}[X]/(\overline{f}_{ij}(X))$, which corresponds to a 'brick' in Fig. 4. The map $\mathsf{CRT}_{t_i, \overline{f}_i}$, whose inverse is again easy to compute, can be thought of as a 'vertical' splitting of R_{t_i}. For simplicity we will usually just write $R_{i,j}$ rather than $R_{t_i, j}$. By analogy with CRT_{t_I}, we introduce the surjective ring homomorphism $\mathsf{CRT}_{t_i, \overline{f}_J}$ from R_{t_i} to $\prod_{j \in J} R_{t_i, j}$ where $J = \{j_1, \ldots, j_c\} \subseteq [r_i]$.

5 Improved SIMD Encoding

In this section we combine the results of Sects. 3 and 4 to derive flexible SIMD encoding and decoding algorithms. Recall that to correctly decode the result of a circuit evaluation $\mathcal{C}(\mathbf{a})$, we require that the bounding box of the plaintext space covers the bounding box of $\mathcal{C}(\mathbf{a})$. We assume that this is indeed the case, and show how to select a minimal number of bricks of R_t to cover the bounding box of $\mathcal{C}(\mathbf{a})$, leaving the other bricks available for doing parallel computations.

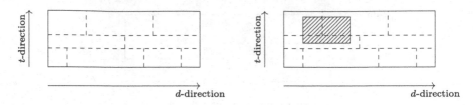

Fig. 4. Decomposition of R_t using factorization of t and $\overline{f}_i$'s

Fig. 5. Encoding of a single Laurent polynomial into R_t.

Recall that each brick corresponds to a ring $R_{i,j}$ in the decomposition

$$R_t \to R_{t_1} \times \cdots \times R_{t_s} \to (R_{1,1} \times \cdots \times R_{1,r_1}) \times \cdots \times (R_{s,1} \times \cdots \times R_{s,r_s}).$$

Each ring $R_{i,j}$ has its own bounding box of size $(d_{ij}, \log_2 t_i)$, where $d_{ij} = \deg \overline{f}_{ij}$. Assuming that the bounding box of $\mathcal{C}(\mathbf{a})$ is given by (w, h), we need to combine enough horizontal slices to cover the height h, and inside each horizontal slice, we need to select enough bricks to cover the width w as illustrated in Fig. 5. Any unused bricks can be used to encode other data values, for instance to compute $\mathcal{C}(\mathbf{b})$ for some other input vector $\mathbf{b}$, immediately resulting in SIMD computations.

We formalize this approach by combining bricks into a block structure: we call a *block* a set of tuples $\mathcal{B} = \{(t_i, \overline{f}_{ij})\}_{i \in I(\mathcal{B}), j \in J(\mathcal{B},i)}$ with index sets $I(\mathcal{B}) \subseteq [s]$ and $J(\mathcal{B}, i) \subseteq [r_i]$, where we recall that r_i is the number of irreducible factors of $\overline{f}_i$. We of course think of this as corresponding to the set of $R_{i,j}$'s with $i \in I(\mathcal{B}), j \in J(\mathcal{B},i)$. Equivalently, through an application of the CRT this corresponds to the set of quotient rings $\{R_{t_i}/(\overline{F}_{i,\mathcal{B}})\}_{i \in I(\mathcal{B})}$ where $\overline{F}_{i,\mathcal{B}} = \prod_{j \in J(\mathcal{B},i)} \overline{f}_{ij}$. Graphically we think of a block as a set of bricks of R_t, which are combined such that the $R_{i,j}$'s with the same index i are glued column-wise and the resulting rows are placed on top of each other (Fig. 6).

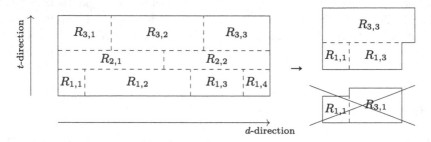

Fig. 6. Example of a block taken from the CRT decomposition of R_t. The bottom combination of 'bricks' is not a block because their first indices do not coincide.

In order for a block $\mathcal{B}$ to be suitable for computing $\mathcal{C}(\mathbf{a})$, whose bounding box we denote by (w, h), we note that the bounding box of $R_{t_i}/(\overline{F}_{i,\mathcal{B}})$ with $i \in I(\mathcal{B})$ is $(w_{i,\mathcal{B}}, \log_2 t_i)$ where

$$w_{i,\mathcal{B}} = \deg \overline{F}_{i,\mathcal{B}} = \sum_{j \in J(\mathcal{B},i)} d_{ij}.$$

If $\min_{i \in I(\mathcal{B})} w_{i,\mathcal{B}} \geq w$ and $\sum_{i \in I(\mathcal{B})} \log_2 t_i \geq h$ then we say that $\mathcal{B}$ *covers the bounding box* (w, h). As we will see $\mathcal{C}(\mathbf{a})$ will be decoded correctly as soon as an encoding block $\mathcal{B}$ is used that covers its bounding box.

Example 4. We decompose $R_t = \mathbb{Z}_{2761}[X]/(f(X))$ where $f(X) = X^{20} + X^{15} + 1$. The plaintext modulus factors into $t_1 = 11$ and $t_2 = 251$ and

$$
\begin{aligned}
f(X) &\equiv f_{1,1}(X) \cdot f_{1,2}(X) \\
&\equiv (X^5 + 3)(X^{15} + 9X^{10} + 6X^5 + 4) \bmod 11, \\
f(X) &\equiv f_{2,1}(X) \cdot f_{2,2}(X) \cdot f_{2,3}(X) \\
&\equiv (X^5 + 18)(X^5 + 120)(X^{10} + 114X^5 + 180) \bmod 251.
\end{aligned}
$$

Accordingly, R_t splits into $(R_{1,1} \times R_{1,2}) \times (R_{2,1} \times R_{2,2} \times R_{2,3})$. Overall we have 5 'bricks' that can be combined into 31 different blocks. For example, one can take a block $\{(11, X^{15}+9X^{10}+6X^5+4), (251, X^5+18), (251, X^5+120)\}$ corresponding to the combination of $R_{1,2}, R_{2,1}$ and $R_{2,2}$ or $\{(11, X^5 + 3), (11, X^{15} + 9X^{10} + 6X^5 + 4)\}$ which simply corresponds to $R_{11} = R_t/(11)$ (see Fig. 7).

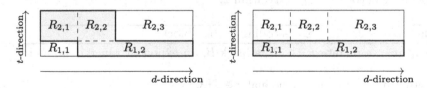

Fig. 7. The block structure of $R_t = \mathbb{Z}[X]/(2651, X^{20} + X^{15} + 1)$ with two blocks colored in gray.

The whole plaintext space can be represented by a block as well

$$\mathcal{P} = \bigcup_{i \in [s]} \bigcup_{j \in [r_i]} \{(t_i, \overline{f}_{ij})\}.$$

Therefore, the SIMD packing problem consists in finding a set of disjoint blocks $S = \{\mathcal{B}_1, \ldots, \mathcal{B}_u\}$ such that $\bigcup_{\mathcal{B} \in S} \mathcal{B} = \mathcal{P}$ and every block covers the maximal bounding box among the corresponding output values.

To a partition S of $\mathcal{P}$ there naturally corresponds a factorization of $\overline{f}_i$ for every $i \in [s]$:

$$\overline{f}_i(X) = \prod_{\mathcal{B} \in S, i \in I(\mathcal{B})} \overline{F}_{i,\mathcal{B}}(X).$$

This induces a family of CRT isomorphisms

$$\mathsf{CRT}_{t_i, \overline{f}_i, S} : R_{t_i} \to \prod_{\mathcal{B} \in S, i \in I(\mathcal{B})} R_{t_i}/(\overline{F}_{i,\mathcal{B}}).$$

Now we have all the ingredients to pack a number of data values into one plaintext as described in Algorithm 1.

Algorithm 1. Plaintext packing.

Input : a set of disjoint blocks $S = \{\mathcal{B}_1, \ldots, \mathcal{B}_u\}$ with corresponding
data values $a_1, \ldots, a_u \in \mathbb{Z}[X^{\pm 1}]$ such that $\bigcup_{k=1}^{u} \mathcal{B}_k = \mathcal{P}$.

Output: $b \in R_t$

1 **for** $k \leftarrow 1$ **to** u **do**
2   **for** $i \in I(\mathcal{B}_k)$ **do**
3           $a_{t_i, \overline{F}_{i,\mathcal{B}_k}} \leftarrow \mathsf{Encd}_{\overline{F}_{i,\mathcal{B}_k}}(a_k)$

4 **for** $i \leftarrow 1$ **to** s **do**
5       $b_i \leftarrow \mathsf{CRT}^{-1}_{t_i, \overline{f}_i, S}(\{a_{t_i, \overline{F}_{i,\mathcal{B}}}\}_{\mathcal{B}, i \in I(\mathcal{B})})$
6 $b \leftarrow \mathsf{CRT}^{-1}_t(b_1, \ldots, b_s)$

After packing one can encrypt the output and feed it to an arithmetic circuit (together with other packings in case the circuit takes more than one argument). The resulting plaintext contains multiple evaluations corresponding to each block that can be decoded using Algorithm 2.

Algorithm 2. Plaintext decoding for one block.

Input : a plaintext $\overline{c} \in R_t$, a block $\mathcal{B}$, an exponent range $[\ell, m]$ and a
coefficient set $A \in \mathbb{Z}$

Output: a Laurent polynomial $a \in \mathbb{Z}[X^{\pm 1}]$

1 $t_I \leftarrow 1$
2 **for** $i \in I(\mathcal{B})$ **do**
3       $t_I \leftarrow t_I \cdot t_i$
4       $\overline{c}_i \leftarrow \overline{c} \bmod t_i$
5       $\overline{c}_i \leftarrow \overline{c}_i \bmod \overline{F}_{i,\mathcal{B}}$
6       $m_i \leftarrow \ell + w_{i,\mathcal{B}} - 1$
7       $c_i \leftarrow \theta_{\overline{F}_{i,\mathcal{B}}, \ell, m_i}(\overline{c}_i)$

8 $a \leftarrow$ coefficient-wise CRT^{-1} of $\{c_i\}_{i \in I(\mathcal{B})}$ to $\mathbb{Z}_{t_I}[X^{\pm 1}]$
9 $a \leftarrow$ selecting coefficient representatives of a from the set A

Algorithm 2 produces correct circuit evaluations for all blocks occurring in Algorithm 1 that satisfy the properties outlined in the next theorem.

Theorem 1. *Let S be a set of disjoint blocks such that $\bigcup_{B \in S} B = P$. Let C be an arithmetic circuit taking v arguments and for each block B let $\mathbf{a}_B = (a_{B,1}, \ldots, a_{B,v})$ be a vector of Laurent polynomials. For each $k = 1, \ldots, v$ let b_k denote the output of Algorithm 1 upon input of $(a_{B,k})_{B \in S}$. Let $\bar{c} = C(b_1, \ldots, b_v)$. Then for each block B we have that if it covers the bounding box of $C(\mathbf{a}_B)$, then upon input of $\bar{c}$ Algorithm 2 produces $C(\mathbf{a}_B)$, for an appropriate choice of ℓ, m and A.*

Proof. By our assumption there are ℓ, m such that $C(\mathbf{a}_B) = \sum_{i=\ell}^{m} \alpha_i X^i$ where

$$\min_{i \in I(B)} w_{i,B} \geq m - \ell + 1 \quad \text{and} \quad \prod_{i \in I(B)} t_i \geq |A|, \quad (3)$$

with $A = \{\min_i \alpha_i, \ldots, \max_i \alpha_i\}$. Let a denote the output of Algorithm 2 upon input of $\bar{c}$ using these ℓ, m, and A. Since this is a Laurent polynomial having coefficients in A, by (3) it suffices to prove that the reductions of a and $C(\mathbf{a}_B)$ modulo t_i are the same for each $i \in I(B)$. Again by (3) these reductions are contained in $\mathbb{Z}_{t_i}[X^{\pm 1}]_{\ell}^{m_i}$ where $m_i = \ell + w_{i,B} - 1$, so by injectivity of $\eta_{\overline{F}_{i,B}}$ it suffices to prove that

$$\mathsf{Encd}_{\overline{F}_{i,B}}(a) = \mathsf{Encd}_{\overline{F}_{i,B}}(C(\mathbf{a}_B)).$$

From Algorithm 2 we see that the left-hand side is just the reduction of $\bar{c}$ into $R_{t_i}/(\overline{F}_{i,B})$, while the right hand side is

$$C(\mathsf{Encd}_{\overline{F}_{i,B}}(a_{B,1}), \ldots, \mathsf{Encd}_{\overline{F}_{i,B}}(a_{B,v}))$$

because of the homomorphic properties of the encoding map. From Algorithm 1 we clearly see that $\mathsf{Encd}_{\overline{F}_{i,B}}(a_{B,k})$ is the reduction of b_k into $R_{t_i}/(\overline{F}_{i,B})$, for all $k = 1, \ldots, v$, so the theorem follows. □

Example 5. Using the CRT decomposition of R_t from Example 4 we cube two Laurent polynomials simultaneously using SIMD, namely $u(X) = 7X^3 + 7X^2$ and $v(X) = 8X^5 + 7X$. To encode u^3 we take the block B_1 with rings $R_{1,1}, R_{2,1}$ and the remaining bricks to build the block B_2 to hold the result v^3.

Since only positive exponents are present in the data, all encoding functions $\mathsf{Encd}_{\overline{F}_{i,B_1}}$ and $\mathsf{Encd}_{\overline{F}_{i,B_2}}$ map $u(X)$ and $v(X)$ identically to the corresponding $R_{i,j}$'s. Then we get

$$a_{11,\overline{F}_{1,B_1}}(X) = 7X^3 + 7X^2 \in R_{1,1} = R_{11}/(X^5 + 3),$$
$$a_{251,\overline{F}_{2,B_1}}(X) = 7X^3 + 7X^2 \in R_{2,1} = R_{251}/(X^5 + 18),$$
$$a_{11,\overline{F}_{1,B_2}}(X) = 8X^5 + 7X \in R_{1,2} = R_{11}/(X^{15} + 9X^{10} + 6X^5 + 4),$$
$$a_{251,\overline{F}_{2,B_2}}(X) = 8X^5 + 7X \in R_{2,2} \times R_{2,3} \cong R_{251}/(X^{15} + 234X^{10} + 55X^5 + 14).$$

Applying $\mathsf{CRT}^{-1}_{t_i,\overline{f}_i,\{B_1,B_2\}}$ for each t_i we find

$$b_1 = X^{18} + X^{17} + 10X^{16} + 5X^{15} + 9X^{13} + 9X^{12}$$
$$+ 2X^{11} + X^{10} + 6X^8 + 6X^7 + 5X^6 + 5X^5 + 4X^3 + 4X^2 + 3X + 9 \in R_{11},$$
$$b_2 = 162X^{18} + 162X^{17} + 89X^{16} + 213X^{15} + 7X^{13} + 7X^{12}$$
$$+ 244X^{11} + 144X^{10} + 125X^8 + 125X^7 + 126X^6 + 177X^5$$
$$+ 9X^3 + 9X^2 + 249X + 221 \in R_{251},$$

which finally leads to the following plaintext via CRT_t^{-1}

$$b = 2421X^{18} + 2421X^{17} + 340X^{16} + 1468X^{15} + 2517X^{13} + 2517X^{12}$$
$$+ 244X^{11} + 144X^{10} + 2635X^8 + 2635X^7 + 126X^6 + 2436X^5$$
$$+ 2017X^3 + 2017X^2 + 751X + 1978 \in R_{2761}.$$

Now we evaluate an arithmetic circuit $z \mapsto z^3$ in b and obtain

$$\bar{c} = 1943X^{19} + 401X^{18} + 745X^{17} + 391X^{16} + 433X^{15}$$
$$+ 2109X^{14} + 1717X^{13} + 2646X^{12} + 2729X^{11} + 2347X^{10}$$
$$+ 2198X^9 + 1724X^8 + 234X^7 + 421X^6 + 2683X^5 + 94X^4$$
$$+ 1188X^3 + 1143X^2 + 1960X + 1906 \in R_{2761},$$

which simultaneously encodes u^3 and v^3.

In order to decode the data we apply Algorithm 2 starting with the block $\mathcal{B}_1$ equipped with the exponent range $[6, 9]$ and the coefficient set $A_{\mathcal{B}_1} = [0, 2760]$. At first, we should reduce $\bar{c}$ modulo $\overline{F}_{i,\mathcal{B}_1}$ and t_i for each $i \in I(\mathcal{B}_1)$. As a result, we find

$$\bar{c}_{1,\mathcal{B}_1} = 5X^4 + 4X^3 + 4X^2 + 5X \in R_{11}/(X^5 + 3),$$
$$\bar{c}_{2,\mathcal{B}_1} = 101X^4 + 52X^3 + 52X^2 + 101X \in R_{251}/(X^5 + 18).$$

To decode into Laurent polynomials we set $\ell_i = 6$ and $m_i = 10$ for every $i \in I(\mathcal{B}_1)$ because $\deg \overline{F}_{1,\mathcal{B}_1} = \deg \overline{F}_{2,\mathcal{B}_1} = 5$. Then we follow the same procedure as in Example 3 to define $\theta_{\overline{F}_1,6,10}$ and $\theta_{\overline{F}_2,6,10}$ via matrices $M_1 = 7 \cdot M$ and $M_2 = 237 \cdot M$ where

$$M = \begin{bmatrix} 0 & 1 & 0 & 0 & 0 \\ 0 & 0 & 1 & 0 & 0 \\ 0 & 0 & 0 & 1 & 0 \\ 0 & 0 & 0 & 0 & 1 \\ 1 & 0 & 0 & 0 & 0 \end{bmatrix}.$$

These linear transformations give us two Laurent polynomials modulo 11 and 251, respectively

$$c_{1,\mathcal{B}_1} = 2X^9 + 6X^8 + 6X^7 + 2X^6 \in \mathbb{Z}_{11}[X^{\pm 1}],$$
$$c_{2,\mathcal{B}_1} = 92X^9 + 25X^8 + 25X^7 + 92X^6 \in \mathbb{Z}_{251}[X^{\pm 1}].$$

Using the coefficient-wise CRT and lifting coefficients in $A_{\mathcal{B}_1}$ we recover the Laurent polynomial

$$a_{\mathcal{B}_1} = 343X^9 + 1029X^8 + 1029X^7 + 343X^6 \in \mathbb{Z}[X^{\pm 1}],$$

which is equal to u^3.

We repeat the same steps for the block $\mathcal{B}_2$ with the exponent range $[3, 15]$ and the same coefficient set A. This block has again the polynomials $\overline{F}_{i,\mathcal{B}_2}$ of the same degree and thus every $m_i = 17$ and $\ell_i = 3$. Executing Algorithm 2 we get the following sequence of calculations

$$\overline{c}_{1,\mathcal{B}_2} = 2X^{11} + X^{10} + 10X^7 + 8X^5 + 2X^3 + 9,$$
$$\overline{c}_{2,\mathcal{B}_2} = 89X^{11} + 170X^{10} + 172X^7 + 203X^5 + 92X^3 + 111,$$
$$\downarrow$$
$$c_{1,\mathcal{B}_2} = 6X^{15} + 2X^{11} + 10X^7 + 2X^3,$$
$$c_{2,\mathcal{B}_2} = 10X^{15} + 89X^{11} + 172X^7 + 92X^3,$$
$$\downarrow$$
$$a_{\mathcal{B}_2} = 512X^{15} + 1344X^{11} + 1176X^7 + 343X^3.$$

The last polynomial is exactly v^3 so we correctly cubed two Laurent polynomials.

Remark 3. The CRT factorization can also be exploited when a homomorphic algorithm needs a bigger plaintext modulus than the upper bound $t_{\mathcal{C}}^{\max}$ discussed above. Let us denote this modulus with a capital T to emphasize direct incompatibility of this parameter with other SHE parameters, namely, $T > t_{\mathcal{C}}^{\max}$. However, one can find a set of natural numbers $\{T_i \leq t_{\mathcal{C}}^{\max}\}$ such that $T \leq T' = \prod_i T_i$. Then $R_{T'}$ splits into smaller quotient rings R_{T_i}. A plaintext $a \in R_{T'}$ then maps to a vector whose ith component lies in R_{T_i}. In that case the plaintext space splits into quotient rings with smaller moduli via CRT such that each ring fits the SHE settings according to the following diagram

$$R_{T'} \xrightarrow{\text{CRT}} \begin{cases} R_{T_1} \xrightarrow{\text{CRT}} \prod_{t'|T_1} \prod_{f'|\overline{f} \bmod t'} R_{t',f'} \xrightarrow{\text{Alg 1}} R_{T_1} \\ \cdots \\ R_{T_s} \xrightarrow{\text{CRT}} \prod_{t'|T_s} \prod_{f'|\overline{f} \bmod t'} R_{t',f'} \xrightarrow{\text{Alg 1}} R_{T_s} \end{cases}$$

A homomorphic circuit evaluation must then be repeated over each CRT factor T_i. Nevertheless, this gives some freedom of choice for T_i's so as to find $R_{T'}$ with a nice CRT decomposition.

6   Parameter Choice

In this section we discuss a set of tools that will allow implementers to benefit from our enhanced SIMD approach as much as possible. There are three parameters that directly affect the packing capacity. We list them below in an order that seems natural for solving any packing problem. Nevertheless, all parameters depend on each other.

Plaintext Modulus. Earlier we defined the range $[t_{\mathcal{C}}^{\min}, t_{\mathcal{C}}^{\max}]$ from which the plaintext modulus t is allowed to be chosen. Additionally, at the end of Sect. 4 we discussed the CRT trick that allows to handle plaintext moduli that are bigger than $t_{\mathcal{C}}^{\max}$. Altogether this gives a designer some freedom to choose t such that it splits into many 'advantageous' t_i's. An 'advantageous' t_i means that the factorization of $\overline{f}_i$ is such that the resulting CRT decomposition can embed as many plaintexts as possible, which is usually facilitated by a finer brick structure as in Fig. 8.

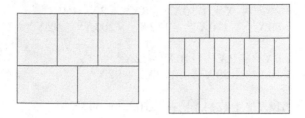

Fig. 8. The CRT decompositions of plaintext spaces corresponding to different t's.

This brick structure is defined by the t_i's and by the degrees of the $\overline{f}_{ij}$'s, namely $d_{i1}, \ldots, d_{ir_i}$ which constitute *a decomposition type* of f modulo t_i. Let G be the Galois group of the splitting field of f over $\mathbb{Q}$. It can be considered as a subgroup of the group S_d of permutations of d elements. Every automorphism σ can be represented as a product of cycle permutations with a corresponding pattern of cycle lengths. Additionally, we say that a set P of prime numbers has density δ if

$$\lim_{x \to \infty} \frac{|\{p \le x : p \in P\}|}{|\{p \le x : p \text{ prime}\}|} = \delta.$$

Then the probability that a desired decomposition type occurs for some random t_i is estimated by the following classical theorem.

Theorem 2 (Frobenius). *The density of the set P of primes modulo which f has a given decomposition type $d_1, d_2, \ldots, d_r$ exists, and it is equal to $1/|G|$ times the number of automorphisms $\sigma \in G$ with cycle pattern $d_1, d_2, \ldots, d_r$.*

An interesting case is where $\overline{f}_i$ splits into linear factors since it gives maximal flexibility to combine blocks. There exists only one $\sigma \in G$ corresponding to such a decomposition which is the identity permutation, so the corresponding probability is $1/|G|$.

Example 6. If $f(X)$ is the nth cyclotomic polynomial then its Galois group G has $d = \phi(n)$ elements and it always splits into irreducible factors of the same degree, i.e. its decomposition type modulo t_i is always $(d', \ldots, d')$ where d' is the order of t_i modulo n; here we implicitly assume that $\gcd(t_i, n) = 1$. Let us take $f(X) = X^{2^k} + 1$. Its Galois group is isomorphic to $\mathbb{Z}_{2^{k+1}}^{\times}$ or to the direct product of two cyclic groups $C_2 \times C_{2^{k-1}}$. It contains 2^k elements with orders shown in the following table:

ord	1	2	4	...	2^{k-1}
$\#\{a \in \mathbb{Z}_{2^{k+1}}^{\times}\}$	1	3	4	...	2^{k-1}

This implies that f splits into $2^{k'}$ irreducible factors of degree $2^{k-k'}$ modulo a random t_i with probability $2^{-k'}$, for any $k' \in \{1, \ldots, k-2, k-1\}$.

In the classical example of a homomorphic application a client encrypts his data and sends it to a third party to perform calculations. Since encryption and decryption are done only on the client side, he therefore has the possibility to tweak the plaintext modulus without re-generation of keys as long as the evaluation (or linearization) key does not depend on t. It is important to note that the plaintext modulus does not affect the security level of an SHE scheme but it does affect the decryption correctness. Hence, t should fit the upper bound $t_C^{\max}$ introduced by the noise growth inside ciphertexts. As a result, one can exploit the same technique as above to find R_t with the most useful decomposition.

Block Set. Recall that the plaintext space can be thought of as a set of bricks $\mathcal{P}$. Every block is then a subset of $\mathcal{P}$. The packing problem consists in finding a partition of $\mathcal{P}$ with the maximal number of blocks where each one satisfies Theorem 1. It is clear that the partition search is highly dependent on the data values and the arithmetic operations being performed homomorphically. Therefore the same plaintext space can be used differently for various applications as shown in Fig. 9. If $r = \sum_{i=1}^{s} r_i$ is the cardinality of $\mathcal{P}$ then the total number of partitions is equal to the r-th Bell number B_r. That number grows exponentially (see [17]) while r is increasing according to

$$\frac{\ln B_r}{r} \simeq \ln r.$$

As a result a system designer has a lot of flexibility to play with the plaintext space partitions to fit data into some block structure. Obviously, the maximal number of blocks cannot be bigger than r, in which case the blocks are just the singletons $\{R_{i,j}\}$. A plaintext space with many CRT factors is usually easier to handle because it is more flexible for block constructions.

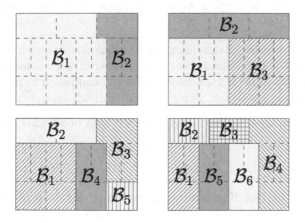

Fig. 9. Different partitions of $\mathcal{P}$.

If one does not find a satisfying partition of all of $\mathcal{P}$, it is of course also possible to leave a couple of bricks unused by packing zeros in them (or even random values).

Encoding Base. Representing data using Laurent polynomials requires a numerical base b which can be a real or a complex number. The size of b affects the length of a representation as well as the size of its coefficients.

In [3] it was shown that non-integral bases taken from the interval $(1, 2)$ have a simple greedy algorithm that, given a real number, produces a base-b expansion with a ternary set of coefficients. This procedure has the property that smaller bases lead to sparser representations and thus smaller coefficient growth but longer expansions. To illustrate this we resort again to the box representation of a Laurent polynomial (Fig. 10).

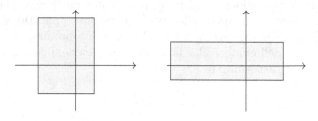

Fig. 10. The examples of bounding boxes corresponding to different encoding bases.

As a result, by changing the encoding base one could play a trade-off game between degree and coefficient size such that the number of plaintexts fitting a block structure is maximal. Furthermore, each block allows to encode data in a different base because neither Algorithm 1 nor Algorithm 2 depends on the choice of b.

Example 7. To illustrate the aforementioned techniques we revisit a medical application of the YASHE homomorphic encryption scheme [4] given in [5]. In this paper the standard logistic function is homomorphically computed to predict the probability of having a heart attack. The algorithm is divided into two steps.

Step 1. One computes the following weighted sum of six encrypted predictive variables

$$z = 0.072 \cdot z_1 + 0.013 \cdot z_2 - 0.029 \cdot z_3 + 0.008 \cdot z_4 - 0.053 \cdot z_5 + 0.021 \cdot z_6,$$

where each $z_i \in [0, 400]$. The multiplicative depth of the corresponding circuit is 1. For this step we take the same YASHE parameters as in [5], i.e. $q \simeq 2^{128}$ and $f(X) = X^{4096} + 1$. Given these parameters we derive $t^{\mathcal{C}}_{max} = 2097152 \simeq 2^{21}$ using [4, Lem. 9]. Running over all primes less than $t^{\mathcal{C}}_{max}$ we find that modulo $t_1 = 257$ and modulo $t_2 = 3583$ our polynomial $f(X)$ can be written as a product of 128 coprime factors of degree 32. With $t = t_1$ the conventional SIMD technique allows then to pack at most 128 values into one plaintext. This capacity can be achieved with base-3 balanced ternary expansions that result in an output bounding box of size $(29, \log_2 53)$. However, our approach supports $t = t_1 \cdot t_2$ so one can pack 256 values using the same encoding method.

Step 2. The output of Step 1 is decrypted, decoded to a real number and encoded again to a plaintext. This 'refreshed' encoding is then encrypted and given as input to the following approximation of the logistic function

$$P(x) = \frac{1}{2} + \frac{1}{4}x - \frac{1}{48}x^3 + \frac{1}{480}x^5 - \frac{17}{80640}x^7.$$

In this step the multiplicative depth is 3, $q \simeq 2^{512}$ and $f(X) = X^{16384} + 1$. These parameters lead to $t_{\max}^C \simeq 2^{50}$. Using the previous SIMD technique the maximal plaintext capacity can be achieved with the plaintext modulus $t \simeq 2^{30.54}$ and base-3 balanced ternary encoding. In this case $f(X)$ splits into 8192 quadratic factors and the output bounding box is of size $(229, 29.54)$. We can thus compose 71 blocks with 115 slots and one block with the remaining slots. As a result, one plaintext can contain at most 71 values.

This capacity can be increased with our SIM^{2}D technique. In particular, one can notice that the ratio between $t_{\max}^C$ and the previously mentioned modulus t is around $2^{19.46}$, which implies some part of the plaintext space remains unfilled. We can fill that space setting the plaintext modulus to $t_1 \cdot t_2$ with $t_1 \simeq 2^{30.54}$ and $t_2 = 675071 \simeq 2^{19.36}$. The polynomial $f(X)$ splits into 128 factors of degree 128 modulo t_2. To fit the modulus t_2 we encoded real values with the non-integral base $b = 1.16391$ and obtained the output bounding box $(1684, 19.36)$. Therefore one block should consist of 14 slots, and we can construct 9 such blocks. As a result, we can combine these blocks with the 71 blocks given by the old SIMD technique, which results in a total plaintext capacity of 80 values.

7   Conclusion

In this paper we presented two techniques that make SIMD operations in the setting of homomorphic encryption more flexible and efficient. Our first technique showed how data values that are naturally represented as Laurent polynomials can be encoded into a plaintext space of the form $\mathbb{Z}_t[X]/(\overline{f}(X))$. Furthermore, we also provided sufficient conditions for correct decoding after evaluation of an arithmetic circuit. Our second technique relied on a fine-grained CRT decomposition of the plaintext space resulting in a much denser and thus more efficient data packing compared to the state of the art. Finally, we provided guidelines on how to choose system parameters in order to find the most efficient packing strategy for a particular task.

References

1. Albrecht, M.R., Player, R., Scott, S.: On the concrete hardness of learning with errors. J. Math. Cryptol. **9**(3), 169–203 (2015)
2. Benhamouda, F.., Lepoint, T., Mathieu, C., Zhou, H.: Optimization of bootstrapping in circuits. In: Klein, P.N. (ed.) Proceedings of the Twenty-Eighth Annual ACM-SIAM Symposium on Discrete Algorithms, pp. 2423–2433. ACM-SIAM (2017)
3. Bonte, C., Bootland, C., Bos, J.W., Castryck, W., Iliashenko, I., Vercauteren, F.: Faster homomorphic function evaluation using non-integral base encoding. In: Fischer, W., Homma, N. (eds.) CHES 2017. LNCS, vol. 10529, pp. 579–600. Springer, Cham (2017). https://doi.org/10.1007/978-3-319-66787-4_28
4. Bos, J.W., Lauter, K., Loftus, J., Naehrig, M.: Improved security for a ring-based fully homomorphic encryption scheme. In: Stam, M. (ed.) IMACC 2013. LNCS, vol. 8308, pp. 45–64. Springer, Heidelberg (2013). https://doi.org/10.1007/978-3-642-45239-0_4
5. Bos, J.W., Lauter, K.E., Naehrig, M.: Private predictive analysis on encrypted medical data. J. Biomed. Inform. **50**, 234–243 (2014)
6. Brakerski, Z.: Fully homomorphic encryption without modulus switching from classical GapSVP. In: Safavi-Naini, R., Canetti, R. (eds.) CRYPTO 2012. LNCS, vol. 7417, pp. 868–886. Springer, Heidelberg (2012). https://doi.org/10.1007/978-3-642-32009-5_50
7. Brakerski, Z., Gentry, C., Vaikuntanathan, V.: (Leveled) fully homomorphic encryption without bootstrapping. In: Goldwasser, S. (ed.) Proceedings of the 3rd Innovations in Theoretical Computer Science Conference ITCS 2012, pp. 309–325. ACM (2012)
8. Brakerski, Z., Vaikuntanathan, V.: Efficient fully homomorphic encryption from (standard) LWE. In: Ostrovsky, R. (ed.), Proceedings of the 2011 IEEE 52nd Annual Symposium on Foundations of Computer Science FOCS, pp. 97–106. IEEE Computer Society Press (2011)
9. Brakerski, Z., Vaikuntanathan, V.: Fully homomorphic encryption from ring-LWE and security for key dependent messages. In: Rogaway, P. (ed.) CRYPTO 2011. LNCS, vol. 6841, pp. 505–524. Springer, Heidelberg (2011). https://doi.org/10.1007/978-3-642-22792-9_29
10. Chen, H., Laine, K., Player, R., Xia, Y.: High-precision arithmetic in homomorphic encryption. In: Smart, N.P. (ed.) CT-RSA 2018. LNCS, vol. 10808. Springer, Heidelberg (2018). To appear
11. Cheon, J.H., Han, K., Kim, D.: Faster bootstrapping of FHE over the integers. Cryptology ePrint Archive, Report 2017/079 (2017). http://eprint.iacr.org/2017/079
12. Cheon, J.H., Jeong, J., Lee, J., Lee, K.: Privacy-preserving computations of predictive medical models with minimax approximation and non-adjacent form. In: Brenner, M., Rohloff, K., Bonneau, J., Miller, A., Ryan, P.Y.A., Teague, V., Bracciali, A., Sala, M., Pintore, F., Jakobsson, M. (eds.) FC 2017. LNCS, vol. 10323, pp. 53–74. Springer, Cham (2017). https://doi.org/10.1007/978-3-319-70278-0_4
13. Chillotti, I., Gama, N., Georgieva, M., Izabachène, M.: Faster fully homomorphic encryption: bootstrapping in less than 0.1 seconds. In: Cheon, J.H., Takagi, T. (eds.) ASIACRYPT 2016. LNCS, vol. 10031, pp. 3–33. Springer, Heidelberg (2016). https://doi.org/10.1007/978-3-662-53887-6_1

14. Costache, A., Smart, N.P., Vivek, S.: Faster homomorphic evaluation of discrete fourier transforms. In: Kiayias, A. (ed.) FC 2017. LNCS, vol. 10322, pp. 517–529. Springer, Cham (2017). https://doi.org/10.1007/978-3-319-70972-7_29

15. Costache, A., Smart, N.P., Vivek, S., Waller, A.: Fixed-Point Arithmetic in SHE Schemes. In: Avanzi, R., Heys, H. (eds.) SAC 2016. LNCS, vol. 10532, pp. 401–422. Springer, Cham (2017). https://doi.org/10.1007/978-3-319-69453-5_22

16. CryptoExperts. FV-NFLlib (2016). https://github.com/CryptoExperts/FV-NFLlib

17. Bruijn, N.G.D.: Asymptotic methods in analysis. Dover, New York (1958)

18. Dowlin, N., Gilad-Bachrach, R., Laine, K., Lauter, K.E., Naehrig, M., Wernsing, J.: Manual for using homomorphic encryption for bioinformatics. Proc. IEEE **105**(3), 552–567 (2017)

19. Oswald, E., Fischlin, M. (eds.): EUROCRYPT 2015. LNCS, vol. 9056. Springer, Heidelberg (2015). https://doi.org/10.1007/978-3-662-46800-5

20. Fan, J., Vercauteren, F.: Somewhat practical fully homomorphic encryption. Cryptology ePrint Archive, Report 2012/144 (2012). http://eprint.iacr.org/2012/144

21. Gentry, C.: Fully homomorphic encryption using ideal lattices. In: Mitzenmacher, M. (ed.) 41st ACM Symposium on Theory of Computing STOC, pp. 169–178. ACM Press (2009)

22. Gentry, C., Halevi, S., Smart, N.P.: Fully homomorphic encryption with polylog overhead. In: Pointcheval, D., Johansson, T. (eds.) EUROCRYPT 2012. LNCS, vol. 7237, pp. 465–482. Springer, Heidelberg (2012). https://doi.org/10.1007/978-3-642-29011-4_28

23. Gentry, C., Halevi, S., Smart, N.P.: Fully homomorphic encryption with polylog overhead. In: Pointcheval, D., Johansson, T. (eds.) EUROCRYPT 2012. LNCS, vol. 7237, pp. 465–482. Springer, Heidelberg (2012). https://doi.org/10.1007/978-3-642-29011-4_28

24. Huang, Z., Jalali, A., Chen, H., Han, K., Laine, K.: Simple encrypted arithmetic library – SEAL (v2.3). Technical report, Technical report, Microsoft Research (2017)

25. Naehrig, M., Lauter, K.E., Vaikuntanathan, V.: Can homomorphic encryption be practical? In: Cachin, C., Ristenpart, T., (eds.) Proceedings of the 3rd ACM Cloud Computing Security Workshop, CCSW 2011, pp. 113–124. ACM (2011)

26. Rivest, R.L., Adleman, L., Dertouzos, M.L.: On data banks and privacy homomorphisms. Found. Secure Comput. **4**(11), 169–180 (1978)

27. Smart, N.P., Vercauteren, F.: Fully homomorphic SIMD operations. Des. Codes Crypt. **71**(1), 57–81 (2014)

Bootstrapping for Approximate Homomorphic Encryption

Jung Hee Cheon[1], Kyoohyung Han[1], Andrey Kim[1], Miran Kim[2],
and Yongsoo Song[1,2(✉)]

[1] Seoul National University, Seoul, Republic of Korea
{jhcheon,satanigh,kimandrik,lucius05}@snu.ac.kr
[2] University of California, San Diego, USA
{mrkim,yongsoosong}@ucsd.edu

Abstract. This paper extends the leveled homomorphic encryption scheme for an approximate arithmetic of Cheon et al. (ASIACRYPT 2017) to a fully homomorphic encryption, i.e., we propose a new technique to refresh low-level ciphertexts based on Gentry's bootstrapping procedure. The modular reduction operation is the main bottleneck in the homomorphic evaluation of the decryption circuit. We exploit a scaled sine function as an approximation of the modular reduction operation and present an efficient evaluation strategy. Our method requires only one homomorphic multiplication for each of iterations and so the total computation cost grows linearly with the depth of the decryption circuit. We also show how to recrypt packed ciphertexts on the RLWE construction with an open-source implementation. For example, it takes 139.8 s to refresh a ciphertext that encrypts 128 numbers with 12 bits of precision, yielding an amortized rate of 1.1 seconds per slot.

Keywords: Homomorphic encryption · Approximate arithmetic
Bootstrapping

1 Introduction

Homomorphic encryption (HE) is a cryptographic scheme that allows us to evaluate an arbitrary arithmetic circuit on encrypted data without decryption. There have been a number of studies [5–9,11,16,19,22,25,30,31] to improve the efficiency of HE cryptosystem after Gentry's blueprint [26]. This cryptographic primitive has a number of prospective real-world applications based on the secure outsourcing of computation in public clouds. For example, HE can be a solution to performing the computation of various algorithms on financial, medical, or genomic data without any information leakage [15,36,38,39,41].

Unfortunately, most of existing HE schemes support the exact arithmetic operations over some discrete spaces (e.g. finite field), so that they are not suitable for many real-world applications which require a floating point operation or real number arithmetic. To be specific, *bitwise encryption* schemes [17,24]

© International Association for Cryptologic Research 2018
J. B. Nielsen and V. Rijmen (Eds.): EUROCRYPT 2018, LNCS 10820, pp. 360–384, 2018.
https://doi.org/10.1007/978-3-319-78381-9_14

can evaluate a boolean gate with bootstrapping in much shorter time, but it is necessary to evaluate a deep circuit with a number of gates to perform a single arithmetic operation (e.g. addition or multiplication) between high-precision numbers. Moreover, a huge expansion rate of ciphertexts is another issue that stands in the way of the practical use of bitwise encryptions. On the other hand, *word encryption* schemes [6,7,25,30] can encrypt multiple high-precision numbers in a single ciphertext but the rounding operation is difficult to be evaluated since it is not expressed as a small-degree polynomial. Therefore, they require either a plaintext space with an exponentially large bit size on the depth of a circuit, or an expensive computation such as rounding operation and extraction of the most significant bits.

Recently, Cheon et al. [14] proposed a HE scheme for an Arithmetic of Approximate Numbers (called HEAAN in what follows) based on the ring learning with errors (RLWE) problem. The main idea is to consider an encryption error as part of a computational error that occurs during approximate computations. For an encryption ct of a message m with a secret key sk, the decryption algorithm $[\langle ct, sk \rangle]_q$ outputs an approximate value $m + e$ of the original message with a small error e. The main advantage of HEAAN comes from the *rescaling* procedure for managing the magnitude of plaintexts. It truncates a ciphertext into a smaller modulus, which leads to an approximate rounding of the encrypted plaintext. As a result, it achieved the first linear growth of the ciphertext modulus on the depth of the circuit being evaluated, against the exponential growth in previous word encryption schemes. In addition, the RLWE-based HEAAN scheme has its own packing strategy to encrypt multiple complex numbers in a single ciphertext and perform a parallel computation. However, HEAAN is a *leveled* HE scheme which can only evaluate a circuit of fixed depth. As homomorphic operations progresses, the ciphertext modulus decreases and finally becomes too small to carry out more computations.

In previous literature, Gentry's bootstrapping is the only known method to construct a fully homomorphic encryption (FHE) scheme which allows us to evaluate an arbitrary circuit. Technically, the bootstrapping method can be understood as a homomorphic evaluation of the decryption circuit to refresh a ciphertext for more computations. The HEAAN scheme does not support the modular arithmetic, however, its decryption circuit $[\langle ct, sk \rangle]_q$ requires the modular reduction operation, which makes its bootstrapping much harder. Therefore, the bootstrapping of HEAAN can be reduced to a problem that represents the modular reduction function $F(t) = [t]_q$ as a polynomial over the integers (or, complex numbers). One may use the polynomial interpolation of this function over the domain of $t = \langle ct, sk \rangle$, but it is a limiting factor for practical implementation due to a huge computational cost of an evaluation.

Our Contributions. We present a methodology to refresh ciphertexts of HEAAN and make it bootstrappable for the evaluation of an arbitrary circuit. We take advantage of its intrinsic characteristic - *approximate computation* on encrypted data. Our bootstrapping procedure aims to evaluate the decryption formula approximately and obtain an encryption of the original message in a large ciphertext modulus. Hence, we find an approximation of the modular reduction function that can be evaluated efficiently using arithmetic operations of HEAAN. The approximation error should be small enough to maintain the precision of an input plaintext.

We first note that the modular reduction function $F(t) = [t]_q$ is the identity nearby zero and periodic with period q. If $t = \langle \mathsf{ct}, \mathsf{sk} \rangle$ is close to a multiple of the ciphertext modulus q (or equivalently, if the encrypted plaintext $m = [t]_q$ is small compared to q), then a trigonometric function can be a good approximation to the modular reduction. Namely, the decryption formula of HEAAN can be represented using the following scaled *sine* function as

$$[\langle \mathsf{ct}, \mathsf{sk} \rangle]_q = \frac{q}{2\pi} \cdot \sin\left(\frac{2\pi}{q} \cdot \langle \mathsf{ct}, \mathsf{sk} \rangle\right) + O(\epsilon^3 \cdot q),$$

when $|[\langle \mathsf{ct}, \mathsf{sk} \rangle]_q| \leq \epsilon \cdot q$. Hence we may use this analytic function instead of the modular reduction in the decryption formula.

Now our goal is to homomorphically evaluate the trigonometric function $\frac{q}{2\pi} \cdot \sin\left(\frac{2\pi}{q} \cdot t\right)$ with an input $t = \langle \mathsf{ct}, \mathsf{sk} \rangle$, which is bounded by Kq for some constant $K = O(\lambda)$ with λ the security parameter. We can consider the Taylor polynomial as an approximation to the trigonometric function, but its degree should be at least $O(Kq)$ to make an error term small enough on the interval $(-Kq, Kq)$. The evaluation of polynomial can be done in $O(\sqrt{Kq})$ homomorphic multiplications with Paterson-Stockmeyer method [40], but this complexity of recryption grows exponentially with the depth $L = O(\log q)$ of the decryption circuit - which is still quite substantial.

We suggest an evaluation strategy of the trigonometric function to reduce its computation cost by exploiting the following double-angle formulas:

$$\begin{cases} \cos(2\theta) = \cos^2 \theta - \sin^2 \theta, \\ \sin(2\theta) = 2 \cos \theta \cdot \sin \theta, \end{cases}$$

which means that we can obtain some approximate values of $\cos(2\theta)$ and $\sin(2\theta)$ from approximate values of $\cos \theta$ and $\sin \theta$. In our bootstrapping process, we first compute the Taylor expansions of $\cos\left(\frac{2\pi}{q} \cdot \frac{t}{2^r}\right)$ and $\sin\left(\frac{2\pi}{q} \cdot \frac{t}{2^r}\right)$ of a small degree $d_0 = O(1)$ for some $r = O(\log(Kq))$. Then we use the doubling-angle formulas r times recursively to get an approximate value of $\sin\left(\frac{2\pi}{q} \cdot t\right)$. In the case of the RLWE-based construction, this evaluation can be even more simplified by encrypting the complex exponentiation $\exp(i\theta) = \cos \theta + i \cdot \sin \theta$ and adapting the identity $\exp(i \cdot 2\theta) = (\exp(i \cdot \theta))^2$.

Results. Our bootstrapping technique for HEAAN is a new cryptographic primitive for FHE mechanisms, which yields the first word encryption scheme for approximate arithmetic. For a ciphertext ct with a modulus q, our bootstrapping procedure generates a ciphertext ct' with a larger ciphertext modulus $Q \gg q$, satisfying the condition $[\langle \mathsf{ct}', \mathsf{sk} \rangle]_Q \approx [\langle \mathsf{ct}, \mathsf{sk} \rangle]_q$ while an error is kept small enough

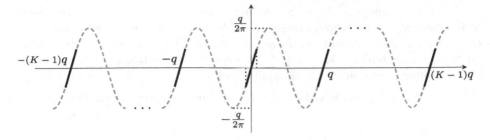

Fig. 1. Modular reduction and scaled sine functions

not to destroy the significant digits of a plaintext. The output ciphertext will have a sufficiently large modulus compared to a plaintext, thereby enabling further computation on the ciphertext. In addition, our approximation to a trigonometric function and efficient evaluation strategy reduce the complexity of the evaluation down to $O(L)$ homomorphic multiplications for the depth $L = O(\log q)$ of the decryption circuit.

We also give an open-source implementation [12] to demonstrate the performance of our bootstrapping method. It contains some optimization techniques including the linear transformation method of [33] for the recryption over the packed ciphertexts. When we want to preserve 12 bits of precision, our bootstrapping on a single-slot ciphertext takes about 26.6 s. We also optimize the linear transforms for sparsely packed ciphertexts and it takes about 139.8 s to recrypt a ciphertext that encrypts 128 complex numbers in plaintext slots, yielding an amortized rate of 1.1 seconds per slot.

Implications of Our Bootstrapping Method. The main feature of approximate arithmetic is that every number contains an error of which could increase during computation. The precision of a number is reduced by approximately one bit after multiplication and finally we may not extract any meaningful information from the computation result if the depth of a circuit is larger than the bit precision of the input data. On the other hand, our bootstrapping procedure is to refresh ciphertexts and then perform further computation on encrypted data. This concept of an *unlimited* computation may seem a contradiction to the property of finite precision in the approximate arithmetic.

However, it turns out to be better for real-world applications that have a property of negative feedback or stability. For example, a cyber-physical system (CPS) is a compromised mechanism of physical and computational components. A computational element commutes with the sensors and every signal contains a small error. One can guarantee the correctness of CPS only when it is stable because an error is reduced by negative feedback to the input. Another example is the gradient descent method, which is the most widely used algorithm to perform optimization. It has a number of applications in machine learning such as logistic regression and neural networks. It computes the gradient of a point and moves it closer to an optimal point, which reduces the effects of perturbations in the output.

As in the examples above, we do not have to worry about the precision of numbers when the overall system is stable. In fact, there are some proof-of-concept implementations about the secure control of CPS [35] and secure logistic regression using biomedical data [37]. We expect that our bootstrapping process can be applied to these real-world applications.

Related Works. There have been several attempts to carry out an approximate arithmetic using HE. Downlin et al. [23] (see also [4,21]) described a method to transform a real number into a polynomial with small and sparse coefficients to reduce the required size of a plaintext modulus. Costache et al. [20] suggested a similar encoding method with [14] to evaluate the discrete Fourier transformation efficiently, but a ciphertext could encrypt only one value. Chen et al. [10] uses a technique of [34] to encrypt a single high-precision number. However, they still have some problems: (1) the coefficients and the degree of encoded polynomial grow exponentially with the depth of a circuit and (2) there is no known result to achieve an FHE scheme because a polynomial should be re-encoded to be represented with a smaller degree and coefficients for more computations and the bootstrapping method of [33] is not enough for this functionality.

The original Gentry's bootstrapping technique was implemented by Gentry and Halevi [27], which took a half hour to recrypt a single bit ciphertext. Gentry et al. [28] represented the decryption circuit of RLWE-based HE with a lower depth circuit using a special modulus space. The Halevi-Shoup FHE implementation [33] reported a recryption time of approximately six minutes per slot. Meanwhile, Ducas and Micciancio [24] proposed the FHEW scheme that bootstraps a single-bit encryption in less than a second based on the framework of [2]. Chillotti et al. [17] obtained a speed up to less than 0.1 s. The following works [3,18] improved the performance by using the evaluation of a look-up table before bootstrapping. However, the size of an input plaintext of bootstrapping is very limited since it is related to the ring dimension of an intermediate Ring GSW scheme. In addition, a huge expansion rate of ciphertexts is still an open problem in bitwise encryption schemes.

The previous FHE schemes evaluate the exact decryption circuit using the structure of a finite field or a polynomial ring in bootstrapping algorithm. The evaluation of an arbitrary polynomial of degree d requires $O(\sqrt{d})$ homomorphic multiplications, but Halevi and Shoup [33] used a polynomial with the lifting property to reduce the computational cost of bootstrapping. They used a recursive algorithm to extract some digits in an encrypted state, so the number of homomorphic multiplications for bootstrapping was reduced down to $O(\log^2 d)$. Contrary to the work of Halevi and Shoup, we find an approximate decryption circuit using a trigonometric function and suggest an even simpler recursive algorithm. As a result, our algorithm only requires $O(\log d)$ number of homomorphic multiplications, which results in an enhanced performance.

Road-map. Section 2 briefly introduces notations and some preliminaries about algebra. We also review the HEAAN scheme of Cheon et al. [14]. Section 3 explains our simplified decryption formula by using a trigonometric function. In Sect. 4, we recall the ciphertext packing method of HEAAN and describe a linear transformation on packed ciphertexts. In Sect. 5, we present our bootstrapping technique with a precise noise estimation. In Sect. 6, we implement the recryption procedure based on the proposed method and discuss the performance results.

2 Preliminaries

The binary logarithm will be simply denoted by $\log(\cdot)$. We denote vectors in bold, e.g. $\mathbf{a}$, and every vector in this paper is a column vector. For a $n_1 \times m$ matrix A_1 and a $n_2 \times m$ matrix A_2, $(A_1; A_2)$ denotes the $(n_1 + n_2) \times m$ matrix obtained by concatenating matrices A_1 and A_2 in a vertical direction.

We denote by $\langle \cdot, \cdot \rangle$ the usual dot product of two vectors. For a real number r, $\lfloor r \rceil$ denotes the nearest integer to r, rounding upwards in case of a tie. For an integer q, we identify $\mathbb{Z} \cap (-q/2, q/2]$ as a representative of $\mathbb{Z}_q$ and use $[z]_q$ to denote the reduction of the integer z modulo q into that interval. We use $x \leftarrow D$ to denote the sampling x according to distribution D. The uniform distribution over a finite set S is denoted by $U(S)$. We let λ denote the security parameter throughout the paper: all known valid attacks against the cryptographic scheme under scope should take $\Omega(2^\lambda)$ bit operations.

2.1 Cyclotomic Ring

For a positive integer M, let $\Phi_M(X)$ be the M-th cyclotomic polynomial of degree $N = \phi(M)$. Let $\mathcal{R} = \mathbb{Z}[X]/(\Phi_M(X))$ be the ring of integers of a number field $\mathbb{Q}[X]/(\Phi_M(X))$. We write $\mathcal{R}_q = \mathcal{R}/q\mathcal{R}$ for the residue ring of $\mathcal{R}$ modulo an integer q. An arbitrary element of the set $\mathcal{P} = \mathbb{R}[X]/(\Phi_M(X))$ will be represented as a polynomial $a(X) = \sum_{j=0}^{N-1} a_j X^j$ of degree strictly less than N and identified with its coefficients vector $\mathbf{a} = (a_0, \ldots, a_{N-1}) \in \mathbb{R}^N$. We define $\|a\|_\infty$ and $\|a\|_1$ by the relevant norms on the coefficients vector $\mathbf{a}$.

Write $\mathbb{Z}_M^* = \{x \in \mathbb{Z}_M : \gcd(x, M) = 1\}$ for the multiplicative group of units in $\mathbb{Z}_M$. Recall that the canonical embedding of $a(X) \in \mathbb{Q}[X]/(\Phi_M(X))$ into $\mathbb{C}^N$ is the vector of evaluations of $a(X)$ at the M-th primitive roots of unity. We use its natural extension σ to $\mathcal{P}$, defined by $\sigma(a) = (a(\zeta^j))_{j \in \mathbb{Z}_M^*}$ for $\zeta = \exp(2\pi i/M)$. Its ℓ_∞-norm is called the *canonical embedding norm*, denoted by $\|a\|_\infty^{\mathsf{can}} = \|\sigma(a)\|_\infty$.

2.2 Homomorphic Encryption for Arithmetic of Approximate Numbers

HE is one of the prospective cryptographic primitives for secure outsourcing computation without information leakage. However, an inefficiency of real number computation is one of the main obstacles to apply HE schemes in real-world applications. Recently Cheon et al. [14] proposed a method to construct the HE scheme for approximate arithmetic, called HEAAN. Their scheme supports an efficient rounding operation of encrypted plaintext as well as basic arithmetic operations. This subsection gives a concrete description of the RLWE-based HEAAN scheme.

For a real $\sigma > 0$, $\mathcal{DG}(\sigma^2)$ denotes a distribution over $\mathbb{Z}^N$ which samples its components independently from the discrete Gaussian distribution of variance σ^2. For an positive integer h, $\mathcal{HWT}(h)$ denotes a uniform distribution over the set of signed binary vectors in $\{\pm 1\}^N$ whose Hamming weight is exactly h. For a real $0 \leq \rho \leq 1$, the distribution $\mathcal{ZO}(\rho)$ draws each entry in the vector from $\{0, \pm 1\}$, with probability $\rho/2$ for each of -1 and $+1$, and probability being zero $1 - \rho$.

- KeyGen(1^λ).
    - For a base p and an integer L, let $q_\ell = p^\ell$ for $\ell = 1, \ldots, L$. Given the security parameter λ, choose a power-of-two M, an integer h, an integer P, and a real number $\sigma > 0$ for an RLWE problem that achieves λ-bit of security level.
    - Sample $s \leftarrow \mathcal{HWT}(h)$, $a \leftarrow U(\mathcal{R}_{q_L})$ and $e \leftarrow \mathcal{DG}(\sigma^2)$. Set the secret key as $\mathsf{sk} \leftarrow (1, s)$ and the public key as $\mathsf{pk} \leftarrow (b, a) \in \mathcal{R}_{q_L}^2$ where $b \leftarrow -as + e$ (mod q_L).
- KSGen$_{\mathsf{sk}}$(s'). For $s' \in \mathcal{R}$, sample $a' \leftarrow U(\mathcal{R}_{P \cdot q_L})$ and $e' \leftarrow \mathcal{DG}(\sigma^2)$. Output the switching key as $\mathsf{swk} \leftarrow (b', a') \in \mathcal{R}_{P \cdot q_L}^2$ where $b' \leftarrow -a's + e' + Ps'$ (mod $P \cdot q_L$).
    - Set the evaluation key as $\mathsf{evk} \leftarrow \mathsf{KSGen}_{\mathsf{sk}}(s^2)$.
- Enc$_{\mathsf{pk}}$(m). For $m \in \mathcal{R}$, sample $v \leftarrow \mathcal{ZO}(0.5)$ and $e_0, e_1 \leftarrow \mathcal{DG}(\sigma^2)$. Output $v \cdot \mathsf{pk} + (m + e_0, e_1)$ (mod q_L).
- Dec$_{\mathsf{sk}}$(ct). For $\mathsf{ct} = (c_0, c_1) \in \mathcal{R}_{q_\ell}^2$, output $m = c_0 + c_1 \cdot s$ (mod q_ℓ).
- Add($\mathsf{ct}_1, \mathsf{ct}_2$). For $\mathsf{ct}_1, \mathsf{ct}_2 \in \mathcal{R}_{q_\ell}^2$, output $\mathsf{ct}_{\mathsf{add}} \leftarrow \mathsf{ct}_1 + \mathsf{ct}_2$ (mod q_ℓ).
- Mult$_{\mathsf{evk}}$($\mathsf{ct}_1, \mathsf{ct}_2$). For $\mathsf{ct}_1 = (b_1, a_1), \mathsf{ct}_2 = (b_2, a_2) \in \mathcal{R}_{q_\ell}^2$, let $(d_0, d_1, d_2) = (b_1 b_2, a_1 b_2 + a_2 b_1, a_1 a_2)$ (mod q_ℓ). Output $\mathsf{ct}_{\mathsf{mult}} \leftarrow (d_0, d_1) + \lfloor P^{-1} \cdot d_2 \cdot \mathsf{evk} \rceil$ (mod q_ℓ).
- RS$_{\ell \to \ell'}$(ct). For a ciphertext $\mathsf{ct} \in \mathcal{R}_{q_\ell}^2$ at level ℓ, output $\mathsf{ct}' \leftarrow \lfloor p^{\ell' - \ell} \cdot \mathsf{ct} \rceil$ (mod $q_{\ell'}$). We will omit the subscript ($\ell \to \ell'$) when $\ell' = \ell - 1$.

The native plaintext space of HEAAN can be understood as the set of polynomials $m(X)$ in $\mathbb{Z}[X]/(\Phi_M(X))$ such that $\|m\|_\infty^{\mathsf{can}} < q/2$. For convenience, we allow an arbitrary element of $\mathcal{P} = \mathbb{R}[X]/(\Phi_M(X))$ as a plaintext polynomial, so that a ciphertext $\mathsf{ct} = (c_0, c_1) \in \mathcal{R}_{q_\ell}^2$ at level ℓ will be called an encryption of $m(X) \in \mathcal{P}$ with an error bound B if it satisfies $\langle \mathsf{ct}, \mathsf{sk} \rangle = m + e$ (mod q_ℓ) for some polynomial $e(X) \in \mathcal{P}$ satisfying $\|e\|_\infty^{\mathsf{can}} \leq B$. The set $\mathcal{P} = \mathbb{R}[X]/(\Phi_M(X))$ can be identified with the complex coordinate space $\mathbb{C}^{N/2}$ using a ring isomorphism. This decoding map allows us to encrypt at most $(N/2)$ numbers in a single ciphertext and carry out parallel operations in a Single Instruction Multiple Data (SIMD) manner. A simple description of the packing method will be described in Sect. 4.1.

We will make the use of the following lemmas from [14] for noise estimation. We adapt some notations from [14], defining the constants B_{ks} and B_{rs}.

Lemma 1 ([14, Lemma 1]). *Let* $\mathsf{ct} \leftarrow \mathsf{Enc}_{\mathsf{pk}}(m)$ *be an encryption of* $m \in \mathcal{R}$. *Then* $\langle \mathsf{ct}, \mathsf{sk} \rangle = m + e$ (mod q_L) *for some* $e \in \mathcal{R}$ *satisfying* $\|e\|_\infty^{\mathsf{can}} \leq B_{\mathsf{clean}}$ *for* $B_{\mathsf{clean}} = 8\sqrt{2}\sigma N + 6\sigma\sqrt{N} + 16\sigma\sqrt{hN}$.

Lemma 2 ([14, Lemma 2]). *Let* $\mathsf{ct}' \leftarrow RS_{\ell \to \ell'}(\mathsf{ct})$ *for a ciphertext* $\mathsf{ct} \in \mathcal{R}_{q_\ell}^2$. *Then* $\langle \mathsf{ct}', \mathsf{sk} \rangle = \frac{q_{\ell'}}{q_\ell} \langle \mathsf{ct}, \mathsf{sk} \rangle + e \pmod{q_{\ell'}}$ *for some* $e \in \mathcal{P}$ *satisfying* $\|e\|_\infty^{can} \leq B_{\mathsf{rs}}$ *for* $B_{\mathsf{rs}} = \sqrt{N/3} \cdot (3 + 8\sqrt{h})$.

Lemma 3 ([14, Lemma 3]). *Let* $\mathsf{ct}_{\mathsf{mult}} \leftarrow Mult_{\mathsf{evk}}(\mathsf{ct}_1, \mathsf{ct}_2)$ *for two ciphertexts* $\mathsf{ct}_1, \mathsf{ct}_2 \in \mathcal{R}_{q_\ell}^2$. *Then* $\langle \mathsf{ct}_{\mathsf{mult}}, \mathsf{sk} \rangle = \langle \mathsf{ct}_1, \mathsf{sk} \rangle \cdot \langle \mathsf{ct}_2, \mathsf{sk} \rangle + e_{\mathsf{mult}} \pmod{q_\ell}$ *for some* $e \in \mathcal{R}$ *satisfying* $\|e_{\mathsf{mult}}\|_\infty^{can} \leq B_{\mathsf{mult}}(\ell)$ *for* $B_{\mathsf{ks}} = 8\sigma N/\sqrt{3}$ *and* $B_{\mathsf{mult}}(\ell) = P^{-1} \cdot q_\ell \cdot B_{\mathsf{ks}} + B_{\mathsf{rs}}$.

A rescaling (rounding) error is the smallest error type of homomorphic operations. The least digits of a plaintext is destroyed by some error after multiplication or rescaling, so its significand should be placed in higher digits not to lose the precision of the resulting plaintext.

3   Decryption Formula over the Integers

The goal of bootstrapping is to refresh a ciphertext and keep computing on encrypted data. Recall that HEAAN supports arithmetic operations on a characteristic zero plaintext space such as $\mathbb{C}$ However, its decryption formula consists of two steps: the inner product $t = \langle \mathsf{ct}, \mathsf{sk} \rangle$ over the integers and the modular reduction $m = [t]_q$. We therefore have to express this decryption formula efficiently using homomorphic operations provided in the HEAAN scheme.

The main difficulty comes from the fact that the reduction modular q function $F(t) = [t]_q$ is not represented as a small-degree polynomial. A naive approach such as the polynomial interpolation causes a huge degree, resulting in a large parameter size and an expensive computational cost for bootstrapping process. Instead, we reduce the required circuit depth and the evaluation complexity by exploiting a polynomial approximation of the decryption formula and taking advantage of approximate arithmetic.

3.1   Approximation of the Modular Reduction Function

Let ct be a ciphertext relative to a secret key sk and a modulus q. Since sk is sampled from a small distribution, the size of its decryption structure $t = \langle \mathsf{ct}, \mathsf{sk} \rangle$ is bounded by Kq for some fixed constant K. So we can say that the decryption formula of HEAAN is defined on the set $\mathbb{Z} \cap (-Kq, Kq)$ and it maps an arbitrary integer $t \in \mathbb{Z} \cap (-Kq, Kq)$ to the reduction modular q.

It is infeasible to find a good approximation of the modular reduction function since it is not continuous. We first assume that a message m of an input ciphertext is still much smaller than a ciphertext modulus q, so that $t = \langle \mathsf{ct}, \mathsf{sk} \rangle$ can be expressed as $qI + m$ for some I and m such that $|I| < K$ and $|m| \ll q$. This assumption is reasonable because one can start the bootstrapping procedure on a ciphertext before its modulus becomes too small. Then the modular reduction $F(t) = [t]_q$ on a restricted domain becomes a *piecewise linear* function (see Fig. 1). We point out that this function is the identity near zero and periodic, so it looks like a part of the scaled sine

$$S(t) = \frac{q}{2\pi} \sin\left(\frac{2\pi t}{q}\right).$$

Note that it gives a good approximation to the piecewise linear function when an input value $t = qI + m$ is close to a multiple of q. Specifically, an error between $F(t)$ and $S(t)$ is bounded by

$$|F(t) - S(t)| = \frac{q}{2\pi}\left|\frac{2\pi m}{q} - \sin\left(\frac{2\pi m}{q}\right)\right| \le \frac{q}{2\pi} \cdot \frac{1}{3!}\left(\frac{2\pi|m|}{q}\right)^3 = O\left(q \cdot \frac{|m|^3}{q^3}\right),$$

which is equivalently $O(1)$ when $m = O(q^{2/3})$.

3.2 Homomorphic Evaluation of the Complex Exponential Function

As discussed before, the scaled sine function $S(t)$ is a good approximation of the reduction modulo q. However, this function cannot be evaluated directly using HE since it is not a polynomial function. The goal of this subsection is to explain how to approximately and efficiently evaluate this trigonometric function based on HEAAN.

We may consider the Taylor polynomial $\frac{q}{2\pi}\sum_{j=0}^{d-1}\frac{(-1)^j}{(2j+1)!}\left(\frac{2\pi t}{q}\right)^{2j+1}$ of $S(t)$. The size of error converges to zero very rapidly as the degree grows, i.e., an error between $S(t)$ and its Taylor polynomial of degree $2d$ is bounded by $\frac{q}{2\pi} \cdot \frac{1}{(2d+1)!}(2\pi K)^{2d+1}$ when $|t| < Kq$, and it becomes small enough when the degree of the Taylor polynomial is $O(Kq)$. However, despite its high precision, this naive method has an ineffective problem in practice. The complexity grows exponentially with the depth of a circuit, e.g. $O(\sqrt{d})$ using the Paterson-Stockmeyer algorithm [40] for an evaluation of a degree-d polynomial.

Instead, we can reduce the computational cost by exploiting the following double-angle formulas: $\cos(2\theta) = \cos^2\theta - \sin^2\theta$ and $\sin(2\theta) = 2\cos\theta \cdot \sin\theta$. From approximate values of trigonometric functions in a small domain, we extend to find good approximations of the sign function on a wider (doubled) range. In particular, the RLWE-based HEAAN scheme can encrypt the complex numbers, so that the evaluation algorithm can be more simplified using the complex exponential function. Specifically, we use the identities

$$\begin{cases} \exp(i\theta) = \cos\theta + i \cdot \sin\theta, \\ \exp(2i\theta) = (\exp(i\theta))^2, \end{cases}$$

and the error growth from squaring can be bounded by about one bit since $(\exp(i\theta) \pm \epsilon)^2 \approx \exp(2i\theta) \pm 2\epsilon$.

We take the Taylor polynomial of a small degree $d_0 \ge 1$ as a high-precision approximation of the complex exponential function within a small range.

Then we perform the squaring operation repeatedly to get an approximation of the complex exponential function over the desirable domain. Note that we multiply a scale factor of Δ to prevent the precision loss and divide the intermediate ciphertexts by a constant Δ using the rescaling procedure of HEAAN.

The use of the complex exponential function has another advantage in error analysis. When we consider the RLWE-based HEAAN scheme, small *complex* errors are added to plaintext slots during encryption, evaluation, rescaling and slot permutation. Therefore, we have only one constraint such that a decryption formula should be tolerant of small complex errors. Another advantage of our method comes from the fact that the complex exponential function is analytic with a bounded derivative over the whole complex plane, and therefore an error does not blow up by the decryption formula. The whole procedure is explicitly described as follows.

A value $t \in (-Kq, Kq)$ is given as an input of the decryption formula.

1. Consider the complex exponential function of $\exp\left(\frac{2\pi it}{2^r \cdot q}\right)$ and compute its (scaled) Taylor expansion as

$$P_0(t) = \Delta \cdot \sum_{k=0}^{d_0} \frac{1}{k!} \left(\frac{2\pi it}{2^r \cdot q}\right)^k$$

of degree $d_0 \geq 1$.
2. For $j = 0, 1, \ldots, r-1$, repeat the squaring $P_{j+1}(t) \leftarrow \Delta^{-1} \cdot (P_j(t))^2$.
3. Return $P_r(t)$.

The degree d_0 of the initial Taylor polynomial, the scaling factor of Δ, and the number r of the iterations (squaring) are determined by the following noise analysis. Since the size of the initial input $(2\pi t)/(2^r \cdot q)$ of the complex exponential function has a small upper bound $(2\pi K/2^r)$, even the Taylor polynomial of a small degree d_0 can be a good approximation to the complex exponential function $\exp\left(\frac{2\pi it}{2^r \cdot q}\right)$. From the above observation, the output $P_r(t)$ is a polynomial of degree $d_r = d_0 \cdot 2^r$ and it is an approximation of $E(t) := \Delta \cdot \exp\left(\frac{2\pi it}{q}\right)$ on a wide interval $t \in (-Kq, Kq)$. After the evaluation of the complex exponential function, we can extract the imaginary (sine) part by conjugation operation (i.e., $2\sin\theta = \exp(i\theta) - \exp(-i\theta)$), which will be described in the next section.

For the estimation of noise, we start from an initial error between $P_0(t)$ and $\Delta \cdot \exp\left(\frac{2\pi it}{2^r \cdot q}\right)$, which is bounded by $\frac{\Delta}{(d_0+1)!}\left|\frac{2\pi K}{2^r}\right|^{d_0+1}$ from the Taylor remainder theorem. As described above, the error bound is almost doubled after each squaring. Therefore, we get a bound from an approximation as follows:

$$|P_r(t) - E(t)| \leq \frac{\Delta \cdot 2^r}{(d_0 + 1)!} \left(\frac{2\pi K}{2^r} \right)^{d_0 + 1}$$

$$\leq \frac{\Delta \cdot 2^r}{\sqrt{2\pi(d_0 + 1)}} \left(\frac{e\pi K}{2^{r-1}(d_0 + 1)} \right)^{d_0 + 1}$$

from Stirling's formula. Asymptotically the choice of parameters $d_0 = O(1)$ and $r = O(\log(Kq))$ gives us a sufficiently small error bound. Note that the complexity of the algorithm is $r = O(\log(Kq))$ homomorphic multiplications and it grows linearly with the depth of the decryption circuit.

4   Linear Transformation on Packed Ciphertexts

In this section, we explain how to homomorphically evaluate the linear transformations over the vector of plaintext slots. We first present a simple description of the packing method of HEAAN. We then explain how to compute the rotation and the complex conjugation over the plaintext slots using the key-switching technique. These functionalities can be applied to the evaluation of a linear transformation over plaintext slots.

4.1   Packing Method

The packing technique of HE schemes allows us to encrypt multiple messages in a single ciphertext and enables a parallel computation in a SIMD manner. Cheon et al. [14] proposed a method to identify a cyclotomic polynomial with real coefficients to a vector of complex numbers. We clarify this encoding method and give a simpler description using the structure of a cyclotomic ring with a power-of-two dimension.

Recall that for a power-of-two integer $M > 4$, we have $N = M/2$ and $\Phi_M(X) = X^N + 1$. The integer 5 has the order of $(N/2)$ modulo M and spans $\mathbb{Z}_M^*$ with the integer "-1". Hence $\{\zeta_j, \overline{\zeta_j} : 0 \leq j < N/2\}$ forms the set of the primitive M-th roots of unity for $\zeta_j := \zeta^{5^j}$ and $0 \leq j < N/2$. We use the notation $\tau : \mathcal{P} = \mathbb{R}[X]/(X^N + 1) \to \mathbb{C}^{N/2}$ to denote a variant of the complex canonical embedding map defined by $\tau : m(X) \mapsto z = (z_j)_{0 \leq j < N/2}$ such that $z_j = m(\zeta_j)$. Note that τ is an isometric ring homomorphism between metric spaces $(\mathcal{P}, \|\cdot\|_\infty^{\mathsf{can}})$ and $(\mathbb{C}^{N/2}, \|\cdot\|_\infty)$. We use this isomorphism τ as the decoding function for packing of $(N/2)$ complex numbers in a single polynomial. By identifying a polynomial $m(X) = \sum_{i=0}^{N-1} m_i X^i \in \mathcal{P}$ with the vector of its coefficients $m = (m_0, \ldots, m_{N-1})$, the decoding algorithm τ can be understood as a linear transformation from $\mathbb{R}^N$ to $\mathbb{C}^{N/2}$. Its matrix representation is given by

$$U = \begin{bmatrix} 1 & \zeta_0 & \zeta_0^2 & \cdots & \zeta_0^{N-1} \\ 1 & \zeta_1 & \zeta_1^2 & \cdots & \zeta_1^{N-1} \\ \vdots & \vdots & \vdots & \ddots & \vdots \\ 1 & \zeta_{\frac{N}{2}-1} & \zeta_{\frac{N}{2}-1}^2 & \cdots & \zeta_{\frac{N}{2}-1}^{N-1} \end{bmatrix}$$

which is the $(N/2) \times N$ Vandermonde matrix generated by $\{\zeta_j : 0 \leq j < N/2\}$.

In order to compute the encoding function, which is the inverse of τ, we first note that the relation $\overline{z} = \overline{U} \cdot m$ is obtained from $z = U \cdot m$ by taking the conjugation. If we write $\mathsf{CRT} = (U; \overline{U})$ as the CRT matrix generated by the set $\{\zeta_j, \overline{\zeta}_j : 0 \leq j < N/2\}$ of M-th primitive roots of unity, we have the identities $(z; \overline{z}) = \mathsf{CRT} \cdot m$ and $\mathsf{CRT}^{-1} = \frac{1}{N}\overline{\mathsf{CRT}}^T$. This implies that the inverse of τ can be computed by

$$m = \frac{1}{N}(\overline{U}^T \cdot z + U^T \cdot \overline{z}).$$

Throughout this paper, we will identify two spaces $\mathcal{P}$ and $\mathbb{C}^{N/2}$ via the map τ, and hence a ciphertext will be called an encryption of $z \in \mathbb{C}^{N/2}$ if it encrypts the corresponding polynomial $m(X) = \tau^{-1}(z)$.

4.2   Rotation and Conjugation

The purpose of the key-switching operation is to convert a ciphertext under a secret s' into a ciphertext of the same message with respect to another secret key $\mathsf{sk} = (1, s)$. The switching key swk can be generated by the procedure of $\mathsf{KSGen}_{\mathsf{sk}}(s')$. Given a ciphertext $\mathsf{ct} = (c_0, c_1)$ at level ℓ, the procedure $\mathsf{KS}_{\mathsf{swk}}(\mathsf{ct})$ proceeds as follows.

- $\underline{\mathsf{KS}_{\mathsf{swk}}(\mathsf{ct})}$. Output the ciphertext $\mathsf{ct}' \leftarrow (c_0, 0) + \lfloor P^{-1} \cdot c_1 \cdot \mathsf{swk}\rceil \pmod{q_\ell}$.

The following lemma shows the correctness of key-switching procedure and estimates a noise bound. It has a similar noise bound $P^{-1} \cdot q \cdot B_{\mathsf{ks}} + B_{\mathsf{rs}} \approx B_{\mathsf{rs}}$ with the rescaling process.

Lemma 4 (Key-switching). *Let* $\mathsf{ct} = (c_0, c_1) \in \mathcal{R}_q^2$ *be a ciphertext with respect to a secret key* $\mathsf{sk}' = (1, s')$ *and let* $\mathsf{swk} \leftarrow \mathsf{KSGen}_{\mathsf{sk}}(s')$. *Then* $\mathsf{ct}' \leftarrow \mathsf{KS}_{\mathsf{swk}}(\mathsf{ct})$ *satisfies* $\langle \mathsf{ct}', \mathsf{sk} \rangle = \langle \mathsf{ct}, \mathsf{sk}' \rangle + e_{\mathsf{ks}} \pmod{q}$ *for some* $e_{\mathsf{ks}} \in \mathcal{R}$ *with* $\|e_{\mathsf{ks}}\|_\infty^{\mathsf{can}} \leq P^{-1} \cdot q \cdot B_{\mathsf{ks}} + B_{\mathsf{rs}}$.

Proof. Let $e' = \langle \mathsf{swk}, \mathsf{sk} \rangle - P \cdot s' \pmod{P \cdot q_L}$ be the inserted error of the switching key swk. It was shown in [14, Lemma 3] that the ciphertext ct' contains an additional error $e'' = c_1 \cdot e'$ in the modulus $P \cdot q$ from the key-switching operation. The key-switching error is the sum of $P^{-1} \cdot e''$ and a rounding error e_{rs} of $P^{-1} \cdot c_1 \cdot \mathsf{swk}$, so its size is bounded by

$$\|e_{\mathsf{ks}}\|_\infty^{\mathsf{can}} \leq P^{-1} \cdot \|e''\|_\infty^{\mathsf{can}} + \|e_{\mathsf{rs}}\|_\infty^{\mathsf{can}} \leq P^{-1} \cdot q \cdot B_{\mathsf{ks}} + B_{\mathsf{rs}},$$

as desired.   □

For an integer k co-prime with M, let $\kappa_k : m(X) \mapsto m(X^k) \pmod{\Phi_M(X)}$ be a mapping defined on the set $\mathcal{P}$. As noted in [29], this transformation can be used to provide more functionalities on plaintext slots. In some more details, given a ciphertext ct of a message m with $\mathsf{sk} = (1, s)$, we denote $\kappa_k(\mathsf{ct})$ the ciphertext which is obtained by applying κ_k to each entry of ct. Then $\kappa_k(\mathsf{ct})$ is a valid encryption of $\kappa_k(m)$ with the secret key $\kappa_k(s)$. The key-switching technique can be applied to the ciphertext $\kappa_k(\text{ct})$ to get an encryption of the same message $\kappa_k(m)$ with respect to the original secret key sk.

Rotation. Suppose that ct is an encryption of a message $m(X)$ with the corresponding plaintext vector $\boldsymbol{z} = (z_j)_{0 \leq j < \frac{N}{2}} \in \mathbb{C}^{N/2}$. For any $0 \leq i, j < N/2$, there is a mapping κ_k which sends an element in the slot of index i to an element in the slot of index j. Let us define $k = 5^{i-j} \pmod{M}$ and $\tilde{m} = \kappa_k(m)$. Denoting $\tilde{\boldsymbol{z}} = (\tilde{z}_j)_{0 \leq j < \frac{N}{2}}$ as the corresponding plaintext vector of $\tilde{m}$, we have

$$\tilde{z}_j = \tilde{m}(\zeta_j) = m(\zeta_j^{5^{i-j}}) = m(\zeta_i) = z_i,$$

so the j-th slot of $\tau(\tilde{m})$ and the i-th entry of $\tau(m)$ have the same value. In general, we may get a ciphertext $\kappa_{5^r}(\text{ct})$ encrypting $\rho(\boldsymbol{z}; r) := (z_r, \ldots, z_{\frac{N}{2}-1}, z_0, \ldots, z_{r-1})$, which is the vector obtained from $\boldsymbol{z}$ by rotation. Below, we describe the rotation procedure including the key-switching operation.

- Generate the rotation key $\text{rk}_r \leftarrow \text{KSGen}_{\text{sk}}(\kappa_{5^r}(s))$.
- $\underline{\text{Rot}(\text{ct}; r)}$. Output the ciphertext $\text{KS}_{\text{rk}_r}(\kappa_{5^r}(\text{ct}))$.

Conjugation. We see that $\kappa_{-1}(\text{ct})$ is a valid encryption of the plaintext vector $\overline{\boldsymbol{z}} = (\overline{z_j})_{0 \leq j < N/2}$ with the secret key $\kappa_{-1}(s)$. It follows from that fact that

$$\overline{z_j} = \overline{m(\zeta_j)} = m(\overline{\zeta_j}) = m(\zeta_j^{-1}).$$

Then the homomorphic evaluation of the conjugation operation over plaintext slots consists of two procedures:

- Generate the conjugation key $\text{ck} \leftarrow \text{KSGen}_{\text{sk}}(\kappa_{-1}(s))$.
- $\underline{\text{Conj}(\text{ct})}$. Output the ciphertext $\text{KS}_{\text{ck}}(\kappa_{-1}(\text{ct}))$.

4.3 Linear Transformations

In general, an arbitrary linear transformation over the vector of plaintext slots in $\mathbb{C}^{N/2}$ can be represented as $\boldsymbol{z} \mapsto A \cdot \boldsymbol{z} + B \cdot \overline{\boldsymbol{z}}$ for some complex matrices $A, B \in \mathbb{C}^{N/2 \times N/2}$. As discussed in [32], it can be efficiently done by handling the matrix in a diagonal order and making use of SIMD computation. Specifically, let $\boldsymbol{u}_j = (A_{0,j}, A_{1,j+1}, \ldots, A_{\frac{N}{2}-j-1, \frac{N}{2}-1}, A_{\frac{N}{2}-j,0}, \ldots, A_{\frac{N}{2}-1,j-1}) \in \mathbb{C}^{N/2}$ denote the shifted diagonal vector of A for $0 \leq j < N/2$. Then we have

$$A \cdot \boldsymbol{z} = \sum_{0 \leq j < N/2} (\boldsymbol{u}_j \odot \rho(\boldsymbol{z}; j)) \tag{1}$$

where $\odot$ denotes the Hadamard (component-wise) multiplication between vectors. Therefore, if the matrix A is given in plaintext and ct is given as an encryption of the vector $\boldsymbol{z}$, the matrix-vector multiplication $A \cdot \boldsymbol{z}$ is expressed as combination of rotations and scalar multiplications. The vector rotation $\rho(\boldsymbol{z}; j)$ can be homomorphically computed by $\mathsf{Rot}(\mathsf{ct}; j)$ and the Hadamard (component wise) scalar multiplication is done by multiplying the polynomial $\tau^{-1}(\boldsymbol{u}_j)$. See Algorithm 1 for an explicit description of the homomorphic matrix multiplication. Similarly, the second term $B \cdot \bar{\boldsymbol{z}}$ can be obtained by multiplying the matrix B after applying the slot-wise conjugation on $\boldsymbol{z}$.

Algorithm 1. Homomorphic evaluation of matrix multiplication

1: **procedure** MATMULT($\mathsf{ct} \in \mathcal{R}_q^2, A \in \mathbb{C}^{N/2 \times N/2}$)
2:     $\mathsf{ct}' \leftarrow \lfloor \tau^{-1}(\boldsymbol{u}_0) \rceil \cdot \mathsf{ct} \pmod{q}$
3:     **for** $j = 1$ to $N/2 - 1$ **do**
4:         $\mathsf{ct}_j \leftarrow \lfloor \tau^{-1}(\boldsymbol{u}_j) \rceil \cdot \mathsf{Rot}(\mathsf{ct}; j) \pmod{q}$
5:         $\mathsf{ct}' \leftarrow \mathsf{Add}(\mathsf{ct}', \mathsf{ct}_j) \pmod{q}$
6:     **end for**
7:     **return** ct'
8: **end procedure**

Algorithm 1 requires $(N/2 - 1)$ rotations and N multiplications with scalar polynomials but the complexity can be reduced down using the idea of Baby-Step Giant-Step algorithm. Let $N_1 = O(\sqrt{N})$ be a divisor of $N/2$ and denote $N_2 = N/2N_1$. It follows from [33] that Eq. (1) can be expressed as

$$A \cdot \boldsymbol{z} = \sum_{0 \leq j < N_2} \sum_{0 \leq i < N_1} (\boldsymbol{u}_{N_1 \cdot j + i} \odot \rho(\boldsymbol{z}; N_1 \cdot j + i))$$

$$= \sum_{0 \leq j < N_2} \rho \left(\sum_{0 \leq i < N_1} \rho(\boldsymbol{u}_{N_1 \cdot j + i}; -N_1 \cdot j) \odot \rho(\boldsymbol{z}; i); N_1 \cdot j \right).$$

For the homomorphic evaluation of the arithmetic circuit, we first compute the ciphertexts of $\rho(\boldsymbol{z}; k)$ for $i = 1, \ldots, N_1 - 1$. For each index j, we perform N_1 scalar multiplications and aggregate the resulting ciphertexts. In total, the matrix multiplication can be homomorphically evaluated with $(N_1 - 1) + (N_2 - 1) = O(\sqrt{N})$ rotations and $N_1 \cdot N_2 = O(N)$ scalar multiplications.

We provide a trade-off between the precision of a plaintext and the size of a ciphertext modulus. The output plaintext contains errors from the rounding operation of scalar polynomial and the homomorphic rotation. We can reduce the relative size of the rounding error by multiplying a scaling factor of $\Delta \geq 1$ to the scalar polynomials, and the relative size of the rotation error can be controlled by placing the significand of an input ciphertext in higher digits. Therefore, the modulus of an output ciphertext is reduced after the evaluation of a linear transformation if we use a scaling factor to get a better precision of plaintexts.

4.4   Sparsely Packed Ciphertext

The packing method described in Sect. 4.1 allows us to encrypt $(N/2)$ complex numbers in a single ciphertext. However, it is sufficient to deal with a small number of slots in some applications of HE. In this section, we explain how to encode a sparse plaintext vector and describe the relation with the ordinary packing method. This idea will be applied to our bootstrapping method and provide a trade-off between the latency time and the amortized time of bootstrapping procedure.

Let $n \geq 2$ be a divisor of N and let $Y = X^{N/n}$. The native plaintext space of HEAAN is the set of small polynomials in $\mathbb{Z}[X]/(X^N + 1)$, and it has a subring $\mathbb{Z}[Y]/(Y^n + 1)$. Note that $\mathbb{Z}[Y]/(Y^n + 1)$ can be identified with the complex coordinate space $\mathbb{C}^{n/2}$ by adapting the idea of ordinary packing method in Sect. 4.1. Specifically, a polynomial $m(Y)$ is mapped to the vector $\boldsymbol{w} = (w_j)_{0 \leq j < n/2}$ where $w_j = m(\xi_j)$, $\xi = \exp(-2\pi i/n) = \zeta^{N/n}$, and $\xi_j = \xi^{5^j}$ for $0 \leq j < n/2$. If we consider a plaintext polynomial $m(Y) \in \mathbb{Z}[Y]/(Y^n + 1)$ as a polynomial $\tilde{m}(X) = m(X^{N/n})$ in X, then the image of $\tilde{m}$ through the ordinary decoding function τ is obtained by $\tau(\tilde{m}) = (z_j)_{0 \leq j < N/2}$ where

$$z_j = \tilde{m}(\zeta_j^{N/n}) = m(\xi^{5^j}) = w_j \pmod{n/2}$$

for $0 \leq j < N/2$. Hence $\boldsymbol{z} = (z_j)_{0 \leq j < N/2}$ can be understood as the vector obtained from $\boldsymbol{w}$ by concatenating itself (N/n) times.

An encryption of a plaintext polynomial $m(Y) \in \mathbb{Z}[Y]/(Y^n + 1)$ with respect to a secret key $\mathsf{sk} = (1, s)$ will be pairs as $\mathsf{ct} \in \mathcal{R}_q^2$ satisfying $\langle \mathsf{ct}, \mathsf{sk} \rangle = m(Y) + e(X) \pmod{q}$ for some small polynomial $e(X) \in \mathcal{R}$. We may employ key-switching technique to get the functionalities of rotation, conjugation, and linear transformation on sparsely packed slots. The main advantage of this method is that it reduces the complexity of an arbitrary linear transformation: the total complexity on $(n/2)$-sparsely packed ciphertexts is bounded by $O(\sqrt{n})$ rotations and $O(n)$ scalar multiplications.

5   Bootstrapping for HEAAN

5.1   Overview of the Recryption Procedure

This section gives a high level structure of the bootstrapping process for the HEAAN scheme. We employ the ciphertext packing method and combine it with our efficient evaluation strategy to achieve a better performance in terms of memory and computation cost. Below we describe all the parts of the recryption procedure in more details. We denote the following five steps by MODRAISE, COEFFTOSLOT, EVALEXP, IMGEXT and SLOTTOCOEFF, respectively. See Fig. 2 for an illustration.

Modulus Raising. Let ct be an input ciphertext of the bootstrapping procedure with a ciphertext modulus q satisfying $m(X) = [\langle \mathsf{ct}, \mathsf{sk} \rangle]_q$. We start with the point that its inner product $t(X) = \langle \mathsf{ct}, \mathsf{sk} \rangle \pmod{X^N + 1}$ is of the form $t = qI + m$ for some small $I(X) \in \mathcal{R}$ with a bound $\|I\|_\infty < K$. Thus ct itself can be viewed as an encryption of $t(X) = t_0 + t_1 X + \cdots + t_{N-1} X^{N-1}$ in a large modulus $Q_0 \gg q$ due to $[\langle \mathsf{ct}, \mathsf{sk} \rangle]_{Q_0} = t(X)$. Our bootstrapping procedure aims to

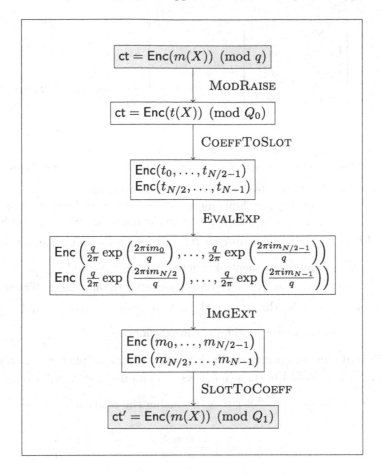

Fig. 2. Pipeline of our bootstrapping process

homomorphically and approximately evaluate the reduction mod q $F(t) = [t]_q$ using arithmetic operations over the integers, and hence we can generate an encryption of the original message $m = [t]_q$ with a ciphertext modulus larger than q.

Putting Polynomial Coefficients in Plaintext Slots. Given the input ciphertext $\mathsf{ct} \in \mathcal{R}_{Q_0}^2$ with a decryption structure $t(X) = \langle \mathsf{ct}, \mathsf{sk} \rangle$, this step aims to put the coefficients $t_0, \ldots, t_{N-1}$ in plaintext slots in order to evaluate the modular reduction function $F(t)$ coefficient-wisely. Let $z' = \tau(t) \in \mathbb{C}^{N/2}$ be the corresponding vector of plaintext slots of the ciphertext ct. Since each ciphertext can store at most $N/2$ plaintext values, we will generate two ciphertexts encrypting the vectors $z'_0 = (t_0, \ldots, t_{\frac{N}{2}-1})$ and $z'_1 = (t_{\frac{N}{2}}, \ldots, t_{N-1})$, respectively.

As mentioned in Sect. 4.1, recall the linear relation between the coefficient vector of a polynomial and its corresponding vector of plaintext slots. If we divide the matrix U into two square matrices

$$U_0 = \begin{bmatrix} 1 & \zeta_0 & \cdots & \zeta_0^{\frac{N}{2}-1} \\ 1 & \zeta_1 & \cdots & \zeta_1^{\frac{N}{2}-1} \\ \vdots & \vdots & \ddots & \vdots \\ 1 & \zeta_{\frac{N}{2}-1} & \cdots & \zeta_{\frac{N}{2}-1}^{\frac{N}{2}-1} \end{bmatrix} \quad \text{and} \quad U_1 = \begin{bmatrix} \zeta_0^{\frac{N}{2}} & \zeta_0^{\frac{N}{2}+1} & \cdots & \zeta_0^{N-1} \\ \zeta_1^{\frac{N}{2}} & \zeta_1^{\frac{N}{2}+1} & \cdots & \zeta_1^{N-1} \\ \vdots & \vdots & \ddots & \vdots \\ \zeta_{\frac{N}{2}-1}^{\frac{N}{2}} & \zeta_{\frac{N}{2}-1}^{\frac{N}{2}+1} & \cdots & \zeta_{\frac{N}{2}-1}^{N-1} \end{bmatrix},$$

then we get an identity $z_k' = \frac{1}{N}(\overline{U_k}^T \cdot z' + U_k^T \cdot \overline{z'})$ for $k = 0, 1$. Therefore, we can generate encryptions of z_0' and z_1' using the linear transformations on the plaintext vector z'. One can apply our general method in Sect. 4.3 to this step.

Evaluation of the Complex Exponential Function. This step takes the results of the previous step and homomorphically computes the reduction mod q function $F(t) = [t]_q$ homomorphically. We use the trigonometric function $S(t) := \frac{q}{2\pi} \sin\left(\frac{2\pi t}{q}\right)$ as an approximation of $F(t)$ and adapt the optimized evaluation strategy for the complex exponential function $E(t) := \frac{q}{2\pi} \exp\left(\frac{2\pi i t}{q}\right)$.

Since the plaintext slots of the input ciphertexts contain the coefficients $t_j = qI_j + m_j$ for $0 \le j < N$, the output ciphertexts will encrypt $\frac{q}{2\pi} \exp\left(\frac{2\pi i t_j}{q}\right) = \frac{q}{2\pi} \exp\left(\frac{2\pi i m_j}{q}\right)$ in the corresponding plaintext slots.

Extraction of the Imaginary Part. We take two input ciphertexts encrypting the values $\frac{q}{2\pi} \exp\left(\frac{2\pi i m_j}{q}\right)$ in their plaintext slots for $0 \le j < N$. We extract their imaginary parts as $\frac{q}{2\pi} \sin\left(\frac{2\pi m_j}{q}\right) \approx m_j$ by using the relation

$$\sin\left(\frac{2\pi m_j}{q}\right) = \frac{1}{2}\left(\exp\left(\frac{2\pi i m_j}{q}\right) - \exp\left(\frac{-2\pi i m_j}{q}\right)\right)$$

and applying the evaluation method of slot-wise conjugation described in Sect. 4.2.

Switching Back to the Coefficient Representation. The final step is to pack all the coefficients m_j in the plaintext slots of two ciphertexts back in a single ciphertext. This procedure is exactly the inverse of the COEFFTOSLOT transformation. That is, when given two ciphertexts that encrypt the vectors $z_0 = (m_0, \ldots, m_{\frac{N}{2}-1})$ and $z_1 = (m_{\frac{N}{2}}, \ldots, m_{N-1})$, we aim to generate an encryption of $m(X)$. Since the plaintext vector $z = \tau(m)$ of $m(X)$ satisfies the identity $z = U \cdot m = U_0 \cdot z_0 + U_1 \cdot z_1$, this transformation is also represented as a linear transformation over the plaintext vectors.

Our bootstrapping process returns an encryption of $m(X)$ with a ciphertext modulus $Q_1 < Q_0$, which is much larger enough than the initial modulus q to allow us to perform further homomorphic operations on the ciphertext.

We can perform the final two steps together by pre-computing the composition of linear transformations. The (inverse) linear transformation step consumes only one level for scalar multiplications but requires a number of slot rotations.

On the other hand, EVALEXP performs homomorphic evaluation of the polynomial $P_r(\cdot)$, which is the most levels consuming part of the recryption but requires relatively small computational cost from our recursive evaluation strategy: linear with the depth.

5.2 Recryption with Sparsely Packed Ciphertexts

As mentioned in Sect. 4.4, the use of sparsely packed ciphertexts has an advantage in some applications, in that it reduces the complexity of linear transformation steps. However, the recryption of sparsely packed ciphertexts requires an additional step before the COEFFTOSLOT step.

Let $n \geq 2$ be a divisor of N and let $Y = X^{N/n}$ as in Sect. 4.4. Assume that we are given an encryption ct of $m(Y) \in \mathbb{Z}[Y]/(Y^n + 1)$ such that $\langle \mathsf{ct}, \mathsf{sk} \rangle \approx q \cdot I(X) + m(Y)$ for some $I(X) = I_0 + I_1 \cdot X + \cdots + I_{N-1} \cdot X^{N-1} \in \mathcal{R}$. Then the MODRAISE step returns an encryption of $q \cdot I(X) + m(Y)$ which is not a polynomial of Y. We aim to generate an encryption of $q \cdot \tilde{I}(Y) + m(Y)$ for some $\tilde{I}(Y) \in \mathbb{Z}[Y]/(Y^n + 1)$ before the next COEFFTOSLOT step. It proceeds as described in Algorithm 2.

Algorithm 2. Homomorphic evaluation of the partial-sum procedure

---

1: **procedure** PARTIALSUM($\mathsf{ct} \in \mathcal{R}_q^2, n|N, n \geq 2$)
2:     $\mathsf{ct}' \leftarrow \mathsf{ct} \pmod{q}$
3:     **for** $j = 0$ to $\log(N/n) - 1$ **do**
4:         $\mathsf{ct}_j \leftarrow \mathsf{Rot}(\mathsf{ct}'; 2^j \cdot (n/2)) \pmod{q}$
5:         $\mathsf{ct}' \leftarrow \mathsf{Add}(\mathsf{ct}', \mathsf{ct}_j) \pmod{q}$
6:     **end for**
7:     **return** ct'
8: **end procedure**

Note that the monomial X^k vanishes by the homomorphism $X \mapsto X - X^{n+1} + X^{2n+1} - \cdots - X^{N-n+1}$ if k is not divisible by (N/n); otherwise, it is multiplied by the constant (N/n). Hence $q \cdot I(X) + m(Y)$ is mapped to $(N/n) \cdot (q \cdot \tilde{I}(Y) + m(Y))$ by this homomorphism where $\tilde{I}(Y) = I_0 + I_{N/n} \cdot Y + \cdots + I_{N-(N/n)} \cdot Y^{n-1}$. For an efficient evaluation of this homomorphism, Algorithm 2 uses the rotation operation repeatedly to fill the same value in the plaintext slots of index j (mod $n/2$) for each $j = 0, \ldots, n/2 - 1$.

As mentioned before, the main advantage of sparsely packed ciphertexts is that the COEFFTOSLOT step can be represented as relatively small matrices multiplications with only a single encryption of the coefficients vector of $t(Y) = q \cdot \tilde{I}(Y) + m(Y)$ while fully-packed slots need two ciphertexts for the COEFFTOSLOT step. In some more details, for the plaintext vector $\boldsymbol{w} \in \mathbb{C}^{n/2}$ of $t(Y)$, the desired ciphertext can be computed by $\frac{1}{n}(\overline{U'}^T \cdot \boldsymbol{w} + U'^T \cdot \overline{\boldsymbol{w}})$ where

$$U' = \begin{bmatrix} 1 & \xi_0 & \xi_0^2 & \cdots & \xi_0^{n-1} \\ 1 & \xi_1 & \xi_1^2 & \cdots & \xi_1^{n-1} \\ \vdots & \vdots & \vdots & \ddots & \vdots \\ 1 & \xi_{\frac{n}{2}-1} & \xi_{\frac{n}{2}-1}^2 & \cdots & \xi_{\frac{n}{2}-1}^{n-1} \end{bmatrix},$$

as in Sect. 4.1. We can either generate two ciphertexts encrypting two plaintext vectors $(w_0, \ldots, w_{n/2-1})$ and $(w_{n/2}, \ldots, w_{n-1})$ by separating U' into two square matrices, or compute a single encryption of $(w_0, \ldots, w_{n-1})$ with n plaintext slots when $n < N$. In the latter case, EVALEXP and IMGEXT perform the same operations as in the fulled-packed ciphertexts, but the memory and the computational cost are reduced by half since we can work on a single ciphertext. The final SLOTTOCOEFF step is also expressed as a linear transformation over n-dimensional vector.

5.3   Noise Estimation of Recryption

In this section, we describe each step of recryption procedure with a noise analysis. We start with an upper bound K of $\|I\|_\infty$. Since each coefficient of a ciphertext $\mathsf{ct} = (c_0, c_1)$ is an element of $\mathbb{Z}_q$ and the signed binary secret key s has exactly h nonzero coefficients, each coefficient of $\langle \mathsf{ct}, \mathsf{sk} \rangle = c_0 + c_1 s$ can be considered to be the sum of $(h+1)$ elements in $\mathbb{Z}_q$, which is bounded by $\frac{q}{2}(h+1)$. Hence all the coefficient of $I(X) = \lfloor \frac{1}{q} \langle \mathsf{ct}, \mathsf{sk} \rangle \rceil$ is bounded by $\frac{1}{2}(h+1) \approx \frac{1}{2}\|s\|_1$. In practice, the coefficients of c_i look like a random variable over the interval $\mathbb{Z}_q$ and a coefficient of $\frac{1}{q}\langle \mathsf{ct}, \mathsf{sk} \rangle$ behaves as the sum of $(h+1)$-numbers of i.i.d. uniform random variables over the interval $(-\frac{1}{2}, \frac{1}{2})$. This heuristic assumption gives us a smaller bound as $K = O(\sqrt{h})$ for $\|I\|_\infty$.

We now consider the error growth during homomorphic evaluation of a linear transformation. As noted in Sect. 4.3, it induces two types of errors such as from the rounding of a scalar polynomial and the key-switching operation. We multiply a sufficiently large scaling factor of $\Delta = O(q)$ to scalar polynomials, so that the precision of rounded polynomials becomes larger than that of an input plaintext. Then we do not need to consider the rounding errors because they have no effect on the precision of the resulting plaintext. The second type of error is added to a plaintext when we apply the key-switching technique for some functionalities such as rotation or conjugation. From Lemma 4, the key-switching error is bounded by $P^{-1} \cdot q \cdot B_{\mathsf{ks}} + B_{\mathsf{rs}} \approx B_{\mathsf{rs}}$ since we set a ciphertext modulus q much smaller than P. During matrix multiplication, key-switching errors are multiplied with diagonal vectors of matrix. Since infinite norms of diagonal vectors of $\frac{1}{N}U_0$ and $\frac{1}{N}U_1$ are exactly one, the total error of second type is bounded by $O(B_{\mathsf{rs}})$.

In the EVALEXP step, we take two ciphertexts to be homomorphically evaluated by the approximate polynomial $P_r(\cdot)$ of the complex exponential polynomial. Each component of the corresponding plaintext slots contains $t_j + e_j$ for some small error e_j such that $|e_j| \leq O(B_{\mathsf{rs}})$. Hence, the error between the desired value $E(t_j)$ and the resulting plaintext of EVALEXP can be measured by

$$|E(t_j) - P_r(t_j + e_j)| \leq |E(t_j) - E(t_j + e_j)| + |E(t_j + e_j) - P_r(t_j + e_j)|$$

$$\leq \frac{\Delta \cdot 2\pi}{q}|e_j| + \frac{\Delta \cdot 2^r}{\sqrt{2\pi(d_0 + 1)}}\left(\frac{e\pi K}{2^{r-1}(d_0 + 1)}\right)^{d_0+1},$$

since $E(\cdot)$ is analytic and $|E'(\cdot)| \leq \Delta \cdot 2\pi/q$. The second term can be bounded by $O(1)$ and so it is negligible when $d_0 = O(1)$ and $r = O(\log(qK))$ as described in Sect. 3.2. By combining the error bound as $|e_j| \leq O(B_{rs})$ of the previous step, we deduce that the output ciphertexts of EVALEXP encrypt $E(t_j) = \Delta \cdot \exp\left(\frac{2\pi i t_j}{q}\right)$ in their plaintext slots with errors bounded by $O(B_{rs})$.

The imaginary part of $E(t)$ is $\frac{\Delta \cdot 2\pi}{q}S(t) = \Delta \cdot \sin\left(\frac{2\pi t}{q}\right)$, which is an approximation of $\frac{\Delta \cdot 2\pi}{q}m$ with an error bounded by $O\left(q \cdot \frac{|m|^3}{q^3}\right)$ for $m = [t]_q$. We set a small bound on an input plaintext (e.g. $|m| \leq q^{2/3}$) so that an approximation error does not destroy the significant digits of the plaintext. Hence the IMGEXT step does not change the magnitude of bootstrapping error.

Finally, in the SLOTTOCOEFF step, the plaintext vectors are multiplied with the matrices U_0 and U_1, and their diagonal vectors have the size of one. The size of error in the resulting plaintext is bounded by $O(N \cdot B_{rs})$ since it is the sum of N errors of size $O(B_{rs})$. In practice, under the heuristic assumption that these errors behaves as independent Gaussian distributions, we get a reduced error bound as $O(\sqrt{N} \cdot B_{rs})$.

In summary, for an input ciphertext $\mathsf{ct} \in \mathcal{R}_q^2$ satisfying $\langle \mathsf{ct}, \mathsf{sk} \rangle = m \pmod{q}$, our bootstrapping process returns a ciphertext ct' such that $\langle \mathsf{ct}', \mathsf{sk} \rangle = m + e \pmod{Q_1}$ for a modulus $Q_1 \gg q$ and some error e with $\|e\|_\infty^{\mathsf{can}} \leq O(\sqrt{N} \cdot B_{rs})$. It consists of the initial/final linear transformations and the evaluation of the complex exponential function, so the total depth (number of levels consumed) for bootstrapping is $O(\log(Kq)) = O(\log \lambda)$. The linear transformations require and $O(\sqrt{N})$ rotations, while the evaluation of the exponential function needs $r = O(\log(Kq)) = O(\log \lambda)$ homomorphic multiplications, which is linear with the depth of the decryption formula. As described above, the total complexity can be significantly reduced when we handle a sparsely packed ciphertext with $(n/2)$ slots: the linear transformations require $O(\sqrt{n})$ rotations and the evaluation complexity of the exponential function is reduced by half.

6   Implementation

In this section, we suggest some parameter sets for our bootstrapping procedure with experimental results. Our implementation is based on the HEAAN library [13] implementing the HE scheme of Cheon et al. [14]. The source code is publicly available at github [12].

6.1   Parameter Selection

We adapt the estimator of Albrecht at el. [1] to guarantee the concrete security of the proposed parameters. All the parameter sets achieve at least 80-bit security level against the known attacks of the LWE problem.

The key-switching keys have the largest modulus in the HEAAN scheme and the modulus Q_0 (after MODRAISE) has half the bit size of the modulus. We use the discrete Gaussian distribution of standard deviation $\sigma = 3.2$ to sample error polynomials and set the Hamming weight $h = 64$ of the secret key $s(X)$. The parameters $d_0 = O(1)$ and $r = O(\log q)$ were chosen asymptotically in the above sections, but in practice, we set the parameter experimentally based on the bootstrapping error. We take the degree 7 Taylor expansion as an initial approximation of the exponential function and choose a sufficiently large number of iterations r to maintain the precision of output plaintexts.

The parameter $\log p$ is the bit size of plaintexts and the plaintext precision denotes the number of significant bits of plaintexts after bootstrapping. The before and after levels L_0, L_1 are obtained by dividing $\log Q_0$ and $\log Q_1$ by $\log p$, respectively. The whole parameter sets are described in Table 1.

Table 1. Parameter sets

Parameter	$\log N$	$\log p$	$\log q$	r	$\log Q_0$	L_0	$\log Q_1$	L_1	Plaintext precision
Set-I	15	23	29	6	620	26	202	8	8 bits
Set-II		27	37	7		22	64	2	12 bits
Set-III	16	31	41	7	1240	40	631	20	16 bits
Set-IV		39	54	9		31	344	8	24 bits

We present some specific examples of input and output plaintexts. For simplicity, we show the real parts of the plaintext values in four slots. All the values are divided by a factor of p for a clear interpretation.

```
Before bootstrapping : [0.777898 0.541580 0.603675 0.822638]
 After bootstrapping : [0.777435 0.541021 0.603023 0.822321]
```

It shows that the error vectors is bounded by $2^{-11} \cdot p$, i.e., the bootstrapping procedure with the parameter set I outputs a ciphertext of 10 bits of precision.

```
Before bootstrapping : [0.516015 0.772621 0.939175 0.345987]
 After bootstrapping : [0.516027 0.772614 0.939172 0.346001]
```

Similarly, the second example shows that the bootstrapping error with the parameter set II is bounded by $2^{-16} \cdot p$ and the output plaintext has 15 bits of precision.

6.2 Experimental Results

We show the performance of our bootstrapping procedure based on the proposed parameter sets. All the experimentations were performed as a single hyperthread on a 2.10 GHz Intel Xeon E5-2620. The experimental results are

Table 2. Bootstrapping timings with parameter sets I to IV

Parameter	Number of slots	Linear trans.	EVALEXP	Total time	Amortized time
Set-I	1	12.3 s	12.3 s	24.6 s	24.6 s
	32	48.6 s		60.9 s	1.9 s
	64	82.8 s		95.1 s	1.5 s
	128	139.2 s		151.5 s	1.2 s
Set-II	1	14.1 s	12.5 s	26.6 s	26.6 s
	32	46.0 s		58.5 s	1.8 s
	64	77.1 s		89.6 s	1.4 s
	128	127.3 s		139.8 s	1.1 s
Set-III	1	64 s	63 s	127 s	127 s
	32	218 s		281 s	8.8 s
	64	343 s		406 s	6.3 s
	128	528 s		591 s	4.6 s
Set-IV	1	58 s	68 s	126 s	126 s
	32	200 s		268 s	8.4 s
	64	307 s		375 s	5.9 s
	128	456 s		524 s	4.1 s

summarized in Table 2. The linear transformation includes the three steps - PARTIALSUM, COEFFTOSLOT, and SLOTTOCOEFF. The amortized time is obtained by dividing the total bootstrapping time by the number of plaintext slots.

The linear transformations take a longer time as the number of plaintext slots grows while the complexity of the EVALEXP step remains stable. Therefore, we can make a trade-off between the latency time and the amortized time by changing the number of slots. Figure 3 illustrates the trend of evaluation timings for each of the bootstrapping phases (parameter set I).

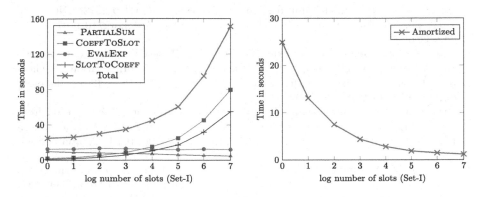

Fig. 3. Tendency of real (left) and amortized (right) bootstrapping timings

7   Conclusion

In this paper, we suggested a method to recrypt a ciphertext of the HEAAN scheme. The performance of our bootstrapping procedure was significantly improved by adapting a trigonometric approximation of the modular reduction function. The linear transformation turns out to be the most time-consuming part, but we used almost the same method as in [28, 33]. It would be an interesting open problem to find an efficient algorithm to evaluate the linear transformations approximately.

Acknowledgments. We would like to thank the anonymous EUROCRYPT'18 reviewers for useful comments. This work was partially supported by Institute for Information & communications Technology Promotion (IITP) grant funded by the Korea government (MSIT) (No. B0717-16-0098, Development of homomorphic encryption for DNA analysis and biometry authentication). M. Kim was supported by the National Institute of Health (NIH) under award number U01EB023685.

References

1. Albrecht, M.R., Player, R., Scott, S.: On the concrete hardness of learning with errors. J. Math. Cryptol. **9**(3), 169–203 (2015)
2. Alperin-Sheriff, J., Peikert, C.: Faster bootstrapping with polynomial error. In: Garay, J.A., Gennaro, R. (eds.) CRYPTO 2014. LNCS, vol. 8616, pp. 297–314. Springer, Heidelberg (2014). https://doi.org/10.1007/978-3-662-44371-2_17
3. Bonnoron, G., Ducas, L., Fillinger, M.: Large fhe gates from tensored homomorphic accumulator. Cryptology ePrint Archive, Report 2017/996 (2017). https://eprint.iacr.org/2017/996
4. Bonte, C., Bootland, C., Bos, J.W., Castryck, W., Iliashenko, I., Vercauteren, F.: Faster homomorphic function evaluation using non-integral base encoding. In: Fischer, W., Homma, N. (eds.) CHES 2017. LNCS, vol. 10529, pp. 579–600. Springer, Cham (2017). https://doi.org/10.1007/978-3-319-66787-4_28
5. Bos, J.W., Lauter, K., Loftus, J., Naehrig, M.: Improved security for a ring-based fully homomorphic encryption scheme. In: Stam, M. (ed.) IMACC 2013. LNCS, vol. 8308, pp. 45–64. Springer, Heidelberg (2013). https://doi.org/10.1007/978-3-642-45239-0_4
6. Brakerski, Z.: Fully homomorphic encryption without modulus switching from classical GapSVP. In: Safavi-Naini, R., Canetti, R. (eds.) CRYPTO 2012. LNCS, vol. 7417, pp. 868–886. Springer, Heidelberg (2012). https://doi.org/10.1007/978-3-642-32009-5_50
7. Brakerski, Z., Gentry, C., Vaikuntanathan, V.: (Leveled) fully homomorphic encryption without bootstrapping. In: Proceedings of the 3rd Innovations in Theoretical Computer Science Conference on - ITCS 2012. ACM Press (2012). https://doi.org/10.1145/2090236.2090262
8. Brakerski, Z., Vaikuntanathan, V.: Efficient fully homomorphic encryption from (standard) LWE. In: Proceedings of the 2011 IEEE 52nd Annual Symposium on Foundations of Computer Science, FOCS 2011, pp. 97–106. IEEE Computer Society (2011)
9. Brakerski, Z., Vaikuntanathan, V.: Fully homomorphic encryption from ring-LWE and security for key dependent messages. In: Rogaway, P. (ed.) CRYPTO 2011. LNCS, vol. 6841, pp. 505–524. Springer, Heidelberg (2011). https://doi.org/10.1007/978-3-642-22792-9_29

10. Chen, H., Laine, K., Player, R., Xia, Y.: High-precision arithmetic in homomorphic encryption. Cryptology ePrint Archive, Report 2017/809 (2017). https://eprint.iacr.org/2017/809
11. Cheon, J.H., Coron, J.-S., Kim, J., Lee, M.S., Lepoint, T., Tibouchi, M., Yun, A.: Batch fully homomorphic encryption over the integers. In: Johansson, T., Nguyen, P.Q. (eds.) EUROCRYPT 2013. LNCS, vol. 7881, pp. 315–335. Springer, Heidelberg (2013). https://doi.org/10.1007/978-3-642-38348-9_20
12. Cheon, J.H., Han, K., Kim, A., Kim, M., Song, Y.: Implementation of boostrapping for HEAAN (2017). https://github.com/kimandrik/HEAANBOOT
13. Cheon, J.H., Kim, A., Kim, M., Song, Y.: Implementation of HEAAN (2016). https://github.com/kimandrik/HEAAN
14. Cheon, J.H., Kim, A., Kim, M., Song, Y.: Homomorphic encryption for arithmetic of approximate numbers. In: Takagi, T., Peyrin, T. (eds.) ASIACRYPT 2017. LNCS, vol. 10624, pp. 409–437. Springer, Cham (2017). https://doi.org/10.1007/978-3-319-70694-8_15
15. Cheon, J.H., Kim, M., Lauter, K.: Homomorphic computation of edit distance. In: Brenner, M., Christin, N., Johnson, B., Rohloff, K. (eds.) FC 2015. LNCS, vol. 8976, pp. 194–212. Springer, Heidelberg (2015). https://doi.org/10.1007/978-3-662-48051-9_15
16. Cheon, J.H., Stehlé, D.: Fully homomophic encryption over the integers revisited. In: Oswald, E., Fischlin, M. (eds.) EUROCRYPT 2015. LNCS, vol. 9056, pp. 513–536. Springer, Heidelberg (2015). https://doi.org/10.1007/978-3-662-46800-5_20
17. Chillotti, I., Gama, N., Georgieva, M., Izabachène, M.: Faster fully homomorphic encryption: bootstrapping in less than 0.1 seconds. In: Cheon, J.H., Takagi, T. (eds.) ASIACRYPT 2016. LNCS, vol. 10031, pp. 3–33. Springer, Heidelberg (2016). https://doi.org/10.1007/978-3-662-53887-6_1
18. Chillotti, I., Gama, N., Georgieva, M., Izabachène, M.: Faster packed homomorphic operations and efficient circuit bootstrapping for TFHE. In: Takagi, T., Peyrin, T. (eds.) ASIACRYPT 2017. LNCS, vol. 10624, pp. 377–408. Springer, Cham (2017). https://doi.org/10.1007/978-3-319-70694-8_14
19. Coron, J.-S., Lepoint, T., Tibouchi, M.: Scale-invariant fully homomorphic encryption over the integers. In: Krawczyk, H. (ed.) PKC 2014. LNCS, vol. 8383, pp. 311–328. Springer, Heidelberg (2014). https://doi.org/10.1007/978-3-642-54631-0_18
20. Costache, A., Smart, N.P., Vivek, S.: Faster homomorphic evaluation of discrete fourier transforms. In: Kiayias, A. (ed.) FC 2017. LNCS, vol. 10322, pp. 517–529. Springer, Cham (2017). https://doi.org/10.1007/978-3-319-70972-7_29
21. Costache, A., Smart, N.P., Vivek, S., Waller, A.: Fixed-point arithmetic in SHE schemes. In: Avanzi, R., Heys, H. (eds.) SAC 2016. LNCS, vol. 10532, pp. 401–422. Springer, Cham (2017). https://doi.org/10.1007/978-3-319-69453-5_22
22. van Dijk, M., Gentry, C., Halevi, S., Vaikuntanathan, V.: Fully homomorphic encryption over the integers. In: Gilbert, H. (ed.) EUROCRYPT 2010. LNCS, vol. 6110, pp. 24–43. Springer, Heidelberg (2010). https://doi.org/10.1007/978-3-642-13190-5_2
23. Dowlin, N., Gilad-Bachrach, R., Laine, K., Lauter, K., Naehrig, M., Wernsing, J.: Manual for using homomorphic encryption for bioinformatics. Proc. IEEE **105**(3), 552–567 (2017)
24. Ducas, L., Micciancio, D.: FHEW: bootstrapping homomorphic encryption in less than a second. In: Oswald, E., Fischlin, M. (eds.) EUROCRYPT 2015. LNCS, vol. 9056, pp. 617–640. Springer, Heidelberg (2015). https://doi.org/10.1007/978-3-662-46800-5_24

25. Fan, J., Vercauteren, F.: Somewhat practical fully homomorphic encryption. IACR Cryptology ePrint Archive 2012:144 (2012)
26. Gentry, C., et al.: Fully homomorphic encryption using ideal lattices. In: STOC, vol. 9, pp. 169–178 (2009)
27. Gentry, C., Halevi, S.: Implementing gentry's fully-homomorphic encryption scheme. In: Paterson, K.G. (ed.) EUROCRYPT 2011. LNCS, vol. 6632, pp. 129–148. Springer, Heidelberg (2011). https://doi.org/10.1007/978-3-642-20465-4_9
28. Gentry, C., Halevi, S., Smart, N.P.: Better bootstrapping in fully homomorphic encryption. In: Fischlin, M., Buchmann, J., Manulis, M. (eds.) PKC 2012. LNCS, vol. 7293, pp. 1–16. Springer, Heidelberg (2012). https://doi.org/10.1007/978-3-642-30057-8_1
29. Gentry, C., Halevi, S., Smart, N.P.: Fully homomorphic encryption with polylog overhead. In: Pointcheval, D., Johansson, T. (eds.) EUROCRYPT 2012. LNCS, vol. 7237, pp. 465–482. Springer, Heidelberg (2012). https://doi.org/10.1007/978-3-642-29011-4_28
30. Gentry, C., Halevi, S., Smart, N.P.: Homomorphic evaluation of the AES circuit. In: Safavi-Naini, R., Canetti, R. (eds.) CRYPTO 2012. LNCS, vol. 7417, pp. 850–867. Springer, Heidelberg (2012). https://doi.org/10.1007/978-3-642-32009-5_49
31. Gentry, C., Sahai, A., Waters, B.: Homomorphic encryption from learning with errors: conceptually-simpler, asymptotically-faster, attribute-based. In: Canetti, R., Garay, J.A. (eds.) CRYPTO 2013. LNCS, vol. 8042, pp. 75–92. Springer, Heidelberg (2013). https://doi.org/10.1007/978-3-642-40041-4_5
32. Halevi, S., Shoup, V.: Algorithms in HElib. In: Garay, J.A., Gennaro, R. (eds.) CRYPTO 2014. LNCS, vol. 8616, pp. 554–571. Springer, Heidelberg (2014). https://doi.org/10.1007/978-3-662-44371-2_31
33. Halevi, S., Shoup, V.: Bootstrapping for HElib. In: Oswald, E., Fischlin, M. (eds.) EUROCRYPT 2015. LNCS, vol. 9056, pp. 641–670. Springer, Heidelberg (2015). https://doi.org/10.1007/978-3-662-46800-5_25
34. Hoffstein, J., Silverman, J.: Optimizations for ntru. In: Public-Key Cryptography and Computational Number Theory, Warsaw, pp. 77–88 (2001)
35. Kim, J., Lee, C., Shim, H., Cheon, J.H., Kim, A., Kim, M., Song, Y.: Encrypting controller using fully homomorphic encryption for security of cyber-physical systems. IFAC-PapersOnLine 49(22), 175–180 (2016)
36. Kim, M., Song, Y., Cheon, J.H.: Secure searching of biomarkers through hybrid homomorphic encryption scheme. BMC Med. Genomics 10(2), 42 (2017)
37. Kim, M., Song, Y., Wang, S., Xia, Y., Jiang, X.: Secure logistic regression based on homomorphic encryption. JMIR Med. Inform. (2018). https://doi.org/10.2196/medinform.8805, (forthcoming)
38. Lauter, K., López-Alt, A., Naehrig, M.: Private computation on encrypted genomic data. In: Aranha, D.F., Menezes, A. (eds.) LATINCRYPT 2014. LNCS, vol. 8895, pp. 3–27. Springer, Cham (2015). https://doi.org/10.1007/978-3-319-16295-9_1
39. Naehrig, M., Lauter, K., Vaikuntanathan, V.: Can homomorphic encryption be practical? In: Proceedings of the 3rd ACM Workshop on Cloud Computing Security Workshop, pp. 113–124. ACM (2011)
40. Paterson, M.S., Stockmeyer, L.J.: On the number of nonscalar multiplications necessary to evaluate polynomials. SIAM J. Comput. 2(1), 60–66 (1973)
41. Wang, S., Zhang, Y., Dai, W., Lauter, K., Kim, M., Tang, Y., Xiong, H., Jiang, X.: Healer: homomorphic computation of exact logistic regression for secure rare disease variants analysis in GWAS. Bioinformatics 32(2), 211–218 (2016)

Permutations

Full Indifferentiable Security of the Xor of Two or More Random Permutations Using the χ^2 Method

Srimanta Bhattacharya[✉] and Mridul Nandi

Indian Statistical Institute, Kolkata, India
mail.srimanta@gmail.com, mridul.nandi@gmail.com

Abstract. The construction XORP (bitwise-xor of outputs of two independent n-bit random permutations) has gained broad attention over the last two decades due to its high security. Very recently, Dai *et al.* (CRYPTO'17), by using a method which they term the *Chi-squared method* (χ^2 method), have shown n-bit security of XORP when the underlying random permutations are kept secret to the adversary. In this work, we consider the case where the underlying random permutations are publicly available to the adversary. The best known security of XORP in this security game (also known as *indifferentiable security*) is $\frac{2n}{3}$-bit, due to Mennink *et al.* (ACNS'15). Later, Lee (IEEE-IT'17) proved a better $\frac{(k-1)n}{k}$-bit security for the general construction XORP[k] which returns the xor of k (≥ 2) independent random permutations. However, the security was shown only for the cases where k is an even integer. In this paper, we improve all these known bounds and prove full, *i.e.*, n-bit (indifferentiable) security of XORP as well as XORP[k] for any k. Our main result is n-bit security of XORP, and we use the χ^2 method to prove it.

Keywords: Random permutation · Indifferentiable security
χ^2 method · XOR construction · Simulator

1 Introduction

The problem to construct *pseudorandom functions* (PRFs) from *pseudorandom permutations* (PRPs) is called "Luby-Rackoff Backwards" [BKR98] (referring to the well known work of Luby and Rackoff who showed how to construct a PRP from a PRF [LR88]). In [BKR98], the authors considered two sequential block cipher calls, where the output of the first call is the key input to the second one. However, this construction achieves security only up to the birthday bound on the output size. Achieving security beyond the birthday bound is somewhat non-trivial. Xoring the outputs of two independent n-bit *random permutations*[1]

---

[1] In this work, we will essentially focus on information theoretic security in the *ideal model*. Therefore, the permutations and functions that we will consider, will be random (and not pseudorandom).

© International Association for Cryptologic Research 2018
J. B. Nielsen and V. Rijmen (Eds.): EUROCRYPT 2018, LNCS 10820, pp. 387–412, 2018.
https://doi.org/10.1007/978-3-319-78381-9_15

is a very simple way to construct *random functions* from random permutations. We call it the *XOR construction* and denote it as XORP. We also consider a generalized version of the XOR construction in which we xor k independent n-bit random permutations, and denote it as XORP[k]. Lucks [Luc00] showed beyond the birthday bound security for XORP[k] for all $k \geq 2$. In particular, he showed that the construction achieves at least $\frac{kn}{k+1}$-bit security. This bound was further improved in a sequence of papers [BI99, CLP14, Pat10, Pat08b]. Very recently, Dai *et al.* [DHT17] have shown n-bit security for XORP. Earlier, Mennink *et al.* [MP15] showed a reduction proving that the security of XORP[k] can be reduced to that of XORP for any $k \geq 3$. Hence, XORP[k] also achieves n-bit security. The XORP (or its general version XORP[k]) construction is important since it has been used to obtain some constructions achieving beyond the birthday bound (or sometimes almost full) security (e.g., CENC [Iwa06, IMV16], PMAC_Plus [Yas11], and ZMAC [IMPS17]).

Moving from secret to public random permutation. While to a certain degree it is possible to view the permutations as secret, there are many reasons to consider the setting where they are public. For example, we sometimes instantiate block ciphers with fixed keys. Moreover, many unkeyed permutations are designed as an underlying primitive of encryption [BDPVA11a], MAC [BDPVA11b], hash functions [BDP+13, RAB+08, Wu11, GKM+09], etc. The CAESAR competition [CAE] received various permutation-based authenticated encryptions, and all of these constructions have been analyzed in the public permutation model.

The security game, in this setting, is clearly different from the standard indistinguishable model due to the public access of the adversary to the underlying permutations. An appropriate notion is the indifferentiability framework, introduced by Maurer *et al.* [MRH04]. Informally, it gives a sufficient condition under which an ideal functionality can be replaced by an indifferentiable-secure construction based on ideal, publicly available underlying primitives. We note that the security game for indifferentiability is also an indistinguishability game in which one has to design a simulator aimed to simulate the underlying primitive. In the past, many constructions were analyzed (e.g., [AMP10, BDPVA08, BMN10]) under this security notion.

Known indifferentiable security bounds of XORP and XORP[k]. In this indifferentiability model, Mandal *et al.* [MPN10] proved $\frac{2n}{3}$-bit security for XORP. Later, Mennink *et al.* [MP15] pointed out a subtle but non-negligible flaw in their proof and fixed the security proof. Recently, Lee [Lee17] has shown improved security for the general construction XORP[k]. In particular, he has proved $\frac{(k-1)n}{k}$-bit security for the general construction XORP[k] when k is an even integer. Table 1 summarizes the state-of-the-art for XORP and XORP[k] in the public permutation setting.

Table 1. A brief comparison of known bounds and our bounds for the constructions XORP and XORP[k]. Here q denotes the total number of queries made by the adversary to all oracles.

Construction	Best known bound	Our bound
XORP	$q^3/2^{2n}$ [MP15]	$\sqrt{q/2^n}$
XORP[k]	$\frac{q^{k+1}}{2^{nk}}$ ($k \geq 4$ even) [Lee17]	$\sqrt{q/2^n}$

Mirror theory and its limitation. Patarin introduced a combinatorial problem motivated from the PRF-security of XORP[k] type constructions. Informally, *mirror theory* (see [Pat10]) provides a suitable lower bound on the number of solutions satisfying a system of linear equations involving exactly two variables at a time. Together with the *H-coefficient technique* [Vau03, Pat08a, IMV16], this leads to a bound on the PRF-distinguishing advantage of XORP. The mirror theory seems to be very powerful as it can be applied to prove optimal security for many constructions such as EDM, EWCDM, etc. [MN17a, MN17b]. However, the proof of the mirror theory is quite complex with some of its steps lacking necessary details. Later, Patarin [CLP14] himself provided a simpler alternative but sub-optimal proof for XORP[k] (which is a trivial corollary of the mirror theory).

One may wonder whether the same technique can be applied to the indifferentiability setup or not. Here, we note that the mirror theory puts a constraint on the system of equations so that no equation in one variable can be generated through linear combination of equations from the system. On the other hand, in the indifferentiable security game, the adversary can make public permutation calls and observe the responses. So, along with the two variables linear equations, we also have to consider several single variable equations. This shows the limitation of the mirror theory in this setup.

Our contribution and the proof technique. Proving full security of XORP in the public permutation model was an open problem so far. The original simulator [MPN10], used in the security proof of XORP, is conjectured to allow for security up to 2^n queries. However, the authors of [MP15] expressed this as a highly nontrivial exercise. In this paper, we resolve this open problem and prove n-bit indifferentiable security of XORP and XORP[k] for all $k \geq 3$. Full indifferentiable security of XORP is our main result which we state and prove in Theorem 2. Subsequently, in Theorem 3, we show full indifferentiability of XORP[k]; for this, we reduce the security of XORP[k]($k \geq 3$), to the security of XORP, and then apply our main result.

The simulator (described in Sect. 3) that we consider in the security proof of XORP follows the same steps as the simulator of [MPN10, Lee17] in the case of forward queries. However, the simulator differs in the responses to the backward queries. In the case of backward queries, the simulator queries the ideal random function repeatedly (about n times) until it succeeds in its goal.

We follow the recently introduced χ^2 method [DHT17] to prove our claim. This method was implicitly used by Stam [Sta78] while proving a bound on the total variation between a *truncated random permutation* and a random function. Though in a purely statistical context, (to the best of our knowledge) Stam's work can be viewed as the origin of the χ^2 method, which led to a bound on the PRF-security of the truncated random permutation construction (see [GG16, GGM17] for recent results and discussion on this construction). In [DHT17], the authors used this method to obtain bounds on the PRF-security of XORP and the EDM construction [CS16a, MN17b]. Also, using this method full PRF-security of variable output length XOR pseudorandom functions has been shown [BN18].

In this paper, we show another application of the χ^2 method in (symmetric-key) cryptography in the context of XORP[k] type construction. Our main result demonstrates the power of this method as the proof of full security of XORP, in the indifferentiability setup, becomes very hard with the existing methods. However, our proof using the χ^2 method is not a straightforward extension of the proof in the indistinguishability framework due to Dai *et al.*; it is somewhat complicated as, unlike in the indistinguishability framework, we will need to consider the primitive queries (*i.e.*, outputs of the individual permutations). Moreover, we will have to handle the backward queries whose analysis is somewhat involved.

Outline of the paper. In the next section, we cover the preliminaries where we discuss the notion of indifferentiability and the χ^2 method. In Sect. 3, we describe the simulator that we consider in the proof of our main result (Theorem 2). In Sect. 4, we state and prove Theorem 2. Some auxilliary proofs, used in the proof of Theorem 2, are given in Sect. 5. Finally, in Sect. 6, we show full indifferentiability of XORP[k].

2   Preliminaries

In this section, we cover the technical preliminaries required to understand our results. We begin with the notational setup. Then we recall the preliminary security notions related to adversary and its advantage in the context of an indistinguishability game. This is to motivate our subsequent discussion on the notion of indifferentiability. Finally, we briefly describe the χ^2 method which is our main tool.

Notational convention. We will use upper case letters to denote random variables and their corresponding lower case letters to denote particular realizations of the variables. Given an integer s we will use the notation X^s to denote the tuple $(X_1, \ldots, X_s)$ of random variables and use x^s to denote the tuple $(x_1, \ldots, x_s)$ of corresponding realizations. Moreover, we write $\{X^s\}$ to denote the set $\{X_i : 1 \leq i \leq s\}$. Given a set $\mathcal{S}$, we will write $X \leftarrow_\$ \mathcal{S}$ to mean that X is sampled uniformly at random from the set $\mathcal{S}$.

2.1 Adversary and Advantage

Here, we recall the notion of adversarial advantage in the context of a generic *indistinguishability game*. An *adversary* $\mathscr{A}$ is an *oracle algorithm* that interacts with an *oracle* $\mathscr{O}$ through queries and responses. Finally, it returns a bit $b \in \{0,1\}$. We express this as $\mathscr{A}^{\mathscr{O}} \rightarrow b$.

In an *indistinguishability game*, $\mathscr{A}$ interacts with two oracles $\mathscr{O}_1$ and $\mathscr{O}_2$. The goal of $\mathscr{A}$ is to distinguish between $\mathscr{O}_1$ and $\mathscr{O}_2$ only from the corresponding queries and responses. The *advantage* of the adversary in this game, denoted $\mathsf{Adv}_{\mathscr{A}}(\mathscr{O}_1, \mathscr{O}_2)$, is given by

$$\mathsf{Adv}^{\mathrm{dist}}_{\mathscr{O}_1,\mathscr{O}_2}(\mathscr{A}) := |\mathbf{Pr}[\mathscr{A}^{\mathscr{O}_1} \rightarrow 1] - \mathbf{Pr}[\mathscr{A}^{\mathscr{O}_2} \rightarrow 1]|,$$

where the probabilities are taken over the random coins of $\mathscr{A}, \mathscr{O}_1$, and $\mathscr{O}_2$.

In this work, we will focus on the information theoretic security of the constructions (XORP and XORP[k]). So, we let $\mathscr{A}$ to be computationally unbounded. Therefore, without loss of any generality, we assume $\mathscr{A}$ to be deterministic (it can always fix its internal coin tosses to those which maximizes its advantage). However, we restrict $\mathscr{A}$ to only q queries. Let the corresponding replies from $\mathscr{O}_1$ and $\mathscr{O}_2$ be $X_1^q = (X_{1,1}, \ldots, X_{1,q})$ and $X_2^q = (X_{2,1}, \ldots, X_{2,q})$ respectively. Note that X_1^q and X_2^q are random variables that capture the randomness of the oracles $\mathscr{O}_1$ and $\mathscr{O}_2$ respectively. Both X_1^q and X_2^q are distributed over the output alphabet $\Omega^q = \Omega \times \cdots \times \Omega$ of the oracles. Then in this setting, it is not difficult to see that

$$\mathsf{Adv}^{\mathrm{dist}}_{\mathscr{O}_1,\mathscr{O}_2}(\mathscr{A}) = |\mathbf{Pr}[\mathscr{A}^{\mathscr{O}_1} \rightarrow 1] - \mathbf{Pr}[\mathscr{A}^{\mathscr{O}_2} \rightarrow 1]|$$
$$\leq \max_{\mathscr{E} \subseteq \Omega^q} \sum_{x^q \in \mathscr{E}} (\mathbf{Pr}[X_1^q = x^q] - \mathbf{Pr}[X_2^q = x^q]). \qquad (1)$$

The quantity on the r.h.s. of (1) is the *statistical distance* or the *total variation distance* between X_1^q and X_2^q. We will consider it slightly more formally in Sect. 2.3. We denote by $\mathsf{Adv}^{\mathrm{dist}}_{\mathscr{O}_1,\mathscr{O}_2}(q)$ the maximum of the distinguishing advantages $\mathsf{Adv}^{\mathrm{dist}}_{\mathscr{O}_1,\mathscr{O}_2}(\mathscr{A})$ among all the adversaries $\mathscr{A}$ making at most q queries.

2.2 Indifferentiability

The notion of indifferentiability was introduced by Maurer *et al.* in [MRH04]. It is a stronger security notion than indistinguishability in the following sense. Informally, let a construction T have oracle access to an ideal primitive F. Then in an indistinguishability game, when T is presented as an oracle to the adversary $\mathscr{A}$, it can only query T in a *black-box* manner, i.e., $\mathscr{A}$ can not query F. Whereas in the indifferentiability game, $\mathscr{A}$ can query both T and F.

As shown in Fig. 1, in the indifferentiability game, in the *real world*, a construction T has oracle access to an *ideal primitive* F. On the other hand, in the *ideal world*, the simulator S has access to another ideal primitive G. $\mathscr{A}$ can query any of these four entities with the goal of distinguishing between the two worlds. In this case, $\mathscr{A}$'s advantage can be written as

$$\mathsf{Adv}_{\mathsf{T}^{\mathsf{F}},\mathsf{G}^{\mathsf{S}}}^{\mathrm{diff}}(\mathscr{A}) = |\mathbf{Pr}[\mathscr{A}^{\mathsf{T},\mathsf{F}} \to 1] - \mathbf{Pr}[\mathscr{A}^{\mathsf{G},\mathsf{S}} \to 1]|.$$

In order to prove indifferentiability of T from G, it is sufficient to construct a simulator S in such a way that $\mathsf{Adv}_{\mathsf{T}^{\mathsf{F}},\mathsf{G}^{\mathsf{S}}}^{\mathrm{diff}}(\mathscr{A})$ becomes negligible for any adversary $\mathscr{A}$. The following definition captures this idea more formally.

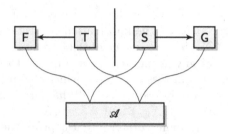

Fig. 1. Indifferentiability game

Definition 1 (Indifferentiability [MRH04]). *A Turing machine* T *with oracle access to an ideal primitive* F *is said to be* $(t, q_{\mathsf{T}}, q_{\mathsf{F}}, \varepsilon)$-*indifferentiable from an ideal primitive* G *if there exists a simulator* S *with oracle access to* G *and running time at most* t*, such that for any adversary* $\mathscr{A}$*, it holds that*

$$\mathsf{Adv}_{\mathsf{T}^{\mathsf{F}},\mathsf{G}^{\mathsf{S}}}^{\mathrm{diff}}(\mathscr{A}) < \varepsilon.$$

$\mathscr{A}$ *makes at most* q_{T} *queries to* T *or* G *and at most* q_{F} *queries to* F *or* S*. Similarly,* T^{F} *is said to be computationally indifferentiable from* G *if the running time of* $\mathscr{A}$ *is bounded above by a polynomial in the security parameter and* ε *is a negligible function of the security parameter.*

Remark 1. For our purpose, we will not consider the parameter t. Also, we will not consider q_{T} and q_{F} separately and focus on their sum $q = q_{\mathsf{T}} + q_{\mathsf{F}}$, which is the total number of queries made by $\mathscr{A}$. Moreover, when F and S are adequately understood we will write the advantage term as $\mathsf{Adv}_{\mathsf{T},\mathsf{G}}^{\mathrm{diff}}(\mathscr{A})$.

We write $\mathsf{Adv}_{\mathsf{T},\mathsf{G}}^{\mathrm{diff}}(q) = \max_{\mathscr{A}} \mathsf{Adv}_{\mathsf{T},\mathsf{G}}^{\mathrm{diff}}(\mathscr{A})$, where maximum is taken over all adversaries making at most q queries to its oracles.

Indifferentiable security of XORP and XORP[k]. We first describe the XORP and XORP[k] constructions. Let Perm denote the set of all permutations over the set $\{0,1\}^n$. Let Π_0 and Π_1 be two independent random permutations, *i.e.*,

$\Pi_0, \Pi_1 \leftarrow_\$ $ Perm. The XORP construction takes an input x from $\{0,1\}^n$ and returns the element $\Pi_0(x) \oplus \Pi_1(x)$. This construction can be further generalized to k permutations. Let $\Pi_0, \ldots, \Pi_{k-1}$ be k independent random permutations. We define

$$\mathsf{XORP}[k](x) = \bigoplus_{i=0}^{k-1} \Pi_i(x). \tag{2}$$

So, $\mathsf{XORP}[2]$ is same as XORP. Now, we describe the setting of indifferentiable security in our context.

Real world. In the real world, the construction XORP has oracle access to the random permutations Π_0 and Π_1. When the adversary $\mathscr{A}$ queries the construction XORP with a value $x \in \{0,1\}^n$, XORP queries the oracles Π_0 and Π_1 with x and receives back $\Pi_0(x)$ and $\Pi_1(x)$ respectively. Finally, it computes $\Pi_0(x) \oplus \Pi_1(x)$ and returns it to $\mathscr{A}$. In addition to querying the XORP construction, $\mathscr{A}$ can directly query the oracles Π_0 and Π_1 and obtain the values of $\Pi_0(y), \Pi_1(y), \Pi_0^{-1}(y)$, and $\Pi_1^{-1}(y)$ for any $y \in \{0,1\}^n$. The queries for $\Pi_0(y)$ and $\Pi_1(y)$ are *forward queries* and the queries $\Pi_0^{-1}(y)$ and $\Pi_1^{-1}(y)$ are *backward queries*.

Ideal world. In the ideal world, $\mathscr{A}$ queries the random function $\$$ and the simulator S. S has oracle access to $\$$. However, S does not have access to (the transcripts of) the interactions between $\mathscr{A}$ and $\$$. The purpose of S is to simulate the output behavior of the oracles Π_0 and Π_1. That is, for $b \in \{0,1\}$, when $\mathscr{A}$ makes a forward query (x, b) with $x \in \{0,1\}^n$, S returns a random variable $V_b \in \{0,1\}^n$. So, for $b \in \{0,1\}$, V_b simulates $\Pi_b(x)$. Similarly, when $\mathscr{A}$ makes a backward query (y, b) (with $y \in \{0,1\}^n$ and $b \in \{0,1\}$) S returns a random variable $V_b \in \{0,1\}^n \cup \{\perp\}$. $V_b \in \{0,1\}^n$ simulates $\Pi_b^{-1}(y)$. The output $V_b = \perp$ indicates that S aborted after a fixed number of iterations. This will be more clear when we will describe the simulator S in Sect. 3. In order to prove that XORP is indifferentiable from $\$$ it is enough to construct simulator S in such a way that no adversary $\mathscr{A}$ can distinguish between the distributions of V_b and Π_b. In other words, advantage of any adversary $\mathscr{A}$, which, in this case, can be written as below,

$$\mathsf{Adv}_{\mathsf{XORP},\$}^{\mathrm{diff}}(\mathscr{A}) = |\mathbf{Pr}[\mathscr{A}^{\mathsf{XOR},(\Pi_0,\Pi_1,\Pi_0^{-1},\Pi_1^{-1})} \to 1] - \mathbf{Pr}[\mathscr{A}^{\$,\mathsf{S}} \to 1]|$$

becomes negligible. In our case, we will restrict $\mathscr{A}$ to q queries and obtain a concrete upper bound on $\mathsf{Adv}_{\mathsf{XORP},\$}^{\mathrm{diff}}(\mathscr{A})$ (in terms of parameters q and n). This will be sufficient to show indifferentiability of XORP with $\$$. For the XORP$[k]$ construction, we obtain similar upper bound on $\mathsf{Adv}_{\mathsf{XORP}[k],\$}^{\mathrm{diff}}(\mathscr{A})$.

2.3 χ^2 Method for Bounding Total Variation

Here, we provide a brief description of the χ^2 method. Given a set Ω, let $X^q := (X_1, \ldots, X_q)$ and $Z^q := (Z_1, \ldots, Z_q)$ be two random vectors distributed over

$\Omega^q = \Omega \times \cdots \times \Omega$ (q times) according to the distributions $\mathbf{P_0}$ and $\mathbf{P_1}$ respectively. Then the *total variation distance* or *statistical distance* between the distributions $\mathbf{P_0}$ and $\mathbf{P_1}$ is defined as

$$\|\mathbf{P_0} - \mathbf{P_1}\| := \frac{1}{2} \sum_{x^q \in \Omega^q} |\mathbf{P_0}(x^q) - \mathbf{P_1}(x^q)| = \max_{\mathscr{E} \in \Omega^q} \left(\sum_{x^q \in \mathscr{E}} \mathbf{P_0}(x^q) - \mathbf{P_1}(x^q) \right).$$

In what follows, we will require the following conditional distributions.

$$\mathbf{P_{0|x^{i-1}}}(x_i) := \mathbf{Pr}[X_i = x_i \mid X_1 = x_1, \ldots, X_{i-1} = x_{i-1}],$$
$$\mathbf{P_{1|x^{i-1}}}(x_i) := \mathbf{Pr}[Z_i = x_i \mid Z_1 = x_1, \ldots, Z_{i-1} = x_{i-1}].$$

When $i = 1$, $\mathbf{P_{0|x^{i-1}}}[x_1]$ represents $\mathbf{Pr}[X_1 = x_1]$. Similarly, for $\mathbf{P_{1|x^{i-1}}}[x_1]$. Let $x^{i-1} \in \Omega^{i-1}$, $i \geq 1$. The χ^2-distance[2] between these two conditional probability distributions is defined as

$$\chi^2(\mathbf{P_{0|x^{i-1}}}, \mathbf{P_{1|x^{i-1}}}) := \sum_{x_i \in \Omega} \frac{(\mathbf{P_{0|x^{i-1}}}(x_i) - \mathbf{P_{1|x^{i-1}}}(x_i))^2}{\mathbf{P_{1|x^{i-1}}}(x_i)}. \tag{3}$$

Note that for the above definition to work, it is required that the support of the distribution $\mathbf{P_{0|x^{i-1}}}$ be contained within the support of the distribution $\mathbf{P_{1|x^{i-1}}}$. Further, when the distributions $\mathbf{P_{0|x^{i-1}}}$ and $\mathbf{P_{1|x^{i-1}}}$ are clear from the context we will use the notation $\chi^2(x^{i-1})$ for $\chi^2(\mathbf{P_{0|x^{i-1}}}, \mathbf{P_{1|x^{i-1}}})$.

In a very recent work [DHT17], Dai *et al.* introduced a new method, which they term the χ^2 method (Chi-squared method), to bound the statistical distance between two joint distributions in terms of the expectations of the χ^2-distances of the corresponding conditional distributions. At the heart of the χ^2 method is the following theorem, stated in our notation and setting.

Theorem 1 ([DHT17]). *Following the notation as above and suppose the support of the distribution $\mathbf{P_{0|x^{i-1}}}$ is contained within the support of the distribution $\mathbf{P_{1|x^{i-1}}}$ for all x^{i-1}, then*

$$\|\mathbf{P_0} - \mathbf{P_1}\| \leq \left(\frac{1}{2} \sum_{i=1}^{q} \mathbf{Ex}[\chi^2(X^{i-1})] \right)^{\frac{1}{2}}, \tag{4}$$

where for each i, the expectation is over the $(i-1)$-th marginal distribution of $\mathbf{P_0}$.

As an aside, we mention that the main ingredients of the proof of Theorem 1 are (*i*) Pinsker's inequality, (*ii*) chain rule of *Kullback-Leibler divergence* (KL divergence), and (*iii*) Jensen's inequality[3]: Pinsker's inequality upper bounds statistical distance between the distributions by the KL divergence between the

---

[2] χ^2-distance has its origin in mathematical statistics dating back to the work of Pearson (see [LV87]). It may be observed that χ^2-distance is not symmetric and hence it is not a metric.

[3] See [CT06] for a background on these topics.

two distributions, chain rule of KL divergence expresses the KL divergence of two joint distributions as the sum of the KL divergences between corresponding conditional distributions, and finally Jensen's inequality is used to upper bound the KL divergence between two distributions by their χ^2-divergence (χ^2-distance).

In [DHT17], Dai *et al.* have applied Theorem 1 to show PRF-security of two well known constructions, namely the xor of two random permutations [Pat08b, Pat10, BI99, Luc00] and the encrypted Davies-Meyer (EDM) construction [CS16a, MN17a]. Subsequently, in [BN18], this method has been applied to prove full PRF-security of the variable output length XOR pseudorandom function construction. This method seems to have potential for further application to obtain better bounds (and simplified proofs) on the PRF-security of other constructions where proofs so far have evaded more classical methods, such as the H-coefficient method [Pat08a]. In fact, much earlier, Stam [Sta78] used this method, implicitly and in a purely statistical context, to obtain a PRF-security bound of the truncated random permutation construction.

3 Simulator and Transcripts

3.1 Description of the Simulator

Here, we describe the simulator S used in the proof of Theorem 2.[4] The goal of the simulator S is to mimic the permutations $\Pi := (\Pi_0(.), \Pi_1(.), \Pi_0^{-1}(.), \Pi_1^{-1}(.))$ in such a way that (XORP, Π) and $(\$, \mathsf{S})$ look indistinguishable. So, S has interfaces corresponding to the forward and backward queries of the random permutations Π_0 and Π_1. Formally, S consists of a pair of stateful randomized algorithms $\mathsf{SIM_{FWD}}$ (which is invoked for the responses to the forward queries) and $\mathsf{SIM_{BCK}}$ (which is invoked for responses to the backward queries). More precisely, for $x \in \{0,1\}^n$ and $b \in \{0,1\}$, when an adversary $\mathscr{A}$ makes a forward query (x, b) to S, S runs the algorithm $\mathsf{SIM_{FWD}}$ and returns a random variable $V_b \in \{0,1\}^n$. So, for $b \in \{0,1\}$, V_b simulates $\Pi_b(x)$. Similarly, when $\mathscr{A}$ makes a backward query (y, b) (with $y \in \{0,1\}^n$ and $b \in \{0,1\}$) to S, S runs the algorithm $\mathsf{SIM_{BCK}}$ and returns a random variable $V_b \in \{0,1\}^n \cup \{\bot\}$. $V_b \in \{0,1\}^n$ simulates $\Pi_b^{-1}(y)$. Note that S has access to the random function $\$$ which returns random elements from $\{0,1\}^n$ on every fresh query. The goal of the simulator S is to simulate the output behavior of $\Pi_0(.), \Pi_1(.), \Pi_0^{-1}(.)$, and $\Pi_1^{-1}(.)$ in the ideal world in such a way that it remains consistent with the XORP construction, which is given by the condition

$$\$(x) = \mathsf{SIM_{FWD}}(x, 0) \oplus \mathsf{SIM_{FWD}}(x, 1) \text{ for } x \in \{0,1\}^n.$$

However, S may fail to maintain the condition. Whenever it fails (which happens only for the backward queries), $\mathsf{SIM_{BCK}}$ returns $\bot$. Before returning $\bot$ it makes several attempts where it interacts with $\$$. If after n attempts it fails to maintain the condition (we will show that would happen with very low probability), it aborts. $V_b = \bot$ indicates that event.

---

[4] We will consider another simulator in the proof of Theorem 3.

Description of the internal state. In order to be consistent with its replies, *i.e.*, to output the same V_b corresponding to the same queries (forward or backward), S is stateful, *i.e.*, it maintains a history of all the previous interactions (*i.e.*, queries and responses) with $\mathscr{A}$. To do this, S internally maintains three sets $\mathscr{D}, \mathscr{R}_0$, and $\mathscr{R}_1$, and also maintains two lists (indexed by elements of $\mathscr{D}$) $\mathscr{L}_0, \mathscr{L}_1$.

The set $\mathscr{D}$ contains all $x \in \{0,1\}^n$ belonging to the forward queries (x, b) made by $\mathscr{A}$ and all $V_b \in \{0,1\}^n$ that the simulator output on a backward query made by $\mathscr{A}$. For $b \in \{0,1\}$, the set $\mathscr{R}_b$ contains all $y \in \{0,1\}^n$ belonging to the backward queries (y, b) made by $\mathscr{A}$ along with all V_b output by S on a forward query made by $\mathscr{A}$. The lists $\mathscr{L}_0, \mathscr{L}_1$ capture the input-output mapping of S. More precisely, for $b \in \{0,1\}, x \in \mathscr{D}, y \in \mathscr{R}_b$, $\mathscr{L}_b(x) = y$ implies either $V_b = y$ was output on a forward query (x, b) or $V_b = x$ was output on a backward query (y, b). More importantly, for all $x \in \mathscr{D}$, the relationship $\mathscr{L}_0(x) \oplus \mathscr{L}_1(x) = \$(x)$ is always satisfied.

Now, we describe how the simulator works via the algorithms $\mathsf{SIM_{FWD}}$ and $\mathsf{SIM_{BCK}}$. Details of the these algorithms are given in Figs. 2 and 3. In the following, we assume that $\mathscr{A}$ always makes fresh queries since otherwise the simulator can repeat the previous responses (as it maintains internal states keeping all previous queries and responses).

Algorithm $\mathsf{SIM_{FWD}}$ (see Fig. 2). On an input $(x \in \{0,1\}^n, b \in \{0,1\})$, S queries $\$$ to obtain $Z = \$(x)$. Then, S samples V_b randomly from the set $\{0,1\}^n \setminus \{\mathscr{R}_b \cup \{Z \oplus \mathscr{R}_{1-b}\}\}$, where $Z \oplus \mathscr{R}_{1-b}$ denotes the set $\{Z \oplus y | y \in \mathscr{R}_{1-b}\}$. Here, it can

---

Algorithm: $\mathsf{SIM_{FWD}}$
Input: $x \in \{0,1\}^n, b \in \{0,1\}$
Output: $V_b \in \{0,1\}^n$

1 :   // check if the query on x is already answered

2 :   **if** $x \in \mathscr{D}$

3 :   **return** $\mathscr{L}_b(x)$

4 :   // otherwise execute the main part

5 : $Z \leftarrow \$(x)$

6 : $V_b \leftarrow_\$ \{0,1\}^n \setminus \{\mathscr{R}_b \cup \{Z \oplus \mathscr{R}_{1-b}\}\}$

7 :   // update the state

8 : $\mathscr{R}_b \leftarrow \mathscr{R}_b \cup \{V_b\}, \mathscr{R}_{1-b} \leftarrow \mathscr{R}_{1-b} \cup \{Z \oplus V_b\}$

9 : $\mathscr{D} \leftarrow \mathscr{D} \cup \{x\}$

10 : $\mathscr{L}_b(x) \leftarrow V_b, \mathscr{L}_{1-b}(x) \leftarrow Z \oplus V_b$

11 :   // **return** the output

12 : **return** V_b

Fig. 2. Description of the simulator for all forward queries.

be observed that the set $\{0,1\}^n \setminus \{\mathscr{R}_b \cup \{Z \oplus \mathscr{R}_{1-b}\}\}$ is non-empty, provided $q < 2^{n-1}$. Therefore, for $q < 2^{n-1}$, the sampling is always possible. Subsequently, S sets V_b and $Z \oplus V_b$ as outputs of $\mathsf{SIM_{FWD}}(x, b)$ and $\mathsf{SIM_{FWD}}(x, 1-b)$ respectively (and hence $\mathsf{SIM_{FWD}}(x, 0) \oplus \mathsf{SIM_{FWD}}(x, 1) = \$(x)$). Before S returns V_b to the adversary, it updates all internal sets accordingly.

Algorithm $\mathsf{SIM_{BCK}}$ (see Fig. 3). Next, we present the algorithm $\mathsf{SIM_{BCK}}$. On an input $(y \in \{0,1\}^n, b \in \{0,1\})$, S samples an element V_b randomly from outside the set $\mathscr{D}$ and then obtains $\$(V_b)$ by querying \$. Now, there is a certain chance that $\$(V_b) \oplus y$ is in the range set $\mathscr{R}_{1-b}$, which would then violate the permutation property of Π_{1-b} that S is simulating. So, S continue with similar attempts until it samples a V_b such that $\$(V_b) \oplus y \notin \mathscr{R}_{1-b}$. It makes at most n such attempts. If it fails after all these n attempts, it returns $\perp$. In all these attempts, the S maintains an auxiliary set $\mathscr{D}'$ which is not a part of its state and only used locally during an execution. At the beginning of the algorithm, $\mathscr{D}'$ is initialized to the current domain $\mathscr{D}$. At the start of each iteration, a fresh V_b is sampled from the set $\{0,1\}^n \setminus \mathscr{D}'$. If the conditions $y \oplus \$(V_b) \in \mathscr{R}_{1-b}$ is satisfied (i.e., the sampled V_b turns out to be a bad choice), then V_b is appended to $\mathscr{D}'$ and the next iteration begins. Note that if $q + n < 2^n$ then the set $\{0,1\}^n \setminus \mathscr{D}'$ is always non-empty so that the sampling of V_b (in Step 6) is possible in every iteration. But $q + n < 2^n$ is trivially satisfied for $n \geq 3$ and $q < 2^{n-1}$. When the condition is not satisfied (i.e., when $y \oplus \$(V_b) \notin \mathscr{R}_{1-b}$) then S returns V_b after appropriately updating all the internal sets.

Remark 2. Here, as an aside, one may notice that there is a chance of collision due to two backward queries made to the two random permutations in the real world or two interfaces in the ideal world. We explain this with the following example. Assume that $\mathscr{A}$ makes backward queries $(y, 0)$ and $(y', 1)$ in the real world. Then it is easy to see that there is a positive probability of getting the same output in both the cases (as Π_0 and Π_1 are sampled independently from the set Perm). On the other hand, in the ideal world, when $\mathscr{A}$ makes the query $(y, 0)$, then if $y \oplus \$(V_0) \notin \mathscr{R}_1$ (which has positive probability) then at Step 13 of $\mathsf{SIM_{BCK}}$ $\mathscr{L}_1(V_0)$ is set to $\$(V_0) \oplus y$ and V_0 is returned to $\mathscr{A}$. Now, if $\mathscr{A}$ makes the query $(\$(V_0) \oplus y, 1)$ then due to the check at Step 2, V_0 is again returned to $\mathscr{A}$. Therefore, there is a positive probability of collision for the queries $(y, 0)$ and $(y', 1)$ in the ideal world as well (as was to be expected since the simulator is simulating the permutations Π_0 and Π_1), where $y' = \$(V_0) \oplus y$ in this case.

3.2 Additional Information to the Adversary

After the adversary $\mathscr{A}$ has finished its interaction in the real/ideal world, i.e., when it has made q queries and received corresponding replies, it is provided with the following additional information. Note that the additional information does not degrade $\mathscr{A}$'s advantage as it is always possible to discard it. Below we assume x, x_i, y are from $\{0,1\}^n$ and b is from $\{0,1\}$.

Algorithm: $\mathsf{SIM_{BCK}}$
Input: $y \in \{0,1\}^n, b \in \{0,1\}$
Output: $V_b \in \{0,1\}^n \cup \{\bot\}$

---

1 :   // check if the query on y is already answered

2 :   **if** $y = \mathscr{L}_b(x)$ **for** $x \in \mathscr{D}$

3 :   **return** x

4 : $\mathscr{D}' \leftarrow \mathscr{D}$

5 :   **repeat** n times

6 : $V_b \leftarrow_\$ \{0,1\}^n \setminus \mathscr{D}'$

7 : $Z \leftarrow \$(V_b)$

8 :   // check if $Z \oplus y$ is outside of the set $\mathscr{R}_{1-b}$

9 :   **if** $Z \oplus y \notin \mathscr{R}_{1-b}$

10 :   // update the state

11 : $\mathscr{D} \leftarrow \mathscr{D} \cup \{V_b\}$

12 : $\mathscr{R}_b \leftarrow \mathscr{R}_b \cup \{y\}, \mathscr{R}_{1-b} \leftarrow \mathscr{R}_{1-b} \cup \{Z \oplus y\}$

13 : $\mathscr{L}_b(V_b) \leftarrow y, \mathscr{L}_{1-b}(V_b) \leftarrow Z \oplus y$

14 :   // **return** the inverse

15 :   **return** V_b

16 :   // otherwise append V_b to the set $\mathscr{D}'$ sothat it is not sampled in the next iteration

17 : $\mathscr{D}' \leftarrow \mathscr{D}' \cup \{V_b\}$

18 :   // abort after t iterations

19 : **return** $\bot$

Fig. 3. Description of the simulator for all backward queries.

1. For each query x made to the construction XORP, $\mathscr{A}$ is given the values $\Pi_0(x)$ and $\Pi_1(x)$. Similarly, for each query x made to the random function $\$$, $\mathscr{A}$ is given the outputs of the simulator S corresponding to the forward queries $(x,0)$ and $(x,1)$.

2. For each forward query (x,b) made to Π_b (*i.e.*, for each value of $\Pi_b(x)$), it is also given $\Pi_{1-b}(x)$. Similarly, for each forward query (x,b) made to S, $\mathscr{A}$ is also given the value corresponding to the forward query $(x, 1-b)$.

3. For each backward query (y,b) made to Π_b (*i.e.*, for each value of $\Pi_b^{-1}(y)$), it is also given $\Pi_{1-b}(\Pi_b^{-1}(y))$. For each backward query (y,b) made to S, $\mathscr{A}$ is also given the value corresponding to the forward query $(x, 1-b)$, where x is the value returned by S on the backward query (y,b).

With access to this extra information, $\mathscr{A}$ knows the tuple $(x_i, \Pi_0(x_i), \Pi_1(x_i))$ corresponding to its i-th query in the real world. Note that from $\Pi_0(x_i)$ and $\Pi_1(x_i)$, $\mathscr{A}$ can always obtain $\Pi_0(x_i) \oplus \Pi_1(x_i)$ (which is, in fact, the output of XORP when queried with x_i). Therefore, we do not include this redundant information in the tuple. When $\Pi_0(x_i)$ and $\Pi_1(x_i)$ are treated as random variables, we will denote $\Pi_0(x_i)$ by $U_{0,i}$ and $\Pi_1(x_i)$ by $U_{1,i}$. So, the tuple $(x_i, U_{0,i}, U_{1,i})$ is a random variable and an arbitrary but fixed value of this random variable will be denoted by $(x_i, u_{0,i}, u_{1,i})$. Similarly, in the ideal world, corresponding to the i-th query, $\mathscr{A}$ knows the tuple $(x_i, V_{0,i}, V_{1,i})$, where for $b \in \{0, 1\}$, $V_{b,i}$ is the reply of S to the forward query (x_i, b). Similar to the previous case, we will denote a fixed value of the random variable $(x_i, V_{0,i}, V_{1,i})$ by $(x_i, v_{0,i}, v_{1,i})$. In the case where the backward query resulted in an abort, $i.e.$, the output was $\perp$, we take $x_i = \perp$ and $v_{0,i}$ and $v_{1,i}$ can be arbitrary (but fixed). In fact, in this case, $v_{0,i}$ and $v_{1,i}$ are purely included to maintain uniformity of presentation and will be disregarded in subsequent calculations. Further, slightly abusing the notation for its simplicity, we will denote any such tuple ($i.e.$, a tuple with $x_i = \perp$) by $\perp$. Note that we did not include the query type ($i.e.$, forward or backward) information in the tuple as, in our calculation, we will consider both the possibilities for a tuple. However, for the sake of completeness, one can assume that $\mathscr{A}$ has this information.

Without loss of any generality we will assume that $\mathscr{A}$ does not repeat its queries as the response will be the same for a repeated query. Also, we will discard any duplicate copy of a tuple that may have occurred due to the extra information supplied to $\mathscr{A}$[5].

(Extended) transcript of the adversary. In the real world, the sequence of random variables $(x_i, U_{0,i}, U_{1,i})$, with $1 \leq i \leq q$, is supported on the set $\mathscr{T}_u$ of sequences $(x_i, u_{0,i}, u_{1,i}), 1 \leq i \leq q$, $x_i, u_{0,i}, u_{1,i} \in \{0, 1\}^n$ and $x_i \neq x_j, u_{0,i} \neq u_{0,j}, u_{1,i} \neq u_{1,j}$ for $1 \leq i < j \leq q$. Whereas in the ideal world the sequence of random variables $(x_i, V_{0,i}, V_{1,i})$, with $1 \leq i \leq q$, is supported on the set $\mathscr{T}_v$ of sequences $(x_i, v_{0,i}, v_{1,i}), 1 \leq i \leq q$, $x_i \in \{0, 1\}^n \cup \{\perp\}$, $v_{0,i}, v_{1,i} \in \{0, 1\}^n$ and $x_i \neq x_j, v_{0,i} \neq v_{0,j}, v_{1,i} \neq v_{1,j}$ for each $1 \leq i < j \leq q$ such that $x_i \neq \perp \neq x_j$. So, we have $\mathscr{T}_u \subset \mathscr{T}_v$. We term elements of $\mathscr{T}_u$ and $\mathscr{T}_v$ *views* of the adversary. In our subsequent treatment, we will solely work with the views from the real and the ideal world, and the fact that $\mathscr{T}_u \subset \mathscr{T}_v$ will be essential for the application of the χ^2 method.

4   Main Result

In this section, we state and prove our main result. We continue in the setup of the previous section. To simplify the presentation we denote 2^n by N. Our main result is the following.

Theorem 2. *Let $N \geq 16$ and $q < \frac{N}{2}$. Then*

$$\mathsf{Adv}^{\mathrm{diff}}_{\mathsf{XORP}, \$}(q) \leq \sqrt{\frac{1.25q}{N}}$$

---

[5] For example, such a duplicate of a tuple $(x_i, u_{0,i}, u_{1,i})$ can occur in the real world if $\mathscr{A}$ queries the XORP with x_i and later makes a backward query to Π_0 with $u_{0,i}$.

Proof. Before presenting the technical details we will provide a brief outline of the proof to help the reader follow the underlying idea and the flow of the proof. But before that we will describe our notational setup for the proof.

Notational setup: To simplify the notation we will denote the random variable $(x_i, U_{0,i}, U_{1,i})$ by S_i and $(x_i, V_{0,i}, V_{1,i})$ by T_i. So, S_i (resp. T_i) follows the distribution of the real (resp. ideal) world which we denote by $\mathsf{p}_0^{\mathsf{fwd}}(.)$ (resp $\mathsf{p}_1^{\mathsf{fwd}}$) when S_i (resp T_i) is a forward query and by $\mathsf{p}_0^{\mathsf{bck}}(.)$ (resp $\mathsf{p}_1^{\mathsf{bck}}$) when S_i (resp T_i) is a backward query. Hence, we denote $\mathbf{Pr}[S_i = s_i]$ by $\mathsf{p}_0^{\mathsf{fwd}}(s_i)$ and $\mathbf{Pr}[T_i = t_i]$ by $\mathsf{p}_1^{\mathsf{fwd}}(t_i)$ when S_i and T_i are forward queries and likewise for backward queries. Further, we will abuse the notation to denote the joint distribution of S^{i-1} by $\mathsf{p}_0^{\mathsf{fwd}}$ when S^{i-1} corresponds to $i-1$ forward queries and by $\mathsf{p}_0^{\mathsf{bck}}$ when S^{i-1} corresponds to $i-1$ backward queries. Moreover, for fixed s_i and s^{i-1}, we denote $\mathbf{Pr}[S_i = s_i \mid S_1 = s_1, \ldots, S_{i-1} = s_{i-1}]$ by $\mathsf{p}_0^{\mathsf{fwd}}(s_i \mid s^{i-1})$ when S_i corresponds to a forward query; likewise for the other cases.

Outline of the proof: The main tool we use in our proof is Theorem 1. Our goal is to evaluate the r.h.s. of (4). In doing so, we calculate $\mathbf{Ex}[\chi^2(S^{i-1})]$ over the real world distributions ($\mathsf{p}_0^{\mathsf{fwd}}$ and $\mathsf{p}_0^{\mathsf{bck}}$). More precisely, we consider the two cases; (*i*) when s_i is a forward query, and (*ii*) when s_i is a backward query. For the forward query case, we first calculate $\chi^2(s^{i-1})$ for fixed s^{i-1}, which is given by the sum of

$$\frac{(\mathsf{p}_0^{\mathsf{fwd}}(s_i \mid s^{i-1}) - \mathsf{p}_1^{\mathsf{fwd}}(s_i \mid s^{i-1}))^2}{\mathsf{p}_1^{\mathsf{fwd}}(s_i \mid s^{i-1})}$$

taken over all possible s_i given s^{i-1}. Here, we note that the support $\mathscr{T}_u$ of real world distributions ($\mathsf{p}_0^{\mathsf{fwd}}$ and $\mathsf{p}_0^{\mathsf{bck}}$) is included in the supports $\mathscr{T}_u$ and $\mathscr{T}_v$ of the ideal world distributions $\mathsf{p}_0^{\mathsf{fwd}}$ and $\mathsf{p}_0^{\mathsf{bck}}$ respectively. Hence, $\chi^2(s^{i-1})$ is well defined. Next, we consider the random variable S^{i-1} in the real world. Each $S_j \in \{S^{i-1}\}$ may correspond to a forward query or a backward query. However, since the distributions $\mathsf{p}_0^{\mathsf{fwd}}$ and $\mathsf{p}_0^{\mathsf{bck}}$ are identical, the distribution of S^{i-1} does not depend on the query type of each individual S_j. So, we treat $\chi^2(S^{i-1})$ as a random variable and take its expectation under the distribution of S^{i-1}. Finally, we take the sum of $\mathbf{Ex}[\chi^2(S^{i-1})]$ for all i in the range $1 \leq i \leq q$, which turns out to be $\frac{8q^3}{N^3}$.

Corresponding steps for the backward query case are exactly similar to the forward query case when $s_i \neq \perp$. The case when $s_i = \perp$ is treated separately. Summing the expectations $\mathbf{Ex}[\chi^2(S^{i-1})]$ for the two subcases (*i.e.*, for $s_i \neq \perp$ and $s_i = \perp$) we obtain the final sum (taken over all i in the range $1 \leq i \leq q$) for the backward query case to be $\frac{2.5q}{N}$. Finally, we get the upper bound on $\mathsf{Adv}_{\mathsf{XORP},\$}^{\mathsf{diff}}(q)$ by applying Theorem 1, where we get an upper bound on the r.h.s. of (4) by taking the maximum of the forward and backward queries for all the q queries (which in this case turns out to be the backward query).

Technical details: Following the above discussion we first consider the case when s_i is a forward query, and then consider the case when it is a backward query. To simplify notation, from here on, we will denote $i - 1$ by r.

Forward query

First, we calculate $\mathsf{p}_0^{\mathsf{fwd}}(s_i \mid s^r)$ and $\mathsf{p}_1^{\mathsf{fwd}}(s_i \mid s^r)$ for fixed s_i and s^r, where $s_i = (x_i, u_{0,i}, u_{1,i})$. $\mathsf{p}_0^{\mathsf{fwd}}(s_i \mid s^r)$ is straightforward to calculate. Since $x_i \notin \{x^r\}$, $\Pi_0(x_i)$ and $\Pi_1(x_i)$ are two independent random samples drawn from outside the sets $\{u_0^r\}$ and $\{u_1^r\}$ respectively. Thus

$$\mathsf{p}_0^{\mathsf{fwd}}(s_i \mid s^r) = \mathbf{Pr}[S_i = (x_i, u_{0,i}, u_{1,i}) \mid S_1 = (x_1, u_{0,1}, u_{1,1}), \ldots,$$
$$S_r = (x_r, u_{0,r}, u_{1,r})]$$
$$= \mathbf{Pr}[\Pi_0(x_i) = u_{0,i} \wedge \Pi_1(x_i) = u_{1,i} \mid \Pi_0(x_j) = u_{0,j} \wedge$$
$$\Pi_1(x_j) = u_{1,j} \forall 1 \leq j \leq r]$$
$$= \frac{1}{(N-r)^2} \tag{5}$$

To calculate $\mathsf{p}_1^{\mathsf{fwd}}(s_i \mid s^r)$, we consider, without loss of any generality, the execution of the algorithm $\mathsf{SIM}_{\mathsf{FWD}}$ algorithm on the forward query $(v_{0,i}, 0)$ (the case when the forward query is $(v_{1,i}, 1)$ is identical). In this case, $\mathscr{D} = \{x^r\}, \mathscr{R}_0 = \{u_0^r\}, \mathscr{R}_1 = \{u_1^r\}$. Then we have

$$\mathsf{p}_1^{\mathsf{fwd}}(s_i \mid s^r) = \mathbf{Pr}[T_i = (x_i, v_{0,i}, v_{1,i}) \mid T_1 = (x_1, v_{0,1}, v_{1,1}), \ldots, T_r = (x_r, v_{0,r}, v_{1,r})]$$
$$= \mathbf{Pr}[\$(x_i) = v_{0,i} \oplus v_{1,i} \wedge V_0 = v_{0,i} \mid \mathscr{D} = \{x^r\}, \mathscr{R}_0 = \{u_0^r\}, \mathscr{R}_1 = \{u_1^r\}]$$
$$= \frac{1}{N} \times \frac{1}{N - |\mathscr{W}_{x_i}|}, \tag{6}$$

where $\mathscr{W}_{x_i} = \mathscr{R}_0 \cup \{\$(x_i) \oplus \mathscr{R}_1\}$. From (5) and (6) we derive the expression for $\chi^2(s^r)$ below.

$$\chi^2(s^r) = \sum_{s_i} \frac{(\mathsf{p}_0^{\mathsf{fwd}}(s_i|s^r) - \mathsf{p}_1^{\mathsf{fwd}}(s_i|s^r))^2}{\mathsf{p}_1^{\mathsf{fwd}}(s_i|s^r)}$$
$$= \sum_{s_i} \frac{\left(\frac{1}{(N-r)^2} - \frac{1}{N(N-|\mathscr{W}_{x_i}|)}\right)^2}{\frac{1}{N(N-|\mathscr{W}_{x_i}|)}}$$
$$= \sum_{s_i} \frac{N\left(|\mathscr{W}_{x_i}| - \frac{2rN-r^2}{N}\right)^2}{(N-|\mathscr{W}_{x_i}|)(N-r)^4}. \tag{7}$$

The sum in (7) is over all possible s_i's given s^r. The number of such number of such s_i's is $(N-r)(N-|\mathscr{W}_{x_i}|)$. Therefore,

$$\chi^2(s^r) = \frac{N\left(|\mathscr{W}_{x_i}| - \frac{2rN-r^2}{N}\right)^2}{(N-r)^3} \tag{8}$$

Let S^r be chosen according to the distribution $\mathsf{p}_0^{\mathsf{fwd}}$. Then $\mathscr{D}, \mathscr{R}_0, \mathscr{R}_1$ are random variables. This, in turn, means $\mathscr{W}_{x_i}$ and $\chi^2(S^r)$ are also random variables (as functions of $\mathscr{D}, \mathscr{R}_0, \mathscr{R}_1$). Our goal is to calculate the expectation of $\chi^2(S^r)$ under the distribution $\mathsf{p}_0^{\mathsf{fwd}}$. For notational simplicity, we denote the random variable $|\mathscr{W}_{x_i}|$ by W (mildly violating our notational convention). So, from (8) we have

$$\mathbf{Ex}[\chi^2(Z^r)] = \mathbf{Ex}\left[\frac{N\left(\mathsf{W} - \frac{2rN - r^2}{N}\right)^2}{(N-r)^3}\right]$$

$$= \frac{N}{(N-r)^3} \times \mathbf{Ex}\left[\left(\mathsf{W} - \frac{2rN - r^2}{N}\right)^2\right]. \tag{9}$$

In the next lemma, whose proof is postponed to Sect. 5, we calculate $\mathbf{Ex}[\mathsf{W}]$.

Lemma 1. *With the above notation*

$$\mathbf{Ex}[\mathsf{W}] = \frac{2rN - r^2}{N}, \ \ and \ \mathbf{Var}[\mathsf{W}] \le \frac{r^2}{N}.$$

Using Lemma 1, (9) can be written as

$$\mathbf{Ex}[\chi^2(S^r)] = \frac{N}{(N-r)^3} \times \mathbf{Ex}\left[(\mathsf{W} - \mathbf{Ex}[\mathsf{W}])^2\right]$$

$$= \frac{N}{(N-r)^3} \times \mathbf{Var}[\mathsf{W}].$$

In Lemma 1, we also showed that $\mathbf{Var}[\mathsf{W}] \le \frac{r^2}{N}$. This leads to the following final expression for the forward query case.

$$\sum_{i=1}^{q} \mathbf{Ex}[\chi^2(S^r)] \le \sum_{i=1}^{q} \frac{r^2}{(N-r)^3}$$

$$\le \frac{8q^3}{N^3}. \tag{10}$$

In (10), we used the fact $r < q$ and $q < \frac{N}{2}$.

Backward query

Let $\mathscr{X}$ be the set of all possible s_i's which are not 'abort', *i.e.*, $s_i \ne \perp$. Then for backward queries we have the following split.

$$\mathbf{Ex}[\chi^2(S^r)] = \mathbf{Ex}\left[\sum_{s_i \in \mathscr{X}} \frac{(\mathsf{p}_0^{\mathsf{bck}}(s_i \mid S^r) - \mathsf{p}_1^{\mathsf{bck}}(s_i \mid S^r))^2}{\mathsf{p}_1^{\mathsf{bck}}(s_i \mid S^r)}\right]$$

$$+ \mathbf{Ex}\left[\frac{(\mathsf{p}_0^{\mathsf{bck}}(\perp \mid S^r) - \mathsf{p}_1^{\mathsf{bck}}(\perp \mid S^r))^2}{\mathsf{p}_1^{\mathsf{bck}}(\perp \mid S^r)}\right] \tag{11}$$

We evaluate the two expectations on the r.h.s. of (11) in the following two cases.

CASE 1: $s_i \in \mathscr{X}$

In this case, we have for fixed s^r,

$$p_0^{\mathsf{bck}}(s_i \mid s^r) = p_0^{\mathsf{fwd}}(s_i \mid s^r) = \frac{1}{(N-r)^2}.$$

Next, we calculate $p_1^{\mathsf{bck}}(s_i \mid s^r)$. For this, we need to consider the execution of the algorithm $\mathsf{SIM}_{\mathsf{BCK}}$. Let the backward query, without loss of any generality, be $(v_{0,i}, 0)$. Further, let us denote by V_0^j the V_0 sampled by the algorithm $\mathsf{SIM}_{\mathsf{BCK}}$ at the j-th iteration, where by j-th iteration we mean j-th repeated execution of the steps 6 to 19 of $\mathsf{SIM}_{\mathsf{BCK}}$. Let us assume that $\mathsf{SIM}_{\mathsf{BCK}}$ succeeds at the ℓ-th iteration for $1 \leq \ell \leq n$, i.e., for $1 \leq j \leq \ell - 1$, $\$(V_0^j) \oplus v_{0,i} \in \mathscr{R}_1$, and $V_0^\ell = x_i$, where $\mathscr{R}_1 = \{v_1^r\}$ (also $\mathscr{R}_0 = \{v_0^r\}, \mathscr{D} = \{x^r\}$). Let us denote by $\mathsf{BAD}_{\ell-1}$ the event $\$(V_0^1) \oplus v_{0,i}, \ldots, \$(V_0^{\ell-1}) \oplus v_{0,i} \in \mathscr{R}_1$ and by E the event $\mathscr{D} = \{x^r\} \wedge \mathscr{R}_0 = \{v_0^r\} \wedge \mathscr{R}_1 = \{v_1^r\}$. Then

$$p_1^{\mathsf{bck}}(s_i \mid s^r) = \mathbf{Pr}[T_i = (x_i, v_{0,i}, v_{1,i}) \mid T_r = (x_r, v_{0,r}, v_{1,r}), \ldots, T_1 = (x_1, v_{0,1}, v_{1,1})]$$

$$= \sum_{\ell=1}^{n} \mathbf{Pr}[\mathsf{BAD}_{\ell-1} \wedge V_0^\ell = x_i \wedge \$(x_i) = v_{0,i} \oplus v_{1,i} \mid \mathsf{E}]$$

$$= \sum_{\ell=1}^{n} \mathbf{Pr}[V_0^\ell = x_i \wedge \$(x_i) = v_{0,i} \oplus v_{1,i} \mid \mathsf{BAD}_{\ell-1}, \mathsf{E}] \times \mathbf{Pr}[\mathsf{BAD}_{\ell-1} \mid \mathsf{E}]$$

Now, $\mathbf{Pr}[\mathsf{BAD}_{\ell-1} \mid \mathsf{E}]$ can be calculated as

$$\mathbf{Pr}[\mathsf{BAD}_{\ell-1} \mid \mathsf{E}] = \prod_{j=1}^{\ell-1} \mathbf{Pr}[\$(V_0^j) \oplus v_{0,i} \in \mathscr{R}_1 = \{v_1^r\} \mid \mathsf{E}]$$

$$= \left(\frac{r}{N}\right)^{\ell-1}. \tag{12}$$

To justify (12) we first note that the distribution $p_0^{\mathsf{bck}}(.)$ is supported on the set of tuples $s^r = (s_1, \ldots, s_r)$ such that none of the s_j, with $1 \leq j \leq r$, is $\perp$. So, in the $\mathsf{SIM}_{\mathsf{BCK}}$ algorithm the set $\mathscr{R}_1$ has size r. Also, at the j-th iteration, with $1 \leq j \leq \ell - 1$ a fresh V_0^j (sampled from outside the set $\mathscr{D}'$) is given to $\$$. Therefore, $\mathsf{BAD}_{\ell-1}$ occurs when $\ell - 1$ independent events each with probability $\frac{r}{N}$ occur, leading to the expression in (12).

Next, at the ℓ-th iteration the set $\mathscr{D}'$ has size $r + \ell - 1$. Since V_0^ℓ is sampled at random from the set $\{0,1\}^n \setminus \mathscr{D}'$, we immediately have

$$\mathbf{Pr}[V_0^\ell = x_i \wedge \$(x_i) = v_{0,i} \oplus v_{1,i} \mid \mathsf{BAD}_{\ell-1}, \mathsf{E}] = \frac{1}{N} \times \frac{1}{N - r - \ell + 1}. \tag{13}$$

By combining (12) and (13) we get

$$p_1^{\mathsf{bck}}(s_i \mid s^r) = \sum_{\ell=1}^{n} \frac{1}{N} \times \frac{1}{N - r - \ell + 1} \times \left(\frac{r}{N}\right)^{\ell-1}. \tag{14}$$

In the following lemma, we derive a lower and an upper bound on $p_1^{\mathsf{bck}}(s_i \mid s^r)$. Proof of the lemma is given in Sect. 5.

Lemma 2. *With the above notation, the following bounds hold for* $\mathsf{p}_1^{\mathsf{bck}}(s_i \mid s^r)$.

$$\frac{1}{(N-r)^2} \times \left(1 - \left(\frac{r}{N}\right)^n\right) \leq \mathsf{p}_1^{\mathsf{bck}}(s_i \mid s^r) \leq \frac{4}{N(N-r)}.$$

Let us denote the lower and upper bounds in Lemma 2 by L and U respectively. Then

$$\frac{\left(\mathsf{p}_0^{\mathsf{bck}}(s_i \mid s^r) - \mathsf{p}_1^{\mathsf{bck}}(s_i \mid s^r)\right)^2}{\mathsf{p}_1^{\mathsf{bck}}(s_i \mid s^r)} \leq \max \left\{ \frac{\left(\mathsf{U} - \frac{1}{(N-r)^2}\right)^2}{\mathsf{U}}, \frac{\left(\mathsf{L} - \frac{1}{(N-r)^2}\right)^2}{\mathsf{L}} \right\}.$$

$$(15)$$

(15) is justified because the function $\frac{\left(y - \frac{1}{(N-r)^2}\right)^2}{y}$ attains its minimum ($= 0$) at $y = \frac{1}{(N-r)^2}$ and is strictly increasing for $y \geq \frac{1}{(N-r)^2}$ and strictly decreasing for $y \leq \frac{1}{(N-r)^2}$. Now,

$$\frac{\left(\mathsf{U} - \frac{1}{(N-r)^2}\right)^2}{\mathsf{U}} = \frac{3N - 4r}{4N(N-r)^3},$$

and

$$\frac{\left(\mathsf{L} - \frac{1}{(N-r)^2}\right)^2}{\mathsf{L}} = \frac{\left(\frac{r}{N}\right)^{2n}}{(N-r)^2 \times \left(1 - \left(\frac{r}{N}\right)^n\right)}.$$

Further, considering that $|\mathscr{L}|$ is at most $(N - |\mathscr{D}|)(N - |\mathscr{R}_1|) = (N-r)^2$, we get

$$\sum_{s_i \in \mathscr{L}} \frac{\left(\mathsf{p}_0^{\mathsf{bck}}(s_i \mid s^r) - \mathsf{p}_1^{\mathsf{bck}}(s_i \mid s^r)\right)^2}{\mathsf{p}_1^{\mathsf{bck}}(s_i \mid s^r)} \leq \max \left\{ \frac{3N - 4r}{4N(N-r)}, \frac{\left(\frac{r}{N}\right)^{2n}}{\left(1 - \left(\frac{r}{N}\right)^n\right)} \right\}.$$

Therefore, when S^r is a random variable that follows the distribution $\mathsf{p}_0^{\mathsf{bck}}$, we obtain the following expectation under the distribution $\mathsf{p}_0^{\mathsf{bck}}$.

$$\mathbf{Ex}\left[\sum_{s_i \in \mathscr{L}} \frac{\left(\mathsf{p}_0^{\mathsf{bck}}(s_i \mid S^r) - \mathsf{p}_1^{\mathsf{bck}}(s_i \mid S^r)\right)^2}{\mathsf{p}_1^{\mathsf{bck}}(s_i \mid S^r)} \right] \leq \max \left\{ \frac{3N - 4r}{4N(N-r)}, \frac{\left(\frac{r}{N}\right)^{2n}}{\left(1 - \left(\frac{r}{N}\right)^n\right)} \right\}.$$

$$(16)$$

CASE 2: $s_i = \bot$
In the real world, there is no abort, so $\mathsf{p}_0^{\mathsf{bck}}(\bot \mid S^r) = 0$. Therefore, similar to (12),

$$\mathbf{Ex}\left[\frac{\left(\mathsf{p}_0^{\mathsf{bck}}(\bot \mid S^r) - \mathsf{p}_1^{\mathsf{bck}}(\bot \mid S^r)\right)^2}{\mathsf{p}_1^{\mathsf{bck}}(\bot \mid S^r)} \right] = \mathbf{Ex}\left[\mathsf{p}_1^{\mathsf{bck}}(\bot \mid S^r) \right]$$

$$= \mathsf{p}_1^{\mathsf{bck}}(\bot)$$

$$= \left(\frac{r}{N}\right)^n.$$

$$(17)$$

From (11), (16), and (17) we derive

$$\sum_{r=0}^{q-1} \mathbf{Ex}[\chi^2(S^r)] \le \sum_{r=0}^{q-1} \max\left\{ \frac{3N-4r}{4N(N-r)}, \frac{\left(\frac{r}{N}\right)^{2n}}{\left(1-\left(\frac{r}{N}\right)^n\right)} \right\} + \left(\frac{r}{N}\right)^n$$

$$\le q \times \left(\max\left\{ \frac{3}{4(N-q)}, \frac{\left(\frac{q}{N}\right)^{2n}}{\left(1-\left(\frac{q}{N}\right)^n\right)} \right\} + \left(\frac{q}{N}\right)^n \right).$$

For $q < \frac{N}{2}$ we have the following bounds,

$$\frac{3}{4(N-q)} < \frac{3}{2N}, \quad \frac{\left(\frac{q}{N}\right)^{2n}}{\left(1-\left(\frac{q}{N}\right)^n\right)} < \frac{1}{N(N-1)}, \quad \text{and} \quad \left(\frac{q}{N}\right)^n < \frac{1}{N}.$$

Hence, we have for the backward query

$$\sum_{r=0}^{q-1} \mathbf{Ex}[\chi^2(S^r)] \le \frac{2.5q}{N}. \tag{18}$$

Finally, we get the following upper bound on $\mathsf{Adv}_{\mathsf{XORP},\$}^{\mathrm{diff}}(q)$.

$$\mathsf{Adv}_{\mathsf{XORP},\$}^{\mathrm{diff}}(q) = \|S^q - T^q\| \tag{19}$$

$$\le \sqrt{\frac{1}{2}\sum_{r=0}^{q-1} \mathbf{Ex}[\chi^2(S^r)]} \tag{20}$$

$$\le \sqrt{\frac{1.25q}{N}}, \tag{21}$$

where (19) is from the definition of $\mathsf{Adv}_{\mathsf{XORP},\$}^{\mathrm{diff}}(q)$. (20) is given by (4) and (21) is given by the maximum of (10) and (18) (which is (18)) for the q queries. □

5   Auxiliary Proofs

In this section we state and prove Lemmas 1 and 2. We begin with Lemma 1 where we work with the same notation and setting of the Forward Query part of the proof of Theorem 2.

5.1   Proof of Lemma 1

Lemma 1. *With the notation of Theorem 2,*

$$\mathbf{Ex}[\mathsf{W}] = \frac{2rN-r^2}{N}, \quad \text{and} \quad \mathbf{Var}[\mathsf{W}] \le \frac{r^2}{N}.$$

Proof. When Z^r is chosen according to the distribution $\mathsf{p}_0^{\mathsf{fwd}}$ the sets $\{U_0^r\}$ and $\{U_1^r\}$ are two random subsets (sampled independently) of $\{0,1\}^n$ of cardinality r. Also, in keeping with the notation of Theorem 2, we assume x_i to be a fixed element of $\{0,1\}^n$. Now, for each $g \in \{0,1\}^n$, we define an indicator random variable I_g as follows.

$$I_g = \begin{cases} 1 \text{ if } g \in \{U_0^r\} \text{ and } g \oplus x_i \in \{U_1^r\} \\ 0 \text{ otherwise.} \end{cases}$$

Therefore,

$$\mathbf{Ex}[I_g] = \mathbf{Pr}[I_g = 1] = \mathbf{Pr}[g \in \{U_0^r\} \wedge g \oplus x_i \in \{U_1^r\}] = \frac{r}{N} \times \frac{r}{N} = \frac{r^2}{N^2}. \quad (22)$$

Also,

$$W = 2r - \sum_{g \in \{0,1\}^n} I_g.$$

Thus,

$$\mathbf{Ex}[W] = 2r - \mathbf{Ex}\left[\sum_{g \in \{0,1\}^n} I_g\right] = 2r - \sum_{g \in \{0,1\}^n} \mathbf{Ex}[I_g] = \frac{2rN - r^2}{N}.$$

Next, to calculate $\mathbf{Var}[W]$ we use the following relationship.

$$\mathbf{Var}[W] = \mathbf{Var}\left[\sum_{g \in \{0,1\}^n} I_g\right] = \sum_{g \in \{0,1\}^n} \mathbf{Var}[I_g] + \sum_{g \neq h \in \{0,1\}^n} \mathbf{Cov}[I_g, I_h].$$

$\mathbf{Var}[I_g]$ is straightforward to calculate from the defintion;

$$\mathbf{Var}[I_g] = \mathbf{Ex}[I_g^2] - \mathbf{Ex}[I_g]^2 = \mathbf{Ex}[I_g](1 - \mathbf{Ex}[I_g])$$
$$= \frac{r^2}{N^2} \times \left(1 - \frac{r^2}{N^2}\right)$$
$$< \frac{r^2}{N^2}.$$

From the definition, $\mathbf{Cov}[I_g, I_h]$ is given by $\mathbf{Cov}[I_g, I_h] = \mathbf{Ex}[I_g I_h] - \mathbf{Ex}[I_g]\mathbf{Ex}[I_h]$. Since $\mathbf{Ex}[I_g] = \mathbf{Ex}[I_h] = \frac{r^2}{N^2}$ is given by (22), the task reduces to the calculation of $\mathbf{Ex}[I_g I_h]$ which we consider below.

$$\mathbf{Ex}[I_g I_h] = \mathbf{Pr}[I_g = 1 \wedge I_h = 1]$$
$$= \mathbf{Pr}[g \in \{U_0^r\} \wedge h \in \{U_0^r\} \wedge g \oplus x_i \in \{U_1^r\} \wedge h \oplus x_i \in \{U_1^r\}]$$
$$= \mathbf{Pr}[g \in \{U_0^r\} \wedge h \in \{U_0^r\}] \times \mathbf{Pr}[g \oplus x_i \in \{U_1^r\} \wedge h \oplus x_i \in \{U_1^r\}]$$
$$= \frac{r(r-1)}{N(N-1)} \times \frac{r(r-1)}{N(N-1)}$$
$$= \left(\frac{r(r-1)}{N(N-1)}\right)^2$$

Therefore, $\mathbf{Cov}[I_g, I_h] = \left(\frac{r(r-1)}{N(N-1)}\right)^2 - \left(\frac{r^2}{N^2}\right)^2 < 0$. This implies that

$$\mathbf{Var}[\mathsf{W}] < N \times \frac{r^2}{N^2} = \frac{r^2}{N}.$$

This finishes the proof of the lemma.   □

5.2   Proof of Lemma 2

Lemma 2. *With the notation of Theorem 2, the following bounds hold for* $\mathsf{p}_1^{bck}(s_i \mid s^r)$.

$$\frac{1}{(N-r)^2} \times \left(1 - \left(\frac{r}{N}\right)^n\right) \leq \mathsf{p}_1^{bck}(z_i \mid z^r) \leq \frac{4}{N(N-r)}.$$

Proof. The lower bound is justified as follows.

$$\mathsf{p}_1^{bck}(z_i \mid z^r) = \sum_{\ell=1}^{n} \frac{1}{N} \times \frac{1}{N-r-\ell+1} \times \left(\frac{r}{N}\right)^{\ell-1}$$

$$\geq \frac{1}{N(N-r)} \times \sum_{\ell=1}^{n} \left(\frac{r}{N}\right)^{\ell-1}$$

$$= \frac{1}{N(N-r)} \times \frac{1 - \left(\frac{r}{N}\right)^n}{1 - \left(\frac{r}{N}\right)}$$

$$= \frac{1}{(N-r)^2} \times \left(1 - \left(\frac{r}{N}\right)^n\right).$$

For the upper bound, we get

$$\mathsf{p}_1^{bck}(z_i \mid z^r) = \sum_{\ell=1}^{n} \frac{1}{N} \times \frac{1}{N-r-\ell+1} \times \left(\frac{r}{N}\right)^{\ell-1} \leq \frac{4}{N^2} \times \sum_{\ell=1}^{\infty} \left(\frac{r}{N}\right)^{\ell-1} \quad (23)$$

$$= \frac{4}{N(N-r)}.$$

The first term on the r.h.s. of (23) follows by noting that $r < q < \frac{N}{2}$ and $\ell < n = \log N \leq \frac{N}{4}$, for $N \geq 16$.   □

6   Extension to the Xor of k Permutations

In this section, we apply our main result (Theorem 2) to show full indifferentiable security of the XORP$[k]$ construction for any k. Following Theorem 2, it is sufficient to consider XORP$[k]$ with $k \geq 3$. In particular, our result is the following.

Theorem 3. *Let $N \geq 16$ and $q < \frac{N}{2}$. Then, there exists a simulator S' for $\mathsf{XORP}[k]$, $k \geq 3$, such that for any adversary $\mathscr{A}'$, there exists an adversary $\mathscr{A}$ with*

$$\mathsf{Adv}^{\mathrm{diff}}_{\mathsf{XORP}[k],\$}(\mathscr{A}') = \mathsf{Adv}^{\mathrm{diff}}_{\mathsf{XORP},\$}(\mathscr{A})$$

and hence, $\mathsf{Adv}^{\mathrm{diff}}_{\mathsf{XORP}[k],\$}(q) \leq \sqrt{\frac{1.25q}{N}}$.

Proof. The indifferentiable security analysis of $\mathsf{XORP}[k]$ follows a reduction technique which is similar to the technique used in [MP15] to prove PRF-security of $\mathsf{XORP}[k]$ in the indistinguishability setting. However, in our case, we additionally need to consider the simulator S'.

Brief description of S'. First, we recall the simulator S for XORP from Sect. 3. The simulator S' works almost the same way as S works. It first samples $(k-2)$ independent random permutations $\Pi_2, \ldots, \Pi_{k-1}$. Note that the sampling can be simulated as a *lazy sampling* in an efficient manner instead of sampling the whole permutations at a time.

In case of a forward or backward query (x, i), with $i \geq 2$, S' responds honestly (*i.e.*, it uses its own sampled random permutation as mentioned above). When $i \in \{0, 1\}$, it behaves exactly in the same way as S except that it computes $\$'(x) = \$(x) \oplus \Pi_2(x) \oplus \cdots \Pi_{k-1}(x)$ and then applies Step 5 of $\mathsf{SIM}_{\mathsf{FWD}}$ (see Fig. 2) in case of a forward query, or Step 7 of $\mathsf{SIM}_{\mathsf{BCK}}$ (see Fig. 3) in case of a backward query.

Next, we describe the reduction for the adversaries. Suppose there is an adversary $\mathscr{A}'$ against $\mathsf{XORP}[k]$ and consider the simulator S' defined above. Now, we construct an adversary $\mathscr{A}$ against XORP and the simulator S. The adversary $\mathscr{A}$ first stores the permutations $\Pi_2, \ldots, \Pi_{k-1}$ (again using lazy sampling to make those efficient). Next, $\mathscr{A}$ runs the algorithm $\mathscr{A}'$ which can make two types of queries, namely (a) primitive or simulator queries and (b) construction or random function queries. Below, we consider these two types of queries.

(a) In case of a primitive or simulator query (x, i) (either forward or backward), $\mathscr{A}$ first checks whether $i \geq 2$ or not. If $i \geq 2$, then $\mathscr{A}$ can simulate the response on its own, *i.e.*, it computes $\Pi_i(x)$ or $\Pi_i^{-1}(x)$, where $\Pi_i \in \{\Pi_2, \ldots, \Pi_{k-1}\}$, and sends the output back to $\mathscr{A}'$. If $i = 0$ or 1, then $\mathscr{A}$ forwards the query to its simulator/primitive oracle and whatever response it gets again forwards to $\mathscr{A}'$; so, it basically relays the queries and responses.

(b) In case of a construction or random function query, $\mathscr{A}$ forwards the query to its corresponding construction/random function oracle. Suppose $\mathscr{A}$ gets Z as a response. Then it computes $Z' = Z \oplus \bigoplus_{i=2}^{k-1} \Pi_i(x)$, and sends Z' back to $\mathscr{A}'$.

Note that $\mathscr{A}$ is actually interacting with $(\mathsf{XORP}, (\Pi_0, \Pi_1, \Pi_0^{-1}, \Pi_1^{-1}))$, whereas the interaction interface of $\mathscr{A}'$ is equivalent to

$$(\mathsf{XORP}[k], (\Pi_0, \ldots, \Pi_{k-1}, \Pi_0^{-1}, \ldots, \Pi_{k-1}^{-1})).$$

Now, assume that $\mathscr{A}$ is interacting with $(\$, \mathsf{S})$, the interaction interface of $\mathscr{A}'$ is then equivalent to $(\$ \oplus \mathsf{XORP}[k-2], \mathsf{S}')$. It is easy to see the correctness of the first oracle as $\mathscr{A}'$ xors his computation of $\mathsf{XORP}[k-2]$ with the output of $\$$. Similarly, one can show the simulator interface of $\mathscr{A}'$ is S'. Note that $\$ \oplus \mathsf{XORP}[k-2]$ is completely independent of $\mathsf{XORP}[k-2]$, and we can consider it as another independent random function $\$'$. Thus, the interface of $\mathscr{A}'$ is equivalent to $(\$', \mathsf{S}')$. So, $\mathscr{A}$ perfectly simulates the real and the ideal world of $\mathscr{A}'$. Therefore, $\mathsf{Adv}^{\text{diff}}_{\mathsf{XORP}[k],\$}(\mathscr{A}') = \mathsf{Adv}^{\text{diff}}_{\mathsf{XORP},\$}(\mathscr{A})$. By Theorem 2, we finally have

$$\mathsf{Adv}^{\text{diff}}_{\mathsf{XORP}[k],\$}(q) \leq \sqrt{\frac{1.25q}{N}}. \qquad \qquad \square$$

7 Conclusion

Proving full security of XORP construction in the secret or public permutation model (*i.e.*, indifferentiable security) is a challenging problem. Recently, Dai et al. introduced a method, called the χ^2 method, using which they were able to obtain full PRF-security of XORP in the secret random permutation model. The full security in the public permutation model for this construction was an open problem. In this paper, we apply the χ^2 method to the XORP construction to prove its full indifferentiable security. We believe this method can also be used for other cryptographic constructions for which the full security is not known.

Here, we also remark that though our bound shows full (*i.e.*, n-bit) indifferentiable security of XORP and XORP[k], in practice (*i.e.*, for realistic setting of parameters), it does not lead to full n-bit security (mainly due to the presence of the square root in the bound). As an immediate goal, it will be interesting to investigate if a more sophisticated application of the χ^2 method can get rid of the square root in our bound.

Acknowledgement. We are indebted to the reviewers for their patient reading and valuable comments which improved the quality of this paper significantly.

This work is supported in part by the WISEKEY project, which we gratefully acknowledge.

References

[AMP10] Andreeva, E., Mennink, B., Preneel, B.: On the indifferentiability of the Grøstl hash function. In: Garay, J.A., De Prisco, R. (eds.) SCN 2010. LNCS, vol. 6280, pp. 88–105. Springer, Heidelberg (2010). https://doi.org/10.1007/978-3-642-15317-4_7

[BDP+13] Bertoni, G., Daemen, J., Peeters, M., Van Assche, G., NIST, G.: Keccak and the SHA-3 Standardization (2013)

[BDPVA08] Bertoni, G., Daemen, J., Peeters, M., Van Assche, G.: On the indifferentiability of the sponge construction. In: Smart, N. (ed.) EUROCRYPT 2008. LNCS, vol. 4965, pp. 181–197. Springer, Heidelberg (2008). https://doi.org/10.1007/978-3-540-78967-3_11

[BDPVA11a] Bertoni, G., Daemen, J., Peeters, M., Van Assche, G.: Duplexing the sponge: single-pass authenticated encryption and other applications. In: Miri, A., Vaudenay, S. (eds.) SAC 2011. LNCS, vol. 7118, pp. 320–337. Springer, Heidelberg (2012). https://doi.org/10.1007/978-3-642-28496-0_19

[BDPVA11b] Bertoni, G., Daemen, J., Peeters, M., Van Assche, G.: On the security of the keyed sponge construction. In: Symmetric Key Encryption Workshop (SKEW 2011) (2011)

[BI99] Bellare, M., Impagliazzo, R.: A tool for obtaining tighter security analyses of pseudorandom function based constructions, with applications to PRP to PRF conversion. IACR Cryptol. ePrint Arch. **1999**, 24 (1999)

[BKR98] Bellare, M., Krovetz, T., Rogaway, P.: Luby-Rackoff backwards: Increasing security by making block ciphers non-invertible. In: Nyberg, K. (ed.) EUROCRYPT 1998. LNCS, vol. 1403, pp. 266–280. Springer, Heidelberg (1998). https://doi.org/10.1007/BFb0054132

[BMN10] Bhattacharyya, R., Mandal, A., Nandi, M.: Security analysis of the mode of JH hash function. In: Hong, S., Iwata, T. (eds.) FSE 2010. LNCS, vol. 6147, pp. 168–191. Springer, Heidelberg (2010). https://doi.org/10.1007/978-3-642-13858-4_10

[BN18] Bhattacharya, S., Nandi, M.: Revisiting variable output length pseudorandom functions. IACR Trans. Symmetric Cryptol. **2018**(1) (2018, to appear)

[CAE] CAESAR: Competition for Authenticated Encryption: Security, Applicability, and Robustness. http://competitions.cr.yp.to/caesar.html/

[CLP14] Cogliati, B., Lampe, R., Patarin, J.: The indistinguishability of the XOR of k permutations. In: Cid, C., Rechberger, C. (eds.) FSE 2014. LNCS, vol. 8540, pp. 285–302. Springer, Heidelberg (2015). https://doi.org/10.1007/978-3-662-46706-0_15

[CS16a] Cogliati, B., Seurin, Y.: EWCDM: an efficient, beyond-birthday secure, nonce-misuse resistant MAC. In: Robshaw, M., Katz, J. (eds.) CRYPTO 2016. LNCS, vol. 9814, pp. 121–149. Springer, Heidelberg (2016). https://doi.org/10.1007/978-3-662-53018-4_5

[CT06] Cover, T.M., Thomas, J.A.: Elements of Information Theory (Wiley Series in Telecommunications and Signal Processing), Wiley-Interscience (2006)

[DHT17] Dai, W., Hoang, V.T., Tessaro, S.: Information-theoretic indistinguishabilityvia the chi-squared method. In: Katz and Shacham [KS17], pp. 497–523 (2017)

[GG16] Gilboa, S., Gueron, S.: The Advantage of Truncated Permutations, CoRR abs/1610.02518 (2016)

[GGM17] Gilboa, S., Gueron, S., Morris, B.: How many queries are needed to distinguish a truncated random permutation from a random function? J. Cryptol. **31**(1), 162–171 (2017)

[GKM+09] Gauravaram, P., Knudsen, L.R., Matusiewicz, K., Mendel, F., Rechberger, C., Schläffer, M., Thomsen, S.S.: Grøstl-a SHA-3 candidate. In: Dagstuhl Seminar Proceedings, Schloss Dagstuhl-Leibniz-Zentrum für Informatik (2009)

[IMPS17] Iwata, T., Minematsu, K., Peyrin, T., Seurin, Y.: ZMAC: a fast tweakable block cipher mode for highly secure message authentication. IACR Cryptol. ePrint Arch. **2017**, 535 (2017)

[IMV16] Iwata, T., Mennink, B., Vizár, D.: CENC is optimally secure. IACR Cryptol. ePrint Arch. **2016**, 1087 (2016)

[Iwa06] Iwata, T.: New blockcipher modes of operation with beyond the birthday bound security. In: Robshaw, M. (ed.) FSE 2006. LNCS, vol. 4047, pp. 310–327. Springer, Heidelberg (2006). https://doi.org/10.1007/11799313_20

[KS17]   Katz, J., Shacham, H. (eds.): CRYPTO 2017. LNCS, vol. 10403. Springer, Cham (2017). https://doi.org/10.1007/978-3-319-63697-9

[Lee17] Lee, J.: Indifferentiability of the sum of random permutations towards optimal security. IEEE Trans. Inf. Theory **63**(6), 4050–4054 (2017)

[LR88]   Luby, M., Rackoff, C.: How to construct pseudorandom permutations from pseudorandom functions. SIAM J. Comput. **17**(2), 373–386 (1988)

[Luc00] Lucks, S.: The sum of PRPs Is a secure PRF. In: Preneel, B. (ed.) EUROCRYPT 2000. LNCS, vol. 1807, pp. 470–484. Springer, Heidelberg (2000). https://doi.org/10.1007/3-540-45539-6_34

[LV87]   Liese, F., Vajda, I.: Convex Statistical Distances. Teubner, Leipzig (1987)

[MN17a] Mennink, B., Neves, S.: Encrypted Davies-Meyer and its dual: Towards optimal security using Mirror theory, Cryptology ePrint Archive, Report 2017/xxx, to be published in CRYPTO 2017 (2017). http://eprint.iacr.org/2017/537

[MN17b] Mennink, B., Neves, S.: Encrypted davies-meyer and its dual: towards optimal security using mirror theory. In: Katz and Shacham [KS17], pp. 556–583 (2017)

[MP15]   Mennink, B., Preneel, B.: On the XOR of multiple random permutations. In: Malkin, T., Kolesnikov, V., Lewko, A.B., Polychronakis, M. (eds.) ACNS 2015. LNCS, vol. 9092, pp. 619–634. Springer, Cham (2015). https://doi.org/10.1007/978-3-319-28166-7_30

[MPN10] Mandal, A., Patarin, J., Nachef, V.: Indifferentiability beyond the birthday bound for the xor of two public random permutations. In: Gong, G., Gupta, K.C. (eds.) INDOCRYPT 2010. LNCS, vol. 6498, pp. 69–81. Springer, Heidelberg (2010). https://doi.org/10.1007/978-3-642-17401-8_6

[MRH04] Maurer, U., Renner, R., Holenstein, C.: Indifferentiability, impossibility results on reductions, and applications to the random oracle methodology. In: Naor, M. (ed.) TCC 2004. LNCS, vol. 2951, pp. 21–39. Springer, Heidelberg (2004). https://doi.org/10.1007/978-3-540-24638-1_2

[Pat08a] Patarin, J.: The "Coefficients H" technique. In: Avanzi, R.M., Keliher, L., Sica, F. (eds.) SAC 2008. LNCS, vol. 5381, pp. 328–345. Springer, Heidelberg (2009). https://doi.org/10.1007/978-3-642-04159-4_21

[Pat08b] Patarin, J.: A proof of security in $O(2^n)$ for the xor of two random permutations. In: Safavi-Naini, R. (ed.) ICITS 2008. LNCS, vol. 5155, pp. 232–248. Springer, Heidelberg (2008). https://doi.org/10.1007/978-3-540-85093-9_22

[Pat10] Patarin, J.: Introduction to Mirror Theory: Analysis of Systems of Linear Equalities and Linear Non Equalities for Cryptography. Cryptology ePrint Archive, Report 2017/287 (2010). http://eprint.iacr.org/2010/287

[RAB+08] Rivest, R.L., Agre, B., Bailey, D.V., Crutchfield, C., Dodis, Y., Fleming, K.E., Khan, A., Krishnamurthy, J., Lin, Y., Reyzin, L., et al.: The MD6 hash function-a proposal to NIST for SHA-3. NIST **2**(3) (2008, submitted)

[Sta78] Stam, A.J.: Distance between sampling with and without replacement. Statistica Neerlandica **32**(2), 81–91 (1978)

[Vau03] Vaudenay, S.: Decorrelation: a theory for block cipher security. J. Cryptol. **16**(4), 249–286 (2003)

[Wu11] Wu, H.: The hash function JH, NIST (round 3), 6 (2011, submitted)

[Yas11] Yasuda, K.: A new variant of PMAC: beyond the birthday bound. In: Rogaway, P. (ed.) CRYPTO 2011. LNCS, vol. 6841, pp. 596–609. Springer, Heidelberg (2011). https://doi.org/10.1007/978-3-642-22792-9_34

An Improved Affine Equivalence Algorithm for Random Permutations

Itai Dinur$^{(\boxtimes)}$

Department of Computer Science, Ben-Gurion University, Beersheba, Israel
dinuri@cs.bgu.ac.il

Abstract. In this paper we study the affine equivalence problem, where given two functions $F, G : \{0,1\}^n \to \{0,1\}^n$, the goal is to determine whether there exist invertible affine transformations A_1, A_2 over $GF(2)^n$ such that $G = A_2 \circ F \circ A_1$. Algorithms for this problem have several well-known applications in the design and analysis of Sboxes, cryptanalysis of white-box ciphers and breaking a generalized Even-Mansour scheme.

We describe a new algorithm for the affine equivalence problem and focus on the variant where F, G are permutations over n-bit words, as it has the widest applicability. The complexity of our algorithm is about $n^3 2^n$ bit operations with very high probability whenever F (or G) is a random permutation. This improves upon the best known algorithms for this problem (published by Biryukov et al. at EUROCRYPT 2003), where the first algorithm has time complexity of $n^3 2^{2n}$ and the second has time complexity of about $n^3 2^{3n/2}$ and roughly the same memory complexity.

Our algorithm is based on a new structure (called a *rank table*) which is used to analyze particular algebraic properties of a function that remain invariant under invertible affine transformations. Besides its standard application in our new algorithm, the rank table is of independent interest and we discuss several of its additional potential applications.

Keywords: Affine equivalence problem · Block cipher
Even-Mansour cipher · Cryptanalysis · Rank table

1 Introduction

In the affine equivalence problem, the input consists of two functions F, G and the goal is to determine whether they are affine equivalent, and if so, output the equivalence relations. More precisely, if there exist invertible affine transformations (over some field) A_1, A_2 such that $G = A_2 \circ F \circ A_1$, output A_1, A_2. Otherwise, assert that F, G are not affine equivalent.

Variants of the affine equivalence problem have been studied in several branches of mathematics and are relevant to both asymmetric and symmetric cryptography. In the context of asymmetric cryptography, the problem was

The author was supported in part by the Israeli Science Foundation through grant No. 573/16.

J. B. Nielsen and V. Rijmen (Eds.): EUROCRYPT 2018, LNCS 10820, pp. 413–442, 2018.
https://doi.org/10.1007/978-3-319-78381-9_16

first formalized by Patarin [17] and referred to as isomorphism of polynomials. In this setting F, G are typically of low algebraic degree (mainly quadratic) over some field.

The focus of this work is on the affine equivalence variant in which F, G map between n-bit words and the affine transformations A_1, A_2 are over $GF(2)^n$. This variant is mostly relevant in several contexts of symmetric-key cryptography. In particular, it is relevant to the classification and analysis of Sboxes (see [6,14]) as affine equivalent Sboxes share differential, linear and several algebraic properties (refer to [7] for recent results on this subject). Moreover, algorithms for the affine equivalence problem were applied in [3] to generate equivalent representations of AES and other block ciphers. These algorithms also have cryptanalytic applications and were used to break white-box ciphers (e.g., in [15]). Additionally, solving the affine equivalence problem can be viewed as breaking a generalization of the Even-Mansour scheme [11], which has received substantial attention from the cryptographic community in recent years. The original scheme builds a block cipher from a public permutation F using two n-bit keys k_1, k_2 and its encryption function is defined as $E(p) = F(p + k_1) + k_2$ (where addition is over $GF(2)^n$). The generalized Even-Mansour scheme replaces the key additions with secret affine mappings and breaking it reduces to solving the affine equivalence problem, as originally described in [3].

The best known algorithms for the affine equivalence problem were presented by Biryukov et al. at EUROCRYPT 2003 [3]. The main algorithm described in [3] has complexity of about $n^3 2^{2n}$ bit operations, while a secondary algorithm has time complexity of about $n^3 2^{3n/2}$, but also uses about the same amount of memory.[1] Besides its high memory consumption, another disadvantage of the secondary algorithm of [3] is that it cannot be used to prove that F and G are not affine equivalent.

In this paper we devise a new algorithm for the affine equivalence problem whose complexity is about $n^3 2^n$ bit operations with very high probability whenever F (or G) is chosen uniformly at random from the set of all permutations on n-bit words. Our algorithm is also applicable without any modification to arbitrary functions (rather than permutations) and seems to perform similarly on random functions. However we focus on permutations as almost all applications actually require solving the affine equivalence problem for permutations. Since our algorithm can be used to prove that F and G are not affine equivalent, it does not share the disadvantage of the secondary algorithm of [3].

As a consequence of our improved algorithm, we solve within several minutes affine equivalence problem instances of size up to $n = 28$ on a single core. Optimizing our implementation and exploiting parallelism would most likely allow solving instances of size at least $n = 40$ using an academic budget. Such instances are out of reach of all previous algorithms for the problem.

---

[1] Biryukov et al. also described a more efficient algorithm of complexity $n^3 2^n$ for the linear equivalence problem, which is a restricted variant of the affine equivalence problem.

Technically, the main algorithm devised in [3] for the affine equivalence problem is a guess-and-determine algorithm (which is related to the "to and fro" algorithm of [18], devised to solve the problem of isomorphism of polynomials) whereas the secondary algorithm is based on collision search (it generalizes Daemen's attack on the original Even-Mansour cipher [8]). On the other hand, algorithms that use algebraic techniques (such as [5]) are mainly known for the asymmetric variant, in which F, G are functions of low degree, and it is not clear how to adapt them to arbitrary functions.

In contrast to previous algorithms, our approach involves analyzing algebraic properties of F, G which are of high algebraic degree. More specifically, we are interested in the polynomial representation (algebraic normal form or ANF) of each of the n output bits of F (and G) as a Boolean function in the n input bits. In fact, we are mainly interested in "truncated" polynomials that include only monomials of degree at least d (in particular, we choose $d = n - 2$). Each such polynomial can be viewed a vector in a vector space (with the standard basis of all monomials of degree at least $d = n - 2$). Therefore we can define the rank of the set of n truncated polynomials for each F, G as the rank of the matrix formed by arranging these polynomials as row vectors. In other words, we associate a rank value (which is an integer between 0 and n) to F (and to G) by computing the rank of its n truncated polynomials (derived from its n output bits) as vectors. We first show that if F, G are affine equivalent, their associated ranks are equal.[2]

To proceed, we analyze F, G independently. We derive from F several functions, each one defined by restricting its 2^n inputs to an affine subspace of dimension $n - 1$. Since each such derived function (restricted to an affine subspace) has an associated rank, we assign to each possible $(n - 1)$-dimensional subspace a corresponding rank. As there are 2^{n+1} possible affine subspaces (such a subspace can be characterized using its orthogonal subspace by a single linear expression over n variables and a free coefficient), we obtain 2^{n+1} rank values for F. These values are collected in the *rank table* of F, where a rank table entry r stores the set of all affine subspaces (more precisely, their compact representations as linear expressions) assigned to rank r.[3]

The main idea of the algorithm is to compute the rank tables of both F and G and then use these tables (and additional more complex structures derived from them) to recover the (unknown) affine transformation A_1, assuming that $G = A_2 \circ F \circ A_1$. In essence, the rank tables allow us to recover *matchings* between $(n - 1)$-dimensional affine subspaces that are defined by A_1: an affine subspace S in matched with S' if A_1 transforms S to the affine subspace S'. Each such matching between S and S' reveals information about A_1 in the form of linear equations. Hence we aim to use the rank tables to recover sufficiently many such matchings and compute A_1 using linear algebra. Once A_1 is derived,

---

[2] The choice of $d = n - 2$ seems arbitrary at this stage. We only note that choosing a larger or smaller value for d typically results in a rank value which is constant for almost all functions, providing no information about them.

[3] The formal definition of a rank table is slightly different.

computing A_2 is trivial. The main property of the rank table that we prove and exploit to recover the matchings is that if S in matched with S', then S appears in the rank table of G in the same entry r as S' appears in the rank table of F.

Since the number of $(n-1)$-dimensional affine subspaces is 2^{n+1}, each containing 2^{n-1} elements, a naive approach to computing the rank table (which works independently on each subspace) has complexity of at least $2^{n+1} \cdot 2^{n-1} = 2^{2n}$. However, using symbolic computation of polynomials, we show how to reduce this complexity to about $n^3 2^n$ bit operations. While this computational step is easy to analyze, this is not the case for the overall algorithm's performance. Indeed, its success probability and complexity depend on the monomials of degree at least $n - 2$ of F and G. In particular, if all n output bits of F and G are functions of degree $n - 3$ or lower, they do not contain any such monomials. As a result, all affine subspaces for F and G are assigned rank zero and the rank tables of these functions contain no useful information, leading to failure of the algorithm.[4]

When F (or G) is chosen uniformly at random from the set of all possible n-bit permutations (or n-bit functions in general), the case that its algebraic degree is less than $n - 2$ is extremely unlikely for $n \geq 8$. Nevertheless, rigorous analysis of the algorithm seems challenging as its performance depends on subtle algebraic properties of random permutations. To deal with this situation, we make a heuristic assumption about the distribution of high degree monomials in random permutations which enables us to use well-known results regarding the rank distribution of random Boolean matrices. Consequently, we derive the distribution of the sizes of the rank table entries for a random permutation. This distribution and additional properties enable us to show that asymptotically the algorithm succeeds with probability close to 1 in complexity of about $n^3 2^n$ bit operations. This heuristic analysis is backed up by thousands of experiments on various problem instances of different sizes. Rigorously analyzing the algorithm and extending it to succeed on all functions (or permutations) with probability 1 in the same complexity remain open problems.

The properties of the rank table and the algorithm for computing it are of independent interest. In particular, we propose methods to build experimental distinguishers for block ciphers based on the rank table and a method to efficiently detect high-order differential distinguishers based on the algorithm for its computation. Furthermore, our techniques are relevant to decomposition attacks on the white-box ASASA block cipher instances proposed by Biryukov et al. [2]. In this application, we adapt the algorithm for computing the rank table in order to improve the complexity of the integral attack on ASASA published in [10] from $2^{3n/2}$ to about 2^n (where n is the block size of the instance).

The rest of the paper is organized as follows. In Sect. 2 we describe some preliminaries and give an overview of the new affine equivalence algorithm in Sect. 3. In Sect. 4 we prove the basic property of rank equality for affine equivalent functions, while in Sect. 5 we define and analyze the matching between

---

[4] In case the algorithm fails, one can try to apply it to F^{-1} and G^{-1} which may be of higher algebraic degree.

$(n - 1)$-dimensional affine subspaces that we use to recover A_1. In Sect. 6 we define the rank table and additional objects used in our algorithm, and describe the relation between these objects for affine equivalent functions. In Sect. 7 we analyze properties of rank tables for random permutations under our heuristic assumption. Then, we describe and analyze the new affine equivalence algorithm in Sect. 8. Next, in Sect. 9, we describe applications of the new algorithm and the rank table structure. Finally, we conclude the paper in Sect. 10.

2 Preliminaries

For a finite set R, denote by $|R|$ its size. Given a vector $u = (u[1], \ldots, u[n]) \in GF(2)^n$, let $wt(u)$ denote its Hamming weight. Throughout this paper, addition between vectors $u_1, u_2 \in GF(2)^n$ is performed bit-wise over $GF(2)^n$.

Multivariate Polynomials. Any Boolean function $F : \{0,1\}^n \rightarrow \{0,1\}$ can be represented as a multivariate polynomial whose algebraic normal form (ANF) is unique and given as $F(x[1], \ldots, x[n]) = \sum_{u=(u[1],\ldots,u[n]) \in \{0,1\}^n} \alpha_u M_u$, where $\alpha_u \in \{0,1\}$ is the coefficient of the monomial $M_u = \prod_{i=1}^n x[i]^{u[i]}$, and the sum is over $GF(2)$. The algebraic degree of the function F is defined as $deg(F) = max\{wt(u) \mid \alpha_u \neq 0\}$.

In several cases it will be more convenient to directly manipulate the representation of F as a multivariate polynomial $P(x[1], \ldots, x[n]) = \sum_{u \in \{0,1\}^n} \alpha_u M_u$. Note that unlike F, the polynomial P is not treated as a function but rather as a symbolic object. $P(x[1], \ldots, x[n])$ can be viewed as a vector in the vector space spanned by the set of all monomials $\{M_u \mid u \in \{0,1\}^n\}$.

Given a multivariate polynomial $P(x[1], \ldots, x[n]) = \sum_{u \in \{0,1\}^n} \alpha_u M_u$ and a positive integer d, define $P_{(\geq d)}$ by taking all the monomials of P of degree at least d, namely, $P_{(\geq d)}(x[1], \ldots, x[n]) = \sum_{u \in \{0,1\}^n \wedge wt(u) \geq d} \alpha_u M_u$. Note that $P_{(\geq d)}$ can be represented using at most $\sum_{i=d}^n \binom{n}{i}$ non-zero coefficients.

Given a function $F : \{0,1\}^n \rightarrow \{0,1\}$ represented by a polynomial $P(x[1], \ldots, x[n])$, define $F_{(\geq d)} : \{0,1\}^n \rightarrow \{0,1\}$ as the function represented by $P_{(\geq d)}$.

Vectorial Functions and Polynomials. Given a vectorial Boolean function $\boldsymbol{F} : \{0,1\}^n \rightarrow \{0,1\}^m$, let $F^{(i)} : \{0,1\}^n \rightarrow \{0,1\}$ denote the Boolean function of its i'th output bit.

We say that a sequence of m polynomials $\boldsymbol{P} = \{P^{(i)}(x[1], \ldots, x[n])\}_{i=1}^m$ represents $\boldsymbol{F}$ if for each $i \in \{1, 2, \ldots, m\}$, the i'th polynomial $P^{(i)}$ represents $F^{(i)}$.

Given a positive integer d, denote $\boldsymbol{P}_{(\geq d)} = \{P_{(\geq d)}^{(i)}(x[1], \ldots, x[n])\}_{i=1}^m$. The vectorial function $\boldsymbol{F}_{(\geq d)} : \{0,1\}^n \rightarrow \{0,1\}^m$ is defined analogously.

The algebraic degree $deg(\boldsymbol{P})$ of $\boldsymbol{P}$ is defined as the maximal degree of its m polynomials. The algebraic degree $deg(\boldsymbol{F})$ is defined analogously.

As each $P^{(i)}$ can be viewed as a vector in a vector space, we define the *symbolic rank* of P as the rank of the m vectors $\{P^{(i)}\}_{i=1}^{m}$. We denote the symbolic rank of P as $SR(P)$. Note that $SR(P) \in \mathbb{Z}_{m+1}$.

Affine Transformations and Affine Equivalence. An affine transformation $A : \{0,1\}^m \rightarrow \{0,1\}^n$ over $GF(2)^m$ is defined using a Boolean matrix $L_{n \times m}$ and a word $a \in \{0,1\}^n$ as $A(x) = L(x) + a$ ($L(x)$ is simply matrix multiplication). The transformation is invertible if $m = n$ and L is an invertible matrix. If $a = 0$, then the A is called a linear transformation (such functions are a subclass of affine functions).

Two functions $F : \{0,1\}^n \rightarrow \{0,1\}^m$, $G : \{0,1\}^n \rightarrow \{0,1\}^m$ are *affine equivalent* is there exist two invertible affine transformations $A_1 : \{0,1\}^n \rightarrow \{0,1\}^n$ and $A_2 : \{0,1\}^m \rightarrow \{0,1\}^m$ such that $G = A_2 \circ F \circ A_1$. It is easy to show that the affine equivalence relation partitions the set of all functions into (affine) equivalence classes. We denote $F \equiv G$ if F is affine equivalent to G.

Symbolic Composition. Given $P = \{P^{(i)}(x[1], \ldots, x[n])\}_{i=1}^{m}$, and an affine function $A_1 : \{0,1\}^{n'} \rightarrow \{0,1\}^n$, the composition $P \circ A_1$ is a sequence of m polynomials in n' variables. For $i \in \{1, 2, \ldots, m\}$, the i'th polynomial in this composition is $P^{(i)} \circ A_1$. It can be computed by substituting each variable $x[j]$ in $P^{(i)}$ with the (affine) symbolic representation of the j'th output bit of A_1 (and simplifying the outcome to obtain the ANF).

For example, given $P(x[1], x[2], x[3]) = x[1]x[2] + x[1]x[3] + x[2] + 1$ and $A_1 : \{0,1\}^2 \rightarrow \{0,1\}^3$ defined by the relations $x[1] = y[1] + y[2] + 1, x[2] = y[2], x[3] = y[1] + y[2]$, then

$$P \circ A_1 = (y[1] + y[2] + 1)(y[2]) + (y[1] + y[2] + 1)(y[1] + y[2]) + y[2] + 1 =$$
$$(y[1]y[2] + y[2] + y[2]) + (y[1] + y[1]y[2] + y[1]y[2] + y[2] + y[1] + y[2]) + y[2] + 1 =$$
$$y[1]y[2] + y[2] + 1.$$

Thus, we compose each monomial M_u with coefficient 1 in P with A_1 to obtain a polynomial expression, add all the expressions and simplify the result. Formally, if we denote P's M_u coefficient by α_u, then

$$P \circ A_1 = \sum_{u \in \{0,1\}^n} \alpha_u \cdot (M_u \circ A_1).$$

Note that composition with an affine function does not increase the algebraic degree of the composed polynomial, namely $deg(P \circ A_1) \leq deg(P)$.

Analogously, given an affine function $A_2 : \{0,1\}^m \rightarrow \{0,1\}^{m'}$, the composition $A_2 \circ P$ is a sequence of m' polynomials in n variables. It can be computed by substituting each variable $x[j]$ in the (affine) symbolic representation of A_2 with $P^{(j)}$. Equivalently, if $A_2(x) = L(x) + a$, then $A_2 \circ P$ can be computed by symbolic matrix multiplication (and addition of a) as $L(P) + a$. In particular, if $m' = 1$ and $a = 0$, then A_2 reduces to a vector $v = (v[1], v[2], \ldots, v[m]) \in \{0,1\}^m$ and $v(P) = \sum_{i=1}^{m} v[i]P^{(i)}$ is a symbolic inner product.

By rules of composition, if F is represented by P, then $P \circ A_1$ represents $F \circ A_1$ (which is a standard composition of functions) and $A_2 \circ P$ represents $A_2 \circ F$.

Half-Space Masks and Coefficients. Let $A : \{0,1\}^{n-1} \to \{0,1\}^n$ be an affine transformation such that $A(x) = L(x) + a$ for a matrix $L_{n \times n-1}$ with linearly independent columns. Then the (affine) range of A is an $(n-1)$-dimensional affine subspace spanned by the columns of L with the addition of a. The subspace orthogonal to the range of A is of dimension 1 and hence spanned by a single non-zero vector $h \in \{0,1\}^n$. Namely, a vector $v \in \{0,1\}^n$ is in the range of A if and only if $h(v + a) = 0$, i.e., v satisfies the linear equation $h(v) + h(a) = 0$.

Since h partitions the space of $\{0,1\}^n$ into two halves, we call h the *half-space mask* (HSM) of A and call the bit $h(a)$ the *half-space free coefficient* (HSC) of A.

We call the linear subspace spanned by the columns of L the *linear range* of A. A vector $v \in \{0,1\}^n$ is in the linear range of A if and only if $h(v) = 0$.

Canonical Affine Transformations. Given non-zero $h \in \{0,1\}^n$ and $c \in \{0,1\}$, there exist many affine transformations whose HSM and HSC are equal to h, c, respectively. We will use the fact (stated formally below) that all affine transformations with an identical affine range are related by composition on the right with an invertible affine transformation.

Fact 1. *The affine transformations $A_1 : \{0,1\}^{n-1} \to \{0,1\}^n$ and $A_2 : \{0,1\}^{n-1} \to \{0,1\}^n$ have the same affine range if and only if there exists an invertible affine transformation $A' : \{0,1\}^{n-1} \to \{0,1\}^{n-1}$ such that $A_1 = A_2 \circ A'$.*

Given A_1, A_2 the matrix A' above can be computed using basic linear algebra.

We now define the *canonical affine transformation* $C_{|h,c} : \{0,1\}^{n-1} \to \{0,1\}^n$ with respect to h, c. Let ℓ denote the index of the first non-zero bit of $h = (h[1], \ldots, h[n])$. Write $C_{|h,c}(x) = L(x) + a$. We define $a = c \cdot e_\ell$ (where e_ℓ is the ℓ'th unit vector) and define $L[i]$ (the i'th column of L) using h and the unit vectors as follows:

$$
L[i] = \begin{cases} e_i & \text{if } i < \ell \\ e_{i+1} + h[i+1]e_\ell & \text{otherwise } (\ell \le i \le n-1) \end{cases}
$$

Thus, on input $(y[1], \ldots, y[n-1])$, the transformation $C_{|h,c}$ is defined by the symbolic form

$$
(x[1], x[2], \ldots, x[n]) = (y[1], \ldots, y[\ell-1], \sum_{i=\ell}^{n-1} h[i+1]y[i] + c, y[\ell], \ldots, y[n-1]).
$$

The motivation behind the definition of $C_{|h,c}$ is that it allows very simple symbolic composition when applied on the right: its main action is to replace the variable $x[\ell]$ with the affine combination that is specified by the coefficients of h and by c. Other variables are just renamed: variables with index $i < \ell$ remain with the same index, while for each variable with index $i > \ell$, its index it reduced by 1.

Remark 1. Note that we have to show that the definition of $C_{|h,c}$ is valid. First, the $n-1$ columns of L are clearly linearly independent. It remains to prove that h,c are indeed the HSM and HSC of $C_{|h,c}$. For this purpose, it suffices to show that for each column $L[i]$, the vector $L[i] + a$ satisfies the equation $h(L[i] + a) + c = 0$. Since $h(a) = h(c \cdot e_\ell) = c \cdot h(e_\ell) = c$ (as $h_\ell = 1$), it remains to show that $h(L[i]) = 0$. Indeed, if $0 \le i < \ell$, then $h(L[i]) = h_i = 0$ (as ℓ is the index of the first non-zero bit of h). Otherwise, $\ell \le i \le n-1$, then $h(L[i]) = h[i+1] + h[i+1]h[\ell] = 0$ (as $h[\ell] = 1$).

3   Overview of the New Affine Equivalence Algorithm

We demonstrate the new algorithm using an example. Although it is oversimplified, this example is sufficient to convey the main ideas of our algorithm.

Definition of Functions. We define the function $\boldsymbol{F} : \{0,1\}^3 \to \{0,1\}^3$ using its symbolic representation $\boldsymbol{P} = \{P^{(i)}(x[1], x[2], x[3])\}_{i=1}^3$,

$$P^{(1)}(x[1], x[2], x[3]) = x[1]x[2] + x[1]x[3] + x[2] + 1$$
$$P^{(2)}(x[1], x[2], x[3]) = x[1]x[2] + x[1] + x[2]$$
$$P^{(3)}(x[1], x[2], x[3]) = x[1]x[3] + x[3].$$

We define $\boldsymbol{G} : \{0,1\}^3 \to \{0,1\}^3$ using 2 affine transformations as $\boldsymbol{G} = A_2 \circ \boldsymbol{F} \circ A_1$, where A_2 is simply the identity and A_1 is defined using the relations

$$x[1] = y[1] + y[3] + 1, \ x[2] = y[1] + y[2], \ x[3] = y[2].$$

Composing $A_2 \circ \boldsymbol{P} \circ A_1$ and simplifying the resultant ANFs, gives the symbolic representation of $\boldsymbol{G}$ as $\boldsymbol{Q} = \{Q^{(i)}(y[1], y[2], y[3])\}_{i=1}^3$, where

$$Q^{(1)}(y[1], y[2], y[3]) = y[1]y[3] + y[1] + y[2] + 1$$
$$Q^{(2)}(y[1], y[2], y[3]) = y[1]y[2] + y[1]y[3] + y[2]y[3] + y[3] + 1$$
$$Q^{(3)}(y[1], y[2], y[3]) = y[1]y[2] + y[2]y[3].$$

The input to our affine equivalence algorithm is $\boldsymbol{F}, \boldsymbol{G}$ defined above and its goal is to recover the (presumably) unknown affine transformation A_1. The first step of the algorithm is to interpolate $\boldsymbol{F}, \boldsymbol{G}$ and obtain $\boldsymbol{P}, \boldsymbol{Q}$, respectively.

Rank Tables and Histograms. The most basic property that we prove in Theorem 1 is that since $\boldsymbol{F}$ and $\boldsymbol{G}$ are affine equivalent, the symbolic ranks of $\boldsymbol{P}$ and $\boldsymbol{Q}$ (as vectors) are equal. Indeed, it is easy to verify that both $\boldsymbol{P}$ and $\boldsymbol{Q}$ have symbolic rank of 3. More significantly, Theorem 1 is stronger and asserts that $SR(\boldsymbol{P}_{(\ge d)}) = SR(\boldsymbol{Q}_{(\ge d)})$ for every $d \ge 1$. Indeed, if we take $d = 2$, we get $\boldsymbol{P}_{(\ge 2)} = \{x[1]x[2] + x[1]x[3], x[1]x[2], x[1]x[3]\}$, which has symbolic rank 2. This is also the symbolic rank of $\boldsymbol{Q}_{(\ge 2)} = \{y[1]y[3], y[1]y[2]+y[1]y[3]+y[2]y[3], y[1]y[2]+y[2]y[3]\}$.

We would like to use this property to recover A_1. Let us examine the 2-dimensional affine subspace defined by the 3-bit HSM $h' = 100$ (whose bits are $h[1] = 1, h[2] = 0, h[3] = 0$) and the single bit HSC $c = 0$. We calculate the symbolic form of $(F \circ C_{|h',0})_{(\geq 1)}$ by evaluating $(P \circ C_{|h',0})_{(\geq 1)}$ (i.e., plugging $x[1] = 0$ into $P_{(\geq 1)}$) and obtain $\{x[2], x[2], x[3]\}$ which has symbolic rank 2. Similarly, for $c = 1$ we calculate $(P \circ C_{|h',1})_{(\geq 1)}$ (i.e., plug $x[1] = 1$ into $P_{(\geq 1)}$) and obtain $\{x[3], 0, 0\}$ which has symbolic rank 1. Hence, we attach the symbolic rank pair $(2, 1)$ to $h' = 100$. We do the same for all 7 non-zero $h' \in \{0, 1\}^3$. The result is a table whose entries are pairs of ranks of the form $(maxR, minR) \in \mathbb{Z}_4 \cdot \mathbb{Z}_4$ (where $maxR \geq minR$), such that entry $(maxR, minR)$ stores the set of HSMs that are associated with this pair of ranks.

$$(3, 2) : \{010, 011, 111, 110\}$$
$$(2, 2) : \{001\}$$
$$(2, 1) : \{100, 101\}$$

This table is called the *rank table* of F (with respect to the degree $d = 1$ as we only considered monomials of degree at least 1). The set of HSMs in an entry $(maxR, minR)$ of the rank table is called a *rank group* (e.g., the rank group with index $(2, 1)$ is $\{100, 101\}$). Similarly, we compute the rank table of G with respect to $d = 1$.

$$(3, 2) : \{100, 001, 110, 011\}$$
$$(2, 2) : \{010\}$$
$$(2, 1) : \{101, 111\}$$

Although the rank tables are different, the size of each rank group $(maxR, minR)$ of F, G is identical. We define the *rank histogram* of F (with respect to d) as a mapping from each $(maxR, minR)$ value to the corresponding rank group size (e.g., the histogram entry for F with index $(2, 1)$ has value $|\{100, 101\}| = 2$). As we show in Lemma 9, that rank histograms of affine equivalent functions (such as F, G) are identical.

To explain this, we look at the HSM $h = 101$ in the rank group of $G = A_2 \circ F \circ A_1$ with index $(2, 1)$ and note that it partitions the space into halves $\{000, 101, 010, 111\}$ and $\{001, 011, 100, 110\}$ (according to whether $x[1] + x[3] = 0$ or $x[1] + x[3] = 1$). After applying A_1, these half-spaces are mapped into $\{100, 101, 110, 111\}$ and $\{000, 001, 010, 011\}$. This is exactly the partition defined by $h' = 100$, which is in the rank group of F with the same index $(2, 1)$. In terms of canonical affine transformations, $C_{|h',c}$ and $A_1 \circ C_{|h,0}$ have the same half-space range of $\{100, 101, 110, 111\}$ (for $c = 1$ in this case) and we define a mapping $h \mapsto_{A_1} h'$ to capture this. In Lemma 3 we show that this mapping is a bijection as A_1 is invertible.

Exploiting Matchings. The central property of the mapping $\mapsto_{A_1}$ is proved in Lemma 6 which asserts that it preserves affine equivalence. Namely, if $F \equiv G$

and $h \mapsto_{A_1} h'$, then $\boldsymbol{F} \circ C_{|h',c} \equiv \boldsymbol{G} \circ C_{|h,0}$ (for some $c \in \{0,1\}$). By flipping the half-space ranges and applying the same argument, we also obtain $\boldsymbol{F} \circ C_{|h',c+1} \equiv \boldsymbol{G} \circ C_{|h,1}$. Combined with Theorem 1 (which states that symbolic rank is an invariant of affine equivalent functions) we obtain that for any $d \geq 1$, $r_0 \triangleq SR((\boldsymbol{F} \circ C_{|h',c})_{(\geq d)}) = SR((\boldsymbol{G} \circ C_{|h,0})_{(\geq d)})$ and $r_1 \triangleq SR((\boldsymbol{F} \circ C_{|h',c+1})_{(\geq d)}) = SR((\boldsymbol{G} \circ C_{|h,1})_{(\geq d)})$. Since the ordered rank pairs for h', h in $\boldsymbol{F}, \boldsymbol{G}$ are equal to $(maxR, minR)$ (for $maxR = max\{r_0, r_1\}, minR = min\{r_0, r_1\}$), they belong to rank groups with the same index $(maxR, minR)$ in the rank tables of $\boldsymbol{F}, \boldsymbol{G}$, respectively. The fact that $\mapsto_{A_1}$ is a bijection leads to Lemma 9 (which assets that the rank histograms of $\boldsymbol{F}, \boldsymbol{G}$ are identical).

The main goal of our affine equivalence algorithm is to recover *matchings* $h \mapsto_{A_1} h'$ for several pairs h, h'. This is useful, as in Lemma 5 we show that each such matching gives n linear equations on the unknown matrix L of $A_1(x) = L(x) + a$. Furthermore, the constant c associated with $h \mapsto_{A_1} h'$ (which determines whether $\boldsymbol{F} \circ C_{|h',0} \equiv \boldsymbol{G} \circ C_{|h,0}$ or $\boldsymbol{F} \circ C_{|h',1} \equiv \boldsymbol{G} \circ C_{|h,0}$) gives a linear equation on a (once again, by Lemma 5). In total, we need to find about n matchings $h \mapsto_{A_1} h'$ along with their associated constants to completely recover A_1.

Going back to the example, the rank group with index $(2,2)$ for $\boldsymbol{G}$ is $\{010\}$, while the rank group with the same index for $\boldsymbol{F}$ is $\{001\}$. Therefore, after computing the rank tables we know that

$$010 \mapsto_{A_1} 001. \tag{1}$$

Remark 2. We matched $010 \mapsto_{A_1} 001$ in the rank group $(maxR, minR) = (2,2)$ and since $maxR = minR$ we cannot derive the constant c associated with this matching (hence we cannot derive a linear equation on a). Such constants can only be derived for matchings $h \mapsto_{(A_1)} h'$ in rank groups where $maxR > minR$, as in such cases we know whether $\boldsymbol{F} \circ C_{|h',0} \equiv \boldsymbol{G} \circ C_{|h,0}$ or $\boldsymbol{F} \circ C_{|h',1} \equiv \boldsymbol{G} \circ C_{|h,0}$ according to the equality $SR((\boldsymbol{F} \circ C_{|h',c})_{(\geq d)}) = SR((\boldsymbol{G} \circ C_{|h,0})_{(\geq d)})$. More precisely, if $maxR > minR$, then either $SR((\boldsymbol{F} \circ C_{|h',0})_{(\geq d)}) = SR((\boldsymbol{G} \circ C_{|h,0})_{(\geq d)})$ or $SR((\boldsymbol{F} \circ C_{|h',1})_{(\geq d)}) = SR((\boldsymbol{G} \circ C_{|h,0})_{(\geq d)})$, but not both. In this sense, it is more useful to recover matchings for HSMs in rank groups $(maxR, minR)$ such that $maxR > minR$.

By applying similar arguments to the rank group $(2,1)$, we know that either $101 \mapsto_{A_1} 100$ or $101 \mapsto_{A_1} 101$ (and similarly $111 \mapsto_{A_1} 100$ or $111 \mapsto_{A_1} 101$). Since we have very few possibilities, we can guess which matchings hold, derive A_1 and test our guess. Unfortunately, for larger n we expect the rank groups to be much bigger and it would be inefficient to exhaustively match HSMs for $\boldsymbol{F}, \boldsymbol{G}$ only based on their ranks. Thus, to narrow down the number of possibilities and eventually uniquely match sufficiently many pairs h, h' such that $h \mapsto_{A_1} h'$, we need to attach more data to each HSM for $\boldsymbol{F}, \boldsymbol{G}$.

HSM Rank Histograms. The main observation that allows attaching more data to each HSM is given in Lemma 4 which shows that the mapping $\mapsto_{A_1}$ is additive.

Consider the two rank groups with index $(3, 2)$ for $\boldsymbol{F}, \boldsymbol{G}$. They are of size 4 and their HSMs cannot be uniquely matched. We first focus on $\boldsymbol{G}$ and examine the rank group $(3, 2)$ which is $\{100, 001, 110, 011\}$. We take $h_1 = 011$ and compute its *HSM rank histogram* with respect to the rank group $(2, 1)$ (which is $\{101, 111\}$). This is done by computing the $(maxR, minR)$ rank pairs for the set defined by adding all elements of rank group $(2, 1)$ to 011, namely $\{011 + 101, 011 + 111\} = \{110, 100\}$. Looking for 110 and 100 in the rank table of $\boldsymbol{G}$, both HSMs have ranks $(3, 2)$. Thus, the HSM rank histogram of $h_1 = 011$ with respect to rank group $(2, 1)$ has a single non-zero entry $(3, 2)$ with the value of 2. We write this HSM rank histogram in short as $[(3, 2) : 2]$.

We now consider the match of $h_1 = 011$ under A_1 which is $h_1' = 110$ (namely, $h_1 \mapsto_{(A_1)} h_1'$). Similarly to h_1, we compute the HSM rank histogram of h_1' with respect to the rank group $(2, 1)$ for $\boldsymbol{F}$ (which is $\{100, 101\}$) and obtain the same HSM rank histogram $[(3, 2) : 2]$. This is a particular case of Lemma 10, which shows that matching HSMs for $\boldsymbol{F}, \boldsymbol{G}$ have identical HSM rank histograms (with respect to a fixed rank group). Lemma 10 is derived using Lemma 4 which asserts that the mapping $\mapsto_{A_1}$ is additive: if $h_1 \mapsto_{(A_1)} h_1'$ and $h_2 \mapsto_{(A_1)} h_2'$, then $(h_1 + h_2) \mapsto_{(A_1)} (h_1' + h_2')$.

Fixing $h_1 = 011$ for $\boldsymbol{G}$ and its match $h_1' = 110$ under A_1, let $h_2^i, h_2'^i$ for $i \in \{1, 2\}$ vary over the 2 elements of the rank groups with index $(2, 1)$ in $\boldsymbol{G}, \boldsymbol{F}$, respectively. Then, as $h_1 \mapsto_{(A_1)} h_1'$ and $h_2^i \mapsto_{(A_1)} h_2'^i$ for $i \in \{1, 2\}$, we get $(h_1 + h_2^i) \mapsto_{(A_1)} (h_1' + h_2'^i)$. By the aforementioned Theorem 1 and Lemma 6 (equating ranks for matching HSMs) we conclude that indeed the HSM rank histograms of 011 and 110 with respect to rank group $(2, 1)$ are identical (which is a special case of Lemma 10).

HSM Rank Histogram Multi-Sets. Since we do not know in advance that $h_1 \mapsto_{(A_1)} h_1'$, we have to compute the HSM rank histograms (with respect to rank group $(2, 1)$) for all HSMs in rank group $(3, 2)$. The outcome is the *HSM rank histogram multi-set* of rank group $(3, 2)$ with respect to rank group $(2, 1)$. It is computed by considering all the HSMs in the rank group $(3, 2)$, namely $\{100, 001, 110, 011\}$ for $\boldsymbol{G}$ and $\{010, 011, 111, 110\}$ for $\boldsymbol{F}$. Lemma 11 (whose proof is based on Lemma 10) asserts that these HSM rank histogram multi-sets are identical as $\boldsymbol{F}, \boldsymbol{G}$ are affine equivalent.

We hope that these multi-sets contain unique HSM rank histograms (with multiplicity 1), which would allow us to derive more matching between HSMs. Unfortunately, the resultant multi-set (for both $\boldsymbol{F}$ and $\boldsymbol{G}$) is $\{[(3, 2) : 2], [(3, 2) : 2], [(3, 2) : 2], [(3, 2) : 2]\}$. It contains 4 identical elements and does not give us any new information about A_1. If the multiplicity of the element $[(3, 2) : 2]$ (calculated above for h_1, h_1') in this multi-set would have been 1, we could have derived the relation $h_1 \mapsto_{(A_1)} h_1'$.

Remark 3. Generally, when n is very small (as in our case), the direct application of the algorithm is more likely to fail to completely recover A_1. As we show later in this paper, for $n \geq 8$ the fraction of instances for which this occurs is very small (and tends to 0 as n grows). In some cases a failure to retrieve A_1 occurs since the affine mappings A_1, A_2 are not uniquely defined. In particular, if there are several solutions for A_1, then we cannot hope to obtain unique matchings that completely define A_1, but we can recover all possible solutions to the affine equivalence problem by enumerating several possibilities for the matchings.

In conclusion, we attached to each HSM in rank group $(3,2)$ for $\boldsymbol{F}, \boldsymbol{G}$ its HSM rank histogram with respect to rank group $(2,1)$ and in general such data may allow us to derive additional matchings $h \mapsto_{(A_1)} h'$. Once we obtain about n matchings, we can recover A_1 by solving a system of linear equations.

4   A Basic Property of Affine Equivalent Functions

Before proving the main result of this section, we state two useful lemmas (the first is proved in the extended version of this paper [9]).

Lemma 1. Let $\boldsymbol{P} = \{P^{(i)}(x[1], \ldots, x[n])\}_{i=1}^m$, let $A_1 : \{0,1\}^{n'} \to \{0,1\}^n$, $A_2 : \{0,1\}^m \to \{0,1\}^{m'}$ be affine functions, and d be a positive integer. Then,

1. $(\boldsymbol{P}_{(\geq d)} \circ A_1)_{(\geq d)} = (\boldsymbol{P} \circ A_1)_{(\geq d)}$
2. $(A_2 \circ (\boldsymbol{P}_{(\geq d)}))_{(\geq d)} = (A_2 \circ \boldsymbol{P})_{(\geq d)}$ and if A_2 is a linear function, $A_2 \circ (\boldsymbol{P}_{(\geq d)}) = (A_2 \circ \boldsymbol{P})_{(\geq d)}$.

Essentially, the lemma states that removing all monomials of degree less than d from $\boldsymbol{P}$ can be done before or after composing it with an affine function and the outcomes are identical.

Note that a potentially simplified first part of the lemma which equates $\boldsymbol{P}_{(\geq d)} \circ A_1$ and $(\boldsymbol{P} \circ A_1)_{(\geq d)}$ is generally incorrect, as the first expression may contain monomials of degree less than d. For example, if $d = 2$ and we compose the affine transformation defined by $x[1] = y[1] + y[2]$ and $x[2] = y[2]$ with the polynomial $x[1]x[2]$, then we get the polynomial $y[1]y[2] + y[2]$ which has a monomial of degree 1.

Lemma 2. Let $\boldsymbol{P} = \{P^{(i)}(x[1], \ldots, x[n])\}_{i=1}^m$, and let $A_1 : \{0,1\}^n \to \{0,1\}^n$ be an invertible affine function. Then, $deg(\boldsymbol{P}) = deg(\boldsymbol{P} \circ A_1)$.

Proof. We show that for $i \in \{1, 2, \ldots, m\}$, $deg(P^{(i)}) = deg(P^{(i)} \circ A_1)$. Observe that $deg(P^{(i)}) \geq deg(P^{(i)} \circ A_1)$ as composition with an affine function cannot increase the algebraic degree of a polynomial. By the same argument and by the invertibility of A_1, we also obtain $deg(P^{(i)} \circ A_1) \geq deg(P^{(i)} \circ A_1 \circ (A_1)^{-1}) = deg(P^{(i)})$. ∎

Theorem 1. Let $\boldsymbol{F} : \{0,1\}^n \to \{0,1\}^m$, $\boldsymbol{G} : \{0,1\}^n \to \{0,1\}^m$ be two affine equivalent functions, represented by $\boldsymbol{P}, \boldsymbol{Q}$, respectively. Then, for every positive integer d, $SR(\boldsymbol{P}_{(\geq d)}) = SR(\boldsymbol{Q}_{(\geq d)})$.

Proof. At a high level, the fact that P and Q have the same symbolic rank follows since rank is preserved by composition with invertible affine transformations. Moreover, this rank equality is preserved after truncating low degree monomials since they cannot affect the high degree monomials when composing with invertible affine transformations. The formal proof is below.

Write $G = A_2 \circ F \circ A_1$, implying that $Q = A_2 \circ P \circ A_1$. Denote $P' = P \circ A_1$ and observe that

$$SR(P'_{(\geq d)}) = SR((A_2 \circ (P'_{(\geq d)}))_{(\geq d)}) = SR((A_2 \circ P')_{(\geq d)}) = SR(Q_{(\geq d)}),$$

where the first equality holds since rank is preserved by invertible linear transformations[5] and the second equality is due to the second part of Lemma 1.

It remains to show that $SR(P'_{(\geq d)}) = SR(P_{(\geq d)})$, or $SR((P \circ A_1)_{(\geq d)}) = SR(P_{(\geq d)})$. We first show that $SR(P_{(\geq d)}) \geq SR((P \circ A_1)_{(\geq d)})$.

If $P_{(\geq d)}$ has full rank of m then the claim is trivial. Otherwise, let $v \in \{0,1\}^m$ be a non-zero vector in the kernel of $P_{(\geq d)}$, namely $v(P_{(\geq d)}) = 0$. Then,

$$v((P \circ A_1)_{(\geq d)}) = (v((P \circ A_1)_{(\geq d)}))_{(\geq d)} = (v(P_{(\geq d)}) \circ A_1)_{(\geq d)} = 0,$$

where the first equality follows from the second part of Lemma 1 and the second equality follows from the first part of this lemma. This implies that v is also in the kernel of $(P \circ A_1)_{(\geq d)}$, as required.

To prove that $SR(P_{(\geq d)}) \leq SR((P \circ A_1)_{(\geq d)})$, observe that if v is a non-zero vector in the kernel of $(P \circ A_1)_{(\geq d)}$, then by the equality above we have $0 = v((P \circ A_1)_{(\geq d)}) = (v(P_{(\geq d)}) \circ A_1)_{(\geq d)}$. This implies that $deg(v(P_{(\geq d)}) \circ A_1) < d$ and since A_1 is invertible, by Lemma 2, $deg(v(P_{(\geq d)})) = deg(v(P_{(\geq d)}) \circ A_1) < d$. This gives $v(P_{(\geq d)}) = 0$, as the polynomial does not contain monomials of degree less than d. Hence, v is in the kernel of $P_{(\geq d)}$ which completes the proof. ∎

5   The Half-Space Mask Bijection and Its Properties

Definition 1. *Let $A : \{0,1\}^n \to \{0,1\}^n$ be an invertible affine transformation. Define a mapping between HSMs using A as follows: $h \in \{0,1\}^n$ is mapped to h' if there exists $c \in \{0,1\}$ such that the affine ranges of $A \circ C_{|h,0}$ and $C_{|h',c}$ are equal. We write $h \mapsto_{(A)} h'$ and say that h and h' match (under A). The bit c is called the associated constant of $h \mapsto_{(A)} h'$.*

Lemma 3. *Let $A : \{0,1\}^n \to \{0,1\}^n$ be an invertible affine transformation. The mapping $\mapsto_{(A)}$ is a bijection and its inverse is given by $\mapsto_{(A^{-1})}$.*

Proof. The proof follows from the invertibility of A. Given that $h \mapsto_{(A)} h'$, there exists $c \in \{0,1\}$ such that the affine ranges of $A \circ C_{|h,0}$ and $C_{|h',c}$ are equal. According to Fact 1, this implies that there exists an invertible affine transformation $A' : \{0,1\}^{n-1} \to \{0,1\}^{n-1}$ such that $A \circ C_{|h,0} = C_{|h',c} \circ A'$.

---

[5] The affine transformation A_2 also adds a constant, but it does contribute to the rank as $d > 0$.

Consequently, $C_{|h,0} \circ (A')^{-1} = A^{-1} \circ C_{|h',c}$ and the affine ranges of $C_{|h,0}$ and $A^{-1} \circ C_{|h',c}$ are equal (again, according to Fact 1). This implies that the affine ranges of $A^{-1} \circ C_{|h',0}$ and $C_{|h,c}$ are equal (flipping the HSC of both sides if $c = 1$), namely $h' \mapsto_{(A^{-1})} h$.   ∎

A property of $\mapsto_{(A)}$ which will be very useful is that it is additive. This is established by the lemma below (proved in the extended version of this paper [9]).

Lemma 4. *Let $A : \{0,1\}^n \to \{0,1\}^n$ be an invertible affine transformation. Let $h_1, h'_1, h_2, h'_2 \in \{0,1\}^n$ be HSMs where $h_1 \neq h_2$ and $h_1 \mapsto_{(A)} h'_1$, $h_2 \mapsto_{(A)} h'_2$ with associated constants c_1, c_2, respectively. Then $(h_1 + h_2) \mapsto_{(A)} (h'_1 + h'_2)$ with the associated constant $c_1 + c_2$.*

The following lemma (proved in the extended version of this paper [9]) shows that the bijection reveals information about the presumably unknown transformation A.

Lemma 5. *Let $A : \{0,1\}^n \to \{0,1\}^n$ be an invertible affine transformation such that $A(x) = L(x) + a$. Let $h, h' \in \{0,1\}^n$ be HSMs such that $h \mapsto_{(A)} h'$ with associated constant c. Then, A satisfies the following constraints.*

1. *For each $i \in \{1, 2, \ldots, n\}$, the i'th column of L, denoted by $L[i]$, satisfies the equation $h'(L[i]) = h[i]$, where $h[i]$ is the i'th bit of h.*
2. *The vector a satisfies the equation $h'(a) = c$.*

The following lemma asserts that affine equivalence is preserved under composition with matching HSMs.

Lemma 6. *Let $\boldsymbol{F} : \{0,1\}^n \to \{0,1\}^m$, $\boldsymbol{G} : \{0,1\}^n \to \{0,1\}^m$ be two affine equivalent functions such that $\boldsymbol{G} = A_2 \circ \boldsymbol{F} \circ A_1$. Let $h, h' \in \{0,1\}^n$ be HSMs such that $h \mapsto_{(A_1)} h'$ with associated constant c. Then, $\boldsymbol{F} \circ C_{|h',c} \equiv \boldsymbol{G} \circ C_{|h,0}$ and $\boldsymbol{F} \circ C_{|h',c+1} \equiv \boldsymbol{G} \circ C_{|h,1}$.*

Proof. Since $h \mapsto_{(A_1)} h'$ with associated constant c, the affine ranges of $A_1 \circ C_{|h,0}$ and $C_{|h',c}$ are equal. According to Fact 1, there exists an invertible affine transformation $A'_1 : \{0,1\}^{n-1} \to \{0,1\}^{n-1}$ such that $A_1 \circ C_{|h,0} = C_{|h',c} \circ A'_1$.

We obtain, $A_2 \circ \boldsymbol{F} \circ C_{|h',c} \circ A'_1 = A_2 \circ \boldsymbol{F} \circ A_1 \circ C_{|h,0} = \boldsymbol{G} \circ C_{|h,0}$, implying that $\boldsymbol{F} \circ C_{|h',c}$ and $\boldsymbol{G} \circ C_{|h,0}$ are affine equivalent.

The claim that $\boldsymbol{F} \circ C_{|h',c+1}$ and $\boldsymbol{G} \circ C_{|h,1}$ are affine equivalent follows by considering the complimentary half-space and observing that the affine ranges of $A_1 \circ C_{|h,1}$ and $C_{|h',c+1}$ are equal. The remainder of the proof is similar.   ∎

Definition 2. *Let $\boldsymbol{F} : \{0,1\}^n \to \{0,1\}^m$ be a function represented by $\boldsymbol{P}$, let d be a positive integer and let $h \in \{0,1\}^n$ be a HSM. Let $r_0 = SR((\boldsymbol{P} \circ C_{|h,0})_{(\geq d)})$, $r_1 = SR((\boldsymbol{P} \circ C_{|h,1})_{(\geq d)})$, $maxR = max\{r_0, r_1\}$ and $minR = min\{r_0, r_1\}$.*

1. *The HSM rank of h (with respect $\boldsymbol{F}, d$) is the ordered pair of integers $(maxR, minR)$, denoted as $R_{\boldsymbol{F},d,h}$,*

2. *The attached constant of h is the value $c \in \{0,1\}$ such that $maxR = SR((\boldsymbol{P} \circ C_{|h,c})_{(\geq d)})$ (if $maxR = minR$, the attached constant is undefined).*

The lemma below states that the HSM ranks of matching HSMs are equal for affine equivalent functions.

Lemma 7. *Let $\boldsymbol{F} : \{0,1\}^n \rightarrow \{0,1\}^m$, $\boldsymbol{G} : \{0,1\}^n \rightarrow \{0,1\}^m$ be two affine equivalent functions and let d be a positive integer. Assume that $\boldsymbol{G} = A_2 \circ \boldsymbol{F} \circ A_1$. Let $h, h' \in \{0,1\}^n$ be HSMs such that $h \mapsto_{(A_1)} h'$. Then $R_{\boldsymbol{G},d,h} = R_{\boldsymbol{F},d,h'}$.*

Proof. Let $c \in \{0,1\}$ be the associated constant of $h \mapsto_{(A_1)} h'$. According to Lemma 6, $\boldsymbol{F} \circ C_{|h',c} \equiv \boldsymbol{G} \circ C_{|h,0}$ and $\boldsymbol{F} \circ C_{|h',c+1} \equiv \boldsymbol{G} \circ C_{|h,1}$.

Assume that $\boldsymbol{F}, \boldsymbol{G}$ are represented by $\boldsymbol{P}, \boldsymbol{Q}$, respectively. Denote $r_0' = SR((\boldsymbol{F} \circ C_{|h',c})_{(\geq d)})$, $r_1' = SR((\boldsymbol{F} \circ C_{|h',c+1})_{(\geq d)})$, $r_0 = SR((\boldsymbol{G} \circ C_{|h,0})_{(\geq d)})$, $r_1 = SR((\boldsymbol{G} \circ C_{|h,1})_{(\geq d)})$.

By the above affine equivalences and Theorem 1, we have $r_0 = r_0'$ and $r_1 = r_1'$. Hence $max(r_0, r_1) = max(r_0', r_1')$ and $min(r_0, r_1) = min(r_0', r_1')$ and the lemma follows. ∎

6 Rank Tables, Rank Histograms and their Properties

Definition 3. *Given a function $\boldsymbol{F} : \{0,1\}^n \rightarrow \{0,1\}^m$ and a positive integer d, define the following mappings.*

1. *The rank table of $\boldsymbol{F}$ with respect to d is a mapping $\mathcal{T}_{\boldsymbol{F},d}$, whose keys (indexes) are integer pairs $(maxR, minR) \in \mathbb{Z}_{m+1} \times \mathbb{Z}_{m+1}$ such that $maxR \geq minR$. It is defined as*

$$\mathcal{T}_{\boldsymbol{F},d}(maxR, minR) = \{h \in \{0,1\}^n \mid R_{\boldsymbol{F},d,h} = (maxR, minR)\}.$$

   Moreover, along with each such HSM h, the table stores its attached constant $c \in \{0,1\}$ (if defined).
   An entry in the rank table $\mathcal{T}_{\boldsymbol{F},d}(maxR, minR)$ (containing all HSMs with this rank) is called a rank group.
2. *The rank histogram of $\boldsymbol{F}$ with respect to d is a mapping $\mathcal{H}_{\boldsymbol{F},d} : \mathbb{Z}_{m+1} \times \mathbb{Z}_{m+1} \rightarrow \mathbb{Z}$ such that $\mathcal{H}_{\boldsymbol{F},d}(maxR, minR) = |\mathcal{T}_{\boldsymbol{F},d}(maxR, minR)|$.*

To simplify our notation, in the following we refer to a HSM rank $(maxR, minR) \in \mathbb{Z}_{m+1} \times \mathbb{Z}_{m+1}$ such that $maxR \geq minR$ using a single symbol $\boldsymbol{r}$.

The lemma below states that if $\boldsymbol{F} \equiv \boldsymbol{G}$, then each HSM with rank $\boldsymbol{r}$ for $\boldsymbol{G}$ is matched in the rank group with the same HSM rank $\boldsymbol{r}$ for $\boldsymbol{F}$.

Lemma 8. *Let $\boldsymbol{F} : \{0,1\}^n \rightarrow \{0,1\}^m$, $\boldsymbol{G} : \{0,1\}^n \rightarrow \{0,1\}^m$ be two affine equivalent functions and let d be a positive integer. Assume that $\boldsymbol{G} = A_2 \circ \boldsymbol{F} \circ A_1$ and let $\boldsymbol{r} \in \mathbb{Z}_{m+1} \times \mathbb{Z}_{m+1}$. Then, for each $h \in \mathcal{T}_{\boldsymbol{G},d}(\boldsymbol{r})$, there exists $h' \in \mathcal{T}_{\boldsymbol{F},d}(\boldsymbol{r})$ such that $h \mapsto_{(A_1)} h'$.*

Proof. By Lemma 7, given $h \in \mathcal{T}_{G,d}(r)$, its match h' under A_1 satisfies $R_{F,d,h'} = R_{G,d,h} = r$ hence $h' \in \mathcal{T}_{F,d}(r)$ as claimed. ∎

Lemma 9. *Let $F : \{0,1\}^n \to \{0,1\}^m$, $G : \{0,1\}^n \to \{0,1\}^m$ be two affine equivalent functions and let d be a positive integer. Then the rank histograms of F and G with respect to d are equal, namely $\mathcal{H}_{F,d} = \mathcal{H}_{G,d}$.*

Proof. Assume that $G = A_2 \circ F \circ A_1$. Given a histogram entry with index r, for each $h \in \mathcal{T}_{G,d}(r)$ let h' be its match $h \mapsto_{(A_1)} h'$. Then, by Lemma 8, $h' \in \mathcal{T}_{F,d}(r)$. Since $h \mapsto_{(A_1)} h'$ is a bijection, this shows that $\mathcal{H}_{G,d}(r) = |\mathcal{T}_{G,d}(r)| \leq |\mathcal{T}_{F,d}(r)| = \mathcal{H}_{F,d}(r)$. On the other hand, as $\mathcal{H}_{G,d}(r) \leq \mathcal{H}_{F,d}(r)$ holds for all histogram entries r and the sum of entries in both histograms is $2^n - 1$, this implies that $\mathcal{H}_{F,d} = \mathcal{H}_{G,d}$. ∎

Definition 4. *Given a function $F : \{0,1\}^n \to \{0,1\}^m$, a positive integer d, a HSM $h_1 \in \{0,1\}^n$ and $r \in \mathbb{Z}_{m+1} \times \mathbb{Z}_{m+1}$, we define the HSM rank histogram of h_1 with respect to (or relative to) the rank group r and denote it by $\mathcal{HG}_{F,d,h_1,r}$. As the standard histogram, it is a mapping $\mathcal{HG}_{F,d,h_1,r} : \mathbb{Z}_{m+1} \times \mathbb{Z}_{m+1} \to \mathbb{Z}$, where*

$$\mathcal{HG}_{F,d,h_1,r}(r') = |\{h_1 + h_2 \mid h_2 \in \{0,1\}^n \wedge h_1 \neq h_2 \wedge R_{F,d,h_2} = r \wedge R_{F,d,h_1+h_2} = r'\}|.$$

Note that unlike the (standard) rank histogram, the HSM rank histogram is defined for a specific HSM with respect to a rank group. We further remark that the HSM rank histogram of h_1 can also be defined with respect to its own the rank group (this is assured by the condition $h_1 \neq h_2$).

The following lemma equates HSM rank histograms for matching HSMs in affine equivalent functions.

Lemma 10. *Let $F : \{0,1\}^n \to \{0,1\}^m$, $G : \{0,1\}^n \to \{0,1\}^m$ be two affine equivalent functions and let d be a positive integer. Assume that $G = A_2 \circ F \circ A_1$. Let $h_1, h_1' \in \{0,1\}^n$ be such that $h_1 \mapsto_{(A_1)} h_1'$. Then, for every $r \in \mathbb{Z}_{m+1} \times \mathbb{Z}_{m+1}$, $\mathcal{HG}_{G,d,h_1,r} = \mathcal{HG}_{F,d,h_1',r}$.*

Proof. The proof follows from the fact that the mapping $\mapsto_{(A_1)}$ preserves HSM ranks for affine equivalent functions (Lemma 7), and by exploiting its additive property (Lemma 4).

Fix a HSM rank histogram entry $r' \in \mathbb{Z}_{m+1} \times \mathbb{Z}_{m+1}$. Define the following two sets:

$$D_1 = \{h_1 + h_2 \mid h_2 \in \{0,1\}^n \wedge h_1 \neq h_2 \wedge R_{G,d,h_2} = r \wedge R_{G,d,h_1+h_2} = r'\}$$

and

$$D_2 = \{h_1' + h_2' \mid h_2' \in \{0,1\}^n \wedge h_1' \neq h_2' \wedge R_{F,d,h_2'} = r \wedge R_{F,d,h_1'+h_2'} = r'\}.$$

To prove the lemma, we need to show that $|D_1| = |D_2|$. Let $h_1 + h_2 \in D_1$ and denote by $\hat{h} \in \{0,1\}^n$ the vector such that $h_1 + h_2 \mapsto_{(A_1)} \hat{h}$. We show that $\hat{h} \in D_2$.

Since $h_1 + h_2 \mapsto_{(A_1)} \hat{h}$, by Lemma 7, $R_{F,d,\hat{h}} = R_{G,d,h_1+h_2} = r'$. Next, write $\hat{h} = h_1' + (h_1' + \hat{h})$. Since $h_1 \mapsto_{(A_1)} h_1'$ and $h_1 + h_2 \mapsto_{(A_1)} \hat{h}$, by Lemma 4, $h_2 \mapsto_{(A_1)} h_1' + \hat{h}$, and by Lemma 7, $R_{F,d,h_1'+\hat{h}} = R_{G,d,h_2} = r$, giving $\hat{h} \in D_2$. Since $\mapsto_{(A_1)}$ is a bijection this implies that $|D_2| \geq |D_1|$.

As $|D_2| \geq |D_1|$ holds for all HSM histogram entries r' and the sum of HSM histogram entries in both $\mathcal{HG}_{G,d,h_1,r}$ and $\mathcal{HG}_{F,d,h_1',r}$ is equal to size of the rank group[6] r (which is equal to $\mathcal{H}_{F,d}(r) = \mathcal{H}_{G,d}(r)$), the equality $|D_1| = |D_2|$ holds.   ∎

Definition 5. *Given a function $F : \{0,1\}^n \to \{0,1\}^m$, a positive integer d, HSMs ranks $r, r' \in \mathbb{Z}_{m+1} \times \mathbb{Z}_{m+1}$, we define the HSM rank histogram multi-set of rank group r with respect to rank group r' as*

$$\mathcal{HM}_{F,d,r,r'} = \{\mathcal{HG}_{F,d,h,r'} \mid R_{F,d,h} = r\}.$$

The HSM rank histogram multi-set collects all the HSM histograms for HSMs in rank group r with respect to the rank group r'. Note that it is possible to have $r = r'$.

The following lemma equates HSM rank histogram multi-set in affine equivalent functions.

Lemma 11. *Let $F : \{0,1\}^n \to \{0,1\}^m$, $G : \{0,1\}^n \to \{0,1\}^m$ be two affine equivalent functions and let d be a positive integer. Then, for every $r, r' \in \mathbb{Z}_{m+1} \times \mathbb{Z}_{m+1}$, $\mathcal{HM}_{F,d,r,r'} = \mathcal{HM}_{G,d,r,r'}$.*

Proof. Fix $r, r' \in \mathbb{Z}_{m+1} \times \mathbb{Z}_{m+1}$. We define a mapping between the elements (HSM histograms) of the multi-sets $\mathcal{HM}_{G,d,r,r'}$ and $\mathcal{HM}_{F,d,r,r'}$. Naturally, the mapping is based on the bijection $\mapsto_{(A_1)}$.

Assume that $G = A_2 \circ F \circ A_1$. Let $h \in \{0,1\}^n$ be such that $R_{G,d,h} = r$ which implies that $\mathcal{HG}_{G,d,h,r'} \in \mathcal{HM}_{G,d,r,r'}$. Let $h' \in \{0,1\}^n$ be the HSM such that $h \mapsto_{(A_1)} h'$. By Lemma 10, $\mathcal{HG}_{G,d,h,r'} = \mathcal{HG}_{F,d,h',r'}$. Furthermore, by Lemma 7 we have $R_{F,d,h'} = R_{G,d,h} = r$, hence $\mathcal{HG}_{G,d,h,r'} = \mathcal{HG}_{F,d,h',r'} \in \mathcal{HM}_{F,d,r,r'}$. Since $\mapsto_{(A_1)}$ is a bijection, we obtain $\mathcal{HM}_{G,d,r,r'} \subseteq \mathcal{HM}_{F,d,r,r'}$ as multi-sets.

On the other hand, the number of elements (HSM histograms) in both multi-sets is equal to the size of the rank group r (which is equal to $\mathcal{H}_{F,d}(r) = \mathcal{H}_{G,d}(r)$), hence $\mathcal{HM}_{G,d,r,r'} = \mathcal{HM}_{F,d,r,r'}$.   ∎

Definition 6. *Let $F : \{0,1\}^n \to \{0,1\}^m$, let d be a positive integer and let $r, r' \in \mathbb{Z}_{m+1} \times \mathbb{Z}_{m+1}$. A HSM $h \in \{0,1\}$ such that $R_{F,d,h} = r$ is called unique (with respect to F, d, r') if $\mathcal{HG}_{F,d,h,r'} \in \mathcal{HM}_{F,d,r,r'}$ has multiplicity 1 in this multi-set.*

The following theorem establishes the importance of unique HSMs in recovering matchings between HSMs for affine equivalent functions.

---

[6] Unless the HSM rank of h_1 is r, in which case the sum of HSM histogram entries is $\mathcal{H}_{F,d}(r) - 1$.

Theorem 2. Let $F : \{0,1\}^n \rightarrow \{0,1\}^m$, $G : \{0,1\}^n \rightarrow \{0,1\}^m$ be two affine equivalent functions and let d be a positive integer. Then for every $r, r' \in \mathbb{Z}_{m+1} \times \mathbb{Z}_{m+1}$, if $h \in \{0,1\}^n$ (such that $R_{G,d,h} = r$) is unique with respect to G, d, r', then the following statements hold:

1. There exists $h' \in \{0,1\}^n$ such that $R_{F,d,h'} = r$ and h' is unique with respect to F, d, r'.
2. $\mathcal{HG}_{G,d,h,r'} = \mathcal{HG}_{F,d,h',r'}$.
3. Assume that $G = A_2 \circ F \circ A_1$. Then, $h \mapsto_{(A_1)} h'$. Moreover, if the attached constants of h, h' are defined and equal to c, c', respectively, then the associated constant of $h \mapsto_{(A_1)} h'$ is $c + c'$.

Proof. By Lemma 11, we have equality of the multi-sets $\mathcal{HM}_{F,d,r,r'} = \mathcal{HM}_{G,d,r,r'}$ which immediately implies the first two statements. Denote by $h'' \in \{0,1\}^n$ the HSM such that $h \mapsto_{(A_1)} h''$. To complete the proof of the third statement we show that $h'' = h'$.

By Lemma 7, we have $R_{F,d,h''} = R_{G,d,h} = r$. Hence $\mathcal{HG}_{F,d,h'',r} \in \mathcal{HM}_{F,d,r,r'}$ (and also $\mathcal{HG}_{F,d,h',r} \in \mathcal{HM}_{F,d,r,r'}$ from the first statement). Since h' is unique with respect to F, d, r', then $\mathcal{HG}_{F,d,h',r'}$ has multiplicity 1 in $\mathcal{HM}_{F,d,r,r'}$. Thus if we show that $\mathcal{HG}_{F,d,h',r'} = \mathcal{HG}_{F,d,h'',r}$, then $h'' = h'$ must hold.

According to Lemma 10, $\mathcal{HG}_{G,d,h,r} = \mathcal{HG}_{F,d,h'',r}$ and by the second statement we obtain $\mathcal{HG}_{F,d,h',r'} = \mathcal{HG}_{G,d,h,r'} = \mathcal{HG}_{F,d,h'',r}$ as required.

Finally, we examine the attached constants c, c' of h, h', respectively (assuming they are defined). If $c = c'$, then the affine ranges of $A_1 \circ C_{|h,0}$ and $C_{|h',0}$ are equal implying that the associated constant of $h \mapsto_{(A_1)} h'$ is $0 = c+c'$. Otherwise $c = c' + 1$ and the affine ranges of $A_1 \circ C_{|h,0}$ and $C_{|h',1}$ are equal implying that the associated constant of $h \mapsto_{(A_1)} h'$ is $1 = c + c'$. ∎

7   Analysis of the Distribution of Rank Histogram Entries for Random Permutations

In this section we analyze the distribution of entries of the rank histogram $\mathcal{H}_{F,d}$ for a random permutation $F : \{0,1\}^n \rightarrow \{0,1\}^n$. The analysis is performed for $d = n-2$, which is the value that we use in our algorithm as explained in detail next.

Assume that F is represented by $P = \{P^{(i)}(x[1], \ldots, x[n])\}_{i=1}^n$. For a given $h \in \{0,1\}^n$, we consider $SR((P \circ C_{|h,0})_{(\geq n-2)})$ and $SR((P \circ C_{|h,1})_{(\geq n-2)})$. For $c \in \{0,1\}$, every one of the n polynomials of $(P \circ C_{|h,c})_{(\geq n-2)}$ has $n-1$ variables (the number of variables in P is reduced by 1 after composition with $C_{|h,c}$). Hence, the number of possible non-zero monomial coefficients in each such polynomial is $\binom{n-1}{n-1} + \binom{n-1}{n-2} = 1+n-1 = n$. Therefore, $(P \circ C_{|h,c})_{(\geq n-2)}$ can be represented by an $n \times n$ Boolean matrix and we are interested in its rank.

Choosing $d = n-1$ would leave at most one non-zero monomial which almost always would be present in $(P \circ C_{|h,c})_{(\geq n-1)}$. Hence, essentially all HSMs would fall into a single rank group and the affine equivalence algorithm would not be able to distinguish and match them. On the other hand, choosing $d \leq n-3$ would leave $\Omega(n^2)$ non-zero monomials and $(\boldsymbol{P} \circ C_{|h,c})_{(\geq r)}$ would almost always have full rank, leading once again to a single rank group. We conclude that $d = n - 1$ is indeed the optimal choice.

Our analysis is based on the following heuristic assumption.

Assumption 1. *For a random permutation* $\boldsymbol{F} : \{0,1\}^n \to \{0,1\}^n$ *represented by* $\boldsymbol{P}$, *for every* $h \in \{0,1\}^n$ *and* $c \in \{0,1\}$, *the entries of the* $n \times n$ *Boolean matrix* $(\boldsymbol{P} \circ C_{|h,c})_{(\geq n-2)}$ *are uniform independent random variables.*

The $n \times n$ Boolean matrix $(\boldsymbol{P} \circ C_{|h,c})_{(\geq n-2)}$ is indeed uniform for a random function $\boldsymbol{F}$ (rather than a random permutation), given any $h \in \{0,1\}^n$ and $c \in \{0,1\}$. However, even for a random function the Boolean matrices obtained for different h, c values are correlated. Nevertheless, these correlations (and the fact that $\boldsymbol{F}$ is a permutation) do not seem to have a noticeable influence on our algorithm in practice (as we demonstrate in Sect. 8.5 and the extended version of this paper [9]).

The rank of random matrices is a well-studied problem. For large n and a non-negative integer $r \leq n$, we denote the probability that a random Boolean $n \times n$ matrix has rank r by β_r. We can lower bound β_r by considering the event where we first select r linearly independent rows to form a subspace of size 2^r (which occurs with constant probability) and then select the remaining $n - r$ rows within this subspace (which occurs with probability $2^{-(r-n)^2}$). This gives a lower bound of $\Omega(2^{-(r-n)^2})$ on β_r. The exact formula is given by the theorem below, taken and adapted from [13].

Theorem 3 ([13], p. 126, adapted). *For* $n \to \infty$, *the probability that a random Boolean* $n \times n$ *matrix has rank* r *is*

$$\beta_r = 2^{-(r-n)^2} \cdot \alpha \cdot \prod_{i=1}^{n-r} (1 - 1/2^i)^{-2}, \tag{2}$$

where $\alpha = \prod_{i=1}^{\infty} (1 - 1/2^i) \approx 0.2888$.

Since $\alpha \leq \alpha \cdot \prod_{i=1}^{n-r} (1 - 1/2^i)^{-2} < 1/\alpha$, the initial probability estimation of $\approx 2^{-(r-n)^2}$ is correct up to a small constant. We also note that (2) is a good estimation even for relatively small values of n (e.g., $n \geq 8$). Let

$$p_{maxR,minR} = \begin{cases} 2\beta_{maxR}\beta_{minR} & \text{if } minR < maxR \\ \beta_{maxR}\beta_{minR} & \text{otherwise } (minR = maxR), \end{cases} \tag{3}$$

where

$$\beta_{maxR}\beta_{minR} = \alpha^2 \cdot 2^{-(maxR-n)^2-(minR-n)^2} \cdot \prod_{i=1}^{n-maxR} (1 - 1/2^i)^{-2} \cdot \prod_{i=1}^{n-minR} (1 - 1/2^i)^{-2}.$$

Then, based on Assumption 1 and Theorem 3, for every $(maxR, minR) \in \mathbb{Z}_{m+1} \times \mathbb{Z}_{m+1}$ such that $maxR \geq minR$, given $h \in \{0,1\}^n$, we have $Pr[R_{F,n-2,h} = (maxR, minR)] \approx p_{maxR,minR}$. Hence, according to Assumption 1, the entries of $\mathcal{H}_{F,n-2}$ are distributed multinomially, with parameter 2^n (the number of HSMs[7]) and probabilities given by $p_{maxR,minR}$. In particular, each individual histogram entry $\mathcal{H}_{F,n-2}(maxR, minR)$ is distributed binomially with parameter 2^n and probability $p_{maxR,minR}$.

Experimental results that support this conclusion are given in the extended version of this paper [9].

Asymptotic Analysis of Specific Histogram Entries. For large n, the binomial variable $\mathcal{H}_{F,n-2}(maxR, minR)$ is with high probability very close to its expectation, which is about

$$2^n \cdot p_{maxR,minR}.$$

If we ignore constant multiplicative factors, we can approximate this expectation by

$$2^n \cdot 2^{-(maxR-n)^2-(minR-n)^2}, \tag{4}$$

as $p_{maxR,minR} \approx 2^{-(maxR-n)^2-(minR-n)^2}$.

We now approximate (up to constant multiplicative factors) the expected values of two specific histogram entries which will be useful for our algorithm. Denote $\gamma_n = \lfloor (n/2)^{1/2} \rfloor$, and let $r_1 = (n+1-\gamma_n, n-\gamma_n)$ and $r_2 = (n, n-\gamma_n)$. Define the random variables $S_1 = \mathcal{H}_{F,d}(r_1)$ and $S_2 = \mathcal{H}_{F,d}(r_2)$. Below, we estimate their expected values according to (4).

Write $\gamma_n = \lfloor (n/2)^{1/2} \rfloor = (n/2)^{1/2} - k$, where $0 \leq k < 1$. Hence, with very high probability we have $S_2 = \mathcal{H}_{F,d}(r_2) = \mathcal{H}_{F,d}(n, n-\gamma_n) \approx 2^n \cdot p_{n,n-\gamma_n} \approx 2^n \cdot 2^{-(\gamma_n)^2} = 2^n \cdot 2^{-((n/2)^{1/2}-k)^2} = 2^n \cdot 2^{-n/2+2k(n/2)^{1/2}-k^2} = 2^{n/2+O(n^{1/2})}$. Therefore, S_2 is close to $2^{n/2}$.

Similarly $S_1 = \mathcal{H}_{F,d}(r_1) = \mathcal{H}_{F,d}(n+1-\gamma_n, n-\gamma_n) \approx 2^n \cdot p_{n+1-\gamma_n,n-\gamma_n} \approx 2^n \cdot 2^{-(\gamma_n-1)^2-(\gamma_n)^2} = 2^n \cdot 2^{-2(\gamma_n)^2+2\gamma_n-1} = 2^n \cdot 2^{-n+(4k+2)(n/2)^{1/2}-2k^2-2k-1} = 2^{\Theta(n^{1/2})}$. Hence S_1 is sub-exponential in n.

8   Details of the New Affine Equivalence Algorithm

In this section we describe and analyze our new affine equivalence algorithm. We start with a description of the auxiliary algorithms it uses.

---

[7] More accurately, the parameter is $2^n - 1$ as HSMs are non-zero.

8.1   The Rank Table and Histogram Algorithm

For $F : \{0,1\}^n \rightarrow \{0,1\}^n$ represented by $P = \{P^{(i)}(x[1], \dots, x[n])\}_{i=1}^n$, the following algorithm computes the rank histogram $\mathcal{H}_{F,d}$ and rank table $\mathcal{T}_{F,d}$ for $d = n - 2$. The algorithm is given as input $P_{\geq(n-2)}$.

---

1. For each non-zero HSM $h \in \{0,1\}^n$:
   (a) Compute $R_{F,n-2,h} = (maxR, minR)$ as follows. Compute $(P_{(\geq n-2)} \circ C_{|h,0})_{(\geq n-2)} = (P \circ C_{|h,0})_{(\geq n-2)}$ and calculate its symbolic rank r_0 using Gaussian elimination. Similarly, compute $(P_{(\geq n-2)} \circ C_{|h,1})_{(\geq n-2)} = (P \circ C_{|h,1})_{(\geq n-2)}$ and its symbolic rank r_1. Let $maxR = max\{r_0, r_1\}$ and $minR = min\{r_0, r_1\}$.
   (b) Insert h into $\mathcal{T}_{F,n-2}(maxR, minR)$, along with the value of the attached constant $c \in \{0,1\}$ such that $maxR = SR((F \circ C_{|h,c})_{(\geq n-2)})$ (if $maxR > minR$). In addition, increment entry $\mathcal{H}_{F,n-2}(maxR, minR)$.

Note that $(P_{(\geq n-2)} \circ C_{|h,0})_{(\geq n-2)} = (P \circ C_{|h,0})_{(\geq n-2)}$ holds according to the first part of Lemma 1.

The time complexity of the algorithm depends on how a polynomial is represented. Here, we represent it using a bit array that specifies the values of its monomial coefficients.

We first analyze the complexity of computing the composition $(P_{(\geq n-2)} \circ C_{|h,c})_{(\geq n-2)}$ in Step 1.(a), which is performed for each non-zero $h \in \{0,1\}^n$ and $c \in \{0,1\}$. Each of the n polynomials of $P_{(\geq n-2)}$ contains at most $\binom{n}{n} + \binom{n}{n-1} + \binom{n}{n-2} < n^2$ non-zero monomials. As described in Sect. 2, computing the composition $P_{(\geq n-2)} \circ C_{|h,c}$ requires substituting one of the n variables with a linear combination of the remaining $n-1$ variables (while renaming the variables of the monomials).

In total, for each polynomial of $P_{(\geq n-2)} \circ C_{|h,c}$, we compose its n^2 monomials with a linear combination of size n, which requires $n^2 \cdot n = n^3$ bit operations. However, as we are only interested in monomials of degree at least $n - 2$, the outcome $(P_{(\geq n-2)} \circ C_{|h,0})_{(\geq n-2)}$ is a polynomial of at most $\binom{n-1}{n-1} + \binom{n-1}{n-2} = 1 + n - 1 = n$ monomials, and the average complexity can be easily reduced to n^2 using low-level optimization techniques.[8]

In conclusion, the average complexity of computing the n polynomials of $(P_{(\geq n-2)} \circ C_{|h,0})_{(\geq n-2)}$ is $n \cdot n^2 = n^3$ and the total time spent on composition is $n^3 \cdot 2^n$ bit operations (up to multiplicative constant factors). Similarly, Gaussian elimination requires n^3 bit operations, hence the total time complexity of the algorithm is $n^3 \cdot 2^n$ bit operations.

---

[8] For example, we can exploit the fact that the composition $(P_{(\geq n-2)} \circ C_{|h,c})_{(\geq n-2)}$ is computed for each $h \in \{0,1\}^n$ and $c \in \{0,1\}$, and the effect of flipping a bit in h on the outcome can be precomputed. Consequently, we iterate over $h \in \{0,1\}^n$ using a Gray code.

8.2   The Unique HSM Algorithm

The following algorithm computes the HSM rank histogram multi-set $\mathcal{HM}_{F,n-2,r,r'}$ and uses it to compute a set of unique HSMs, denoted by U_F. This set contains triplets of the form $(h, c, \mathcal{HG}_{F,n-2,h,r'})$, where $h \in \mathcal{T}_{F,n-2}(r)$ is unique with respect to $F, n-2, r'$ and $c \in \{0, 1\}$ is its attached constant. Note that for the attached constant to be defined, we must have $maxR > minR$, where $(maxR, minR) = r$.

The algorithm is given as input the rank table $\mathcal{T}_{F,n-2}$ and rank group indexes r, r'.

---

1. For each $h \in \mathcal{T}_{F,n-2}(r)$, compute $\mathcal{HG}_{F,n-2,h,r'}$ as follows:
   (a) for each $h' \in \mathcal{T}_{F,n-2}(r')$:
       i. Compute $h + h'$, find its rank $r'' = R_{F,n-2,h+h'}$ in $\mathcal{T}_{F,n-2}$ and increment $\mathcal{HG}_{F,n-2,h,r'}(r'')$.
   (b) Insert $\mathcal{HG}_{F,n-2,h,r'}$ along with h and its attached constant c into the multi-set $\mathcal{HM}_{F,n-2,r,r'}$.
2. For each unique HSM h in $\mathcal{HM}_{F,n-2,r,r'}$, add the triplet $(h, c, \mathcal{HG}_{F,n-2,h,r'})$ to U_F.

The time complexity of the algorithm is the product of sizes of the rank groups $|\mathcal{T}_{F,n-2}(r)| \cdot |\mathcal{T}_{F,n-2}(r')| = \mathcal{H}_{F,n-2}(r) \cdot \mathcal{H}_{F,n-2}(r')$.

Since the goal of the affine equivalence algorithm will be to find n linearly independent unique HSMs, it is useful to estimate their number. In the extended version of this paper [9] we lower bound the expected number of unique HSMs in $\mathcal{HM}_{F,n-2,r,r'}$ asymptotically (ignoring constant factors) based on Assumption 1, given that F is a random permutation. More specifically, we obtain the lower bound of $S - S^2/\sqrt{S'}$, where $S = \mathcal{H}_{F,n-2}(r)$, and $S' = \mathcal{H}_{F,n-2}(r')$.

8.3   The Affine Transformation A_1 Recovery Algorithm

Assume that we have affine equivalent functions F and G such that $G = A_2 \circ F \circ A_1$ and $A_1(x) = L(x) + a$.

The following algorithm recovers A_1 using sets of unique HSMs U_F and U_G, computed with the previous algorithm of Sect. 8.2 (where its invocations for F and G use the same parameters values of r, r'). Since F and G are affine equivalent, the HSM rank histograms of the HSMs in these sets have to match according to Theorem 2. Each equal HSM histogram pair reveals the matching $h \mapsto_{A_1} h'$ and its associated constant is revealed by adding the attached constants of h, h' (which are defined in case $maxR > minR$, where $(maxR, minR) = r$), again by Theorem 2.

Each matching $h \mapsto_{A_1} h'$ and its associated constant give linear equations on the columns of L and on a (respectively) according to Lemma 5. Assuming that U_F and U_G contain n linearly independent unique HSMs, A_1 is recovered by linear algebra.

1. Allocate $n+1$ linear equation systems $\{E_i\}_{i=1}^{n+1}$, each of dimension $n \times n$: the first n equation systems are on the columns $L[i]$ of L and the final equation system E_{n+1} is on a.
2. Locate n linearly independent HSMs in U_G. For each such HSM h:
   (a) Recover the triplet $(h, c, \mathcal{HG}_{G,n-2,h,r'})$ from U_G.
   (b) Search U_F for a triplet $(h', c', \mathcal{HG}_{F,n-2,h',r'})$ such that $\mathcal{HG}_{F,n-2,h',r'} = \mathcal{HG}_{G,n-2,h,r'}$. If no match exists, return "Not Equivalent".
   (c) Based on Lemma 5, for $i = 1, 2, \ldots, n$ add equation $h'(L[i]) = h[i]$ to E_i.
   (d) Based on Lemma 5, add equation $h'(a) = c + c'$ to E_{n+1}.
3. Solve each one of $\{E_i\}_{i=1}^{n+1}$, recover A_1 and return its matrix L and vector a.

The complexity of the algorithm is about $n \cdot n^3 = n^4$ bit operations, which is polynomial in n. Since we solve the same linear equation (with coefficients given by the h' vectors) $n + 1$ times with different constants, the complexity can be reduced to n^3 by inverting the matrix which defines the linear equations.

8.4   The New Affine Equivalence Algorithm

We describe the new affine equivalence algorithm below. Let $r_1 = (n+1-\gamma_n, n-\gamma_n)$ and $r_2 = (n, n - \gamma_n)$ for $\gamma_n = \lfloor (n/2)^{1/2} \rfloor$, as defined in Sect. 7.

1. Given $F : \{0,1\}^n \rightarrow \{0,1\}^n$, $G : \{0,1\}^n \rightarrow \{0,1\}^n$, compute their corresponding ANF representations $P_{\geq(n-2)}$ and $Q_{\geq(n-2)}$.
2. Run the algorithm of Section 8.1 to compute the rank table $\mathcal{T}_{F,n-2}$ and rank histogram $\mathcal{H}_{F,n-2}$ for F using $P_{\geq(n-2)}$, and similarly compute $\mathcal{T}_{G,n-2}$ and $\mathcal{H}_{G,n-2}$ for G. If $\mathcal{H}_{F,n-2} \neq \mathcal{H}_{G,n-2}$, return "Not Equivalent".
3. Run the unique HSM algorithm of Section 8.2 for F on inputs $\mathcal{T}_{F,n-2}$ and r_1, r_2 defined above, and obtain the set U_F. Similarly, obtain the set for U_G by running this algorithm on inputs $\mathcal{T}_{G,n-2}$ and r_1, r_2.
   If $|U_F| \neq |U_G|$, return "Not Equivalent". Otherwise, if U_F does not contain n linearly independent HSMs, return "Fail".
4. Run the affine transformation recovery algorithm of Section 8.3 on inputs U_F and U_G. If it returns "Not Equivalent", return the same output. Otherwise, it returns a candidate for A_1.
5. Recover a candidate for $A_2 = L_2(x) + a_2$ by evaluating inputs $v \in \{0,1\}^n$ to $F \circ A_1$ and G: each input v gives n linear equations on L_2 and a_2. Hence, after a bit more than n evaluations, we expect the linear equation system to have a single solution which gives a candidate for A_2.
6. Test the candidates A_1, A_2 by equating the evaluations of G and $A_2 \circ F \circ A_1$ on all 2^n possible inputs. If $G(v) \neq A_2 \circ F \circ A_1(v)$ for some $v \in \{0,1\}^n$, return "Not Equivalent". Otherwise, return A_1, A_2.

The correctness of the algorithm follows from the correctness of the sub-procedures is executes and from the results obtained so far. In particular,

Step 2 is correct according to Lemma 9, while the correctness of Step 3 is based on Theorem 2. The correctness of the final step is trivial.

Step 3 is the most complex to analyze in terms of success probability and complexity (which is the product $\mathcal{H}_{F,n-2}(r_1) \cdot \mathcal{H}_{F,n-2}(r_2)$). We first focus on the time complexity analysis of the other steps.

The complexities of steps 4, 5 are at most polynomial in n and can be neglected. Step 1 interpolates the 2^n ANF coefficients for each of the n output bits of F, G. Each such interpolation can be performed in $n2^n$ bit operations using the Moebius transform [12]. Hence this step requires $n^2 2^n$ bit operations and 2^n function evaluations in total. The complexity of Step 2 was shown to be $n^3 2^n$ bit operations, while the complexity of Step 6 is 2^n function evaluations.

In total, the time complexity of the algorithm is at most $n^3 2^n$ bit operations and 2^n function evaluations, assuming that the complexity of Step 3 does not dominate the algorithm (as we show below).

The memory complexity is 2^n words of n bits, but it can be significantly reduced in some cases as described in Sect. 8.6.

Asymptotic Analysis of the Unique HSM Algorithm. As in Sect. 7, denote $S_1 = \mathcal{H}_{F,d}(r_1)$ and $S_2 = \mathcal{H}_{F,d}(r_2)$ and recall that their expected values are $2^{\Theta(n^{1/2})}$ and $2^{n/2+O(n^{1/2})}$, respectively.

The expected asymptotic complexity of the unique HSM algorithm is therefore at most

$$S_1 \cdot S_2 = 2^{n/2+O(n^{1/2})} \ll 2^n.$$

Hence, the complexity of Step 3 is negligible compared to the complexity of the remaining steps of the affine equivalence algorithm described above.

According to the analysis of the unique HSM algorithm given in the extended version of this paper [9], the asymptotic lower bound on the expected number of unique HSMs in $\mathcal{HM}_{F,n-2,r,r'}$ is

$$S_1 - (S_1)^2/\sqrt{S_2} > 2^{\Theta(n^{1/2})} - 2^{\Theta(n^{1/2})}/2^{n/4+O(n^{1/4})} = 2^{\Theta(n^{1/2})} \gg n.$$

Out of these unique HSMs, n are very likely to be linearly independent. This shows that asymptotically the algorithm succeeds with overwhelming probability.

We remark that there are many possible ways to select the rank group indexes r_1 and r_2 that give similar results.

8.5   Experimental Results

In practice we do not pre-fix the rank groups of F, G for which we run the unique HSM algorithm. Instead, we select a *reference rank group* r' such that $|\mathcal{T}_{F,n-2}(r')| \approx 2^{n/2}$ (as r_2 defined above). We then iterate over the rank groups r from the smallest to the largest, while collecting unique HSMs using repeated executions of the unique HSM algorithm with inputs r, r'. We stop once we collect n linearly independent unique HSMs. This practical variant is more flexible and succeeds given that the variant above succeeds.

We implemented the algorithm and tested it for various values of $8 \leq n \leq 28$. In each trial we first selected the permutation F uniformly at random. We then chose invertible affine mappings A_1, A_2 uniformly at random and defined $G = A_2 \circ F \circ A_1$. After calculating these inputs, we executed the algorithm and verified that it correctly recovered A_1, A_2.

Following this initial verification, our goal was to collect statistics that support the asymptotic complexity analysis of the unique HSM algorithm above. For this purpose, we selected the permutation F at random and calculated the success rate and complexity of Step 3, which executes the unique HSM algorithm on F (after running steps 1, 2). If Step 3 succeeds to return n linearly independent unique HSMs with a certain complexity for F, then it would succeed with identical complexity on any linearly equivalent G, hence analyzing a single permutation is sufficient for the purpose of gathering statistics.

Our results for $n \in \{8, 12, 16, 20, 24, 28\}$ are summarized in Table 1. This table shows that all the trials for the various choices of n were successful. In terms of complexity, for $n = 8$, the unique HSM algorithm had to iterate over $2^{10.5} > 2^8$ HSMs on average in order to find 8 unique linearly independent HSMs. This relatively high complexity is due to the fact that our asymptotic analysis ignores constants whose effect is more pronounced for smaller values of n. Nevertheless, the complexity of the algorithm for $n = 8$ in terms of bit operations remains roughly $8^3 \cdot 2^8$, as the unique HSM algorithm does not perform linear algebra.

For $n \geq 12$, the average complexity of the unique HSM algorithm is below 2^n, and this gap increases as n grows (as predicted by the asymptotic analysis). Note that the complexity drops in two cases (between $n = 16$ and $n = 20$ and between $n = 24$ and $n = 28$) since for larger n we have more non-empty rank groups of various sizes and hence more flexibility in the algorithm (which happens to be quite substantial for $n = 20$ and $n = 28$). Finally, we note that we did not optimize the index r' of the reference rank group and better options that improve the complexities are likely to exist. However, since the unique HSM algorithm does not dominate the overall complexity, such improvements would have negligible effect.

In addition to the experiments on random permutations, we also performed simulations on random functions and obtained similar results.

8.6   Additional Variants of the New Affine Equivalence Algorithm

We describe several variants of the affine equivalence algorithm.

Using Rank Group Sums. The first additional variant we describe uses the rank tables of F, G to directly recover several matchings in the initial stage of the algorithm. It is based on the observation that for each non-empty rank group r, the HSM obtained by summing of all HSMs in $\mathcal{T}_{G,n-2}(r)$ has to match (under $\mapsto_{(A_1)}$) the HSM obtained by summing of all HSMs in $\mathcal{T}_{F,n-2}(r)$ due to the additive property of the HSM bijection. Simple analysis (based on Assumption 1 and backed up by experimental results) shows that the number of non-empty rank

Table 1. Experimental results for the unique HSM algorithm

n	Number of trials	Number of successful trials	Average complexity
8	1000	1000	$1486 \approx 2^{10.5}$
12	1000	1000	$3229 \approx 2^{11.7}$
16	1000	1000	$25599 \approx 2^{14.6}$
20	1000	1000	$15154 \approx 2^{13.9}$
24	1000	1000	$126777 \approx 2^{17}$
28	100	100	$40834 \approx 2^{15.3}$

groups for a random permutation F is at least $n/4$ with very high probability. Hence we can initially recover at least $n/4$ matchings using this approach. There are several ways to recover the remaining matchings by exploiting the fact that we have essentially reduced the size of the problem from 2^n to at most $2^{3n/4}$. We can also continue in a similar way, further exploiting additive properties of the bijection: we take a uniquely matched HSM pair h, h'. For G, we compute the *HSM rank table* for h with respect to some rank group r' by adding it to all HSMs in this group. We do the same for F by computing the HSM rank table of h' with respect to r'. As in the initial observation, the sum of HSMs in each non-empty rank group of these smaller tables for F, G match under $\mapsto_{(A_1)}$, revealing additional matchings. We repeat this process for several uniquely matched HSM pairs (computing additional HSM rank tables) until we identify the required n linearly independent matchings.

Reducing the Memory Complexity. The memory complexity of the algorithm is about 2^n words of n bits. If the functions F, G are given as truth tables, then the memory complexity cannot be reduced by much. However, if we are given access to F, G via oracles (e.g., they are implemented by block ciphers with a fixed key), then we can significantly reduce the memory complexity with no substantial effect on the time complexity.

First, instead of using the Moebius transform in Step 1 in order to interpolate all the coefficients of F, G, we simply interpolate each of the relevant $\approx n^2$ coefficients of degree at least $n - 2$ independently, increasing the complexity of Step 1 by a factor of about n. Next, in Step 2 we do not store the entire rank table, but only the relevant rank groups with indexes r_1 and r_2. As a result, we now have to recompute the ranks of $S_1 \cdot S_2$ HSMs in Step 3, but this requires much lower complexity than 2^n.

Overall, the memory complexity of this low-memory variant is dominated by the size of largest rank group stored in memory S_2, which is bit more than $2^{n/2}$. Finally, by a different choice of rank groups of indexes r_1 and r_2, it is possible reduce the memory to be sub-exponential in n.

Multiple Solutions to the Affine Equivalence Problem. Consider the case where there are two or more solutions of the form $(A_1^{(i)}, A_2^{(i)})$ to an instance of the affine equivalence problem F, G. This may occur (for example) if F is self-affine equivalent, namely, there exist (A_1, A_2) (that are not both identities) such that $F = A_2 \circ F \circ A_1$. We note that this case is extremely unlikely if F is chosen uniformly at random for $n \geq 8$, but it may occur for specific choices (e.g., the AES Sbox is self-affine equivalent).

In case of multiple solutions, a straightforward application of the affine equivalence algorithm would fail, as a HSM h would most likely match a different $h'^{(i)}$ for each solution $A_1^{(i)}$, namely $h \mapsto_{A_1^{(i)}} h'^{(i)}$. Consequently, we would not be able to find sufficiently many unique HSMs in Step 4. However, we can tweak the algorithm to deal with this case by working on each match $h \mapsto_{A_1^{(i)}} h'^{(i)}$ separately. More specifically, according to Lemma 6 we know that $F \circ C_{|h'^{(i)}, c} \equiv G \circ C_{|h, 0}$ and we can apply the algorithm recursively on these functions.

Affine Equivalences Among a Set of Functions. We consider a generalization of the affine equivalence problem that was described in [3]. Given a set of K functions $\{F_i\}_{i=1}^{K}$, our goal is to partition them into groups of affine equivalent functions. The naive approach is to run the affine equivalence algorithm on each pair of functions, which results in complexity of $K^2 \cdot n^3 2^n$.

We can improve this complexity by noticing that up to Step 4 of the affine equivalence algorithm the functions F, G are analyzed independently. In particular, we can compute the rank histogram $\mathcal{H}_{F_i, n-2}$ for each function F_i independently (as done in Step 2) in time $n^3 2^n$, and then sort the functions and classify them according to their rank histograms.[9] This reduces the time complexity to about $K \cdot n^3 2^n + \tilde{O}(K^2)$ (where $\tilde{O}$ hides a small polynomial factor in n), improving upon the time complexity of $K \cdot n^3 2^{2n} + \tilde{O}(K^2)$, obtained in [3].

9 Applications

We describe applications of the affine equivalence algorithm and then focus on additional applications of the new objects and algorithms defined in this paper.

9.1 Applications of the New Affine Equivalence Algorithm

Algorithms for the affine equivalence problem are useful in several contexts such as classification of Sboxes [6,14], producing equivalent representations of block ciphers [3] and attacking white-box ciphers [15]. In all of these contexts, if the goal is to apply the algorithm a few times to functions with a small domain size n, then the main algorithm of Biryukov et al. [3] is already practical and there is little to be gained by using our algorithm.

---

[9] We can also attach more data to each function by computing HSM histogram multisets between groups $\mathcal{HM}_{F_i, n-2, r, r'}$.

On the other hand, our algorithm may provide an advantage if the goal is to solve the affine equivalence problem on functions with a larger domain sizes (e.g., the domain size of the CAST Sbox [1] is $n = 32$). Furthermore, our algorithm may be beneficial if we need to solve the affine equivalence problem for many functions with domain size $n \geq 8$. For example, if we want to classify a large set of 8-bit Sboxes produced based on some design criteria, we can use the variant that searches for affine equivalences among a set of functions (described in Sect. 8.6).

An additional application (which is also described in [3]) is cryptanalysis of a generalization of the Even-Mansour scheme. The original scheme [11] builds a block cipher using a public permutation $F : \{0,1\}^n \rightarrow \{0,1\}^n$ and a pair of n-bit keys k_1, k_2 by defining the encryption function on a plaintext $p \in \{0,1\}^n$ as $E(p) = F(p+k_1)+k_2$. Breaking the scheme may be considered as a special case of solving the affine equivalence problem where the linear matrices are identities. Thus, in the generalized scheme, arbitrary affine transformations A_1, A_2 are used as the key and the encryption function is defined as $E(p) = A_2 \circ F \circ A_1(p)$. Clearly, breaking the generalized Even-Mansour scheme reduces to solving the affine equivalence problem. The currently best know attack on this scheme (given in [3]) requires about $2^{3n/2}$ time and memory. It uses a birthday paradox based approach that generalizes Daemen's attack on the original Even-Mansour cipher [8]. Hence, we improve the complexity of the best known attack on the generalized Even-Mansour cipher from about $2^{3n/2}$ to 2^n.

9.2   Additional Applications

We describe additional applications of the rank table and histogram objects defined in this paper, and the algorithm used to compute them.

Application to Decomposition of the ASASA Construction. The ASASA construction is an SP-network that consists of three secret affine layers (A) interleaved with two secret Sbox layers (S). At ASIACRYPT 2014, Biryukov et al. [2] proposed several concrete ASASA block cipher designs as candidates for white-box cryptography, whose security was based on the alleged difficulty of recovering their internal components. These designs were subsequently broken in [16] and [10].

Of particular interest is the integral attack of [10]. While the full details of this attack are out of the scope of this paper, we focus on its heaviest computational step that consists of summing over about 2^n affine subspaces of dimension slightly less than n (where n is the block size of the scheme). This step was performed in [10] in complexity of about $2^{3n/2}$. We can improve the complexity of this step (and the complexity of the full attack) to about 2^n by using a symbolic algorithm which is similar to the one used for computing the rank table.

Application to Distinguishers on Sboxes and Block Ciphers. In [4] Biryukov and Perrin considered the problem of reverse-engineering Sboxes and proposed techniques to check whether a given Sbox was selected at random or was designed according to some unknown criteria. These techniques are based on the linear approximation table (LAT) and difference distribution table (DDT) of the Sbox. Here, we provide another method based on the distribution of entries in the rank histogram of the Sbox. More specifically, an Sbox would be considered suspicious if its rank histogram entry sizes differ significantly from their expected values according to the distribution derived in Sect. 7 (supported by the experimental results of the extended version of this paper [9]).

An advantage of our proposal is that the LAT and DDT require about 2^{2n} time and memory to compute and store, whereas the rank histogram can be computed in time of about 2^n. Hence, our proposal can be used to analyze larger Sboxes. We can also use additional properties of HSM rank histogram multi-sets (such as the number of unique HSMs) as possible distinguishing techniques.

In a related application, the rank table (and additional structures defined in this paper) can be used to experimentally construct distinguishers on block ciphers with a small block size (e.g., 32 bits). This is done be selecting a few keys for the block cipher at random and detecting consistent deviations from random among the resultant permutations. In particular, if there is a linear combination of the output bits that is a low-degree function of some $(n-1)$-dimensional input subspace, then we can detect it in time complexity of about 2^n. Since there are 2^{n+1} possible $(n-1)$-dimensional affine subspaces and 2^n linear combinations of output bits, we search over a space of 2^{2n+1} possible distinguishers in about 2^n time. This can be viewed as an improvement over known experimental methods [19] that search a much smaller space containing about n^2 potential high-order differential distinguishers in similar complexity (these methods only consider the input and output bits, but not their linear combinations). Finally, the technique can also be used on block ciphers with larger block sizes by considering linear subspaces of the input domain and output range.

10 Conclusions and Open Problems

In this paper we described an improved algorithm for the affine equivalence problem, focusing on randomly chosen permutations. The main open problem is to further improve the algorithm's complexity and applicability. An additional future work item is to find more applications for the rank table and related structures defined in this paper.

References

1. Adams, C.M.: Constructing symmetric ciphers using the CAST design procedure. Des. Codes Cryptogr. **12**(3), 283–316 (1997)
2. Biryukov, A., Bouillaguet, C., Khovratovich, D.: Cryptographic schemes based on the ASASA structure: black-box, white-box, and public-key (extended abstract). In: Sarkar, P., Iwata, T. (eds.) ASIACRYPT 2014. LNCS, vol. 8873, pp. 63–84. Springer, Heidelberg (2014). https://doi.org/10.1007/978-3-662-45611-8_4

3. Biryukov, A., De Cannière, C., Braeken, A., Preneel, B.: A toolbox for cryptanalysis: linear and affine equivalence algorithms. In: Biham, E. (ed.) EUROCRYPT 2003. LNCS, vol. 2656, pp. 33–50. Springer, Heidelberg (2003). https://doi.org/10.1007/3-540-39200-9_3
4. Biryukov, A., Perrin, L.: On reverse-engineering S-boxes with hidden design criteria or structure. In: Gennaro, R., Robshaw, M. (eds.) CRYPTO 2015. LNCS, vol. 9215, pp. 116–140. Springer, Heidelberg (2015). https://doi.org/10.1007/978-3-662-47989-6_6
5. Bouillaguet, C., Fouque, P.-A., Véber, A.: Graph-theoretic algorithms for the "Isomorphism of Polynomials" problem. In: Johansson, T., Nguyen, P.Q. (eds.) EUROCRYPT 2013. LNCS, vol. 7881, pp. 211–227. Springer, Heidelberg (2013). https://doi.org/10.1007/978-3-642-38348-9_13
6. Brinkmann, M., Leander, G.: On the classification of APN functions up to dimension five. Des. Codes Cryptogr. **49**(1–3), 273–288 (2008)
7. Canteaut, A., Roué, J.: On the behaviors of affine equivalent sboxes regarding differential and linear attacks. In: Oswald, E., Fischlin, M. (eds.) EUROCRYPT 2015. LNCS, vol. 9056, pp. 45–74. Springer, Heidelberg (2015). https://doi.org/10.1007/978-3-662-46800-5_3
8. Daemen, J.: Limitations of the Even-Mansour construction. In: Imai, H., Rivest, R.L., Matsumoto, T. (eds.) ASIACRYPT 1991. LNCS, vol. 739, pp. 495–498. Springer, Heidelberg (1993). https://doi.org/10.1007/3-540-57332-1_46
9. Dinur, I.: An improved affine equivalence algorithm for random permutations. IACR Cryptology ePrint Archive 2018, p. 115 (2018)
10. Dinur, I., Dunkelman, O., Kranz, T., Leander, G.: Decomposing the asasa block cipher construction. IACR Cryptology ePrint Archive 2015, p. 507 (2015)
11. Even, S., Mansour, Y.: A construction of a cipher from a single pseudorandom permutation. J. Cryptol. **10**(3), 151–162 (1997)
12. Joux, A.: Algorithmic Cryptanalysis. Chapman & Hall, London (2009)
13. Kolchin, V.F.: Random Graphs. Cambridge University Press, Cambridge (1999)
14. Leander, G., Poschmann, A.: On the classification of 4 bit S-boxes. In: Carlet, C., Sunar, B. (eds.) WAIFI 2007. LNCS, vol. 4547, pp. 159–176. Springer, Heidelberg (2007). https://doi.org/10.1007/978-3-540-73074-3_13
15. Michiels, W., Gorissen, P., Hollmann, H.D.L.: Cryptanalysis of a generic class of white-box implementations. In: Avanzi, R.M., Keliher, L., Sica, F. (eds.) SAC 2008. LNCS, vol. 5381, pp. 414–428. Springer, Heidelberg (2009). https://doi.org/10.1007/978-3-642-04159-4_27
16. Minaud, B., Derbez, P., Fouque, P.-A., Karpman, P.: Key-recovery attacks on ASASA. In: Iwata, T., Cheon, J.H. (eds.) ASIACRYPT 2015. LNCS, vol. 9453, pp. 3–27. Springer, Heidelberg (2015). https://doi.org/10.1007/978-3-662-48800-3_1
17. Patarin, J.: Hidden Fields Equations (HFE) and Isomorphisms of Polynomials (IP): two new families of asymmetric algorithms. In: Maurer, U. (ed.) EUROCRYPT 1996. LNCS, vol. 1070, pp. 33–48. Springer, Heidelberg (1996). https://doi.org/10.1007/3-540-68339-9_4
18. Patarin, J., Goubin, L., Courtois, N.: Improved algorithms for isomorphisms of polynomials. In: Nyberg, K. (ed.) EUROCRYPT 1998. LNCS, vol. 1403, pp. 184–200. Springer, Heidelberg (1998). https://doi.org/10.1007/BFb0054126
19. Wang, Q., Liu, Z., Varıcı, K., Sasaki, Y., Rijmen, V., Todo, Y.: Cryptanalysis of reduced-round SIMON32 and SIMON48. In: Meier, W., Mukhopadhyay, D. (eds.) INDOCRYPT 2014. LNCS, vol. 8885, pp. 143–160. Springer, Cham (2014). https://doi.org/10.1007/978-3-319-13039-2_9

Galois Counter Mode

Optimal Forgeries Against
Polynomial-Based MACs and GCM

Atul Luykx[1]([✉]) and Bart Preneel[2]

[1] Visa Research, Palo Alto, USA
aluykx@visa.com
[2] imec-COSIC, KU Leuven, Leuven, Belgium
bart.preneel@esat.kuleuven.be

Abstract. Polynomial-based authentication algorithms, such as GCM
and Poly1305, have seen widespread adoption in practice. Due to their
importance, a significant amount of attention has been given to understanding and improving both proofs and attacks against such schemes.
At EUROCRYPT 2005, Bernstein published the best known analysis of
the schemes when instantiated with PRPs, thereby establishing the most
lenient limits on the amount of data the schemes can process per key.
A long line of work, initiated by Handschuh and Preneel at CRYPTO
2008, finds the best known attacks, advancing our understanding of the
fragility of the schemes. Yet surprisingly, no known attacks perform as
well as the predicted worst-case attacks allowed by Bernstein's analysis, nor has there been any advancement in proofs improving Bernstein's
bounds, and the gap between attacks and analysis is significant. We settle
the issue by finding a novel attack against polynomial-based authentication algorithms using PRPs, and combine it with new analysis, to show
that Bernstein's bound, and our attacks, are optimal.

Keywords: Forgery · Wegman-Carter · Authenticator · MAC
GCM · Universal hash · Polynomial

1 Introduction

Polynomial-based universal hash functions [dB93, Tay93, BJKS93] are simple and
fast. They map inputs to polynomials, which are then evaluated on keys to produce output. When used to provide data authenticity as Message Authentication
Code (MAC) algorithms or in Authenticated Encryption (AE) schemes, they
often take the form of Wegman-Carter (WC) authenticators [WC81], which add
the polynomial output to randomly generated values.

Part of the appeal of such polynomial-based WC authenticators is that if the
polynomial keys and random values are generated independently and uniformly
for each message, then information-theoretic security is achieved, as initially
explored by Gilbert, MacWilliams, and Sloane [GMS74], following pioneering

© International Association for Cryptologic Research 2018
J. B. Nielsen and V. Rijmen (Eds.): EUROCRYPT 2018, LNCS 10820, pp. 445–467, 2018.
https://doi.org/10.1007/978-3-319-78381-9_17

work by Simmons as described in [Sim91]. However, in the interest of speed and practicality, tweaks were introduced to WC authenticators, seemingly not affecting security.

Wegman and Carter [WC81] introduced one of the first such tweaks[1], by holding polynomial keys constant across messages, which maintained security as long as the polynomial outputs are still added to fresh random values each time. Further work then instantiated the random values via a pseudorandom number generator [Bra82], pseudorandom function (PRF), and then pseudorandom permutation (PRP) outputs [Sho96], the latter being dubbed Wegman-Carter-Shoup (WCS) authenticators by Bernstein [Ber05b]. Uniqueness of the PRF and PRP outputs is guaranteed using a nonce. With m the message and n the nonce, the resulting constructions take the form $(n, m) \mapsto \pi(n) + \rho(m)$, with π the PRF or PRP, and ρ the universal hash function.

The switch to using PRFs and PRPs means that information-theoretic is replaced by complexity-theoretic security. Furthermore, switching to PRPs in WCS authenticators results in security bound degradation, impacting the amount of data that can be processed per key (as, for example, exploited by the Sweet32 attacks [BL16]). Naïve analysis uses the fact that PRPs are indistinguishable from PRFs up to the birthday bound, however this imposes stringent limits. Shoup [Sho96], and then Bernstein [Ber05b] improve this analysis significantly using advanced techniques, yet do not remove the birthday bound limit. Regardless, despite the data limits, the use of PRPs enables practical and fast instantiations of MAC and AE algorithms, such as Poly1305-AES [Ber05c] and GCM [MV04a, MV04b], the latter of which has seen widespread adoption in practice [VM06, SMC08, IS09].

As a result of the increased significance of WCS authenticators schemes like GCM, more recent work has focused on trying to understand their fragility when deployed in the real-world. The history of attacks against WC and WCS authenticators consists of work exploring the consequences of fixing the polynomial key across all messages—once the polynomial key is known, all security is lost.

Joux [Jou] and Handschuh and Preneel [HP08] exhibit attacks which recover the polynomial key the moment a nonce is repeated. Ferguson [Fer05] explores attacks when tags are too short, further improved by Mattson and Westerlund [MW16]. A long line of work initiated by Handschuh and Preneel [HP08], illustrates how to efficiently exploit verification attempts to eliminate false keys, by systematically narrowing the set of potential polynomial keys and searching for so-called "weak" keys [Saa12, PC15, ABBT15, ZW17, ZTG13].

However, interestingly, in the case of polynomial-based WCS authenticators, none of the nonce-respecting attacks match the success of the predicted worst-case attacks by Bernstein [Ber05b]. Furthermore, the gap in success between the predicted worst-case and best-known attacks grows quadratically in the number of queries made to the authenticator. Naturally, one is led to question whether

---

[1] Strictly speaking, Wegman and Carter did not tweak the constructions pioneered by Simmons, as the connection between the two works was made only later by Stinson [Sti91].

Bernstein's analysis is in fact the best one can do, or whether there actually is an attack, forcing us to abide by the data limits.

1.1 Contributions

We exhibit novel nonce-respecting attacks against polynomial-based WCS authenticators (Sect. 3), and show how they naturally arise from a new, simplified proof (Sect. 4). We prove that both our attack and Bernstein's bound [Ber05b] are optimal, by showing they match (Sect. 5).

Unlike other birthday bound attacks, our attacks work by establishing quadratically many polynomial systems of equations from the tagging queries. It applies to polynomial-based WCS authenticators such as Poly1305-AES, as well as GCM and the variant SGCM [Saa11]. We achieve optimality in a chosen-plaintext setting, however the attacks can be mounted passively, using just known plaintext for MACs and ciphertext for AE schemes.

1.2 Related Work

Our introduction provides only a narrow view of the history of universal hash functions, targeted to ones based on polynomials. Bernstein [Ber05c] provides a genealogy of polynomial-based universal hash functions and Wegman-Carter authenticators, and both Procter and Cid [PC15,PC13] and Abdelraheem et al. [ABBT15] provide detailed overviews of the past attacks against polynomial-based Wegman-Carter MACs and GCM.

Zhu, Tan, and Gong [ZTG13] and Ferguson [Fer05] have pointed out that non-96-bit nonce GCM suffers from birthday bound attacks which lead to immediate recovery of the polynomial key. Such attacks use the fact that the nonce is processed by the universal hash function before being used, resulting in block cipher call collisions. These attacks are not applicable to the most widely deployed version of GCM, which uses 96 bit nonces, nor to polynomial-based WCS authenticators in general.

Iwata et al. [IOM12] identify and correct issues with GCM's original analysis [MV04a]. Niwa et al. find further improvements in GCM's bounds [NOMI15]. Their proofs do not improve over Bernstein's analysis [Ber05b].

New constructions using universal hash functions like EWCDM [CS16] achieve full security [MN17] in the nonce-respecting setting, and maintain security during nonce-misuse.

McGrew and Fluhrer [MF05] and Black and Cochran [BC09] explore how easy it is to find multiple forgeries once a single forgery has been performed.

A long line of research seeks attacks and proofs of constructions which match each other, such as the generic attack by Preneel and van Oorschot [PvO99], tight analysis for CBC-MAC [BPR05,Pie06], keyed sponges and truncated CBC [GPT15], and HMAC [GPR14], and new attacks for PMAC [LPSY16, GPR16].

2 Preliminaries

2.1 Basic Definitions and Notation

The notation used throughout the paper is summarized in Appendix C. Unless specified otherwise, all sets are assumed to be finite. Vectors are denoted $x \in X^q$, with corresponding components $(x_1, x_2, \ldots, x_q)$. Given a set X, $X^{\leq \ell}$ denotes the set of non-empty sequences of elements of X with length not greater than ℓ.

A *random function* $\rho : M \to T$ is a random variable distributed over the set of all functions from M to T. A *uniformly distributed random permutation* (URP) $\varphi : N \to N$ is a random variable distributed over the set of all permutations on N, where N is assumed to be finite. When we write $\varphi : N \to T$ is a URP, we implicitly assume that $N = T$.

The symbol $\mathbb{P}$ denotes a probability measure, and $\mathbb{E}$ expected value.

We make the following simplifications when discussing the algorithms. We analyze block cipher-based constructions by replacing each block cipher call with a URP call. This commonly used technique allows us to focus on the constructions' security without worrying about the underlying block cipher's quality. See for example [Ber05b]. Furthermore, although our analysis uses information-theoretic adversaries, the attacks we describe are efficient, but require large storage.

We also implicitly include key generation as part of the oracles. For example, consider a construction $E : K \times M \to T$, where E is stateless and deterministic, and K is its "key" input. In the settings we consider, E-queries are only actually made to $E(k, \cdot)$, where the key input is fixed to some random variable k chosen uniformly at random from K. Hence, rather than each time talking of $E(k, \cdot)$, we simplify notation by considering the random function $\rho(m) \stackrel{\text{def}}{=} E(k, m)$, with the uniform random variable k implicitly part of ρ's description.

2.2 Polynomial-Based WCS Authenticators

Although not necessary, for simplicity we fix tags to lie in a commutative group. The following definition is from Bernstein [Ber05b].

Definition 2.1 (WCS Authenticator). *Let* T *be a commutative group with operation* $+$. *Let* $\pi : N \to T$ *be a URP, and* $\rho : M \to T$ *a random function. The* Wegman-Carter-Shoup (WCS) authenticator *maps elements* $(n, m) \in N \times M$ *to* $\pi(n) + \rho(m)$.

We take the following definition from Procter and Cid [PC15].

Definition 2.2 (Polynomial-Based Universal Hash). *Let* X *be a field and* ℓ *a positive integer. Given* $x = (x_1, x_2, \ldots, x_l) \in X^{\leq \ell}$, *define the polynomial* $p_x(\alpha)$ *by*

$$p_x(\alpha) \stackrel{\text{def}}{=} \sum_{i=1}^{l} x_i \cdot \alpha^i. \tag{1}$$

Then the polynomial-based universal hash function $\rho : X^{\leq \ell} \to X$ *is the random function* $\rho(\boldsymbol{x}) \stackrel{\text{def}}{=} p_{\boldsymbol{x}}(\kappa)$, *where* κ *is a uniform random variable over* X, *and* $\boldsymbol{x} \in X^{\leq \ell}$.

We say that the input messages $X^{\leq \ell}$ to the polynomial-based universal hash consist of *blocks*, with the *block length* of the messages being at most ℓ.

When a WCS authenticator uses a polynomial-based universal hash function, we call the resulting construction a *polynomial-based WCS authenticator*.

Let $\gamma : N \times M \to T$ be a WCS authenticator. An adversary **A** interacting with γ is said to be *nonce-respecting* if it never repeats N-input to γ. Furthermore, the *verification oracle* associated to γ, $V : N \times M \times T \to \{0, 1\}$, is defined as

$$V(n, m, t) = \begin{cases} 1 & \text{if } \gamma(n, m) = t \\ 0 & \text{otherwise} \end{cases}. \qquad (2)$$

Nonce-respecting adversaries may repeat nonce-input to V.

Definition 2.3 (Authenticity Advantage). *Let* **A** *be a nonce-respecting adversary interacting with WCS authenticator* $\gamma : N \times M \to T$ *and associated verification oracle* V. *Then* **A**'s authenticity advantage, *denoted* $\mathsf{Auth}_\gamma(\mathbf{A})$, *is the probability that* **A** *makes a* V-*query* (n^*, m^*, t^*) *resulting in* V *outputting* 1 *and* $\gamma(n^*, m^*) = t^*$ *was not a previous query-response from* γ.

In our analysis we will also need the following definition.

Definition 2.4 (Single-Forgery Advantage). *Let* **A** *be a nonce-respecting adversary interacting with WCS authenticator* $\gamma : N \times M \to T$, *resulting in queries* $\gamma(n_i, m_i) = t_i$ *for* $i = 1, \ldots, q$. *Say that* **A** *outputs* (n^*, m^*, t^*) *after its interaction. Then* **A**'s single-forgery advantage *is*

$$\mathsf{sAuth}_\gamma(\mathbf{A}) \stackrel{\text{def}}{=} \mathbb{P}\left[\gamma(n^*, m^*) = t^*, (n^*, m^*, t^*) \neq (n_i, m_i, t_i) \; \forall i\right]. \qquad (3)$$

The maximum over all adversaries making at most q *queries is denoted* $\mathsf{sAuth}_\gamma(q)$.

Bernstein connects Auth and sAuth as follows.

Theorem 2.1 ([Ber05a]). *Let* **A** *be an authenticity adversary making at most* q γ *queries and* v *verification queries, then*

$$\mathsf{Auth}_\gamma(\mathbf{A}) \leq v \cdot \mathsf{sAuth}_\gamma(q). \qquad (4)$$

Bellare et al. prove a similar result for different constructions [BGM04].

2.3 GCM

We present those details of GCM [MV04a, MV04b] necessary to describe our attacks. GCM takes nonce, associated data, and plaintext input. It operates by first encrypting the plaintext using CTR mode [Nat80] into a ciphertext c. Then it processes the ciphertext and associated data using a WCS authenticator into a tag.

GCM only uses one key, namely a block cipher key; as explained before, we view the keyed block cipher as a URP π over the set of 128-bit strings, hence the block cipher key is implicit in our description. An authentication key L is computed as the output of π under the all-zero string, which we denote 0: $L \stackrel{\text{def}}{=} \pi(0)$.

GCM's WCS authenticator views the set of 128-bit strings as a finite field with 2^{128} elements. Once the ciphertext c has been computed using CTR mode, its length is encoded in a 64-bit string and the ciphertext is padded with zeros to have length a multiple of 128 bits. The associated data is processed in the same way. Let $a_1, a_2, \ldots, a_l$ and $c_1, c_2, \ldots, c_{l'}$ denote the padded associated data and ciphertext, respectively, where the length of all blocks a_i and c_i is 128 bits. Let x_0 denote the concatenation of the encoded lengths of the associated data and ciphertext. Then, if $x = (x_0, a_l, a_{l-1}, \ldots, a_1, c_{l'}, \ldots, c_1)$, GCM computes its tag as

$$p_x(L) + \pi(n) \quad \text{with } L = \pi(0), \tag{5}$$

where n is a value deduced from the nonce.

All π-input in GCM can be derived from the nonce and L, and no two π-inputs are the same, unless some unlikely event happens, in which case GCM loses all security [Jou, Fer05, ZTG13]. In more detail, the nonce is converted into distinct counters for CTR mode, as well as an additional, distinct input, which is used for the URP input in GCM's WCS authenticator, denoted n in 5. In 96-bit nonce GCM, n is equal to the nonce concatenated with a string consisting of 31 zeroes, followed by a 1, and the counters used in CTR mode increment the last 32 bits of n.

In our attacks and analysis below we mostly focus on plain WCS authenticators, however everything translates nearly verbatim over to GCM's WCS authenticator.

3   Key Recovery Attacks

Most of the previously published attacks aim to recover the polynomial key of the WCS authenticator in order to be able to construct arbitrary forgeries. All known key recovery attacks focus either on reducing the set of candidate keys $\mathcal{T}$, which contains the actual key, or, equivalently, increasing $\mathcal{T}$'s complement $\mathcal{F}$, the set of "false" keys. The former can be achieved through nonce misuse [Jou, HP08], which allows one to obtain a polynomial for which the key is a root, thereby reducing $\mathcal{T}$ to the set of all roots of the polynomial. Although nonce misuse attacks are important to understand the fragility of the schemes, we focus on attacks which stay in the nonce-respecting model.

In contrast, the nonce-respecting attacks reduce $\mathcal{T}$ via repeated verification attempts [HP08, PC15, ABBT15]. Their goal is to construct a *forgery polynomial* which evaluates to zero on the key. Then the forgery polynomial is combined with a previous tagging query into a verification attempt in such a way that if the verification attempt fails, then one knows that the key is not one of the roots of the forgery polynomial. If the forgery polynomial has degree ℓ, then at most ℓ faulty keys can be removed for each verification attempt, resulting in a success probability of at most

$$\frac{1}{|T| - v\ell}, \tag{6}$$

where v is the number of verification attempts.

Our attacks differ from the previous nonce-respecting attacks in two ways: they do not require verification attempts in order to increase $\mathcal{F}$, and $\mathcal{F}$ increases quadratically as a function of the number of tagging queries, q, giving a success probability of roughly

$$\frac{1}{|T| - q^2}. \tag{7}$$

We describe chosen-plaintext attacks which perfectly match the bounds for both polynomial-based WCS MACs and GCM. The attacks can also be applied passively, where adversaries do not have chosen-plaintext control. Success then depends in a non-trivial way on the message distribution, which in turn depends on the application in consideration; we leave further detailed analysis of the known-plaintext attacks for future work. In Sect. 5 we show that our chosen-plaintext attacks are optimal.

3.1 WCS Authenticator Attacks

Constructing the False-Key Set. Let $\gamma(n, m) = \pi(n) + \rho(m)$ be a polynomial-based WCS authenticator, with π a URP and ρ a polynomial-based universal hash function. Say that we somehow know that the queries $\gamma(n_i, m_i) = t_i$ for $i = 1, \ldots, q$ were made. This means

$$\pi(n_i) + \rho(m_i) = t_i \quad \text{or} \quad \pi(n_i) = t_i - \rho(m_i), \quad \text{for } i = 1, \ldots, q. \tag{8}$$

Since π is a permutation, this means

$$t_i - \rho(m_i) \neq t_j - \rho(m_j), \quad \text{for } i \neq j. \tag{9}$$

In particular, we know that the real key κ does *not* satisfy the polynomial equations

$$\rho(m_i) - \rho(m_j) + t_j - t_i = 0, \quad \text{for } i \neq j. \tag{10}$$

Therefore, each query to γ might allow us to increase the set of false keys. In fact, the jth query to γ gives an additional $j - 1$ equations which can be used to discard keys.

Known-plaintext Attack. Given (n_i, m_i, t_i) for $i = 1, \ldots, q$, perform the following:

1. Construct

$$\mathcal{F} \stackrel{\text{def}}{=} \{k \mid p_{m_i}(k) - p_{m_j}(k) + t_j - t_i = 0, i \neq j\}. \tag{11}$$

2. Pick any $k^* \notin \mathcal{F}$, output k^*.

Analysis of the known-plaintext attack is complicated by the choice of distribution for the messages m_i. We focus instead on analyzing the chosen-plaintext attack below.

Chosen-Plaintext Attack. Choose q distinct messages of length one block, m_1, $m_2, \ldots, m_q$, and q nonces $n_1, n_2, \ldots, n_q$. For example, one could pick $m_i = n_i = i$, for some encoding of i. Then conclude with the known-plaintext attack described above. The resulting false-key set is

$$\mathcal{F} = \left\{ \frac{t_i - t_j}{m_i - m_j}, i \neq j \right\}. \tag{12}$$

The following proposition establishes the expected size of $\mathcal{F}$ for this attack. In Sect. 5 we connect the expected size of $\mathcal{F}$ with the success of key recovery attacks and forgeries.

Proposition 3.1. *Let* $N = |\mathsf{T}|$, *and say that* $q \leq \sqrt{N-3}$, *then*

$$\mathbb{E}(|\mathcal{F}|) \geq \frac{q(q-1)}{4}, \tag{13}$$

where $\mathcal{F}$ *is from (12) and* $|\mathcal{F}|$ *denotes its cardinality.*

Proof. Let κ denote the real key, then

$$\mathcal{F} = \left\{ \frac{\pi(n_i) - \pi(n_j) + \kappa m_i - \kappa m_j}{m_i - m_j}, i \neq j \right\} \tag{14}$$

$$= \left\{ \frac{\pi(n_i) - \pi(n_j) + \kappa(m_i - m_j)}{m_i - m_j}, i \neq j \right\} \tag{15}$$

$$= \left\{ \frac{\pi(n_i) - \pi(n_j)}{m_i - m_j} + \kappa, i \neq j \right\}. \tag{16}$$

Let $S = \{(\pi(n_i) - \pi(n_j))/(m_i - m_j), i \neq j\}$, so that $|S| = |\mathcal{F}|$.

By Markov's inequality,

$$\mathbb{E}(|S|) \geq \mathbb{P}\left[|S| \geq \frac{q(q-1)}{2} \right] \cdot \frac{q(q-1)}{2}, \tag{17}$$

and $|S| \geq q(q-1)/2$ only if none of the $(\pi(n_i) - \pi(n_j))/(m_i - m_j)$ collide. By applying a union bound we know that the probability there is such a collision is at most $q(q-1)/(2(N-3))$, hence

$$\mathbb{P}\left[|S| \geq \frac{q(q-1)}{2} \right] \geq 1 - \frac{q(q-1)}{2(N-3)}. \tag{18}$$

If $q \leq \sqrt{N-3}$, then

$$1 - \frac{q(q-1)}{2(N-3)} \geq \frac{1}{2}, \tag{19}$$

and we have our desired bound. □

3.2   GCM Attacks

With a known-plaintext attack against GCM it is possible to increase $\mathcal{F}$ without resorting to verification attempts or polynomial equations. Since we know that the authentication key is computed as $\pi(0)$, and all inputs to π are distinct, each URP output from CTR mode reduces the set of valid keys, which you can compute easily if you know the plaintext. However, such an attack still requires known plaintext, potentially making it more difficult to implement in practice.

In contrast, if we apply our WCS authenticator attacks described above to GCM, by replacing messages with ciphertexts, then we arrive at an attack which potentially only requires ciphertext. In a passive setting, the steps are identical: create a false-key set $\mathcal{F}$ as in Eq. 11, except the polynomials are replaced by GCM's, from (5).

The optimal chosen-plaintext attack changes slightly for GCM, since we need to deal with the encoded lengths of the ciphertexts in the polynomials of Eq. 5. Instead of choosing q distinct plaintexts m_i, we now set all plaintexts to be the all-zero string of length one block. This results in polynomials

$$xL + c_i L^2, \tag{20}$$

where x is the encoding of the length of a one-block length ciphertext, and the c_i are the ciphertexts, all distinct from each other. The resulting false-key set is as follows:

$$\left\{ \sqrt{\frac{t_i - t_j}{c_i - c_j}}, i \neq j \right\}. \tag{21}$$

Since the square root is bijective in finite fields of characteristic two, we have that the above set contains the same number of elements as

$$\left\{ \frac{t_i - t_j}{c_i - c_j}, i \neq j \right\}, \tag{22}$$

and the analysis made for WCS authenticators holds with little modification.

4   Bounding Authenticity with Key Recovery

4.1   Bernstein's Analysis

Bernstein analyzes a generalization of Wegman-Carter and WCS MACs, namely those of the form $(n, m) \mapsto \rho(m) + \varphi(n)$, where $\rho : \mathsf{M} \to \mathsf{T}$ and $\varphi : \mathsf{N} \to \mathsf{T}$ are independent random functions. Wegman-Carter authenticators fix φ to be a uniformly distributed random function, and WCS authenticators fix φ to be a URP. As part of his analysis, Bernstein uses *differential probability* [Ber05b], more commonly known as ϵ-almost (XOR) universal, given by

$$\Delta_\rho \stackrel{\text{def}}{=} \max_{\substack{m \neq m' \\ t \in \mathsf{T}}} \mathbb{P}\left[\rho(m) = \rho(m') + t\right]. \tag{23}$$

Various papers [dB93, Tay93, BJKS93] establish that for a polynomial-based universal hash function $\rho : \mathsf{M} \to \mathsf{T}$, $\Delta_\rho \leq \ell / |\mathsf{T}|$, where $\mathsf{M} = \mathsf{T}^{\leq \ell}$.

Bernstein also introduces the concept of *interpolation probabilities* of a random function φ, which is the probability that $\varphi(x_i) = y_i$ for some values $x_1, \ldots, x_q$ and $y_1, \ldots, y_q$. Bernstein establishes that $\rho(m) + \varphi(n)$ is secure if ρ's differential and φ's interpolation probabilities are small. Ultimately when applied to polynomial-based WCS authenticators, we get the following.

Theorem 4.1. *Let* $\gamma : \mathsf{N} \times \mathsf{M} \to \mathsf{T}$ *be a polynomial-based WCS authenticator with* $\mathsf{M} = \mathsf{T}^{\leq \ell}$ *and let* **A** *be a nonce-respecting adversary against* γ *making at most* q γ *queries and* v *verification queries, then*

$$\mathsf{Auth}_\gamma(\mathbf{A}) \leq v \cdot \frac{\ell}{|\mathsf{T}|} \cdot \left(1 - \frac{q}{|\mathsf{T}|}\right)^{-\frac{q+1}{2}}. \tag{24}$$

4.2   Reshaping Authenticity Advantage

Although Bernstein's analysis is general and applies to more than just polynomial-based WCS MACs, a targeted analysis will elucidate the gap between currently known attacks and the bound given by Bernstein.

Whereas Bernstein proves bounds for $\varphi(n) + \rho(m)$ in terms of φ's interpolation and ρ's differential probability, we instead rework the bounds to φ's unpredictability (Sect. 4.3) and key recovery against ρ (Sect. 4.4), the latter only applying to polynomial-based MACs. The concepts introduced in this section will allow us to prove that the CPA attacks introduced in Sect. 3 are in fact optimal.

Instrumental to our analysis is the fact that an adversary's single-forgery advantage can be split in two, according to whether its attempted forgery (n^*, m^*, t^*) uses a nonce n^* that was never used before, or not. We let $\mathsf{sAuth}_\gamma^{\text{new}}(\mathbf{A})$ denote the probability that **A** forges and uses a new nonce, and $\mathsf{sAuth}_\gamma^{\text{old}}(\mathbf{A})$ the probability that **A** forges and uses an old nonce. By basic probability theory,

$$\mathsf{sAuth}_\gamma(\mathbf{A}) \leq \max\left\{\mathsf{sAuth}_\gamma^{\text{new}}(\mathbf{A}), \mathsf{sAuth}_\gamma^{\text{old}}(\mathbf{A})\right\}. \tag{25}$$

Letting KR denote polynomial key recovery advantage (see Definition 4.2), we establish the following result.

Corollary 4.1. *Let* $\gamma : (n, m) \mapsto \rho(m) + \pi(n)$ *be a polynomial-based WCS authenticator with* $\rho : \mathsf{M} \to \mathsf{T}$ *a random function, and* $\pi : \mathsf{N} \to \mathsf{T}$ *an independent URP. Let* $\mathbf{A}$ *be an authenticity adversary against* γ *making at most q queries of length at most ℓ. Then*

$$\mathsf{Auth}_\gamma(\mathbf{A}) \le v \cdot \max\left\{\ell \cdot \mathsf{KR}_\gamma(q), \frac{1}{|\mathsf{T}| - q}\right\}. \tag{26}$$

The proof can be found in Appendix A, which relies on results developed in the next sections.

4.3   Unpredictability

We show how any attempted forgery using a new nonce against a WCS authenticator has low success probability. This means if authenticity adversaries want to achieve significant advantage, then they must re-use nonces during forgeries. We state the result more generally than for only polynomial-based WCS authenticators.

Definition 4.1 (Unpredictability). *Let* $\mathbf{A}$ *be an adversary interacting with random function* $\varphi : \mathsf{X} \to \mathsf{Y}$. *Say that* $\mathbf{A}$ *produces the sequence* $\boldsymbol{x} \in \mathsf{X}^q$ *and* φ *responds with outputs* $\boldsymbol{y} \in \mathsf{Y}^q$. *Let* (x^*, y^*) *be* $\mathbf{A}$'s *output, then* $\mathbf{A}$'s *unpredictability advantage against* φ *is*

$$\mathsf{Unpred}_\varphi(\mathbf{A}) \overset{\text{def}}{=} \mathbb{P}\left[\varphi(x^*) = y^*, x^* \ne x_i, i = 1, \dots, q\right], \tag{27}$$

where the probability is taken over the randomness of $\mathbf{A}$ *and* φ.

Let $\gamma : (n, m) \mapsto \rho(m) + \pi(n)$ be any Wegman-Carter-style MAC using random functions $\rho : \mathsf{M} \to \mathsf{T}$ and $\varphi : \mathsf{N} \to \mathsf{T}$ which are independent of each other. Let $\mathbf{A}$ be an authenticity adversary against γ. We construct an unpredictability adversary $\mathbf{B}\langle\mathbf{A}\rangle$ against φ as follows.

1. $\mathbf{B}$ runs $\mathbf{A}$.
2. $\mathbf{B}$ simulates ρ using its own randomness; call it ρ'.
3. Every γ-query made by $\mathbf{A}$ is reconstructed by $\mathbf{B}$ using ρ' and the φ-oracle $\mathbf{B}$ interacts with. Concretely, every $\gamma(n, m)$ made by $\mathbf{A}$ gets forwarded as $\varphi(n)$, and $\mathbf{B}$ returns $\varphi(n) + \rho'(m)$.
4. $\mathbf{B}$ receives $\mathbf{A}$'s final output, (n^*, m^*, t^*), and finally outputs $(n^*, t^* - \rho'(m^*))$.

Proposition 4.1
$$\mathsf{sAuth}_\gamma^{new}(\mathbf{A}) \le \mathsf{Unpred}_\varphi(\mathbf{B}\langle\mathbf{A}\rangle). \tag{28}$$

Proof. First note that $\mathbf{B}$ perfectly reconstructs $\mathbf{A}$'s authenticity game since ρ' is independent of φ. Then, if $\mathbf{A}$ wins its authenticity game, $\gamma(n^*, m^*) = t^*$, or in other words, $\varphi(n^*) + \rho(m^*) = t^*$. In particular, $\varphi(n^*) = t^* - \rho(m^*)$. If n^* has never been queried to φ before, $t^* - \rho(m^*)$ would correctly predict φ's output on an unknown input, hence $\mathbf{B}\langle\mathbf{A}\rangle$ would win its unpredictability game. $\square$

Lemma 4.1. *Let* $\pi : \mathsf{N} \to \mathsf{T}$ *be a URP and* $\mathbf{B}$ *an adversary making at most* q *queries, then*

$$\mathsf{Unpred}_\pi(\mathbf{B}) \leq \frac{1}{|\mathsf{T}| - q}. \tag{29}$$

4.4   Bounding Forgeries with Key Recovery

Having set aside adversaries which use new nonces for forgeries, we can focus on those that re-use nonces. This section applies only to polynomial-based WCS authenticators.

Definition 4.2 (Polynomial Key Recovery). *Let* $\mathbf{A}$ *be a nonce-respecting adversary interacting with polynomial-based WCS authenticator* γ *using URP* π *and polynomial-based universal hash* ρ, *with* κ *denoting the random variable representing the key underlying* ρ. *Say that* $\mathbf{A}$ *outputs an element* $k^* \in \mathsf{K}$, *then* $\mathbf{A}$*'s polynomial key recovery advantage against* γ *is*

$$\mathsf{KR}_\gamma(\mathbf{A}) \overset{\text{def}}{=} \mathbb{P}\left[k^* = \kappa\right], \tag{30}$$

where the randomness is taken over $\mathbf{A}$ *and* γ. *We let* $\mathsf{KR}_\gamma(q)$ *denote the maximum of* $\mathsf{KR}_\gamma(\mathbf{A})$ *over all adversaries* $\mathbf{A}$ *making at most* q *queries.*

Forgeries can be used to recover authentication keys. We construct a polynomial key recovery adversary $\mathbf{C}\langle\mathbf{A}\rangle$ against γ.

1. $\mathbf{C}$ runs $\mathbf{A}$.
2. Every (n, m) query by $\mathbf{A}$ gets forwarded to $\mathbf{C}$'s oracle, and $\mathbf{C}$ returns the output $\gamma(n, m)$ to $\mathbf{A}$.
3. When $\mathbf{A}$ outputs (n^*, m^*, t^*), then $\mathbf{C}$ checks to see if $n^* = n_i$ for some previous query $\gamma(n_i, m_i) = t_i$. If this is not the case, then $\mathbf{C}$ aborts. Otherwise $\mathbf{C}$ computes the roots of the polynomial[2] $p_{m^*}(\alpha) - p_{m_i}(\alpha) - t^* + t_i = 0$, and chooses a key uniformly at random from the set of roots.

Proposition 4.2. *Let* $\mathbf{A}$ *be an adversary making queries of length at most* ℓ. *The probability that* $\mathbf{A}$ *wins its authenticity game and outputs* (n^*, m^*, t^*) *where* $n^* = n_i$ *for some previous query* (n_i, m_i) *to* γ, *is bounded above by*

$$\ell \cdot \mathsf{KR}_\gamma(\mathbf{C}\langle\mathbf{A}\rangle). \tag{31}$$

Proof. If $\mathbf{A}$ wins with $n^* = n_i$, then

$$\gamma(n^*, m^*) = \gamma(n_i, m^*) = \varphi(n_i) + \rho(m^*) = t^*, \tag{32}$$

and

$$\gamma(n_i, m_i) = \varphi(n_i) + \rho(m_i) = t_i, \tag{33}$$

therefore $\rho(m^*) - \rho(m_i) - t^* + t_i = 0$. We know that the key used by ρ is in the set of roots of the polynomial $p_{m^*}(\alpha) - p_{m_i}(\alpha) - t^* + t_i$, which has size at most $\max\{|m^*|, |m_i|\}$. Picking an element uniformly at random from this set, we have that $\mathbf{C}$ wins with probability at least $1/\max\{|m^*|, |m_i|\}$.   □

---

[2] Finding roots of polynomials over a finite field is computationally efficient using Berlekamp's algorithm [Ber70] or the Cantor-Zassenhaus algorithm [CZ81].

5   Using Key Recovery to Mount Forgeries

The previous section discussed how to convert authenticity attacks into key recovery attacks to reshape the upper bounds on forgery attacks. Here we discuss the opposite, namely how to use key recovery adversaries to mount forgeries. This will allow us to not only show that the analysis of Sect. 4 is tight, but also that the attacks of Sect. 3 are optimal, using Bernstein's analysis.

5.1   Key-Set Recovery

The obvious way to convert a key recovery attack into an authenticity attack is to run the key recovery adversary and use the output of the key recovery adversary to mount a forgery. We explain this formally in Appendix B. However, this method constructs authenticity adversaries which are about as successful as key recovery adversaries.

In contrast, as seen in Sect. 4.4, Proposition 4.2, authenticity adversaries might improve over key recovery adversaries by up to a factor of ℓ. Intuitively, given a key recovery adversary, one could try to do this by taking the candidate key k^* output by the key recovery adversary, and finding a polynomial of degree ℓ which contains k^* as a root, and then construct a forgery using this polynomial. The problem with this approach is that most of the roots of the polynomial chosen by the resulting authenticity adversary could be useless, as they could, for example, lie in some false-key set determined by the key recovery adversary. Without any further information about the key recovery adversary it does not seem possible to improve the authenticity adversary.

However, if we instead look at *key-set recovery adversaries*, we can improve our chances of constructing forgeries. We will show that key-set recovery and key-recovery adversaries are in fact very similar, allowing us to prove tight bounds on the connection between key-recovery and forgeries.

Definition 5.1 (Polynomial Key-Set Recovery). *Let* $\mathbf{A}$ *be a nonce-respecting adversary interacting with polynomial-based WCS authenticator γ using URP π and polynomial-based universal hash ρ, with κ denoting the random variable representing the key underlying ρ. Say that $\mathbf{A}$ outputs a set $K^* \subset \mathsf{K}$, and let 1_{K^*} denote the random variable which equals one if $\kappa \in K^*$ and zero otherwise. Then $\mathbf{A}$'s polynomial key-set recovery advantage against γ is*

$$\mathsf{KS}_\gamma(\mathbf{A}) \overset{\text{def}}{=} \mathbb{E}\left(\frac{1_{K^*}}{|K^*|}\right), \tag{34}$$

where the randomness is taken over $\mathbf{A}$ and γ. We let $\mathsf{KS}_\gamma(q)$ denote the maximum of $\mathsf{KS}_\gamma(\mathbf{A})$ taken over all adversaries making at most q queries.

Let $\mathbf{C}$ be a key-set recovery adversary. Once $\mathbf{C}$ has made all its queries, it is possible to compute $\mathcal{F}_{\mathbf{C}}$, the random set of false keys given by Eq. (11), and $\mathcal{T}_{\mathbf{C}}$ its complement. Then it is straightforward to construct key-set adversary $\mathbf{D}\langle\mathbf{C}\rangle$ which runs $\mathbf{C}$, and then returns $\mathcal{T}_{\mathbf{C}}$. We argue that $\mathbf{C}$'s advantage is not greater than $\mathbf{D}$'s.

Lemma 5.1. *Let* **C** *and* **D** $\langle$**C**$\rangle$ *be defined as above, then*

$$\mathsf{KS}_\gamma(\mathbf{C}) \leq \mathsf{KS}_\gamma(\mathbf{D}\,\langle\mathbf{C}\rangle). \tag{35}$$

Proof. First note that κ, the key underlying the polynomial-based universal hash, must be in $\mathcal{T}_\mathbf{C}$, since by definition it cannot satisfy any of the equations given in (11). Therefore, if **C**'s output, denoted K^*, contains elements not in $\mathcal{T}_\mathbf{C}$, then it is possible to improve **C**'s advantage by having **C** output $K^* \cap \mathcal{T}_\mathbf{C}$, since that would reduce **C**'s output set size without affecting the probability that κ is in the set. Therefore without loss of generality we assume that $K^* \subset \mathcal{T}_\mathbf{C}$.

Then, given any sequence of q queries that **C** makes, $\mathcal{T}_\mathbf{C}$ describes exactly those keys which satisfy the transcript, and in particular κ is uniformly distributed over $\mathcal{T}_\mathbf{C}$. Therefore, if $K^* \subset \mathcal{T}_\mathbf{C}$, then **C**'s advantage is the same as **D**:

$$\mathbb{E}\left(\frac{1_{K^*}}{|K^*|}\right) = \sum_n \frac{1}{n} \sum_m \mathbb{P}\left[\kappa \in K^*, |K^*| = n, |\mathcal{T}_\mathbf{C}| = m\right] \tag{36}$$

$$= \sum_n \frac{1}{n} \sum_m \mathbb{P}\left[\kappa \in K^* \,\middle|\, |K^*| = n, |\mathcal{T}_\mathbf{C}| = m\right] \mathbb{P}\left[|K^*| = n, |\mathcal{T}_\mathbf{C}| = m\right] \tag{37}$$

$$= \sum_n \frac{1}{n} \sum_m \frac{n}{m} \cdot \mathbb{P}\left[|K^*| = n, |\mathcal{T}_\mathbf{C}| = m\right] \tag{38}$$

$$= \sum_m \frac{1}{m} \mathbb{P}\left[|\mathcal{T}_\mathbf{C}| = m\right] \tag{39}$$

$$= \mathbb{E}\left(\frac{1_{\mathcal{T}_\mathbf{C}}}{|\mathcal{T}_\mathbf{C}|}\right). \tag{40}$$

$\square$

Since our focus is on optimal, information-theoretic adversaries, without loss of generality we assume that all key-set recovery adversaries return $\mathcal{T}$.

Given such a key-set recovery adversary **D**, we construct single-forgery adversary **A** $\langle$**D**$\rangle$ as follows:

1. **A** runs **D**, and responds to any **D**-query (n, m) with $\gamma(n, m)$.
2. When **D** outputs the candidate set $\mathcal{T}_\mathbf{D}$, **A** picks ℓ distinct elements uniformly at random from $\mathcal{T}_\mathbf{D}$ and constructs a polynomial p_{m^*} with those elements as roots.
3. **A** picks any previous query $\gamma(n, m) = t$ made by **D**, adds m^* to m componentwise to get $m' = (m_1 + m_1^*, m_2 + m_2^*, \ldots)$, and submits the forgery attempt (n, m', t).

Naturally this reduction becomes void if the size of $\mathcal{T}_\mathbf{D}$ is less than ℓ, however as we will see in Sect. 5.2, this can only happen if q nearly as large as the number of nonces the adversary can query. We capture this limit on q with M_γ, which is defined to be

$$M_\gamma \stackrel{\text{def}}{=} \max\left\{q \,\middle|\, \min_{\substack{m_1,\ldots,m_q, \\ t_1,\ldots,t_q}} |\mathsf{T}| \geq \ell\right\}. \tag{41}$$

The following proposition shows that one can construct better forgeries using key-set recovery adversaries.

Proposition 5.1. *Let* $q \leq M_\gamma$, *then*

$$\ell \cdot \mathsf{KS}_\gamma(\mathbf{D}) \leq \mathsf{sAuth}_\gamma^{old}(\mathbf{A}\langle \mathbf{D} \rangle). \tag{42}$$

Proof. Let L denote the ℓ elements that $\mathbf{A}$ picks from $\mathcal{T}_\mathbf{D}$. Adversary $\mathbf{A}$ wins if $\kappa \in L$, since then $p_{m^*}(\kappa) = 0$ and so $p_{m+m^*}(\kappa) + \pi(n) = t$.

$$\mathbb{P}\left[\kappa \in L\right] = \sum_n \mathbb{P}\left[\kappa \in L \mid |\mathcal{T}_\mathbf{D}| = n, \kappa \in \mathcal{T}_\mathbf{D}\right] \mathbb{P}\left[\kappa \in \mathcal{T}_\mathbf{D}, |\mathcal{T}_\mathbf{D}| = n\right] \tag{43}$$

$$= \sum_n \frac{\ell}{n} \cdot \mathbb{P}\left[\kappa \in \mathcal{T}_\mathbf{D}, |\mathcal{T}_\mathbf{D}| = n\right] \tag{44}$$

$$= \ell \cdot \mathbb{E}\left(\frac{1_{\mathcal{T}_\mathbf{D}}}{|\mathcal{T}_\mathbf{D}|}\right) = \ell \cdot \mathsf{KS}_\gamma(\mathbf{D}). \tag{45}$$

□

Furthermore, there is little real difference between key-recovery and key-set recovery advantage.

Proposition 5.2

$$\mathsf{KS}_\gamma(q) = \mathsf{KR}_\gamma(q). \tag{46}$$

Proof. If the output set size of a key-set recovery adversary is always one, then key-set recovery advantage is identical to key-recovery advantage. Since any key-recovery adversary can be converted into a key-set recovery adversary with output set size one, we have that $\mathsf{KR}_\gamma(q) \leq \mathsf{KS}_\gamma(q)$.

Given any key-set recovery adversary $\mathbf{C}$, we convert it into a key-recovery adversary $\mathbf{C}'$ by picking a candidate key k^* uniformly at random from the output set K^*. Then

$$\mathsf{KR}_\gamma(\mathbf{C}') = \mathbb{P}\left[\kappa = k^*\right] \tag{47}$$

$$= \sum_n \mathbb{P}\left[\kappa = k^* \mid \kappa \in K^*, |K^*| = n\right] \mathbb{P}\left[\kappa \in K^*, |K^*| = n\right] \tag{48}$$

$$= \sum_n \frac{1}{n} \mathbb{P}\left[\kappa \in K^*, |K^*| = n\right] = \mathsf{KS}_\gamma(\mathbf{C}). \tag{49}$$

□

Propositions 4.2, 5.1 and 5.2 establish the following result, confirming that the analysis of Sect. 4.4 is tight.

Corollary 5.1. *Let* $q \leq M_\gamma$, *then*

$$\ell \cdot \mathsf{KR}_\gamma(q) = \mathsf{sAuth}_\gamma^{old}(q). \tag{50}$$

5.2   Attack Success Probability and Optimality

Our chosen-plaintext attack only uses messages of length one block, which is reflected in the fact that $|\mathcal{F}|$ only grows as a function of q. Intuitively one would expect to be able to increase $\mathcal{F}$ as well by taking advantage of longer messages and the fact that polynomials of higher degree have more roots. However, here we show that this is impossible.

The success probability of the key recovery attacks from Sect. 3 is given as follows, which results from the observation that the real key cannot be in $\mathcal{F}$ by definition.

Proposition 5.3. *Let* **A** *denote the chosen-plaintext attack from Sect. 3, then*

$$\mathrm{KR}_\gamma(\mathbf{A}) \geq \frac{1}{|\mathsf{T}| - \mathbb{E}\left(|\mathcal{F}|\right)}. \tag{51}$$

Combining this result with Bernstein's result, we have the following.

Theorem 5.1. *Let* $\mathcal{F}$ *be defined as in Sect. 3, then*

$$\mathbb{E}\left(|\mathcal{F}|\right) \leq \frac{q(q+1)}{2}. \tag{52}$$

Proof. Using Theorem 4.1, Corollary 5.1, and Proposition 5.3, we have

$$\ell \cdot \frac{1}{|\mathsf{T}| - \mathbb{E}\left(|\mathcal{F}|\right)} \leq \frac{\ell}{|\mathsf{T}|} \cdot \left(1 - \frac{q}{|\mathsf{T}|}\right)^{-\frac{q+1}{2}}. \tag{53}$$

Letting x denote $\mathbb{E}(|\mathcal{F}|)$ and $N = |\mathsf{T}|$, we have

$$\frac{1}{N-x} \leq \frac{1}{N}\left(1 - \frac{q}{N}\right)^{-\frac{q+1}{2}} \tag{54}$$

$$x \leq N\left[1 - \left(1 - \frac{q}{N}\right)^{\frac{q+1}{2}}\right]. \tag{55}$$

We apply Bernoulli's inequality, namely that $(1+x)^r \geq 1 + rx$ if $r \geq 1$ and $x \geq -1$, which holds in our case when $1 \leq q \leq N$, to get

$$\left(1 - \frac{q}{N}\right)^{\frac{q+1}{2}} \geq 1 - \frac{q+1}{2} \cdot \frac{q}{N}, \tag{56}$$

hence

$$x \leq \frac{q(q+1)}{2}. \tag{57}$$

$\square$

6 Conclusions, Limitations, and Open Problems

Using new analysis and attacks we have shown that, without further restrictions on the adversaries, Bernstein's analysis is in fact optimal. We can therefore conclude that the data limits imposed by Bernstein's bounds are necessary.

Our attacks illustrate for the first time how to maximally take advantage of tagging queries without needing verification queries in order to attack WCS authenticators. However, there are limitations on the applicability of the attacks.

As implied by the introduction, our attacks only work against polynomial-based WCS authenticators when they re-use the polynomial key, and is therefore not applicable to, for example, SNOW 3G [3GP17] or Poly1305 as used in NaCl [Ber09, Ber09].

The attacks work best when tags are not truncated, since the underlying PRP behaves more like a PRF with increased truncation [GGM18, HWKS98]. However, as pointed out by Ferguson [Fer05] and Mattsson and Westerlund [MW16], one must take care when truncating tags in WCS authenticators. In some cases standards mandate that tags not be truncated [VM06, SMC08, IS09].

The attacks are not directly applicable to constructions which do not follow the WCS authenticator structure of mapping (n, m) to $\pi(n) + \rho(m)$. A few different constructions are discussed by Bernstein [Ber05c] and Handschuh and Preneel [HP08]. In particular, if a PRF instead of a PRP is used to hide the polynomial output, or if multiple PRP calls are XORed together as with CWC [KVW04] and GCM/2^+ [AY12], then the attacks are not applicable; it remains an open problem whether the analyses of the latter constructions are tight.

WCS authenticators can also be instantiated using non-polynomial-based universal hash functions, [BHK+99, HK97, EPR99, Joh97, KYS05, Kro06, BHK+99]. We expect that similar attacks are applicable to these functions.

As shown by Luykx et al. [LMP17], the attacks' success probability will not improve in the multi-key setting.

Finally, although our attacks show that one should abide by Bernstein's bounds, implementing the attacks seems to require a large amount of storage to achieve significant success probability. It is unclear whether there is a compact way of representing the set of false keys. Alternatively, if one were able to prove lower bounds on the storage requirements for any attacker, one could possibly afford to use keys beyond the data limits recommended by Bernstein's analysis, assuming adversaries have bounded storage capabilities.

Acknowledgments. The authors would like to thank Guy Barwell, Dan Bernstein, Bart Mennink, Scott Fluhrer, and the anonymous reviewers for their comments, as well as Mridul Nandi for pointing out an error in a previous version of the manuscript.

A   Proof of Corollary 4.1

We re-use the notation and definitions from Sects. 4.3 and 4.4.

Corollary. Let $\gamma : (n, m) \mapsto \rho(m) + \pi(n)$ be a polynomial-based WCS authenticator with $\rho : M \to T$ a random function, and $\pi : N \to T$ an independent URP. Let A be an authenticity adversary against γ making at most q queries of length at most ℓ. Then A's advantage against γ is bounded by

$$v \cdot \max \left\{ \ell \cdot \mathsf{KR}_\gamma(C \langle A \rangle), \frac{1}{|T| - q} \right\}. \tag{58}$$

Proof. We restrict our attention to single-forgery adversaries, and use Theorem 2.1 to generalize to any authenticity adversary.

Let E denote the event that n^* does not equal a previous query to φ. By Proposition 4.1, the probability that A wins and E occurs is bounded above by the probability that $B \langle A \rangle$ wins, which is at most $1/(|T| - q)$ by Lemma 4.1. By Proposition 4.2, the probability that A wins and E does not occur is bounded above by ℓ times the probability that $C \langle A \rangle$ wins.   □

B   From Key Recovery to Forgeries

Let C be a polynomial authenticator key recovery adversary against γ, then we construct an authenticity adversary $A \langle C \rangle$ against γ as follows:

1. A runs C.
2. Every (n, m) query by C gets forwarded to A's oracle, and A returns the output $\gamma(n, m)$ to C.
3. When C outputs k^*, A uses it to compute $y^* = \gamma(n_1, m_1) - p_{m_1}(k^*)$, where (n_1, m_1) is the first query made by C. Then A picks a message m^*, and attempts the forgery $(n_1, m^*, y^* + p_{m^*}(k^*))$.

Proposition B.1

$$\mathsf{KR}_\gamma(C) \leq \mathsf{Auth}_\gamma(A \langle C \rangle). \tag{59}$$

Proof. If C wins its game, then $k^* = k$, the key used by the polynomial hash. Then we have

$$\gamma(n_1, m^*) = \pi(n_1) + p_{m^*}(k) \tag{60}$$
$$= \gamma(n_1, m_1) - p_{m_1}(k) + p_{m^*}(k) \tag{61}$$
$$= \gamma(n_1, m_1) - p_{m_1}(k^*) + p_{m^*}(k^*), \tag{62}$$

which is exactly the tag submitted by A.   □

C   Notation

Table 1. List of notation.

Symbol	Description
Quantities	
v	Number of verification queries
q	Number of tagging queries
ℓ	Maximum message length
N	Size of T
Random variables	
φ	Random function
π	URP
γ	Authenticator
ρ	Polynomial-based universal hash
κ	Key of a polynomial hash
Sets	
$x \in \mathsf{X}$	Domain, block
$y \in \mathsf{Y}$	Range
$k \in \mathsf{K}$	Key set
$n \in \mathsf{N}$	Nonce set
$m \in \mathsf{M}$	Message space
$t \in \mathsf{T}$	Tag space
$\mathcal{F}$	Faulty keys output by attacks
$\mathcal{T}$	Complement of $\mathcal{F}$, i.e. $\mathsf{K}\backslash\mathcal{F}$
Adversaries	
A	Adversary (generic or authenticity)
B	Unpredictability adversary
C	Key recovery or key set recovery adversary
D	Optimal key recovery or key set recovery adversary
Miscellaneous	
$\boldsymbol{x}$	Vector of elements
$p_m(k)$	Polynomial defined by m evaluated at k

References

[3GP17] Specification of the 3GPP: Confidentiality and Integrity Algorithms UEA2 & UIA2; Document 2: SNOW 3G specification (2017). https://portal.3gpp.org/desktopmodules/Specifications/SpecificationDetails.aspx?specificationId=2396

[ABBT15] Abdelraheem, M.A., Beelen, P., Bogdanov, A., Tischhauser, E.: Twisted polynomials and forgery attacks on GCM. In: Oswald, E., Fischlin, M. (eds.) EUROCRYPT 2015. LNCS, vol. 9056, pp. 762–786. Springer, Heidelberg (2015). https://doi.org/10.1007/978-3-662-46800-5_29

[AY12] Aoki, K., Yasuda, K.: The security and performance of "GCM" when short multiplications are used instead. In: Kutyłowski, M., Yung, M. (eds.) Inscrypt 2012. LNCS, vol. 7763, pp. 225–245. Springer, Heidelberg (2013). https://doi.org/10.1007/978-3-642-38519-3_15

[BC09] Black, J., Cochran, M.: MAC reforgeability. In: Dunkelman, O. (ed.) FSE 2009. LNCS, vol. 5665, pp. 345–362. Springer, Heidelberg (2009). https://doi.org/10.1007/978-3-642-03317-9_21

[Ber70] Berlekamp, E.R.: Factoring polynomials over large finite fields. Math. Comput. 24(111), 713–735 (1970)

[Ber05a] Bernstein, D.J.: Stronger security bounds for permutations (2005). http://cr.yp.to/papers.html#permutations. Accessed 9 April 2015

[Ber05b] Bernstein, D.J.: Stronger security bounds for Wegman-Carter-Shoup authenticators. In: Cramer, R. (ed.) EUROCRYPT 2005. LNCS, vol. 3494, pp. 164–180. Springer, Heidelberg (2005). https://doi.org/10.1007/11426639_10

[Ber05c] Bernstein, D.J.: The Poly1305-AES message-authentication code. In: Gilbert, H., Handschuh, H. (eds.) FSE 2005. LNCS, vol. 3557, pp. 32–49. Springer, Heidelberg (2005). https://doi.org/10.1007/11502760_3

[Ber09] Bernstein, D.J.: Cryptography in NaCl (2009). http://cr.yp.to/papers.html#naclcrypto. Accessed 14 Sept 2017

[BGM04] Bellare, M., Goldreich, O., Mityagin, A.: The power of verification queries in message authentication and authenticated encryption. IACR Cryptology ePrint Archive 2004, p. 309 (2004)

[BHK+99] Black, J., Halevi, S., Krawczyk, H., Krovetz, T., Rogaway, P.: UMAC: fast and secure message authentication. In: Wiener [Wie99], pp. 216–233

[BJKS93] Bierbrauer, J., Johansson, T., Kabatianskii, G., Smeets, B.: On families of hash functions via geometric codes and concatenation. In: Stinson [Sti94], pp. 331–342

[BL16] Bhargavan, K., Leurent, G.: On the practical (in-)security of 64-bit block ciphers: collision attacks on HTTP over TLS and OpenVPN. In: Weippl, E.R., Katzenbeisser, S., Kruegel, C., Myers, A.C., Halevi, S. (eds.) Proceedings of the 2016 ACM SIGSAC Conference on Computer and Communications Security, 24–28 October 2016, Vienna, Austria, pp. 456–467. ACM (2016)

[BPR05] Bellare, M., Pietrzak, K., Rogaway, P.: Improved security analyses for CBC MACs. In: Shoup, V. (ed.) CRYPTO 2005. LNCS, vol. 3621, pp. 527–545. Springer, Heidelberg (2005). https://doi.org/10.1007/11535218_32

[Bra82] Brassard, G.: On computationally secure authentication tags requiring short secret shared keys. In: Chaum, D., Rivest, R.L., Sherman, A.T. (eds.) Advances in Cryptology, pp. 79–86. Springer, Boston (1983). https://doi.org/10.1007/978-1-4757-0602-4_7

[CS16]   Cogliati, B., Seurin, Y.: EWCDM: an efficient, beyond-birthday secure, nonce-misuse resistant MAC. In: Robshaw, M., Katz, J. (eds.) CRYPTO 2016. LNCS, vol. 9814, pp. 121–149. Springer, Heidelberg (2016). https://doi.org/10.1007/978-3-662-53018-4_5

[CZ81]   Cantor, D.G., Zassenhaus, H.: A new algorithm for factoring polynomials over finite fields. Math. Comput. **36**(154), 587–592 (1981)

[dB93]   den Boer, B.: A simple and key-economical unconditional authentication scheme. J. Comput. Secur. **2**, 65–72 (1993)

[EPR99] Etzel, M., Patel, S., Ramzan, Z.: Square hash: fast message authentication via optimized universal hash functions. In: Wiener [Wie99], pp. 234–251

[Fer05] Ferguson, N.: Authentication weaknesses in GCM. Comments submitted to NIST Modes of Operation Process (2005)

[GGM18] Gilboa, S., Gueron, S., Morris, B.: How many queries are needed to distinguish a truncated random permutation from a random function? J. Cryptol. **31**(1), 162–171 (2018)

[GMS74] Gilbert, E.N., MacWilliams, F.J., Sloane, N.J.A.: Codes which detect deception. Bell Syst. Tech. J. **53**(3), 405–424 (1974)

[GPR14] Gaži, P., Pietrzak, K., Rybár, M.: The exact PRF-security of NMAC and HMAC. In: Garay, J.A., Gennaro, R. (eds.) CRYPTO 2014. LNCS, vol. 8616, pp. 113–130. Springer, Heidelberg (2014). https://doi.org/10.1007/978-3-662-44371-2_7

[GPR16] Gazi, P., Pietrzak, K., Rybár, M.: The exact security of PMAC. IACR Trans. Symmetric Cryptol. **2016**(2), 145–161 (2016)

[GPT15] Gaži, P., Pietrzak, K., Tessaro, S.: The exact PRF security of truncation: tight bounds for keyed sponges and truncated CBC. In: Gennaro, R., Robshaw, M. (eds.) CRYPTO 2015. LNCS, vol. 9215, pp. 368–387. Springer, Heidelberg (2015). https://doi.org/10.1007/978-3-662-47989-6_18

[HK97]   Halevi, S., Krawczyk, H.: MMH: software message authentication in the Gbit/second rates. In: Biham, E. (ed.) FSE 1997. LNCS, vol. 1267, pp. 172–189. Springer, Heidelberg (1997). https://doi.org/10.1007/BFb0052345

[HP08]   Handschuh, H., Preneel, B.: Key-recovery attacks on universal hash function based MAC algorithms. In: Wagner, D. (ed.) CRYPTO 2008. LNCS, vol. 5157, pp. 144–161. Springer, Heidelberg (2008). https://doi.org/10.1007/978-3-540-85174-5_9

[HWKS98] Hall, C., Wagner, D., Kelsey, J., Schneier, B.: Building PRFs from PRPs. In: Krawczyk, H. (ed.) CRYPTO 1998. LNCS, vol. 1462, pp. 370–389. Springer, Heidelberg (1998). https://doi.org/10.1007/BFb0055742

[IOM12] Iwata, T., Ohashi, K., Minematsu, K.: Breaking and repairing GCM security proofs. In: Safavi-Naini, R., Canetti, R. (eds.) CRYPTO 2012. LNCS, vol. 7417, pp. 31–49. Springer, Heidelberg (2012). https://doi.org/10.1007/978-3-642-32009-5_3

[IS09]   Igoe, K., Solinas, J.: AES Galois Counter Mode for the secure shell transport layer protocol. RFC 5647, August 2009

[Joh97] Johansson, T.: Bucket hashing with a small key size. In: Fumy, W. (ed.) EUROCRYPT 1997. LNCS, vol. 1233, pp. 149–162. Springer, Heidelberg (1997). https://doi.org/10.1007/3-540-69053-0_12

[Jou]   Joux, A.: Comments on the draft GCM specification - authentication failures in NIST version of GCM. http://csrc.nist.gov/groups/ST/toolkit/BCM/documents/comments/800-38_Series-Drafts/GCM/Joux_comments.pdf

[Kro06]   Krovetz, T.: Message authentication on 64-bit architectures. In: Biham, E., Youssef, A.M. (eds.) SAC 2006. LNCS, vol. 4356, pp. 327–341. Springer, Heidelberg (2007). https://doi.org/10.1007/978-3-540-74462-7_23

[KVW04]   Kohno, T., Viega, J., Whiting, D.: CWC: a high-performance conventional authenticated encryption mode. In: Roy, B., Meier, W. (eds.) FSE 2004. LNCS, vol. 3017, pp. 408–426. Springer, Heidelberg (2004). https://doi.org/10.1007/978-3-540-25937-4_26

[KYS05]   Kaps, J.-P., Yüksel, K., Sunar, B.: Energy scalable universal hashing. IEEE Trans. Comput. **54**(12), 1484–1495 (2005)

[LMP17]   Luykx, A., Mennink, B., Paterson, K.G.: Analyzing multi-key security degradation. In: Takagi, T., Peyrin, T. (eds.) ASIACRYPT 2017. LNCS, vol. 10625, pp. 575–605. Springer, Cham (2017). https://doi.org/10.1007/978-3-319-70697-9_20

[LPSY16] Luykx, A., Preneel, B., Szepieniec, A., Yasuda, K.: On the influence of message length in PMAC's security bounds. In: Fischlin, M., Coron, J.-S. (eds.) EUROCRYPT 2016. LNCS, vol. 9665, pp. 596–621. Springer, Heidelberg (2016). https://doi.org/10.1007/978-3-662-49890-3_23

[MF05]   McGrew, D.A., Fluhrer, S.R.: Multiple forgery attacks against message authentication codes. Cryptology ePrint Archive, Report 2005/161 (2005). http://eprint.iacr.org/2005/161

[MN17]   Mennink, B., Neves, S.: Encrypted Davies-Meyer and its dual: towards optimal security using mirror theory. In: Katz, J., Shacham, H. (eds.) CRYPTO 2017. LNCS, vol. 10403, pp. 556–583. Springer, Cham (2017). https://doi.org/10.1007/978-3-319-63697-9_19

[MV04a]   McGrew, D.A., Viega, J.: The security and performance of the Galois/Counter Mode (GCM) of operation. In: Canteaut, A., Viswanathan, K. (eds.) INDOCRYPT 2004. LNCS, vol. 3348, pp. 343–355. Springer, Heidelberg (2004). https://doi.org/10.1007/978-3-540-30556-9_27

[MV04b]   McGrew, D.A., Viega, J.: The security and performance of the Galois/Counter Mode of operation (Full Version). IACR Cryptology ePrint Archive 2004, p. 193 (2004)

[MW16]   Mattsson, J., Westerlund, M.: Authentication key recovery on Galois/Counter Mode (GCM). In: Pointcheval, D., Nitaj, A., Rachidi, T. (eds.) AFRICACRYPT 2016. LNCS, vol. 9646, pp. 127–143. Springer, Cham (2016). https://doi.org/10.1007/978-3-319-31517-1_7

[Nat80]   National Institute of Standards and Technology. DES Modes of Operation. FIPS 81, December 1980

[NOMI15] Niwa, Y., Ohashi, K., Minematsu, K., Iwata, T.: GCM security bounds reconsidered. In: Leander, G. (ed.) FSE 2015. LNCS, vol. 9054, pp. 385–407. Springer, Heidelberg (2015). https://doi.org/10.1007/978-3-662-48116-5_19

[PC13]   Procter, G., Cid, C.: On weak keys and forgery attacks against polynomial-based MAC schemes. In: Moriai, S. (ed.) FSE 2013. LNCS, vol. 8424, pp. 287–304. Springer, Heidelberg (2014). https://doi.org/10.1007/978-3-662-43933-3_15

[PC15]   Procter, G., Cid, C.: On weak keys and forgery attacks against polynomial-based MAC schemes. J. Cryptol. **28**(4), 769–795 (2015)

[Pie06]   Pietrzak, K.: A tight bound for EMAC. In: Bugliesi, M., Preneel, B., Sassone, V., Wegener, I. (eds.) ICALP 2006. LNCS, vol. 4052, pp. 168–179. Springer, Heidelberg (2006). https://doi.org/10.1007/11787006_15

[PvO99] Preneel, B., van Oorschot, P.C.: On the security of iterated message authentication codes. IEEE Trans. Inf. Theor. **45**(1), 188–199 (1999)

[Saa11] Saarinen, M.-J.O.: SGCM: The Sophie Germain Counter Mode. Cryptology ePrint Archive, Report 2011/326 (2011). http://eprint.iacr.org/2011/326

[Saa12] Saarinen, M.-J.O.: Cycling attacks on GCM, GHASH and other polynomial MACs and Hashes. In: Canteaut, A. (ed.) FSE 2012. LNCS, vol. 7549, pp. 216–225. Springer, Heidelberg (2012). https://doi.org/10.1007/978-3-642-34047-5_13

[Sho96] Shoup, V.: On fast and provably secure message authentication based on universal hashing. In: Koblitz, N. (ed.) CRYPTO 1996. LNCS, vol. 1109, pp. 313–328. Springer, Heidelberg (1996). https://doi.org/10.1007/3-540-68697-5_24

[Sim91] Simmons, G.J.: A survey of information authentication. In: Simmons, G.J. (ed.) Contemporary Cryptology: The Science of Information Integrity, pp. 381–419. IEEE Press, New York (1991)

[SMC08] Salowey, J.A., McGrew, D.A., Choudhury, A.: AES Galois Counter Mode (GCM) Cipher Suites for TLS. RFC 5288, August 2008

[Sti91] Stinson, D.R.: Universal hashing and authentication codes. In: Feigenbaum, J. (ed.) CRYPTO 1991. LNCS, vol. 576, pp. 74–85. Springer, Heidelberg (1992). https://doi.org/10.1007/3-540-46766-1_5

[Sti94] Stinson, D.R. (ed.): CRYPTO 1993. LNCS, vol. 773. Springer, Heidelberg (1994). https://doi.org/10.1007/3-540-48329-2

[Tay93] Taylor, R.: An integrity check value algorithm for stream ciphers. In: Stinson [Sti94], pp. 40–48

[VM06] Viega, J., McGrew, D.A.: The use of Galois Message Authentication Code (GMAC) in IPsec ESP and AH. RFC 4543, May 2006

[WC81] Wegman, M.N., Carter, L.: New hash functions and their use in authentication and set equality. J. Comput. Syst. Sci. **22**(3), 265–279 (1981)

[Wie99] Wiener, M. (ed.): CRYPTO 1999. LNCS, vol. 1666. Springer, Heidelberg (1999). https://doi.org/10.1007/3-540-48405-1

[ZTG13] Zhu, B., Tan, Y., Gong, G.: Revisiting MAC forgeries, weak keys and provable security of galois/counter mode of operation. In: Abdalla, M., Nita-Rotaru, C., Dahab, R. (eds.) CANS 2013. LNCS, vol. 8257, pp. 20–38. Springer, Cham (2013). https://doi.org/10.1007/978-3-319-02937-5_2

[ZW17] Zheng, K., Wang, P.: A uniform class of weak keys for universal hash functions. Cryptology ePrint Archive, Report 2017/436 (2017). http://eprint.iacr.org/2017/436

Revisiting AES-GCM-SIV: Multi-user Security, Faster Key Derivation, and Better Bounds

Priyanka Bose[1(✉)], Viet Tung Hoang[2], and Stefano Tessaro[1]

[1] Department of Computer Science, University of California, Santa Barbara, USA
priyanka@cs.ucsb.edu
[2] Department of Computer Science, Florida State University, Tallahassee, USA

Abstract. This paper revisits the multi-user (mu) security of symmetric encryption, from the perspective of delivering an analysis of the AES-GCM-SIV AEAD scheme. Our end result shows that its mu security is comparable to that achieved in the single-user setting. In particular, even when instantiated with short keys (e.g., 128 bits), the security of AES-GCM-SIV is not impacted by the collisions of two user keys, as long as each individual nonce is not re-used by too many users. Our bounds also improve existing analyses in the single-user setting, in particular when messages of variable lengths are encrypted. We also validate security against a general class of key-derivation methods, including one that *halves* the complexity of the final proposal.

As an intermediate step, we consider mu security in a setting where the data processed by every user is bounded, and where user keys are generated according to arbitrary, possibly correlated distributions. This viewpoint generalizes the currently adopted one in mu security, and can be used to analyze re-keying practices.

Keywords: Multi-user security · AES-GCM-SIV
Authenticated encryption · Concrete security

1 Introduction

This work continues the study of the *multi-user (mu) security* of symmetric cryptography, the setting where the adversary distributes its resources to attack multiple instances of a cryptosystem, with the end goal of compromising *at least one of them*. This attack model was recently the object of extensive scrutiny [2, 9,21,22,26,29,35], and its relevance stems from the *en masse* deployment of symmetric cryptography, e.g., within billions of daily TLS connections. The main goal is to study the degradation in security as the number of users increases.

OUR CONTRIBUTIONS. This paper will extend this line of work in different ways. The most tangible contribution is a complete analysis in the mu setting of the AES-GCM-SIV [18] scheme by Gueron, Langley, and Lindell, an AES-based scheme for *authenticated encryption with associated data* (AEAD) which

© International Association for Cryptologic Research 2018
J. B. Nielsen and V. Rijmen (Eds.): EUROCRYPT 2018, LNCS 10820, pp. 468–499, 2018.
https://doi.org/10.1007/978-3-319-78381-9_18

is meant to resist nonce misuse. Our main result will show that the scheme's security does not degrade in the mu setting, in a sense much stronger than what was claimed in the previous mu analyses. Also, we abstract the requirement needed for AES-GCM-SIV's key-derivation step, and show that a very simple KDF is sufficient for high security. Beyond this, our analysis also delivers conceptual and technical insights of wider interest.

Concretely, our result will highlight the benefit of ensuring limited nonce reuse across different users (e.g., by choosing nonces randomly). We show that in this setting AES-GCM-SIV does *not* suffer any impact from key-collisions, in particular allowing security to go beyond the Birthday barrier (wrt the key length) even in the multi-user setting. The resulting analysis is particularly involved, and calls for a precise understanding of the power of verification queries (for which nonce re-use across multiple users *cannot* be restricted). Previous analyses of AE schemes (specifically, those of [9]) do not ensure security when two users have the same key, thus forcing either an increase of the key length or a worse security guarantee.

On the way, we analyze the building blocks of AES-GCM-SIV in a refined model of mu security where the amount of data processed by each user is bounded, and where keys come from arbitrary distributions. These results could be of independent interest.

We now continue with a more detailed overview of our results.

MULTI-USER SECURITY. *Multi-user* (mu) security was introduced by Bellare, Boldyreva and Micali [3] in the public-key setting as an explicit security target, although in the symmetric setting the notion had already been targeted in attacks [10,11], and was used implicitly as a technical tool in [4].

For example, in the mu definition of encryption security under chosen-plaintext attacks, each user i is assigned a secret key K_i, and the attacker's encryption queries $\text{ENC}(i, M)$ result in either an encryption of M under K_i (in the real world), or an equally long random ciphertext (in the ideal world). The goal is to distinguish the real from the ideal-world.

Assessing security in this model is interesting and non-trivial. Take for example randomized counter-mode encryption (CTR), based on a block cipher with key length k and block length n. The advantage of any *single-user* adversary encrypting, in total, L blocks of data and making p queries to the cipher (which we model as ideal) is upper bounded by $\epsilon_{su}(L, p) \leq \frac{L^2}{2^n} + \frac{p}{2^k}$ (cf. e.g. [5]). If the attacker now adaptively distributes its queries across u users, a hybrid argument shows that the bound is $\epsilon_{mu}(L, p, u) \leq u \cdot \epsilon_{su}(L, p + L) \leq \frac{2uL^2}{2^n} + \frac{u(p+L)}{2^k}$.

Usually, we do not want to fix u, and allow the adversary to encrypt its budget of L blocks *adaptively* across as many users as it sees fit. In particular, the adversary could (1) query one message only with length L, or (2) query L messages with length 1, each to a different user. Thus, in the worst case, the bound becomes $\epsilon_{mu}(L, p) \leq \frac{2L^3}{2^n} + \frac{Lp + L^2}{2^k}$. A number of recent works [2, 21, 22, 29, 35] have shown that this is overly pessimistic, and the security loss can be much smaller; in fact, often $\epsilon_{mu}(L, p) \approx \epsilon_{su}(L, p)$ holds.

BOUNDING THE PER-USER DATA COMPLEXITY. Note that even if $\epsilon_{mu}(L, p) \approx \epsilon_{su}(L, p)$ above, the matching attack could be a single-user attack, requiring a single honest user to encrypt $L \approx 2^{n/2}$ blocks under the same key. For $k = n = 128$, this would require a single honest user to willingly encrypt multiple exabytes of data, and there are many scenarios where we can easily enforce this not to happen. If we enforce a per-user upper bound B on the number of encrypted blocks, an L-block adversary would be forced to spread its effort across at least L/B users, and the advantage could become even *smaller*. Indeed, tightening existing bounds, we show below that for CTR, the advantage of such an attacker is at most

$$\frac{LB}{2^n} + \frac{L^2}{2^{n+k}} + \frac{ap}{2^k}.$$

for some constant a. This bound shows that the fewer blocks we encrypt per user, the higher the security: Beyond-birthday security is possible, e.g., for $k = n = 128$ and $B = 2^{32}$, the bound is of the order $L/2^{96} + p/2^{128}$. Also, the bound is independent of the number of users, and in particular the role of off-line computation – captured here by p – is also independent of L. Note that most previous results on mu security target deterministic security games, such as PRFs/PRPs [2, 21, 22, 29, 35] or deterministic AE [9, 26], and security falls apart when more than $2^{k/2}$ users are present, and their keys collide. Here, key-collisions are irrelevant, and security well beyond $2^{k/2}$ users is possible.

AES-GCM-SIV: OVERVIEW AND BOUNDS. The above viewpoint generalizes that of Abdalla and Bellare [1], who were first to observe, in a simpler model, that re-keying after encrypting B blocks increases security. The fewer data we encrypt per key, the higher the security.

AES-GCM-SIV adapts the re-keying idea to the AEAD setting, making it in particular *nonce based* – i.e., to encrypt a message M with a nonce N, we use a key-derivation function (KDF) KD to derive a key $K_N \leftarrow \mathsf{KD}(K, N)$ from the master secret key K and the nonce N, and then encrypt the message M with the nonce N under the key K_N using a base AE scheme AE. Now, the keys K_N can be thought as belonging to different (virtual) users. Existing analyses [20, 24] show indeed that, assuming KD is a good PRF, a mu security bound for AE can be lifted to a bound on the end scheme in the *single-user* setting, where now B is a bound on the amount of data encrypted *per nonce*, rather than per user. If nonces are not re-used, B is the maximum block length of an encrypted message.

Concretely, in AES-GCM-SIV, the underlying AE is GCM-SIV$^+$, a slight modification of GCM-SIV [19]. This relies in turn on SIV ("synthetic IV") [34], an AEAD scheme which combines a PRF F and an encryption scheme SE (only meant to be CPA secure) to achieve nonce-misuse resistance. For message M, nonce N, and associated data A, the encryption of SIV results into a ciphertext C obtained as

$$\mathsf{IV} \leftarrow \mathsf{F}(K_\mathsf{F}, (M, N, A)), \quad C \leftarrow \mathsf{SE.E}(K_\mathsf{E}, M; \mathsf{IV}),$$

where K_F and K_E are the two components of the secret key, and $\mathsf{SE.E}(K_\mathsf{E}, M; \mathsf{IV})$ is the deterministic encryption function of SE run with IV IV.

In GCM-SIV$^+$, SE is counter mode, and F is what we call GMAC$^+$, a Wegman-Carter MAC [38] similar to, but different from, the one used in GCM [28]. It composes an xor-universal hash function with n-bit key, with a block cipher of block length n and key length k. GMAC$^+$'s total key length is hence $k + n$ bits. (As we target AES, $n = 128$ and $k \in \{128, 256\}$.) A difference from the original SIV scheme is that the same block cipher key is used across GMAC$^+$ and counter-mode, but an appropriate domain separation is used.

For *nonce-misuse resistance* (so-called mrae security), the best published bound for AES-GCM-SIV with key length 128 bits is of order

$$\frac{QB^2}{2^{128}} + \frac{\ell_{\max}QR}{2^{128}} + \frac{p}{2^{128}} + \epsilon(Q),$$

for any adversary that makes at most p ideal-cipher queries, encrypts at most B blocks *per nonce*, uses at most $Q < 2^{64}$ nonces in encryption/verification queries, where R is the maximum number of repetition of a nonce, and $\ell_{\max}$ is the maximal length of a verification query. Here, $\epsilon(Q)$ is the PRF advantage of KD against Q queries, and it is $Q/2^{96}$ for the considered instantiation.

OUR BOUNDS IN THE MU SETTING. The analysis of AES-GCM-SIV *uses* mu security as a tool, but still only gives su security bounds. A valid question is whether its security substantially degrades in the mu setting or not.

We answer this question, and show that for a large class of suitable instantiations of KD, *multi*-user mrae security of AES-GCM-SIV is of order

$$\frac{LB}{2^{128}} + \frac{d(p + L)}{2^{128}},$$

where L, B, and d are upper bounds, respectively, of the overall number of encrypted/verified blocks, of the number of blocks encrypted per user-nonce pair, and of the number of users that re-use a particular nonce value.

This shows a number of things: First off, our bound is an improvement even in the single-user case, as $d = 1$ vacuously holds, and even if we use the KDF considered in the previous works. (Note in particular that the PRF advantage term $\epsilon(Q)$ disappears from the bound.) The term $\frac{LB}{2^{128}}$ can be much smaller than $\frac{QB^2}{2^{128}}$, as in many settings Q and L can be quite close (e.g., if most messages are very short). In fact, the point is slightly more subtle, and we elaborate on it at the end of the introduction. Second, if d is constant (which we can safely assume if nonces are randomly chosen), security does not degrade as the number of users increases. In particular, the security is unaffected by key collisions. If d cannot be bounded, we necessarily need to increase the key length to 256 bits, and in this case the second term becomes $\frac{d(p+L)}{2^{256}}$. Finally, we have no assumption on the data amount of verification queries per user-nonce pair (other than the overall bound L), whereas the bounds in prior works can become weak if there is a very long verification query, and the adversary uses only a single nonce among verification queries.

The rest of the introduction will explain some ideas behind the bound and the techniques, which we believe to be more broadly applicable.

CHALLENGES. On the way to our end result, we give a number of results of independent interest. Interestingly, while we will recycle ideas on the way, the approach is less modular than one expects. First off, we analyze CTR and GMAC$^+$ in a regime where the amount of data processed by each user is bounded. We will then obtain an analysis of the mu security GCM-SIV$^+$. Here, due to the key re-use, the technique for generic composition used in the original SIV scheme fails, but we will be able to recycle many low-level parts of the proofs for CTR and GMAC$^+$.

At this point, however, it is unclear whether nonce-based key derivation achieves its purpose in the mu setting, where B is now a bound on the number of blocks encrypted per user-nonce pair. Indeed, say the master secret key K has length $k = 128$. Then, should the number of users exceed $2^{k/2} = 2^{64}$, with high probability two users will end up with *identical* keys. If we treat KD as a PRF, like [20,24] do, all security will vanish at this point. Indeed, the existing mu analysis of GCM succumbs to this problem [9], and the problem seems unavoidable here too, since we are considering a deterministic security game.

BOUNDED NONCE RE-USE ACROSS USERS. The way out from this problem is to assume every nonce is re-used by at most d users. Consider the canonical attack to break privacy of the scheme: Fix a sufficiently long message M and a nonce N, and re-use them over and over in encryption queries for different users, and if the same ciphertext appears twice after roughly $2^{k/2}$ queries, we are likely to be in the real world, as ciphertexts are random and independent in the ideal world. This however requires us to *re-use* the same nonce across $2^{k/2}$ users. A first interesting point we observe is that the security of KD as PRF degrades gracefully with the number of users d that can re-use the same input/nonce.

Unfortunately, this is not enough. The catch is that a bound d on the number of users re-using a nonce is only meaningful for encryption queries, e.g., if nonces are chosen randomly. For authenticity, an attacker would attempt to issue verification queries for as many users as it wishes, and we cannot restrict the choice of nonces. In particular, we cannot prevent that $2^{k/2}$ verification queries for different users with the same nonce may end up using colliding user keys. The question is how far this is an issue.

To get some intuition, consider the security of KD as a MAC, i.e., the adversary issues, in a first stage, queries (i, N), producing output $\mathsf{KD}(K_i, N)$ (where K_i is the key of the i-th user), but respecting the constraint that no nonce is used more than d times across different i's, where d is relatively small. Then, in a second stage, the adversary gets to ask unrestricted verification queries with input (i, N, T), except for the obvious requirement that (i, N) must be previously un-queried. The adversary wins if $\mathsf{KD}(K_i, N) = T$ for one of these verification queries. At first glance, a collision $K_i = K_j$ could help if we have queried (i, N) in the first stage, learnt T, and now can submit (j, N, T) in the second. The caveat is that we need to be able to have *detected* such collisions. This is hard to do during the first stage, even with many queries, due to the constraint of reusing N only d times. Thus, the only obvious way to exploit this would be to try, for each of the q first-stage queries (i, N) with corresponding output T, to query (j, N, T) for many $j \neq i$. This would however require roughly 2^k trials to succeed. Finally, note that while it may be that we ask two verification queries (i, N, T) and (j', N', T') where $K_i = K_j$, this does not seem to give any help in succeeding, because a verification query does not reveal the actual output of KD on that input.

Confirming this intuition is *not* simple. We will do so for a specific class of natural KD constructions outlined below, and point out that the setting of AE is harder than studying the security of KD itself as a MAC. Indeed, our KD is used to derive keys for GMAC$^+$ and CTR *at the same time*, and we need to prove unpredictability of the overall encryption scheme on a new pair (N, i) which was previously unqueried, while producing a bound which does not depend on key collisions. This is the most technically involved part of the paper.

A SIMPLER KDF. Finally, let us address *how* we instantiate KD. The construction of KD from [18] is truncation based, and makes 4 (for $k = 128$), respectively 6 (for $k = 256$) calls to a block cipher to derive a key. A recent proposal [24] suggests using the so-called XOR construction to achieve higher security, as multiple analyses [7, 14, 25, 31, 33] confirm better bounds than for truncation [16]. Still, the resulting KD would need 4 resp. 6 calls. They also consider a faster construction, based on CENC [23], which would require 3 resp. 4 calls. All of these constructions are required to be good PRFs in existing analyses.

Rather than studying concrete constructions, we apply our result to a general class of KDFs which includes in particular all of these proposals, but also simpler ones. For instance, our bounds apply to the following simple KDF, a variant of which was in the initial AES-GCM-SIV proposal, but was discarded due to security concerns. Namely, given the underlying block cipher E, the KDF outputs

$$\mathsf{KD}(K, N) = E(K, \mathsf{pad}(N, 0)) \parallel E(K, \mathsf{pad}(N, 1)) \tag{1}$$

for $k = n$ and N an nl-bit string, with nl $\leq n - 2$, and, analogously, for $k = 2n$, one can extend this by additionally concatenating $E(K, \mathsf{pad}(N, 2))$. Here, pad is a mapping with the property that the sets $\{\mathsf{pad}(N, 0), \mathsf{pad}(N, 1), \mathsf{pad}(N, 2)\}$ defined by each N are disjoint. This approach seems to contradict common sense which was adopted in the new KDF variants for AES-GCM-SIV, because the derived keys are not truly random. However, a crucial point of our analyses is that we do not prove PRF security of these KDFs. Rather, we study the distributions on keys they induce, and then (implicitly) rely on the security of the underlying components using keys obtained from (slightly) non-uniform distributions.[1]

---

[1] This key-derivation scheme is also used to derive sub-keys from tweaks in the setting of FPE within the DFF construction [37]. DFF is a replacement for FF2 [36], a scheme proposed to NIST for standardization but eventually rejected due to a birthday-bound key-recovery attack [15]. The security of DFF is formalized and studied in [6], but their analysis is still in the su setting, namely there is only one master key for KD.

In platforms that support AES hardware acceleration, the difference in performance between the KDF in Eq. (1) and the current one in AES-GCM-SIV is not important, as demonstrated via the experiments in [18]. Still, we believe it is important for schemes to be minimal, and thus to understand the security of the simplest possible instantiations of the KDF.

SUB-OPTIMALITY OF POLYVAL. We also observe that the universal hash POLYVAL within GMAC$^+$ is somewhat suboptimal. That is, if both the message and the associated data are the empty string, then their hash image under POLYVAL is always 0^{128}, regardless of the hash key. This does not create any issue in the single-user setting, but substantially weakens the mu security of GCM-SIV$^+$ and GMAC$^+$ to $\frac{LB}{2^{128}} + \frac{d(p+L)}{2^{128}}$, despite their use of 256-bit keys. Had the padding in POLYVAL ensured that the hash image of empty strings under a random key has a uniform distribution, the security of GCM-SIV$^+$ and GMAC$^+$ could be improved to $\frac{LB}{2^{128}} + \frac{Lp}{2^{256}}$, meaning this bound is independent of the number d of users that reuse any particular nonce. While this issue does not affect the concrete security bound of AES-GCM-SIV, this change becomes necessary if GCM-SIV$^+$ or GMAC$^+$ are used as standalone schemes.

RELATION TO EXISTING WORKS. We elaborate further on our improvements in the su setting over recent analyses [20, 24]. As mentioned above, their bound contains a term of the order $QB^2/2^n$, which we improve to $LB/2^n$. The fact that the latter is better is not quite obvious. Indeed, it is not hard to improve the term $QB^2/2^n$ in [20, 24] to $\sum_{i=1}^{Q} B_i^2/2^n$, where B_i is a bound on the number of blocks encrypted with the i-th nonce. This seems to address the point that different amounts of data can be encrypted for different nonces.

The crucial point is that we capture a far more general class of attacks by only limiting the adversary in terms of L, p, and d. For instance, for a parameter L, consider the following single-user adversary using $Q = L/2$ nonces. It will select a random subset of the Q nonces, of size $L/(2B)$, for which it encrypts B blocks of data, and for the remaining $L/2 - L/(2B)$ nonces, it only encrypts one block of data. In our bound, we still get a term $LB/2^n$. In contrast, with the parametrization adopted by [20, 24], we can only set $Q = L/2$ and $B_i = B$ for all $i \in [Q]$, because any of the nonces can, a priori, be used to encrypt B blocks. This ends up giving a term of magnitude $LB^2/2^n$, however, which is much larger. For $B = 2^{32}$, the difference between $L/2^{64}$ and $L/2^{96}$ is enormous.

Switching to the type of bounds is non-trivial: The adversary can adopt an arbitrarily adaptive attack pattern. Handling such adversaries was the object of recent works in the mu regime [2, 21, 22, 26, 29, 35].

STANDARD VS IDEAL-MODEL. We also note that the bound of [24] is expressed in the standard model, and contains a term $Q\epsilon$, where ϵ is the advantage of a PRF adversary $\mathcal{A}'$ against the cipher E, making B queries. The catch is that ϵ is very sensitive to the time complexity of $\mathcal{A}'$, which we approximate with the number of ideal-cipher queries p. Thus, $Q\epsilon$ is of order $Q(B^2/2^n + p/2^k)$. While [24] argues that $QB^2/2^n$ is the largest term, the ideal model makes it evident that the hidden term $Qp/2^k$ is likely to be far more problematic in the case $n = k$. Indeed, $p \geq Q$ and $B^2 \leq Q$ are both plausible (the attacker can more easily invest in local computation than obtaining honest encryptions under equal nonces), and this becomes $\frac{Q^2}{2^k}$. This shows security is bounded by $2^{k/2}$. The work of [26] on classical GCM also seemingly focuses on the standard model and thus seems to fail to capture such hidden terms. In contrast, [20] handles this properly.

We stress that we share the sentiment that ideal-model analysis may oversimplify some security issues. However, we find them a necessary evil when trying to capture the influence of local computation in multi-user attacks, which is a fundamental part of the analysis.

OUTLINE OF THIS PAPER. We introduce basic notions and security definitions in the multi-user setting in Sect. 3. Then, in Sect. 4, we study the security of our basic building blocks, CTR and GMAC$^+$, in the multi-user setting. In Sect. 5, we analyze the SIV composition when keys are re-used across encryption and PRF, and observe this to work in particular for the setting of GCM-SIV. Finally, Sect. 6 studies our variant of AES-GCM-SIV with more general key derivation.

2   Preliminaries

NOTATION. Let ε denote the empty string. For a finite set S, we let $x \leftarrow_\$ S$ denote the uniform sampling from S and assigning the value to x. Let $|x|$ denote the length of the string x, and for $1 \leq i < j \leq |x|$, let $x[i, j]$ (and also $x[i : j]$) denote the substring from the ith bit to the jth bit (inclusive) of x. If A is an algorithm, we let $y \leftarrow A(x_1, \ldots; r)$ denote running A with randomness r on inputs $x_1, \ldots$ and assigning the output to y. In the context that we use a blockcipher $E : \{0,1\}^k \times \{0,1\}^n \rightarrow \{0,1\}^n$, the block length of a string x, denoted $|x|_n$, is $\max\{1, \lceil |x|/n \rceil\}$.

SYSTEMS AND TRANSCRIPTS. Following the notation from [21] (which was in turn inspired by Maurer's framework [27]), it is convenient to consider interactions of a distinguisher A with an abstract system $\mathbf{S}$ which answers A's queries. The resulting interaction then generates a transcript $\tau = ((X_1, Y_1), \ldots, (X_q, Y_q))$ of query-answer pairs. It is well known that $\mathbf{S}$ is entirely described by the probabilities $\mathsf{p_S}(\tau)$ that if we make queries in τ to system $\mathbf{S}$, we will receive the answers as indicated in τ.

We will generally describe systems informally, or more formally in terms a set of oracles they provide, and only use the fact that they define corresponding probabilities $\mathsf{p_S}(\tau)$ without explicitly giving these probabilities.

THE H-COEFFICIENT TECHNIQUE. We now describe the H-coefficient technique of Patarin [13, 32]. Generically, it considers a deterministic distinguisher $\mathcal{A}$, interacting with system $\mathbf{S}_0$ or with system $\mathbf{S}_1$. Let $\mathcal{X}_0$ and $\mathcal{X}_1$ be random variables for the transcripts defined by these interactions with $\mathbf{S}_0$ and $\mathbf{S}_1$, and a bound on the distinguishing advantage of $\mathcal{A}$ is given by the statistical distance $\mathsf{SD}(\mathcal{X}_0, \mathcal{X}_1)$.

Lemma 1. [13,32] *Supposed we can partition transcripts into good and bad transcripts. Further, suppose that there exists $\epsilon \geq 0$ such that $1 - \frac{\mathsf{ps}_0(\tau)}{\mathsf{ps}_1(\tau)} \leq \epsilon$ for every good transcript τ such that $\mathsf{ps}_1(\tau) > 0$. Then,*

$$\mathsf{SD}(\mathcal{X}_1, \mathcal{X}_0) \leq \epsilon + \Pr[\mathcal{X}_1 \text{ is bad}].$$

3   Multi-user Security of Symmetric Primitives

We revisit security definitions for basic symmetric primitives in the multi-user setting. We will in particular extend existing security definitions to impose overall bounds on the volume of data processed by each user, however we will relegate this matter to theorem statements restricting the considered adversaries, rather than hard-coding these bounds in the definitions.

3.1   Symmetric and Authenticated Encryption

We define AE syntax here, as well as natural multi-user generalizations of classical security notions for confidentiality and integrity. Since this paper will deal both with probabilistic and deterministic schemes, we define both, following the treatment of Namprempre, Rogaway, and Shrimpton [30]. Our notational conventions are similar to those from [9].

IV-BASED ENCRYPTION. An *IV-based symmetric encryption* scheme SE consists of two algorithms, the *randomized* encryption algorithm SE.E and the deterministic *decryption algorithm* SE.D, and is associated with a corresponding key length SE.kl $\in \mathbb{N}$ and initialization-vector (IV) length SE.vl $\in \mathbb{N}$. Here, SE.E takes as input a secret key $K \in \{0,1\}^{\mathsf{SE.kl}}$ and a plaintext $M \in \{0,1\}^*$. It then samples IV $\leftarrow_\$ \{0,1\}^{\mathsf{SE.vl}}$, deterministically computes a ciphertext core C' from K, M and IV, and returns $C \leftarrow \mathsf{IV} \parallel C'$. We often write $C \leftarrow_\$ \mathsf{SE.E}_K(M)$ or $C \leftarrow_\$ \mathsf{SE.E}(K, M)$. If we want to force the encryption scheme to run on a specific initialization vector IV, then we write $\mathsf{SE.E}(K, M; \mathsf{IV})$. The corresponding decryption algorithm SE.D takes as input a key $K \in \{0,1\}^{\mathsf{SE.kl}}$ and a ciphertext $C \in \{0,1\}^*$, returns either a plaintext $M \in \{0,1\}^*$, or an error symbol $\bot$. For correctness, we require that if C is output by $\mathsf{SE.E}_K(M)$, then $\mathsf{SE.D}_K(C)$ returns M. We allow all algorithms to make queries to an ideal primitive Π, in which case this will be made explicit when not clear from the context, e.g., by writing $\mathsf{SE}[\Pi]$ in lieu of SE.

CHOSEN-PLAINTEXT SECURITY FOR IV-BASED ENCRYPTION. We re-define the traditional security notion of ind-security for the multi-user setting. Our definition will however incorporate a general, stateful *key-generation* algorithm KeyGen which is invoked every time a new user is spawned via a call to the NEW oracle. KeyGen is a parameter of the game, and it takes additionally some input string aux which is supplied by the adversary. The traditional mu security setting would have KeyGen simply output a random string, and ignore aux, but we will consider a more general setting to lift mu bounds to the key-derivation setting.

The game is further generalized to handle an arbitrary ideal primitive (an ideal cipher, a random oracle, or a combination thereof) via an oracle PRIM.[2] Also note that the oracle PRIM can simply trivially provide no functionality, in which case we revert to the standard-model definition. We note that the key-generation algorithm KeyGen does not have access to the oracle PRIM.

Given an adversary $\mathcal{A}$, the resulting game is $\mathbf{G}^{\text{mu-ind}}_{\text{SE,KeyGen},\Pi}(\mathcal{A})$, and is depicted at the top of Fig. 1. The associated advantage is

$$\mathsf{Adv}^{\text{mu-ind}}_{\text{SE,KeyGen},\Pi}(\mathcal{A}) = 2 \cdot \Pr\left[\mathbf{G}^{\text{mu-ind}}_{\text{SE,KeyGen},\Pi}(\mathcal{A})\right] - 1.$$

Whenever we use the canonical KeyGen which outputs a random string regardless of its input, we will often omit it, and just write $\mathsf{Adv}^{\text{mu-ind}}_{\text{SE},\Pi}(\mathcal{A})$ instead.

AUTHENTICATED ENCRYPTION SCHEME. An authenticated encryption scheme AE with associated data (also referred to as an AEAD scheme), the algorithms AE.E and AE.D are both deterministic. In particular, AE.E takes as input a secret key $K \in \{0,1\}^{\text{AE.kl}}$, a *nonce* $N \in \{0,1\}^{\text{AE.nl}}$, a plaintext $M \in \{0,1\}^*$, and the *associated data* A, and returns the ciphertext $C \leftarrow \mathsf{AE.E}(K, N, M, A)$. The corresponding decryption algorithm AE.D takes as input a key $K \in \{0,1\}^{\text{AE.kl}}$, the nonce N, the ciphertext $C \in \{0,1\}^*$, and the associated data A, and returns either a plaintext $M \in \{0,1\}^*$, or an error symbol $\perp$. We require that if C is output by $\mathsf{AE.E}_K(M, N, A)$, then $\mathsf{AE.D}_K(C, N, A)$ returns M.

Our security notion for AE is nonce-misuse-resistant: Ciphertexts produced by encryptions with the same nonce are pseudorandom *as long as* the encryptions are on different messages or associated data, even if they are for the same nonce. Our formalization of AE multi-user security in terms of $\mathbf{G}^{\text{mu-mrae}}_{\text{AE,KeyGen},\Pi}(\mathcal{A})$ is that of Bellare and Tackmann [9], with the addition of a KeyGen algorithm to handle arbitrary correlated key distributions. It is depicted in Fig. 1, at the bottom.

Given an adversary $\mathcal{A}$ and a key-generation algorithm KeyGen, we are then going to define

$$\mathsf{Adv}^{\text{mu-mrae}}_{\text{AE,KeyGen},\Pi}(\mathcal{A}) = 2 \cdot \Pr\left[\mathbf{G}^{\text{mu-mrae}}_{\text{AE,KeyGen},\Pi}(\mathcal{A})\right] - 1.$$

As above, KeyGen is omitted if it is the canonical one.

We say that an adversary is *d-repeating* if among the encryption queries, an adversary only uses each nonce for at most d users. We stress that we make no assumption on how the adversary picks nonces for the verification queries, and for each individual user, the adversary can repeat nonces in encryption queries as often as it wishes. If nonces are chosen arbitrarily then d can be as big as the number of encryption queries. If nonces are picked at random then d is a small constant.

A KEY-COLLISION ATTACK. We now show that for any AE scheme AE that uses the canonical KeyGen, if an adversary can choose nonces arbitrarily then there

---

[2] If PRIM is meant to handle multiple primitives, we assume they can be accessed through the same interface by pre-pending to the query a prefix indicating which primitive is meant to be queried.

Game $\mathbf{G}^{\text{mu-ind}}_{\text{SE,KeyGen},\Pi}(\mathcal{A})$	$\text{ENC}(i, M)$
$\text{st}_0 \leftarrow \varepsilon;\ v \leftarrow 0;\ b \leftarrow\!\!\text{\textdollar}\ \{0,1\}$ $b' \leftarrow\!\!\text{\textdollar}\ \mathcal{A}^{\text{NEW,ENC,PRIM}}$ Return $(b' = b)$	If $i \notin \{1, \ldots, v\}$ then return $\bot$ $C_1 \leftarrow\!\!\text{\textdollar}\ \text{SE.E}^{\text{PRIM}}(K_i, M)$ $C_0 \leftarrow\!\!\text{\textdollar}\ \{0,1\}^{|C_1|}$ Return C_b
$\underline{\text{NEW}(\text{aux})}$ $v \leftarrow v+1$ $(K_v, \text{st}_v) \leftarrow\!\!\text{\textdollar}\ \text{KeyGen}(\text{st}_{v-1}, \text{aux})$	

Game $\mathbf{G}^{\text{mu-mrae}}_{\text{AE,KeyGen},\Pi}(\mathcal{A})$	$\underline{\text{NEW}(\text{aux})}$
$\text{st}_0 \leftarrow \varepsilon;\ v \leftarrow 0;\ b \leftarrow\!\!\text{\textdollar}\ \{0,1\}$ $b' \leftarrow\!\!\text{\textdollar}\ \mathcal{A}^{\text{NEW,ENC,VF,PRIM}}$ Return $(b' = b)$	$v \leftarrow v+1$ $(K_v, \text{st}_v) \leftarrow\!\!\text{\textdollar}\ \text{KeyGen}(\text{st}_{v-1}, \text{aux})$
$\underline{\text{VF}(i, N, C, A)}$	$\underline{\text{ENC}(i, N, M, A)}$
If $i \notin \{1, \ldots, v\}$ then return $\bot$ If $(i, N, C, A) \in V[i]$ then return true If $b = 0$ then return false $M \leftarrow \text{AE.D}^{\text{PRIM}}(K_i, N, C, A)$ Return $(M \neq \bot)$	If $i \notin \{1, \ldots, v\}$ then return $\bot$ If $(i, N, M, A) \in U[i]$ then return $\bot$ $C_1 \leftarrow \text{AE.E}^{\text{PRIM}}(K_i, N, M, A)$ $C_0 \leftarrow\!\!\text{\textdollar}\ \{0,1\}^{|C_1|}$ $U[i] \leftarrow U[i] \cup \{(i, N, M, A)\}$ $V[i] \leftarrow V[i] \cup \{(i, N, C_b, A)\}$ Return C_b

Fig. 1. Security definitions for chosen-plaintext security of IV-based encryption (top), as well as nonce-misuse resistance for authenticated encryption (bottom). We assume (without making this explicit) that PRIM implements the ideal-primitive Π.

is an attack, using q encryption queries and no verification query, that achieves advantage $q(q-1)/2^{\text{AE.kl}+3}$.

Suppose that under AE, a ciphertext is always at least as long as the corresponding plaintext. Fix an arbitrary message M such that $|M| \geq \text{AE.kl} + 2$. Fix a nonce N and associated data A. The adversary $\mathcal{A}$ attacks q users, and for each user i, it queries $\text{ENC}(i, N, M, A)$ to get answer C_i. If there are distinct i and j such that $C_i = C_j$ then it outputs 1, hoping that users i and j have the same key. For analysis, we need the following well-known result; see, for example, [17, Chapter 5.8] for a proof.

Lemma 2 (Lower bound for birthday attack). *Let $q, N \geq 1$ be integers such that $q \leq \sqrt{2N}$. Suppose that we throw q balls at random into N bins. Then the chance that there is a bin of at least two balls is at least $\frac{q(q-1)}{4N}$.*

From Lemma 2 above, in the real world, the adversary will output 1 if two users have the same key, which happens with probability at least $q(q-1)/2^{\text{AE.kl}+2}$. In contrast, since the ciphertexts are at least $|M|$-bit long, in the ideal world, it outputs 1 with probability at most $q(q-1)/2^{|M|+1} \leq q(q-1)/2^{\text{AE.kl}+3}$. Hence

$$\mathsf{Adv}_{\mathsf{AE},\varPi}^{\mathsf{mu\text{-}mrae}}(\mathcal{A}) \geq \frac{q(q-1)}{2^{\mathsf{AE.kl}+2}} - \frac{q(q-1)}{2^{\mathsf{AE.kl}+3}} = \frac{q(q-1)}{2^{\mathsf{AE.kl}+3}}.$$

3.2 Multi-user PRF Security

We consider keyed functions $\mathsf{F} : \{0,1\}^{\mathsf{F.kl}} \times \{0,1\}^{\mathsf{F.il}} \to \{0,1\}^{\mathsf{F.ol}}$, possibly making queries to an ideal primitive $\varPi$. Here, note that we allow $\mathsf{F.il} = *$, indicating a variable-input-length function. We define a variant of the standard multi-user version of PRF security from [4] using (as in the previous section) a general algorithm KeyGen to sample possibly correlated keys.

Concretely, let $\mathsf{Func}(\mathsf{il},\mathsf{ol})$ be the set of all functions $\{0,1\}^{\mathsf{il}} \to \{0,1\}^{\mathsf{ol}}$, where, once again, $\mathsf{il} = *$ is allowed. We give the multi-user PRF security game in Fig. 2. There, F's access to $\varPi$ is modeled by having oracle access to PRIM. For any adversary $\mathcal{A}$, and key-generation algorithm KeyGen, we define

$$\mathsf{Adv}_{\mathsf{F},\mathsf{KeyGen},\varPi}^{\mathsf{mu\text{-}prf}}(\mathcal{A}) = 2 \cdot \Pr\left[\mathbf{G}_{\mathsf{F},\mathsf{KeyGen},\varPi}^{\mathsf{mu\text{-}prf}}(\mathcal{A})\right] - 1.$$

As usual, we will omit KeyGen when it is the canonical key generator outputting independent random keys.

Game $\mathbf{G}_{\mathsf{F},\mathsf{KeyGen},\varPi}^{\mathsf{mu\text{-}prf}}(\mathcal{A})$	NEW(aux)	EVAL(i, M)
$v \leftarrow 0;\ \mathsf{st}_0 \leftarrow \emptyset$	$v \leftarrow v + 1$	If $i \notin \{1,\dots,v\}$ return $\perp$
$b \leftarrow\!\!\$\ \{0,1\}$	$(K_v,\mathsf{st}_v) \leftarrow\!\!\$\ \mathsf{KeyGen}(\mathsf{st}_{v-1},\mathsf{aux})$	$Y_1 \leftarrow \mathsf{F}^{\mathrm{PRIM}}(K_i, M)$
$b' \leftarrow\!\!\$\ \mathcal{A}^{\mathrm{NEW},\mathrm{EVAL},\mathrm{PRIM}}$	$\rho_v \leftarrow\!\!\$\ \mathsf{Func}(\mathsf{F.il},\mathsf{F.ol})$	$Y_0 \leftarrow\!\!\$\ \rho_i(M)$
Return $(b' = b)$		Return Y_b

Fig. 2. Definition of multi-user PRF security. Again, PRIM implements the ideal primitive $\varPi$.

3.3 Decomposing AE Security

While the notion mu-mrae is very strong, it might be difficult to prove that an AE scheme, say AES-GCM-SIV meets this notion, if one aims for beyond-birthday bounds. We therefore decompose this notion into separate privacy and authenticity notions, as defined below.

PRIVACY. Consider the game $\mathbf{G}_{\mathsf{AE},\mathsf{KeyGen},\varPi}^{\mathsf{mu\text{-}priv}}(\mathcal{A})$ in Fig. 3 that defines the (misuse-resistant) privacy of an AE scheme AE, with respect to a key-generation algorithm KeyGen, and an ideal primitive $\varPi$. Define

$$\mathsf{Adv}_{\mathsf{AE},\mathsf{KeyGen},\varPi}^{\mathsf{mu\text{-}priv}}(\mathcal{A}) = 2\Pr[\mathbf{G}_{\mathsf{AE},\mathsf{KeyGen},\varPi}^{\mathsf{mu\text{-}priv}}(\mathcal{A})] - 1.$$

Under this notion, the adversary is given access to an encryption oracle that either implements the true encryption or returns a random string of appropriate length, but there is no decryption oracle. If the adversary repeats a prior encryption query then this query will be ignored.

Game $\mathbf{G}^{\text{mu-priv}}_{\text{AE,KeyGen},\Pi}(\mathcal{A})$	Game $\mathbf{G}^{\text{mu-auth}}_{\text{AE,KeyGen},\Pi}(\mathcal{A})$
$v \leftarrow 0;\, \mathsf{st}_0 \leftarrow \varepsilon;\, b \leftarrow\!\!{\scriptstyle\$}\, \{0,1\}$ $b' \leftarrow\!\!{\scriptstyle\$}\, \mathcal{A}^{\text{New,Enc,Prim}}$ Return $(b' = b)$	$v \leftarrow 0;\, \mathsf{st}_0 \leftarrow \varepsilon;\, b \leftarrow 0$ $\mathcal{A}^{\text{New,Enc,Vf,Prim}}$ Return $(b = 1)$
$\underline{\text{New}(\text{aux})}$ $v \leftarrow v+1$ $(K_v, \mathsf{st}_v) \leftarrow\!\!{\scriptstyle\$}\, \mathsf{KeyGen}(\mathsf{st}_{v-1}, \mathsf{aux})$	$\underline{\text{New}(\text{aux})}$ $v \leftarrow v+1$ $(K_v, \mathsf{st}_v) \leftarrow\!\!{\scriptstyle\$}\, \mathsf{KeyGen}(\mathsf{st}_{v-1}, \mathsf{aux})$
$\underline{\text{Enc}(i, N, M, A)}$ If $i \notin \{1, \dots, v\}$ then return $\bot$ If $(i, N, M, A) \in U[i]$ then return $\bot$ $C_1 \leftarrow \mathsf{AE.E}^{\text{Prim}}(K_i, N, M, A)$ $C_0 \leftarrow\!\!{\scriptstyle\$}\, \{0,1\}^{|C_1|}$ $U[i] \leftarrow U[i] \cup \{(i, N, M, A)\}$ Return C_b	$\underline{\text{Enc}(i, N, M, A)}$ If $i \notin \{1, \dots, v\}$ then return $\bot$ $C \leftarrow \mathsf{AE.E}^{\text{Prim}}(K_i, N, M, A)$ $V[i] \leftarrow V[i] \cup \{(i, N, C, A)\}$ Return C $\underline{\text{Vf}(i, N, C, A)}$ If $i \notin \{1, \dots, v\}$ then return $\bot$ If $(i, N, C, A) \notin V[i]$ then $\quad M \leftarrow \mathsf{AE.D}^{\text{Prim}}(K_i, N, C, A)$ $\quad$ If $(M \neq \bot)$ then $b \leftarrow 1$

Fig. 3. Games to define privacy (left), and authenticity (right) of an AE scheme AE with respect to a key-generation algorithm $\mathsf{KeyGen} : \mathcal{K} \times \mathcal{N} \to \{0,1\}^{\text{AE.kl}}$. The oracle Prim implements the ideal primitive Π. In the authenticity notion, queries to Vf must be performed *after* all queries to Enc.

Authenticity. Consider the game $\mathbf{G}^{\text{mu-auth}}_{\text{AE,KeyGen},\Pi}(\mathcal{A})$ in Fig. 3 that defines the (misuse-resistant) authenticity of an AE scheme AE, with respect to a key-generation algorithm KeyGen, and an ideal primitive Π. Define

$$\mathsf{Adv}^{\text{mu-auth}}_{\text{AE,KeyGen},\Pi}(\mathcal{A}) = 2\Pr[\mathbf{G}^{\text{mu-auth}}_{\text{AE,KeyGen},\Pi}(\mathcal{A})] - 1.$$

Under this notion, initially a bit b is set to 0 and the adversary is given an encryption oracle that always implements the true encryption, and a verification oracle. We require that the verification queries be made *after* all the evaluation queries. On a verification (i, N, C, A), if there is a prior encryption query (i, N, M, A) for an answer C, then the oracle ignores this query. Otherwise, the oracle sets $b \leftarrow 1$ if $\mathsf{AE.D}^{\text{Prim}}(K_i, N, C, A)$ returns a non-$\bot$ answer. The goal of the adversary is to set $b = 1$.

Relations. Note that in the mrae notion, the adversary can perform encryption and verification queries in an arbitrary order. In contrast, in the authenticity notion, the adversary can only call the verification oracle after it finishes querying the encryption oracle. Still, in Proposition 1 below, we show that authenticity and privacy tightly implies mrae security. The proof is in the full version of this paper [12].

Proposition 1. *Let* AE *be an AE scheme associated with a key-generation algorithm* KeyGen *and an ideal primitive* Π. *Suppose that a ciphertext in* AE *is always* *at least n-bit longer than the corresponding plaintext. For any adversary $\mathcal{A}_0$ that makes q_v verification queries, we can construct adversaries $\mathcal{A}_1$ and $\mathcal{A}_2$ such that*

$$\mathsf{Adv}^{\mathrm{mu\text{-}mrae}}_{\mathsf{AE},\mathsf{KeyGen},\Pi}(\mathcal{A}_0) \leq \mathsf{Adv}^{\mathrm{mu\text{-}priv}}_{\mathsf{AE},\mathsf{KeyGen},\Pi}(\mathcal{A}_1) + \mathsf{Adv}^{\mathrm{mu\text{-}auth}}_{\mathsf{AE},\mathsf{KeyGen},\Pi}(\mathcal{A}_2) + \frac{2q_v}{2^n}.$$

Any query of $\mathcal{A}_1$ or $\mathcal{A}_2$ is produced directly from $\mathcal{A}_0$. If $\mathcal{A}_0$ is d-repeating then so are $\mathcal{A}_1$ and $\mathcal{A}_2$.

4   Multi-user Security of Basic Symmetric Schemes

4.1   Security of Counter-Mode Encryption

We study the mu security of counter mode encryption, or CTR for short. While this is interesting on its own right (we are not aware of any analysis achieving a comparable bound in the literature), we will also use Theorem 1 below to obtain security results for AES-GCM-SIV. For this reason, we introduce some extra notions to handle the degree of generality needed for our proof. Also, our result is general enough to suggest an efficient solution to the re-keying problem first studied by Abdalla and Bellare [1].

GENERAL IVs. We will consider a general IV-increasing procedure add, which is associated with some maximal message length of $L_{\max}$ blocks, and a block length n. In particular, add takes an n-bit string IV and an offset $i \in \{0, \ldots, L_{\max} - 1\}$ as inputs, and is such that $\mathsf{add}(\mathsf{IV}, i)$ returns an n-bit string, and for all IV, the strings $\mathsf{add}(\mathsf{IV}, 0), \ldots, \mathsf{add}(\mathsf{IV}, L_{\max} - 1)$ are distinct. We also say that add has *min-entropy h* if for a random n-bit IV, and every $i \in \mathbb{Z}_{L_{\max}}$, $\mathsf{add}(\mathsf{IV}, i)$ takes any value with probability at most 2^{-h}, i.e., its min-entropy is at least h.

For example, the canonical IV addition is such that $\mathsf{add}(\mathsf{IV}, i) = \mathsf{IV} + i \pmod{2^n}$, where we identify n-bit strings with integers in $\mathbb{Z}_{2^n}$. Here, $L_{\max} = 2^n$. In contrast, AES-GCM-SIV will use CTR with $L_{\max} = 2^{32}$, $n = 128$, and $\mathsf{add}(\mathsf{IV}, i) = 1 \parallel \mathsf{IV}[2, 96] \parallel (\mathsf{IV}[97, 128] + i \pmod{2^{32}})$. Clearly, here, the min-entropy is 127 bits, due to the first bit being set to one.

CTR ENCRYPTION. Let $E : \{0,1\}^k \times \{0,1\}^n \to \{0,1\}^n$ be a block cipher, i.e., $E(K, \cdot)$ is a permutation for all k-bit K. We denote $E(K, \cdot) = E_K(\cdot)$, and E_K^{-1} is the inverse of E_K. Further, let add be a general IV-increasing procedure with maximal block length $L_{\max}$. We define the IV-based encryption scheme CTR = CTR$[E, \mathsf{add}]$ with CTR.kl $= k$, and where encryption operates as follows (where we use $\xleftarrow{n}$ to denote some function which pads a message M into n-bit blocks).

$\underline{\mathsf{CTR}.\mathsf{E}(K, M):}$

$C[0] \leftarrow \mathsf{IV} \leftarrow_\$ \{0,1\}^n, M[1], \ldots, M[\ell] \xleftarrow{n} M$
If $\ell > L_{\max}$ then return $\bot$
For $i = 1$ to ℓ do $C[i] \leftarrow E_K(\mathsf{add}(\mathsf{IV}, i - 1)) \oplus M[i]$
Return $C[0] \parallel C[1] \parallel \cdots \parallel C[\ell]$

Decryption CTR.D re-computes the masks $E_K(\mathsf{add}(\mathsf{IV}, i-1))$ using $C[0] = \mathsf{IV}$, and then retrieves the message blocks by xoring the masks to the ciphertext. Here, we assume without loss of generality messages are padded (e.g., PKCS#7), so that they are split uniquely into full-length n-bit blocks. Our result extends easily to the more common padding-free variant where the last block is allowed to be shorter than n bits, and the output of $E_K(\mathsf{add}(\mathsf{IV}, \ell-1))$ is truncated accordingly, since an adversary can simulate the padding-free version by removing the appropriate number of bits from the received ciphertexts.

SECURITY OF CTR. We establish the (CPA) security of randomized CTR in the ideal-cipher model for an arbitrary key-generation algorithm KeyGen which produces keys that collide with small probability. In particular, we say that KeyGen is α-*smooth* if for a sequence of keys $(K_1, \ldots, K_u)$ output by an arbitrary interaction with NEW, we have $\Pr[K_i = K] \leq \alpha$ for all i and $K \in \{0,1\}^k$, and $\Pr[K_i = K_j] \leq \alpha$ for all $i \neq j$. The canonical KeyGen is α-smooth for $\alpha = 2^{-k}$. See the full version of this paper [12] for the proof.

Theorem 1. *Let E be modeled as an ideal cipher, add have min-entropy h, and KeyGen be α-smooth. Further, let $L, B \geq 1$ such that $L \leq 2^{(1-\epsilon)h-1}$, for some $\epsilon \in (0,1]$, and let $\mathcal{A}$ be an adversary that queries ENC for at most L n-bit blocks, and at most B blocks for each user, and makes p PRIM queries. Then,*

$$\mathsf{Adv}^{\mathrm{mu\text{-}ind}}_{\mathsf{CTR}[E,\mathsf{add}],\mathsf{KeyGen},E}(\mathcal{A}) \leq 2^{-n/2} + \left(LB + L^2\alpha\right) \cdot \left(\frac{1}{2^n} + \frac{1}{2^h}\right) + ap\alpha,$$

where $a := \left\lceil \frac{1.5n}{\epsilon h} \right\rceil - 1$.

The bound highlights the benefits when each user only encrypts B blocks. In particular, assume $h = n$, $\alpha = 1/2^k$. If $B = 2^b$, then the number L of blocks encrypted overall by the scheme can be as high as 2^{n-b}. (The second term has L^2 in the numerator, but the denominator is much larger, i.e., 2^{n+k}.) Another interesting feature is that the contribution of PRIM queries to the bound is independent of the number of users and L.

MORE ON THE BOUND. Previous works [20,24] implicitly give mu security bounds for CTR, but adopt a different model, where the adversary is a-priori constrained in (1) the number of queries q, (2) a bound B_i on the number of blocks encrypted per user $i \in [u]$. The resulting bounds contain a leading term $\sum_{i=1}^{u} B_i^2/2^n$, assuming no primitive queries are made (adding primitive queries p only degrades the bound). This is essentially what one can obtain by applying a naïve hybrid argument to the single-user analysis. We discussed the disadvantage of such a bound in the introduction already.

RE-KEYING, REVISITED. Also, in contrast to the previous works, the above result holds for an arbitrary KeyGen, and only requires *very weak* randomness from it. This suggests a new and efficient solutions for the re-keying problem of [1]. Let $H : \{0,1\}^k \times \{0,1\}^* \to \{0,1\}^k$ be a hash function, and let KeyGen, on input aux $\in \{0,1\}^*$, simply output $H(K, \mathsf{aux})$ for some master secret key K, and this KeyGen is α-smooth if H is for example POLYVAL from AES-GCM-SIV, where $\alpha = \ell/2^k$, and ℓ is an upper bound on the length of aux. We can assume ℓ to be fixed to something short, even 1. Indeed, aux could be a counter, or some other short string. The resulting bound (when $h = n$) would be $2^{-n/2} + \frac{2LB}{2^n} + \frac{2L^2}{2^{n+k}} + ap/2^k$. Note that this solution heavily exploits the ideal-cipher model — clearly, we are indirectly assuming some form of related-key security on E implicitly, and one should carefully assess the security of E in this setting.

The results in the model of Abdalla and Bellare [1] are weaker in that they only study more involved key-derivation methods (but with the benefit of a standard-model security reduction), in a more constrained model, where the adversary sequentially queries B blocks on a key, before moving to the next key. Our model, however, is adaptive, as the adversary can distribute queries as it pleases across users. But difference is not only qualitative, as quantitative bounds in [1] are obtained via naïve hybrid arguments.

4.2   Security of GMAC⁺

This section deals with an abstraction of GMAC⁺, the PRF used within the AES-GCM-SIV mode of operation. We show good mu bounds for this construction. The ideas extend similarly to various Wegman-Carter type MACs [38], but we focus here on GMAC⁺.

THE GMAC⁺ CONSTRUCTION. The construction relies on a hash function H : $\{0,1\}^n \times \{0,1\}^* \times \{0,1\}^* \to \{0,1\}^n$, which is meant to satisfy the following properties. (We employ the shorthand $H_K(M, A) = H(K, M, A)$.)

Definition 1. *Let $H : \{0,1\}^n \times \{0,1\}^* \times \{0,1\}^* \to \{0,1\}^n$. We say that H is c-almost XOR universal if for all $(M, A) \neq (M', A')$, and all $\Delta \in \{0,1\}^n$, and $K \leftarrow_\$ \{0,1\}^n$,*

$$\Pr[H_K(M, A) \oplus H_K(M', A') = \Delta] \leq \frac{c \cdot \max\{|M|_n + |A|_n, |M'|_n + |A'|_n\}}{2^n},$$

where $|X|_n = \max\{1, \lceil |X|/n \rceil\}$ is the block length of string X, as defined in Sect. 2. Further, we say it is c-regular if for all $Y \in \{0,1\}^n$, $M, A \in \{0,1\}^*$, and $K \leftarrow_\$ \{0,1\}^n$,

$$\Pr[H_K(M, A) = Y] \leq \frac{c \cdot (|M|_n + |A|_n)}{2^n}.$$

We say it is weakly *c-regular if this is only true for $(M, A) \neq (\varepsilon, \varepsilon)$, and $H_K(\varepsilon, \varepsilon) = 0^n$ for all K.*

Remark 1. Note that for POLYVAL as used in AES-GCM-SIV, we can set $c = 1.5$ *provided* that we exclude the empty string as input. This is because the empty string results in POLYVAL outputting 0^n regardless of the key, and thus POLYVAL is only *weakly* c-regular. It is easy to fix POLYVAL so that this does not happen (as the input is padded with its length, it is sufficient to ensure that the length padding of the empty string contains at least one bit with value 1). See the full version of this paper [12] for more details.

We also consider a generic function xor : $\{0,1\}^n \times \{0,1\}^{nl} \rightarrow \{0,1\}^n$, for $nl < n$, which is meant to add a nonce to a string. In particular, we require: (1) λ-*regularity:* For every $N \in \{0,1\}^{nl}$ and $Z \in \{0,1\}^n$, there are at most λ strings $Y \in \{0,1\}^n$ such that $\mathsf{xor}(Y,N) = Z$, (2) *injectivity:* For every Y, $\mathsf{xor}(Y, \cdot)$ is injective, and (3) *linearity:* For every Y, Y', N, N', we have $\mathsf{xor}(Y,N) \oplus \mathsf{xor}(Y',N') = \mathsf{xor}(Y \oplus Y', N \oplus N')$.

Example 1. In GCM-SIV and AES-GCM-SIV, one uses

$$\mathsf{xor}(Y,N) = 0 \parallel (Y \oplus 0^{n-nl}N)[2:n].$$

This is clearly 2-regular, injective, and linear. Note that here it is important to prepend 0's to the nonce N; if one instead appends 0's to N then injectivity of xor will be destroyed.

Given H and xor, as well as a block cipher $E : \{0,1\}^k \times \{0,1\}^n \rightarrow \{0,1\}^n$, we define $\mathsf{GMAC}^+ = \mathsf{GMAC}^+[H,E,\mathsf{xor}] : \{0,1\}^{k+n} \times (\{0,1\}^* \times \{0,1\}^* \times \{0,1\}^{nl}) \rightarrow \{0,1\}^n$ such that

$$\mathsf{GMAC}^+(K_{\mathsf{in}} \parallel K_{\mathsf{out}}, (M,A,N)) = E_{K_{\mathsf{out}}}(\mathsf{xor}(H_{K_{\mathsf{in}}}(M,A), N)). \tag{2}$$

MU-PRF SECURITY OF GMAC^+. We upper bound the mu-prf advantage for GMAC^+. We stress here that the adversary's EVAL queries have form (i, M, A, N), and the length of such queries is implicitly defined as $|M|_n + |A|_n$.

We also consider an arbitrary KeyGen algorithm, which outputs pairs of keys $(K_{\mathsf{in}}^i, K_{\mathsf{out}}^i) \in \{0,1\}^n \times \{0,1\}^k$. We will only require these keys to be pairwise-close to uniform, i.e., we say that KeyGen is β-*pairwise almost uniform (AU)* if for every $i \neq j$, the distribution of $(K_{\mathsf{in}}^i, K_{\mathsf{out}}^i), (K_{\mathsf{in}}^j, K_{\mathsf{out}}^j)$ is such that every pair of $(n+k)$-bit strings appears with probability at most $\beta \frac{1}{2^{2(n+k)}}$. Clearly, the canonical KeyGen satisfies this with $\beta = 1$, but we will be for instance interested later on in cases where $\beta = 1 + \epsilon$ for some small constant $\epsilon > 0$.

The proof of the following theorem is in the full version of this paper [12].

Theorem 2 (Security of GMAC^+). *Let $H : \{0,1\}^n \times \{0,1\}^* \times \{0,1\}^* \rightarrow \{0,1\}^n$ be c-almost xor universal and c-regular, KeyGen be β-pairwise AU, xor be injective, linear, and λ-regular, and let $E : \{0,1\}^k \times \{0,1\}^n \rightarrow \{0,1\}^n$ be a block cipher, which we model as an ideal cipher. Then, for any adversary $\mathcal{A}$ making q EVAL queries of at most L n-bit blocks (with at most B blocks queries per user), as well as p ideal-cipher queries,*

$$\mathsf{Adv}^{\mathsf{mu\text{-}prf}}_{\mathsf{GMAC}^+[H,E,\mathsf{xor}],B,E}(\mathcal{A}) \leq \frac{(1+C)qB}{2^n} + \frac{CL(p+q) + \beta q^2}{2^{n+k}}, \tag{3}$$

where $C := c \cdot \lambda \cdot \beta$.

Here, parameters are even better than in the case of counter-mode, but this is in part due to the longer key. In particular, this being PRF security, it is unavoidable that security is compromised when more than $2^{(k+n)/2}$ users are involved. The interesting fact is that *partial* key collisions (i.e., a collision in the hash keys or in the cipher keys) alone do not help.

For example, take $k = n = 128$, $C = \beta = 1$, $B = 2^{32}$, $L = qB$, $q \leq 2^{95}$, then the bound becomes roughly $q/2^{95} + p/2^{128}$, and note that this is when processing up to 2^{128} blocks of data.

WEAK REGULARITY. We also provide a version of Theorem 2 for the case where H is only weakly c-regular. We stress that the security loss is substantial here (and thus if using GMAC$^+$ alone, one should rather make sure H is c-regular), but nonetheless the security is preserved in the case where a nonce N is reused across a sufficiently small number d of users. A proof sketch is in the full version of this paper [12].

Theorem 3 (Security of GMAC$^+$, weak regularity). *Let* $H : \{0,1\}^n \times \{0,1\}^* \times \{0,1\}^* \rightarrow \{0,1\}^n$ *be* c-almost xor universal and weakly c-regular, KeyGen *be* β-pairwise AU, xor *be injective, linear, and* λ-regular, and let $E : \{0,1\}^k \times \{0,1\}^n \rightarrow \{0,1\}^n$ *be a block cipher, which we model as an ideal cipher. Then, for any adversary* $\mathcal{A}$ *making* q EVAL *queries of at most* L n-bit blocks (with at most B blocks queries per user), as well as p ideal-cipher queries,

$$\mathsf{Adv}^{\mathsf{mu\text{-}prf}}_{\mathsf{GMAC}^+[H,E,\mathsf{xor}],B,E}(\mathcal{A}) \leq \frac{(1+C)qB}{2^n} + \frac{CL(p+2q) + \beta q^2}{2^{n+k}} + \frac{d(p+q)}{2^k}, \quad (4)$$

where $C := c \cdot \lambda \cdot \beta$, *and* d *is a bound on the number of users re-using any given nonce.*

5 SIV Composition with Key Reuse

SIV WITH KEY REUSE. Let $E : \{0,1\}^k \times \{0,1\}^n \rightarrow \{0,1\}^n$ be a blockcipher that we will model as an ideal cipher. Let $\mathsf{F} : \{0,1\}^{\mathsf{F.kl}} \times \mathcal{N} \times \{0,1\}^* \times \{0,1\}^* \rightarrow \{0,1\}^*$ be a keyed function, with $\mathsf{F.kl} \geq k$. Let $\mathsf{SE} : \{0,1\}^k \times \{0,1\}^* \rightarrow \{0,1\}^*$ be an IV-based encryption scheme of IV length n. Both F and SE are built on top of E. In a generic SIV composition, the key $K_{\mathsf{in}} \parallel K_{\mathsf{out}}$ of F and the key J of SE will be chosen independently. However, for efficiency, it would be convenient if one can reuse $K_{\mathsf{out}} = J$, which GCM-SIV$^+$ does. Formally, let $\mathsf{AE} = \mathsf{SIV}[\mathsf{F}, \mathsf{SE}]$ be the AE scheme as defined in Fig. 4.

RESULTS. We consider security of the SIV construction for $\mathsf{F} = \mathsf{GMAC}^+$ and $\mathsf{SE} = \mathsf{CTR}$. We assume that GMAC$^+$ and CTR use functions xor and add, respectively, such that (1) xor is 2-regular, injective, and linear, and $\mathsf{xor}(X, N) \in 0\{0,1\}^{n-1}$ for every string $X \in \{0,1\}^n$ and every nonce $N \in \{0,1\}^{\mathsf{nl}}$, and (2) add has min-entropy $n-1$, and $\mathsf{add}(\mathsf{IV}, \ell) \in 1\{0,1\}^{n-1}$ for every $\mathsf{IV} \in \{0,1\}^n$ and every $\ell \in \mathbb{N}$. (Those notions for add and xor can be found in Sects. 4.1 and 4.2 respectively.) This assumption holds for the design choice of AES-GCM-SIV. We thus only write $\mathsf{CTR}[E]$ or $\mathsf{GMAC}^+[H, E]$ instead of $\mathsf{CTR}[E, \mathsf{add}]$ or $\mathsf{GMAC}^+[H, E, \mathsf{xor}]$. Below, we show the mu-mrae security of $\mathsf{SIV}[\mathsf{GMAC}^+[H, E], \mathsf{CTR}[E]]$, with respect to a pairwise AU KeyGen, and a c-regular, c-AXU hash function H; the notion of pairwise AU for key-generation algorithms can be found in Sect. 4.2. See the full version of this paper [12] for the proof.

$\mathsf{AE.E}(K_{\mathsf{in}} \parallel K_{\mathsf{out}}, N, M, A)$	$\mathsf{AE.D}(K_{\mathsf{in}} \parallel K_{\mathsf{out}}, N, C, A)$
$\mathsf{IV} \leftarrow \mathsf{F}(K_{\mathsf{in}} \parallel K_{\mathsf{out}}, N, M, A)$	$\mathsf{IV} \parallel C' \leftarrow C; \; M \leftarrow \mathsf{SE.D}^E(K_{\mathsf{out}}, C)$
$C \leftarrow \mathsf{SE.E}^E(K_{\mathsf{out}}, M; \mathsf{IV})$	$T \leftarrow \mathsf{F}^E(K_{\mathsf{in}} \parallel K_{\mathsf{out}}, N, M, A)$
Return C	If $T \neq \mathsf{IV}$ then return $\perp$ else return M

Fig. 4. The SIV construction (with key reuse) $\mathsf{AE} = \mathsf{SIV}[\mathsf{F}, \mathsf{SE}]$ that is built on top of an ideal cipher E.

Theorem 4 (Security of SIV). *Let* $E : \{0,1\}^k \times \{0,1\}^n \to \{0,1\}^n$ *be a blockcipher that we will model as an ideal cipher. Fix* $0 < \epsilon < 1$. *Let* $H : \{0,1\}^n \times \{0,1\}^* \times \{0,1\}^* \to \{0,1\}^*$ *be a c-regular, c-AXU hash. Let* $\mathsf{AE} \leftarrow \mathsf{SIV}[\mathsf{GMAC}^+[H, E], \mathsf{CTR}[E]]$. *Then for any* β-*pairwise AU* KeyGen *and for any adversary* $\mathcal{A}$ *that makes at most* q *encryption/verification queries whose total block length is at most* $L \leq 2^{(1-\epsilon)n-4}$, *and encryption queries of at most* B *blocks per user, and* $p \leq 2^{(1-\epsilon)n-4}$ *ideal-cipher queries,*

$$\mathsf{Adv}^{\mathsf{mu\text{-}mrae}}_{\mathsf{AE}, \mathsf{KeyGen}, E}(\mathcal{A}) \leq \frac{1}{2^{n/2}} + \frac{\beta a p}{2^k} + \frac{(3\beta c + 7\beta)L^2 + 4\beta c L p}{2^{n+k}}$$
$$+ \frac{(4c\beta + 0.5\beta + 6.5)LB}{2^n},$$

where $a = \lceil 1.5n/(n-1)\epsilon \rceil - 1$.

REMARKS. The proof of Theorem 4 only needs to know that the mu-ind proof of CTR and the mu-prf proof of GMAC^+ follow some high-level structure that we will describe below. We do not need to know any other specific details about those two proofs. This saves us the burden of repeating the entire prior proofs in Sects. 4.1 and 4.2. The mu-ind proof of CTR uses the H-coefficient technique and follows this canonical structure:

(i) When the adversary finishes querying, we grant it all the keys. Note that in the ideal world, the keys are still created but not used.

(ii) For each ideal-cipher query $E_K(X)$ for answer Y, the transcript correspondingly stores an entry $(\mathtt{prim}, K, X, Y, +)$. Likewise, for each query $E^{-1}(K, Y)$ for answer X, the transcript stores an entry $(\mathtt{prim}, K, X, Y, -)$. For each query $\mathrm{ENC}(i, M)$ with answer C, we store an entry $(\mathtt{enc}, i, M, C)$.

(iii) When the adversary finishes querying, for each entry $(\mathtt{enc}, i, M, C)$, in the real world, we grant it a table that stores all triples $(K_i, X, E(K_i, X))$ for all queries $E(K_i, X)$ that $\mathsf{CTR.E}[E](K_i, M; T)$ makes, where K_i is the key of user i and T is the IV of C. In the ideal world, the proof generates a corresponding fake table as follows. If we consider the version of CTR in which messages are padded (e.g., PKCS#7), then one can first parse $\mathsf{IV} \parallel C_1 \parallel \cdots \parallel C_m \xleftarrow{n} C$ and $M_1 \parallel \cdots \parallel M_m \xleftarrow{n} M$ and then return $(K_i, X_1, C_1 \oplus M_1), \ldots, (K_i, X_m, C_m \oplus M_m)$, where $X_i = \mathsf{add}(\mathsf{IV}, i-1)$ and we use $\xleftarrow{n}$ to denote some function that pads a message into n-bit blocks. If one uses the well-known padding-free version of CTR where the last block of the message is allowed to be shorter than n-bit, then one first pads C with random bits so that the last fragmentary block becomes n-bit long, and likewise pads M with 0's so that the last fragmentary block becomes n-bit long, and then proceeds as above. (This step can be optionally omitted for the padding version since the adversary can generate the table by itself.)

(iv) Consider a transcript τ. If there are two tables $\mathcal{T}_1$ and $\mathcal{T}_2$ in τ that contain triples (K, X, Y) and (K, X', Y') respectively, and either $X = X'$, or $Y = Y'$, then τ must be considered bad. If there is a table $\mathcal{T}$ that contains triples (K, X, Y) and (K, X', Y') such that either $X = X'$, or $Y = Y'$, then τ is also considered bad. In addition, if there is a table $\mathcal{T}$ that contains a triple (K, X, Y), and there is an entry $(\mathtt{prim}, K, X', Y', \cdot)$, and either $X = X'$ or $Y = Y'$, then τ is considered bad. The proof may define some other criteria for badness of transcripts.

We say that a CTR transcript is CTR-*bad* if it is bad according to the criteria defined by the proof of Theorem 1. (Note that although not all of those criteria are specified in the structure above, it is enough for our purpose, as our proof of Theorem 4 does not need to know those specific details.) The proof of GMAC$^+$ also follows a similar high-level structure. We say that a GMAC$^+$ transcript is GMAC$^+$-*bad* if it is bad according to the criteria defined by the proof of Theorem 2.

WEAK REGULARITY. We also provide a version of Theorem 4 for the case where H is only weakly c-regular. Again, the security loss is substantial here, but security is preserved if each nonce is reused across a sufficiently small number d of users. A proof sketch is given in the full version of this paper [12].

Theorem 5 (Security of SIV, weak regularity). *Let $E : \{0,1\}^k \times \{0,1\}^n \to \{0,1\}^n$ be a blockcipher that we will model as an ideal cipher. Fix $0 < \epsilon < 1$. Let $H : \{0,1\}^n \times \{0,1\}^* \times \{0,1\}^* \to \{0,1\}^*$ be a weakly c-regular, c-AXU hash. Let $\mathsf{AE} \leftarrow \mathsf{SIV}[\mathsf{GMAC}^+[H, E], \mathsf{CTR}[E]]$. Then for any β-pairwise AU KeyGen and for any adversary $\mathcal{A}$ that makes at most q encryption/verification queries whose total block length is at most $L \le 2^{(1-\epsilon)n-4}$, and encryption queries of at most B blocks per user, and $p \le 2^{(1-\epsilon)n-4}$ ideal-cipher queries,*

$$\mathsf{Adv}^{\mathsf{mu\text{-}mrae}}_{\mathsf{AE}, \mathsf{KeyGen}, E}(\mathcal{A}) \le \frac{1}{2^{n/2}} + \frac{\beta ap}{2^k} + \frac{(3\beta c + 7\beta)L^2 + 4\beta cLp}{2^{n+k}}$$
$$+ \frac{(4c\beta + 0.5\beta + 6.5)LB}{2^n} + \frac{dp + (2d + a)L}{2^k},$$

where $a = \lceil 1.5n/(n-1)\epsilon \rceil - 1$, and d is a bound on the number of users re-using any given nonce.

6 AES-GCM-SIV with a Generic Key Derivation

In this section we consider the mu-mrae security of AES-GCM-SIV with respect to a quite generic class of key-derivation functions. This class includes the current

KDF KD_0 of AES-GCM-SIV, but it contains another KDF KD_1 that is not only simpler but also twice faster. This KD_1 was the original KDF in AES-GCM-SIV, but then subsequently replaced by KD_0. Our multi-user bound is even better than the single-user bound of Gueron and Lindell [20]. In this section, we assume that $GMAC^+$ and CTR use functions xor and add, respectively, such that (1) xor is 2-regular, injective, and linear, and $xor(X, N) \in 0\{0,1\}^{n-1}$ for every string $X \in \{0,1\}^n$ and every nonce $N \in \mathcal{N} = \{0,1\}^{nl}$, and (2) add has min-entropy $n-1$, and $add(IV, \ell) \in 1\{0,1\}^{n-1}$ for every $IV \in \{0,1\}^n$ and every $\ell \in \mathbb{N}$. (Those notions for add and xor can be found in Sects. 4.1 and 4.2 respectively.) This assumption holds for the design choice of AES-GCM-SIV. We thus only write $CTR[E]$ or $GMAC^+[H, E]$ instead of $CTR[E, add]$ or $GMAC^+[H, E, xor]$.

Below, we will formalize the Key-then-Encrypt transform that captures the way AES-GCM-SIV generates session keys for every encryption/decryption. We then describe our class of KDFs.

THE KtE TRANSFORM. Let AE be an AE scheme of nonce space $\mathcal{N}$ and let $KD : \mathcal{K} \times \mathcal{N} \rightarrow \{0,1\}^{AE.kl}$ be a key-derivation function. Given KD and AE, the Key-then-Encrypt (KtE) transform constructs another AE scheme $\overline{AE} = KtE[KD, AE]$ as shown in Fig. 5.

$\overline{AE}.E(K, N, M, A)$	$\overline{AE}.D(K, N, C, A)$
$J \leftarrow KD(K, N); C \leftarrow AE.E(J, N, M, A)$	$J \leftarrow KD(K, N); M \leftarrow AE.D(J, N, C, A)$
Return C	Return M

Fig. 5. The AE scheme $\overline{AE} = KtE[KD, AE]$ constructed from an AE scheme AE and a key-derivation function KD, under the KtE transform.

NATURAL KDFs. Let $n \geq 1$ be an integer and let $k \in \{n, 2n\}$. Let $E : \{0,1\}^k \times \{0,1\}^n \rightarrow \{0,1\}^n$ be a blockcipher that we will model as an ideal cipher. Let $pad : \mathcal{N} \times \{0, \ldots, 5\} \rightarrow \{0,1\}^n$ be a padding mechanism such that $pad(N_0, s_0) \neq pad(N_1, s_1)$ for every distinct pairs $(N_0, s_0), (N_1, s_1) \in \mathcal{N} \times \{0, \ldots, 5\}$. Let $KD[E] : \{0,1\}^k \times \mathcal{N} \rightarrow \{0,1\}^{n+k}$ be a KDF that is associated with a deterministic algorithm $KD.Map : (\{0,1\}^n)^6 \rightarrow \{0,1\}^{n+k}$. We say that $KD[E]$ is *natural* if on input (K, N), $KD[E]$ first calls $R_0 \leftarrow E(K, pad(N, 0)), \ldots, R_5 \leftarrow E(K, pad(N, 5))$, and then returns $KD.Map(R_0, \ldots, R_5)$.

It might seem arbitrary to limit the number of blockcipher calls of a natural KDF to six. However, note that since $k \leq 2n$, the block length of each $(k + n)$-bit derived key is at most three. All known good constructions, which we list below, use at most six blockcipher calls. Using more would simply make the performance and even the bounds worse. We therefore define a natural KDF to use at most six blockcipher calls.

The current KDF $KD_0[E]$ of AES-GCM-SIV, as shown in the left panel of Fig. 6, is natural; it is defined for even n only. For $k = n$, it can be implemented using four blockcipher calls, but for $k = 2n$ it needs six blockcipher calls. Consider the KDF $KD_1[E]$ on the right panel of Fig. 6. For $k = n$ it can be implemented using two blockcipher calls, and $k = 2n$ it needs three blockcipher calls. This KDF is also simpler to implement than KD_0. Iwata and Seurin [24] propose to use either the XOR construction [8,14] or the CENC construction [23]. Both the XOR and CENC constructions are natural; the former uses four blockcipher calls for $k = n$ and six blockcipher calls for $k = 2n$, and the latter uses three and four blockcipher calls respectively.

$KD_0[E](K, N)$	$KD_1[E](K, N)$
For $s = 0$ to 5 do $R_s \leftarrow E_K(\mathsf{pad}(N, s))$	For $s = 0$ to 5 do $R_i \leftarrow E_K(\mathsf{pad}(N, s))$
For $i = 0$ to 2 do	Return $(R_0 \parallel R_1 \parallel R_2)[1 : n + k]$
$\quad V_i \leftarrow R_{2i}[1 : n/2] \parallel R_{2i+1}[1 : n/2]$	
Return $(V_0 \parallel V_1 \parallel V_2)[1 : n + k]$	

Fig. 6. Key-derivation functions KD_0 (left) and KD_1 (right).

For a natural key-derivation function $KD[E]$, we say that it is γ-*unpredictable* if for any subset $S \subseteq \{0, 1\}^n$ of size at least $\frac{15}{16} \cdot 2^n$ and any $s \in \{0, 1\}^{n+k}$, if the random variables $R_0, \ldots, R_5$ are sampled uniformly without replacement from S then $\Pr[KD.\mathsf{Map}(R_0, \ldots, R_5) = s] \leq \gamma/2^{n+k}$. Lemma 3 below shows that both $KD_0[E]$ and $KD_1[E]$ are 2-unpredictable; see the full version of this paper [12] for the proof. One might also show that both the XOR and CENC constructions are 2-unpredictable. Therefore, in the remainder of this section, we only consider natural, 2-unpredictable KDFs.

Lemma 3. *Let $n \geq 128$ be an even integer and let $k \in \{n, 2n\}$. Let $E : \{0, 1\}^k \times \{0, 1\}^n \to \{0, 1\}^n$ be a blockcipher that we will model as an ideal cipher. Then both $KD_0[E]$ and $KD_1[E]$ are 2-unpredictable.*

IDEAL COUNTERPART OF NATURAL KDF. For a natural KDF $KD[E]$, consider its following ideal version $KD[k]$. The key space of $KD[k]$ is the entire set $\mathsf{Perm}(n)$. It takes as input a permutation $\pi \in \mathsf{Perm}(n)$ and a string $N \in \mathcal{N}$, computes $R_s \leftarrow \pi(\mathsf{pad}(N, s))$ for all $s \in \{0, \ldots, 5\}$, and returns $KD.\mathsf{Map}(R_0, \ldots, R_5)$. Of course $KD[k]$ is impractical since its key length is huge, but it will be useful in studying the security of the KtE transform. The following bounds the privacy and authenticity of $KtE[KD[k], AE]$ via the mu-mrae security of the AE scheme AE; the proof is in the full version of this paper [12]. In light of that, in the subsequent subsections, we will analyze the difference between security of $KtE[KD[E], AE]$ and that of $KtE[KD[k], AE]$.

Proposition 2. *Let $n \geq 8$ be an integer and let $k \in \{n, 2n\}$. Let $E : \{0, 1\}^k \times \{0, 1\}^n \to \{0, 1\}^n$ be a blockcipher that we will model as an ideal cipher. Let $KD[E]$ be a natural KDF. Let AE be an AE scheme of key length $k + n$. Let $\overline{AE} = KtE[KD[k], AE]$. Then for any adversaries $\overline{A}_1$ and $\overline{A}_2$, we can construct*

KeyGen(st, aux)
———————————————
$(N, i) \leftarrow \text{aux}; (\pi_1, S_1, \ldots, \pi_m, S_m) \leftarrow \text{st}$
If $(i \in \{1, \ldots, m\}$ and $N \in S_i)$ or $(i \notin \{1, \ldots, m+1\})$ then
 //Unexpected input, return a random key anyway
  $K \leftarrow_{\$} \{0, 1\}^{k+n}$; return (K, st)
If $i \in \{1, \ldots, m\}$ then $S_i \leftarrow S_i \cup \{N\}$; $\text{st} \leftarrow (\pi_1, S_1, \ldots, \pi_m, S_m)$
If $i = m+1$ then $\pi_{m+1} \leftarrow_{\$} \text{Perm}(n); S_{m+1} \leftarrow \{N\}$; $\text{st} \leftarrow (\pi_1, S_1, \ldots, \pi_{m+1}, S_{m+1})$
Return $(\text{KD}[k](\pi_i, N), \text{st})$

Fig. 7. Key-generation algorithm KeyGen corresponding to $\text{KD}[k]$.

a key-generation algorithm KD.KeyGen *as shown in Fig. 7, and an adversary* $\mathcal{A}$ *such that*

$$\text{Adv}_{\overline{\text{AE}}, E}^{\text{mu-priv}}(\overline{\mathcal{A}}_1) + \text{Adv}_{\overline{\text{AE}}, E}^{\text{mu-auth}}(\overline{\mathcal{A}}_2) \leq 3 \, \text{Adv}_{\text{AE}, \text{KeyGen}, E}^{\text{mu-mrae}}(\mathcal{A}).$$

For any type of queries, the number of $\mathcal{A}$*'s queries is at most the maximum of that of* $\overline{\mathcal{A}}_1$ *and* $\overline{\mathcal{A}}_2$*, and the similar claim holds for the total block length of the encryption/verification queries. Moreover, the maximum of total block length of encryption queries per user of* $\mathcal{A}$ *is at most the maximum of that per (user, nonce) pair of* $\overline{\mathcal{A}}_1$ *and* $\overline{\mathcal{A}}_2$.

The following lemma says that if $\text{KD}[E]$ is 2-unpredictable then the constructed KeyGen in the theorem statement of Proposition 2 is 4-pairwise AU; the notion of pairwise AU for key-generation algorithms can be found in Sect. 4.2. The proof is in the full version of this paper [12].

Lemma 4. *Let* $n \geq 8$ *be an integer and let* $k \in \{n, 2n\}$. *Let* $E : \{0, 1\}^k \times \{0, 1\}^n \rightarrow \{0, 1\}^n$ *be a blockcipher that we will model as an ideal cipher. Let* $\text{KD}[E]$ *be a natural, 2-unpredictable KDF. Then the corresponding key-generation algorithm* KeyGen *in Fig. 7 is 4-pairwise AU.*

INDISTINGUISHABILITY OF $\text{KD}[E]$. For an adversary $\mathcal{A}$, define

$$\text{Adv}_{\text{KD}[E]}^{\text{dist}}(\mathcal{A}) = 2 \Pr[\mathbf{G}_{\text{KD}[E]}^{\text{dist}}(\mathcal{A})] - 1$$

as the advantage of $\mathcal{A}$ in distinguishing a natural KDF $\text{KD}[E]$ and its ideal counterpart $\text{KD}[k]$ in the multi-user setting, where game $\mathbf{G}_{\text{KD}[E]}^{\text{dist}}(\mathcal{A})$ is defined in Fig. 8. Under this notion, the adversary is given access to both E and E^{-1}, an oracle NEW() to initialize a new user v with a truly random master key K_v and a secret ideal permutation π_v, and an evaluation oracle EVAL that either implements $\text{KD}[E]$ or $\text{KD}[k]$. We say that an adversary $\mathcal{A}$ is d-*repeating* if among its evaluation queries, a nonce is used for at most d users.

Lemma 5 below bounds the indistinguishability advantage between $\text{KD}[E]$ and $\text{KD}[k]$. The proof is in the full version of this paper [12].

Game $\mathbf{G}^{\text{dist}}_{\text{KD}[E]}(\mathcal{A})$	$\text{EVAL}(i, N)$
$v \leftarrow 0;\ b \leftarrow\!\!\$ \{0,1\};\ b' \leftarrow\!\!\$ \mathcal{A}^{\text{NEW},\text{EVAL},E,E^{-1}}$	If $i > v$ then return $\perp$
Return $(b' = b)$	If $b = 1$ then return $\text{KD}[E](K_i, N)$
	Else return $\text{KD}[k](\pi_i, N)$
Procedure $\text{NEW}()$	
$v \leftarrow v + 1;\ K_v \leftarrow\!\!\$ \{0,1\}^k;\ \pi_v \leftarrow\!\!\$ \text{Perm}(n)$	

Fig. 8. Game to distinguish $\text{KD}[E]$ and its ideal counterpart $\text{KD}[k]$.

Lemma 5. *Fix $0 < \epsilon < 1$. Let $n \geq 16$ be an integer and let $k \in \{n, 2n\}$. Let $E : \{0,1\}^k \times \{0,1\}^n \rightarrow \{0,1\}^n$ be a blockcipher that we will model as an ideal cipher. Let $\text{KD}[E]$ be a natural KDF. For any d-repeating adversary $\mathcal{A}$ that makes at most $p \leq 2^{n-4}$ ideal-cipher queries, and $q \leq 2^{(1-\epsilon)n-4}$ evaluation queries,*

$$\text{Adv}^{\text{dist}}_{\text{KD}[E]}(\mathcal{A}) \leq \frac{1}{2^{n/2}} + \frac{24pq + 18q^2}{2^{k+n}} + \frac{ap + d(p + 3q)}{2^k}$$

where $a = \lceil 1.5/\epsilon \rceil - 1$. The theorem statement still holds if we grant the adversary the master keys when it finishes querying.

6.1 Privacy Analysis

Lemma 6 below reduces the privacy security of $\text{KtE}[\text{KD}[E], \text{AE}]$ for a generic AE scheme AE, to that of $\text{KtE}[\text{KD}[k], \text{AE}]$; the proof relies crucially on Lemma 5.

Lemma 6. *Fix $0 < \epsilon < 1$. Let $n \geq 16$ be an integer and let $k \in \{n, 2n\}$. Let $E : \{0,1\}^k \times \{0,1\}^n \rightarrow \{0,1\}^n$ be a blockcipher that we will model as an ideal cipher. Let $\text{KD}[E]$ be a natural KDF. Let AE be an AE scheme of key length $k+n$, and let $\overline{\text{AE}} = \text{KtE}[\text{KD}[E], \text{AE}]$. Consider a d-repeating adversary $\mathcal{A}$ that makes $p \leq 2^{n-5}$ ideal-cipher queries and $q \leq 2^{(1-\epsilon)n-4}$ encryption queries. Suppose that using AE to encrypt $\mathcal{A}$'s encryption queries would need to make $L \leq 2^{n-5}$ ideal-cipher queries. Then*

$$\text{Adv}^{\text{mu-priv}}_{\overline{\text{AE}},E}(\mathcal{A}) \leq \text{Adv}^{\text{mu-priv}}_{\text{KtE}[\text{KD}[k],\text{AE}],E}(\mathcal{A}) + \frac{2}{2^{n/2}} + \frac{48(L + p)q + 36q^2}{2^{k+n}}$$
$$+ \frac{2a(L + p) + 2d(L + p + 3q)}{2^k},$$

where $a = \lceil 1.5/\epsilon \rceil - 1$.

Proof. We first construct an adversary $\overline{\mathcal{A}}$ that tries to distinguish $\text{KD}[E]$ and $\text{KD}[k]$. Adversary $\overline{\mathcal{A}}$ simulates game $\mathbf{G}^{\text{mu-priv}}_{\overline{\text{AE}},E}(\mathcal{A})$, but each time it needs to generate a session key, it uses its EVAL oracle instead of $\text{KD}[E]$. However, if $\overline{\mathcal{A}}$ previously queried $\text{EVAL}(i, N)$ for an answer K, next time it simply uses K without querying. Finally, adversary $\overline{\mathcal{A}}$ outputs 1 only if the simulated game returns true. Let b be the challenge bit in game $\mathbf{G}^{\text{dist}}_{\text{KD}[E]}(\overline{\mathcal{A}})$. Then

$$\Pr[\mathbf{G}^{\mathsf{dist}}_{\mathsf{KD}[E]}(\overline{\mathcal{A}}) \Rightarrow \mathsf{true} \mid b = 1] = \Pr[\mathbf{G}^{\mathsf{mu\text{-}priv}}_{\overline{\mathsf{AE}},E}(\mathcal{A})], \text{ and}$$

$$\Pr[\mathbf{G}^{\mathsf{dist}}_{\mathsf{KD}[E]}(\overline{\mathcal{A}}) \Rightarrow \mathsf{false} \mid b = 0] = \Pr[\mathbf{G}^{\mathsf{mu\text{-}priv}}_{\mathsf{KtE}[\mathsf{KD}[k],\mathsf{AE}],E}(\mathcal{A})].$$

Subtracting, we get

$$\mathsf{Adv}^{\mathsf{dist}}_{\mathsf{KD}[E]}(\overline{\mathcal{A}}) = \frac{1}{2}\left(\mathsf{Adv}^{\mathsf{mu\text{-}priv}}_{\overline{\mathsf{AE}},E}(\mathcal{A}_1) - \mathsf{Adv}^{\mathsf{mu\text{-}priv}}_{\mathsf{KtE}[\mathsf{KD}[k],\mathsf{AE}],E}(\mathcal{A}_1)\right).$$

Note that $\overline{\mathcal{A}}$ makes at most $p + L \leq 2^{n-4}$ ideal-cipher queries, and q EVAL queries. Moreover, $\overline{\mathcal{A}}$ is also d-repeating. Hence using Lemma 5,

$$\mathsf{Adv}^{\mathsf{dist}}_{\mathsf{KD}[E],\mathsf{KD}[k]}(\overline{\mathcal{A}}) \leq \frac{1}{2^{n/2}} + \frac{24(L+p)q + 18q^2}{2^{k+n}} + \frac{a(L+p) + d(L+p+3q)}{2^k}.$$

Putting this all together,

$$\mathsf{Adv}^{\mathsf{mu\text{-}priv}}_{\overline{\mathsf{AE}},E}(\mathcal{A}) \leq \mathsf{Adv}^{\mathsf{mu\text{-}priv}}_{\mathsf{KtE}[\mathsf{KD}[k],\mathsf{AE}],E}(\mathcal{A}) + \frac{2}{2^{n/2}} + \frac{48(L+p)q + 36q^2}{2^{k+n}}$$
$$+ \frac{2a(L+p) + 2d(L+p+3q)}{2^k}.$$

This concludes the proof. ☐

6.2   Authenticity Analysis

In Sect. 6.1, we bound the privacy advantage by constructing a d-repeating adversary distinguishing $\mathsf{KD}[E]$ and $\mathsf{KD}[k]$, and then using Lemma 5. This method does not work for authenticity: the constructed adversary might be q-repeating, because there is no restriction of the nonces in verification queries, and one would end up with an inferior term $q(L+p+q)/2^k$. We instead give a dedicated analysis.

RESTRICTING TO SIMPLE ADVERSARIES. We say that an adversary is *simple* if for any nonce N and user i, if the adversary uses N for an encryption query of user i, then it will never use nonce N on verification queries for user i. Lemma 7 below reduces the authenticity advantage of a general adversary against $\mathsf{KtE}[\mathsf{KD}[E],\mathsf{AE}]$ to that of a simple adversary; the proof is in the full version of this paper [12], and is based on the idea of splitting the cases of where the adversary forges on a fresh (N, i) pair and where it does not, and the latter can be handled using Lemma 5 above. Handling the former is the harder part, which we deal with below. We discuss the bound however below, and give an overview of the proof.

Lemma 7. *Let $n \geq 16$ be an integer and let $k \in \{n, 2n\}$. Let $E : \{0,1\}^k \times \{0,1\}^n \to \{0,1\}^n$ be a blockcipher that we will model as an ideal cipher. Let $\mathsf{KD}[E]$ be a natural KDF. Let AE be an AE scheme of key length $n + k$, and let $\overline{\mathsf{AE}} = \mathsf{KtE}[\mathsf{KD}[E], \mathsf{AE}]$. Let $\mathcal{A}_0$ be a d-repeating adversary that makes at most $q \leq 2^{(1-\epsilon)n-4}$ encryption/verification queries and $p \leq 2^{n-5}$ ideal-cipher queries.*

Suppose that using AE *to encrypt* $\mathcal{A}_0$ *'s encryption queries and decrypt its verifi-cation queries would need to make* $L \leq 2^{n-5}$ *ideal-cipher queries. Then, we can construct an adversary* $\mathcal{A}_1$ *and a simple adversary* $\mathcal{A}_2$*, both d-repeating, such that*

$$\mathsf{Adv}^{\text{mu-auth}}_{\overline{\mathsf{AE}},E}(\mathcal{A}_0) \leq \mathsf{Adv}^{\text{mu-auth}}_{\mathsf{KtE}[\mathsf{KD}[k],\mathsf{AE}],E}(\mathcal{A}_1) + \mathsf{Adv}^{\text{mu-auth}}_{\overline{\mathsf{AE}},E}(\mathcal{A}_2)$$

$$+ \frac{2}{2^{n/2}} + \frac{48(L+p)q + 36q^2}{2^{n+k}} + \frac{2(a+d)L + 2(a+d)p + 6dq}{2^k},$$

where $a = \lceil 1.5/\epsilon \rceil - 1$*. Any query of* $\mathcal{A}_1$ *or* $\mathcal{A}_2$ *is also a query of* $\mathcal{A}_0$*.*

HANDLING SIMPLE ADVERSARIES. Lemma 8 below shows that the AE scheme $\mathsf{KtE}[\mathsf{KD}[E], \mathsf{SIV}[\mathsf{GMAC}^+[H, E], \mathsf{CTR}[E]]]$ has good authenticity against simple adversaries, for any 2-unpredictable, natural KDF $\mathsf{KD}[E]$. See the full version [12] for the proof. Note that here we can handle both regular and weakly regular hash functions. (If we instead consider just regular hash functions, we can slightly improve the bound, but the difference is inconsequential.)

Lemma 8. *Fix* $0 < \epsilon < 1$ *and let* $a = \lceil 1.5/\epsilon \rceil - 1$*. Let* $n \geq 128$ *be an integer, and let* $k \in \{n, 2n\}$*. Let* $E : \{0,1\}^k \times \{0,1\}^n \rightarrow \{0,1\}^n$ *be a blockcipher that we will model as an ideal cipher. Let* $H : \{0,1\}^n \times \{0,1\}^* \times \{0,1\}^* \rightarrow \{0,1\}^n$ *be a hash function that is either c-regular or weakly c-regular. Let* $\mathsf{KD}[E]$ *be a natural, 2-unpredictable KDF. Let* $\mathsf{AE} = \mathsf{SIV}[\mathsf{GMAC}^+[H, E], \mathsf{CTR}[E]]$ *and* $\overline{\mathsf{AE}} = \mathsf{KtE}[\mathsf{KD}[E], \mathsf{AE}]$*. Let* $\mathcal{A}$ *be a d-repeating, simple adversary that makes at most* $p \leq 2^{(1-\epsilon)n-8}$ *ideal-cipher queries, and* $q \leq 2^{(1-\epsilon)n-8}$ *encryption/verification queries whose total block length is at most* $L \leq 2^{(1-\epsilon)n-8}$*. Then*

$$\mathsf{Adv}^{\text{mu-auth}}_{\overline{\mathsf{AE}},E}(\mathcal{A}) \leq \frac{3}{2^{n/2}} + \frac{11q}{2^n} + \frac{288(L+p)q + 36q^2 + 48c(L+p+q)L}{2^{n+k}}$$

$$+ \frac{(8a + 7a^2 + 3d)q}{2^k} + \frac{(na + 6a + 6d)L + 6(a+d)p}{2^k}.$$

DISCUSSION. The bound in Lemma 8 consists of three important terms $\frac{q}{2^n}, \frac{pd}{2^k}$, and $\frac{naL}{2^k}$, each corresponding to an actual attack. Let us revisit these, as this will be helpful in explaining the proof below. First, since the IV length is only n-bit long, even if an adversary simply outputs q verification queries in a random fashion, it would get an advantage about $\frac{q}{2^n}$. Next, for the term $\frac{pd}{2^k}$, consider an adversary that picks a long enough message M and then makes encryption queries $(1, N, M, A), \ldots, (d, N, M, A)$ of the same nonce N and asso-ciated data, for answers $C_1, \ldots, C_d$ respectively. (Recall that the adversary is d-repeating, so it cannot use the nonce N in encryption queries for more than d users.) By picking p candidate master keys $K_1, \ldots, K_p$ and comparing C_i with $\overline{\mathsf{AE}}.E(K_j, N, M, A)$ for all $i \leq d$ and $j \leq p$, the adversary can recover one master key with probability about $\frac{pd}{2^k}$.

Finally, for the term $\frac{naL}{2^k}$, consider the following attack. The adversary first picks a nonce N and p candidate keys $K_1, \ldots, K_p$, and then queries

$R_{0,j} \leftarrow E_K(K_j, \mathsf{pad}(N,0)), \ldots, R_{5,j} \leftarrow E(K_j, \mathsf{pad}(N,5))$ for every $j \leq p$. Let $K_{\mathsf{in}}^j \parallel K_{\mathsf{out}}^j \leftarrow \mathsf{KD.Map}(R_{0,j}, \ldots, R_{5,j})$. Now, if some K_j is the master key of some user i then $K_{\mathsf{in}}^j \parallel K_{\mathsf{out}}^j$ will be the session key of that user i for nonce N. The adversary then picks an arbitrary ciphertext C, and then computes $M_j \leftarrow \mathsf{CTR}[E].\mathsf{D}(K_j, C)$ and $V_j \leftarrow E^{-1}(K_{\mathsf{out}}^j, T)$ for each $j \leq p$, where T is the IV of C. The goal of the adversary is to make a sequence of q verification queries $(1, N, C, A), \ldots, (q, N, C, A)$, for an ℓ-block associated data A that it will determine later. (Recall that in verification queries, the adversary can reuse a nonce across as many users as it likes.) To maximize its chance of winning, the adversary will iterate through every possible string A^* of block length ℓ, and let $count(A^*)$ denote the number of j's that $\mathsf{xor}(H(K_{\mathsf{in}}^j, M_j, A^*), N) = V_j$. Then it picks A as the string to maximize $count(A)$. The proof of Lemma 8 essentially shows that with very high probability, we have $count(A) \leq na(\ell + |C|_n) \leq \frac{naL}{q}$, and thus the advantage of this attack is bounded by $\frac{naL}{2^k}$.

PROOF IDEAS. We now sketch some ideas in the proof of Lemma 8. First consider an adversary that does not use the encryption oracle. Assume that the adversary does not repeat a prior ideal-cipher query, or make redundant ideal-cipher queries. For each query $E_K(Y)$ of answer Y, create an entry $(\mathsf{prim}, K, X, Y, +)$. Likewise, for each query $E_K^{-1}(Y)$ of answer X, create an entry $(\mathsf{prim}, K, X, Y, -)$. Consider a verification query (i, N, C, A). Let K_i be the secret master key of user i, and let $K_{\mathsf{in}} \parallel K_{\mathsf{out}}$ be the session key of user i for nonce N. Let T be the IV of C. The proof examines several cases, but here we only discuss a few selective ones. If there is no entry $(\mathsf{prim}, K_i, X, Y, \cdot)$ such that $X \in \{\mathsf{pad}(N,0), \ldots, \mathsf{pad}(N,5)\}$ then given the view of the adversary, the session key $K_{\mathsf{in}} \parallel K_{\mathsf{out}}$ still has at least $k + n - 1$ bits of (conditional) min-entropy. In this case, the chance that $\mathsf{AE.D}(K_{\mathsf{in}} \parallel K_{\mathsf{out}}, N, C, M)$ returns a non-$\perp$ answer is roughly $1/2^n$. Next, suppose that there is an entry $(\mathsf{prim}, K, X, Y, -)$ such that $K = K_i$ and $X \in \{\mathsf{pad}(N,0), \ldots, \mathsf{pad}(N,5)\}$. By using some balls-into-bins analysis,[3] we can argue that it is very likely that there are at most $6a$ entries $(\mathsf{prim}, K^*, X^*, Y^*, -)$ such that $X^* \in \{\mathsf{pad}(N,0), \ldots, \mathsf{pad}(N,5)\}$. Hence the chance this case happens is at most $6a/2^k$.

Now consider the case that there are entries $(\mathsf{prim}, K_i, \mathsf{pad}(N,0), R_0, +), \ldots,$ $(\mathsf{prim}, K_i, \mathsf{pad}(N,5), R_5, +)$, and $(\mathsf{prim}, K_{\mathsf{out}}, V, T, -)$, with $V \in 0\{0,1\}^{n-1}$ and $K_{\mathsf{in}} \parallel K_{\mathsf{out}} \leftarrow \mathsf{KD.Map}(R_0, \ldots, R_5)$. This corresponds to the last attack in the discussion above. We need to bound $\Pr[\mathsf{Bad}]$, where Bad is the the event (i) this case happens, and (ii) $V = \mathsf{xor}(H(K_{\mathsf{in}}, M, A), N)$, where $M \leftarrow \mathsf{CTR}[E].\mathsf{D}(K_{\mathsf{out}}, C)$. This is highly non-trivial because somehow the adversary already sees the keys K_i and $K_{\mathsf{in}} \parallel K_{\mathsf{out}}$, and can *adaptively* pick (C, A), as shown in the third attack above.

To deal with this, we consider a *fixed* (i^*, N^*, C^*, A^*). There are at most p septets $\mathcal{T}$ of entries $(\mathsf{prim}, K, \mathsf{pad}(N^*, 0), R_0^*, +), \ldots, (\mathsf{prim}, K, \mathsf{pad}(N^*, 5), R_5^*, +)$

---

[3] We note that this is not the classic balls-into-bins setting, because the balls are thrown in an inter-dependent way. In the full version [12], we give an analysis of this biased balls-into-bins setting.

and $(\texttt{prim}, J, U, T^*, -)$, with $U \in 0\{0,1\}^{n-1}$ and $J' \parallel J \leftarrow \mathsf{KD.Map}(R_0^*, \dots, R_5^*)$. We then show that the chance that there are $n\ell a$ such septets T such that $\mathsf{xor}(H(J'(T)), M^*(T), A^*), N^*) = U(T)$ is at most $2^{1-(3\ell n+2n)}$, where $\ell = |C^*|_n + |A^*|_n \geq 2$ and $M^*(T) \leftarrow \mathsf{CTR}[E].\mathsf{D}(J(T), C^*)$. Hence, regardless of how the adversary picks (i, N, C, A) from all possible choices of (i^*, N^*, C^*, A^*), the chance that there are $na(|C|_n + |A|_n)$ septets T such that $\mathsf{xor}(H(J'(T)), M(T), A), N) = U(T)$, where $M(T) \leftarrow \mathsf{CTR}[E].\mathsf{D}(J(T), C)$, is at most

$$\sum_{\ell=2}^{\infty} \sum_{\substack{(i^*, N^*, C^*, A^*) \\ |C^*|_n + |A^*|_n = \ell}} 2^{1-(3n\ell+2n)} \leq \sum_{\ell=2}^{\infty} 2^{2n\ell+2n} \cdot 2^{1-(3n\ell+2n)} = \sum_{\ell=2}^{\infty} \frac{2}{2^{n\ell}} \leq \frac{1}{2^n}.$$

Thus $\Pr[\mathsf{Bad}] \leq \frac{1}{2^n} + \frac{na \cdot \mathbf{E}[|A|_n + |C|_n]}{2^k}$.

Now we consider the general case where the adversary $\mathcal{A}$ might use the encryption oracle. Clearly if for each encryption query (i, N, M, A), we grant the adversary the session key $\mathsf{KD}[E](K_i, N)$, where K_i is the master key of user i, then it only helps the adversary. Recall that here the adversary is simple, so it cannot query $\mathrm{ENC}(i, N, M, A)$ and later query $\mathrm{VF}(i, N, C', A')$. We also let the adversary compute up to $L + p$ ideal-cipher queries, so that the encryption oracle does not have to give the ciphertexts to the adversary. Effectively, we can view that $\mathcal{A}$ is in the following game G_0. It is given access to E/E^{-1} and an oracle $\mathrm{EVAL}(i, N)$ that generates $\mathsf{KD}[E](i, N)$. Then it has to generate a list of verification queries. The game then tries to decrypt those, and returns true only if some gives a non-$\perp$ answer.

To remove the use of the EVAL oracle, it is tempting to consider the variant G_1 of game G_0 where EVAL instead implements $\mathsf{KD}[k]$, and then bound the gap between G_0 and G_1 by constructing a d-repeating adversary $\overline{\mathcal{A}}$ distinguishing $\mathsf{KD}[E]$ and $\mathsf{KD}[k]$. However, this approach does not work because it is impossible for $\overline{\mathcal{A}}$ to correctly simulate the processing of the verification queries. Instead, we define game G_1 as follows. Its EVAL again implements $\mathsf{KD}[k]$, but after the adversary produces its verification queries, the game tries to *program* E so that the outputs of EVAL are consistent with $\mathsf{KD}[E]$ on random master keys $K_1, K_2, \cdots \leftarrow_\$ \{0,1\}^{n+k}$. (But E still has to remain consistent with its past ideal-cipher queries.) Of course it is not always possible, because the fake EVAL might have generated some inconsistency. In this case, the game returns false, meaning that the adversary *loses*. If there is no inconsistency, then after the programming, the game processes the verification queries as in G_0.

To bound the gap between G_0 and G_1, we will construct a d-repeating adversary $\overline{\mathcal{A}}$ distinguishing $\mathsf{KD}[E]$ and $\mathsf{KD}[k]$, but additionally, it wants to be granted the master keys after it finishes querying. Note that Lemma 5 applies to this key-revealing setting. Now, after the adversary $\overline{\mathcal{A}}$ finishes querying, it is granted the master keys and checks for inconsistency between the outputs of EVAL and the ideal-cipher queries. If there is inconsistency then $\overline{\mathcal{A}}$ outputs 0, indicating that it has been dealing with $\mathsf{KD}[k]$. Otherwise, it has to simulate the processing of the verification queries. However, although it knows the keys now, it can no longer queries E. Instead, $\overline{\mathcal{A}}$ tries to sample an *independent* blockcipher $\tilde{E}$, subject to (1) $\tilde{E}$ and E agree on the outputs of the past ideal-cipher queries, and the outputs of EVAL are consistent with $\mathsf{KD}[\tilde{E}]$ on the master keys $K_1, K_2, \ldots$. It then processes the verification queries using this blockcipher $\tilde{E}$ instead of E.

Although the game G_1 above does not completely remove the use of the EVAL oracle, it still creates some sort of independence between the sampling of the master keys, and the outputs that the adversary $\mathcal{A}$ receives, allowing us to repeat several proof ideas above.

HANDLING GENERAL ADVERSARIES. Combining Lemmas 7 and 8, we immediately obtain the following result.

Lemma 9. *Fix $0 < \epsilon < 1$ and let $a = \lceil 1.5/\epsilon \rceil - 1$. Let $n \geq 128$ be an integer, and let $k \in \{n, 2n\}$. Let $E : \{0,1\}^k \times \{0,1\}^n \to \{0,1\}^n$ be a blockcipher that we will model as an ideal cipher. Let $H : \{0,1\}^n \times \{0,1\}^* \times \{0,1\}^* \to \{0,1\}^n$ be a hash function that is either c-regular hash or weakly c-regular. Let $\mathsf{KD}[E]$ be a natural, 2-unpredictable KDF. Let $\mathsf{AE} = \mathsf{SIV}[\mathsf{GMAC}^+[H, E], \mathsf{CTR}[E]]$ and $\overline{\mathsf{AE}} = \mathsf{KtE}[\mathsf{KD}[E], \mathsf{AE}]$. Let $\mathcal{A}$ be a d-repeating adversary that makes at most $p \leq 2^{(1-\epsilon)n-8}$ ideal-cipher queries, and $q \leq 2^{(1-\epsilon)n-8}$ encryption/verification queries whose total block length is at most $L \leq 2^{(1-\epsilon)n-8}$. Then we can construct a d-repeating adversary $\overline{\mathcal{A}}$ such that*

$$\mathsf{Adv}_{\overline{\mathsf{AE}}, E}^{\mathsf{mu\text{-}auth}}(\mathcal{A}) \leq \mathsf{Adv}_{\mathsf{KtE}[\mathsf{KD}[k], \mathsf{AE}], E}^{\mathsf{mu\text{-}auth}}(\overline{\mathcal{A}}) + \frac{5}{2^{n/2}} + \frac{11q}{2^n} + \frac{336(L+p)q + 72q^2}{2^{n+k}}$$
$$+ \frac{48c(L+p+q)L}{2^{n+k}} + \frac{(8a + 7a^2 + 9d)q + (na + 8a + 8d)L + 8(a+d)p}{2^k}.$$

Moreover, any query of $\overline{\mathcal{A}}$ is also a query of $\mathcal{A}$.

6.3 Unwinding Mu-Mrae Security

The following Theorem 6 concludes the mu-mrae security of AE scheme $\overline{\mathsf{AE}} = \mathsf{KtE}[\mathsf{KD}[E], \mathsf{SIV}[\mathsf{GMAC}^+[H, E], \mathsf{CTR}[E]]]$; the proof is in the full version of this paper [12]. Note that here we can handle both regular and weakly regular hash functions. (If we instead consider just regular hash functions, we can slightly improve the bound, but the difference is inconsequential.)

Theorem 6 (Security of AES-GCM-SIV). *Let $n \geq 128$ be an integer, and let $k \in \{n, 2n\}$. Fix $0 < \epsilon < 1$ and let $a = \lceil 1.5n/(n-1)\epsilon \rceil - 1$. Let $E : \{0,1\}^k \times \{0,1\}^n \to \{0,1\}^n$ be a blockcipher that we will model as an ideal cipher. Let $H : \{0,1\}^n \times \{0,1\}^* \times \{0,1\}^* \to \{0,1\}^n$ be a c-AXU hash function. Moreover, either H is c-regular, or weakly c-regular. Let $\mathsf{KD}[E]$ be a natural, 2-unpredictable KDF. Let $\mathsf{AE} = \mathsf{SIV}[\mathsf{GMAC}^+[H, E], \mathsf{CTR}[E]]$ and $\overline{\mathsf{AE}} = \mathsf{KtE}[\mathsf{KD}[E], \mathsf{AE}]$. Let $\mathcal{A}$ be a d-repeating adversary that makes at most $p \leq 2^{(1-\epsilon)n-8}$ ideal-cipher queries, and $q \leq 2^{(1-\epsilon)n-8}$ encryption/verification queries whose total block length is at most $L \leq 2^{(1-\epsilon)n-8}$ and encryption queries of at most B blocks per (user, nonce) pair. Then,*

$$\mathsf{Adv}^{\mathsf{mu\text{-}mrae}}_{\mathsf{AE},E}(\mathcal{A}) \leq \frac{10}{2^{n/2}} + \frac{(17a + 4a^2 + 24d + na)L + (22a + 13d)p}{2^k}$$
$$+ \frac{(48c + 30)LB}{2^n} + \frac{(303 + 108c)L^2 + (192 + 96c)Lp}{2^{n+k}}.$$

We note that one way that d can be kept small is by choosing nonces randomly, or at least with sufficient entropy. Then, by a classical balls-into-bins analysis, if q is quite smaller than 2^{nl}, where nl is the nonce length, which holds in practice for $nl = 96$, then the value d is bounded by a constant with high probability. We also point out that if d cannot be bounded, then our security bound still gives very meaningful security guarantees if $k = 2n$ (i.e., this would have us use AES-256). As there is a matching attack in the unbounded d case, which just exploits key collisions, this suggests the need to increase the key length to 256 bits in the multi-user case. However, many uses in practice will have d bounded, and for these 128-bit keys will suffice.

Acknowledgments. We thank Mihir Bellare, Shay Gueron, Yehuda Lindell, and anonymous CRYPTO reviewers for insightful feedback.

Priyanka Bose and Stefano Tessaro were supported by NSF grants CNS-1553758 (CAREER), CNS-1423566, CNS-1719146, CNS-1528178, and IIS-1528041, and by a Sloan Research Fellowship. Viet Tung Hoang was supported in part by NSF grant CICI-1738912 and the First Year Assistant Professor Award of Florida State University.

References

1. Abdalla, M., Bellare, M.: Increasing the lifetime of a key: a comparative analysis of the security of re-keying techniques. In: Okamoto, T. (ed.) ASIACRYPT 2000. LNCS, vol. 1976, pp. 546–559. Springer, Heidelberg (2000). https://doi.org/10.1007/3-540-44448-3_42

2. Bellare, M., Bernstein, D.J., Tessaro, S.: Hash-function based PRFs: AMAC and its multi-user security. In: Fischlin, M., Coron, J.-S. (eds.) EUROCRYPT 2016. LNCS, vol. 9665, pp. 566–595. Springer, Heidelberg (2016). https://doi.org/10.1007/978-3-662-49890-3_22

3. Bellare, M., Boldyreva, A., Micali, S.: Public-key encryption in a multi-user setting: security proofs and improvements. In: Preneel, B. (ed.) EUROCRYPT 2000. LNCS, vol. 1807, pp. 259–274. Springer, Heidelberg (2000). https://doi.org/10.1007/3-540-45539-6_18

4. Bellare, M., Canetti, R., Krawczyk, H.: Pseudorandom functions revisited: the cascade construction and its concrete security. In: 37th FOCS, pp. 514–523. IEEE Computer Society Press, October 1996

5. Bellare, M., Desai, A., Jokipii, E., Rogaway, P.: A concrete security treatment of symmetric encryption. In: 38th FOCS, pp. 394–403. IEEE Computer Society Press, October 1997

6. Bellare, M., Hoang, V.T.: Identity-based format-preserving encryption. In: CCS 2017, pp. 1515–1532 (2017)

7. Bellare, M., Impagliazzo, R.: A tool for obtaining tighter security analyses of pseudorandom function based constructions, with applications to PRP to PRF conversion. Cryptology ePrint Archive, Report 1999/024 (1999). http://eprint.iacr.org/1999/024

8. Bellare, M., Krovetz, T., Rogaway, P.: Luby-rackoff backwards: increasing security by making block ciphers non-invertible. In: Nyberg, K. (ed.) EUROCRYPT 1998. LNCS, vol. 1403, pp. 266–280. Springer, Heidelberg (1998). https://doi.org/10.1007/BFb0054132

9. Bellare, M., Tackmann, B.: The multi-user security of authenticated encryption: AES-GCM in TLS 1.3. In: Robshaw, M., Katz, J. (eds.) CRYPTO 2016. LNCS, vol. 9814, pp. 247–276. Springer, Heidelberg (2016). https://doi.org/10.1007/978-3-662-53018-4_10

10. Biham, E.: How to forge DES-encrypted messages in 2^{28} steps. Technical Report CS0884, Technion - Israel Institute of Technology (1996)

11. Biham, E.: How to decrypt or even substitute DES-encrypted messages in 2^{28} steps. Inf. Process. Lett. **84**(3), 117–124 (2002)

12. Bose, P., Hoang, V.T., Tessaro, S.: Revisiting AES-GCM-SIV: Multi-user security, faster key derivation, and better bounds. Cryptology ePrint Archive, Report 2018/136 (2018). https://eprint.iacr.org/2018/136

13. Chen, S., Steinberger, J.: Tight security bounds for key-alternating ciphers. In: Nguyen, P.Q., Oswald, E. (eds.) EUROCRYPT 2014. LNCS, vol. 8441, pp. 327–350. Springer, Heidelberg (2014). https://doi.org/10.1007/978-3-642-55220-5_19

14. Dai, W., Hoang, V.T., Tessaro, S.: Information-theoretic indistinguishability via the chi-squared method. In: Katz, J., Shacham, H. (eds.) CRYPTO 2017. LNCS, vol. 10403, pp. 497–523. Springer, Cham (2017). https://doi.org/10.1007/978-3-319-63697-9_17

15. Dworkin, M., Perlner, R.: Analysis of VAES3 (FF2). Cryptology ePrint Archive, Report 2015/306 (2015). http://eprint.iacr.org/2015/306

16. Gilboa, S., Gueron, S.: Distinguishing a truncated random permutation from a random function. Cryptology ePrint Archive, Report 2015/773 (2015). http://eprint.iacr.org/2015/773

17. Goldwasser, S., Bellare, M.: Lecture Notes on Cryptography. Summer Course "Cryptography and Computer Security". MIT (1999)

18. Gueron, S., Langley, A., Lindell, Y.: AES-GCM-SIV: Specification and analysis. Cryptology ePrint Archive, Report 2017/168 (2017). http://eprint.iacr.org/2017/168

19. Gueron, S., Lindell, Y.: GCM-SIV: Full nonce misuse-resistant authenticated encryption at under one cycle per byte. In: Ray, I., Li, N., Kruegel, C. (eds.) ACM CCS 2015, pp. 109–119. ACM Press, October 2015

20. Gueron, S., Lindell, Y.: Better bounds for block cipher modes of operation via nonce-based key derivation. In: CCS 2017, pp. 1019–1036 (2017)

21. Hoang, V.T., Tessaro, S.: Key-alternating ciphers and key-length extension: exact bounds and multi-user security. In: Robshaw, M., Katz, J. (eds.) CRYPTO 2016. LNCS, vol. 9814, pp. 3–32. Springer, Heidelberg (2016). https://doi.org/10.1007/978-3-662-53018-4_1

22. Hoang, V.T., Tessaro, S.: The multi-user security of double encryption. In: Coron, J.-S., Nielsen, J.B. (eds.) EUROCRYPT 2017. LNCS, vol. 10211, pp. 381–411. Springer, Cham (2017). https://doi.org/10.1007/978-3-319-56614-6_13

23. Iwata, T.: New blockcipher modes of operation with beyond the birthday bound security. In: Robshaw, M. (ed.) FSE 2006. LNCS, vol. 4047, pp. 310–327. Springer, Heidelberg (2006). https://doi.org/10.1007/11799313_20

24. Iwata, T., Seurin, Y.: Reconsidering the security bound of AES-GCM-SIV. IACR Trans. Symm. Cryptol. **2017**(4), 240–267 (2017)

25. Lucks, S.: The sum of PRPs is a secure PRF. In: Preneel, B. (ed.) EUROCRYPT 2000. LNCS, vol. 1807, pp. 470–484. Springer, Heidelberg (2000). https://doi.org/10.1007/3-540-45539-6_34

26. Luykx, A., Mennink, B., Paterson, K.G.: Analyzing multi-key security degradation. In: Takagi, T., Peyrin, T. (eds.) ASIACRYPT 2017. LNCS, vol. 10625, pp. 575–605. Springer, Cham (2017). https://doi.org/10.1007/978-3-319-70697-9_20

27. Maurer, U.: Indistinguishability of Random Systems. In: Knudsen, L.R. (ed.) EUROCRYPT 2002. LNCS, vol. 2332, pp. 110–132. Springer, Heidelberg (2002). https://doi.org/10.1007/3-540-46035-7_8

28. McGrew, D.A., Viega, J.: The security and performance of the galois/counter mode (GCM) of operation. In: Canteaut, A., Viswanathan, K. (eds.) INDOCRYPT 2004. LNCS, vol. 3348, pp. 343–355. Springer, Heidelberg (2004). https://doi.org/10.1007/978-3-540-30556-9_27

29. Mouha, N., Luykx, A.: Multi-key security: the even-mansour construction revisited. In: Gennaro, R., Robshaw, M. (eds.) CRYPTO 2015. LNCS, vol. 9215, pp. 209–223. Springer, Heidelberg (2015). https://doi.org/10.1007/978-3-662-47989-6_10

30. Namprempre, C., Rogaway, P., Shrimpton, T.: Reconsidering generic composition. In: Nguyen, P.Q., Oswald, E. (eds.) EUROCRYPT 2014. LNCS, vol. 8441, pp. 257–274. Springer, Heidelberg (2014). https://doi.org/10.1007/978-3-642-55220-5_15

31. Patarin, J.: A proof of security in $O(2^n)$ for the xor of two random permutations. In: Safavi-Naini, R. (ed.) ICITS 08. LNCS, vol. 5155, pp. 232–248. Springer, Heidelberg (2008). https://doi.org/10.1007/978-3-540-85093-9_22

32. Patarin, J.: The "coefficients H" technique (invited talk). In: Avanzi, R.M., Keliher, L., Sica, F. (eds.) SAC 2008. LNCS, vol. 5381, pp. 328–345. Springer, Heidelberg (2009). https://doi.org/10.1007/978-3-642-04159-4_21

33. Patarin, J.: Introduction to mirror theory: Analysis of systems of linear equalities and linear non equalities for cryptography. Cryptology ePrint Archive, Report 2010/287 (2010). http://eprint.iacr.org/2010/287

34. Rogaway, P., Shrimpton, T.: A provable-security treatment of the key-wrap problem. In: Vaudenay, S. (ed.) EUROCRYPT 2006. LNCS, vol. 4004, pp. 373–390. Springer, Heidelberg (2006). https://doi.org/10.1007/11761679_23

35. Tessaro, S.: Optimally secure block ciphers from ideal primitives. In: Iwata, T., Cheon, J.H. (eds.) ASIACRYPT 2015. LNCS, vol. 9453, pp. 437–462. Springer, Heidelberg (2015). https://doi.org/10.1007/978-3-662-48800-3_18

36. Vance, J.: VAES3 scheme for FFX: An addendum to "The FFX mode of operation for Format Preserving Encryption". Submission to NIST, May 2011

37. Vance, J., Bellare, M.: Delegatable Feistel-based Format Preserving Encryption mode. Submission to NIST, November 2015

38. Wegman, M.N., Carter, L.: New hash functions and their use in authentication and set equality. J. Comput. Syst. Sci. 22, 265–279 (1981)

Attribute-Based Encryption

Unbounded ABE via Bilinear Entropy Expansion, Revisited

Jie Chen[1], Junqing Gong[2(✉)], Lucas Kowalczyk[3], and Hoeteck Wee[3,4]

[1] East China Normal University, Shanghai, China
s080001@e.ntu.edu.sg
[2] ENS de Lyon, Laboratoire LIP (U. Lyon, CNRS, ENSL, INRIA, UCBL),
Lyon, France
junqing.gong@ens-lyon.fr
[3] Columbia University, New York, USA
luke@cs.columbia.edu
[4] CNRS, ENS, Paris, France
wee@di.ens.fr

Abstract. We present simpler and improved constructions of unbounded attribute-based encryption (ABE) schemes with constant-size public parameters under static assumptions in bilinear groups. Concretely, we obtain:

- a simple and adaptively secure unbounded ABE scheme in composite-order groups, improving upon a previous construction of Lewko and Waters (Eurocrypt '11) which only achieves selective security;
- an improved adaptively secure unbounded ABE scheme based on the k-linear assumption in prime-order groups with shorter ciphertexts and secret keys than those of Okamoto and Takashima (Asiacrypt '12);
- the first adaptively secure unbounded ABE scheme for arithmetic branching programs under static assumptions.

J. Chen—School of Computer Science and Software Engineering. Supported by the National Natural Science Foundation of China (Nos. 61472142, 61632012, U1705264) and the Young Elite Scientists Sponsorship Program by CAST (2017QNRC001). Homepage: http://www.jchen.top.

J. Gong—Supported in part by the French ANR ALAMBIC Project (ANR-16-CE39-0006).

L. Kowalczyk—Supported in part by an NSF Graduate Research Fellowship DGE-16-44869; The Leona M. & Harry B. Helmsley Charitable Trust; ERC Project aSCEND (H2020 639554); the Defense Advanced Research Project Agency (DARPA) and Army Research Office (ARO) under Contract W911NF-15-C-0236; and NSF grants CNS-1445424, CNS-1552932 and CCF-1423306. Any opinions, findings and conclusions or recommendations expressed are those of the authors and do not necessarily reflect the views of the Defense Advanced Research Projects Agency, Army Research Office, the National Science Foundation, or the U.S. Government.

H. Wee—Supported in part by ERC Project aSCEND (H2020 639554), H2020 FENTEC and NSF Award CNS-1445424.

J. B. Nielsen and V. Rijmen (Eds.): EUROCRYPT 2018, LNCS 10820, pp. 503–534, 2018.
https://doi.org/10.1007/978-3-319-78381-9_19

At the core of all of these constructions is a "bilinear entropy expansion" lemma that allows us to generate any polynomial amount of entropy starting from constant-size public parameters; the entropy can then be used to transform existing adaptively secure "bounded" ABE schemes into unbounded ones.

1   Introduction

Attribute-based encryption (ABE) [13,25] is a generalization of public-key encryption to support fine-grained access control for encrypted data. Here, ciphertexts and keys are associated with descriptive values which determine whether decryption is possible. In a key-policy ABE (KP-ABE) scheme for instance, ciphertexts are associated with attributes like '(author:Waters), (inst:UT), (topic:PK)' and keys with access policies like '((topic:MPC) OR (topic:Qu)) AND (NOT(inst:CWI))', and decryption is possible only when the attributes satisfy the access policy. A ciphertext-policy (CP-ABE) scheme is the dual of KP-ABE with ciphertexts associated with policies and keys with attributes.

Over past decade, substantial progress has been made in the design and analysis of ABE schemes, leading to a large families of schemes that achieve various trade-offs between efficiency, security and underlying assumptions. Meanwhile, ABE has found use as a tool for providing and enhancing privacy in a variety of settings from electronic medical records to messaging systems and online social networks.

As institutions grow and with new emerging and more complex applications for ABE, it became clear that we need ABE schemes that can readily accommodate the addition of new roles, entities, attributes and policies. This means that the ABE set-up algorithm should put no restriction on the length of the attributes or the size of the policies that will be used in the ciphertexts and keys. This requirement was introduced and first realized in the work of Lewko and Waters [21] under the term *unbounded ABE*. Their constructions have since been improved and extended in several subsequent works [1–3,5,12,17,18,23,24] (cf. Figs. 1 and 2).

In this work, we put forth new ABE schemes that simultaneously:

(1) are unbounded (the set-up algorithm is independent of the length of the attributes or the size of the policies);
(2) can be based on faster asymmetric prime-order bilinear groups;
(3) achieve adaptive security;
(4) rely on simple hardness assumptions in the standard model.

All four properties are highly desirable from both a practical and theoretical stand-point and moreover, properties (1)–(3) are crucial for many real-world applications of ABE. Indeed, properties (2), (3) and (4) are by now standard cryptographic requirements pertaining to speed and efficiency, strong security guarantees under realistic and natural attack models, and minimal hardness

reference	adaptive	assumption	standard model
OT12 [23]	✓	2-Lin	✓
RW13 [24]		q-type	✓
Att16 [3]	✓	q-type + k-Lin	✓
AC17 [1]	✓	k-Lin, $k \geq 2$	
ours	✓	k-Lin, $k \geq 1$ ✓	✓

Fig. 1. Summary of unbounded KP-ABE schemes for monotone span programs from *prime-order* groups with $O(1)$-size mpk.

reference	|mpk|	adaptive	assumption
LW11 [21]	$O(1)$✓		static ✓
Att14 [2]	$O(1)$✓	✓	q-type
KL15 [17]	$O(\log n)$	✓	static ✓
ours	$O(1)$✓	✓	static ✓

Fig. 2. Summary of unbounded KP-ABE schemes for monotone span programs with n-bit attributes (i.e. universe $[n]$) from *composite-order* groups.

assumptions. Property (2) is additionally motivated by the fact that pairing-based schemes are currently more widely implemented and deployed than lattice-based ones. There is now a vast body of works (e.g. [2,3,6,19,22,27]) showing how to achieve properties (2)–(4) for "bounded" ABE where the set-up time and public parameters grow with the attributes or policies, culminating in unifying frameworks that provide a solid understanding of the design and analysis of these schemes. Unbounded ABE, on the other hand, has received comparatively much less attention in the literature; this is in part because the schemes and proofs remain fairly complex and delicate. Amongst these latter works, only the work of Okamato and Takashima (OT) [23] simultaneously achieved (1)–(4).

Our Results. We present simpler and more modular constructions of unbounded ABE that realize properties (1)–(4) with better efficiency and expressiveness than was previously known.

(i) We present new adaptively secure, unbounded KP-ABE schemes for monotone span programs –which capture access policies computable by monotone Boolean formulas– whose ciphertexts are 42% smaller and our keys are 8% smaller than the state-of-the-art in [23] (with even more substantial savings with our SXDH-based scheme), as well as CP-ABE schemes with similar savings, cf. Fig. 3.

(ii) Our constructions generalize to the larger class of arithmetic span programs [15], which capture many natural computational models, such as monotone Boolean formulas, as well as Boolean and arithmetic branching programs; this yields the first adaptively secure, unbounded KP-ABE for arithmetic span programs. Prior to this work, we do not even know any selectively secure, unbounded KP-ABE for arithmetic span programs.

Moreover, our constructions generalize readily to the k-Lin assumption.

At the core of all of these constructions is a "bilinear entropy expansion" lemma [17] that allows us to generate any polynomial amount of entropy starting

reference	$|\mathsf{mpk}|$	$|\mathsf{sk}|$	$|\mathsf{ct}|$	assumption
KP-ABE OT12 [23]	$79|G_1| + |G_T|$	$14n + 5$	$14n + 5$	DLIN
Ours	$9|G_1| + |G_T|$	$8n$	$5n + 3$	SXDH
	$28|G_1| + 2|G_T|$	$13n$	$8n + 5$	DLIN
	$(5k^2 + 4k)|G_1| + k|G_T|$	$(5k + 3)n$	$(3k + 2)n + 2k + 1$	k-LIN
CP-ABE OT12 [23]	$79|G_1| + |G_T|$	$14n + 5$	$14n + 5$	DLIN
Ours	$11|G_1| + |G_T|$	$5n + 5$	$7n + 3$	SXDH
	$32|G_1| + 2|G_T|$	$9n + 9$	$12n + 6$	DLIN
	$(7k^2 + 4)|G_1| + k|G_T|$	$(4k + 1)(n + 1)$	$(5k + 2)n + 3k$	k-LIN

Fig. 3. Summary of adaptively secure, unbounded ABE schemes for read-once monotone span programs with n-bit attributes (i.e. universe $[n]$) from *prime-order* groups. The columns $|\mathsf{sk}|$ and $|\mathsf{ct}|$ refer to the number of group elements in G_2 and G_1 respectively (minus a $|G_T|$ contribution in ct).

from constant-size public parameters; the entropy can then be used to transform existing adaptively secure bounded ABE schemes into unbounded ones in a *single* shot. The fact that we only need to invoke our entropy expansion lemma *once* yields both quantitative and qualitative advantages over prior works [17,23]: (i) we achieve security loss $O(n + Q)$ for n-bit attributes (i.e. universe $[n]$) and Q secret key queries, improving upon $O(n \cdot Q)$ in [23] and $O(\log n \cdot Q)$ in [17] and (ii) there is clear delineation between entropy expansion and the analysis of the underlying bounded ABE schemes, whereas prior works interweave both techniques in a more complex nested manner.

Following the recent literature on adaptively secure bounded ABE, we first describe our constructions in the simpler setting of composite-order bilinear groups, and then derive our final prime-order schemes by building upon and extending previous frameworks in [6,7,11]. Along the way, we also present a simple adaptively secure unbounded KP-ABE scheme in composite-order groups whose hardness relies on standard, static assumptions (cf. Fig. 2).

1.1   Technical Overview

We will start with asymmetric composite-order bilinear groups (G_N, H_N, G_T) whose order N is the product of three primes p_1, p_2, p_3. Let g_i, h_i denote generators of order p_i in G_N and H_N, for $i = 1, 2, 3$.

Warm-Up. We begin with the LOSTW KP-ABE for monotone span programs [19]; this is a bounded, adaptively secure scheme that uses composite-order groups.

Here, ciphertexts $\mathsf{ct_x}$ are associated with attribute vector[1] $\mathbf{x} \in \{0,1\}^n$ and keys $\mathsf{sk_M}$ with read-once monotone span programs $\mathbf{M}$.[2]

$$\mathsf{mpk} := (g_1, g_1^{v_1}, \ldots, g_1^{v_n}, e(g_1, h_1)^\alpha) \tag{1}$$

$$\mathsf{ct_x} := (g_1^s, \{g_1^{sv_j}\}_{x_j=1}, e(g_1, h_1)^{\alpha s} \cdot m)$$

$$\mathsf{sk_M} := (\{h_1^{\alpha_j + r_j v_j}, h_1^{r_j}\}_{j \in [n]})$$

where $\alpha_1, \ldots, \alpha_n$ are shares of α w.r.t. the span program $\mathbf{M}$; the shares satisfy the requirement that for any $\mathbf{x} \in \{0,1\}^n$, the shares $\{\alpha_j\}_{x_j=1}$ determine α if $\mathbf{x}$ satisfies $\mathbf{M}$, and reveal nothing about α otherwise. For decryption, observe that we can compute $\{e(g_1, h_1)^{\alpha_j s}\}_{x_j=1}$, from which we can compute the blinding factor $e(g_1, h_1)^{\alpha s}$. The proof of security relies on Waters' dual system encryption methodology [2,20,26,27], in the most basic setting at the core of which is an information-theoretic statement about α_j, v_j.

Towards Our Unbounded ABE. The main challenge in building an unbounded ABE lies in "compressing" $g_1^{v_1}, \ldots, g_1^{v_n}$ in mpk down to a constant number of group elements. The first idea following [21,23] is to generate $\{v_j\}_{j \in [n]}$ via a pairwise-independent hash function as $w_0 + j \cdot w_1$, as in the Lewko-Waters IBE. Simply replacing v_j with $w_0 + j \cdot w_1$ leads to natural malleability attacks on the ciphertext, and instead, we would replace sv_j with $s_j(w_0 + j \cdot w_1)$, where $s_1, \ldots, s_n$ are fresh randomness used in encryption. Next, we need to bind the $s_j(w_0 + j \cdot w_1)$'s together via some common randomness s; it suffices to use $sw + s_j(w_0 + j \cdot w_1)$ in the ciphertext. That is, we start with the scheme in (1) and we perform the substitutions (*) for each $j \in [n]$:

$$\begin{array}{ll} \text{ciphertext:} & (s, sv_j) \mapsto (s, sw + s_j(w_0 + j \cdot w_1), s_j) \\ \text{secret key:} & (\alpha_j + v_j r_j, r_j) \mapsto (\alpha_j + r_j w, r_j, r_j(w_0 + j \cdot w_1)) \end{array} \tag{*}$$

This yields the following scheme:

$$\mathsf{mpk} := (g_1, g_1^w, g_1^{w_0}, g_1^{w_1}, e(g_1, h_1)^\alpha) \tag{2}$$

$$\mathsf{ct_x} := (g_1^s, \{g_1^{sw+s_j(w_0+j \cdot w_1)}, g_1^{s_j}\}_{x_j=1}, e(g_1, h_1)^{\alpha s} \cdot m)$$

$$\mathsf{sk_M} := (\{h_1^{\alpha_j + r_j w}, h_1^{r_j}, h_1^{r_j(w_0+j \cdot w_1)}\}_{j \in [n]})$$

As a sanity check for decryption, observe that we can compute $\{e(g_1, h_1)^{\alpha_j s}\}_{x_j=1}$ and then $e(g_1, h_1)^{\alpha s}$ as before. We note that the ensuing scheme is similar to

---

[1] Some works associate ciphertexts with a set $S \subseteq [n]$ where $[n]$ is referred to as the attribute universe, in which case $\mathbf{x} \in \{0,1\}^n$ corresponds to the characteristic vector of S.

[2] All known adaptively secure ABE for monotone span programs under static assumptions in the standard model (even in the bounded setting and even with composite-order groups) have a read-once restriction [2,3,6,19,22,27].

Attrapadung's unbounded KP-ABE in [2, Sect. 7.1], except the latter requires q-type assumptions.[3]

Our Proof Strategy. To analyze our scheme in (2), we follow a very simple and natural proof strategy: we would "undo" the substitutions described in (*) to recover ciphertext and keys similar to those in the LOSTW KP-ABE, upon which we could apply the analysis for the bounded setting from the prior works. That is, we want to computationally replace each $w_0 + j \cdot w_1$ with a fresh u_j:

$$\left\{ \begin{array}{l} g_1^s, \{g_1^{sw+s_j(w_0+j \cdot w_1)}, \ g_1^{s_j}\}_{j \in [n]} \\ \{h_1^{\alpha_j+r_j w}, \ h_1^{r_j}, \ h_1^{r_j(w_0+j \cdot w_1)}\}_{j \in [n]} \end{array} \right\} \stackrel{\text{hopefully}}{\approx_c} \left\{ \begin{array}{l} g_1^s, \{g_1^{sw+s_j u_j}, \ g_1^{s_j}\}_{j \in [n]} \\ \{h_1^{\alpha_j+r_j w}, \ h_1^{r_j}, \ h_1^{r_j u_j}\}_{j \in [n]} \end{array} \right\} \quad (3)$$

Unfortunately, once we give out $g_1^{w_0}, g_1^{w_1}$ in mpk, the above distributions are trivially distinguishable by using the relation $e(g_1, h_1^{r_j(w_0+j \cdot w_1)}) = e(g_1^{w_0+j \cdot w_1}, h_1^{r_j})$. Furthermore, the above statement does not yield a scheme similar to LOSTW when applied to our scheme in (2); for that, we would need to also replace w on the RHS in (3) with fresh v_j as described by

$$(g_1^{sw+s_j u_j}, h_1^{\alpha_j+r_j w}) \mapsto (g_1^{\underline{sv_j}+s_j u_j}, h_1^{\alpha_j+r_j \underline{v_j}})$$

in order to match up with the LOSTW KP-ABE in (1).

1.2   Bilinear Entropy Expansion

The core of our analysis is a *(bilinear) entropy expansion lemma* [17] that captures the spirit of the above statement in (3), namely, it allows us to generate fresh independent randomness starting from the correlated randomness, albeit in a new subgroup of order p_2 generated by g_2, h_2.

More formally, given public parameters $(g_1, g_1^w, g_1^{w_0}, g_1^{w_1}, h_1, h_1^w, h_1^{w_0}, h_1^{w_1})$, we show that

$$\left\{ \begin{array}{l} g_1^s, \{g_1^{sw+s_j(w_0+j \cdot w_1)}, \ g_1^{s_j}\}_{j \in [n]} \\ \{h_1^{r_j w}, \ h_1^{r_j}, \ h_1^{r_j(w_0+j \cdot w_1)}\}_{j \in [n]} \end{array} \right\} \approx_c - \cdot \boxed{\left\{ \begin{array}{l} g_2^s, \{g_2^{sv_j+s_j u_j}, \ g_2^{s_j}\}_{j \in [n]} \\ \{h_2^{r_j v_j}, \ h_2^{r_j}, \ h_2^{r_j u_j}\}_{j \in [n]} \end{array} \right\}} \quad (4)$$

where "—" is short-hand for duplicating the terms on the LHS, so that the g_1, h_1-components remain unchanged. That is, starting with the LHS, we replaced (i) $w_0 + j \cdot w_1$ with fresh u_j, and (ii) w with fresh v_j, both in the p_2-subgroup. We also omitted the α_j's from (3). We clarify that the trivial distinguisher on (3) fails here because $e(g_1, h_2) = 1$.

---

[3] Attrapadung's unbounded KP-ABE does have the advantage that there is no read-once restriction on the span programs, but even with the read-once restriction, the proof still requires q-type assumptions.

Prior Work. We clarify that the name "bilinear entropy expansion" was introduced in the prior work of Kowalczyk and Lewko (KL) [17], which also proved a statement similar to (3), with three notable differences: (i) our entropy expansion lemma starts with 3 units of entropy (w, w_0, w_1) whereas KL uses $O(\log n)$ units of entropy; (ii) the KL statement does not account for the public parameters, and therefore (unlike our lemma) cannot serve as an immediate bridge from the unbounded ABE to the bounded variant; (iii) our entropy expansion lemma admits an analogue in prime-order groups, which in turn yields an unbounded ABE scheme in prime-order groups, whereas the composite-order ABE scheme in KL does not have an analogue in prime-order setting (an earlier prime-order construction was retracted on June 1, 2016). In fact, the "consistent randomness amplification" techniques used in the unbounded ABE schemes of Okamoto and Takashima (OT) [23] also seem to yield an entropy expansion lemma with $O(1)$ units of entropy in prime-order groups. As noted earlier in the introduction, our approach is also different from both KL and OT in the sense that we only need to invoke our entropy expansion lemma once when proving security of the unbounded ABE.

Proof Overview. We provide a proof overview of our entropy expansion lemma in (4). The proof proceeds in two steps: (i) replacing $w_0 + j \cdot w_1$ with fresh u_j, and then (ii) replacing w with fresh v_j.

(i) We replace $w_0 + j \cdot w_1$ with fresh u_j; that is,

$$\left\{ \begin{array}{l} \{g_1^{s_j(w_0+j\cdot w_1)},\ g_1^{s_j}\}_{j\in[n]} \\ \{h_1^{r_j},\ h_1^{r_j(w_0+j\cdot w_1)}\}_{j\in[n]}, \end{array} \right\} \approx_c -\cdot \boxed{\left\{ \begin{array}{l} \{g_2^{s_j u_j},\ g_2^{s_j}\}_{j\in[n]} \\ \{h_2^{r_j},\ h_2^{r_j u_j}\}_{j\in[n]} \end{array} \right\}} \quad (5)$$

where we suppressed the terms involving w; moreover, this holds even given $g_1, g_1^{w_0}, g_1^{w_1}$. Our first observation is that we can easily adapt the proof of Lewko-Waters IBE [8,20] to show that for each $i \in [n]$,

$$\left\{ \begin{array}{l} g_1^{s_i(w_0+i\cdot w_1)},\ g_1^{s_i} \\ \{h_1^{r_j},\ h_1^{r_j(w_0+j\cdot w_1)}\}_{j\neq i} \end{array} \right\} \approx_c -\cdot \boxed{\left\{ \begin{array}{l} g_2^{s_i u_i},\ g_2^{s_i} \\ \{h_2^{r_j},\ h_2^{r_j u_j}\}_{j\neq i} \end{array} \right\}} \quad (6)$$

The idea is that the first term on the LHS corresponds to an encryption for the identity i, and the next $n-1$ terms correspond to secret keys for identities $j \neq i$; on the right, we have the corresponding "semi-functional entities". At this point, we can easily handle $(h_2^{r_i}, h_2^{r_i(w_0+i\cdot w_1)})$ via a statistical argument, thanks to the entropy in $w_0 + i \cdot w_1 \bmod p_2$. Next, we need to get from a single $(g_1^{s_i(w_0+i\cdot w_1)},\ g_1^{s_i})$ on the LHS in (6) to n such terms on the LHS in (5). This requires a delicate "two slot" hybrid argument over $i \in [n]$ and the use of an additional subgroup; similar arguments also appeared in [14,23]. This is where we used the fact that N is a product of three primes, whereas the Lewko-Waters IBE and the statement in (6) works with two primes in the asymmetric setting.

(ii) Next, we replace w with fresh v_j; that is,

$$\left\{ \begin{array}{l} g_2^s, \{g_2^{sw+s_ju_j}, g_2^{s_j}\}_{j\in[n]} \\ \{h_2^{r_jw}, h_2^{r_j}, h_2^{r_ju_j}\}_{j\in[n]} \end{array} \right\} \approx_c \left\{ \begin{array}{l} g_2^s, \{g_2^{sv_j+s_ju_j}, g_2^{s_j}\}_{j\in[n]} \\ \{h_2^{r_jv_j}, h_2^{r_j}, h_2^{r_ju_j}\}_{j\in[n]} \end{array} \right\}$$

Intuitively, this should follow from the DDH assumption in the p_2-subgroup, which says that $(h_2^{r_jw}, h_2^{r_j}) \approx_c (h_2^{r_jv_j}, h_2^{r_j})$. The actual proof is more delicate since w also appears on the other side of the pairing as $g_2^{sw+s_ju_j}$; fortunately, we can treat u_j as a one-time pad that masks w.

Completing the Proof of Unbounded ABE. We return to a proof sketch of our unbounded ABE in (2). Let us start with the simpler setting where the adversary makes only a single key query. Upon applying our entropy expansion lemma[4], we have that the ciphertext/key pair $(\mathsf{ct}_\mathsf{x}, \mathsf{sk}_\mathsf{M})$ satisfies

$$\left\{ \begin{array}{l} g_1^s, \{g_1^{sw+s_j(w_0+j\cdot w_1)}, g_1^{s_j}\}_{x_j=1} \\ \{h_1^{\alpha_j+r_jw}, h_1^{r_j}, h_1^{r_j(w_0+j\cdot w_1)}\}_{j\in[n]} \end{array} \right\} \approx_c - \cdot \boxed{\left\{ \begin{array}{l} g_2^s, \{g_1^{sv_j+s_ju_j}, g_2^{s_j}\}_{x_j=1} \\ \{h_2^{\alpha_j+r_jv_j}, h_2^{r_j}, h_2^{r_ju_j}\}_{j\in[n]} \end{array} \right\}}$$

with $e(g_1, h_1)^{\alpha s} \cdot m$ omitted. Note that the boxed term on the RHS is *exactly* the LOSTW KP-ABE ciphertext/key pair in (1) over the p_2-subgroup, once we strip away the terms involving u_j, s_j.

Finally, to handle the general setting where the ABE adversary makes Q key queries, we simply observe that thanks to self-reducibility, our entropy expansion lemma extends to a Q-fold setting (with Q copies of $\{r_j\}_{j\in[n]}$) without any additional security loss:

$$\left\{ \begin{array}{l} g_1^s, \{g_1^{sw+s_j(w_0+j\cdot w_1)}, g_1^{s_j}\}_{j\in[n]} \\ \{h_1^{r_{j,1}w}, h_1^{r_{j,1}}, h_1^{r_{j,1}(w_0+j\cdot w_1)}\}_{j\in[n]} \\ \vdots \\ \{h_1^{r_{j,Q}w}, h_1^{r_{j,Q}}, h_1^{r_{j,Q}(w_0+j\cdot w_1)}\}_{j\in[n]} \end{array} \right\} \approx_c - \cdot \left\{ \begin{array}{l} g_2^s, \{g_2^{sv_j+s_ju_j}, g_2^{s_j}\}_{j\in[n]} \\ \{h_2^{r_{j,1}v_j}, h_2^{r_{j,1}}, h_2^{r_{j,1}u_j}\}_{j\in[n]} \\ \vdots \\ \{h_2^{r_{j,Q}v_j}, h_2^{r_{j,Q}}, h_2^{r_{j,Q}u_j}\}_{j\in[n]} \end{array} \right\}$$

At this point, we can rely on the (adaptive) security for the LOSTW KP-ABE for the setting with a single challenge ciphertext and Q key queries.

1.3   Our Prime-Order Scheme

To obtain prime-order analogues of our composite-order schemes, we build upon and extend the previous framework of Chen et al. [6, 11] for simulating composite-order groups in prime-order ones. Along the way, we present a more general framework that provides prime-order analogues of the static assumptions used in the security proof for our composite-order ABE. Moreover, we show that these prime-order analogues follow from the standard k-Linear assumption (and more generally, the MDDH assumption [9]) in prime-order bilinear groups.

---

[4] And a subgroup assumption to introduce the $h_2^{\alpha_j}$'s.

Our KP-ABE. Let (G_1, G_2, G_T) be a bilinear group of prime order p. Following [6,11], we start with our composite-order KP-ABE scheme in (2), sample $\mathbf{A}_1 \leftarrow_R \mathbb{Z}_p^{(2k+1) \times k}, \mathbf{B} \leftarrow_R \mathbb{Z}_p^{(k+1) \times k}$, and carry out the following substitutions:

$$
\begin{aligned}
g_1 &\mapsto [\mathbf{A}_1]_1, & h_1 &\mapsto [\mathbf{B}]_2 \\
\alpha &\mapsto \mathbf{k} \in \mathbb{Z}_p^{2k+1} & w, w_0, w_1 &\mapsto \mathbf{W}, \mathbf{W}_0, \mathbf{W}_1 \in \mathbb{Z}_p^{(2k+1) \times (k+1)} \\
s, s_j &\mapsto \mathbf{s}, \mathbf{s}_j \in \mathbb{Z}_p^k, & r_j &\mapsto \mathbf{r}_j \in \mathbb{Z}_p^k \\
g_1^s &\mapsto [\mathbf{s}^\top \mathbf{A}_1^\top]_1, & h_1^{r_j} &\mapsto [\mathbf{Br}_j]_2 \\
g_1^{ws} &\mapsto [\mathbf{s}^\top \mathbf{A}_1^\top \mathbf{W}]_1, & h_1^{wr_j} &\mapsto [\mathbf{WBr}_j]_2
\end{aligned} \tag{7}
$$

where $[\cdot]_1, [\cdot]_2$ correspond respectively to exponentiations in the prime-order groups G_1, G_2. This yields the following prime-order KP-ABE scheme for monotone span programs:

$$
\begin{aligned}
\mathsf{mpk} &:= (\ [\mathbf{A}_1^\top]_1, [\mathbf{A}_1^\top \mathbf{W}]_1, [\mathbf{A}_1^\top \mathbf{W}_0]_1, [\mathbf{A}_1^\top \mathbf{W}_1]_1,\ e([\mathbf{A}_1^\top]_1, [\mathbf{k}]_2)\ ), \\
\mathsf{ct_x} &:= (\ [\mathbf{s}^\top \mathbf{A}_1^\top]_1, \{[\mathbf{s}^\top \mathbf{A}_1^\top \mathbf{W} + \mathbf{s}_j^\top \mathbf{A}_1^\top (\mathbf{W}_0 + j \cdot \mathbf{W}_1)]_1, [\mathbf{s}_j^\top \mathbf{A}_1^\top]_1\}_{x_j=1}, \\
&\qquad\qquad\qquad\qquad\qquad\qquad\qquad e([\mathbf{s}^\top \mathbf{A}_1^\top]_1, [\mathbf{k}]_2) \cdot m\ ) \\
\mathsf{sk_M} &:= (\ \{[\mathbf{k}_j + \mathbf{WBr}_j]_2, [\mathbf{Br}_j]_2, [(\mathbf{W}_0 + j \cdot \mathbf{W}_1)\mathbf{Br}_j]_2\}_{j \in [n]}\ )
\end{aligned}
$$

where $\mathbf{k}_j$ is the j'th share of $\mathbf{k}$. Decryption proceeds as before by first computing $\{e([\mathbf{s}^\top \mathbf{A}_1^\top]_1, [\mathbf{k}_j]_2)\}_{x_j=1}$ and relies on the associativity relations $\mathbf{A}_1^\top \mathbf{W} \cdot \mathbf{B} = \mathbf{A}_1^\top \cdot \mathbf{WB}$ (ditto $\mathbf{W}_0 + j \cdot \mathbf{W}_1$) [7].

Dimensions of $\mathbf{A}_1, \mathbf{B}$. It is helpful to compare the dimensions of $\mathbf{A}_1, \mathbf{B}$ to those of the CGW prime-order analogue of LOSTW in [6]; once we fix the dimensions of $\mathbf{A}_1, \mathbf{B}$, the dimensions of $\mathbf{W}, \mathbf{W}_0, \mathbf{W}_1$ are also fixed. In all of these constructions, the width of $\mathbf{A}_1, \mathbf{B}$ is always k, for constructions based on the k-linear assumption. CGW uses a shorter $\mathbf{A}_1$ of dimensions $(k+1) \times k$, and a $\mathbf{B}$ of the same dimensions $(k+1) \times k$. Roughly speaking, increasing the height of $\mathbf{A}_1$ by k plays the role of adding a subgroup in our composite-order scheme; in particular, the LOSTW KP-ABE uses a group of order $p_1 p_2$ in the asymmetric setting, whereas our unbounded ABE uses a group of order $p_1 p_2 p_3$.

We note that the direct adaptation of the prior techniques in [11] would yield $\mathbf{A}_1$ of height $3k$ and $\mathbf{B}$ of height $k+1$, and reducing the height of $\mathbf{A}_1$ down to $2k+1$ is the key to our efficiency improvements over the prime-order unbounded KP-ABE scheme in [23]. To accomplish this, we need to optimize on the static assumptions used in the composite-order bilinear entropy expansion lemma, and thereafter, carefully transfer these optimizations to the prime-order setting, building upon and extending the recent prime-order IBE schemes in [11].

Bilinear Entropy Expansion Lemma. In the rest of this overview, we motivate the prime-order analogue of our bilinear entropy expansion lemma in (4), and defer a more accurate treatment to Sect. 6. Upon our substitutions in (7), we expect to prove a statement of the form:

$$\left\{ \begin{array}{l} [\mathbf{s}^\top \mathbf{A}_1^\top]_1,\ \{[\mathbf{s}^\top \mathbf{A}_1^\top \mathbf{W} + \mathbf{s}_j^\top \mathbf{A}_1^\top (\mathbf{W}_0 + j \cdot \mathbf{W}_1)]_1,\ [\mathbf{s}_j^\top \mathbf{A}_1^\top]_1\}_{j \in [n]} \\ \{[\mathbf{W}\mathbf{Br}_j]_2,\ [\mathbf{Br}_j]_2,\ [(\mathbf{W}_0 + j \cdot \mathbf{W}_1)\mathbf{Br}_j]_2\}_{j \in [n]} \end{array} \right\} \tag{8}$$

$$\underset{\approx_c}{\text{roughly}} \quad \boxed{\left\{ \begin{array}{l} [\hat{\mathbf{s}}^\top \mathbf{A}_2^\top]_1,\ \{[\hat{\mathbf{s}}^\top \mathbf{A}_2^\top \mathbf{V}_j + \hat{\mathbf{s}}_j^\top \mathbf{A}_2^\top \mathbf{U}_j]_1,\ [\hat{\mathbf{s}}_j^\top \mathbf{A}_2^\top]_1\}_{j \in [n]} \\ \{[\mathbf{V}_j \mathbf{Br}_j]_2,\ [\mathbf{0}]_2,\ [\mathbf{U}_j \mathbf{Br}_j]_2\}_{j \in [n]} \end{array} \right\}}$$

given also the public parameters $[\mathbf{A}_1^\top]_1, [\mathbf{A}_1^\top \mathbf{W}]_1, [\mathbf{A}_1^\top \mathbf{W}_0]_1, [\mathbf{A}_1^\top \mathbf{W}_1]_1$. Here, $\mathbf{A}_2 \leftarrow_\mathrm{R} \mathbb{Z}_p^{(2k+1)\times k}$ is an additional matrix that plays the role of g_2, whereas $\mathbf{U}_j, \mathbf{V}_j$ play the roles of the fresh entropy u_j, v_j. Note that we do not introduce additional terms that correspond to those involving h_2 on the RHS, and can therefore keep $\mathbf{B}$ of dimensions $(k+1) \times k$. To prevent a trivial distinguishing attack based on the associativity relation $\mathbf{A}_1^\top \mathbf{W} \cdot \mathbf{B} = \mathbf{A}_1^\top \cdot \mathbf{W}\mathbf{B}$, we need to sample random $\mathbf{U}_j, \mathbf{V}_j$ subject to the constraints $\mathbf{A}_1^\top \mathbf{U}_j = \mathbf{A}_1^\top \mathbf{V}_j = \mathbf{0}$. In the proof of the entropy expansion lemma, we will show that the k-Lin assumption implies

$$(\mathbf{A}_1, \mathbf{A}_1^\top \mathbf{W}, \{[\mathbf{W}\mathbf{Br}_j]_2, [\mathbf{Br}_j]_2\}_{j \in [n]}) \approx_c (\mathbf{A}_1, \mathbf{A}_1^\top \mathbf{W}, \{[(\mathbf{W} + \boxed{\mathbf{U}_j})\mathbf{Br}_j]_2, [\mathbf{Br}_j]_2\}_{j \in [n]}).$$

To complete the proof of the unbounded ABE, we proceed as before in the composite-order setting, and observe that the boxed term in (8) above (once we strip away the terms involving $\mathbf{U}_j$ and $\hat{\mathbf{s}}_j$) correspond to the prime-order variant of the LOSTW KP-ABE in CGW, as given by:

$$\mathsf{ct}_{\mathbf{x}} := (\ [\hat{\mathbf{s}}^\top \mathbf{A}_2^\top]_1, \{[\hat{\mathbf{s}}^\top \mathbf{A}_2^\top \mathbf{V}_j]_1\}_{x_j = 1}, e([\hat{\mathbf{s}}^\top \mathbf{A}_2^\top]_1, [\mathbf{k}]_2) \cdot m\ )$$
$$\mathsf{sk}_{\mathsf{M}} := (\ \{[\mathbf{k}_j + \mathbf{V}_j \mathbf{Br}_j]_2, [\mathbf{Br}_j]_2\}_{j \in [n]}\ )$$

As in the composite-order setting, we need to first extend our bilinear entropy expansion lemma to a Q-fold setting via random self-reducibility. We may then carry out the analysis in CGW to complete the proof of our unbounded ABE.

1.4 Extensions

Due to lack of space, we briefly sketch two extensions: CP-ABE for monotone span programs, and KP-ABE for arithmetic span programs.

CP-ABE. Here, we start with the LOSTW CP-ABE for monotone span programs [19], which basically reverses the structures of the ciphertexts and keys. This means that we will need a variant of our entropy expansion lemma that accommodates a similar reversal. The statement adapts naturally to this setting, and so does the proof, except we need to make some changes to step two, which requires that we start with a taller $\mathbf{A}_1 \in \mathbb{Z}_q^{3k \times k}$. This gives rise to the following prime-order CP-ABE:

$$\mathsf{mpk} := (\ [\mathbf{A}_1^\top]_1, [\mathbf{A}_1^\top \mathbf{W}]_1, [\mathbf{A}_1^\top \mathbf{W}_0]_1, [\mathbf{A}_1^\top \mathbf{W}_1]_1, [\mathbf{A}_1^\top \mathbf{U}_0]_1\ e([\mathbf{A}_1^\top]_1, [\mathbf{k}]_2)\ ),$$
$$\mathsf{ct}_{\mathsf{M}} := (\ [\mathbf{s}^\top \mathbf{A}_1^\top]_1, \{\ [\mathbf{c}_{0,j}^\top + \mathbf{s}_j^\top \mathbf{A}_1^\top \mathbf{W}]_1, [\mathbf{s}_j^\top \mathbf{A}_1^\top]_1, [\mathbf{s}_j^\top \mathbf{A}_1^\top (\mathbf{W}_0 + j \cdot \mathbf{W}_1)]_1\ \}_{j \in [n]},$$
$$e([\mathbf{s}^\top \mathbf{A}_1^\top]_1, [\mathbf{k}]_2) \cdot m\ )$$
$$\mathsf{sk}_{\mathbf{x}} := (\ [\mathbf{k} + \mathbf{U}_0 \mathbf{Br}]_2, [\mathbf{Br}]_2, \{\ [\mathbf{W}\mathbf{Br} + (\mathbf{W}_0 + j \cdot \mathbf{W}_1)\mathbf{Br}_j]_2, [\mathbf{Br}_j]_2\ \}_{x_j = 1}\ )$$

where $c_{0,j}$ is the j'th share of $c_0 := s^\top A_1^\top U_0$ w.r.t. M. Decryption proceeds by first computing $\{e([c_{0,j}^\top]_1, [Br]_2)\}_{x_j=1}$ and then $e([c_0^\top]_1, [Br]_2)$.

Arithmetic Span Programs. In arithmetic span programs, the attributes x come from $\mathbb{Z}_p^n$ instead of $\{0,1\}^n$, which enable richer and more expressive arithmetic computation. The analogue of the LOSTW KP-ABE for arithmetic span programs [6,15] will then have ciphertexts:

$$\mathsf{ct_x} := (g_1^s, \{g_1^{(v_j+x_jv_j')s}\}_{j\in[n]}, e(g_1, h_1)^{\alpha s}\cdot m).$$

That is, we replaced $g_1^{x_jv_js}$ in (1) with $g_1^{(v_j+x_jv_j')s}$. In the unbounded setting, we will need to generate $\{v_j\}_{j\in[n]}$ and $\{v_j'\}_{j\in[n]}$ via two different pairwise-independent hash functions, given by $w_0 + j\cdot w_1$ and $w_0' + j\cdot w_1'$ respectively. Our entropy expansion lemma generalizes naturally to this setting.

2 Preliminaries

Notation. We denote by $s \leftarrow_R S$ the fact that s is picked uniformly at random from a finite set S. By PPT, we denote a probabilistic polynomial-time algorithm. Throughout this paper, we use 1^λ as the security parameter. We use lower case boldface to denote (column) vectors and upper case boldcase to denote matrices. We use $\equiv$ to denote two distributions being identically distributed, and $\approx_c$ to denote two distributions being computationally indistinguishable. For any two finite sets (also including spaces and groups) S_1 and S_2, the notation "$S_1 \approx_c S_2$" means the uniform distributions over them are computationally indistinguishable.

2.1 Monotone Span Programs

We define (monotone) span programs [16].

Definition 1 (span programs [4,16]). *A (monotone) span program for attribute universe $[n]$ is a pair (M, ρ) where M is a $\ell \times \ell'$ matrix over $\mathbb{Z}_p$ and $\rho: [\ell] \to [n]$. Given $x = (x_1, \ldots, x_n) \in \{0,1\}^n$, we say that*

$$x \text{ satisfies } (M, \rho) \text{ iff } 1 \in \mathsf{span}\langle M_x\rangle,$$

Here, $1 := (1, 0, \ldots, 0)^\top \in \mathbb{Z}^{1\times\ell'}$ is a row vector; M_x denotes the collection of vectors $\{M_j : x_{\rho(j)} = 1\}$ where M_j denotes the j'th row of M; and span refers to linear span of collection of (row) vectors over $\mathbb{Z}_p$.

That is, x satisfies (M, ρ) iff there exists constants $\omega_1, \ldots, \omega_\ell \in \mathbb{Z}_p$ such that

$$\sum_{j:x_{\rho(j)}=1} \omega_j M_j = 1 \tag{9}$$

Observe that the constants $\{\omega_j\}$ can be computed in time polynomial in the size of the matrix $\mathbf{M}$ via Gaussian elimination. Like in [6, 19], we need to impose a one-use restriction, that is, ρ is a permutation and $\ell = n$. By re-ordering the rows of $\mathbf{M}$, we may assume WLOG that ρ is the identity map, which we omit in the rest of this section.

Lemma 1 (statistical lemma [6, Appendix A.6]**).** *For any* $\mathbf{x}$ *that does not satisfy* $\mathbf{M}$, *the distributions*

$$(\{v_j\}_{j:x_j=1}, \{\mathbf{M}_j\left(\begin{smallmatrix} \alpha \\ \mathbf{u} \end{smallmatrix}\right) + r_j v_j, \ r_j\}_{j\in[n]})$$

perfectly hide α, *where the randomness is taken over* $v_j \leftarrow_{\mathrm{R}} \mathbb{Z}_p, \mathbf{u} \leftarrow_{\mathrm{R}} \mathbb{Z}_p^{\ell'-1}$, *and for any fixed* $r_j \neq 0$.

2.2 Attribute-Based Encryption

An attribute-based encryption (ABE) scheme for a predicate $\mathsf{P}(\cdot, \cdot)$ consists of four algorithms $(\mathsf{Setup}, \mathsf{Enc}, \mathsf{KeyGen}, \mathsf{Dec})$:

$\mathsf{Setup}(1^\lambda, \mathcal{X}, \mathcal{Y}, \mathcal{M}) \to (\mathsf{mpk}, \mathsf{msk})$. The setup algorithm gets as input the security parameter λ, the attribute universe $\mathcal{X}$, the predicate universe $\mathcal{Y}$, the message space $\mathcal{M}$ and outputs the public parameter mpk, and the master key msk.

$\mathsf{Enc}(\mathsf{mpk}, x, m) \to \mathsf{ct}_x$. The encryption algorithm gets as input mpk, an attribute $x \in \mathcal{X}$ and a message $m \in \mathcal{M}$. It outputs a ciphertext ct_x. Note that x is public given ct_x.

$\mathsf{KeyGen}(\mathsf{mpk}, \mathsf{msk}, y) \to \mathsf{sk}_y$. The key generation algorithm gets as input msk and a value $y \in \mathcal{Y}$. It outputs a secret key sk_y. Note that y is public given sk_y.

$\mathsf{Dec}(\mathsf{mpk}, \mathsf{sk}_y, \mathsf{ct}_x) \to m$. The decryption algorithm gets as input sk_y and ct_x such that $\mathsf{P}(x, y) = 1$. It outputs a message m.

Correctness. We require that for all $(x, y) \in \mathcal{X} \times \mathcal{Y}$ such that $\mathsf{P}(x, y) = 1$ and all $m \in \mathcal{M}$,

$$\Pr[\mathsf{Dec}(\mathsf{mpk}, \mathsf{sk}_y, \mathsf{Enc}(\mathsf{mpk}, x, m)) = m] = 1,$$

where the probability is taken over $(\mathsf{mpk}, \mathsf{msk}) \leftarrow \mathsf{Setup}(1^\lambda, \mathcal{X}, \mathcal{Y}, \mathcal{M})$, $\mathsf{sk}_y \leftarrow \mathsf{KeyGen}(\mathsf{mpk}, \mathsf{msk}, y)$, and the coins of Enc.

Security Definition. For a stateful adversary $\mathcal{A}$, we define the advantage function

$$\mathsf{Adv}_{\mathcal{A}}^{\mathrm{ABE}}(\lambda) := \Pr\left[b = b' : \begin{array}{l} (\mathsf{mpk}, \mathsf{msk}) \leftarrow \mathsf{Setup}(1^\lambda, \mathcal{X}, \mathcal{Y}, \mathcal{M}); \\ (x^*, m_0, m_1) \leftarrow \mathcal{A}^{\mathsf{KeyGen}(\mathsf{msk}, \cdot)}(\mathsf{mpk}); \\ b \leftarrow_{\mathrm{R}} \{0, 1\}; \mathsf{ct}_{x^*} \leftarrow \mathsf{Enc}(\mathsf{mpk}, x^*, m_b); \\ b' \leftarrow \mathcal{A}^{\mathsf{KeyGen}(\mathsf{msk}, \cdot)}(\mathsf{ct}_{x^*}) \end{array} \right] - \frac{1}{2}$$

with the restriction that all queries y that $\mathcal{A}$ makes to $\mathsf{KeyGen}(\mathsf{msk}, \cdot)$ satisfies $\mathsf{P}(x^*, y) = 0$ (that is, sk_y does not decrypt ct_{x^*}). An ABE scheme is *adaptively secure* if for all PPT adversaries $\mathcal{A}$, the advantage $\mathsf{Adv}_{\mathcal{A}}^{\mathrm{ABE}}(\lambda)$ is a negligible function in λ.

Unbounded ABE. An ABE scheme is *unbounded* [21] if the running time of Setup only depends on λ; otherwise, we say that it is bounded.

3   Bilinear Entropy Expansion, Revisited

3.1   Composite-Order Bilinear Groups and Computational assumptions

A generator $\mathcal{G}$ takes as input a security parameter λ and outputs $\mathbb{G} := (G_N, H_N, G_T, e)$, where N is product of three primes p_1, p_2, p_3 of $\Theta(\lambda)$ bits, G_N, H_N and G_T are cyclic groups of order N and $e : G_N \times H_N \to G_T$ is a non-degenerate bilinear map. We require that the group operations in G_N, H_N and G_T as well the bilinear map e are computable in deterministic polynomial time with respect to λ. We assume that a random generator g (resp. h) of G_N (resp. H_N) is always contained in the description of bilinear groups. For every divisor n of N, we denote by G_n the subgroup of G_N of order n. We use g_1, g_2, g_3 to denote random generators of the subgroups $G_{p_1}, G_{p_2}, G_{p_3}$ respectively. We define h_1, h_2, h_3 random generators of the subgroups $H_{p_1}, H_{p_2}, H_{p_3}$ analogously.

Computational Assumptions. We review two static computational assumptions in the composite-order group, used e.g. in [8, 20].

Assumption 1 ($\mathrm{SD}^{G_N}_{p_1 \mapsto p_1 p_2}$). *We say that $(p_1 \mapsto p_1 p_2)$-subgroup decision assumption, denoted by $SD^{G_N}_{p_1 \mapsto p_1 p_2}$, holds if for all PPT adversaries $\mathcal{A}$, the following advantage function is negligible in λ.*

$$\mathsf{Adv}_{\mathcal{A}}^{\mathrm{SD}^{G_N}_{p_1 \mapsto p_1 p_2}}(\lambda) := \big| \Pr[\mathcal{A}(\mathbb{G}, D, T_0) = 1] - \Pr[\mathcal{A}(\mathbb{G}, D, T_1) = 1] \big|$$

where

$$D := (g_1, g_2, g_3, h_1, h_3, h_{12}), \quad h_{12} \leftarrow_{\mathrm{R}} H_{p_1 p_2}$$
$$T_0 \leftarrow_{\mathrm{R}} \boxed{G_{p_1}}, \quad T_1 \leftarrow_{\mathrm{R}} \boxed{G_{p_1 p_2}}.$$

Assumption 2 ($\mathrm{DDH}^{H_N}_{p_1}$). *We say that p_1-subgroup Diffie-Hellman assumption, denoted by $DDH^{H_N}_{p_1}$, holds if for all PPT adversaries $\mathcal{A}$, the following advantage function is negligible in λ.*

$$\mathsf{Adv}_{\mathcal{A}}^{\mathrm{DDH}^{H_N}_{p_1}}(\lambda) := \big| \Pr[\mathcal{A}(\mathbb{G}, D, T_0) = 1] - \Pr[\mathcal{A}(\mathbb{G}, D, T_1) = 1] \big|$$

where

$$D := (g_1, g_2, g_3, h_1, h_2, h_3),$$
$$T_0 := (h_1^x, h_1^y, \boxed{h_1^{xy}}), \quad T_1 := (h_1^x, h_1^y, \boxed{h_1^{xy+z}}), \quad x, y, z \leftarrow_{\mathrm{R}} \mathbb{Z}_N.$$

By symmetry, one may permute the indices for subgroups and/or exchange the roles of G_N and H_N, and define $\mathrm{SD}^{G_N}_{p_1 \mapsto p_1 p_3}$, $\mathrm{SD}^{G_N}_{p_3 \mapsto p_3 p_2}$, $\mathrm{SD}^{H_N}_{p_1 \mapsto p_1 p_2}$, $\mathrm{SD}^{H_N}_{p_1 \mapsto p_1 p_3}$ and $\mathrm{DDH}^{H_N}_{p_2}, \mathrm{DDH}^{H_N}_{p_3}$ analogously.

3.2   Lemma in Composite-Order Groups

We state our entropy expansion lemma in composite-order groups as follows.

Lemma 2 (Bilinear entropy expansion lemma). *Under the $SD_{p_1 \mapsto p_1 p_2}^{H_N}$, $SD_{p_1 \mapsto p_1 p_3}^{H_N}$, $SD_{p_1 \mapsto p_1 p_2}^{G_N}$, $DDH_{p_2}^{H_N}$, $SD_{p_1 \mapsto p_1 p_3}^{G_N}$, $DDH_{p_3}^{H_N}$, $SD_{p_3 \mapsto p_3 p_2}^{G_N}$ assumptions, we have*

$$\approx_c \left\{ \begin{array}{l} \mathsf{aux} : g_1, g_1^w, g_1^{w_0}, g_1^{w_1} \\ \mathsf{ct} : g_1^s, \{g_1^{sw+s_j(w_0+j\cdot w_1)}, \ g_1^{s_j}\}_{j\in[n]} \\ \mathsf{sk} : \{h_1^{r_j w}, \ h_1^{r_j}, \ h_1^{r_j(w_0+j\cdot w_1)}\}_{j\in[n]} \end{array} \right.$$

$$\approx_c \left\{ \begin{array}{l} \mathsf{aux} : g_1, g_1^w, g_1^{w_0}, g_1^{w_1} \\ \mathsf{ct} : g_1^s \cdot \boxed{g_2^s}, \{g_1^{sw+s_j(w_0+j\cdot w_1)} \cdot \boxed{g_2^{sv_j+s_j u_j}}, \ g_1^{s_j} \cdot \boxed{g_2^{s_j}}\}_{j\in[n]} \\ \mathsf{sk} : \{h_1^{r_j w} \cdot \boxed{h_2^{r_j v_j}}, \ h_1^{r_j} \cdot \boxed{h_2^{r_j}}, \ h_1^{r_j(w_0+j\cdot w_1)} \cdot \boxed{h_2^{r_j u_j}}\}_{j\in[n]} \end{array} \right.$$

where

$$w, w_0, w_1 \leftarrow_{\mathrm{R}} \mathbb{Z}_N, \ v_j, u_j \leftarrow_{\mathrm{R}} \mathbb{Z}_N, \ s, s_j \leftarrow_{\mathrm{R}} \mathbb{Z}_N, \ r_j \leftarrow_{\mathrm{R}} \mathbb{Z}_N.$$

Concretely, the distinguishing advantage $\mathsf{Adv}_{\mathcal{A}}^{\mathrm{ExpLem}}(\lambda)$ is at most

$$\mathsf{Adv}_{\mathcal{B}}^{\mathrm{SD}_{p_1 \mapsto p_1 p_2}^{H_N}}(\lambda) + \mathsf{Adv}_{\mathcal{B}'}^{\mathrm{SD}_{p_1 \mapsto p_1 p_3}^{H_N}}(\lambda) + \mathsf{Adv}_{\mathcal{B}''}^{\mathrm{SD}_{p_1 \mapsto p_1 p_2}^{G_N}}(\lambda) + \mathsf{Adv}_{\mathcal{B}'''}^{\mathrm{SD}_{p_1 \mapsto p_1 p_3}^{H_N}}(\lambda)$$

$$+ \mathsf{Adv}_{\mathcal{B}_0}^{\mathrm{DDH}_{p_2}^{H_N}}(\lambda) + n \cdot \left(\mathsf{Adv}_{\mathcal{B}_1}^{\mathrm{SD}_{p_1 \mapsto p_1 p_3}^{G_N}}(\lambda) + \mathsf{Adv}_{\mathcal{B}_2}^{\mathrm{DDH}_{p_3}^{H_N}}(\lambda) + \mathsf{Adv}_{\mathcal{B}_4}^{\mathrm{SD}_{p_3 \mapsto p_3 p_2}^{G_N}}(\lambda) \right.$$

$$\left. + \mathsf{Adv}_{\mathcal{B}_6}^{\mathrm{DDH}_{p_3}^{H_N}}(\lambda) + \mathsf{Adv}_{\mathcal{B}_7}^{\mathrm{SD}_{p_1 \mapsto p_1 p_3}^{G_N}} \right)(\lambda) + \mathsf{Adv}_{\mathcal{B}_8}^{\mathrm{DDH}_{p_2}^{H_N}}(\lambda)$$

where $\mathsf{Time}(\mathcal{B})$, $\mathsf{Time}(\mathcal{B}')$, $\mathsf{Time}(\mathcal{B}'')$, $\mathsf{Time}(\mathcal{B}''')$, $\mathsf{Time}(\mathcal{B}_0)$, $\mathsf{Time}(\mathcal{B}_1)$, $\mathsf{Time}(\mathcal{B}_2)$, $\mathsf{Time}(\mathcal{B}_4)$, $\mathsf{Time}(\mathcal{B}_6)$, $\mathsf{Time}(\mathcal{B}_7)$, $\mathsf{Time}(\mathcal{B}_8) \approx \mathsf{Time}(\mathcal{A})$.

We will prove the lemma in two main steps (cf. Sect. 1.2), which are formulated via the following two lemmas.

Lemma 3 (Bilinear entropy expansion lemma (step one)). *Under the $DDH_{p_2}^{H_N}$, $SD_{p_1 \mapsto p_1 p_3}^{G_N}$, $DDH_{p_3}^{H_N}$, $SD_{p_3 \mapsto p_3 p_2}^{G_N}$ assumptions, we have*

$$\left\{ \begin{array}{l} \mathsf{aux} : g_1, g_1^{w_0}, g_1^{w_1}, g_2 \\ \mathsf{ct} : \{g_1^{s_j(w_0+j\cdot w_1)}, \ g_1^{s_j}\}_{j\in[n]} \\ \mathsf{sk} : \{h_{123}^{r_j}, \ h_{123}^{r_j(w_0+j\cdot w_1)}\}_{j\in[n]} \end{array} \right\} \approx_c \left\{ \begin{array}{l} \mathsf{aux} : g_1, g_1^{w_0}, g_1^{w_1}, g_2 \\ \mathsf{ct} : \{g_1^{s_j(w_0+j\cdot w_1)} \cdot \boxed{g_2^{s_j u_j}}, \ g_1^{s_j} \cdot \boxed{g_2^{s_j}}\}_{j\in[n]} \\ \mathsf{sk} : \{h_{13}^{r_j} \cdot \boxed{h_2^{r_j}}, \ h_{13}^{r_j(w_0+j\cdot w_1)} \cdot \boxed{h_2^{r_j u_j}}\}_{j\in[n]} \end{array} \right\}$$

where

$$w_0, w_1 \leftarrow_{\mathrm{R}} \mathbb{Z}_N, \ u_j \leftarrow_{\mathrm{R}} \mathbb{Z}_N, \ s_j \leftarrow_{\mathrm{R}} \mathbb{Z}_N, \ r_j \leftarrow_{\mathrm{R}} \mathbb{Z}_N.$$

Concretely, the distinguishing advantage $\mathsf{Adv}_{\mathcal{A}}^{\mathrm{Step1}}(\lambda)$ is at most

$$\mathsf{Adv}_{\mathcal{B}_0}^{\mathrm{DDH}_{p_2}^{H_N}}(\lambda) + n \cdot \left(\mathsf{Adv}_{\mathcal{B}_1}^{\mathrm{SD}_{p_1 \mapsto p_1 p_3}^{G_N}}(\lambda) + \mathsf{Adv}_{\mathcal{B}_2}^{\mathrm{DDH}_{p_3}^{H_N}}(\lambda) + \mathsf{Adv}_{\mathcal{B}_4}^{\mathrm{SD}_{p_3 \mapsto p_3 p_2}^{G_N}}(\lambda) \right.$$

$$\left. + \mathsf{Adv}_{\mathcal{B}_6}^{\mathrm{DDH}_{p_3}^{H_N}}(\lambda) + \mathsf{Adv}_{\mathcal{B}_7}^{\mathrm{SD}_{p_1 \mapsto p_1 p_3}^{G_N}}(\lambda) \right)$$

where $\mathsf{Time}(\mathcal{B}_0)$, $\mathsf{Time}(\mathcal{B}_1)$, $\mathsf{Time}(\mathcal{B}_2)$, $\mathsf{Time}(\mathcal{B}_4)$, $\mathsf{Time}(\mathcal{B}_6)$, $\mathsf{Time}(\mathcal{B}_7) \approx \mathsf{Time}(\mathcal{A})$.

Note that sk in the LHS of this lemma has an extra h_{23}-component, which we may introduce using the $\mathsf{SD}^{H_N}_{p_1 \mapsto p_1 p_2}$ and $\mathsf{SD}^{H_N}_{p_1 \mapsto p_1 p_3}$ assumption. The proof of this lemma is fairly involved, and we defer the proof to Sect. 3.3.

Lemma 4 (Bilinear entropy expansion lemma (step two)). *Under the* $DDH^{H_N}_{p_2}$ *assumption, we have*

$$
\left\{
\begin{array}{l}
\mathsf{aux} : g_1, g_1^w, h_1, h_1^w \\
\mathsf{ct} : g_2^s, \{\, g_2^{s\,w} \cdot g_2^{s_j u_j}, \, g_2^{s_j} \,\}_{j \in [n]} \\
\mathsf{sk} : \{ h_2^{r_j w}, \, h_2^{r_j}, \, h_2^{r_j u_j} \}_{j \in [n]}
\end{array}
\right\}
\approx_c
\left\{
\begin{array}{l}
\mathsf{aux} : g_1, g_1^w, h_1, h_1^w \\
\mathsf{ct} : g_2^s, \{\, \boxed{g_2^{s v_j}} \cdot g_2^{s_j u_j}, \, g_2^{s_j} \,\}_{j \in [n]} \\
\mathsf{sk} : \{ \boxed{h_2^{r_j v_j}}, \, h_2^{r_j}, \, h_2^{r_j u_j} \}_{j \in [n]}
\end{array}
\right\}
$$

where

$$ w \leftarrow_{\mathsf{R}} \mathbb{Z}_N, \; v_j, u_j \leftarrow_{\mathsf{R}} \mathbb{Z}_N, \; s, s_j \leftarrow_{\mathsf{R}} \mathbb{Z}_N, \; r_j \leftarrow_{\mathsf{R}} \mathbb{Z}_N. $$

Concretely, the distinguishing advantage $\mathsf{Adv}^{\mathrm{STEP2}}_{\mathcal{A}}(\lambda)$ *is at most* $\mathsf{Adv}^{DDH^{H_N}_{p_2}}_{\mathcal{B}_8}(\lambda)$ *where* $\mathsf{Time}(\mathcal{B}_8) \approx \mathsf{Time}(\mathcal{A})$.

Proof. This follows from the $DDH^{H_N}_{p_2}$ assumption, which tells us that

$$ \{ h_2^{r_j}, h_2^{r_j w} \}_{j \in [n]} \approx_c \{ h_2^{r_j}, \boxed{h_2^{r_j v_j}} \}_{j \in [n]}. $$

The adversary $\mathcal{B}_8$ on input $\{ h_2^{r_j}, T_j \}_{j \in [n]}$ along with g_1, g_2, h_1, h_2, picks $\tilde{w}, s, s_j, \tilde{u}_j \leftarrow_{\mathsf{R}} \mathbb{Z}_N$ (and implicitly sets $u_j = \frac{1}{s_j}(\tilde{u}_j - sw)$), then runs $\mathcal{A}$ on input

$$
\left\{
\begin{array}{l}
\mathsf{aux} : g_1, g_1^{\tilde{w}}, h_1, h_1^{\tilde{w}} \\
\mathsf{ct} : g_2^s, \{\, g_2^{\tilde{u}_j}, \, g_2^{s_j} \,\}_{j \in [n]} \\
\mathsf{sk} : \{ T_j, \, h_2^{r_j}, \, (h_2^{r_j})^{\frac{\tilde{u}_j}{s_j}} \cdot T_j^{-\frac{s}{s_j}} \}_{j \in [n]}
\end{array}
\right\}.
$$

By the Chinese Remainder Theorem, we have $(g_1^w, h_1^w, g_2^w, h_2^w) \equiv (g_1^{\tilde{w}}, h_1^{\tilde{w}}, g_2^w, h_2^w)$, where $w, \tilde{w} \leftarrow_{\mathsf{R}} \mathbb{Z}_N$. Next, observe that

- When $T_j = g^{r_j w}$ and if we write $r_j u_j = r_j \cdot \frac{\tilde{u}_j}{s_j} + r_j w \cdot (-\frac{s}{s_j})$, then $\tilde{u}_j = sw + s_j u_j$ and the distribution we feed to $\mathcal{A}$ is exactly that of the left distribution.
- When $T_j = g^{r_j v_j}$ and if we write $r_j u_j = r_j \cdot \frac{\tilde{u}_j}{s_j} + r_j v_j \cdot (-\frac{s}{s_j})$, then $\tilde{u}_j = sv_j + s_j u_j$ and the distribution we feed to $\mathcal{A}$ is exactly that of the right distribution.

This completes the proof. □

3.3   Entropy Expansion Lemma: Step One

Proof Overview. First, we note that we can adapt the proof of the Lewko-Waters IBE [8,20][5] to show that under $\mathsf{SD}^{G_N}_{p_1 \mapsto p_1 p_3}$ and $DDH^{H_N}_{p_3}$ assumptions,

---

[5] With two main differences: (i) we are in the selective setting which allows for a much simpler proof, (ii) we allow $j = i$ in sk.

we have that for each $i \in [n]$:

$$\left\{\begin{array}{l} \mathsf{aux}: g_1, g_1^{w_0}, g_1^{w_1} \\ \mathsf{ct}: \{g_1^{s_i(w_0+i\cdot w_1)},\ g_1^{s_i}\} \\ \mathsf{sk}: \{h_{13}^{r_j},\ h_{13}^{r_j(w_0+j\cdot w_1)}\}_{j\in[n]} \end{array}\right\} \approx_c \left\{\begin{array}{l} \mathsf{aux}: g_1, g_1^{w_0}, g_1^{w_1} \\ \mathsf{ct}: \{g_1^{s_i(w_0+i\cdot w_1)}\cdot \boxed{g_3^{s_iu_i}},\ g_1^{s_i}\cdot \boxed{g_3^{s_i}}\} \\ \mathsf{sk}: \{h_1^{r_j}\cdot \boxed{h_3^{r_j}},\ h_1^{r_j(w_0+j\cdot w_1)}\cdot \boxed{h_3^{r_ju_j}}\}_{j\in[n]} \end{array}\right\}.$$

We can then use the $\mathrm{SD}_{p_3 \mapsto p_2 p_3}^{G_N}$ assumption to argue that

$$(g_3^{s_i}, g_3^{s_iu_i}) \approx_c (g_3^{s_i}\cdot \boxed{g_2^{s_i}}, g_3^{s_iu_i}\cdot \boxed{g_2^{s_iu_i}})$$

Roughly speaking, we will then repeat the above argument n times for each $i \in [n]$ (see Sub-Game$_{i,1}$ through Sub-Game$_{i,4}$ below). Here, there is an additional complication arising from the fact that in order to invoke the $\mathrm{SD}_{p_1 \mapsto p_1 p_3}^{G_N}$ assumption, we need to simulate sk given only h_1, h_{13}, h_2. To do this, we need to switch sk back to $\{h_{13}^{r_j}, h_{13}^{r_j(w_0+j\cdot w_1)}\}_{j\in[n]}$, which we do in Sub-Game$_{i,5}$ through Sub-Game$_{i,7}$.

At this point, we are almost done, except we still need to introduce a $(h_2^{r_j}, h_2^{r_ju_j})$-component into sk. We will handle this at the very beginning of the proof (cf. Game$_{0'}$). Fortunately, we can carry out the above argument even with the extra $(h_2^{r_j}, h_2^{r_ju_j})$-component in sk.

Actual Proof. We prove step one of the entropy expansion lemma in Lemma 3 via the following game sequence. Each claim will be followed by a proof sketch but a formal proof is omitted. By ct$_j$ (resp. sk$_j$), we denote the j'th tuple of ct (resp. sk).

Game$_0$. This is the left distribution in Lemma 3:

$$\left\{\begin{array}{l} \mathsf{aux}: g_1, g_1^{w_0}, g_1^{w_1}, g_2 \\ \mathsf{ct}: \{g_1^{s_j(w_0+j\cdot w_1)},\ g_1^{s_j}\}_{j\in[n]} \\ \mathsf{sk}: \{h_{123}^{r_j},\ h_{123}^{r_j(w_0+j\cdot w_1)}\}_{j\in[n]} \end{array}\right\}.$$

Game$_{0'}$. We modify sk as follows:

$$\mathsf{sk}: \{h_{13}^{r_j}\cdot \boxed{h_2^{r_j}},\ h_{13}^{r_j(w_0+j\cdot w_1)}\cdot \boxed{h_2^{r_ju_j}}\}_{j\in[n]}$$

where $u_1,\ldots,u_n \leftarrow_{\mathrm{R}} \mathbb{Z}_N$. We claim that Game$_0 \approx_c$ Game$_{0'}$. This follows from the $\mathrm{DDH}_{p_2}^{H_N}$ assumption, which tells us that

$$\{h_2^{r_j}, h_2^{r_jw_0}\}_{j\in[n]} \approx_c \{h_2^{r_j}, \boxed{h_2^{r_ju_j'}}\}_{j\in[n]} \text{ given } g_1, g_2, h_{13}$$

where $u_j' \leftarrow_{\mathrm{R}} \mathbb{Z}_N$ and we will then implicitly set $u_j = u_j' + j \cdot w_1$ for all $j \in [n]$. In the security reduction, we use the fact that aux, ct leak no information about $w_0 \bmod p_2$.

$\underline{\text{Game}_i}(i = 1, \ldots, n+1)$. We modify ct as follows:

$$\text{ct} : \{g_1^{s_j(w_0 + j \cdot w_1)} \cdot \boxed{g_2^{s_j u_j}}, \; g_1^{s_j} \cdot \boxed{g_2^{s_j}} \}_{j < i}$$
$$\{g_1^{s_j(w_0 + j \cdot w_1)}, \qquad\qquad g_1^{s_j} \qquad\quad \}_{j \geq i}$$

where $u_1, \ldots, u_{i-1}$ are defined as in $\text{Game}_{0'}$. It is easy to see that $\text{Game}_{0'} \equiv \text{Game}_1$. To show that $\text{Game}_i \approx_c \text{Game}_{i+1}$, we will require another sequence of sub-games.

$\underline{\text{Sub-Game}_{i,1}}$. Identical to Game_i except that we modify ct_i as follows:

$$\text{ct}_i : \{g_1^{s_i(w_0 + i \cdot w_1)} \cdot \boxed{g_3^{s_i(w_0 + i \cdot w_1)}}, \; g_1^{s_i} \cdot \boxed{g_3^{s_i}} \}$$

We claim that $\text{Game}_i \approx_c \text{Sub-Game}_{i,1}$. This follows from the $\text{SD}_{p_1 \mapsto p_1 p_3}^{G_N}$ assumption, which tells us that

$$g_1^{s_i} \approx_c g_1^{s_i} \cdot \boxed{g_3^{s_i}} \text{ given } g_1, g_2, h_{13}, h_2$$

In the reduction, we will sample $w_0, w_1, u_j \leftarrow_R \mathbb{Z}_N$ and use g_1, g_2 to simulate aux, $\{\text{ct}_j\}_{j \neq i}$ and h_{13}, h_2 to simulate sk.

$\underline{\text{Sub-Game}_{i,2}}$. We modify the distribution of sk_j for all $j \neq i$ (while keeping sk_i unchanged):

$$\text{sk}_j \; (j \neq i) : h_1^{r_j} \cdot h_2^{r_j} \cdot \boxed{h_3^{r_j}}, \; h_1^{r_j(w_0 + j \cdot w_1)} \cdot h_2^{r_j u_j} \cdot \boxed{h_3^{r_j u_j}}$$

We claim that $\text{Sub-Game}_{i,1} \approx_c \text{Sub-Game}_{i,2}$. This follows from the $\text{DDH}_{p_3}^{H_N}$ assumption, which tells us that

$$\{h_3^{r_j}, h_3^{r_j w_1}\}_{j \neq i} \approx_c \{h_3^{r_j}, \boxed{h_3^{r_j u_j'}}\}_{j \neq i} \text{ given } g_1, g_2, g_3, h_1, h_2, h_3.$$

where $u_j' \leftarrow_R \mathbb{Z}_N$. In the reduction, we will program $w_0 := \tilde{w}_0 - i \cdot w_1 \bmod p_3$ with $\tilde{w}_0 \leftarrow_R \mathbb{Z}_N$ so that we can simulate $g_3^{s_i(w_0 + i \cdot w_1)}$ in ct_i, and then implicitly set $u_j = \tilde{w}_0 + (j - i) \cdot u_j' \bmod p_3$ for all $j \neq i$.

$\underline{\text{Sub-Game}_{i,3}}$. We modify the distribution of ct_i and sk_i simultaneously:

$$\text{ct}_i : g_1^{s_i(w_0 + i \cdot w_1)} \cdot \boxed{g_3^{s_i u_i}}, \; g_1^{s_i} \cdot g_3^{s_i}$$
$$\text{sk}_i : h_1^{r_i} \cdot h_2^{r_i} \cdot h_3^{r_i}, \; h_1^{r_i(w_0 + i \cdot w_1)} \cdot h_2^{r_i u_i} \cdot \boxed{h_3^{r_i u_i}}$$

We claim that $\text{Sub-Game}_{i,2} \equiv \text{Sub-Game}_{i,3}$. This follows from the fact that for all $j \neq i$, the quantity $w_0 + j \cdot w_1 \bmod p_3$ leaked in sk_j is masked by u_j and therefore $\{w_0 + i \cdot w_1 \bmod p_3\} \equiv \{u_i \bmod p_3\}$.

$\underline{\text{Sub-Game}_{i,4}}$. We modify the distribution of ct_i as follows:

$$\text{ct}_i : g_1^{s_i(w_0 + i \cdot w_1)} \cdot \boxed{g_2^{s_i u_i}} \cdot g_3^{s_i u_i}, \; g_1^{s_i} \cdot \boxed{g_2^{s_i}} \cdot g_3^{s_i}$$

We claim that $\mathsf{Sub\text{-}Game}_{i,3} \approx_c \mathsf{Sub\text{-}Game}_{i,4}$. This follows from the $\mathsf{SD}^{G_N}_{p_3 \mapsto p_3 p_2}$ assumption, which tells us that

$$g_3^{s_i} \approx_c \boxed{g_2^{s_i}} \cdot g_3^{s_i} \quad \text{given} \quad g_1, g_2, h_1, h_{23}.$$

In the reduction, we will sample $w_0, w_1, u_j \leftarrow_{\mathrm{R}} \mathbb{Z}_N$ and use g_1, g_2 to simulate $\mathsf{aux}, \{\mathsf{ct}_j\}_{j \neq i}$. In addition, we will use generator h_{23} to sample $\{h_2^{r_j} \cdot h_3^{r_j}, h_2^{r_j u_j} \cdot h_3^{r_j u_j}\}_{j \in [n]}$ in sk.

$\underline{\mathsf{Sub\text{-}Game}_{i,5}}$. We modify the distribution of ct_i and sk_i:

$$\mathsf{ct}_i : g_1^{s_i(w_0+i \cdot w_1)} \cdot g_2^{s_i u_i} \cdot \boxed{g_3^{s_i(w_0+i \cdot w_1)}}, \; g_1^{s_i} \cdot g_2^{s_i} \cdot g_3^{s_i}$$

$$\mathsf{sk}_i : h_1^{r_i} \cdot h_2^{r_i} \cdot h_3^{r_i}, \; h_1^{r_i(w_0+i \cdot w_1)} \cdot h_2^{r_i u_i} \cdot \boxed{h_3^{r_i(w_0+i \cdot w_1)}}$$

We claim that $\mathsf{Sub\text{-}Game}_{i,4} \equiv \mathsf{Sub\text{-}Game}_{i,5}$. The proof is completely analogous to that of $\mathsf{Sub\text{-}Game}_{i,2} \equiv \mathsf{Sub\text{-}Game}_{i,3}$.

$\underline{\mathsf{Sub\text{-}Game}_{i,6}}$. We modify the distribution of sk_j for all $j \neq i$:

$$\mathsf{sk}_j \; (j \neq i) : h_1^{r_j} \cdot h_2^{r_j} \cdot \boxed{h_3^{r_j}}, \; h_1^{r_j(w_0+j \cdot w_1)} \cdot h_2^{r_j u_j} \cdot \boxed{h_3^{r_j(w_0+j \cdot w_1)}}$$

We claim that $\mathsf{Sub\text{-}Game}_{i,5} \approx_c \mathsf{Sub\text{-}Game}_{i,6}$. The proof is completely analogous to that of $\mathsf{Sub\text{-}Game}_{i,1} \approx_c \mathsf{Sub\text{-}Game}_{i,2}$.

$\underline{\mathsf{Sub\text{-}Game}_{i,7}}$. We modify the distribution of ct_i:

$$\mathsf{ct}_i : g_1^{s_i(w_0+i \cdot w_1)} \cdot g_2^{s_i u_i} \cdot \cancel{g_3^{s_i(w_0+i \cdot w_1)}}, \; g_1^{s_i} \cdot g_2^{s_i} \cdot \cancel{g_3^{s_i}}$$

We claim that $\mathsf{Sub\text{-}Game}_{i,6} \approx_c \mathsf{Sub\text{-}Game}_{i,7}$. The proof is completely analogous to that of $\mathsf{Game}_i \approx_c \mathsf{Sub\text{-}Game}_{i,1}$. Furthermore, observe that $\mathsf{Sub\text{-}Game}_{i,7}$ is actually identical to Game_{i+1}.

$\underline{\mathsf{Game}_{n+1}}$. In Game_{n+1}, we have:

$$\left\{ \begin{array}{l} \mathsf{aux} : g_1, g_1^{w_0}, g_1^{w_1}, g_2 \\ \mathsf{ct} : \{g_1^{s_j(w_0+j \cdot w_1)} \cdot \boxed{g_2^{s_j u_j}}, \; g_1^{s_j} \cdot \boxed{g_2^{s_j}}\}_{j \in [n]} \\ \mathsf{sk} : \{h_{13}^{r_j} \cdot \boxed{h_2^{r_j}}, \; h_{13}^{r_j(w_0+j \cdot w_1)} \cdot \boxed{h_2^{r_j u_j}}\}_{j \in [n]} \end{array} \right\}.$$

This is exactly the right distribution of Lemma 3.

4   KP-ABE for Monotone Span Programs in Composite-Order Groups

In this section, we present our adaptively secure, unbounded KP-ABE for monotone span programs based on static assumptions in composite-order groups (cf. Sect. 3.1).

4.1   Construction

Setup($1^\lambda, 1^n$): On input ($1^\lambda, 1^n$), sample $\mathbb{G} := (N = p_1p_2p_3, G_N, H_N, G_T, e) \leftarrow \mathcal{G}(1^\lambda)$ and select random generators g_1, h_1 and h_{123} of G_{p_1}, H_{p_1} and H_N, respectively. Pick

$$w, w_0, w_1 \leftarrow_R \mathbb{Z}_N, \ \alpha \leftarrow_R \mathbb{Z}_N,$$

a pairwise independent hash function $\mathsf{H} : G_T \rightarrow \{0,1\}^\lambda$, and output the master public and secret key pair

$$\mathsf{mpk} := (\ (N, G_N, H_N, G_T, e); \ g_1, \ g_1^w, \ g_1^{w_0}, \ g_1^{w_1}, \ e(g_1, h_{123})^\alpha; \ \mathsf{H}\ )$$
$$\mathsf{msk} := (h_{123}, h_1, \alpha, w, w_0, w_1\ ).$$

Enc($\mathsf{mpk}, \mathbf{x}, m$): On input an attribute vector $\mathbf{x} := (x_1, \ldots, x_n) \in \{0,1\}^n$ and $m \in \{0,1\}^\lambda$, pick $s, s_j \leftarrow_R \mathbb{Z}_N$ for all $j \in [n]$ and output

$$\mathsf{ct}_\mathbf{x} := \left(\begin{array}{c} C_0 := g_1^s, \ \{\ C_{1,j} := g_1^{sw + s_j(w_0 + j \cdot w_1)}, \ C_{2,j} := g_1^{s_j}\ \}_{j:x_j=1}, \\ C := \mathsf{H}(e(g_1, h_{123})^{\alpha s}) \cdot m \\ \in G_N^{2n+1} \times \{0,1\}^\lambda. \end{array} \right)$$

KeyGen($\mathsf{mpk}, \mathsf{msk}, \mathbf{M}$): On input a monotone span program $\mathbf{M} \in \mathbb{Z}_N^{n \times \ell'}$, pick $\mathbf{u} \leftarrow_R \mathbb{Z}_N^{\ell'-1}$ and $r_j \leftarrow_R \mathbb{Z}_N$ for all $j \in [n]$, and output

$$\mathsf{sk}_\mathbf{M} := (\{\ K_{0,j} := h_{123}^{\mathbf{M}_j\binom{\alpha}{\mathbf{u}}} \cdot h_1^{r_j w}, \ K_{1,j} := h_1^{r_j}, \ K_{2,j} := h_1^{r_j(w_0 + j \cdot w_1)}\ \}_{j \in [n]}) \in H_N^{3n}.$$

Dec($\mathsf{mpk}, \mathsf{sk}_\mathbf{M}, \mathsf{ct}_\mathbf{x}$): If $\mathbf{x}$ satisfies $\mathbf{M}$, compute $\omega_1, \ldots, \omega_n \in \mathbb{Z}_p$ such that

$$\sum_{j:x_j=1} \omega_j \mathbf{M}_j = \mathbf{1}.$$

Then, compute

$$K \leftarrow \prod_{j:x_j=1} \left(e(C_0, K_{0,j}) \cdot e(C_{1,j}, K_{1,j})^{-1} \cdot e(C_{2,j}, K_{2,j}) \right)^{\omega_j},$$

and recover the message as $m \leftarrow C/\mathsf{H}(K) \in \{0,1\}^\lambda$.

It is direct to prove the correctness and we omit the detail here.

4.2   Proof of Security

We prove the following theorem:

Theorem 1. *Under the subgroup decision assumptions and the subgroup Diffie-Hellman assumptions (cf. Sect. 3.1), the unbounded KP-ABE scheme described in this section (cf. Sect. 4.1) is adaptively secure (cf. Sect. 2.2).*

Main Technical Lemma. We prove the following technical lemma. Our proof consists of two steps. We first apply the entropy expansion lemma (see Lemma 2) and obtain a copy of the LOSTW KP-ABE (variant there-of) in the p_2-subgroup. We may then carry out the classic dual system methodology used for establishing adaptive security of the LOSTW KP-ABE in the p_2-subgroup with the p_3-subgroup as the semi-functional space.

Lemma 5. *For any adversary $\mathcal{A}$ that makes at most Q key queries against the unbounded KP-ABE scheme, there exist adversaries $\mathcal{B}_0, \mathcal{B}_1, \mathcal{B}_2, \mathcal{B}_2$ such that:*

$$\mathsf{Adv}_{\mathcal{A}}^{\mathrm{ABE}}(\lambda) \leq \mathsf{Adv}_{\mathcal{B}_0}^{\mathrm{EXPLEM}}(\lambda) + \mathsf{Adv}_{\mathcal{B}_1}^{\mathrm{SD}_{p_2 \mapsto p_2 p_3}^{G_N}}(\lambda) + Q \cdot \mathsf{Adv}_{\mathcal{B}_2}^{\mathrm{SD}_{p_2 \mapsto p_2 p_3}^{H_N}}(\lambda) + Q \cdot \mathsf{Adv}_{\mathcal{B}_3}^{\mathrm{SD}_{p_2 \mapsto p_2 p_3}^{H_N}}(\lambda)$$

where $\mathsf{Time}(\mathcal{B}_0), \mathsf{Time}(\mathcal{B}_1), \mathsf{Time}(\mathcal{B}_2), \mathsf{Time}(\mathcal{B}_3) \approx \mathsf{Time}(\mathcal{A})$. *In particular, we achieve security loss* $O(n + Q)$ *based on the* $SD_{p_1 \mapsto p_1 p_2}^{H_N}$, $SD_{p_1 \mapsto p_1 p_3}^{H_N}$, $SD_{p_1 \mapsto p_1 p_2}^{G_N}$, $DDH_{p_2}^{H_N}$, $SD_{p_1 \mapsto p_1 p_3}^{G_N}$, $DDH_{p_3}^{H_N}$, $SD_{p_3 \mapsto p_3 p_2}^{G_N}$, $SD_{p_2 \mapsto p_2 p_3}^{G_N}$, $SD_{p_2 \mapsto p_2 p_3}^{H_N}$ *assumptions.*

The proof follows a series of games based on the dual system methodology (see Fig. 4). We first define the auxiliary distributions, upon which we can describe the games.

Game	CT	SK			Justification
		$\kappa < i$	$\kappa = i$	$\kappa > i$	
0	Normal		Normal		real game
0'	E-normal		E-normal		entropy expansion lemma, Lemma 2
i	SF	SF	E-normal	E-normal	$SD_{p_2 \mapsto p_2 p_3}^{G_N}$, $\mathsf{Game}_i = \mathsf{Game}_{i-1,3}$
$i,1$	—	—	P-normal	—	$SD_{p_2 \mapsto p_2 p_3}^{H_N}$
$i,2$	—	—	P-SF	—	statistical lemma, Lemma 1
$i,3$	—	—	SF	—	$SD_{p_2 \mapsto p_2 p_3}^{H_N}$
Final	random m		SF		statistical hiding

Fig. 4. Game sequence for our composite-order unbounded KP-ABE.

Auxiliary Distributions. We define various forms of a ciphertext (of message m under attribute vector $\mathbf{x}$):

- Normal: Generated by Enc.
- E-normal: Same as a normal ciphertext except that a copy of normal ciphertext is created in G_{p_2} and then we use the substitution:

$$w \mapsto v_j \bmod p_2 \text{ in } j\text{'th component} \quad \text{and} \quad w_0 + j \cdot w_1 \mapsto u_j \bmod p_2 \quad (10)$$

where $v_j, u_j \leftarrow_{\mathrm{R}} \mathbb{Z}_N$. Concretely, an E-normal ciphertext is of the form

$$\mathsf{ct_x} := \left(g_1^s \cdot \boxed{g_2^s}, \; \{ g_1^{sw+s_j(w_0+j\cdot w_1)} \cdot \boxed{g_2^{sv_j+s_ju_j}}, \; g_1^{s_j} \cdot \boxed{g_2^{s_j}} \}_{j:x_j=1}, \; \mathsf{H}(e(g_1^s \cdot \boxed{g_2^s}, h_{123}^\alpha)) \cdot m \right)$$

where g_2 is a random generator of G_{p_2} and $s, s_j \leftarrow_{\mathrm{R}} \mathbb{Z}_N$.

- SF: Same as E-normal ciphertext except that we copy all entropy from G_{p_2} to G_{p_3}. Concretely, an SF ciphertext is of the form

$$\mathsf{ct_x} := \left(\begin{array}{l} g_1^s \cdot g_2^s \cdot \boxed{g_3^s}, \\ \{ g_1^{sw+s_j(w_0+j\cdot w_1)} \cdot g_2^{sv_j+s_ju_j} \cdot \boxed{g_3^{sv_j+s_ju_j}}, \; g_1^{s_j} \cdot g_2^{s_j} \cdot \boxed{g_3^{s_j}} \}_{j:x_j=1}, \\ \mathsf{H}(e(g_1^s \cdot g_2^s \cdot \boxed{g_3^s}, h_{123}^\alpha)) \cdot m \end{array} \right)$$

where g_3 is a random generator of G_{p_3} and $s, s_j \leftarrow_{\mathrm{R}} \mathbb{Z}_N$.

Then we pick $\hat{\alpha} \leftarrow_{\mathrm{R}} \mathbb{Z}_N$ and define various forms of a key (for span program $\mathbf{M}$):

- Normal: Generated by KeyGen.
- E-normal: Same as a normal key except that a copy of $\{h_1^{r_jw}, h_1^{r_j}, h_1^{r_j(w_0+j\cdot w_1)}\}_{j\in[n]}$ is created in H_{p_2} and use the same substitution as in (10). Concretely, an E-normal key is of the form

$$\mathsf{sk_M} := \left(\; \{ h_{123}^{\mathbf{M}_j\left(\begin{smallmatrix}\alpha\\\mathbf{u}\end{smallmatrix}\right)} \cdot h_1^{r_jw} \cdot \boxed{h_2^{r_jv_j}}, \; h_1^{r_j} \cdot \boxed{h_2^{r_j}}, \; h_1^{r_j(w_0+j\cdot w_1)} \cdot \boxed{h_2^{r_ju_j}} \}_{j\in[n]} \; \right)$$

where h_{123}, h_1 and h_2 are respective random generators of H_N, H_{p_1} and H_{p_2}, $\mathbf{u} \leftarrow_{\mathrm{R}} \mathbb{Z}_N^{\ell'-1}$ and $r_j \leftarrow_{\mathrm{R}} \mathbb{Z}_N$.

- P-normal: Same as E-normal key except that a copy of $\{h_2^{r_jv_j}, h_2^{r_j}, h_2^{r_ju_j}\}_{j\in[n]}$ is created in H_{p_3}. Concretely, a P-normal key is of the form

$$\mathsf{sk_M} := \left(\left\{ \begin{array}{l} h_{123}^{\mathbf{M}_j\left(\begin{smallmatrix}\alpha\\\mathbf{u}\end{smallmatrix}\right)} \cdot h_1^{r_jw} \cdot h_2^{r_jv_j} \cdot \boxed{h_3^{r_jv_j}}, \\ h_1^{r_j} \cdot h_2^{r_j} \cdot \boxed{h_3^{r_j}}, \; h_1^{r_j(w_0+j\cdot w_1)} \cdot h_2^{r_ju_j} \cdot \boxed{h_3^{r_ju_j}} \end{array} \right\}_{j\in[n]} \right)$$

where h_3 is a random generator of H_{p_3}, $\mathbf{u} \leftarrow_{\mathrm{R}} \mathbb{Z}_N^{\ell'-1}$ and $r_j \leftarrow_{\mathrm{R}} \mathbb{Z}_N$.

- P-SF: Same as P-normal key except that $\hat{\alpha}$ is introduced in H_{p_3}. Concretely, a P-SF key is of the form

$$\mathsf{sk_M} := \left(\left\{ \begin{array}{l} h_{123}^{\mathbf{M}_j\left(\begin{smallmatrix}\alpha\\\mathbf{u}\end{smallmatrix}\right)} \cdot \boxed{h_3^{\mathbf{M}_j\left(\begin{smallmatrix}\hat{\alpha}\\\mathbf{0}\end{smallmatrix}\right)}} \cdot h_1^{r_jw} \cdot h_2^{r_jv_j} \cdot h_3^{r_jv_j}, \\ h_1^{r_j} \cdot h_2^{r_j} \cdot h_3^{r_j}, \; h_1^{r_j(w_0+j\cdot w_1)} \cdot h_2^{r_ju_j} \cdot h_3^{r_ju_j} \end{array} \right\}_{j\in[n]} \right)$$

where $\mathbf{u} \leftarrow_{\mathrm{R}} \mathbb{Z}_N^{\ell'-1}$ and $r_j \leftarrow_{\mathrm{R}} \mathbb{Z}_N$.

- SF: Same as P-SF key except that $\{h_3^{r_j v_j}, h_3^{r_j}, h_3^{r_j u_j}\}_{j \in [n]}$ is removed. Concretely, a SF key is of the form

$$\mathsf{sk_M} := \left(\left\{ \begin{array}{l} h_{123}^{\mathbf{M}_j\left(\begin{smallmatrix}\alpha\\\mathbf{u}\end{smallmatrix}\right)} \cdot h_3^{\mathbf{M}_j\left(\begin{smallmatrix}\hat{\alpha}\\\mathbf{0}\end{smallmatrix}\right)} \cdot h_1^{r_j w} \cdot h_2^{r_j v_j} \cdot h_3^{\cancel{r_j v_j}}, \\ h_1^{r_j} \cdot h_2^{r_j} \cdot h_3^{\cancel{r_j}}, \; h_1^{r_j(w_0 + j \cdot w_1)} \cdot h_2^{r_j u_j} \cdot h_3^{\cancel{r_j u_j}} \end{array} \right\}_{j \in [n]} \right)$$

where $\mathbf{u} \leftarrow_{\mathrm{R}} \mathbb{Z}_N^{\ell'-1}$ and $r_j \leftarrow_{\mathrm{R}} \mathbb{Z}_N$.

Here E, P, SF means "expanded", "pesudo", "semi-functional", respectively.

Games. We describe the game sequence in detail. For each following claim, we omit its formal proof but provide a proof sketch instead.

Game$_0$. The real security game (cf. Sect. 2.2) where keys and ciphertext are normal.

Game$_0'$. Identical to Game$_0$ except that all keys and the challenge ciphertext are E-normal. We claim that Game$_0 \approx_c$ Game$_{0'}$. This follows from the entropy expansion lemma (see Lemma 2). In the reduction, on input

$$\left\{ \begin{array}{l} \mathsf{aux} : g_1, g_1^w, g_1^{w_0}, g_1^{w_1} \\ \mathsf{ct} : C_0, \{C_{1,j}, C_{2,j}\}_{j \in [n]} \\ \mathsf{sk} : \{K_{0,j}, K_{1,j}, K_{2,j}\}_{j \in [n]} \end{array} \right\},$$

we select a random generator h_{123} of H_N, sample $\alpha \leftarrow_{\mathrm{R}} \mathbb{Z}_N$, $\mathbf{u}_\kappa \leftarrow_{\mathrm{R}} \mathbb{Z}_N^{\ell'-1}$, $\tilde{r}_{j,\kappa} \leftarrow_{\mathrm{R}} \mathbb{Z}_N$ for $j \in [n]$ and $\kappa \in [Q]$, and simulate the game with

$$\left\{ \begin{array}{l} \mathsf{mpk} : \mathsf{aux}, \; e(g_1, h_{123})^\alpha \\ \mathsf{ct_{x^*}} : \{C_0, C_{1,j}, C_{2,j}\}_{j:x_j^*=1}, e(C_0, h_{123}^\alpha) \cdot m_b \\ \mathsf{sk_M^\kappa} : \{ h_{123}^{\mathbf{M}_j\left(\begin{smallmatrix}\alpha\\\mathbf{u}_\kappa\end{smallmatrix}\right)} \cdot K_{0,j}^{\tilde{r}_{j,\kappa}}, K_{1,j}^{\tilde{r}_{j,\kappa}}, K_{2,j}^{\tilde{r}_{j,\kappa}} \}_{j \in [n]} \end{array} \right\}.$$

Game$_i$. Identical to Game$_{0'}$ except that the first $i - 1$ keys and the challenge ciphertext is SF. We claim that Game$_{0'} \approx_c$ Game$_1$. This follows from the $\mathsf{SD}_{p_2 \mapsto p_2 p_3}^{G_N}$ assumption, which asserts that

$$\left(g_2^s, \{g_2^{s_j}\}_{j \in [n]} \right) \approx_c \left(g_2^s \cdot \boxed{g_3^s}, \{g_2^{s_j} \cdot \boxed{g_3^{s_j}}\}_{j \in [n]} \right) \text{ given } g_1, h_1, h_2.$$

In the reduction, we sample $w, w_0, w_1, v_j, u_j \leftarrow_{\mathrm{R}} \mathbb{Z}_N, h_{123} \leftarrow_{\mathrm{R}} H_N, \alpha \leftarrow_{\mathrm{R}} \mathbb{Z}_N$ and simulate $\mathsf{mpk}, \mathsf{sk_M^\kappa}$ honestly. To show that Game$_i \approx_c$ Game$_{i+1}$, we will require another sequence of sub-games.

Game$_{i,1}$. Identical to Game$_i$ except that the i'th key is P-normal. We claim that Game$_i \approx_c$ Game$_{i,1}$. This follows from $\mathsf{SD}_{p_2 \mapsto p_2 p_3}^{H_N}$ assumption which asserts that

$$\{h_2^{r_j}\}_{j \in [n]} \approx_c \{h_2^{r_j} \cdot h_3^{r_j}\}_{j \in [n]} \text{ given } g_1, g_{23}, h_1, h_2, h_3$$

In the reduction, we sample $w, w_0, w_1, v_j, u_j, \alpha, \hat{\alpha} \leftarrow_{\mathrm{R}} \mathbb{Z}_N$ and select a random generator h_{123} of H_N, and simulate $\mathsf{mpk}, \mathsf{ct}, \{\mathsf{sk_M^\kappa}\}_{\kappa \neq i}$ honestly.

$\mathsf{Game}_{i,2}$. Identical to Game_i except that the i'th key is P-SF. We claim that $\overline{\mathsf{Game}_{i,1}} \equiv \mathsf{Game}_{i,2}$. This follows from Lemma 1 in Sect. 2 which ensures that for any $\mathbf{x}$ that does not satisfy $\mathbf{M}$,

$$
\begin{array}{c}
\overbrace{(\ h_2, \{h_2^{v_j}\}_{j\in[n]}, \alpha, \hat{\alpha};}^{\kappa\text{'th sk},\ \kappa \neq i}\ \overbrace{\{g_2, g_2^{v_j}, g_3, g_3^{v_j}\}_{j:x_j=1};}^{\mathsf{SF\ ct}}\ \overbrace{\{h_{123}^{\mathbf{M}_j(\overset{\alpha}{\mathbf{u}})} \cdot h_3^{r_j v_j}, h_3^{r_j}\}_{j\in[n]}\ )}^{\text{P-normal }i\text{'th sk}} \\[2ex]
\equiv (\ h_2, \{h_2^{v_j}\}_{j\in[n]}, \alpha, \hat{\alpha};\ \{g_2, g_2^{v_j}, g_3, g_3^{v_j}\}_{j:x_j=1};\ \underbrace{\{h_{123}^{\mathbf{M}_j(\overset{\alpha}{\mathbf{u}})} \cdot \boxed{h_3^{\mathbf{M}_j(\overset{\hat{\alpha}}{\mathbf{0}})}} \cdot h_3^{r_j v_j}, h_3^{r_j}\}_{j\in[n]}\ )}_{\text{P-SF }i\text{'th sk}}
\end{array}
$$

where $v_j \leftarrow_R \mathbb{Z}_N$ and $\mathbf{u} \leftarrow_R \mathbb{Z}_N^{\ell'-1}$, and for all α, $\hat{\alpha}$, and $r_j \neq 0 \bmod p_3$. It is straight-forward to compute the remaining terms in mpk, the challenge cipher-text and the Q secret keys by sampling $g_1, w, w_0, w_1, u_j, s, s_j$ ourselves.

$\mathsf{Game}_{i,3}$. Identical to Game_i except that the i'th key is SF. We claim that $\overline{\mathsf{Game}_{i,2}} \approx_c \mathsf{Game}_{i,3}$. The proof is completely analogous to that of $\mathsf{Game}_i \approx_c \mathsf{Game}_{i,1}$. Furthermore, observe that $\mathsf{Game}_{i,3}$ is actually identical to Game_{i+1}.

Game_{Final}. Identical to Game_{Q+1} except that the challenge ciphertext is a SF one for a random message in G_T. We claim that $\mathsf{Game}_{Q+1} \equiv \mathsf{Game}_{Final}$. This follows from the fact that

$$
(\ \overbrace{e(g_1, h_{123}^\alpha)}^{\mathsf{mpk}},\ \overbrace{h_{123}^\alpha \cdot h_3^{\hat{\alpha}}}^{\mathsf{SF\ sk}},\ \overbrace{e(g_{123}^s, h_{123}^\alpha)}^{\mathsf{SF\ ct}}\ ) \equiv (\ e(g_1, h_{123}^\alpha),\ h_{123}^\alpha,\ e(g_{123}^s, h_{123}^\alpha \cdot \boxed{h_3^{\hat{\alpha}}})\ )
$$

where g_{123}, h_{123} and h_3 are respective random generators of G_N, H_N and H_{p_3}, $\alpha, \hat{\alpha} \leftarrow_R \mathbb{Z}_N$. The message m_b is statistically hidden by $e(g_{123}^s, h_3^{\hat{\alpha}})$. In Game_{Final}, the view of the adversary is statistically independent of the challenge bit b. Hence, $\mathsf{Adv}_{Final} = 0$.

5 Simulating Composite-Order Groups in Prime-Order Groups

We build upon and extend the previous framework of Chen et al. [6,11] for simulating composite-order groups in prime-order ones. We provide prime-order analogues of the static assumptions $\mathrm{SD}_{p_1 \mapsto p_1 p_2}^{G_N}, \mathrm{DDH}_{p_1}^{H_N}$ used in the previous sections. Moreover, we show that these prime-order analogues follow from the standard k-Linear assumption (and more generally, the MDDH assumption [9]) in prime-order bilinear groups.

Additional Notation. Let $\mathbf{A}$ be a matrix over $\mathbb{Z}_p$. We use $\mathsf{span}(\mathbf{A})$ to denote the column span of $\mathbf{A}$, and we use $\mathsf{span}^\ell(\mathbf{A})$ to denote matrices of width ℓ where each column lies in $\mathsf{span}(\mathbf{A})$; this means $\mathbf{M} \leftarrow_R \mathsf{span}^\ell(\mathbf{A})$ is a random matrix of width ℓ where each column is chosen uniformly from $\mathsf{span}(\mathbf{A})$. We use $\mathsf{basis}(\mathbf{A})$

to denote a basis of span($\mathbf{A}$), and we use ($\mathbf{A}_1 \mid \mathbf{A}_2$) to denote the concatenation of matrices $\mathbf{A}_1, \mathbf{A}_2$. If $\mathbf{A}$ is a m-by-n matrix with $m > n$, we use $\overline{\mathbf{A}}$ to denote the sub-matrix consisting of the first n rows and $\underline{\mathbf{A}}$ the sub-matrix with remaining $m - n$ rows. We let $\mathbf{I}_n$ be the n-by-n identity matrix and $\mathbf{0}$ be a zero matrix whose size will be clear from the context.

5.1 Prime-Order Groups and Matrix Diffie-Hellman Assumptions

A generator $\mathcal{G}$ takes as input a security parameter λ and outputs a description $\mathbb{G} := (p, G_1, G_2, G_T, e)$, where p is a prime of $\Theta(\lambda)$ bits, G_1, G_2 and G_T are cyclic groups of order p, and $e : G_1 \times G_2 \to G_T$ is a non-degenerate bilinear map. We require that the group operations in G_1, G_2 and G_T as well the bilinear map e are computable in deterministic polynomial time with respect to λ. Let $g_1 \in G_1$, $g_2 \in G_2$ and $g_T = e(g_1, g_2) \in G_T$ be the respective generators. We employ the *implicit representation* of group elements: for a matrix $\mathbf{M}$ over $\mathbb{Z}_p$, we define $[\mathbf{M}]_1 := g_1^{\mathbf{M}}, [\mathbf{M}]_2 := g_2^{\mathbf{M}}, [\mathbf{M}]_T := g_T^{\mathbf{M}}$, where exponentiation is carried out component-wise. Also, given $[\mathbf{A}]_1, [\mathbf{B}]_2$, we let $e([\mathbf{A}]_1, [\mathbf{B}]_2) = [\mathbf{AB}]_T$.

We define the matrix Diffie-Hellman (MDDH) assumption on G_1 [9]:

Assumption 3 (MDDH$_{k,\ell}^m$ Assumption). *Let $\ell > k \geq 1$ and $m \geq 1$. We say that the MDDH$_{k,\ell}^m$ assumption holds if for all PPT adversaries $\mathcal{A}$, the following advantage function is negligible in λ.*

$$\mathsf{Adv}_{\mathcal{A}}^{\mathrm{MDDH}_{k,\ell}^m}(\lambda) := \big| \Pr[\mathcal{A}(\mathbb{G}, [\mathbf{M}]_1, [\mathbf{MS}]_1) = 1] - \Pr[\mathcal{A}(\mathbb{G}, [\mathbf{M}]_1, [\mathbf{U}]_1) = 1] \big|$$

where $\mathbf{M} \leftarrow_{\mathrm{R}} \mathbb{Z}_p^{\ell \times k}$, $\mathbf{S} \leftarrow_{\mathrm{R}} \mathbb{Z}_p^{k \times m}$ and $\mathbf{U} \leftarrow_{\mathrm{R}} \mathbb{Z}_p^{\ell \times m}$.

The MDDH assumption on G_2 can be defined in an analogous way. Escala *et al.* [9] showed that

$$k\text{-Lin} \Rightarrow \mathrm{MDDH}_{k,k+1}^1 \Rightarrow \mathrm{MDDH}_{k,\ell}^m \ \forall \ell > k, m \geq 1$$

with a tight security reduction. Henceforth, we will use MDDH$_k$ to denote MDDH$_{k,k+1}^1$.

5.2 Basis Structure

We want to simulate composite-order groups whose order is the product of three primes. Fix parameters $\ell_1, \ell_2, \ell_3, \ell_W \geq 1$. Pick random

$$\mathbf{A}_1 \leftarrow_{\mathrm{R}} \mathbb{Z}_p^{\ell \times \ell_1}, \mathbf{A}_2 \leftarrow_{\mathrm{R}} \mathbb{Z}_p^{\ell \times \ell_2}, \mathbf{A}_3 \leftarrow_{\mathrm{R}} \mathbb{Z}_p^{\ell \times \ell_3}$$

where $\ell := \ell_1 + \ell_2 + \ell_3$. Let $(\mathbf{A}_1^{\|} \mid \mathbf{A}_2^{\|} \mid \mathbf{A}_3^{\|})^{\top}$ denote the inverse of $(\mathbf{A}_1 \mid \mathbf{A}_2 \mid \mathbf{A}_3)$, so that $\mathbf{A}_i^{\top} \mathbf{A}_i^{\|} = \mathbf{I}$ (known as *non-degeneracy*) and $\mathbf{A}_i^{\top} \mathbf{A}_j^{\|} = \mathbf{0}$ if $i \neq j$ (known as *orthogonality*), as depicted in Fig. 5. This generalizes the constructions in [10,11] where $\ell_1 = \ell_2 = \ell_3 = k$.

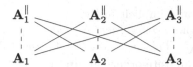

Fig. 5. Basis relations. Solid lines mean orthogonal, dashed lines mean non-degeneracy. Similar relations hold in composite-order groups with (g_1, g_2, g_3) in place of $(\mathbf{A}_1, \mathbf{A}_2, \mathbf{A}_3)$ and (h_1, h_2, h_3) in place of $(\mathbf{A}_1^{\|}, \mathbf{A}_2^{\|}, \mathbf{A}_3^{\|})$.

Correspondence. We have the following correspondence with composite-order groups:

$$g_i \mapsto [\mathbf{A}_i]_1, \qquad g_i^s \mapsto [\mathbf{A}_i \mathbf{s}]_1$$
$$w \in \mathbb{Z}_N \mapsto \mathbf{W} \in \mathbb{Z}_p^{\ell \times \ell_w}, \quad g_i^w \mapsto [\mathbf{A}_i^\top \mathbf{W}]_1$$

The following statistical lemma is analogous to the Chinese Remainder Theorem, which tells us that $w \bmod p_2$ is uniformly random given g_1^w, g_3^w, where $w \leftarrow_{\mathrm{R}} \mathbb{Z}_N$:

Lemma 6 (statistical lemma). *With probability* $1 - 1/p$ *over* $\mathbf{A}_1, \mathbf{A}_2, \mathbf{A}_3$, $\mathbf{A}_1^{\|}, \mathbf{A}_2^{\|}, \mathbf{A}_3^{\|}$, *the following two distributions are statistically identical.*

$$\{ \mathbf{A}_1^\top \mathbf{W}, \mathbf{A}_3^\top \mathbf{W}, \boxed{\mathbf{W}} \} \quad and \quad \{ \mathbf{A}_1^\top \mathbf{W}, \mathbf{A}_3^\top \mathbf{W}, \boxed{\mathbf{W} + \mathbf{U}^{(2)}} \}$$

where $\mathbf{W} \leftarrow_{\mathrm{R}} \mathbb{Z}_p^{\ell \times \ell_w}$ *and* $\mathbf{U}^{(2)} \leftarrow_{\mathrm{R}} \mathsf{span}^{\ell_w}(\mathbf{A}_2^{\|})$.

5.3 Basic Assumptions

We first describe the *prime-order* $(\mathbf{A}_1 \mapsto \mathbf{A}_1, \mathbf{A}_2)$-*subgroup decision assumption*, denoted by $\mathrm{SD}_{\mathbf{A}_1 \mapsto \mathbf{A}_1, \mathbf{A}_2}^{G_1}$. This is analogous to the subgroup decision assumption in composite-order groups $\mathrm{SD}_{p_1 \mapsto p_1 p_2}^{G_N}$ which asserts that $G_{p_1} \approx_c G_{p_1 p_2}$ given h_1, h_3, h_{12} along with g_1, g_2, g_3. By symmetry, we can permute the indices for $\mathbf{A}_1, \mathbf{A}_2, \mathbf{A}_3$.

Lemma 7 ($\mathrm{MDDH}_{\ell_1, \ell_1 + \ell_2} \Rightarrow \mathrm{SD}_{\mathbf{A}_1 \mapsto \mathbf{A}_1, \mathbf{A}_2}^{G_1}$). *Under the* $\mathrm{MDDH}_{\ell_1, \ell_1 + \ell_2}$ *assumption in* G_1, *there exists an efficient sampler outputting random* $([\mathbf{A}_1]_1, [\mathbf{A}_2]_1, [\mathbf{A}_3]_1)$ *(as described in Sect. 5.2) along with base* $\mathsf{basis}(\mathbf{A}_1^{\|})$, $\mathsf{basis}(\mathbf{A}_3^{\|})$, $\mathsf{basis}(\mathbf{A}_1^{\|}, \mathbf{A}_2^{\|})$ *(of arbitrary choice) such that the following advantage function is negligible in* λ.

$$\mathsf{Adv}_{\mathcal{A}}^{\mathrm{SD}_{\mathbf{A}_1 \mapsto \mathbf{A}_1, \mathbf{A}_2}^{G_1}}(\lambda) := \big| \Pr[\mathcal{A}(D, [\mathbf{t}_0]_1) = 1] - \Pr[\mathcal{A}(D, [\mathbf{t}_1]_1) = 1] \big|$$

where

$$D := ([\mathbf{A}_1]_1, [\mathbf{A}_2]_1, [\mathbf{A}_3]_1, \mathsf{basis}(\mathbf{A}_1^{\|}), \mathsf{basis}(\mathbf{A}_3^{\|}), \mathsf{basis}(\mathbf{A}_1^{\|}, \mathbf{A}_2^{\|})),$$
$$\mathbf{t}_0 \leftarrow_{\mathrm{R}} \mathsf{span}(\mathbf{A}_1), \quad \mathbf{t}_1 \leftarrow_{\mathrm{R}} \mathsf{span}(\mathbf{A}_1, \mathbf{A}_2).$$

Similar statements were also implicit in [10, 11].

We then formalize the *prime-order* $\mathbf{A}_1$-*subgroup Diffie-Hellman assumption*, denoted by $\mathrm{DDH}_{\mathbf{A}_1}^{G_2}$. This is analogous to the subgroup Diffie-Hellman assumption in the composite-order group $\mathrm{DDH}_{p_1}^{H_N}$ which ensures that $\{h_1^{r_j w}, h_1^{r_j}\}_{j \in [Q]} \approx_c \{h_1^{r_j w} \cdot h_1^{u_j}, h_1^{r_j}\}_{j \in [Q]}$ given $g_1, g_2, g_3, h_1, h_2, h_3$ for $Q = \mathrm{poly}(\lambda)$. One can permute the indices for $\mathbf{A}_1, \mathbf{A}_2, \mathbf{A}_3$.

Lemma 8 ($\mathrm{MDDH}_{\ell_W, Q}^{\ell_1} \Rightarrow \mathrm{DDH}_{\mathbf{A}_1}^{G_2}$). *Fix $Q = \mathrm{poly}(\lambda)$ with $Q > \ell_W \geq 1$. Under the $\mathrm{MDDH}_{\ell_W, Q}^{\ell_1}$ assumption in G_2, the following advantage function is negligible in λ*

$$\mathsf{Adv}_{\mathcal{A}}^{\mathrm{DDH}_{\mathbf{A}_1}^{G_2}}(\lambda) := \big| \Pr[\mathcal{A}(D, T_0) = 1] - \Pr[\mathcal{A}(D, T_1) = 1] \big|$$

where

$$D := (\; \mathbf{A}_1, \mathbf{A}_2, \mathbf{A}_3, \mathbf{A}_1^{\|}, \mathbf{A}_2^{\|}, \mathbf{A}_3^{\|};\; \mathbf{A}_2^{\top}\mathbf{W}, \mathbf{A}_3^{\top}\mathbf{W}\;),$$
$$T_0 := ([\mathbf{WD}]_2, [\mathbf{D}]_2), \quad T_1 := ([\mathbf{WD} + \mathbf{R}^{(1)}]_2, [\mathbf{D}]_2),$$

and $\mathbf{W} \leftarrow_{\mathrm{R}} \mathbb{Z}_p^{\ell \times \ell_W}$, $\mathbf{D} \leftarrow_{\mathrm{R}} \mathbb{Z}_p^{\ell_W \times Q}$, $\mathbf{R}^{(1)} \leftarrow_{\mathrm{R}} \mathrm{span}^Q(\mathbf{A}_1^{\|})$.

6   KP-ABE for Monotone Span Programs in Prime-Order Groups

In this section, we present our adaptively secure, unbounded KP-ABE for monotone span programs programs based on the k-Lin assumption in prime-order groups.

6.1   Construction

Setup$(1^\lambda, 1^n)$: On input $(1^\lambda, 1^n)$, sample $\mathbf{A}_1 \leftarrow_{\mathrm{R}} \mathbb{Z}_p^{(2k+1) \times k}, \mathbf{B} \leftarrow_{\mathrm{R}} \mathbb{Z}_p^{(k+1) \times k}$ and

$$\mathbf{W}, \mathbf{W}_0, \mathbf{W}_1 \leftarrow_{\mathrm{R}} \mathbb{Z}_p^{(2k+1) \times (k+1)}, \quad \mathbf{k} \leftarrow_{\mathrm{R}} \mathbb{Z}_p^{2k+1}$$

and output the master public and secret key pair

$$\mathsf{mpk} := (\; [\mathbf{A}_1^{\top}, \mathbf{A}_1^{\top}\mathbf{W}, \mathbf{A}_1^{\top}\mathbf{W}_0, \mathbf{A}_1^{\top}\mathbf{W}_1]_1,\; e([\mathbf{A}_1^{\top}]_1, [\mathbf{k}]_2)\;)$$
$$\mathsf{msk} := (\; \mathbf{k},\; \mathbf{B},\; \mathbf{W},\; \mathbf{W}_0,\; \mathbf{W}_1\;).$$

Enc$(\mathsf{mpk}, \mathbf{x}, m)$: On input an attribute vector $\mathbf{x} := (x_1, \ldots, x_n) \in \{0, 1\}^n$ and $m \in G_T$, pick $\mathbf{c}, \mathbf{c}_j \leftarrow_{\mathrm{R}} \mathrm{span}(\mathbf{A}_1)$ for all $j \in [n]$ and output

$$\mathsf{ct}_{\mathbf{x}} := \begin{pmatrix} C_0 := [\mathbf{c}^{\top}]_1, \\ \{\; C_{1,j} := [\mathbf{c}^{\top}\mathbf{W} + \mathbf{c}_j^{\top}(\mathbf{W}_0 + j \cdot \mathbf{W}_1)]_1, C_{2,j} := [\mathbf{c}_j^{\top}]_1 \;\}_{j:x_j=1}, \\ C := e([\mathbf{c}^{\top}]_1, [\mathbf{k}]_2) \cdot m \end{pmatrix}$$
$$\in G_1^{2k+1} \times (G_1^{k+1} \times G_1^{2k+1})^n \times G_T.$$

KeyGen(mpk, msk, $\mathbf{M}$): On input a monotone span program $\mathbf{M} \in \mathbb{Z}_p^{n \times \ell'}$, pick $\mathbf{K}' \leftarrow_R \mathbb{Z}_p^{(2k+1) \times (\ell'-1)}$, $\mathbf{d}_j \leftarrow_R \text{span}(\mathbf{B})$ for all $j \in [n]$, and output

$$\text{sk}_{\mathbf{M}} := \left(\left\{ \begin{array}{c} K_{0,j} := [(\mathbf{k} \| \mathbf{K}') \mathbf{M}_j^\top + \mathbf{W} \mathbf{d}_j]_2, \ K_{1,j} := [\mathbf{d}_j]_2, \\ K_{2,j} := [(\mathbf{W}_0 + j \cdot \mathbf{W}_1) \mathbf{d}_j]_2 \end{array} \right\}_{j \in [n]} \right)$$
$$\in (G_2^{2k+1} \times G_2^{k+1} \times G_2^{2k+1})^n.$$

Dec(mpk, $\text{sk}_{\mathbf{M}}$, $\text{ct}_{\mathbf{x}}$): If $\mathbf{x}$ satisfies $\mathbf{M}$, compute $\omega_1, \ldots, \omega_n \in \mathbb{Z}_p$ such that

$$\sum_{j:x_j=1} \omega_j \mathbf{M}_j = \mathbf{1}.$$

Then, compute

$$K \leftarrow \prod_{j:x_j=1} \left(e(C_0, K_{0,j}) \cdot e(C_{1,j}, K_{1,j})^{-1} \cdot e(C_{2,j}, K_{2,j}) \right)^{\omega_j},$$

and recover the message as $m \leftarrow C/K \in G_T$.

The proof of correctness is direct and we omit it here.

6.2 Entropy Expansion Lemma in Prime-Order Groups

With $\mathbf{A}_1, \mathbf{A}_2, \mathbf{A}_3, \mathbf{A}_1^\|, \mathbf{A}_2^\|, \mathbf{A}_3^\|$ defined as in Sect. 5.2, our prime-order entropy expansion lemma is stated as follows. The proof is analogous to that for composite-order entropy expansion lemma (Lemma 2) shown in Sect. 3.2.

Lemma 9 (prime-order entropy expansion lemma). *Suppose $\ell_1, \ell_3, \ell_W \geq k$. Then, under the $MDDH_k$ assumption, we have*

$$\left\{ \begin{array}{l} \text{aux} : [\mathbf{A}_1^\top]_1, [\mathbf{A}_1^\top \mathbf{W}]_1, [\mathbf{A}_1^\top \mathbf{W}_0]_1, [\mathbf{A}_1^\top \mathbf{W}_1]_1 \\ \text{ct} : [\mathbf{c}^\top]_1, \ \left\{ [\mathbf{c}^\top \mathbf{W} + \mathbf{c}_j^\top (\mathbf{W}_0 + j \cdot \mathbf{W}_1)]_1, \ [\mathbf{c}_j^\top]_1 \right\}_{j \in [n]} \\ \text{sk} : \left\{ [\mathbf{W} \mathbf{D}_j]_2, \ [\mathbf{D}_j]_2, \ [(\mathbf{W}_0 + j \cdot \mathbf{W}_1) \mathbf{D}_j]_2 \right\}_{j \in [n]} \end{array} \right\}$$

$$\approx_c \left\{ \begin{array}{l} \text{aux} : [\mathbf{A}_1^\top]_1, [\mathbf{A}_1^\top \mathbf{W}]_1, [\mathbf{A}_1^\top \mathbf{W}_0]_1, [\mathbf{A}_1^\top \mathbf{W}_1]_1 \\ \text{ct} : [\boxed{\mathbf{c}}^\top]_1, \ \left\{ [\boxed{\mathbf{c}}^\top (\mathbf{W} + \boxed{\mathbf{V}_j^{(2)}}) + \boxed{\mathbf{c}_j}^\top (\mathbf{W}_0 + j \cdot \mathbf{W}_1 + \boxed{\mathbf{U}_j^{(2)}})]_1, \ [\boxed{\mathbf{c}_j}^\top]_1 \right\}_{j \in [n]} \\ \text{sk} : \left\{ [(\mathbf{W} + \boxed{\mathbf{V}_j^{(2)}}) \mathbf{D}_j]_2, \ [\mathbf{D}_j]_2, \ [(\mathbf{W}_0 + j \cdot \mathbf{W}_1 + \boxed{\mathbf{U}_j^{(2)}}) \mathbf{D}_j]_2 \right\}_{j \in [n]} \end{array} \right\}$$

where $\mathbf{W}, \mathbf{W}_0, \mathbf{W}_1 \leftarrow_R \mathbb{Z}_p^{\ell \times \ell_W}$, $\mathbf{V}_j^{(2)}, \mathbf{U}_j^{(2)} \leftarrow_R \text{span}^{\ell_W}(\mathbf{A}_2^\|)$, $\mathbf{D}_j \leftarrow_R \mathbb{Z}_p^{\ell_W \times \ell_W}$, and $\mathbf{c}, \mathbf{c}_j \leftarrow_R \text{span}(\mathbf{A}_1)$ in the left distribution while $\mathbf{c}, \mathbf{c}_j \leftarrow_R \text{span}(\mathbf{A}_1, \mathbf{A}_2)$ in the right distribution. Concretely, the distinguishing advantage $\text{Adv}_{\mathcal{A}}^{\text{ExpLem}}(\lambda)$ is at most

$$\text{Adv}_{\mathcal{B}}^{\text{SD}_{\mathbf{A}_1 \mapsto \mathbf{A}_1, \mathbf{A}_2}^{G_1}}(\lambda) + \text{Adv}_{\mathcal{B}_0}^{\text{DDH}_{\mathbf{A}_2}^{G_2}}(\lambda) + n \cdot \left(\text{Adv}_{\mathcal{B}_1}^{\text{SD}_{\mathbf{A}_1 \mapsto \mathbf{A}_1, \mathbf{A}_3}^{G_1}}(\lambda) + \text{Adv}_{\mathcal{B}_2}^{\text{DDH}_{\mathbf{A}_3}^{G_2}}(\lambda) \right.$$
$$\left. + \text{Adv}_{\mathcal{B}_4}^{\text{SD}_{\mathbf{A}_3 \mapsto \mathbf{A}_3, \mathbf{A}_2}^{G_1}}(\lambda) + \text{Adv}_{\mathcal{B}_6}^{\text{DDH}_{\mathbf{A}_3}^{G_2}}(\lambda) + \text{Adv}_{\mathcal{B}_7}^{\text{SD}_{\mathbf{A}_1 \mapsto \mathbf{A}_1, \mathbf{A}_3}^{G_1}}(\lambda) \right) + \text{Adv}_{\mathcal{B}_8}^{\text{DDH}_{\mathbf{A}_2}^{G_2}}(\lambda)$$

where $\text{Time}(\mathcal{B})$, $\text{Time}(\mathcal{B}_0)$, $\text{Time}(\mathcal{B}_1)$, $\text{Time}(\mathcal{B}_2)$, $\text{Time}(\mathcal{B}_4)$, $\text{Time}(\mathcal{B}_6)$, $\text{Time}(\mathcal{B}_7)$, $\text{Time}(\mathcal{B}_8) \approx \text{Time}(\mathcal{A})$.

Remark 1 (Differences from overview in Sect. 1.3). We stated our prime-order expansion lemma for general ℓ_1, ℓ_2, ℓ_3; for our KP-ABE, it suffices to set $(\ell_1, \ell_2, \ell_3) = (k, 1, k)$. Compared to the informal statement (8) in Sect. 1.3, we use $\mathbf{A}_2 \in \mathbb{Z}_p^{2k+1}$ instead of $\mathbf{A}_2 \in \mathbb{Z}_p^{(2k+1) \times k}$, and we introduced extra $\mathbf{A}_2$-components corresponding to $\mathbf{A}_2^\top \mathbf{W}, \mathbf{A}_2^\top (\mathbf{W}_0 + j \cdot \mathbf{W}_1)$ in ct on the RHS. We have $\mathbf{D}_j$ in place of $\mathbf{B}r_j$ in the above statement, though we will introduce $\mathbf{B}$ later on in Lemma 10. We also picked $\mathbf{D}_j$ to be square matrices to enable random self-reducibility of the sk-terms. Finally, $\mathbf{V}_j^{(2)}, \mathbf{U}_j^{(2)}$ correspond to $\mathbf{V}_j, \mathbf{U}_j$ in the informal statement, and in particular, we have $\mathbf{A}_1^\top \mathbf{V}_j^{(2)} = \mathbf{A}_1^\top \mathbf{U}_j^{(2)} = \mathbf{0}$.

6.3 Proof of Security

We prove the following theorem:

Theorem 2. *Under the $MDDH_k$ assumption in prime-order groups (cf. Sect. 5.1), the unbounded KP-ABE scheme for monotone span programs described in this Section (cf. Sect. 6.1) is adaptively secure (cf. Sect. 2.2).*

Bilinear Entropy Expansion Lemma, Revisited. With the additional basis $\mathbf{B} \in \mathbb{Z}_p^{(k+1) \times k}$, we need a variant of the entropy expansion lemma in Lemma 9 with $(\ell_1, \ell_2, \ell_3, \ell_W) = (k, 1, k, k+1)$ where the columns of $\mathbf{D}_j$ are drawn from $\mathrm{span}(\mathbf{B})$ instead of $\mathbb{Z}_p^{k+1}$ (see Lemma 10).

Lemma 10 (prime-order entropy expansion lemma, revisited). *Pick* $(\mathbf{A}_1, \mathbf{a}_2, \mathbf{A}_3) \leftarrow_R \mathbb{Z}_p^{(2k+1) \times (k+1)} \times \mathbb{Z}_p^{2k+1} \times \mathbb{Z}_p^{(2k+1) \times (k+1)}$ *and define its dual* $(\mathbf{A}_1^\parallel, \mathbf{a}_2^\parallel, \mathbf{A}_3^\parallel)$ *as in Sect. 5.2. With* $\mathbf{B} \leftarrow_R \mathbb{Z}_p^{(k+1) \times k}$, *we have*

$$
\left\{
\begin{aligned}
\mathrm{aux} &: [\mathbf{A}_1^\top]_1, [\mathbf{A}_1^\top \mathbf{W}]_1, [\mathbf{A}_1^\top \mathbf{W}_0]_1, [\mathbf{A}_1^\top \mathbf{W}_1]_1 \\
\mathrm{ct} &: [\mathbf{c}^\top]_1, \{[\mathbf{c}^\top \mathbf{W} + \mathbf{c}_j^\top(\mathbf{W}_0 + j \cdot \mathbf{W}_1)]_1, [\mathbf{c}_j^\top]_1\}_{j \in [n]} \\
\mathrm{sk} &: \{[\mathbf{W}\mathbf{D}_j]_2, [\mathbf{D}_j]_2, [(\mathbf{W}_0 + j \cdot \mathbf{W}_1)\mathbf{D}_j]_2\}_{j \in [n]}
\end{aligned}
\right\}
$$

$$
\approx_c
\left\{
\begin{aligned}
\mathrm{aux} &: [\mathbf{A}_1^\top]_1, [\mathbf{A}_1^\top \mathbf{W}]_1, [\mathbf{A}_1^\top \mathbf{W}_0]_1, [\mathbf{A}_1^\top \mathbf{W}_1]_1 \\
\mathrm{ct} &: [\boxed{\mathbf{c}}^\top]_1, \{[\boxed{\mathbf{c}}^\top(\mathbf{W} + \boxed{\mathbf{V}_j^{(2)}}) + \boxed{\mathbf{c}_j}^\top(\mathbf{W}_0 + j \cdot \mathbf{W}_1 + \boxed{\mathbf{U}_j^{(2)}})]_1, [\boxed{\mathbf{c}_j}^\top]_1\}_{j \in [n]} \\
\mathrm{sk} &: \{[(\mathbf{W} + \boxed{\mathbf{V}_j^{(2)}})\mathbf{D}_j]_2, [\mathbf{D}_j]_2, [(\mathbf{W}_0 + j \cdot \mathbf{W}_1 + \boxed{\mathbf{U}_j^{(2)}})\mathbf{D}_j]_2\}_{j \in [n]}
\end{aligned}
\right\}
$$

where $\mathbf{W}, \mathbf{W}_0, \mathbf{W}_1 \leftarrow_R \mathbb{Z}_p^{(2k+1) \times (k+1)}$, $\mathbf{V}_j^{(2)}, \mathbf{U}_j^{(2)} \leftarrow_R \mathrm{span}^{k+1}(\mathbf{a}_2^\parallel)$, $\mathbf{D}_j \leftarrow_R \mathrm{span}^{k+1}(\mathbf{B})$, *and* $\mathbf{c}, \mathbf{c}_j \leftarrow_R \mathrm{span}(\mathbf{A}_1)$ *in the left distribution while* $\mathbf{c}, \mathbf{c}_j \leftarrow_R \mathrm{span}(\mathbf{A}_1, \mathbf{a}_2)$ *in the right distribution. We let* $\mathsf{Adv}_{\mathcal{A}}^{\mathrm{EXPLEMREV}}(\lambda)$ *denote the distinguishing advantage.*

We claim that the lemma follows from the basic entropy expansion lemma (Lemma 9) and the $MDDH_k$ assumption, which tells us that

$$
\{[\mathbf{D}_j \leftarrow_R \mathbb{Z}_p^{(k+1) \times (k+1)}]_2\}_{j \in [n]} \approx_c \{[\mathbf{D}_j \leftarrow_R \mathrm{span}^{k+1}(\mathbf{B})]_2\}_{j \in [n]}.
$$

Concretely, for all $\mathcal{A}$, we can construct $\mathcal{B}_0$ and $\mathcal{B}_1$ with $\mathsf{Time}(\mathcal{B}_0), \mathsf{Time}(\mathcal{B}_1) \approx \mathsf{Time}(\mathcal{A})$ such that

$$\mathsf{Adv}_{\mathcal{A}}^{\mathrm{ExpLemRev}}(\lambda) \leq \mathsf{Adv}_{\mathcal{B}_0}^{\mathrm{ExpLem}}(\lambda) + 2 \cdot \mathsf{Adv}_{\mathcal{B}_1}^{\mathrm{MDDH}_{k,k+1}^{n(k+1)}}(\lambda).$$

The proof is straight-forward by demonstrating that the left (resp. right) distributions in Lemmas 9 and 10 are indistinguishable under the MDDH_k assumption and then applying Lemma 9. In the reduction, we sample $\mathbf{W}, \mathbf{W}_0, \mathbf{W}_1 \leftarrow_{\mathrm{R}} \mathbb{Z}_p^{(2k+1) \times (k+1)}$ (and $\mathbf{V}_j^{(2)}, \mathbf{U}_j^{(2)} \leftarrow_{\mathrm{R}} \mathsf{span}^{k+1}(\mathbf{a}_2^{\parallel})$ for the right distributions) and simulate $\mathsf{aux}, \mathsf{ct}$ honestly.

Main Technical Lemma. We prove the following technical lemma. As with the composite-order scheme in Sect. 4, we first apply the new entropy expansion lemma in Lemma 10 and obtain a copy of the CGW KP-ABE (variant-thereof) in the $\mathbf{a}_2$-subspace. We may then carry out the classic dual system methodology used for establishing adaptive security of the CGW KP-ABE.

Lemma 11. *For any adversary $\mathcal{A}$ that makes at most Q key queries against the unbounded KP-ABE scheme, there exist adversaries $\mathcal{B}_0, \mathcal{B}_1, \mathcal{B}_2$ such that:*

$$\mathsf{Adv}_{\mathcal{A}}^{\mathrm{ABE}}(\lambda) \leq \mathsf{Adv}_{\mathcal{B}_0}^{\mathrm{ExpLemRev}}(\lambda) + Q \cdot \mathsf{Adv}_{\mathcal{B}_1}^{\mathrm{MDDH}_{k,k+1}^{n}}(\lambda) + Q \cdot \mathsf{Adv}_{\mathcal{B}_2}^{\mathrm{MDDH}_{k,k+1}^{n}}(\lambda) + O(1/p).$$

where $\mathsf{Time}(\mathcal{B}_0), \mathsf{Time}(\mathcal{B}_1), \mathsf{Time}(\mathcal{B}_2) \approx \mathsf{Time}(\mathcal{A})$. In particular, we achieve security loss $O(n + Q)$ based on the MDDH_k assumption.

The proof follows the same game sequence as shown in Sect. 4.2 except that the adversary is given an E-normal challenge ciphertext instead of a SF one in $\mathsf{Game}_i, \mathsf{Game}_{i,1}, \mathsf{Game}_{i,2}, \mathsf{Game}_{i,3}$ (in fact, we do not need to define SF ciphertexts) and the auxiliary distributions are defined as follows.

Auxiliary Distributions. We define various forms of ciphertext (of message m under attribute vector $\mathbf{x}$):

- Normal: Generated by Enc; in particular, $\mathbf{c}, \mathbf{c}_j \leftarrow_{\mathrm{R}} \mathsf{span}(\mathbf{A}_1)$.
- E-normal: Same as a normal ciphertext except that $\mathbf{c}, \mathbf{c}_j \leftarrow_{\mathrm{R}} \mathsf{span}(\mathbf{A}_1, \mathbf{a}_2)$ and we use the substitution:

$$\mathbf{W} \mapsto \mathbf{W} + \mathbf{V}_j^{(2)} \quad \text{in } j\text{'th component}$$

$$\text{and} \quad \mathbf{W}_0 + j \cdot \mathbf{W}_1 \mapsto \mathbf{W}_0 + j \cdot \mathbf{W}_1 + \mathbf{U}_j^{(2)} \tag{11}$$

where $\mathbf{U}_j^{(2)}, \mathbf{V}_j^{(2)} \leftarrow_{\mathrm{R}} \mathsf{span}^{k+1}(\mathbf{a}_2^{\parallel})$.

Then we pick $\alpha \leftarrow_{\mathrm{R}} \mathbb{Z}_p$ and define various forms of key (for span program $\mathbf{M}$):

- Normal: Generated by KeyGen.
- E-normal: Same as a normal key except that we use the same substitution as in (11).

- P-normal: Sample $\mathbf{d}_j \leftarrow_R \mathbb{Z}_p^{k+1}$ in an E-normal key.
- P-SF: Replace $\mathbf{k}$ with $\mathbf{k} + \alpha \mathbf{a}_2^{\parallel}$ in a P-normal key.
- SF: Sample $\mathbf{d}_j \leftarrow_R \mathsf{span}(\mathbf{B})$ in a P-SF key.

Acknowledgments. We greatly thank Katsuyuki Takashima for insightful and constructive feedback. We also thank all anonymous reviewers for their helpful comments.

References

1. Agrawal, S., Chase, M.: FAME: fast attribute-based message encryption. In: ACM CCS (2017)
2. Attrapadung, N.: Dual system encryption via doubly selective security: framework, fully secure functional encryption for regular languages, and more. In: Nguyen, P.Q., Oswald, E. (eds.) EUROCRYPT 2014. LNCS, vol. 8441, pp. 557–577. Springer, Heidelberg (2014). https://doi.org/10.1007/978-3-642-55220-5_31
3. Attrapadung, N.: Dual system encryption framework in prime-order groups via computational pair encodings. In: Cheon, J.H., Takagi, T. (eds.) ASIACRYPT 2016, Part II. LNCS, vol. 10032, pp. 591–623. Springer, Heidelberg (2016). https://doi.org/10.1007/978-3-662-53890-6_20
4. Beimel, A.: Secure Schemes for Secret Sharing and Key Distribution. Ph.D., Technion - Israel Institute of Technology (1996)
5. Brakerski, Z., Vaikuntanathan, V.: Circuit-ABE from LWE: unbounded attributes and semi-adaptive security. In: Robshaw, M., Katz, J. (eds.) CRYPTO 2016, Part III. LNCS, vol. 9816, pp. 363–384. Springer, Heidelberg (2016). https://doi.org/10.1007/978-3-662-53015-3_13
6. Chen, J., Gay, R., Wee, H.: Improved dual system ABE in prime-order groups via predicate encodings. In: Oswald, E., Fischlin, M. (eds.) EUROCRYPT 2015, Part II. LNCS, vol. 9057, pp. 595–624. Springer, Heidelberg (2015). https://doi.org/10.1007/978-3-662-46803-6_20
7. Chen, J., Wee, H.: Fully, (Almost) tightly secure IBE and dual system groups. In: Canetti, R., Garay, J.A. (eds.) CRYPTO 2013, Part II. LNCS, vol. 8043, pp. 435–460. Springer, Heidelberg (2013). https://doi.org/10.1007/978-3-642-40084-1_25
8. Chen, J., Wee, H.: Semi-adaptive attribute-based encryption and improved delegation for boolean formula. In: Abdalla, M., De Prisco, R. (eds.) SCN 2014. LNCS, vol. 8642, pp. 277–297. Springer, Cham (2014). https://doi.org/10.1007/978-3-319-10879-7_16
9. Escala, A., Herold, G., Kiltz, E., Ràfols, C., Villar, J.: An algebraic framework for Diffie-Hellman assumptions. In: Canetti, R., Garay, J.A. (eds.) CRYPTO 2013, Part II. LNCS, vol. 8043, pp. 129–147. Springer, Heidelberg (2013). https://doi.org/10.1007/978-3-642-40084-1_8
10. Gay, R., Hofheinz, D., Kiltz, E., Wee, H.: Tightly CCA-secure encryption without pairings. In: Fischlin, M., Coron, J.-S. (eds.) EUROCRYPT 2016, Part I. LNCS, vol. 9665, pp. 1–27. Springer, Heidelberg (2016). https://doi.org/10.1007/978-3-662-49890-3_1
11. Gong, J., Dong, X., Chen, J., Cao, Z.: Efficient IBE with tight reduction to standard assumption in the multi-challenge setting. In: Cheon, J.H., Takagi, T. (eds.) ASIACRYPT 2016, Part II. LNCS, vol. 10032, pp. 624–654. Springer, Heidelberg (2016). https://doi.org/10.1007/978-3-662-53890-6_21

12. Goyal, R., Koppula, V., Waters, B.: Semi-adaptive security and bundling functionalities made generic and easy. In: Hirt, M., Smith, A. (eds.) TCC 2016, Part II. LNCS, vol. 9986, pp. 361–388. Springer, Heidelberg (2016). https://doi.org/10.1007/978-3-662-53644-5_14

13. Goyal, V., Pandey, O., Sahai, A., Waters, B.: Attribute-based encryption for fine-grained access control of encrypted data. In: Juels, A., Wright, R.N., Vimercati, S. (eds.) ACM CCS 2006, pp. 89–98. ACM Press, October/November 2006. Available as Cryptology ePrint Archive Report 2006/309

14. Hofheinz, D., Jager, T., Knapp, E.: Waters signatures with optimal security reduction. In: Fischlin, M., Buchmann, J., Manulis, M. (eds.) PKC 2012. LNCS, vol. 7293, pp. 66–83. Springer, Heidelberg (2012). https://doi.org/10.1007/978-3-642-30057-8_5

15. Ishai, Y., Wee, H.: Partial garbling schemes and their applications. In: Esparza, J., Fraigniaud, P., Husfeldt, T., Koutsoupias, E. (eds.) ICALP 2014, Part I. LNCS, vol. 8572, pp. 650–662. Springer, Heidelberg (2014). https://doi.org/10.1007/978-3-662-43948-7_54

16. Karchmer, M., Wigderson, A.: On span programs. In: Structure in Complexity Theory Conference, pp. 102–111 (1993)

17. Kowalczyk, L., Lewko, A.B.: Bilinear entropy expansion from the decisional linear assumption. In: Gennaro, R., Robshaw, M. (eds.) CRYPTO 2015, Part II. LNCS, vol. 9216, pp. 524–541. Springer, Heidelberg (2015). https://doi.org/10.1007/978-3-662-48000-7_26

18. Lewko, A.: Tools for simulating features of composite order bilinear groups in the prime order setting. In: Pointcheval, D., Johansson, T. (eds.) EUROCRYPT 2012. LNCS, vol. 7237, pp. 318–335. Springer, Heidelberg (2012). https://doi.org/10.1007/978-3-642-29011-4_20

19. Lewko, A., Okamoto, T., Sahai, A., Takashima, K., Waters, B.: Fully secure functional encryption: attribute-based encryption and (hierarchical) inner product encryption. In: Gilbert, H. (ed.) EUROCRYPT 2010. LNCS, vol. 6110, pp. 62–91. Springer, Heidelberg (2010). https://doi.org/10.1007/978-3-642-13190-5_4

20. Lewko, A., Waters, B.: New techniques for dual system encryption and fully secure HIBE with short ciphertexts. In: Micciancio, D. (ed.) TCC 2010. LNCS, vol. 5978, pp. 455–479. Springer, Heidelberg (2010). https://doi.org/10.1007/978-3-642-11799-2_27

21. Lewko, A., Waters, B.: Unbounded HIBE and attribute-based encryption. In: Paterson, K.G. (ed.) EUROCRYPT 2011. LNCS, vol. 6632, pp. 547–567. Springer, Heidelberg (2011). https://doi.org/10.1007/978-3-642-20465-4_30

22. Okamoto, T., Takashima, K.: Fully secure functional encryption with general relations from the decisional linear assumption. In: Rabin, T. (ed.) CRYPTO 2010. LNCS, vol. 6223, pp. 191–208. Springer, Heidelberg (2010). https://doi.org/10.1007/978-3-642-14623-7_11

23. Okamoto, T., Takashima, K.: Fully secure unbounded inner-product and attribute-based encryption. In: Wang, X., Sako, K. (eds.) ASIACRYPT 2012. LNCS, vol. 7658, pp. 349–366. Springer, Heidelberg (2012). https://doi.org/10.1007/978-3-642-34961-4_22

24. Rouselakis, Y., Waters, B.: Practical constructions and new proof methods for large universe attribute-based encryption. In: Sadeghi, A.-R., Gligor, V.D., Yung, M. (eds.) ACM CCS 2013, pp. 463–474. ACM Press, November 2013

25. Sahai, A., Waters, B.: Fuzzy identity-based encryption. In: Cramer, R. (ed.) EUROCRYPT 2005. LNCS, vol. 3494, pp. 457–473. Springer, Heidelberg (2005). https://doi.org/10.1007/11426639_27

26. Waters, B.: Dual system encryption: realizing fully secure IBE and HIBE under simple assumptions. In: Halevi, S. (ed.) CRYPTO 2009. LNCS, vol. 5677, pp. 619–636. Springer, Heidelberg (2009). https://doi.org/10.1007/978-3-642-03356-8_36
27. Wee, H.: Dual system encryption via predicate encodings. In: Lindell, Y. (ed.) TCC 2014. LNCS, vol. 8349, pp. 616–637. Springer, Heidelberg (2014). https://doi.org/10.1007/978-3-642-54242-8_26

Anonymous IBE, Leakage Resilience and Circular Security from New Assumptions

Zvika Brakerski[1]([envelope]), Alex Lombardi[2], Gil Segev[3], and Vinod Vaikuntanathan[2]

[1] Weizmann Institute of Science, Rehovot, Israel
zvika.brakerski@weizmann.ac.il
[2] MIT, Cambridge, USA
[3] Hebrew University of Jerusalem, Jerusalem, Israel

Abstract. In *anonymous* identity-based encryption (IBE), ciphertexts not only hide their corresponding messages, but also their target identity. We construct an anonymous IBE scheme based on the Computational Diffie-Hellman (CDH) assumption in general groups (and thus, as a special case, based on the hardness of factoring Blum integers).

Our approach extends and refines the recent tree-based approach of Cho et al. (CRYPTO '17) and Döttling and Garg (CRYPTO '17). Whereas the tools underlying their approach do not seem to provide any form of anonymity, we introduce two new building blocks which we utilize for achieving anonymity: *blind garbled circuits* (which we construct based on any one-way function), and *blind batch encryption* (which we construct based on CDH).

We then further demonstrate the applicability of our newly-developed tools by showing that batch encryption implies a public-key encryption scheme that is both resilient to leakage of a $(1-o(1))$-fraction of its secret key, and KDM secure (or circular secure) with respect to all linear functions of its secret key (which, in turn, is known to imply KDM security for bounded-size circuits). These yield the first high-rate leakage-resilient encryption scheme and the first KDM-secure encryption scheme based on the CDH or Factoring assumptions.

Finally, relying on our techniques we also construct a batch encryption scheme based on the hardness of the Learning Parity with Noise (LPN) problem, albeit with very small noise rate $\Omega(\log^2(n)/n)$. Although this batch encryption scheme is not blind, we show that it still implies standard (i.e., non-anonymous) IBE, leakage resilience and KDM security. IBE and high-rate leakage resilience were not previously known from LPN, even with extremely low noise.

1 Introduction

Identity Based Encryption (IBE) is a form of public key encryption where a user's public key is just his name. Specifically, an authority holding a master secret key

---

The full version of this paper [BLSV17] is available on ePrint.

© International Association for Cryptologic Research 2018
J. B. Nielsen and V. Rijmen (Eds.): EUROCRYPT 2018, LNCS 10820, pp. 535–564, 2018.
https://doi.org/10.1007/978-3-319-78381-9_20

msk can generate individual secret keys for users sk_{id} according to their identity id, and encryption is performed using a master public key (mpk) and the identity of the recipient. The notion of IBE was proposed by Shamir [Sha84] but first realized only over 15 years later [BF03, Coc01]. Aside from the obvious utility of using IBE for the purpose for which it was intended, it has also proved to be a useful building block to achieve other cryptographic tasks (e.g. chosen-ciphertext secure encryption [BCHK07]) as well as an inspiration for defining more expressive forms of encryption schemes with *access control*. Most generally, the latter refers to schemes where multiple secret keys can be generated, but each key can only recover encrypted information if some predefined condition holds. The most natural generalization is to attribute based encryption (ABE) [SW05, GPSW06] where secret keys sk_f correspond to policies f, and encryptions are with respect to attributes x, so that the message is decryptable only if $f(x) = 1$. IBE is a special case where f is a point function (i.e. $f_a(x) = 1$ if and only if $x = a$).

Very recently, a beautiful work of Döttling and Garg [DG17a] proposed a new *tree based* approach for IBE and showed that it implies a candidate IBE scheme from the computational Diffie-Hellman assumption (CDH), which was previously unknown. Their main building blocks were garbled circuits and a special form of encryption called Chameleon Encryption. In a follow-up work [DG17b] they showed that tree based constructions can also be used to amplify the properties of IBE schemes.

An important variant of IBE is one where it is also required that a cipher-text for recipient id does not expose id to an unauthorized decryptor. This property is called *anonymity*. Anonymous IBE is quite useful, e.g. for search-able encryption [BCOP04], and analogously to the connection between IBE and ABE, anonymous IBE is a special case of *attribute hiding* ABE (e.g., as in [KSW08]). The latter has raised much interest recently in the cryptographic lit-erature due to its connection to *functional encryption schemes*. Anonymous IBE schemes can be constructed from pairings [BCOP04, ABC+08, BW06, Gen06], lattices [GPV08, ABB10, CHKP12] and quadratic residuosity [BGH07] (the last one in the random oracle model).

The [DG17a, DG17b] constructions are not anonymous for a fundamental reason. Their construction is based on an implicit exponential-size prefix tree representing the entire space of identities. The encryption operation considers a path from the root to the leaf representing the target id and constructs a sequence of garbled circuits, each respective to a node along this path. At decryption time, the garbled circuits are evaluated from root to leaf, where the output of each garbled circuit is used to generate the input labels for the next garbled circuit along the path. Therefore, if one tries to decrypt a ciphertext intended for id using a key for id′, the decryption process will succeed up to the node of divergence between sk and sk′, at which point the $sk_{id'}$ decryptor will not be able to decode the labels that correspond to the next garbled circuit. Thus, this process necessarily reveals the common prefix of id and (a known) id′.

1.1   Our Results

In this work, we present new primitives and techniques showing how to get significantly more mileage out of the tree-based approach. First and most importantly, we build on the tree-based approach using new tools that we call *blind batch encryption* and *blind garbled circuits* to construct anonymous IBE schemes. Secondly, we show that our building blocks can be constructed from assumptions not previously known to imply IBE at all, in particular, the learning parity with noise (LPN) assumption with extremely small noise. Finally, we show that our building blocks can be used to achieve cryptographic capabilities that are apparently unrelated to IBE, namely leakage resilience and KDM security. We elaborate on all of these contributions below.

Batch Encryption and New Constructions of IBE. The recent work of Döttling and Garg [DG17b] show an amplification between notions of identity based encryption. Namely, they show how to go from any selective IBE scheme to a fully secure IBE scheme. We notice that their construction can be repurposed to do something very different. Namely, we show how to start from an IBE scheme which only supports *polynomially many identities* but with short master public key, and construct a full-fledged IBE scheme. In particular, the scheme should support $T = T(\lambda)$ identities with a master public key of size $S = S(\lambda) = T^{1-\epsilon} \cdot \mathsf{poly}(\lambda)$ for some constant $\epsilon > 0$ and a fixed polynomial poly; we call this a weakly compact IBE scheme. We remind the reader that non-compact IBE schemes, namely ones that support T identities and have a master public key that grows linearly with T, in fact follow quite easily from any public-key encryption scheme (see, e.g., [DKXY02]).

Weakly compact IBE turns out to be easier to construct using the techniques of [DG17a], and in particular it does not require the full power of their Chameleon Encryption. We show that it is sufficient to start from a building block that we call *batch encryption*. In particular, whereas Chameleon Encryption is required to have a trapdoor, a batch encryption scheme has no trapdoors. Indeed, looking ahead, we remark that this feature of requiring no trapdoors is what enables our IBE construction from the extremely-low-noise LPN assumption. The batch encryption definition takes after the laconic oblivious transfer primitive presented by Cho, Döttling, Garg, Gupta, Miao and Polychroniadou [CDG+17] (a definition that preceded Chameleon Encryption).

A batch encryption scheme is a public key encryption scheme in which key generation is a projection (i.e. the key generation algorithm takes the secret key as input and outputs a shorter string as the public key). For secret keys of length n, a batch encryption scheme encrypts an array of $n \times 2$ messages at a time. At decryption, only one out of each pair of messages is recovered, depending on the value of the respective secret key bit. We require that we can instantiate the scheme for any n without increasing the length of the public key. Indeed, batch encryption is very similar to laconic oblivious transfer [CDG+17] and the two are essentially existentially equivalent. The formal definition varies slightly in that laconic OT can more efficiently handle situations where only a subset of the n message pairs are encrypted. Another formal difference is that the laconic OT formulation allows for a randomized receiver message, however since receiver privacy is not a requirement for this primitive this is not actually needed and therefore the analogous component in batch encryption is deterministic. The formulation of batch encryption is more useful for our applications, but our constructions can be seen as simply constructing laconic OT.

We show that batch encryption implies weakly compact IBE (as defined above) and that weakly compact IBE can be bootstrapped to a full-fledged IBE scheme.

Batch Encryption from CDH and Extremely-Low-Noise LPN. Batch encryption can be constructed from CDH, using the methods of [DG17a]; it can also be constructed from the Learning with Errors (LWE) assumption in a straightforward manner without using lattice trapdoors. Thus we observe that LWE-based IBE does not require lattice trapdoors, even though they are used by all previous constructions. We note that the resulting IBE scheme is greatly inefficient, quite probably much less efficient than a trapdoor based construction, however the conceptual difference here could be of interest.

We take an additional step forward and show that even the learning parity with noise (LPN) assumption is sufficient to instantiate batch encryption, although we must rely on LPN with very extreme parameters. The LPN assumption with a constant noise rate implies one-way functions; with a noise rate of $1/\sqrt{n}$ (where n is the dimension of the LPN secret), it implies public-key encryption [Ale11]; and with the extremely low noise rate of $\log^2 n/n$, it implies collision-resistant hash functions [BLVW17, YZW+17]. The latter parameter setting is insecure against quasi-polynomial adversaries, but given the state of the art in algorithms for LPN, presumably secure against polynomial-time adversaries. Indeed, it is ill advised to base cryptographic hardness on the gap between polynomial time adversaries and quasi-polynomial time hardness and we see this result mainly as proof of concept showing that batch encryption can be based on structures that were not considered to imply IBE so far.

The Blinding Technique and Anonymous IBE. Our main contribution is a construction of anonymous IBE from the CDH assumption.

To construct anonymous IBE we present techniques that allow us to walk down the identity-space tree at decryption time *blindly*. Namely, in a way that does not reveal to the decryptor whether they are on the correct path until the very end of the process. This allows us to overcome the aforementioned basic obstacle. We present a variety of blind primitives that help us in achieving this goal.

The first building block we introduce is *blind garbled circuits*. Recall that a standard circuit garbling scheme takes a circuit C as input, and outputs a garbled version of the circuit $\widehat{C}$ together with pairs of labels $\mathsf{lab}_{i,b}$ for the input wires. Given $\widehat{C}, \mathsf{lab}_{i,x_i}$, the value $C(x)$ can be computed. For security, there is a simulator that takes $y = C(x)$ and produces a garbled circuit and a set of input labels that are indistinguishable from the original. We augment this definition with a *blindness* property, requiring that the simulated garbled circuit and labels are *completely uniform* when starting with a completely uniform y that is unknown to the distinguisher (indeed, the latter condition is necessary since an attempt to evaluate the simulated garbled circuit should output y). We show that blind garbled circuits can be constructed by properly instantiating the "point-and-permute" construction [BMR90, Rog91], based on any one way function. Interestingly, as far as we know, the point-and-permute construction has been used to achieve more efficient garbled circuits, but has never been used to achieve stronger security properties.

We then introduce *blind batch encryption*, which is the blind version of the aforementioned batch encryption primitive. The use of batch encryption in IBE constructions is as a way to encrypt labels for a garbled circuit so that only one label per input wire can be decrypted (i.e. the one corresponding to the batch encryption secret key). Blind batch encryption is a "blindness preserving" counterpart for blind garbled circuits as follows. We require that if a random message is encrypted using a blind batch encryption scheme, then the resulting ciphertext is completely random as well.[1] This combines very naturally with a blind garbling scheme: if we batch encrypt labels to a blind garbled circuit with a random output, then by simulation security this is indistinguishable from encrypting random labels that are independent of the garbled circuit. Therefore, we are guaranteed that the batch ciphertext itself is random as well. At a very high level, this will allow us to propagate the randomness (blindness) property along the leaf-root path in the tree, and avoid revealing any information via partial decryption.

We show that blind batch encryption can be constructed based on CDH by introducing a modification to the CDH based Chameleon Encryption construction from [DG17a]. Unfortunately, our construction based on extremely low noise LPN is not blind.

We apply these building blocks to anonymize the aforementioned IBE construction from batch encryption. We present a blindness property for IBE that is analogous to the one for batch encryption, requiring that an encryption of random message is indistinguishable from random *even to a user who is permitted to decrypt it*. We show that this notion implies anonymous IBE, and furthermore, the construction of full-fledged IBE from a weakly compact scheme, and a construction of the weakly compact scheme from a batch encryption scheme both preserve blindness (if we use blind garbled circuits). In fact, formally, to avoid redundancy we only present the reduction in the blind setting, and the non-blind variant follows as a special case.

We find it intriguing that even though we only require anonymous IBE at the end, we have to go through the (apparently stronger) primitive of blind IBE. Roughly speaking, the difference is that anonymous IBE only requires hiding of the identities in settings where the adversary cannot decrypt (namely, he only obtains secret keys for identities id different from either of the challenge identities id_0 and id_1) while blind IBE requires hiding of the identities even in settings where the adversary can decrypt. Morally, we think of this as the difference

---

[1] We actually allow a slight relaxation of this condition.

between weak attribute-hiding and strong attribute-hiding in predicate encryption (although the details are somewhat different). We also note that weakly compact anonymous IBE can be constructed generically from any weakly compact IBE scheme. Thus, had we been able to bootstrap from a weakly compact anonymous IBE scheme into a full-fledged anonymous IBE, we would have a generic construction of anonymous IBE scheme from any IBE scheme.

Batch Encryption Implies Leakage Resilience and KDM Security. We show that the utility of batch encryption schemes go beyond IBE, thus expanding [CDG+17] who showed a variety of applications of laconic OT, mostly in the context of multi-party computation. We show that batch encryption naturally gives rise to a public key encryption scheme with desirable properties such as resilience to high rate $(1 - o(1))$ key leakage [AGV09, NS12] and security for key dependent messages [BRS02] (KDM, also known as circular security). This allows us to present constructions from assumptions such as CDH, Factoring and extremely-low-noise LPN that were not known before [AGV09, NS12, BHHO08, ACPS09, BG10, HLWW16]. Note that from [CDG+17] it was not even clear that the (nearly) equivalent notion of laconic OT even implies plain public key encryption (without assuming "receiver privacy"; with receiver privacy, we know that any 2 message OT implies PKE). This further strengthens our impression that batch encryption is a notion worthy of further exploration.

The basic idea is quite straightforward. Recall that a batch encryption scheme encrypts an array of $n \times 2$ bits, and decryption only recovers one out of two pairs. Therefore, if the secret key is $\mathbf{x} \in \{0,1\}^n$ and the encrypted message is $\mathbf{M} \in \{0,1\}^{n \times 2}$, then the decrypted message is equal to $m = \sum_i (M_{i,0}(1 \oplus x_i) \oplus M_{i,1} x_i) = \sum_i M_{i,0} \oplus \sum_i (M_{i,1} \oplus M_{i,0}) x_i$. Denote $\alpha_0 = \sum_i M_{i,0}$, $\alpha_i = M_{i,1} \oplus M_{i,0}$. Note that it is sufficient that one out of each pair $M_{i,0}, M_{i,1}$ is random to make all $\{\alpha_i\}_{i>0}$ completely random, this property will be useful for us. To encrypt, we will n-out-of-n secret share our message $m = \sum_i \mu_i$ and set $M_{i,0} = M_{i,1} = \mu_i$. Decryption follows by decrypting the batch ciphertext and reconstructing m. For security, we notice that the batch security means that we can convert one out of each pair $M_{i,0}, M_{i,1}$ to random (this will be unnoticed even to a distinguisher who has the key x). At this point, we recall that x is in fact information theoretically unknown to the adversary who only sees the projected public key (recall that the projection key generation function is shrinking). Thus the value $\sum_i \alpha_i x_i$ extracts from the remaining entropy in x and is statistically close to uniform (indeed one has to prove that there is no additional usable information in the ciphertext other than the output message m). This argument naturally extends to leakage resilience, since we can allow additional leakage on x so long as sufficient information remains to allow for extraction. It appears that security against computationally (sub-exponentially) hard to invert unbounded length leakage ("auxiliary input resilience" [DGK+10]) should follow in a similar manner, however we do not provide a proof.

For KDM security, we notice that for any linear function of x of the form $\alpha_0 \oplus \sum_i \alpha_i x_i$ the above shows how to simulate a ciphertext that decrypts to this message (in fact, how to sample a random such ciphertext). Indeed this ciphertext is not honestly generated but we can show that it is indistinguishable from one. This is the basis for KDM security. We recall that as shown in [BHHI10, App11], KDM security with respect to linear functions can be amplified to KDM security for bounded polynomial functions of the key. Interestingly, this amplification approach also involves batch encrypting labels for a garbled circuit. For lack of space, we refer the reader to our full version [BLSV17] for the details on the leakage-resilience and KDM security constructions.

1.2 Concurrent Work

In concurrent and independent work, Döttling, Garg, Hajiabadi, and Masny [DGHM18] construct (non-anonymous) IBE from a subexponential assumption on constant-noise LPN (similar in spirit to our assumption). In another concurrent and independent work, Kitagawa and Tanaka [KT18] construct *KDM-secure IBE* from any IBE along with any KDM-secure secret key encryption scheme. Since we construct both IBE and KDM-secure PKE from Batch Encryption, combining [KT18] with our work yields KDM-secure IBE from Batch Encryption (and hence constructions from CDH/Factoring and from $\log^2(n)/n$-noise LPN).

1.3 Our Techniques

The rest of the paper is organized as follows. In Sect. 3, we define the notion of (blind) batch encryption and construct it from the CDH assumption. We also provide a construction of the (non-blind) batch encryption from the extremely low noise LPN assumption. We then introduce the notion of blind garbled circuits and construct it in Sect. 4. Then, in Sect. 5, we show how to use (blind) batch encryption to construct a weakly compact (blind) IBE scheme. In Sect. 6, we bootstrap the weakly compact (blind) IBE scheme into a full-fledged (blind) IBE scheme. The applications to leakage resilience and KDM security, as well as many details in the following sections, are deferred to the full version of our paper [BLSV17].

We first provide an overview of the last step of our anonymous IBE construction, namely our bootstrapping theorem for blind IBE, and then the construction of weakly compact IBE from batch encryption.

Bootstrapping Blind IBE. We start with bootstrapping a regular IBE scheme, and then describe the additional techniques required to handle blindness.

Suppose we have a blind IBE scheme $\mathcal{WIBE}$ that supports $T = T(\lambda)$ identities and has a master public key whose size is $S = S(\lambda) = T^{1-\epsilon} \cdot p(\lambda)$ for some absolute constant $\epsilon > 0$ and a fixed polynomial p. To keep our exposition simple, assume that the ciphertexts in this scheme are truly pseudorandom. We remark that without the restriction on the master public key length, there are generic ways of constructing such schemes from any public-key encryption scheme, resulting in master public key of length $O(T \cdot \lambda)$; see, e.g., [DKXY02].

The key leverage we have in $\mathcal{WIBE}$ is that the master public key grows *sublinearly* with the number of identities the scheme supports.

We will show how to construct another (blind) IBE scheme $\mathcal{WIBE}'$ that supports $2T$ identities without growing the master public key at all. This will not be enough by itself to prove the full bootstrapping theorem by induction because the ciphertext and secret key sizes grow significantly in the transformation. Nevertheless, all of the necessary ideas for the full bootstrapping theorem are in this toy example already.

We start by picking T to be sufficiently large so that the size of the master public key $T^{1-\epsilon} \cdot p(\lambda)$ is at most $T/4$. The master public key of $\mathcal{WIBE}'$ is a single master public key of $\mathcal{WIBE}$; we will denote it by $\mathsf{mpk}^{(\epsilon)}$ and associate it with the root of a depth-2 tree with branching factor 2 in the first level and T in the second. We will also pick two other master public keys $\mathsf{mpk}^{(0)}$ and $\mathsf{mpk}^{(1)}$, but *will not publish it* as part of the $\mathcal{WIBE}'$ master public key. The master secret key in $\mathcal{WIBE}'$ will, however, include $\mathsf{msk}^{(\epsilon)}$ as well as $\mathsf{mpk}^{(i)}, \mathsf{msk}^{(i)}$.

The two questions we address next is (a) how to encrypt a message m for an identity $\mathsf{id} \| \mathsf{id}'$ where $\mathsf{id} \in \{0, 1\}$ and $\mathsf{id}' \in \{0, \ldots, T - 1\}$ and (b) how to generate identity secret keys.

Let us address the question of secret keys first. The secret key for an identity $\mathsf{id} \| \mathsf{id}'$ where $\mathsf{id} \in \{0, 1\}$ and $\mathsf{id}' \in \{0, \ldots, T - 1\}$ will include as part of it $\mathsf{sk}_{\mathsf{id}'}^{(\mathsf{id})}$, namely the secret key for the identity id' generated with respect to the master public key $\mathsf{mpk}^{(\mathsf{id})}$. Thus, it makes sense to encrypt a message m under the identity $\mathsf{id} \| \mathsf{id}'$ by encrypting it with respect to the identity id' under the master public key $\mathsf{mpk}^{(\mathsf{id})}$. If the encryptor could do this, decryption indeed works and we are done! However, the big problem here is that the encryptor does not know $\mathsf{mpk}^{(0)}$ or $\mathsf{mpk}^{(1)}$. How can the encryptor generate a ciphertext without knowing the master public key?

It is here that we use the technique of *deferred encryption* similarly to [GKW16] and the aforementioned [DG17a]. That is, instead of having to generate an encryption of m under an unknown master public key, the encryptor simply constructs a circuit $C[m, \mathsf{id}']$ which has the message m and the identity id' hardcoded. The circuit $C[m, \mathsf{id}']$, on input an mpk, produces an encryption of m under mpk with identity id'. (The circuit also has the encryption randomness r hardcoded).

The encryptor now does two things. It first garbles this circuit to produce $\widehat{C}$, the garbled circuit, together with $2S$ labels $\mathsf{lab}_{i,b}$ for $i \in [S]$ and $b \in \{0, 1\}$. It then encrypts each label $\mathsf{lab}_{i,b}$ using the identity (id, i, b) under the master public key $\mathsf{mpk}^{(\epsilon)}$. It is here that we use compactness of $\mathcal{WIBE}$ in a crucial way: since $\mathcal{WIBE}$ can support $T > 4S$ identities, it can indeed be used to encrypt these labels.

The identity secret key for $\mathsf{id} \| \mathsf{id}'$ now contains two things. As before, it contains the secret key for the identity id' under the master public key $\mathsf{mpk}^{(\mathsf{id})}$. It also contains the secret keys for the S identities $(\mathsf{id}, i, \mathsf{mpk}^{(\mathsf{id})}[i])$ under the master public key $\mathsf{mpk}^{(\epsilon)}$.

Decryption proceeds by first using the secret keys for the S identities to unlock half the labels for the garbled circuit $\widehat{C}$, namely, the labels corresponding to the input $\mathsf{mpk}^{(\mathsf{id})}$. It then decodes the garbled circuit to produce an encryption of m with identity id' under the master public key $\mathsf{mpk}^{(\mathsf{id})}$. The first part of the secret key is now precisely what is necessary to decrypt and obtain the message m.

We first argue semantic security (IND-ID-CPA security), then show the barriers to achieving blindness/anonymity and how our new techniques overcome them. Let the challenge identity be $\mathsf{id}\|\mathsf{id}'$. A ciphertext of a message m under $\mathsf{id}\|\mathsf{id}'$ contains the garbled circuit $\widehat{C}$ and encryptions of the labels $L_{i,b}$ under identities (id, i, b) with respect to the master public key $\mathsf{mpk}^{(\epsilon)}$. Notice first that secret keys for identities that begin with the bit $(1 - \mathsf{id})$ are completely useless in unlocking any of the labels of the garbled circuit. Only secret keys for identities that begin with the bit id are useful. Even they can only ever unlock half the labels of the garbled circuit. Indeed, this is crucial since otherwise we will not be able to invoke the security of the garbled circuit at all!

The secret keys for identities that begin with the (matching) bit id unlock the garbled labels corresponding to the input $\mathsf{mpk}^{(\mathsf{id})}$. One now invokes the security of the garbled circuit which says that the only thing revealed by these labels together with the garbled circuit is the encryption of m under the identity id' generated with the master public key $\mathsf{mpk}^{(\mathsf{id})}$. Now, since the adversary never obtains the secret key for the challenge identity, she never gets the secret key for id' under $\mathsf{mpk}^{(\mathsf{id})}$. Thus, the semantic security of $\mathcal{WIBE}$ tells us that the message m remains hidden.

As described in the introduction, this construction does not lead to an anonymous IBE scheme. Indeed, given a ciphertext with respect to the identity $\mathsf{id}_1\|\mathsf{id}'_1$ and a secret key for $\mathsf{id}_2\|\mathsf{id}'_2 \neq \mathsf{id}_1\|\mathsf{id}'_1$, one can easily tell if $\mathsf{id}_1 = \mathsf{id}_2$ or not, simply by seeing if the first decryption step succeeds. Worse, it is unclear if the anonymity of the underlying $\mathcal{WIBE}$ scheme helps here at all. If $\mathsf{id}_1 = \mathsf{id}_2$, the secret keys are authorized to decrypt half the encrypted labels ("first level ciphertexts"), and if $\mathsf{id}_1 \neq \mathsf{id}_2$, the secret keys do not decrypt any of them. Thus, it seems at first glance that we are doomed: one can seemingly always recover the first bit of the identity in any tree-based scheme.

Our *key observation* is that even in the "partly-authorized case", the ciphertexts are encryptions of fresh random labels. (In reality, these labels do appear again in the garbled circuits; in the proof, this is handled by doing the hybrids in the reverse order from the current presentation where pseudorandomness at the leaves comes from the adversary not having the final secret key corresponding to the target identity.) Thus, if the $\mathcal{WIBE}$ scheme is *blind*, the adversary can still not tell the difference between whether she had an authorized key or not. In both cases, the output of the decryption is a bunch of uniformly random strings! Our troubles, unfortunately, do not stop there. The next line of defense, the garbled circuit, could also help the adversary distinguish whether she obtained the right labels in the first step or just random strings. Blindness again comes to the rescue: this time, we use our blind garbled circuits in conjunction with the fact that the output of the circuit we are garbling is actually pseudorandom.

This concludes a sketch of our toy construction and its security proof.

Of course, there was no reason a-priori to have only one level of garbled circuits. One can garble the "inner $\mathcal{WIBE}$" encryptions and do so for every level in the tree. The inputs to each such garbled circuit is a single master public key, so the input labels to this new garbled circuit will be no larger than the previous level's input labels. We can thus build an IBE scheme corresponding to a tree of any poly(λ) depth, allowing us to support exponentially many identities: a full IBE scheme. Of course, we cannot generate exponentially many $\mathcal{WIBE}$ master public keys (one for each node of the tree), but we can implicitly generate them using a PRF.

For full details on our bootstrapping theorem, see Sect. 6.

From Batch Encryption to Weakly Compact IBE. We now provide a high level overview of how to construct weakly compact IBE from batch encryption. Formally, we construct a scheme that supports any polynomial number T of identities with public key size λ. We focus on the vanilla (non-blind) variant as the blind one follows via a similar construction. We note that batch encryption schemes go hand-in-hand with garbled circuits (a connection that is extensively used in [CDG+17, DG17a]). Consider a batch encryption scheme with secret key x of length $n \gg \lambda$ and public key length λ. Then we can encrypt an array of $n \times 2$ elements, specifically we can encrypt labels for an n-input garbled circuit. The holder of the secret key will be able to evaluate said garbled circuit on the labels that correspond to his secret key. In other words, batch encryption allows us to specify a circuit $C : \{0,1\}^n \rightarrow \{0,1\}^m$ and generate a ciphertext that will reveal only $C(x)$, even to an adversary that holds the secret key.

Recall that the only requirement we want from the resulting IBE is short master public key. All other parameters can depend polynomially on the size of the identity space. We will therefore generate a sequence of T key pairs for a standard public key encryption scheme $(\mathsf{pke.pk}_1, \mathsf{pke.sk}_1), \ldots, (\mathsf{pke.pk}_T, \mathsf{pke.sk}_T)$. For simplicity assume $|\mathsf{pke.pk}_i| = \lambda$. Then we instantiate the batch encryption scheme with $n = T \cdot \lambda$ and generate a batch public key, a projection of $x = \mathsf{pke.pk}_1 \| \cdots \| \mathsf{pke.pk}_T$. The batch public key will serve as mpk of the weakly compact IBE scheme, and indeed its length is λ, independent of T.

To encrypt a ciphertext to target identity $\mathsf{id} \in [T]$, we generate a garbled circuit that expects as input a sequence of T public keys, and takes the id-th of them and uses it to encrypt the message. The IBE secret key for identity id will contain the entire sequence $x = \mathsf{pke.pk}_1 \| \cdots \| \mathsf{pke.pk}_T$, indeed in this case the batch encryption secret key is not secret at all! In addition, the IBE secret key for id will contain $\mathsf{pke.sk_{id}}$. Given a ciphertext, a decryptor will first use x to evaluate the garbled circuit and recover $C(x)$, which in this case is just a public-key encryption ciphertext with respect to $\mathsf{pke.pk_{id}}$. The next step is to just use $\mathsf{pke.sk_{id}}$ to decrypt this ciphertext and recover the message.

Security follows from the security of batch encryption (which conveniently applies also when the batch secret key x is known) and the security of the public key encryption scheme.

2   Preliminaries and Definitions

2.1   (Anonymous) Identity-Based Encryption

Definition 1 (Identity Based Encryption). *An identity based encryption (IBE) scheme consists of five PPT algorithms* (Params, Setup, Keygen, Enc, Dec) *with the following syntax.*

1. Params$(1^\lambda, 1^t)$ *takes as input the security parameter* 1^λ *and an identity length* 1^t. *It returns public parameters* pp *(which can be reused to generate multiple master public key/master secret key pairs).*
2. Setup(pp) *takes as input public parameters* pp *and returns a master public key* mpk *and master secret key* msk.
3. Keygen(pp, msk, id) *takes as input public parameters* pp *and the master secret key* msk. *It outputs a secret key* sk_{id} *associated to* id.
4. Enc(pp, mpk, id, m) *encrypts a message* m *to a specified identity* id. *It outputs a ciphertext* ct.
5. Dec(pp, sk, ct) *decrypts a ciphertext* ct *with secret key* sk, *outputting a plaintext message* m'.

We require that an IBE scheme satisfy the following two properties.

– *Correctness: with probability 1 over the randomness of* (Params, Setup, Keygen, Enc, Dec), *we have that* Dec(pp, sk_{id}, Enc(pp, mpk, id, m)) $= m$ *where* (mpk, msk) ← Setup(pp) *and* sk_{id} ← Keygen(msk, id).
– *IND-ID-CPA Security: a PPT adversary* $\mathcal{A}$ *cannot win the following security game with probability greater than* $\frac{1}{2} + \mathrm{negl}(\lambda)$:
   1. pp ← Params$(1^\lambda, 1^t)$
   2. (mpk, msk) ← Setup(pp)
   3. (id^*, m_0, m_1, st) ← $\mathcal{A}^{\mathsf{Keygen(pp, msk, \cdot)}}$(mpk)
   4. $b \xleftarrow{\$} \{0, 1\}$
   5. ct ← Enc(pp, mpk, id^*, m_b)
   6. b' ← $\mathcal{A}^{\mathsf{Keygen(pp, msk, \cdot)}}$(st, ct)
   7. $\mathcal{A}$ *wins if and only if* $b' = b$ *and* id^* *was never queried by* $\mathcal{A}$ *to its* Keygen *oracle.*

Definition 2 (Anonymous IBE). *An anonymous IBE scheme also has the syntax (*Params, Setup, Keygen, Enc, Dec*) of an IBE scheme. It satisfies the same correctness property as IBE, and has the following stronger notion of security:*

– *IND-ANON-ID-CPA Security: A PPT adversary* $\mathcal{A}$ *cannot with the following security game with probability greater than* $\frac{1}{2} + \mathrm{negl}(\lambda)$:
   1. pp ← Params$(1^\lambda, 1^t)$
   2. (mpk, msk) ← Setup(pp)
   3. $(id_0, id_1, m_0, m_1, st)$ ← $\mathcal{A}^{\mathsf{Keygen(pp, msk, \cdot)}}$(mpk)
   4. $b \xleftarrow{\$} \{0, 1\}$
   5. ct ← Enc(pp, mpk, id_b, m_b)
   6. b' ← $\mathcal{A}^{\mathsf{Keygen(pp, msk, \cdot)}}$(st, ct)
   7. $\mathcal{A}$ *wins if and only if* $b' = b$ *and* id_0, id_1 *were never queried by* $\mathcal{A}$ *to its* Keygen *oracle.*

2.2   Computational Diffie-Hellman (CDH)

Let g be an element of some group $\mathbb{G}$. We say that q is a ϵ-randomizer for g if the statistical distance between g^a for $a \leftarrow \mathbb{Z}_q$ and $h \leftarrow \langle g \rangle$ is at most ϵ. We note that any $q \geq \mathsf{ord}(g) \cdot \lceil 1/\epsilon \rceil$ is an ϵ-randomizer, so it is sufficient to have an upper bound on the order of g in order to compute a randomizer for any ϵ.

A (possibly randomized) group sampler is a ppt algorithm $\mathcal{G}$ that on input the security parameter outputs a tuple $(\mathbb{G}, g, q) \leftarrow \mathcal{G}(1^\lambda)$ which defines a $\mathbb{G}$ by providing a $\mathrm{poly}(\lambda)$-bit representation for group elements, and a polynomial time algorithm for computing the group operation and inversion (and thus also exponentiation), together with an element $g \in \mathbb{G}$ and a $\mathrm{negl}(\lambda)$-randomizer q for $\langle g \rangle$.

The Computational Diffie-Hellman (CDH) assumption with respect to $\mathcal{G}$, denoted $\mathsf{CDH}_\mathcal{G}$, is that for every ppt algorithm $\mathcal{A}$ it holds that

$$\mathrm{Adv}_{\mathsf{CDH}_\mathcal{G}}[\mathcal{A}](\lambda) = \Pr_{\substack{(\mathbb{G},g,q) \leftarrow \mathcal{G}(1^\lambda) \\ a_1, a_2 \leftarrow \mathbb{Z}_q}} [\mathcal{A}(1^\lambda, (\mathbb{G}, g, q), g^{a_1}, g^{a_2}) = g^{a_1 a_2}] = \mathrm{negl}(\lambda).$$

We sometimes omit the indication of $\mathcal{G}$ when it is clear from the context.

We note that there exists a randomized group sampler such that the hardness of factoring Blum integers reduces to the hardness of the CDH problem [Shm85, McC88, BBR99].

2.3   Learning Parity with Noise (LPN)

For all $n \in \mathbb{N}$, row vector $\mathbf{s} \in \{0,1\}^n$ and real value $\epsilon \in [0, 1/2]$, define a randomized oracle $A_{\mathbf{s},\epsilon}$ to be s.t. for every call to $A_{\mathbf{s},\epsilon}$, the oracle samples $\mathbf{a} \leftarrow \{0,1\}^n$, $e \leftarrow \mathsf{Ber}_\epsilon$ (where Ber is the Bernoulli distribution), and outputs $(\mathbf{a}, \mathbf{s} \cdot \mathbf{a} + e)$ where arithmetics are over the binary field. Note that $A_{\mathbf{s},1/2}$ outputs completely uniform entries for every call.

The Learning Parity with Noise assumption $\mathsf{LPN}_{n,\epsilon}$, for a polynomial function $n : \mathbb{N} \to \mathbb{N}$ and a function $\epsilon : \mathbb{N} \to [0, 1/2]$ is that for every ppt oracle algorithm $\mathcal{A}$ it holds that

$$\mathrm{Adv}_{\mathsf{LPN}_{n,\epsilon}}[\mathcal{A}](\lambda) = \left| \Pr_{\mathbf{s} \leftarrow \{0,1\}^n} [\mathcal{A}^{A_{\mathbf{s},\epsilon}}(1^\lambda)] - \Pr[\mathcal{A}^{A_{\mathbf{0},1/2}}(1^\lambda)] \right| = \mathrm{negl}(\lambda),$$

where $n = n(\lambda)$, $\epsilon = \epsilon(\lambda)$.

We note that if $\epsilon = \log n/n$ then LPN is solvable in polynomial time, but no polynomial time algorithm is known for $\epsilon = \Omega(\log^2 n/n)$.

The Collision Resistant Hash Family of [BLVW17]. It is shown in [BLVW17] how to create Collision Resistant Hash functions based on the hardness of $\mathsf{LPN}_{n,\epsilon}$ for any polynomial n, $\epsilon = \Omega(\log^2 n/n)$. Since this construction is the basis for our LPN-based batch encryption construction, let us elaborate a little on it here.

The key to the hash function is a random matrix $\mathbf{A} \in \{0,1\}^{n \times (2n^2/\log n)}$. To apply the hash function on an input $x \in \{0,1\}^{2n}$, they first preprocess it as follows. Interpret x as a collection of $2n/\log n$ blocks, each containing $\log n$ bits. Then interpret each block as a number in $\{1, \ldots, n\}$ using the usual mapping, so $x \in [n]^{2n/\log n}$. Then define a vector $\hat{\mathbf{x}} \in \{0,1\}^{2n^2/\log n}$ as a concatenation of $2n/\log n$ blocks of n-bits, such that each block is a $\{0,1\}^n$ indicator vector of the respective entry in x (i.e. have a single bit equal 1 in the location corresponding to the value of the entry in x). Finally output $\mathbf{A}\hat{\mathbf{x}}$. This is shrinking from $2n$ to n bits, and CRH follows since a collision implies a low norm vector $\mathbf{v}$ s.t. $\mathbf{Av} = 0$. The argument of security for our batch encryption scheme is similar to their proof of security of CRH, however we do not use it as black box.

2.4   One-Time Encryption Using Goldreich-Levin Hard-Core Bit

We show the following one time encryption scheme based on the Goldreich-Levin hard-core bit [GL89].

Definition 3. *Define* gl-enc(x, μ) *as a randomized function that on input* $x \in \{0,1\}^\ell$, $\mu \in \{0,1\}$ *samples* $\alpha \in \{0,1\}^\ell$ *and outputs* $(\alpha, \langle \alpha, x \rangle \oplus \mu)$, *where the inner product is over the binary field. Define* gl-dec$(x, (\alpha, \sigma))$ *be the function that takes* $x \in \{0,1\}^\ell$ *and* $(\alpha, \sigma) \in \{0,1\}^{\ell+1}$ *and outputs* $\sigma \oplus \langle \alpha, x \rangle$.

By definition, for all x, μ it holds that gl-dec$(x,$ gl-enc$(x, \mu)) = \mu$ with probability 1. Furthermore, the Goldreich-Levin Theorem asserts that given an ensemble of joint distributions $\{(X_\lambda, Z_\lambda)\}_\lambda$ s.t. for any polynomial time algorithm $\mathcal{A}$, $\Pr_{(x,z)\leftarrow(X,Z),\mathcal{A}}[\mathcal{A}(1^\lambda, z) = x] = \mathrm{negl}(\lambda)$, then $(z,$ gl-enc$(x, \mu))$ is computationally indistinguishable from $(z, U_{\ell+1})$ for any μ (possibly dependent on z). We furthermore note that if μ is random and unknown to the distinguisher then gl-enc(x, μ) is uniformly random regardless of x.

3   Blind Batch Encryption and Instantiations

3.1   Defining Batch Encryption

A Batch Encryption scheme is an encryption scheme whose key generation is a *projection* function (or a hash function) taking as input a string x to be used as secret key, and outputting a hash value h to be used as public key. The batch encryption scheme is parameterized by a *block size* B. The aforementioned string x should be parsed as $x \in [B]^n$. Batch encryption uses the public key h to encrypt an $n \times B$ matrix $\mathbf{M}$ such that a decryptor with secret key x can obtain exactly M_{i,x_i} for all $i \in [n]$; that is, exactly one matrix element from each row of $\mathbf{M}$. Note that when $B = 2$ we can think of x as a bit vector $x \in \{0,1\}^n$ with the natural translation between $\{0,1\}$ and $\{1,2\}$.

In more detail, the syntax of the batch encryption scheme is as follows, where we think of the function $B = B(\lambda, n)$ as a global parameter of the construction.

1. Setup$(1^\lambda, 1^n)$. Takes as input the security parameter λ and key length n, and outputs a common reference string crs.
2. Gen(crs, x). Using the common reference string, project the secret key $x \in [B]^n$ to a public key h.

3. Enc(crs, h, **M**). Takes as input a common reference string crs, the public key h, and a matrix **M** $\in \{0,1\}^{n \times B}$ and outputs a ciphertext ct. For the purpose of defining the blinding property below, the ciphertext ct can be written as a concatenation of two parts ct = (subct$_1$, subct$_2$).
4. Dec(crs, x, ct). Given a ciphertext ct, output a message vector **m**.

Additionally, a batch encryption scheme supports two *optional* functions.

5. SingleEnc(crs, h, i, **m**). Takes as input a common reference string crs, the public key h, an index $i \in [n]$, and a message **m** $\in \{0,1\}^B$ and outputs a ciphertext ct. As above, the ciphertext ct can be written as a concatenation of two parts ct = (subct$_1$, subct$_2$) for blindness purposes to be defined below.
6. SingleDec(crs, x, i, ct$_i$). Takes as input a common reference string crs, the secret key x, an index $i \in [n]$, and a ciphertext ct$_i$ and outputs a message $m \in \{0,1\}$.

Whenever SingleEnc and SingleDec are defined, we require that Enc(crs, h, **M**) = (ct$_i$)$_{i\in[n]}$ for ct$_i$ $\leftarrow$ SingleEnc(crs, h, i, **m**$_i$), where **m**$_i$ denotes the ith row of **M**. Similarly, we require that for ct = (ct$_i$)$_{i\in[n]}$, the decryption algorithm computes $m_i \leftarrow$ SingleDec(crs, x, i, ct$_i$) for all $i \in [n]$ and outputs their concatenation.

Correctness of Batch Encryption. We define two notions of correctness of a batch encryption scheme, the first stronger than the second.

Definition 4 (Batch Correctness). *Letting* crs = Setup(1^λ, 1^n), *then for all* x, **M**, *it holds that taking* h = Gen(crs, x), ct = Enc(crs, h, **M**), **m**$'$ = Dec(crs, x, ct), *it holds that* **m**$'_i$ = **M**$_{i,x_i}$ *for all* i *with probability at least* $1 - 2^\lambda$ *over the randomness of* Enc.

Definition 5 (δ-Pointwise-Correctness for SingleEnc). *Letting* crs = Setup(1^λ, 1^n), *then for all* x, i, **m**, *taking* h = Gen(crs, x), ct$_i$ = SingleEnc (crs, h, i, **m**), m' = SingleDec(crs, x, i, ct$_i$), *it holds that* $m' = m_{x_i}$ *with probability at least* $1/2 + \delta$ *over the randomness of* SingleEnc.

Note that $1/\text{poly}(\lambda)$-pointwise-correctness implies batch correctness via repetition.

Succinctness of Batch Encryption

Definition 6. *A batch encryption scheme is α-succinct if for* crs = Setup(1^λ, 1^n) *and* h = Gen(crs, x) *for some* $x \in [B]^n$, *it holds that* $|h| \leq \alpha n \log B$.

Definition 7. *A batch encryption scheme is fully succinct if for* crs = Setup(1^λ, 1^n) *and* h = Gen(crs, x) *for some* $x \in [B]^n$, *it holds that* $|h| \leq p(\lambda)$ *for some fixed polynomial* $p(\lambda)$.

Semantic Security of Batch Encryption

Definition 8 (Batch Encryption Security). *The security of a batch encryption scheme is defined using the following game between a challenger and adversary.*

1. *The adversary takes 1^λ as input, and sends 1^n, $x \in [B]^n$ to the challenger.*
2. *The challenger generates $\mathsf{crs} = \mathsf{Setup}(1^\lambda, 1^n)$ and sends crs to the adversary.*
3. *The adversary generates $\mathbf{M}^{(0)}, \mathbf{M}^{(1)} \in \{0,1\}^{n \times B}$ such that $\mathbf{M}^{(0)}_{i,x_i} = \mathbf{M}^{(1)}_{i,x_i}$ for all $i \in [n]$ and sends them to the challenger.*
4. *The challenger computes $h = \mathsf{Gen}(\mathsf{crs}, x)$ and encrypts $\mathsf{ct} = \mathsf{Enc}(\mathsf{crs}, h, M^{(\beta)})$ for a random bit $\beta \in \{0, 1\}$. It sends ct to the adversary.*
5. *The adversary outputs a bit β' and wins if $\beta' = \beta$.*

The batch encryption scheme is secure if no polynomial time adversary can win the above game with probability $\geq 1/2 + 1/poly(\lambda)$.

By a standard hybrid argument, the above definition is implied by the following security property for SingleEnc.

Definition 9 (SingleEnc Security). *We say that a batch encryption scheme satisfies SingleEnc-security if no polynomial time adversary can win the following game with probability $\geq 1/2 + 1/poly(\lambda)$:*

1. *The adversary takes 1^λ as input, and sends 1^n, $x \in [B]^n$, $i \in [n]$ to the challenger.*
2. *The challenger generates $\mathsf{crs} = \mathsf{Setup}(1^\lambda, 1^n)$ and sends crs to the adversary.*
3. *The adversary generates $\mathbf{m}^{(0)}, \mathbf{m}^{(1)} \in \{0,1\}^B$ s.t. $\mathbf{m}^{(0)}_{x_i} = \mathbf{m}^{(1)}_{x_i}$ and sends them to the challenger.*
4. *The challenger computes $h = \mathsf{Gen}(\mathsf{crs}, x)$ and encrypts $\mathsf{ct} = \mathsf{SingleEnc}(\mathsf{crs}, h, i, \mathbf{m}^{(\beta)})$ for a random bit $\beta \in \{0, 1\}$. It sends ct to the adversary.*
5. *The adversary outputs a bit β' and it wins if $\beta' = \beta$.*

Relation to Chameleon Encryption and Laconic Oblivious Transfer. For readers familiar with the notions of chameleon encryption [DG17a] and laconic oblivious transfer [CDG+17], we compare the notion of batch encryption to these objects.

First, we note that the notion of batch encryption is a significant weakening of the notion of a *chameleon encryption scheme* defined in [DG17a] in the following two ways. Most significantly, we do not require a trapdoor which supports finding collisions (namely, the "chameleon" part of chameleon encryption); this is crucial because our construction from LPN does not seem to have an associated trapdoor. Nevertheless, we show that batch encryption is sufficient to construct IBE. As well, our security definition is selective in the input x rather than adaptive (that is, the adversary picks x before seeing the crs), which means that batch encryption does not obviously imply collision resistant hash functions (CRHF), but rather only target collision-resistance. In contrast, the hash function implicit in chameleon encryption is a CRHF).

On the other hand, batch encryption is essentially equivalent to laconic oblivious transfer as defined in [CDG+17], as long as you restrict the first message of the OT protocol to be a deterministic function of the crs and database D (however, since receiver privacy is not required for laconic OT, any laconic OT scheme can be modified to have this property). Our transformations show that batch encryption (or laconic OT) is the right primitive from which to bootstrap and obtain IBE. Additionally, our new blindness property also has an interpretation in the language of laconic OT.

3.2   Defining Blind Batch Encryption

Next, we define the additional *blindness property* of a batch encryption scheme, which asserts that when encrypting a random message that is not known to the distinguisher, the ciphertext is "essentially" indistinguishable from uniform. More specifically, we allow a part of the ciphertext to not be indistinguishable from uniform so long as it does not reveal any information on h or on the encrypted message.

Definition 10 (Blindness). *Let $\mathcal{BBENC} = $ (Setup, Gen, Enc, Dec) be a batch encryption scheme. Furthermore, suppose that*

$$\mathsf{Enc}(\mathsf{crs}, h, \mathbf{M}; r) = E_1(\mathsf{crs}, h, \mathbf{M}; r) \| E_2(\mathsf{crs}, h, \mathbf{M}; r)$$

is some decomposition of $\mathsf{Enc}(\cdot)$ *into two parts. We say that $\mathcal{BBENC}$ is blind if (1) the function $E_1(\mathsf{crs}, h, \mathbf{M}; r) = E_1(\mathsf{crs}; r)$ does not depend on the public key h or message $\mathbf{M}$, and (2) no polynomial time adversary can win the following game with probability $\geq \frac{1}{2} + 1/poly(\lambda)$.*

1. *The adversary takes 1^λ as input, and sends $1^n, x \in [B]^n$ to the challenger.*
2. *The challenger generates $\mathsf{crs} = \mathsf{Setup}(1^\lambda, 1^n)$ and computes $h = \mathsf{Gen}(\mathsf{crs}, x)$. It samples a random $\beta \leftarrow \{0,1\}$, a random message matrix $\mathbf{M} \leftarrow \{0,1\}^{n \times B}$, and encrypts $(\mathsf{subct}_1, \mathsf{subct}_2) \leftarrow \mathsf{Enc}(\mathsf{crs}, h, \mathbf{M})$. It then generates ct as follows.*
   - *If $\beta = 0$ then $\mathsf{ct} = (\mathsf{subct}_1, \mathsf{subct}_2)$.*
   - *If $\beta = 1$ then sample a random bit string subct_2' of the same length as subct_2. Set $\mathsf{ct} = (\mathsf{subct}_1, \mathsf{subct}_2')$.*

   The challenger sends $\mathsf{crs}, \mathsf{ct}$ to the adversary (note that $\mathbf{M}$ is not sent to the adversary).
3. *The adversary outputs a bit β' and it wins if $\beta' = \beta$.*

Again, the above definition of blindness is implied by an analogous blindness property for SingleEnc via a standard hybrid argument. If $\mathcal{BBENC}$ is a blind batch encryption scheme, we call $\mathsf{Enc} = E_1 \| E_2$ the *blind decomposition* of Enc and adopt the notation that outputs of E_1 are denoted by subct_1 and outputs of E_2 are denoted by subct_2.

From Block Size B to Block Size 2. Although our construction of batch encryption itself from LPN constructs a scheme with large block size, the lemma below shows that we can work with block size 2, without loss of generality. The proof of the lemma is in the full version [BLSV17].

Lemma 1. *Suppose that there is an α-succinct (blind) batch encryption scheme with block size B. Then, there is an α-succinct (blind) batch encryption scheme with block size 2.*

From α-Succinct to Fully Succinct (Blind) Batch Encryption. We show that fully succinct (blind) batch encryption can be built from $1/2$-succinct (blind) batch encryption. The construction and proof are similar to the laconic OT bootstrapping theorem of Cho et al. [CDG+17]. However, to preserve blindness, we make use of blind garbled circuits (defined in Sect. 4), similar to its use in Sects. 5 and 6. We state the lemma below and provide the proof in the full version [BLSV17].

Lemma 2. *Suppose that there is a $1/2$-succinct (blind) batch encryption scheme with block size $B = 2$ and a (blind) garbling scheme. Then, there is a fully succinct (blind) batch encryption scheme with block size $B = 2$.*

3.3 Blind Batch Encryption from CDH

In this section, we construct blind batch encryption from the CDH assumption. The scheme has perfect correctness, is *fully succinct*, and has block size $B = 2$. This construction is inspired by the Chameleon Encryption construction in [DG17a] but does not require a trapdoor. Let $\mathcal{G}$ be a group sampler as described in Sect. 2.2. Recall the Goldreich-Levin encoding/decoding procedure as per Sect. 2.4. The blind batch encryption scheme is as follows.

1. CDH-BE.Setup($1^\lambda, 1^n$). Sample $(\mathbb{G}, g, q) \leftarrow \mathcal{G}(1^\lambda)$. Sample $\alpha_{i,b} \leftarrow \mathbb{Z}_q$ for $i \in [n]$, $b \in \{0, 1\}$. Define $g_{i,b} = g^{\alpha_{i,b}}$. Output $\mathsf{crs} = ((\mathbb{G}, g, q), \{g_{i,b}\}_{i,b})$.
2. CDH-BE.Gen(crs, x). Output $h = \prod_i g_{i,x_i}$.
3. CDH-BE.SingleEnc($\mathsf{crs}, h, i, \mathbf{m}$). Sample $r \leftarrow \mathbb{Z}_q$. For all $j \neq i$ and for all $b \in \{0, 1\}$ compute: $\hat{g}_{j,b} = g^r_{j,b}$. Compute $\hat{g}_{i,b} = h^r g^{-r}_{i,b}$, and let $\mu_{i,b} = \mathsf{gl\text{-}enc}(\hat{g}_{i,b}, \mathbf{m}_b)$. Output

$$\mathsf{ct} = \left(\mathsf{subct}_1 = \{\hat{g}_{j,b}\}_{j \neq i, b \in \{0,1\}}, \mathsf{subct}_2 = \{\mu_{i,b}\}_{b \in \{0,1\}}\right).$$

4. CDH-BE.SingleDec($\mathsf{crs}, x, i, \mathsf{ct}$). Given $\mathsf{ct} = \left(\{\hat{g}_{j,b}\}_{j \neq i, b \in \{0,1\}}, \{\mu_{i,b}\}_{b \in \{0,1\}}\right)$. Compute $\hat{g}_{i,x_i} = \prod_{j \neq i} = \hat{g}_{i,x_i}$. Output $m = \mathsf{gl\text{-}dec}(\hat{g}_{i,x_i}, \mu_{i,x_i})$.

Correctness follows immediately by definition. Moreover, we note that this scheme is *fully succinct* (see Definition 7; note that $h \in \mathbb{G}$ has a fixed poly(λ) size representation by assumption).

Lemma 3. *The scheme CDH-BE is secure under the $\mathsf{CDH}_{\mathcal{G}}$ assumption.*

Proof. Consider the following game between a challenger and an adversary.

1. The adversary takes 1^λ as input, and sends $1^n, x \in \{0, 1\}^n, i \in [n]$, to the challenger.
2. The challenger generates $\mathsf{crs} = \mathsf{CDH\text{-}BE.Setup}(1^\lambda, 1^n)$, i.e. a group $(\mathbb{G}, g, q)$ and collection of $g_{j,b}$. It computes $h = \mathsf{CDH\text{-}BE.Gen}(x)$. It then samples $r \leftarrow \mathbb{Z}_q$ and computes $\hat{g}_{j,b} = g^r_{j,b}$ for all $j \neq i$, $b \in \{0, 1\}$, as well as $\hat{g}_{i,x_i} = h^r g^{-r}_{i,x_i}$. It sends crs and the computed $\hat{g}$ values to the adversary.
3. The adversary returns g'.
4. The challenger declares that the adversary wins if $g' = h^r g^{-r}_{i,1-x_i}$.

We will prove that all polynomial time adversaries have negligible advantage in the above game. By the Goldreich-Levin theorem (see Sect. 2.4), this implies the security of the scheme as per Definition 9.

To see that the above holds, an adversary against the above game, and consider an input to the $\mathsf{CDH}_\mathcal{G}$ problem consisting of $(\mathbb{G}, g, q), g^{a_1}, g^{a_2}$. We will show how to produce a challenger for the above game, so that when the adversary succeeds, the value $g^{a_1 a_2}$ can be computed. The challenger, upon receiving $1^n, x, i$ will do the following. Generate $\alpha_{j,b} \leftarrow \mathbb{Z}_q$ for all $j \neq i$, $b \in \{0, 1\}$, and also $\alpha_{i, 1-x_i}$. Conceptually, we will associate a_1 with the value r to be generated by the challenger, and a_2 with the difference $(\alpha_{i, x_i} - \alpha_{i, 1-x_i})$.[2]

Following this intuition, the challenger will generate $g_{i,b} = g^{\alpha_{j,b}}$ for all $j \neq i$, $b \in \{0, 1\}$ as well as for $(j, b) = (i, 1 - x_i)$. Then generate $g_{i,x_i} = g_{i, 1-x_i} \cdot g^{a_2}$. Generate $\hat{g}_{j,b} = (g^{a_1})^{\alpha_{j,b}}$ for all $j \neq i$, $b \in \{0, 1\}$. We are left with generating $\hat{g}_{i,x_i} = h^r g_{i,x_i}^{-r} = \prod_{j \neq i} g_{j,x_j}^r = \prod_{j \neq i} \hat{g}_{j,x_j}$, which can be derived from previously computed values. Note that the computed values are within negligible statistical distance of their distribution in the real experiment. If the adversary manages to compute $g' = h^r g_{i, 1-x_i}^{-r} = \left(\prod_{j \neq i} \hat{g}_{j,x_j} \right) \cdot (g_{i,x_i} / g_{i, 1-x_i})^r = \left(\prod_{j \neq i} \hat{g}_{j,x_j} \right) \cdot g^{a_1 a_2}$, then the product $\prod_{j \neq i} \hat{g}_{j,x_j}$ can be canceled out and a solution to $\mathsf{CDH}_\mathcal{G}$ is achieved.

Lemma 4. *The scheme* CDH-BE *is blind under the* $\mathsf{CDH}_\mathcal{G}$ *assumption.*

Proof. Consider the game in Definition 10. We first of all note that in our scheme, subct_1 is independent of $h, \mathbf{m}$ and therefore the marginal distribution of subct_1 is identical regardless of the value of β. From the properties of gl-enc (see Sect. 2.4), if $\mathbf{m}$ is uniform then the $\mu_{i,b}$ values are uniformly distributed. It follows that any adversary will have exactly $1/2$ probability to win the blindness game.

3.4   Batch Encryption from LPN

In this section we present a candidate construction from LPN with noise rate $\Omega(\log^2(n)/n)$. Specifically, we will show an LPN based construction which has δ-pointwise correctness for $\delta = 1/\text{poly}(n)$, is $\frac{1}{2}$-*succinct*, and has block size $B = n$. Our construction is based on a collision resistant hash function construction of [BLVW17]. See Sect. 2.3 for details about the assumption and the CRH candidate. Unfortunately, we are unable to prove blindness for this candidate. As explained above, the δ point-wise correctness can be amplified, however this amplification does not preserve the blindness property. Therefore, even though our δ-point-wise correct candidate is blind, we cannot amplify it to have batch correctness without giving up blindness.

We introduce the following notation. For any number $j \in [B]$ we define $\mathsf{ind}(j) \in \{0, 1\}^B$ to be the vector with 1 in the j-th coordinate and 0 in all other coordinates. Note that for a matrix $\mathbf{A} \in \{0, 1\}^{k \times B}$ (for arbitrary k) it holds that $\mathbf{A} \cdot \mathsf{ind}(j)$ is exactly the j-th column of $\mathbf{A}$.

---

[2] In fact, this correspondence only needs to hold in the exponent. Specifically, note that both $g^{(\alpha_{i,x_i} - \alpha_{i,1-x_i})}$ and g^{a_2} are statistically indistinguishable from uniform in $\langle g \rangle$ and therefore from each other.

1. LPN-BE.Setup($1^\lambda, 1^n$). Recall that $B = n$, and assume w.l.o.g that $\lambda \leq n$ (otherwise redefine $n = \lambda$ and proceed with the new value, which only strengthens the constructed object). We define $\tilde{n} = \frac{n \log B}{2} = \frac{n \log n}{2}$ and a parameter $\epsilon = \log n / n = \Omega(\log^2(\tilde{n})/\tilde{n})$ to be used below. Sample $\mathbf{A}_1, \ldots, \mathbf{A}_n \leftarrow \{0,1\}^{\tilde{n} \times B}$ (we will also denote $\mathbf{A} = [\mathbf{A}_1 \| \cdots \| \mathbf{A}_n]$). Output $\mathsf{crs} = \{\mathbf{A}_i\}_{i \in [n]}$.
2. LPN-BE.Gen(crs, x). Output $\mathbf{h} = \sum_{i \in [n]} \mathbf{A}_i \cdot \mathsf{ind}(x_i)$.
3. LPN-BE.SingleEnc($\mathsf{crs}, \mathbf{h}, i, \mathbf{m}$). Define $\mathbf{A}_{-i} = [\mathbf{A}_1 \| \cdots \| \mathbf{A}_{i-1} \| \mathbf{A}_{i+1} \| \cdots \| \mathbf{A}_n]$. For all $j \in [B]$ sample $\mathbf{s}^{(j)} \leftarrow \{0,1\}^{\tilde{n}}$ and $\mathbf{e}^{(j)} \leftarrow \mathsf{Ber}_\epsilon^{(n-1)B+1}$. Compute

$$\mathbf{v}^{(j)} = \mathbf{s}^{(j)}[\mathbf{A}_{-i} \| \mathbf{A}_i \cdot \mathsf{ind}(j) - \mathbf{h}] + \mathbf{e}^{(j)} + [0, \ldots, 0, \mathbf{m}_j].$$

   Output $\mathsf{ct} = \mathsf{subct}_2 = \{\mathbf{v}^{(j)}\}_{j \in [B]}$.
4. LPN-BE.SingleDec($\mathsf{crs}, x, i, \mathsf{ct}$). Given $\mathsf{ct} = \{\mathbf{v}^{(j)}\}_{j \in [B]}$, define

$$\hat{\mathbf{x}}_{-i} = [\mathsf{ind}(x_1) \| \cdots \| \mathsf{ind}(x_{i-1}) \| \mathsf{ind}(x_{i+1}) \| \cdots \| \mathsf{ind}(x_n) \| 1]^\dagger,$$

   where $\dagger$ represents vector transpose. Output $m = \mathbf{v}^{(x_i)} \cdot \hat{\mathbf{x}}_{-i}$.

Lemma 5. *The scheme* LPN-BE *is* $1/\mathsf{poly}(n)$*-pointwise correct.*

Proof. Let $\mathsf{crs}, x, i, \mathbf{m}$ be arbitrary, and consider computing $h = \mathsf{Gen}(\mathsf{crs}, x)$, $\mathsf{ct} = \mathsf{SingleEnc}(\mathsf{crs}, h, i, \mathbf{m})$ and $m' = \mathsf{SingleDec}(\mathsf{crs}, x, i, \mathsf{ct})$. Then, by definition

$$\begin{aligned} m' &= \left(\mathbf{s}^{(j)}[\mathbf{A}_{-i} \| \mathbf{A}_i \cdot \mathsf{ind}(x_i) - \mathbf{h}] + \mathbf{e}^{(x_i)} + [0, \ldots, 0, \mathbf{m}_j]\right) \hat{\mathbf{x}}_{-i} \\ &= \mathbf{m}_j + \mathbf{e}^{(x_i)} \cdot \hat{\mathbf{x}}_{-i}, \end{aligned}$$

but since $\mathbf{e}^{(x_i)}$ is Bernoulli with parameter ϵ, and the hamming weight of $\hat{\mathbf{x}}_{-i}$ is exactly n by definition, then $\mathbf{e}^{(x_i)} \cdot \hat{\mathbf{x}}_{-i}$ is Bernoulli with parameter $\epsilon' \leq 1/2 - e^{-2\epsilon n}$. Since we set $\epsilon = \log n / n$, pointwise correctness follows.

Lemma 6. *The scheme* LPN-BE *is secure under the* $\mathsf{LPN}_{\tilde{n}, \epsilon}$ *assumption (we recall that* $\epsilon = \Omega(\log^2(\tilde{n})/\tilde{n})$*).*

Proof. We consider the SingleEnc security game in Definition 9 (recall that this is sufficient for full batch security). We will prove that the view of the adversary is computationally indistinguishable from one where all $\mathbf{v}^{(j)}$ are uniformly random for all $j \neq x_i$. Security will follow.

Consider a challenger that receives an LPN challenge of the form $\mathbf{A}'_1, \ldots, \mathbf{A}'_n \in \{0,1\}^{\tilde{n} \times B}, \{\mathbf{b}_{j,k}\}_{j \in [B] \setminus \{x_i\}, k \in [n]}$, where $\mathbf{b}_{j,k}$ are either all uniform or are of the form $\mathbf{b}_{j,k} = \mathbf{s}^{(j)} \mathbf{A}'_k + \mathbf{e}_{j,k}$. (Note that the challenge does not actually depend on x_i, we can just take $j \in [B - 1]$ and map the values to $[B] \setminus \{x_i\}$ after the fact.)

Upon receiving x, i from the adversary, the challenger computes $\mathbf{h} = \sum_{i \in [n]} \mathbf{A}'_i \cdot \mathsf{ind}(x_i)$. Then, for all $k \neq i$ it sets $\mathbf{A}_k = \mathbf{A}'_k$, and then sets $\mathbf{A}_i$ as follows. Set $\mathbf{A}_i \cdot \mathsf{ind}(x_i) = \mathbf{A}'_i \cdot \mathsf{ind}(x_i)$ (recall that multiplying by $\mathsf{ind}(j)$ is equivalent to selecting the j-th column), and for all $j \neq x_i$ set $\mathbf{A}_i \cdot \mathsf{ind}(j) = \mathbf{A}'_i \cdot \mathsf{ind}(j) + \mathbf{h}$. Note that since $\mathbf{A}_i \cdot \mathsf{ind}(x_i) = \mathbf{A}'_i \cdot \mathsf{ind}(x_i)$ it holds that $\mathbf{h} = \sum_{i \in [n]} \mathbf{A}_i \cdot \mathsf{ind}(x_i)$, and indeed $\mathsf{crs} = \{\mathbf{A}_1, \ldots, \mathbf{A}_n\}, \mathbf{h}, x, i$ are distributed identically to the original game.

The challenger sends crs, **h** to the adversary and receives the message vectors. It then samples $\mathbf{s}^{(x_i)}$, $\mathbf{e}^{(x_i)}$ itself and generates $\mathbf{v}^{(x_i)}$ properly. For all $j \neq x_i$ generate

$$\mathbf{v}^{(j)} = [\mathbf{b}_{j,[n]\setminus\{i\}} \| \mathbf{b}_{j,i} \cdot \mathsf{ind}(j)] + [0,\ldots,0,\mathbf{m}_j],$$

where $\mathbf{b}_{j,[n]\setminus\{i\}}$ is the concatenation of all $\mathbf{b}_{j,k}$ for $k \neq i$ in order. We notice that if the vectors $\{\mathbf{b}_{j,k}\}$ were generated from an LPN distribution, then $\mathbf{v}^{(j)}$ has the correct distribution. This is because we defined $\mathbf{A}_i \cdot \mathsf{ind}(j) - \mathbf{h} = \mathbf{A}'_i \cdot \mathsf{ind}(j)$. On the other hand, if $\{\mathbf{b}_{j,k}\}$ are uniform then all $\mathbf{v}^{(j)}$, $j \neq x_i$ are uniform. Security thus follows.

4   Blind Garbled Circuits

In this section, we define the notion of a blind garbled circuit and show a construction assuming only one-way functions. Indeed, we observe that the widely used "point-and-permute" garbled circuit construction [BMR90, Rog91] is in fact blind. We start with the definition of standard garbled circuits and proceed to define and construct blind garbled circuits.

Definition 11 (Garbled Circuits). *A garbling scheme consists of three algorithms* (Garble, Eval, Sim) *where:*

1. Garble$(1^\lambda, 1^n, 1^m, C)$ *is a PPT algorithm that takes as input the security parameter λ and a circuit $C : \{0,1\}^n \to \{0,1\}^m$, and outputs a garbled circuit $\widehat{C}$ along with input labels $(\mathsf{lab}_{i,b})_{i\in[n],b\in\{0,1\}}$ where each label $\mathsf{lab}_{i,b} \in \{0,1\}^\lambda$.*
2. Eval$(1^\lambda, \widehat{C}, \widehat{L})$ *is a deterministic algorithm that takes as input a garbled circuit $\widehat{C}$ along with a set of n labels $\widehat{L} = (\mathsf{lab}_i)_{i\in[n]}$, and outputs a string $y \in \{0,1\}^m$.*
3. Sim$(1^\lambda, 1^{|C|}, 1^n, y)$ *is a PPT algorithm that takes as input the security parameter, the description length of C, an input length n and a string $y \in \{0,1\}^m$, and outputs a simulated garbled circuit $\widetilde{C}$ and labels $\widetilde{L}$.*

We often omit the first input to these algorithms (namely, 1^λ) when it is clear from the context. We require that the garbling scheme satisfies two properties:

1. *Correctness: For all circuits C, inputs x, and all $(\widehat{C}, (\mathsf{lab}_{i,b})_{i,b}) \leftarrow$ Garble(C, x) and $\widehat{L} = (\mathsf{lab}_{i,x_i})_{i\in[n]}$, we have that Eval$(\widehat{C}, \widehat{L}) = C(x)$.*
2. *Simulation Security: for all circuits $C : \{0,1\}^n \to \{0,1\}^m$ and all inputs $x \in \{0,1\}^n$, the following two distributions are computationally indistinguishable:*

$$\left\{ (\widehat{C}, \widehat{L}) : (\widehat{C}, \mathsf{lab}_{i,b})_{i,b} \leftarrow \mathsf{Garble}(C, x), \widehat{L} = (\mathsf{lab}_{i,x_i})_{i\in[n]} \right\}$$
$$\approx_c \left\{ (\widetilde{C}, \widetilde{L}) : (\widetilde{C}, \widetilde{L}) \leftarrow \mathsf{Sim}(1^\lambda, 1^{|C|}, 1^n, C(x)) \right\}.$$

The traditional notion of security of a garbled circuit requires that the garbling $\widehat{C}$ of a circuit C and the garbled labels $\widehat{L}$ corresponding to an input x together reveal $C(x)$ and nothing more (except the size of the circuit C and the input x). Formally, this is captured by a simulation definition which requires that a simulator who is given only $C(x)$ can faithfully simulate the joint distribution of $\widehat{C}$ and $\widehat{L}$. Blindness requires that the simulator's output is *uniformly random*. Of course, this is simply unachievable if the distinguisher is given the circuit C and the input x, or if the distribution of $C(x)$ is not uniformly random. However, blindness only refers to the setting where the distribution of $C(x)$ is uniformly random.

Definition 12 (Blind Garbled Circuits). *A garbling scheme* (Garble, Eval, Sim) *is called* blind *if the distribution* $\mathsf{Sim}(1^\lambda, 1^c, 1^n, U_m)$, *representing the output of the simulator on a completely uniform output, is indistinguishable from a completely uniform bit string. (Note that the distinguisher must not know the random output value that was used for the simulation.)*

Using a construction essentially identical to the point-and-permute garbled circuits of [BMR90, Rog91], we prove the following result.

Lemma 7. *Assuming the existence of one-way functions, there exists a blind garbling scheme.*

We refer the reader to the full version [BLSV17] for details.

5 Weakly Compact Blind IBE

5.1 Defining Weakly Compact Blind IBE

We now begin our construction of anonymous IBE from blind batch encryption and blind garbled circuits; along the way, we will also construct IBE from batch encryption. As noted earlier, we construct anonymous IBE as a consequence of building a stronger object which we call *blind IBE*. Similar in nature to the blindness property of batch encryption (Definition 10), we say that an IBE scheme is blind if, when encrypting (under some identity id^*) a random message that is not known to the distinguisher, the ciphertext is "essentially" indistinguishable from uniform, even given any polynomial number of secret keys $\{\mathsf{sk_{id}}\}$ possibly including $\mathsf{sk_{id^*}}$.

Definition 13 (Blind IBE). *An IBE scheme satisfies IND-BLIND-ID-CPA security if (1) it satisfies IND-ID-CPA security and (2) the function* $\mathsf{Enc}(\mathsf{pp}, \mathsf{mpk}, \mathsf{id}, m; r)$ *can be expressed as a concatenation* $E_1(\mathsf{pp}; r) || E_2$ ($\mathsf{pp}, \mathsf{mpk}, \mathsf{id}, m; r$) *such that no PPT adversary $\mathcal{A}$ can win the following game with probability greater than* $\frac{1}{2} + \mathrm{negl}(\lambda)$:

1. $\mathsf{pp} \leftarrow \mathsf{Params}(1^\lambda || 1^t)$
2. $(\mathsf{mpk}, \mathsf{msk}) \leftarrow \mathsf{Setup}(\mathsf{pp})$
3. $(\mathsf{id}^*, \mathsf{st}) \leftarrow \mathcal{A}^{\mathsf{Keygen}(\mathsf{pp}, \mathsf{msk}, \cdot)}(\mathsf{mpk})$
4. $m \xleftarrow{\$} \mathcal{M}$
5. $(\mathsf{subct_1}, \mathsf{subct_2}) \leftarrow \mathsf{Enc}(\mathsf{pp}, \mathsf{mpk}, \mathsf{id}^*, m) = (E_1(\mathsf{pp}; r), E_2(\mathsf{pp}, \mathsf{mpk}, \mathsf{id}^*, m; r))$
6. $\beta \xleftarrow{\$} \{0, 1\}$. If $\beta = 1$, $\mathsf{subct_2} \xleftarrow{\$} \{0, 1\}^{|\mathsf{subct_2}|}$
7. $\beta' \leftarrow \mathcal{A}^{\mathsf{Keygen}(\mathsf{pp}, \mathsf{msk}, \cdot)}(\mathsf{st}, (\mathsf{subct_1}, \mathsf{subct_2}))$
8. $\mathcal{A}$ *wins if and only if* $\beta' = \beta$.

We call $\mathsf{Enc} = E_1 || E_2$ *the blind decomposition of* Enc.

Lemma 8. *Any blind IBE scheme is also an anonymous IBE scheme.*

Proof. Consider an adversary $\mathcal{A}$ playing the IND-ANON-ID-CPA security game; $\mathcal{A}$ is eventually given a challenge $\mathsf{ct} \leftarrow \mathsf{Enc}(\mathsf{pp}, \mathsf{mpk}, \mathsf{id}_b, m_b)$ where (id_0, m_0) and (id_1, m_1) are the challenge id-message pairs chosen by $\mathcal{A}$. For each $b \in \{0, 1\}$, it is certainly the case that $\mathcal{A}$ cannot distinguish whether it was given $\mathsf{ct}_{\mathsf{id}_b, m_b} \leftarrow \mathsf{Enc}(\mathsf{pp}, \mathsf{mpk}, \mathsf{id}_b, m_b)$ or $\mathsf{ct}_{\mathsf{id}_b, m} \leftarrow \mathsf{Enc}(\mathsf{pp}, \mathsf{mpk}, \mathsf{id}_b, m)$ where $m \xleftarrow{\$} \mathcal{M}$ is a uniformly random message; this follows from ordinary IBE security. Additionally, by blind IBE security, $\mathcal{A}$ also cannot distinguish whether it is given $\mathsf{ct}_{\mathsf{id}_b, m}$ as above or $\tilde{\mathsf{ct}}_{\mathsf{id}_b, m} \leftarrow E_1(\mathsf{pp}; r) || C$ for $C \xleftarrow{\$} \{0, 1\}^{|E_2(\mathsf{pp}, \mathsf{mpk}, \mathsf{id}_b, m; r)|}$. But $\tilde{\mathsf{ct}}_{\mathsf{id}_0, m}$ and $\tilde{\mathsf{ct}}_{\mathsf{id}_1, m}$ are drawn from identical distributions, so we conclude that $\mathcal{A}$ cannot distinguish whether it was given $\mathsf{ct}_{\mathsf{id}_0, m_0}$ or $\mathsf{ct}_{\mathsf{id}_1, m_1}$, as desired.

Our overall goal is to construct (blind) IBE from (blind) batch encryption; this will be done in two steps. In this section, we construct what we call *weakly compact (blind) IBE*, which is intuitively an IBE scheme for any $T = \mathrm{poly}(\lambda)$ identities which is at least slightly more efficient than the trivial "IBE scheme" consisting of T independent PKE schemes (one for each identity), which has $|\mathsf{mpk}| = T \cdot \mathrm{poly}(\lambda)$. Indeed, all we require is that $|\mathsf{mpk}|$ grows sublinearly with T. In Sect. 6, we show that full (blind) IBE can be bootstrapped from weakly compact (blind) IBE.

Definition 14 (Weakly Compact IBE). *A weakly compact IBE scheme consists of five PPT algorithms* (Params, Setup, Keygen, Enc, Dec) *with the same syntax as an IBE scheme. What distinguishes a weakly compact IBE scheme from a full IBE scheme is the following weakened efficiency requirements:*

- Params *now takes as input* $1^\lambda || 1^T$ *where* $T = 2^t$ *is the number of identities. This means that all five algorithms now run in time* $\mathrm{poly}(T, \lambda)$ *rather than* $\mathrm{poly}(\log T, \lambda)$.[3]
- Weak Compactness: *we require that* $|\mathsf{mpk}| = O(T^{1-\epsilon} \mathrm{poly}(\lambda))$ *for some* $\epsilon > 0$.
- *Security still holds with respect to adversaries running in time* $\mathrm{poly}(\lambda)$, *not* $\mathrm{poly}(\lambda, T)$. /See Footnote 3/

Definition 15. *A weakly compact blind IBE scheme is a weakly compact IBE scheme satisfying IND-BLIND-ID-CPA security.*

We will construct weakly compact (blind) IBE from the following building blocks: (1) (blind) batch encryption, (2) (blind) garbled circuits, and (3) (blind) public key encryption, where blind PKE is defined as follows.

Definition 16 (Blind Public Key Encryption). *An blind public key encryption scheme (with public parameters) is a public key encryption scheme*

---

[3] This is only a technical difference, since we only consider weakly compact IBE schemes with $T = \mathrm{poly}(\lambda)$.

(Params, Gen, Enc, Dec) *which is IND-CPA secure and satisfies the following additional security property: the function* Enc(pp, pk, m; r) *can be expressed as a concatenation* E_1(pp; r)$\|E_2$(pp, pk, m; r) *such that the distribution*

$$\left\{ \text{pp} \leftarrow \text{Params}(1^\lambda), (\text{pk}, \text{sk}) \leftarrow \text{Gen(pp)}, m \xleftarrow{\$} \{0,1\}^n : (\text{pp}, \text{pk}, \text{sk}, \text{Enc(pp}, \text{pk}, m)) \right\}$$

is computationally indistinguishable from the distribution

$$\left\{ \text{pp} \leftarrow \text{Params}(1^\lambda), (\text{pk}, \text{sk}) \leftarrow \text{Gen(pp)}, m \xleftarrow{\$} \mathcal{M}, L = |E_2(\text{pp}, \text{pk}, m; r)|, \right.$$

$$\left. \text{subct}_2 \xleftarrow{\$} \{0,1\}^L : (\text{pp}, \text{pk}, \text{sk}, E_1(\text{pp}; r)\|\text{subct}_2) \right\}.$$

That is, encryptions of random messages are pseudorandom (along with some function independent of the public key) even given the secret key.

We note here that blind public key encryption can be constructed generically from blind batch encryption; indeed, blind batch encryption can be used to build a blind PKE scheme satisfying stronger security notions such as *leakage resilience* and *key-dependent message (KDM) security*.

5.2 The Construction

The construction of our weakly compact blind IBE scheme $\mathcal{WBIBE}$ uses three ingredients:

- A blind public-key encryption scheme

$$\mathcal{BPKE} = (\text{BPKE.Params}, \text{BPKE.Gen}, \text{BPKE.Enc}, \text{BPKE.Dec})$$

where the encryption algorithm can be decomposed into BPKE.$E_1\|$BPKE.E_2 as in Definition 16;
- A blind garbling scheme

$$\mathcal{BGBL} = (\text{BGC.Garble}, \text{BGC.Eval}, \text{BGC.Sim}); \text{ and}$$

- A blind batch encryption scheme

$$\mathcal{BBENC} = (\text{Batch.Setup}, \text{Batch.Gen}, \text{Batch.Enc}, \text{Batch.Dec})$$

where the encryption algorithm can be decomposed into Batch.$E_1\|$Batch.E_2 as in Definition 10. Moreover, we assume that $\mathcal{BBENC}$ is fully succinct.

The construction works as follows.

1. WBIBE.Params(1^T): Given a bound T on the number of identities, the parameter generation algorithm Params first obtains blind public-key encryption parameters bpke.pp $\leftarrow$ BPKE.Params(1^λ). Letting n be the length of the public keys generated by BPKE.Gen, it then obtains a common reference string batch.crs $\leftarrow$ Batch.Setup($1^\lambda, 1^{nT}$). The output is

$$\text{wbibe.pp} = (\text{bpke.pp}, \text{batch.crs})$$

2. WBIBE.Setup(wbibe.pp): On input the public parameters, the setup algorithm first obtains T key pairs $(\text{bpke.pk}_i, \text{bpke.sk}_i) \leftarrow \text{BPKE.Gen}(\text{bpke.pp})$. Secondly, it compresses the sequence of $\mathcal{BPKE}$ public keys into a $\mathcal{BBENC}$ public key:

$$h \leftarrow \text{Batch.Gen}(\text{batch.crs}, (\text{bpke.pk}_0, \text{bpke.pk}_1, \ldots, \text{bpke.pk}_{T-1})).$$

The output is the pair $(\text{wbibe.mpk}, \text{wbibe.msk})$ where

$\text{wbibe.mpk} = h$ and

$\text{wbibe.msk} = (\text{bpke.pk}_0, \ldots, \text{bpke.pk}_{T-1}, \text{bpke.sk}_0, \ldots, \text{bpke.sk}_{T-1})$

3. WBIBE.Keygen(wbibe.pp, wbibe.msk, id): On input the public parameters, the master secret key and an identity $\text{id} \in \{0, 1, \ldots, T-1\}$, the key generation algorithm outputs

$$\text{wbibe.sk}_\text{id} = (\text{id}, \text{bpke.pk}_0, \text{bpke.pk}_1, \ldots, \text{bpke.pk}_{T-1}, \text{bpke.sk}_\text{id}).$$

4. WBIBE.Enc(wbibe.pp, wbibe.mpk, id, m): On input the public parameters, a master public key, an identity id and a message m, the encryption algorithm does the following.

First, sample a uniformly random string r and compute

$$\text{ct}_0 = \text{BPKE}.E_1(\text{bpke.pp}; r).$$

Secondly, let $C[\text{bpke.pp}, m, r]$ be a circuit with public parameters bpke.pp (contained as part of wbibe.pp), the message m and the random string r hardcoded. C takes as input a blind public key and outputs the encryption of m under the public key using randomness r. That is,

$$C[\text{bpke.pp}, m, r](\text{bpke.pk}) = \text{BPKE}.E_2(\text{bpke.pp}, \text{bpke.pk}, m; r)$$

Compute

$$(\widehat{C}, \overline{\text{lab}}) \leftarrow \text{BGC.Garble}(1^\lambda, 1^n, 1^\ell, C[\text{bpke.pp}, m, r])$$

where $\overline{\text{lab}} \in (\{0, 1\}^\lambda)^{n \times 2}$ and ℓ is defined to be the output length of C. Set $\text{ct}_1 := \widehat{C}$.

Finally, let $\mathbf{M} \in (\{0, 1\}^\lambda)^{nT \times 2}$ be a uniformly random nT-by-2 matrix and then *redefine* $\mathbf{M}[\text{id} \cdot n + j, b] = \overline{\text{lab}}[j, b]$ for all $1 \le j \le n, b \in \{0, 1\}$. Compute

$$(\text{ct}_2, \text{ct}_2') \leftarrow \text{Batch.Enc}(\text{batch.crs}, h, \mathbf{M}).$$

Output the ciphertext $\text{wbibe.ct} = (\text{ct}_0, \text{ct}_1, \text{ct}_2, \text{ct}_2')$.

5. WBIBE.Dec(wbibe.pp, wbibe.sk, wbibe.ct): On input the public parameters, a secret key and a ciphertext, the decryption algorithm parses the secret key as $\text{wbibe.sk} = (\text{id}, \text{bpke.pk}_0, \ldots, \text{bpke.pk}_{T-1}, \text{bpke.sk}_\text{id})$, and parses the ciphertext as $\text{wbibe.ct} = (\text{ct}_0, \text{ct}_1, \text{ct}_2, \text{ct}_2')$. It then does three things.

First, it computes

$$\mathbf{m} \leftarrow \mathsf{Batch.Dec}(\mathsf{batch.crs}, (\mathsf{bpke.pk}_0, \mathsf{bpke.pk}_1, \ldots, \mathsf{bpke.pk}_{T-1}), \mathsf{ct}_2 \| \mathsf{ct}_2'),$$

Secondly, it defines $\widehat{L} = (L_j)_{j \in [n]} \in (\{0,1\}^\lambda)^n$ by $L_j = \mathbf{m}[\mathsf{id} \cdot n + j]$ and computes $\mathsf{ct}_0' \leftarrow \mathsf{BGC.Eval}(\mathsf{ct}_1, \widehat{L})$. Finally, it computes and outputs

$$m \leftarrow \mathsf{BPKE.Dec}(\mathsf{bpke.pp}, \mathsf{bpke.sk}_{\mathsf{id}}, \mathsf{ct}_0 \| \mathsf{ct}_0').$$

We show that this scheme is a weakly compact blind IBE scheme.

Theorem 1. *Suppose $\mathcal{BPKE}$ is a blind public-key encryption scheme, $\mathcal{BBENC}$ is a blind batched encryption scheme, and $\mathcal{BGBL}$ is a blind garbling scheme. Then, $\mathcal{WBIBE}$ is a weakly compact blind IBE scheme.*

We defer the reader to the full version [BLSV17] for details.

6 Bootstrapping (Blind) IBE

Our bootstrapping theorem converting a weakly compact (blind) IBE scheme into a full-fledged (blind) IBE scheme follows the ideas of [DG17a, DG17b] and is essentially a way to achieve *domain extension* of the space of identities. The bootstrapping scheme is described in Sect. 6.1 and analyzed in the full version of our paper [BLSV17]. Recall that a high level overview was provided in the introduction (Sect. 1.3).

6.1 The Bootstrapping Theorem

Let $\mathcal{WBIBE}$ denote a weakly compact blind IBE scheme supporting $T = T(\lambda)$ identities with a master public key of size $S = S(\lambda)$ bits. By compactness, we may choose $T = \mathsf{poly}(\lambda)$ large enough so that $S < T/4$. Additionally, let $\mathcal{BGBL} = (\mathsf{BGC.Garble}, \mathsf{BGC.Eval}, \mathsf{BGC.Sim})$ denote a blind garbling scheme. We construct a full-fledged blind IBE scheme $\mathcal{BIBE}$ as follows.

– BIBE.Params$(1^\lambda, 1^n)$: On input the length n of the identities supported by the system, the parameter generation algorithm generates parameter wbibe.pp $\leftarrow$ WBIBE.Params$(1^\lambda, 1^T)$ and outputs bibe.pp $= (1^n, \mathsf{wbibe.pp})$.
– BIBE.Setup(bibe.pp): On input the public parameters, the setup algorithm chooses a seed s for a PRF family $f_s : \{0,1\}^{\leq n} \to \{0,1\}^r$ where r is the number of random bits used by the Setup algorithm of $\mathcal{WBIBE}$. BIBE.Setup then obtains

$$(\mathsf{wbibe.mpk}^{(\epsilon)}, \mathsf{wbibe.msk}^{(\epsilon)}) \leftarrow \mathsf{WBIBE.Setup}(\mathsf{wbibe.pp}; f_s(\epsilon))$$

where ϵ denotes the empty string. The output is

$$\mathsf{bibe.mpk} = \mathsf{wbibe.mpk}^{(\epsilon)} \quad \text{and} \quad \mathsf{bibe.msk} = s.$$

- BIBE.Keygen(bibe.pp, bibe.msk, id): On input the public parameters, the master secret key and an n-bit identity id $=$ id$_1||$id$_2||...$id$_n$, the key generation algorithm does the following.

  First, for each prefix id$[\leq i] =$ id$_1||$id$_2|| \ldots ||$id$_i \in \{0,1\}^i$, compute the master public key wbibe.mpk$^{(\leq i)}$ and the master secret key wbibe.msk$^{(\leq i)}$:

$$(\text{wbibe.mpk}^{(\leq i)}, \text{wbibe.msk}^{(\leq i)}) \leftarrow \text{WBIBE.Setup}(\text{wbibe.pp}; f_s(\text{id}[\leq i])).$$

(By convention, id$[\leq 0] = \epsilon$)

  For each $0 \leq i \leq n - 1$ and $j \in [S]$, define id$'_{i,j} :=$ id$_{i+1}||j||b_{i+1,j} \in \{0,1\} \times [S] \times \{0,1\}$, where $b_{i+1,j} :=$ wbibe.mpk$^{(\leq i+1)}[j]$. Compute

$$\text{sk}_{i,j} \leftarrow \text{WBIBE.Keygen}(\text{wbibe.pp}, \text{wbibe.msk}^{(\leq i)}, \text{id}'_{i,j}).$$

Finally, compute

$$\text{sk}_{\text{leaf}} \leftarrow \text{WBIBE.Keygen}(\text{wbibe.pp}, \text{wbibe.msk}^{(\leq n)}, \text{id}_{\text{null}}),$$

where id$_{\text{null}} = 0^T$ is a default identity, and output

$$\text{bibe.sk}_{\text{id}} = \left((\text{wbibe.mpk}^{(\leq i)})_{0 \leq i \leq n}, (\text{sk}_{i,j})_{j \in [S], 0 \leq i \leq n-1}, \text{sk}_{\text{leaf}} \right).$$

- BIBE.Enc(bibe.pp, bibe.mpk, id, m): On input the public parameters, the master public key, an n-bit identity id, and a message m, the encryption algorithm does the following.

  Let $C[\text{wbibe.pp}, \eta, \overline{\text{lab}}, \overline{\mathbf{r}}]$ be a circuit that computes the function

$$\left(\text{WBIBE}.E_2(\text{wbibe.pp}, \text{wbibe.mpk}, \eta||j||b, \text{lab}_{j,b}; r_{j,b}) \right)_{j \in [S], b \in \{0,1\}}$$

on input wbibe.mpk, where $\overline{\mathbf{r}}$ is the collection of all $r_{j,b}$ and $\overline{\text{lab}}$ is the collection of all $\text{lab}_{j,b}$. Let $C'[\text{wbibe.pp}, m, r]$ be a circuit that computes the function

$$\text{WBIBE}.E_2(\text{wbibe.pp}, \text{wbibe.mpk}, \text{id}_{\text{null}}, m; r)$$

on input wbibe.mpk. Choose random strings $r, \overline{\mathbf{r}}^{(1)}, \ldots, \overline{\mathbf{r}}^{(n)}$.

  Compute $(\widehat{C}_n, \overline{\text{lab}}^{(n)}) \leftarrow \text{BGC.Garble}(C'[\text{wbibe.pp}, m, r])$. For $i = n - 1$ to 0, compute

$$(\widehat{C}_i, \overline{\text{lab}}^{(i)}) \leftarrow \text{BGC.Garble}(C[\text{wbibe.pp}, \text{id}_{i+1}, \overline{\text{lab}}^{(i+1)}, \overline{\mathbf{r}}^{(i+1)}])$$

Compute $\text{ct}_{n+1} \leftarrow \text{WBIBE}.E_1(\text{wbibe.pp}; r)$, and for $i = 1$ to n, compute

$$\text{ct}_{i,j,b} \leftarrow \text{WBIBE}.E_1(\text{wbibe.pp}; r_{j,b}^{(i)}),$$

and let $\text{ct}_i := (\text{ct}_{i,j,b})_{j,b}$.
Output the following as the ciphertext:

$$\text{bibe.ct} = \left(\widehat{C}_0, \ldots, \widehat{C}_{n-1}, \widehat{C}_n, \text{ct}_1, \ldots, \text{ct}_n, \text{ct}_{n+1}, \overline{\text{lab}}^{(0)}[\text{wbibe.mpk}^{(\epsilon)}] \right),$$

where $\overline{\text{lab}}^{(0)}[\text{wbibe.mpk}^{(\epsilon)}]$ is short-hand for $(\text{lab}_{j,b_{0,j}}^{(0)})_{j \in [S]}$.

- BIBE.Dec(bibe.pp, bibe.sk$_{\text{id}}$, bibe.ct): On input the public parameters, an identity secret key and a ciphertext, the decryption algorithm does the following.

  Let $\widehat{L}^{(0)} := \left(\mathsf{lab}_{j,b_{0,j}}^{(0)}\right)_{j\in[S]}$. For $1 \leq i \leq n$, do the following steps one after the other.

  - Compute $\mathsf{ct}_i' \leftarrow \mathsf{Eval}(\widehat{C}_{i-1}, \widehat{L}^{(i-1)})$ which itself consists of ciphertexts $\mathsf{ct}_{i,j,b}'$ for $j \in [S]$ and $b \in \{0,1\}$.
  - Compute $L_j^{(i)} \leftarrow \mathsf{WBIBE.Dec}(\mathsf{wbibe.pp}, \mathsf{sk}_{i,j}, \mathsf{ct}_{i,j,b_{i,j}} \| \mathsf{ct}_{i,j,b_{i,j}}')$ for all $j \in [S]$ and $b_{i,j} = \mathsf{wbibe.mpk}^{(\leq i)}[j]$. Let $\widehat{L}^{(i)}$ denote the collection of all $L_j^{(i)}$.

  Finally, compute $\mathsf{ct}_{n+1}' \leftarrow \mathsf{Eval}(\widehat{C}_n, \widehat{L}^{(n)})$ and output

$$m' \leftarrow \mathsf{WBIBE.Dec}(\mathsf{wbibe.pp}, \mathsf{sk}_{\text{leaf}}, \mathsf{ct}_{n+1} \| \mathsf{ct}_{n+1}').$$

- The blind decomposition of BIBE.Enc is as follows: $\mathsf{BIBE}.E_1(\mathsf{bibe.pp};$ $\mathbf{R})$ is defined to be the collection $(\mathsf{ct}_1, \mathsf{ct}_2, \ldots, \mathsf{ct}_{n+1})$, while $\mathsf{BIBE}.E_2(\mathsf{bibe.pp}, \mathsf{bibe.mpk},\ \mathsf{id}, m; \mathbf{R})$ is defined to be the collection $\left(\widehat{C}_0, \ldots, \widehat{C}_n, \mathsf{lab}^{(0)}[\mathsf{bibe.mpk}]\right)$.

Theorem 2. *Suppose that* $\mathcal{WBIBE}$ *is a weakly compact blind IBE scheme and that* $\mathcal{BGBL}$ *is a blind garbling scheme. Then,* $\mathcal{BIBE}$ *is a blind IBE scheme. Additionally, even without the blindness assumptions,* $\mathcal{BIBE}$ *is an IBE scheme.*

We refer the reader to the full version [BLSV17] for details.

Acknowledgments. The first author was supported by the Israel Science Foundation (Grant No. 468/14), Binational Science Foundation (Grants No. 2016726, 2014276), ERC Project 756482 REACT and European Union PROMETHEUS Project (Horizon 2020 Research and Innovation Program, Grant 780701). The third author was supported by the European Union's 7th Framework Program (FP7) via a Marie Curie Career Integration Grant (Grant No. 618094), by the European Union's Horizon 2020 Framework Program (H2020) via an ERC Grant (Grant No. 714253), by the Israel Science Foundation (Grant No. 483/13), by the Israeli Centers of Research Excellence (I-CORE) Program (Center No. 4/11), by the US-Israel Binational Science Foundation (Grant No. 2014632), and by a Google Faculty Research Award. The second and fourth authors were supported by NSF Grants CNS-1350619 and CNS-1414119, Alfred P. Sloan Research Fellowship, Microsoft Faculty Fellowship, the NEC Corporation, a Steven and Renee Finn Career Development Chair from MIT and by the Defense Advanced Research Projects Agency (DARPA) and the U.S. Army Research Office under contracts W911NF-15-C-0226 and W911NF-15-C-0236. The second author was in addition supported by an Akamai Presidential Fellowship.

References

[ABB10] Agrawal, S., Boneh, D., Boyen, X.: Efficient lattice (H)IBE in the standard model. In: Gilbert, H. (ed.) EUROCRYPT 2010. LNCS, vol. 6110, pp. 553–572. Springer, Heidelberg (2010). https://doi.org/10.1007/978-3-642-13190-5_28

[ABC+08] Abdalla, M., Bellare, M., Catalano, D., Kiltz, E., Kohno, T., Lange, T., Malone-Lee, J., Neven, G., Paillier, P., Shi, H.: Searchable encryption revisited: consistency properties, relation to anonymous IBE, and extensions. J. Cryptol. **21**(3), 350–391 (2008)

[ACPS09] Applebaum, B., Cash, D., Peikert, C., Sahai, A.: Fast cryptographic primitives and circular-secure encryption based on hard learning problems. In: Halevi, S. (ed.) CRYPTO 2009. LNCS, vol. 5677, pp. 595–618. Springer, Heidelberg (2009). https://doi.org/10.1007/978-3-642-03356-8_35

[AGV09] Akavia, A., Goldwasser, S., Vaikuntanathan, V.: Simultaneous hardcore bits and cryptography against memory attacks. In: Reingold, O. (ed.) TCC 2009. LNCS, vol. 5444, pp. 474–495. Springer, Heidelberg (2009). https://doi.org/10.1007/978-3-642-00457-5_28

[Ale11] Alekhnovich, M.: More on average case vs approximation complexity. Comput. Complex. **20**(4), 755–786 (2011)

[App11] Applebaum, B.: Key-dependent message security: generic amplification and completeness. In: Paterson, K.G. (ed.) EUROCRYPT 2011. LNCS, vol. 6632, pp. 527–546. Springer, Heidelberg (2011). https://doi.org/10.1007/978-3-642-20465-4_29

[BBR99] Biham, E., Boneh, D., Reingold, O.: Breaking generalized Diffie-Hellmann modulo a composite is no easier than factoring. Inf. Process. Lett. **70**(2), 83–87 (1999)

[BCHK07] Boneh, D., Canetti, R., Halevi, S., Katz, J.: Chosen-ciphertext security from identity-based encryption. SIAM J. Comput. **36**(5), 1301–1328 (2007)

[BCOP04] Boneh, D., Di Crescenzo, G., Ostrovsky, R., Persiano, G.: Public key encryption with keyword search. In: Cachin, C., Camenisch, J.L. (eds.) EUROCRYPT 2004. LNCS, vol. 3027, pp. 506–522. Springer, Heidelberg (2004). https://doi.org/10.1007/978-3-540-24676-3_30

[BF03] Boneh, D., Franklin, M.K.: Identity-based encryption from the weil pairing. SIAM J. Comput. **32**(3), 586–615 (2003)

[BG10] Brakerski, Z., Goldwasser, S.: Circular and leakage resilient public-key encryption under subgroup indistinguishability. In: Rabin, T. (ed.) CRYPTO 2010. LNCS, vol. 6223, pp. 1–20. Springer, Heidelberg (2010). https://doi.org/10.1007/978-3-642-14623-7_1

[BGH07] Boneh, D., Gentry, C., Hamburg, M.: Space-efficient identity based encryption without pairings. In: Proceedings of 48th Annual IEEE Symposium on Foundations of Computer Science (FOCS 2007), Providence, RI, USA, 20–23 October 2007, pp. 647–657 (2007)

[BHHI10] Barak, B., Haitner, I., Hofheinz, D., Ishai, Y.: Bounded key-dependent message security. In: Gilbert, H. (ed.) EUROCRYPT 2010. LNCS, vol. 6110, pp. 423–444. Springer, Heidelberg (2010). https://doi.org/10.1007/978-3-642-13190-5_22

[BHHO08] Boneh, D., Halevi, S., Hamburg, M., Ostrovsky, R.: Circular-secure encryption from decision Diffie-Hellman. In: Wagner, D. (ed.) CRYPTO 2008. LNCS, vol. 5157, pp. 108–125. Springer, Heidelberg (2008). https://doi.org/10.1007/978-3-540-85174-5_7

[BLSV17] Brakerski, Z., Lombardi, A., Segev, G., Vaikuntanathan, V.: Anonymous IBE, leakage resilience and circular security from new assumptions. Cryptology ePrint Archive, Report 2017/967 (2017). https://eprint.iacr.org/2017/967

[BLVW17] Brakerski, Z., Lyubashevsky, V., Vaikuntanathan, V., Wichs, D.: Cryptographic hashing and worst-case hardness for LPN via code smoothing. Personal Communication (2017)

[BMR90] Beaver, D., Micali, S., Rogaway, P.: The round complexity of secure protocols (extended abstract). In: Proceedings of the 22nd Annual ACM Symposium on Theory of Computing, Baltimore, Maryland, USA, 13–17 May 1990, pp. 503–513 (1990)

[BRS02] Black, J., Rogaway, P., Shrimpton, T.: Encryption-scheme security in the presence of key-dependent messages. In: Nyberg, K., Heys, H. (eds.) SAC 2002. LNCS, vol. 2595, pp. 62–75. Springer, Heidelberg (2003). https://doi.org/10.1007/3-540-36492-7_6

[BW06] Boyen, X., Waters, B.: Anonymous hierarchical identity-based encryption (without random oracles). In: Dwork, C. (ed.) CRYPTO 2006. LNCS, vol. 4117, pp. 290–307. Springer, Heidelberg (2006). https://doi.org/10.1007/11818175_17

[CDG+17] Cho, C., Döttling, N., Garg, S., Gupta, D., Miao, P., Polychroniadou, A.: Laconic oblivious transfer and its applications. In: Katz, J., Shacham, H. (eds.) CRYPTO 2017. LNCS, vol. 10402, pp. 33–65. Springer, Cham (2017). https://doi.org/10.1007/978-3-319-63715-0_2

[CHKP12] Cash, D., Hofheinz, D., Kiltz, E., Peikert, C.: Bonsai trees, or how to delegate a lattice basis. J. Cryptol. 25(4), 601–639 (2012)

[Coc01] Cocks, C.: An identity based encryption scheme based on quadratic residues. In: Proceedings of the 8th IMA International Conference on Cryptography and Coding (2001)

[DG17a] Döttling, N., Garg, S.: Identity-based encryption from the Diffie-Hellman assumption. In: Katz, J., Shacham, H. (eds.) CRYPTO 2017. LNCS, vol. 10401, pp. 537–569. Springer, Cham (2017). https://doi.org/10.1007/978-3-319-63688-7_18

[DG17b] Döttling, N., Garg, S.: From selective IBE to full IBE and selective HIBE. In: Theory of Cryptography Conference (2017, to appear)

[DGHM18] Döttling, N., Garg, S., Hajiabadi, M., Masny, D.: New constructions of identity-based and key-dependent message secure encryption schemes. In: IACR International Workshop on Public Key Cryptography. Springer (2018). https://eprint.iacr.org/2017/978

[DGK+10] Dodis, Y., Goldwasser, S., Tauman Kalai, Y., Peikert, C., Vaikuntanathan, V.: Public-key encryption schemes with auxiliary inputs. In: Micciancio, D. (ed.) TCC 2010. LNCS, vol. 5978, pp. 361–381. Springer, Heidelberg (2010). https://doi.org/10.1007/978-3-642-11799-2_22

[DKXY02] Dodis, Y., Katz, J., Xu, S., Yung, M.: Key-insulated public key cryptosystems. In: Knudsen, L.R. (ed.) EUROCRYPT 2002. LNCS, vol. 2332, pp. 65–82. Springer, Heidelberg (2002). https://doi.org/10.1007/3-540-46035-7_5

[Gen06] Gentry, C.: Practical identity-based encryption without random oracles. In: Vaudenay, S. (ed.) EUROCRYPT 2006. LNCS, vol. 4004, pp. 445–464. Springer, Heidelberg (2006). https://doi.org/10.1007/11761679_27

[GKW16] Goyal, R., Koppula, V., Waters, B.: Semi-adaptive security and bundling functionalities made generic and easy. In: Hirt, M., Smith, A. (eds.) TCC 2016. LNCS, vol. 9986, pp. 361–388. Springer, Heidelberg (2016). https://doi.org/10.1007/978-3-662-53644-5_14

[GL89] Goldreich, O., Levin, L.A.: A hard-core predicate for all one-way functions. In: STOC, pp. 25–32. ACM (1989)

[GPSW06] Goyal, V., Pandey, O., Sahai, A., Waters, B.: Attribute-based encryption for fine-grained access control of encrypted data. In: Proceedings of the 13th ACM Conference on Computer and Communications Security, CCS 2006, Alexandria, VA, USA, 30 October–3 November 2006, pp. 89–98 (2006)

[GPV08] Gentry, C., Peikert, C., Vaikuntanathan, V.: Trapdoors for hard lattices and new cryptographic constructions. In: Proceedings of the 40th Annual ACM Symposium on Theory of Computing, Victoria, British Columbia, Canada, 17–20 May 2008, pp. 197–206 (2008)

[HLWW16] Hazay, C., López-Alt, A., Wee, H., Wichs, D.: Leakage-resilient cryptography from minimal assumptions. J. Cryptol. **29**(3), 514–551 (2016)

[KSW08] Katz, J., Sahai, A., Waters, B.: Predicate encryption supporting disjunctions, polynomial equations, and inner products. In: Smart, N. (ed.) EUROCRYPT 2008. LNCS, vol. 4965, pp. 146–162. Springer, Heidelberg (2008). https://doi.org/10.1007/978-3-540-78967-3_9

[KT18] Kitagawa, F., Tanaka, K.: Key dependent message security and receiver selective opening security for identity-based encryption. In: IACR International Workshop on Public Key Cryptography. Springer (2018). https://eprint.iacr.org/2017/987

[McC88] McCurley, K.S.: A key distribution system equivalent to factoring. J. Cryptol. **1**(2), 95–105 (1988)

[NS12] Naor, M., Segev, G.: Public-key cryptosystems resilient to key leakage. SIAM J. Comput. **41**(4), 772–814 (2012)

[Rog91] Rogaway, P.: The round-complexity of secure protocols. Ph.D. thesis, MIT (1991)

[Sha84] Shamir, A.: Identity-based cryptosystems and signature schemes. In: Blakley, G.R., Chaum, D. (eds.) CRYPTO 1984. LNCS, vol. 196, pp. 47–53. Springer, Heidelberg (1985). https://doi.org/10.1007/3-540-39568-7_5

[Shm85] Shmuely, Z.: Composite Diffie-Hellman public-key generating systems are hard to break, Technion Technical Report (1985). http://www.cs.technion.ac.il/users/wwwb/cgi-bin/tr-get.cgi/1985/CS/CS0356.pdf

[SW05] Sahai, A., Waters, B.: Fuzzy identity-based encryption. In: Cramer, R. (ed.) EUROCRYPT 2005. LNCS, vol. 3494, pp. 457–473. Springer, Heidelberg (2005). https://doi.org/10.1007/11426639_27

[YZW+17] Yu, Y., Zhang, J., Weng, J., Guo, C., Li, X.: Learning parity with noise implies collision resistant hashing. https://eprint.iacr.org/2017/1260.pdf

Secret Sharing

Towards Breaking the Exponential Barrier for General Secret Sharing

Tianren Liu[1]($\boxtimes$), Vinod Vaikuntanathan[1], and Hoeteck Wee[2]

[1] MIT, Cambridge, USA
liutr@mit.edu, vinodv@csail.mit.edu
[2] CNRS and ENS, Paris, France
wee@di.ens.fr

Abstract. A secret-sharing scheme for a monotone Boolean (access) function $F : \{0,1\}^n \to \{0,1\}$ is a randomized algorithm that on input a secret, outputs n shares $s_1, \ldots, s_n$ such that for any $(x_1, \ldots, x_n) \in \{0,1\}^n$, the collection of shares $\{s_i : x_i = 1\}$ determine the secret if $F(x_1, \ldots, x_n) = 1$ and reveal nothing about the secret otherwise. The best secret sharing schemes for general monotone functions have shares of size $\Theta(2^n)$. It has long been conjectured that one cannot do much better than $2^{\Omega(n)}$ share size, and indeed, such a lower bound is known for the restricted class of linear secret-sharing schemes.

In this work, we *refute* two natural strengthenings of the above conjecture:

- First, we present secret-sharing schemes for a family of $2^{2^{n/2}}$ monotone functions over $\{0,1\}^n$ with sub-exponential share size $2^{O(\sqrt{n} \log n)}$. This *unconditionally* refutes the stronger conjecture that circuit size is, within polynomial factors, a lower bound on the share size.
- Second, we disprove the analogous conjecture for non-monotone functions. Namely, we present "non-monotone secret-sharing schemes" for *every access function* over $\{0,1\}^n$ with shares of size $2^{O(\sqrt{n} \log n)}$.

Our construction draws upon a rich interplay amongst old and new problems in information-theoretic cryptography: from secret-sharing, to multi-party computation, to private information retrieval. Along the way, we also construct the first *multi-party* conditional disclosure of secrets (CDS) protocols for general functions $F : \{0,1\}^n \to \{0,1\}$ with communication complexity $2^{O(\sqrt{n} \log n)}$.

T. Liu—Research supported in part by NSF Grants CNS-1350619 and CNS-1414119.
V. Vaikuntanathan—Research supported in part by NSF Grants CNS-1350619 and CNS-1414119, Alfred P. Sloan Research Fellowship, Microsoft Faculty Fellowship, the NEC Corporation, a Steven and Renee Finn Career Development Chair from MIT. This work was also sponsored in part by the Defense Advanced Research Projects Agency (DARPA) and the U.S. Army Research Office under contracts W911NF-15-C-0226 and W911NF-15-C-0236.
H. Wee—Research supported in part by ERC Project aSCEND (H2020 639554) and NSF Award CNS-1445424.

J. B. Nielsen and V. Rijmen (Eds.): EUROCRYPT 2018, LNCS 10820, pp. 567–596, 2018.
https://doi.org/10.1007/978-3-319-78381-9_21

1   Introduction

Secret sharing [Sha79, Bla79] is a powerful cryptographic technique that allows a dealer to distribute shares of a secret to n parties such that certain authorized subsets of parties, and only they, can recover the secret. The original definition of secret sharing is what we now call an (n, t)-threshold secret sharing scheme, where any set of t or more parties can recover the secret, and no subset of fewer than t parties can learn any information about the secret whatsoever.

Later on, this was generalized in [ISN89] to the notion of a secret-sharing scheme realizing a monotone function $F : \{0, 1\}^n \to \{0, 1\}$. This is simply a randomized algorithm that on input a secret, outputs n shares $s_1, \ldots, s_n$ such that for any $(x_1, \ldots, x_n) \in \{0, 1\}^n$, the collection of shares $\{s_i : x_i = 1\}$ determine the secret if $F(x_1, \ldots, x_n) = 1$ and reveal nothing about the secret otherwise.[1] It is easy to see that (n, t)-threshold secret sharing corresponds to the special case where F is the (monotone) threshold function that outputs 1 if and only if at least t of the n input bits are 1.

While the landscape of threshold secret sharing is relatively well-understood, even very basic information-theoretic questions about the more general notion of secret sharing remain embarrassingly open. It is simple to construct a secret sharing scheme realizing any monotone function $F : \{0, 1\}^n \to \{0, 1\}$ where each share is at most 2^n bits; the share size can be improved to $O(2^n/\sqrt{n})$ bits [ISN89]. We also know that there is an (explicit) monotone function $F : \{0, 1\}^n \to \{0, 1\}$ that requires a total share size of $\Omega(n^2/\log n)$ bits [Csi97], a far cry from the upper bound. No better lower bounds are known (except for the restricted class of linear secret-sharing schemes, cf. Sect. 1.3), even in a non-explicit sense.

Closing the exponential gap between the aforementioned upper bound or lower bounds is a long-standing open problem in cryptography. The general consensus appears to be that the upper bound is almost tight, as formalized in a conjecture of Beimel [Bei11]:

Conjecture (Main). There exists a family of monotone functions $\{F_n : \{0, 1\}^n \to \{0, 1\}\}_{n \in \mathbb{N}}$ s.t. the total share size of any secret sharing scheme realizing F_n is $2^{\Omega(n)}$ bits.

Note that this is a purely information-theoretic statement with no reference to the computational complexity of sharing or reconstruction.

1.1   Our Results

In this work, we *refute* two natural strengthenings of the main conjecture by presenting new secret sharing schemes. The first variant of the main conjecture

---

[1] The typical formulation of secret-sharing refers to a dealer that holds a secret distributing shares to n parties, such that only certain subsets of parties —described by a so-called access structure— can reconstruct the secret. In our formulation, the randomized algorithm corresponds to the dealer, s_i corresponds to the share given to party i, $x_i \in \{0, 1\}$ indicates whether party i is present in a subset, and F corresponds to the access structure.

considers a lower bound on share size that depends on the representation size of the function F as is the case for the state-of-the-art upper bounds, and the second variant considers a natural generalization of secret-sharing to non-monotone functions.

The Representation Size Barrier. To construct a secret-sharing scheme for a function F, we would need some representation of the function F, e.g., as a boolean formula or as a circuit or as a span program [KW93]. The most general approach we have for constructing secret-sharing schemes with small share sizes yields share sizes that are linear in the size of the monotone Boolean formula (more generally, the monotone span program) [BL88]. Recall that there are most $2^{O(s \log s)}$ circuits or formulas or span programs of size s. The following conjecture then captures the intuition that any secret-sharing scheme must yield share sizes that is polynomial in the representation size:

Conjecture 1. For any collection F_n of monotone functions over $\{0,1\}^n$ such that $|F_n| \geq 2^{\omega(n \log n)}$, the total share size of any secret-sharing scheme realizing F_n is at least $(\log |F_n|)^{\Omega(1)}$.[2]

Note that there are $2^{\Theta(2^n/\sqrt{n})}$ monotone functions over $\{0,1\}^n$, so the main conjecture is a special case of Conjecture 1. In addition, by a counting argument, the number of unbounded-fan-in circuits with s gates is no more than $2^{O(s(n+s))}$, thus the collection F_n contains functions whose circuit (and thus formula) size at least $\Omega(\sqrt{\log |F_n|} - n)$. This means that if our intuition that the share size is polynomial in the representation size (as a formula or even a circuit) is correct, then the share size for F_n must be at least $(\log |F_n|)^{\Omega(1)}$, as captured by the conjecture. Indeed, we refute this conjecture.

Theorem 1 (Informal). For any $s = s(n) \leq 2^{n/2}$, there is a collection $\hat{F}_{n,s}$ of $2^{s(n)}$ monotone functions over $\{0,1\}^n$ and a secret-sharing scheme for $\hat{F}_{n,s}$ where each share is $2^{O(\sqrt{\log s} \log \log s)} = (\log |\hat{F}_{n,s}|)^{o(1)}$ bits.

In particular, Theorem 1 has the following, we believe surprising, consequences.

First, our result implies that there are secret sharing schemes whose share size is much better than what the "representation size intuition" would suggest. In one extreme case, taking $s(n) = 2^{n/2}$, our result implies a family $\hat{F}_{n,2^{n/2}}$ of $2^{2^{n/2}} = 2^{2^{\Omega(n)}}$ monotone functions and a secret sharing scheme for $\hat{F}_{n,2^{n/2}}$ with share size only $2^{\tilde{O}(\sqrt{n})}$. Whereas, by simple counting arguments, there must be a function in this class with circuits (or formulas or monotone span programs or essentially every other natural computational model we can think of) of size $2^{\Omega(n)}$.

---

[2] The same secret-sharing algorithm can be used to realizing as many as $n!$ different access functions by permuting the parties. This trick comes from the nature of secret sharing, thus two access functions is equivalent if one is the composition of a permutation and the other, and Conjecture 1 should be stated on the number of equivalence classes in F_n. Assuming $|F_n| \gg n!$ has essentially the same effect.

As another reference point, taking $s(n) = n^{\log n}$, it follows that there exists monotone functions over $\{0, 1\}^n$ that require quasi-polynomial (in n) size circuits (and formulas and monotone span programs), but which admit secret-sharing schemes with polynomial share size. This in particular implies that existing secret-sharing schemes with share sizes linear in the formula size are far from optimal.

Second, our result implies that "non-linear reconstruction" totally dominates "linear reconstruction". Secret-sharing schemes with linear reconstruction are known to be equivalent to monotone span programs [KW93], whereas the scheme from Theorem 1 has a *non-linear* reconstruction algorithm. In particular, our results shows that for share size $\text{poly}(n)$ (resp., $2^{\sqrt{n}}$), there are $2^{\text{quasipoly}(n)}$ (resp., $2^{2^{\Omega(n)}}$) access structures that can be realized by secret sharing schemes with non-linear reconstruction, compared to $2^{\text{poly}(n)}$ (resp., $2^{2^{\Omega(\sqrt{n})}}$) by linear schemes.

Prior to this work, such a statement was only known under intractability assumptions pertaining to number-theoretic and combinatorial problems like quadratic residuosity and graph isomorphism [BI01, VV15], whereas our result is *unconditional*.

Non-monotone Secret Sharing. A further generalization called *non-monotone* secret sharing was defined in the work of Beimel and Ishai [BI01] and further studied in [BIKK14, VV15]; this is a natural extension of secret sharing to any arbitrary, possibly non-monotone F. A non-monotone secret-sharing scheme for a function $F : \{0, 1\}^n \to \{0, 1\}$ is a randomized algorithm that on input a secret, outputs n *pairs* of shares $(s_{i,0}, s_{i,1})_{i \in [n]}$ such that for any $(x_1, \ldots, x_n)$, the n shares $(s_{1,x_1}, \ldots, s_{n,x_n})$ determine the secret if $F(x_1, \ldots, x_n) = 1$ and reveal nothing about the secret otherwise. Standard monotone secret-sharing correspond to the special case where F is monotone and $s_{1,0} = \cdots = s_{n,0} = \bot$. Non-monotone secret sharing schemes are natural candidates for use in advanced cryptographic schemes such as attribute-based encryption [GPSW06, OSW07].

It is easy to see that we can construct non-monotone secret-sharing schemes for *all functions* on n bits starting from standard secret-sharing for all monotone functions on $2n$ bits, with a small polynomial blow-up in the share size. This might suggest that the best share sizes for non-monotone secret-sharing and standard secret-sharing are polynomially related, motivating the following strengthening of the main conjecture that we formalize below:

Conjecture 2. There exists a family of functions $\{F_n : \{0, 1\}^n \to \{0, 1\}\}_{n \in \mathbb{N}}$ such that the total share size in any non-monotone secret sharing scheme for F_n is $2^{\Omega(n)}$ bits.

Indeed, we also refute this conjecture:

Theorem 2 (Informal). There is a non-monotone secret-sharing for the family of all functions $F : \{0, 1\}^n \to \{0, 1\}$ where each share is $2^{\tilde{O}(\sqrt{n})}$ bits.

1.2   Overview of Our Constructions

We derive both Theorems 1 and 2 from the construction of a more general cryptographic primitive, namely that of conditional disclosure of secrets (CDS)

[GIKM00], which is a generalization of non-monotone secret-sharing to general, non-boolean inputs. Informally, conditional disclosure of secrets allows a set of parties to disclose a secret to an external party Charlie, subject to a given condition on their joint inputs. Concretely, we consider $(k + 1)$-party CDS for **INDEX**$_N$, where Alice holds $\mathbf{D} \in \{0,1\}^N$, parties $P_1, \ldots, P_k$ "jointly" hold an index $i \in [N]^3$, and all of them hold a secret $\mu \in \{0,1\}$, and Charlie knows $\mathbf{D}, i$ and should learn μ iff $\mathbf{D}[i] = 1$. To enable this, Alice and all the parties should share randomness that is hidden from Charlie (akin to the random coins used in a secret-sharing scheme). Our goal is to minimize the communication complexity in CDS, that is, the total number of bits sent by Alice and the k parties to Charlie.

Our main result is as follows:

Theorem (Main). For any $1 \leq k \leq \log N$, there is a $(k+1)$-party CDS for **INDEX**$_N$ where the total communication complexity is $2^{O(\sqrt{\log N} \cdot \log \log N)}$ bits.

Previously, such a result was only known for $k = 1$ [LVW17]. Before describing our $(k+1)$-party CDS, we briefly explain how Theorems 1 and 2 follow from the CDS.

Our non-monotone secret-sharing scheme for all functions $F : \{0,1\}^n \rightarrow \{0,1\}$ in Theorem 2 follows from the special case $k = \log N$, where $N = 2^n$. Concretely, the non-monotone secret-sharing scheme for F is derived from the $(n + 1)$-party CDS for **INDEX**$_{2^n}$ as follows: Alice holds the truth table $\mathbf{D} \in \{0,1\}^{2^n}$ of a function $F : \{0,1\}^n \rightarrow \{0,1\}$ and each party $P_i, i = 1, \ldots, n$ holds a single bit of the index $i \in [2^n]$, and the messages sent by P_i in the CDS corresponds to the shares. Going from the $(n+1)$-party CDS to Theorem 1 requires an additional folklore transformation which transforms a *non-monotone* secret-sharing scheme for any $F : \{0,1\}^n \rightarrow \{0,1\}$ into a *monotone* secret-sharing scheme with roughly the same share size for a related function F'; see Sect. 6.3 for details.

A General Framework for CDS. We proceed to provide an overview for our $(k+1)$-party CDS protocol. We begin with a general framework for constructing CDS protocols, and then sketch how to instantiate the underlying building blocks to obtain our main result.

The LVW17 Framework. We begin by sketching the 2-party CDS protocol, i.e. $k = 1$, from [LVW17] (which in turn builds upon [BIKK14,DG15]). The starting point of their protocol is a notion of (N, ℓ)-PIR encoding, which encodes $i \in [N]$ as vector $\mathbf{u}_i \in \mathbb{Z}_6^\ell$ and $\mathbf{D}$ as a function $H_\mathbf{D} : \mathbb{Z}_6^\ell \rightarrow \mathbb{Z}_6^\ell$ such that for all $i, \mathbf{D}, \mathbf{w}$, we have

$$\mathbf{D}[i] = \langle H_\mathbf{D}(\mathbf{u}_i + \mathbf{w}), \mathbf{u}_i \rangle - \langle H_\mathbf{D}(\mathbf{w}), \mathbf{u}_i \rangle.$$

---

[3] We will make the precise sense of how the parties "jointly" hold the index clear in a little bit, but roughly speaking, the reader should imagine that each party holds $\lceil (\log N)/k \rceil$ bits of the index.

This immediately implies that for all $\mu \in \{0, 1\}$, we have

$$\mu \mathbf{D}[i] = \langle H_{\mathbf{D}}(\mu \mathbf{u}_i + \mathbf{w}), \mathbf{u}_i \rangle - \langle H_{\mathbf{D}}(\mathbf{w}), \mathbf{u}_i \rangle. \tag{1}$$

[LVW17] constructed a two-party CDS protocol with communication $O(\ell)$ starting from any (N, ℓ)-PIR encoding, and also gave a construction of a $(N, 2^{\tilde{O}(\sqrt{\log N})})$-PIR encoding. The two-party CDS protocol is as follows:

- Alice and P_1 share randomness $\mathbf{w}, \mathbf{r}$ (hidden from Charlie);
- Alice sends $\mathbf{m}_A^1 := H_{\mathbf{D}}(\mathbf{w}) + \mathbf{r}$.
- P_1 sends $\mathbf{m}_B^1 := \mu \mathbf{u}_i + \mathbf{w}$ and $m_B^2 := \langle \mathbf{u}_i, \mathbf{r} \rangle$.
- Charlie can now compute $\mu \mathbf{D}[i]$ (and thus μ) given $\mathbf{D}, i, (\mathbf{m}_A^1, \mathbf{m}_B^1, m_B^2)$ using the relation

$$\mu \mathbf{D}[i] = \langle \underbrace{H_{\mathbf{D}}(\mu \mathbf{u}_i + \mathbf{w})}_{\mathbf{m}_B^1}, \mathbf{u}_i \rangle - \langle \underbrace{H_{\mathbf{D}}(\mathbf{w}) + \mathbf{r}}_{\mathbf{m}_A^1}, \mathbf{u}_i \rangle + \underbrace{\langle \mathbf{r}, \mathbf{u}_i \rangle}_{m_B^2}$$

which follows readily from (1).

It is easy to see that the total communication is $O(\ell)$. Privacy follows fairly readily from the fact that the joint distribution of $(\mathbf{m}_A^1, \mathbf{m}_B^1)$ is uniformly random, and that m_B^2 is completely determined given $(\mathbf{m}_A^1, \mathbf{m}_B^1)$ and $\mu \mathbf{D}[i]$ along with $\mathbf{D}, i$.

The Multi-party Setting. We show how to extend the [LVW17] construction to the multi-party setting with k parties, for any $k \geq 1$. Here, the index i is distributed across k parties $P_1, \ldots, P_k$. The key idea is to have these k parties jointly emulate P_1 in the two-party CDS via secure multi-party computation (MPC); in fact, because of communication constraints, we will use a *private simultaneous messages* protocol [FKN94, IK97] in this setting, where each of the k parties sends a single message to Charlie. That is, these k parties jointly hold inputs $i, \mathbf{w}, \mathbf{r}, \mu$ and they will run an MPC protocol with Charlie so that Charlie learns

$$(\mu \mathbf{u}_i + \mathbf{w}, \langle \mathbf{u}_i, \mathbf{r} \rangle), \tag{2}$$

upon which Charlie can proceed as in the two-party CDS to recover $\mu \mathbf{D}[i]$. Moreover, security of the MPC protocol ensures that what Charlie learns in the k-party CDS is the same as that in the two-party CDS. Correctness and security then follow readily from those of the two-party CDS and the MPC protocol.

Recall that our goal is to obtain a protocol with total communication complexity $2^{o(n)}$, and we need to make sure that the MPC protocol does not blow up the communication by too much. The key insight is that the total size of the inputs for the MPC protocol is $O(\log N + \ell)$ and is in particular independent of $\mathbf{D}$. Therefore, it suffices to design an MPC protocol for computing (2) with polynomial communication for the $(N, 2^{\tilde{O}(\sqrt{\log N})})$-PIR encoding in [LVW17], upon which we will derive a k-party CDS protocol with total communication $\text{poly}(\ell) = 2^{\tilde{O}(\sqrt{\log N})}$.

Minimizing Communication via Decomposability. Prior works on MPC tells us that the communication cost for securely computing (2) is essentially dominated by the cost of (non-securely) computing the k-ary functionality

$$i = (i_1, \ldots, i_k) \mapsto \mathbf{u}_i.$$

In fact, it suffices to construct PIR-encodings where $\mathbf{u}_i \in \mathbb{Z}_6^\ell$ may be derived by applying a simple function to vectors $\mathbf{u}_{i_1}, \ldots, \mathbf{u}_{i_k} \in \mathbb{Z}_6^\ell$, each of which is derived from some local (and possibly complex) computation on $i_1, \ldots, i_k$ respectively. In this work, we consider $\mathbf{u}_i$ that are given by

$$\mathbf{u}_i = \mathbf{u}_{1,i_1} \circ \cdots \circ \mathbf{u}_{k,i_k}$$

where $\circ$ corresponds to point-wise product of vectors. We refer to this property as k-*decomposability*.

Using k-decomposable $\mathbf{u}_i$, the computation in (2) can be written as

$$(i_1, \ldots, i_k, \mathbf{w}, \mathbf{r}, \mu) \mapsto (\mu \mathbf{u}_{1,i_1} \circ \cdots \circ \mathbf{u}_{k,i_k} + \mathbf{w}, \langle \mathbf{u}_{1,i_1} \circ \cdots \circ \mathbf{u}_{k,i_k}, \mathbf{r} \rangle)$$

which is essentially a degree $k+1$ computation over the inputs; concretely, it can be written as the sum of a small number of monomials over $(\mathbf{u}_{i_1}, \ldots, \mathbf{u}_{i_k}, \mathbf{w}, \mathbf{r}, \mu)$. Following [IK00, IK02, CFIK03], such a computation admits a non-interactive MPC protocol satisfying perfect, information-theoretic security, and total communication polynomial in $\ell, k, \log N$.

This brings us to the final building block: a $(N, 2^{\tilde{O}(\sqrt{\log N})})$-PIR encoding that is k-decomposable.

PIR-Encodings from Matching Vector (MV) Families. The key tool in the $(N, 2^{\tilde{O}(\sqrt{\log N})})$-PIR encoding in [LVW17] is matching vector (MV) families, first constructed by Grolmusz [Gro00] and introduced to cryptography in the context of private information retrieval [Yek08, Efr12, DGY11, DG15].

MV Families. A (mod 6) MV family is an explicit collection of vectors $\{(\mathbf{u}_i, \mathbf{v}_i)\}_{i \in [N]}$ such that $\mathbf{u}_i, \mathbf{v}_i \in \mathbb{Z}_6^\ell$ where $\ell = 2^{O(\sqrt{\log N} \cdot \log \log N)}$ and:

$$\langle \mathbf{u}_i, \mathbf{v}_i \rangle = 0,$$
$$\langle \mathbf{u}_i, \mathbf{v}_j \rangle \in \{1, 3, 4\} \quad \text{for } i \neq j.$$

where all computations are done mod 6.

At this point, it may seem like we are abusing notation as we are using $\mathbf{u}_i$ to denote both the vectors in a MV family and those in a PIR encoding. Fortunately, they are essentially the same thing in the [LVW17] construction, and therefore, it suffices to construct MV families where the underlying $\mathbf{u}_i$'s are k-decomposable.

Prior Constructions. We begin with an overview of Grolmusz's MV families [Gro00]. Fix any integers h, w so that $\binom{h}{w} \geq N$. Pick *any* distinct $\mathbf{x}_1, \ldots, \mathbf{x}_N \in \{0,1\}^h$ of Hamming weight exactly w. Observe that for all $i, j \in [N]$:

$$\|\mathbf{x}_i \circ \mathbf{x}_j\|_1 \begin{cases} =w & \text{if } i = j \\ <w & \text{if } i \neq j \end{cases}$$

The vectors $\mathbf{u}_i$ will have length $\ell = h^{O(\sqrt{w})}$ indexed by subsets S of $[h]$ of size at most $O(\sqrt{w})$ and defined as follows:

$$\mathbf{u}_i[S] = \prod_{i' \in S} \mathbf{x}_i[i']$$

The reason for defining $\mathbf{u}_i$ this way is that for any fixed polynomial f of degree $O(\sqrt{w})$ from $\{0,1\}^h \to \mathbb{Z}_6$, we can write $f(\mathbf{x}_i \circ \mathbf{x}_j)$ as $\langle \mathbf{u}_i, \mathbf{v}_j \rangle$ where $\mathbf{v}_j$ depends only on f and $\mathbf{x}_j$. (Roughly speaking, the polynomial f checks whether the Hamming weight of its input is equal to or less than w.) We can then set $h = 2 \log N, w = \log N$, which yields $\ell = 2^{O(\sqrt{\log N} \cdot \log \log N)}$.

Our Construction. In our setting, the index $i = (i_1, \ldots, i_k)$ is divided equally amongst k parties and want $\mathbf{u}_i$ to be of the form

$$\mathbf{u}_i = \mathbf{u}_{1,i_1} \circ \cdots \circ \mathbf{u}_{k,i_k}.$$

To achieve this property, it suffices to modify the choices of $\mathbf{x}_1, \ldots, \mathbf{x}_N$ in the prior construction. Concretely, we pick w, h to be multiples of k such that $\binom{h/k}{w/k}^k \geq N$. Then, we choose $\mathbf{x}_1, \ldots, \mathbf{x}_N$ so that each $\mathbf{x}_i$ can be decomposed into k blocks $(\mathbf{x}_{i_1} \| \cdots \| \mathbf{x}_{i_k})$ each of weight exactly w/k. Recall that each entry of $\mathbf{u}_i$ is a product of $O(\sqrt{w})$ entries of $\mathbf{x}_i$, which means we can write $\mathbf{u}_i = \mathbf{u}_{1,i_1} \circ \cdots \circ \mathbf{u}_{k,i_k}$ where each $\mathbf{u}_{1,i_1}, \ldots, \mathbf{u}_{k,i_k}$ depends on $\mathbf{x}_{i_1}, \ldots, \mathbf{x}_{i_k}$ respectively. We can still set $h = 2 \log N, w = \log N$ as before, since $\binom{h/k}{w/k}^k \geq ((\frac{h}{w})^{w/k})^k = N$ for any $1 \leq k \leq \log N$.

1.3   Related Work

A linear secret-sharing scheme is one where the secret sharing algorithm computes a linear function of the secret and its randomness. Most secret-sharing schemes in the literature are linear secret-sharing schemes, and many cryptographic applications also require the linearity property. For linear secret-sharing schemes, the existing upper bounds (namely, linear in formula or span program size) are essentially optimal, due to their connection to a computational model called monotone span programs defined by Karchmer and Wigderson [KW93].

Using this connection, we know the following results about linear secret-sharing schemes. We know there exist access functions that need $2^{\Omega(n)}$ share size for linear secret sharing via a counting argument [KW93, Bei11]. As for explicit functions, we have $n^{\log n / \log \log n}$ lower bounds for linear secret sharing realizing some explicit access functions [BGP95,BGW99]. Quite recently, this was improved by Pitassi and Robere [RPRC16,PR17] who showed an *exponential* lower bound for monotone span programs realizing some explicit access functions. Therefore, to beat these exponential bounds, as we do in this work, we need to turn to general, *non-linear* secret-sharing schemes.

1.4 Discussion

Our work highlights new connections and exploits the rich interplay amongst old and new problems in information-theoretic cryptography: from secret-sharing (70s), to multi-party computation (80s), to private information retrieval (90s), and brings forth strong evidence against the conjectured optimality of the classic constructions for monotone secret-sharing.

While we do not construct secret-sharing schemes with sub-exponential share sizes for all monotone functions over $\{0,1\}^n$, as would be necessary to refute the main conjecture, we do achieve sub-exponential $2^{O(\sqrt{n} \log n)}$ share sizes for a large number of these functions, namely $2^{2^{n/2}}$ out of $2^{2^{n-O(\log n)}}$ of them. There are several very exciting new research directions at this point, and we highlight two specific questions related to the main conjecture:

- Does there exist a family containing 1% of all monotone functions over $\{0,1\}^n$ that admit a monotone secret-sharing scheme of total share size $2^{o(n)}$? Here, 1% can be replaced by any constant. One way to resolve this question would be to extend our construction to a *larger* set of monotone functions.
- Does there exist a family of $2^{2^{\Omega(n)}}$ functions over $\{0,1\}^n$ that admit a monotone secret-sharing scheme of total share size $2^{o(\sqrt{n})}$? One way to resolve this question would be to *improve* the communication complexity of 2-party CDS, which in turn seems closely related to the problem of improving the communication complexity of the state-of-the-art 2-server private information retrieval.

On another thread, we note that the work of Beimel et al. [BIKK14] showed ways to use improved PIR schemes to obtain protocols for various information-theoretic multiparty tasks improving their communication complexity or randomness complexity. Our work continues this line of thought. We mention that the problem of improving the communication complexity of private simultaneous messages (PSM) protocols, a generalization of CDS, for general functions to a sub-exponential number remains wide open.

1.5 Organization

We start with Sect. 3 which describes the framework of the multiparty CDS construction. That is, a multi-party CDS scheme for the **INDEX**predicate can be constructed from: (a) a "PIR-encoding", and (b) a private simultaneous messages (PSM) protocol computing a special functionality related to the PIR-encoding (see Theorem 3.1). The following two sections construct these two building blocks. Section 4 shows there exists a succinct PIR-encoding that is also decomposable (Theorem 4.9). Section 5 shows there exists an efficient PSM for the special functionality if the PIR-encoding is decomposable (Theorem 5.1). An immediate combination of the above results gives an good multi-party CDS protocol (Theorem 6.1, which is a restatement of the main theorem in the introduction). Section 6 also presents our applications to secret sharing. A good multi-party CDS protocol implies non-monotone secret sharing for all functions (Theorem 6.5, which is a restatement of Theorem 2 in the introduction). It also implies monotone secret sharing for a large class of functions (Theorem 6.7, which is a restatement of Theorem 1 in the introduction). In addition, Sect. 7 shows similar results for linear CDS and linear secret sharing.

2   Preliminaries and Definitions

We start with some notation that we will use throughout the paper.

- Let $\mathcal{R}$ denote a generic commutative ring, $\mathbb{Z}$ denote the integer ring, and $\mathbb{Z}_m$ denote the ring of integers modulo m. Let $\mathbb{F}$ denote a generic finite field. When q is a prime power, let $\mathbb{F}_q$ denote the finite field of size q. For an integer m, let $[m] := \{1, \ldots, m\}$.
- We will let boldface letters, such as $\mathbf{x}$, denote vectors. When $\mathbf{x}$ is a vector, let $\mathbf{x}[i]$ denote its i-th element.
- For vectors $\mathbf{x} \in \mathcal{R}^\ell, \mathbf{y} \in \mathcal{R}^{\ell'}$, let $\mathbf{x}\|\mathbf{y} \in \mathcal{R}^{\ell+\ell'}$ denote their concatenation.
- Call a vector $\mathbf{x} \in \mathcal{R}^\ell$ a *zero-one vector* if its entries are either 0 or 1. For a zero-one vector $\mathbf{x}$, let $\|\mathbf{x}\|_1$ denote the number of 1's in $\mathbf{x}$.

Definition 2.1 (Point-wise product). *For any two vectors* $\mathbf{x}, \mathbf{y} \in \mathcal{R}^\ell$, *their point-wise product, denoted by* $\mathbf{x} \circ \mathbf{y}$, *is a vector in the same linear space whose i-th element is the product of the i-th elements of* $\mathbf{x}, \mathbf{y}$, *i.e.* $(\mathbf{x} \circ \mathbf{y})[i] = \mathbf{x}[i] \cdot \mathbf{y}[i]$.

This is also known in the literature as the Hadamard product or Schur product, typically used in the context of matrices.

2.1 k-party Conditional Disclosure of Secrets (CDS)

In a k-party CDS scheme, there are k parties who know a secret message μ and jointly hold input x. These parties cannot communicate with each other, but instead they have access to a common random string (CRS). Their goal is to send a single message to the CDS referee Charlie, at the end of which Charlie, who already knows x, should learn μ if and only if $\mathsf{P}(x) = 1$, for a fixed predicate P.

Definition 2.2 (Conditional disclosure of secrets (CDS) [GIKM00]). *Let input spaces* $\mathcal{X}_1, \ldots, \mathcal{X}_k$, *secret space* $\mathcal{M}$ *and randomness space* $\mathcal{W}$ *be finite sets.*

Fix a predicate $\mathsf{P} : \mathcal{X}_1 \times \mathcal{X}_2 \times \ldots \times \mathcal{X}_k \to \{0,1\}$. *A* cc*-conditional disclosure of secrets* (CDS) *protocol for* P *is a tuple of deterministic functions* $(\mathsf{B}_1, \ldots, \mathsf{B}_k, \mathsf{C})$

$$\text{Transmitting functions } \mathsf{B}_i : \mathcal{M} \times \mathcal{X}_i \times \mathcal{W} \to \{0,1\}^{\text{cc}}$$

$$\text{Reconstruction function } \mathsf{C} : \mathcal{X}_1 \times \ldots \times \mathcal{X}_k \times \{0,1\}^{\text{cc} \times k} \to \mathcal{M}$$

satisfying the following properties:

(reconstruction). *For all* $(x_1, \ldots, x_k) \in \mathcal{X}_1 \times \ldots \times \mathcal{X}_k$ *such that* $\mathsf{P}(x_1, \ldots, x_k) = 1$, *for all* $w \in \mathcal{W}$, *and for all* $\mu \in \mathcal{M}$:

$$\mathsf{C}(x_1, \ldots, x_k, \mathsf{B}_1(\mu, x_1; w), \ldots, \mathsf{B}_k(\mu, x_k; w)) = \mu.$$

(privacy). *There exists a randomized algorithm* S *such that for all input tuple* $(x_1, \ldots, x_k) \in \mathcal{X}_1 \times \ldots \times \mathcal{X}_k$ *satisfying* $\mathsf{P}(x_1, \ldots, x_k) = 0$, *the joint distribution of* $(\mathsf{B}_1(\mu, x_1; w), \ldots, \mathsf{B}_k(\mu, x_k; w))$ *is perfectly indistinguishable from* $\mathsf{S}(x_1, \ldots, x_k)$, *where the randomness are taken over* $w \xleftarrow{\text{R}} \mathcal{W}$ *and the coin tosses of* S.

Predicates. We consider the following predicates:

- Index $\mathbf{INDEX}_N^{k+1}$: An index $i \in [N]$ is distributed amongst the first k parties. Let $\mathcal{X}_1 = \cdots = \mathcal{X}_k := [\sqrt[k]{N}]$, $\mathcal{X}_{k+1} = \{0,1\}^N$ and under the natural mapping $[N] \ni i \mapsto (i_1, \ldots, i_k) \in ([\sqrt[k]{N}])^k$,

$$\mathsf{P}_{\mathbf{INDEX}}(i_1, \ldots, i_k, \mathbf{D}) = 1 \text{ iff } \mathbf{D}[i] = 1$$

  Note that $\mathbf{D}$ can also be interpreted as the characteristic vector of a subset of $[N]$.
- All ("worst") predicates $\mathbf{ALL}_N^k$: An index $i \in [N]$ is distributed among the k parties as before and the predicate is specified by a truth table. Let $\mathcal{X}_1 = \ldots = \mathcal{X}_k := [\sqrt[k]{N}]$ and there is a fixed public function $F : [N] \to \{0,1\}$. under the natural mapping $i \in [N] \mapsto (i_1, \ldots, i_k) \in [\sqrt[k]{N}]^k$,

$$\mathsf{P}_{\mathbf{ALL}}(i_1, \ldots, i_k) := F(i).$$

$\mathbf{ALL}_N^k$ is an easier predicate (family) than $\mathbf{INDEX}_N^{k+1}$, as any CDS protocol for $\mathbf{INDEX}_N^{k+1}$ implies a CDS protocol for $\mathbf{ALL}_N^k$ with the same total communication complexity (e.g. [LVW17, Sect. 2.3]).

The definitions of both predicates inherently require N to be a perfect k-th power. In the case where N is not a k-th power, we can pad N to the nearest larger k-th power. When $k \leq \log N$, it's guaranteed that the nearest larger k-th power is no greater than N^2. For the sake of our result, a square blowup on N doesn't matter.

2.2 k-party Private Simultaneous Messages (PSM)

In a k-party PSM scheme, there are k parties who jointly have input x, and they cannot communicate with each other, but have access to a common random string (CRS), as in the case of CDS. There is also the PSM referee Charlie who wants to learn $F(x)$, for a fixed functionality F. In the PSM scheme, every party sends a single message to Charlie based on its piece of input and the CRS. Given these messages, Charlie should be able to learn $F(x)$ and nothing else about x.

Definition 2.3 (Private simultaneous message (PSM)). *Let $\mathcal{X}_t$ be the input space of the t-th party, let $\mathcal{X} \subseteq \mathcal{X}_1 \times \ldots \times \mathcal{X}_k$ be the input space, and let $\mathcal{M}$ be the output space. Fix a functionality $F : \mathcal{X} \to \mathcal{M}$. A cc-bits private simultaneous message (PSM) protocol for F is a tuple of deterministic functions $(B_1, \ldots ; B_k, C)$:*

$$\text{Transmitting functions } B_i : \mathcal{X}_i \times \mathcal{W} \to \{0,1\}^{cc},$$
$$\text{Reconstruction function } C : \{0,1\}^{cc \times k} \to \mathcal{M}$$

satisfying the following properties:

(reconstruction). *For all $(x_1, \ldots, x_k) \in \mathcal{X}$:*

$$C(B_1(x_1; w), \ldots, B_k(x_k; w)) = F(x_1, \ldots, x_k)$$

(privacy). *There exists a randomized simulator S, such that for any input tuple $(x_1, \ldots, x_k) \in \mathcal{X}$, the joint distribution $(B_1(x_1; w), \ldots, B_k(x_k; w))$ is perfectly indistinguishable from $S(F(x_1, \ldots, x_k))$, where the distributions are taken over $w \xleftarrow{R} \mathcal{W}$ and the coin tosses of S.*

Common Input. In the PSM functionalities we care about in this work, a part of the input is shared among all parties. That is, the input of the t-th party is of the form $x'_t = (x_t, y)$, where x_t is t-th party's exclusive input, and y is shared input known by all parties but the referee Charlie. This can be formalized by letting $\mathcal{X}'_t = \mathcal{X}_t \times \mathcal{Y}$ as the t-th party's input space and by defining the global input space $\mathcal{X}$ as consisting of vectors $((x_1, y_1), \ldots, (x_k, y_k))$ where $y_1 = \ldots = y_k$. For notational simplicity, let $F(x_1, \ldots, x_k; y)$ denotes $F((x_1, y), \ldots, (x_k, y))$ and let the transmission functions be denoted as $B_i(x_i; y; w)$.

Functionality. We consider the following functionalities of interest:

- Affine functions $\mathbf{AFFINE}^k$: For vectors $\mathbf{x}_1, \ldots, \mathbf{x}_k \in \mathcal{R}^n$ and any affine function $f : \mathcal{R}^{kn} \to \mathcal{R}$

$$F_{\mathbf{AFFINE}}(\mathbf{x}_1, \ldots, \mathbf{x}_k; f) = f(\mathbf{x}_1, \ldots, \mathbf{x}_k).$$

- Branching Program $\mathbf{BP}^k_m$: A mod-$\mathcal{R}$ branching program is a directed acyclic graph with a source s and a sink t, where every edge is labeled with an affine function. Given an input vector $\mathbf{x}$, the value of an edge is the value of its label function when applied to $\mathbf{x}$; the value of an (s,t)-path is the product of the values on its edges; and the value of the branching program is the sum of the values of all (s,t)-paths.

To formalize the branching program problem as a PSM functionality: let the branching program be the shared input; split input vector $\mathbf{x}$ among the parties as their exclusive input. More precisely, in $\mathbf{BP}_m^k$, there are k parties and a branching program with m nodes. Let $f_{i,j}$ $(1 \leq i < j \leq m)$ denote the affine function assigned to edge (i,j). The t-th party's input is $\mathbf{x}_t \in \mathcal{R}^n$ and $\{f_{i,j}\}_{i<j}$. The functionality $\mathbf{BP}_m^k$ is defined as

$$\mathsf{F}_{\mathbf{BP}}(\mathbf{x}_1, \ldots, \mathbf{x}_k; \{f_{i,j}\}_{i<j}) = \sum_{s\text{-}t \text{ path } p} \prod_{\substack{\text{edge } (i,j) \\ \text{in } p}} f_{i,j}(\mathbf{x}_1, \ldots, \mathbf{x}_k).$$

The affine function functionality $\mathbf{AFFINE}^k$ is the special case of $\mathbf{BP}^k$ when the associated graph has only two nodes and one edge.

3   A Framework for Multi-party CDS

In this section, we describe a framework for constructing multi-party conditional disclosure of information (CDS) protocols. Our framework relies on vector families that satisfy two properties described below. The first is the property of being a "PIR encoding", satisfied by matching vector families and was used in [LVW17] to construct two-party CDS. The second is the existence of a communication-efficient private simultaneous messages (PSM) protocol for a functionality associated to the PIR encoding scheme.

PIR Encoding. We define an (N, ℓ)-PIR encoding as a family of N vectors $(\mathbf{u}_i)_{i \in [N]}$ where each $\mathbf{u}_i \in \mathbb{Z}_6^\ell$, along with a mapping $H_{\mathbf{D}} : \mathbb{Z}_6^\ell \to \mathbb{Z}_6^\ell$ for every table (string) $\mathbf{D} \in \{0,1\}^N$. The PIR encoding property requires that for any index $i \in [N]$ and any vector $\mathbf{w} \in \mathbb{Z}_6^\ell$,

$$\langle \mathbf{u}_i, H_{\mathbf{D}}(\mathbf{w} + \mathbf{u}_i) - H_{\mathbf{D}}(\mathbf{w}) \rangle = \mathbf{D}[i] \cdot \phi_{\mathbf{w},i} \tag{3}$$

where $\phi_{\mathbf{w},i} \neq 0$.

As the name suggests, any PIR encoding scheme can be used to construct a 2-server information theoretic private information retrieval (PIR) scheme. Indeed, the 2-server PIR schemes of [CKGS98, WY05, DG15] can all be viewed as instances of this paradigm.

A Special-Purpose PSM Protocol. A PIR encoding scheme can also be used to construct the following CDS protocol as described in [LVW17]. Alice holds a database $\mathbf{D}$, and Bob holds an index i and a secret message μ. Bob sends the pair $(\mathbf{w} + \mu\mathbf{u}_i, \langle \mathbf{u}_i, \mathbf{r} \rangle)$, and Alice sends $H_{\mathbf{D}}(\mathbf{w}) + \mathbf{r}$ to Charlie. Here, $\mathbf{w}$ and $\mathbf{r}$ come from the CRS. Charlie now has enough information to compute

$$\langle \mathbf{u}_i, H_{\mathbf{D}}(\mathbf{w} + \mu\mathbf{u}_i) - H_{\mathbf{D}}(\mathbf{w}) \rangle = \mathbf{D}[i] \cdot \mu \cdot \phi_{\mathbf{w},i}$$

which reveals the secret μ if and only if $\mathbf{D}[i] \neq 0$.

In $(k + 1)$-party CDS for $\mathbf{INDEX}_N$, the index $i = (i_1, \ldots, i_k)$ is divided equally among the first k parties and the $(k + 1)^{th}$ party holds the database $\mathbf{D}$. Our plan is to have the first k parties simulate what Bob did in the 2-party CDS. We describe our CDS protocol in Fig. 1, which assumes a PSM protocol for computing the k-party functionality

$$F_{\mathsf{aux}} : [\sqrt[k]{N}] \times \ldots \times [\sqrt[k]{N}] \times (\{0,1\} \times \mathbb{Z}_6^\ell \times \mathbb{Z}_6^\ell) \to \mathbb{Z}_6^\ell \times \mathbb{Z}_6 \tag{4}$$
$$\text{where } F_{\mathsf{aux}}(i_1, \ldots, i_k; (\mu, \mathbf{w}, \mathbf{r})) \mapsto (\mathbf{w} + \mu\mathbf{u}_i, \langle \mathbf{u}_i, \mathbf{r} \rangle)$$

Here, $\mathbf{w}, \mathbf{r}$ and μ are common inputs and the index i is divided equally among the first k parties. This, in particular, will enable the k parties to simulate the Bob in the 2-party CDS setting, and jointly send $\mathbf{w} + \mu\mathbf{u}_i$ and $\langle \mathbf{u}_i, \mathbf{r} \rangle$ to Charlie without revealing any extra information. More precisely, our construction requires a PIR encoding such that there is a PSM for this functionality with communication complexity $\mathsf{cc}_{\mathsf{PSM}}(N, \ell, k)$ that is as small as possible.

Theorem 3.1. *Assume that there is an (N, ℓ)-PIR encoding scheme $(\mathbf{u}_i)_{i \in [N]}$ and a PSM for F_{aux} in (4) with communication complexity $\mathsf{cc}_{\mathsf{PSM}}(N, \ell, k)$, then there is a $(k+1)$-party CDS protocol for $\mathbf{INDEX}_N^{k+1}$ (Fig. 1) with communication complexity $\ell + \mathsf{cc}_{PSM}(N, \ell, k)$.*

CDS for $\mathbf{INDEX}_N^{k+1}$

Building blocks: An (N, ℓ)-PIR encoding scheme $(\mathbf{u}_i)_{i=1}^N$
    and a PSM protocol $(\mathsf{B}_1, \ldots, \mathsf{B}_{k+1}, \mathsf{C})$ for F_{aux} in (4).
Input of P_t $(1 \leq t \leq k)$: $i_t \in [N^{1/k}]$
Input of P_{k+1} (Alice): $\mathbf{D} \in \{0,1\}^N$.
Shared Randomness: $\mathbf{w}, \mathbf{r} \in \mathbb{Z}_6^\ell$, $\mu \in \{0,1\}$ and randomness s for the PSM.

The protocol proceeds as follows.

- For $1 \leq t \leq k$, the t-th party sends $\mathbf{m}_{\mathsf{psm},t} := \mathsf{B}_t(i_t; (\mu, \mathbf{w}, \mathbf{r}); s)$.
- P_{k+1} (Alice) sends $\mathbf{m}_A := \mathbf{r} + H_{\mathbf{D}}(\mathbf{w})$.
- Charlie computes $(\mathbf{m}_B^1, \mathbf{m}_B^2) := \mathsf{C}(\mathbf{m}_{\mathsf{psm},1}, \ldots, \mathbf{m}_{\mathsf{psm},k})$.
- Charlie outputs 1 if $\langle \mathbf{u}_i, H_{\mathbf{D}}(\mathbf{m}_B^1) \rangle - \langle \mathbf{u}_i, \mathbf{m}_A \rangle + m_B^2 \neq 0$, and outputs 0 otherwise.

Fig. 1. $(k + 1)$-party CDS for $\mathbf{INDEX}_N^{k+1}$ from PIR encodings and PSM.

Proof. By the definition of PIR encoding, for any table $\mathbf{D} \in \{0,1\}^N$, there exists a mapping $H_{\mathbf{D}} : \mathbb{Z}_6^\ell \to \mathbb{Z}_6^\ell$ that satisfies Eq. (3). We now show correctness, privacy and efficiency of the protocol.

Correctness. The correctness of the PSM protocol tells us that $\mathbf{m}_B^1 = \mathbf{w} + \mu \mathbf{u}_i$ and $m_B^2 = \langle \mathbf{u}_i, \mathbf{r} \rangle$. Equation (3) then directly implies that

$$\langle \mathbf{u}_i, H_{\mathbf{D}}(\mathbf{w} + \mu \mathbf{u}_i) \rangle - \langle \mathbf{u}_i, H_{\mathbf{D}}(\mathbf{w}) \rangle = \mathbf{D}[i] \cdot \mu \cdot \phi_{\mathbf{w},i}$$

where $\mu \in \{0, 1\}$ is the secret message. Charlie learns

$$
\begin{aligned}
\langle \mathbf{u}_i, H_{\mathbf{D}}(\mathbf{m}_B^1) \rangle - \langle \mathbf{u}_i, \mathbf{m}_A \rangle + m_B^2 \\
= \langle \mathbf{u}_i, H_{\mathbf{D}}(\mathbf{w} + \mu \mathbf{u}_i) \rangle - \langle \mathbf{u}_i, \mathbf{r} + H_{\mathbf{D}}(\mathbf{w}) \rangle + \langle \mathbf{u}_i, \mathbf{r} \rangle \\
= \langle \mathbf{u}_i, H_{\mathbf{D}}(\mathbf{w} + \mu \mathbf{u}_i) \rangle - \langle \mathbf{u}_i, H_{\mathbf{D}}(\mathbf{w}) \rangle \\
= \mathbf{D}[i] \cdot \mu \phi_{\mathbf{w},i},
\end{aligned}
$$

which, since $\phi_{\mathbf{w},i} \neq 0$, gives μ if and only if $\mathbf{D}[i] = 1$.

Privacy. Privacy follows by putting the following observations together.

- First, the joint distribution of $\mathbf{m}_B^1$ and $\mathbf{m}_A$ is uniformly random, since we are using $(\mathbf{w}, \mathbf{r})$ as one-time pads;
- Secondly, when $\mathbf{D}[i] = 0$, we have $\langle \mathbf{u}_i, H_{\mathbf{D}}(\mathbf{w} + \mu \mathbf{u}_i) \rangle - \langle \mathbf{u}_i, H_{\mathbf{D}}(\mathbf{w}) \rangle = 0$. This means that $m_B^2 = \langle \mathbf{u}_i, \mathbf{m}_A \rangle - \langle \mathbf{u}_i, H_{\mathbf{D}}(\mathbf{m}_B^1) \rangle$ and thus can be simulated knowing only $\mathbf{m}_A$ and $\mathbf{m}_B^1$ (and, of course, $\mathbf{D}$ and i);
- Finally, the joint distribution of $\mathbf{m}_{\mathsf{psm},1}, \ldots, \mathbf{m}_{\mathsf{psm},k}$ can be perfectly simulated from $(\mathbf{m}_B^1, m_B^2)$, due to the privacy of the PSM.

Efficiency. Each party except Alice sends a PSM message of size at most $\mathsf{cc_{PSM}}(N, \ell, k)$. Alice sends a vector of size ℓ. The communication complexity of a party is no more than $\ell + \mathsf{cc_{PSM}}(N, \ell, k)$.   □

4   PIR Encoding from Decomposable Matching Vectors

These two ingredients used to construct multi-party CDS, namely PIR encodings and the special-purpose PSM protocol are connected by the property of *decomposability*.

Definition 4.1 (k-decomposability). *Let $N' := \sqrt[k]{N}$. A family of vectors $(\mathbf{u}_i)_{i=1}^N$ is k-decomposable if there exist vector families $(\mathbf{u}_{1,i})_{i=1}^{N'}, \ldots, (\mathbf{u}_{k,i})_{i=1}^{N'}$ such that under the natural mapping $i \mapsto (i_1, \ldots, i_k) \in [N']^k$*

$$\mathbf{u}_i = \mathbf{u}_{1,i_1} \circ \ldots \circ \mathbf{u}_{k,i_k}$$

for all $i \in [N]$. Here, $\circ$ denotes the component-wise multiplication operation. A family of vectors $(\mathbf{u}_i)_{i \in [N]}$ is called a k-decomposable PIR encoding if it is a PIR encoding and it is k-decomposable.

In this section, we construct a k-decomposable (N, ℓ)-PIR encoding with $\ell = 2^{O(\sqrt{\log N} \log \log N)}$. In Sect. 5, we show an efficient PSM for functionality (4) as long as $(\mathbf{u}_i)_{i \in [N]}$ is k-decomposable. Put together, they fulfill the assumptions in Theorem 3.1 and give us a communication-efficient multiparty CDS protocol.

4.1   PIR Encodings from Matching Vector Families

First, we define matching vector families and show that they give rise to PIR encodings. Our exposition here follows [LVW17] and uses techniques from [DG15].

Definition 4.2 (Matching vector family). *For integers N, ℓ, a collection of N pairs of vectors $\{(\mathbf{u}_i, \mathbf{v}_i)\}_{i=1}^{N}$ where all vectors are in $\mathbb{Z}_6^{\ell}$, is an (N, ℓ)-matching vector family if*

- *for any $i \in [N]$, $\langle \mathbf{u}_i, \mathbf{v}_i \rangle = 0$,*
- *for any $i \neq j \in [N]$, $\langle \mathbf{u}_i, \mathbf{v}_j \rangle \in \{1, 3, 4\}$.*

where all operations are done over $\mathbb{Z}_6$.

This definition is a specialization of the one from [Yek08, Efr12], and is sufficient for our purposes. The magical fact about matching vector families is that they exist for values of ℓ that are significantly less than N. (In contrast, if one replaces $\mathbb{Z}_6$ with $\mathbb{Z}_p$ for a prime p, we know that $\ell \geq N^{1/(p-1)}$ [BF98, BDL12]. It is thus a surprise that one can do much better when the modulus is a (small) composite number.

Lemma 4.3 ([Gro00]). *For every integer N, there exists an (N, ℓ)-matching vector family $\{(\mathbf{u}_i, \mathbf{v}_i)\}_{i \in [N]}$ of length $\ell = 2^{O(\sqrt{\log N \log \log N})} = N^{o(1)}$.*

We now show that any MV family gives rise to a PIR encoding scheme. This lemma was observed in the current form in [LVW17] and is implicit in the 2-server PIR protocol of [DG15].

Lemma 4.4 ([LVW17]). *If $\{(\mathbf{u}_i, \mathbf{v}_i)\}_{i=1}^{N}$ is an (N, ℓ)-matching vector family, then the family of vectors $\{1 \| \mathbf{u}_i\}_{i=1}^{N}$ is an $(N, \ell + 1)$-PIR encoding.*

Proof. Define $\hat{\mathbf{u}}_i = 1 \| \mathbf{u}_i \in \mathbb{Z}_6^{\ell+1}$ and $\hat{\mathbf{v}}_i = 0 \| \mathbf{v}_i \in \mathbb{Z}_6^{\ell+1}$ for all $i \in [N]$. Then $\{(\hat{\mathbf{u}}_i, \hat{\mathbf{v}}_i)\}_{i=1}^{N}$ remains an $(N, \ell + 1)$-matching vector family since

$$\langle \hat{\mathbf{u}}_i, \hat{\mathbf{v}}_i \rangle = \langle \mathbf{u}_i, \mathbf{v}_i \rangle.$$

Given a database $\mathbf{D} \in \{0, 1\}^N$, an index $i \in [N]$, and randomness $\mathbf{w} \in \mathbb{Z}_6^{\ell}$, define auxiliary functions $G, G' : \{0, 1\} \to \mathbb{Z}_6$, Following [DG15]. ($G$ and G' implicitly depend on i, $\mathbf{D}$ and $\mathbf{w}$). G and G' are defined as follows.

$$G(\mu) := \sum_{j \in [N]} \mathbf{D}[j] \cdot (-1)^{\langle \mu \hat{\mathbf{u}}_i + \mathbf{w}, \hat{\mathbf{v}}_j \rangle}, \text{ and}$$

$$G'(\mu) := \sum_{j \in [N]} \langle \hat{\mathbf{u}}_i, \hat{\mathbf{v}}_j \rangle \cdot \mathbf{D}[j] \cdot (-1)^{\langle \mu \hat{\mathbf{u}}_i + \mathbf{w}, \hat{\mathbf{v}}_j \rangle}. \tag{5}$$

A straightforward computation shows the following (see [LVW17, Theorem 4.2]):

$$G'(1) - G(1) + G'(0) - G(0) = \mathbf{D}[i] \cdot (-1)^{\langle \mathbf{w}, \hat{\mathbf{v}}_i \rangle}. \tag{6}$$

For each $\mathbf{D} \in \{0,1\}^N$, define $H_{\mathbf{D}} : \mathbb{Z}_6^\ell \to \mathbb{Z}_6^\ell$ by

$$H_{\mathbf{D}}(\mathbf{z}) := (-1)^{\mathbf{z}[1]} \cdot \sum_{j \in [N]} (\hat{\mathbf{v}}_j - \mathbf{e}_1) \cdot \mathbf{D}[j] \cdot (-1)^{\langle \mathbf{z}, \hat{\mathbf{v}}_j \rangle}$$

where $\mathbf{e}_1 = (1, 0, \ldots, 0)$ is the first vector in the standard basis. Recalling that $\hat{\mathbf{u}}_i[1]$, the first bit of $\hat{\mathbf{u}}_i$, equals 1, we have

$$\begin{aligned}
\langle \hat{\mathbf{u}}_i, H_{\mathbf{D}}(\mu \hat{\mathbf{u}}_i + \mathbf{w}) \rangle &= (-1)^{\mu \hat{\mathbf{u}}_i[1] + \mathbf{w}[1]} \cdot \sum_{j \in [N]} \langle \hat{\mathbf{u}}_i, \hat{\mathbf{v}}_j - \mathbf{e}_1 \rangle \cdot \mathbf{D}[j] \cdot (-1)^{\langle \mu \hat{\mathbf{u}}_i + \mathbf{w}, \hat{\mathbf{v}}_j \rangle} \\
&= (-1)^{\mu + \mathbf{w}[1]} \cdot \sum_{j \in [N]} (\langle \hat{\mathbf{u}}_i, \hat{\mathbf{v}}_j \rangle - 1) \cdot \mathbf{D}[j] \cdot (-1)^{\langle \mu \hat{\mathbf{u}}_i + \mathbf{w}, \hat{\mathbf{v}}_j \rangle} \\
&= (-1)^{\mu + \mathbf{w}[1]} \cdot \Big(G'(\mu) - G(\mu)\Big).
\end{aligned}$$

Combined with Eq. (6), we see that:

$$\begin{aligned}
\langle \hat{\mathbf{u}}_i, H_{\mathbf{D}}(\hat{\mathbf{u}}_i + \mathbf{w}) \rangle - \langle \hat{\mathbf{u}}_i, H_{\mathbf{D}}(\mathbf{w}) \rangle &= (-1)^{1 + \mathbf{w}[1]} \cdot (G'(1) - G(1) + G'(0) - G(0)) \\
&= (-1)^{1 + \mathbf{w}[1]} \cdot \mathbf{D}[i] \cdot (-1)^{\langle \mathbf{w}, \hat{\mathbf{v}}_i \rangle} \\
&= \mathbf{D}[i] \cdot \phi_{\mathbf{w}, i}.
\end{aligned}$$

where $\phi_{\mathbf{w}, i} := (-1)^{1 + \mathbf{w}[1] + \langle \mathbf{w}, \hat{\mathbf{v}}_i \rangle}$. This completes the proof. $\square$

4.2   Decomposable Matching Vector (DMV) Families

The main contribution of this section is the definition and construction of a decomposable matching vector family and thus, decomposable PIR encoding schemes.

Definition 4.5 (Decomposable Matching Vector Family). *For integers N, ℓ and $k \leq \log N$, a collection of vectors $\mathbf{u}_1, \ldots, \mathbf{u}_N, \mathbf{v}_1, \ldots, \mathbf{v}_N \in \mathbb{Z}_6^\ell$ is a k-decomposable (N, ℓ)-matching vector family if it is an (N, ℓ)-matching vector family and $(\mathbf{u}_i)_{i=1}^N$ is k-decomposable (as in Definition 4.1).*

First, we show that decomposable matching vector families imply decomposable PIR encodings, extending Lemma 4.4.

Lemma 4.6. *For integers N, ℓ and $k \leq \log N$, if $\{(\mathbf{u}_i, \mathbf{v}_i)\}_{i=1}^N$ is a k-decomposable (N, ℓ)-matching vector family, then the family $\{1 \| \mathbf{u}_i\}_{i=1}^N$ is a k-decomposable $(N, \ell + 1)$-PIR encoding.*

Proof. By Lemma 4.4, $\{1 \| \mathbf{u}_i\}_{i=1}^N$ is a $(N, \ell + 1)$-PIR encoding.

Let $N' = \sqrt[k]{N}$ and let $\{\mathbf{u}_{1,i}\}_{i=1}^{N'}, \ldots, \{\mathbf{u}_{k,i}\}_{i=1}^{N'}$ be the k-decomposition of $\{\mathbf{u}_i\}_{i=1}^N$, then $\{1 \| \mathbf{u}_i\}_{i=1}^N$ is a k-decomposable and $\{1 \| \mathbf{u}_{1,i}\}_{i=1}^{N'}, \ldots, \{1 \| \mathbf{u}_{k,i}\}_{i=1}^{N'}$ is a k-decomposition of it. $\square$

Thus, our main goal is to construct a decomposable matching vector family (whose parameters are slightly worse than that of the [Gro00] matching vector family). To this end, we build on the construction of MV families from the work of Grolmusz [Gro00].

Lemma 4.7 (Implicit in [Gro00]). *For integers h, w, N and any distinct vectors $\mathbf{x}_1, \ldots, \mathbf{x}_N \in \{0,1\}^h$ of Hamming weight exactly w. Then, there exists a matching vector family of N vectors where the vectors have length $\ell = h^{O(\sqrt{w})}$. In particular, the vectors $\mathbf{u}_i$ are indexed by subsets S of $[h]$ of size at most $O(\sqrt{w})$ and defined as follows:*

$$\mathbf{u}_i[S] = \prod_{j \in S} \mathbf{x}_i[j]$$

Proof (Sketch). Observe that for all $i, j \in [N]$:

$$\|\mathbf{x}_i \circ \mathbf{x}_j\|_1 \begin{cases} = w & \text{if } i = j \\ < w & \text{if } i \neq j \end{cases}$$

Let $\mathsf{thres}_w : \{0,1\}^h \to \mathbb{Z}_6$ denote the function which maps 0–1 vectors of Hamming weight exactly w to 0, and those of weight less than w to $\{1, 3, 4\}$. This means that

$$\mathsf{thres}_w(\mathbf{x}_i \circ \mathbf{x}_i) = 0$$
$$\mathsf{thres}_w(\mathbf{x}_i \circ \mathbf{x}_j) \in \{1, 3, 4\} \quad \text{if } i \neq j$$

The choices of $\mathbb{Z}_6$ and $\{1, 3, 4\}$ come from the work of Barrington et al. [BBR94] which tells us that thres_w can be computed by a multilinear polynomial over $\mathbb{Z}_6$ of total degree $O(\sqrt{w})$.

Next, we will construct the vectors $\mathbf{u}_i$ and $\mathbf{v}_j$ of length $\ell = h^{O(\sqrt{w})}$ from $\mathbf{x}_i$ and $\mathbf{x}_j$ respectively so that $\langle \mathbf{u}_i, \mathbf{v}_j \rangle = \mathsf{thres}_w(\mathbf{x}_i \circ \mathbf{x}_j)$ for all i, j. The bound on ℓ comes from the fact that we can write the evaluation of a multilinear polynomial of total degree $O(\sqrt{w})$ in h variables as the inner product of two vectors of length $\ell = h^{O(\sqrt{w})}$. In particular, $\mathbf{u}_i$ will be defined as above and $\mathbf{v}_j$ will be the coefficient vector of the degree $O(\sqrt{w})$ multilinear polynomial f_j which maps $\mathbf{x} \mapsto \mathsf{thres}_w(\mathbf{x} \circ \mathbf{x}_j)$. $\qquad \square$

In our setting, the index $i = (i_1, \ldots, i_k) \in [\sqrt[k]{N}] \times \cdots \times [\sqrt[k]{N}]$ is divided amongst k players, as described in Sect. 2.1.

Lemma 4.8. *For integers N and $k \leq \log N$, there exists a k-decomposable (N, ℓ)-matching vectors family where $\ell = 2^{O(\sqrt{\log N} \cdot \log \log N)}$.*

Proof. Let h, w be multiples of k such that $\binom{h/k}{w/k}^k \geq N$. Let $\mathbf{y}_1, \ldots, \mathbf{y}_{\sqrt[k]{N}}$ be distinct vectors in $\{0,1\}^{h/k}$, each of Hamming weight w/k. For each $i = (i_1, \ldots, i_k) \in [N]$, we define

$$\mathbf{x}_i := \mathbf{y}_{i_1} \| \cdots \| \mathbf{y}_{i_k} \in \{0,1\}^h$$

Clearly, $\mathbf{x}_1, \ldots, \mathbf{x}_N$ is a collection of distinct vectors with Hamming weight w, and by Lemma 4.7, there exists a matching vector family $\{(\mathbf{u}_i, \mathbf{v}_i)\}_{i \in [N]}$, where the $\mathbf{u}_i$'s are indexed by subsets $S \subseteq [h]$ of size at most $O(\sqrt{w})$ and satisfies

$$\mathbf{u}_i[S] = \prod_{j \in S} \mathbf{x}_i[j] = \prod_{t=1}^{k} \prod_{j \in S_t} \mathbf{x}_i[j].$$

where $S_t := S \cap \{(t-1) \cdot \frac{h}{k} + 1, \ldots, t \cdot \frac{h}{k}\}$. Now, if we define

$$\mathbf{u}_{t,i_t}[S] := \prod_{j \in S_t} \mathbf{x}_i[j] = \prod_{j \in S_t} \mathbf{y}_{i_t}[j - (t-1) \cdot \tfrac{h}{k}],$$

we have $\mathbf{u}_i = \mathbf{u}_{1,i_1} \circ \ldots \circ \mathbf{u}_{k,i_k}$, giving us a k-decomposition.

Set $w = \lceil \log N/k \rceil \cdot k$ and $h = 2w$ (so that h, w are multiples of k). Then,

$$\binom{h/k}{w/k}^k \geq \left((h/w)^{w/k}\right)^k = 2^w \geq N$$

Also, we have $\ell = h^{O(\sqrt{w})} = 2^{O(\sqrt{\log N} \cdot \log \log N)}$. $\qquad\square$

As a result, we have the main theorem of this section:

Theorem 4.9. *For integers N and $k \leq \log N$, there exists a k-decomposable (N, ℓ)-PIR encoding where $\ell = 2^{O(\sqrt{\log N} \cdot \log \log N)}$.*

5   A Special-Purpose PSM Protocol

Given the decomposable PIR encoding from Sect. 4.2, the final piece required to instantiate the framework in Sect. 3 is a special purpose PSM protocol for the functionality described in Eq. 4.

Theorem 5.1. *For integers N, ℓ and $k \leq \log N$, if $(\mathbf{u}_i)_{i=1}^{N}$ is k-decomposable, then there is a PSM for the functionality*

$$F_{\mathsf{aux}} : [\sqrt[k]{N}] \times \ldots \times [\sqrt[k]{N}] \times (\{0,1\} \times \mathbb{Z}_6^\ell \times \mathbb{Z}_6^\ell) \to \mathbb{Z}_6^\ell \times \mathbb{Z}_6 \tag{4}$$

where $F_{\mathsf{aux}}(i_1, \ldots, i_k; (\mu, \mathbf{w}, \mathbf{r})) \mapsto (\mathbf{w} + \mu \mathbf{u}_i, \langle \mathbf{u}_i, \mathbf{r} \rangle)$

with communication complexity $O(\ell k^2)$ per party.

In order to construct a efficient PSM protocol for this specialized functionality, we show that (a) this functionality can be written as an affine mod-6 branching program of size $O(k \cdot \ell)$ and (b) use the fact that there are efficient PSM protocols that compute affine branching programs over rings [IK00, IK02, CFIK03] where the total communication is polynomial in the size of the branching program.

Lemma 5.2. *There is a PSM protocol for the k-party affine branching program functionality $\mathbf{BP}_m^k$ over the ring $\mathbb{Z}_6$ with communication complexity $O(m^2)$ per party.*

Proof. A mod-$\mathbb{Z}_6$ branching program can be locally decomposed into a mod-$\mathbb{F}_2$ branching program and a mod-$\mathbb{F}_3$ branching program using Chinese remainder theorem. [IK02] show a $\mathsf{cc} = O(m^2 \log p)$ PSM protocol for mod-$\mathbb{F}_p$ branching program when p is a prime.

Moreover, [IK02] can be immediately extended to branching programs over any commutative rings; see Appendix A for more details. It is further extended by [CFIK03] to branching programs over any rings. □

Proof (Proof of Theorem 5.1). Let $\{\mathbf{u}_{1,i}\}_{i=1}^{\sqrt[k]{N}}, \ldots, \{\mathbf{u}_{k,i}\}_{i=1}^{\sqrt[k]{N}} \subseteq \mathbb{Z}_6^\ell$ be the decomposition, then under the natural mapping $i \mapsto (i_1, \ldots, i_N) \in [\sqrt[k]{N}]^k$, we have

$$\mathbf{u}_i = \mathbf{u}_{1,i_1} \circ \ldots \circ \mathbf{u}_{k,i_k}$$

for all $i \in [N]$.

For the functionality

$$F_{\mathsf{aux}}^{(1)}(i_1, \ldots, i_k; \mu, \mathbf{w}) = \mu \mathbf{u}_i + \mathbf{w} \in \mathbb{Z}_6^\ell,$$

its j-th output bit can be written as

$$(\mu \mathbf{u}_i + \mathbf{w})[j] = \mu \mathbf{u}_{1,i_1}[j]\mathbf{u}_{2,i_2}[j]\ldots \mathbf{u}_{k,i_k}[j] + \mathbf{w}[j].$$

Since the t-th party $(1 \leq t \leq k)$ can compute $\mathbf{u}_{t,i_t}$ locally, the j-th bit of $\mu \mathbf{u}_i + \mathbf{w}$ can be computed by a mod-$\mathbb{Z}_6$ branching program with $O(k)$ nodes. By Lemma 5.2, there exists a PSM scheme evaluating the j-th output bit using $O(k^2)$ communication per party. Thus, all ℓ outputs can be computed with communication complexity $O(\ell k^2)$ per party.

For the functionality

$$F_{\mathsf{aux}}^{(2)}(i_1, \ldots, i_k; \mathbf{r}) = \langle \mathbf{u}_i, \mathbf{r} \rangle \in \mathbb{Z}_6,$$

let $\mathbf{s} \in \mathbb{Z}_6^\ell$ be a random vector sampled from CRS such that $\sum_{j=1}^\ell \mathbf{s}[j] = 0$. Then, instead of computing the functionality, the parties compute together the functionality

$$F_{\mathsf{aux}}^{(2)'}(i_1, \ldots, i_k; \mathbf{r}, \mathbf{s}) = \mathbf{u}_i \circ \mathbf{r} + \mathbf{s} \in \mathbb{Z}_6^\ell,$$

It is easy to see that $\mathbf{u}_i \circ \mathbf{r} + \mathbf{s}$ reveals $\langle \mathbf{u}_i, \mathbf{r} \rangle$ and nothing more. The j-th bit of $\mathbf{u}_i \circ \mathbf{r} + \mathbf{s}$ can be written as

$$(\mathbf{u}_i \circ \mathbf{r} + \mathbf{s})[j] = \mathbf{u}_{1,i_1}[j]\mathbf{u}_{2,i_2}[j]\ldots \mathbf{u}_{k,i_k}[j]\mathbf{r}[j] + \mathbf{s}[j],$$

which can again be computed by a simple affine branching program with $O(k)$ nodes. By Lemma 5.2, there exists a PSM scheme evaluating this with $O(k^2)$ bits of communication per party. Thus, to compute all the bits, we need $O(\ell k^2)$ bits of communication.

Clearly, once Charlie learns $\mathbf{u}_i \circ \mathbf{r} + \mathbf{s}$, he can add up all the bits to get $\langle \mathbf{u}_i, \mathbf{r} \rangle$. □

6 Putting Together

We finally put together all the pieces to construct a multiparty CDS scheme, and various types of secret sharing schemes.

6.1 Multi-party CDS for $\mathbf{INDEX}_N^{k+1}$

We obtain the multiparty CDS protocol by instantiating our general framework in Sect. 3 with the decomposable PIR encodings in Sect. 4.2 and the PSM protocol in Sect. 5.

Theorem 6.1. *For $1 \leq k \leq \log N$, there is a $(k+1)$-party CDS protocol for* $\mathbf{INDEX}_N^{k+1}$ *whose communication complexity is* $2^{O(\sqrt{\log N} \log \log N)}$.

Proof. Theorem 3.1 gives us a $k+1$-party CDS protocol for $\mathbf{INDEX}_N^{k+1}$ assuming a k-decomposable PIR encoding scheme and a PSM protocol for the associated functionality F_{aux}. Theorem 4.9 constructs such a k-decomposable (N, ℓ)-PIR encoding scheme with $\ell = 2^{O(\sqrt{\log N} \log \log N)}$ and Theorem 5.1 constructs a PSM protocol for the associated functionality F_{aux} with communication complexity $O(k^2 \cdot \ell)$. Put together, applying Theorem 3.1, we get a $(k+1)$-party CDS protocol with communication complexity $O(\ell + k^2 \ell) = 2^{O(\sqrt{\log N} \log \log N)}$. □

Because CDS for $\mathbf{ALL}_N^k$ is easier than CDS for $\mathbf{INDEX}_N^{k+1}$, we also get:

Corollary 6.2. *For $1 \leq k \leq \log N$, there is a k-party CDS protocol for* $\mathbf{ALL}_N^k$ *whose communication complexity is* $2^{O(\sqrt{\log N} \log \log N)}$.

6.2 From CDS to Non-monotone Secret Sharing

A non-monotone secret-sharing scheme for access function F is a randomized algorithm that on input a secret bit, outputs n pairs of shares $(s_{i,0}, s_{i,1})_{i \in [n]}$ such that for any $(x_1, \ldots, x_n) \in \{0,1\}^n$, the n shares $(s_{1,x_1}, \ldots, s_{n,x_n})$ determine the secret if $F(x_1, \ldots, x_n) = 1$ and reveal nothing about the secret otherwise.

Definition 6.3 (Non-monotone Secret Sharing). *Given a function $F : \{0,1\}^n \to \{0,1\}$, a non-monotone secret-sharing scheme for access function F is a randomized algorithm*

$$\mathsf{nmSS} : \mathcal{M} \times \mathcal{W} \to (\{0,1\}^{\mathsf{cc}})^{2n}$$

that on input a secret bit, outputs n pairs of shares $s_{1,0}, s_{1,1}, \ldots, s_{n,0}, s_{n,1} \in \{0,1\}^{\mathsf{cc}}$ satisfying the following properties:

(correctness). *There exists a reconstruction algorithm $\mathsf{C} : \{0,1\}^n \times (\{0,1\}^{\mathsf{cc}})^n \to \mathcal{M}$ such that for all $(x_1, \ldots, x_n) \in \{0,1\}^n$ that $F(x_1, \ldots, x_n) = 1$ and for all $\mu \in \mathcal{M}, w \in \mathcal{W}$,*

$$\mathsf{nmSS}(\mu; w) = (s_{1,0}, s_{1,1}, \ldots, s_{n,0}, s_{n,1}) \implies \mathsf{C}(x_1, \ldots, x_n, s_{1,x_1}, \ldots, s_{n,x_n}) = \mu.$$

(privacy). *There exists a simulator* S *such that for all* $(x_1, \ldots, x_n) \in \{0,1\}^n$
satisfying $F(x_1, \ldots, x_n) = 0$, *the joint distribution of* $(s_{1,x_1}, \ldots, s_{n,x_n})$ *is perfectly indistinguishable from* $S(x_1, \ldots, x_k)$, *where* $(s_{1,0}, s_{1,1}, \ldots, s_{n,0}, s_{n,1}) :=$ nmSS$(\mu; w)$ *and the randomness are taken over* $w \xleftarrow{\text{R}} \mathcal{W}$ *and the coin tosses of* S.

Standard monotone secret-sharing correspond to the special case where F is monotone and $s_{1,0} = \cdots = s_{n,0} = \bot$. In such case, let $s_1, \ldots, s_n$ denotes $s_{1,1}, \ldots, s_{n,1}$ respectively.

Let $N := 2^n$. It is not hard to see that non-monotone secret sharing for all n-party access functions F is the same as n-party CDS for $\mathbf{ALL}_N^n$. This connection is almost syntactic, and can be formalized as follows.

Lemma 6.4. *For any access function* $F : \{0,1\}^n \to \{0,1\}$, *there is a non-monotone secret-sharing scheme for* F *with share size* cc *if and only if there is a CDS scheme for* $\mathbf{ALL}_{2^n}^n$ *with communication complexity* cc *when the predicate is* F.

Proof. Assume $(\mathsf{B}_1, \ldots, \mathsf{B}_n)$ is a CDS for $\mathbf{ALL}_{2^n}^n$ with predicate function F, a non-monotone secret sharing scheme for F can be defined as

$$\mathsf{nmSS}(\mu; w) = (\mathsf{B}_1(\mu, 0; w), \mathsf{B}_1(\mu, 1; w), \ldots, \mathsf{B}_n(\mu, 0; w), \mathsf{B}_n(\mu, 1; w)).$$

Assume nmSS is a non-monotone secret sharing scheme for access function F. Then $(\mathsf{B}_1, \ldots, \mathsf{B}_n)$ is a CDS for $\mathbf{ALL}_{2^n}^n$ with predicate function F if $\mathsf{B}_t(x_t; w)$ outputs s_{t,x_t} and s_{t,x_t} is defined as $(s_{1,0}, s_{1,1}, \ldots, s_{n,0}, s_{n,1}) := \mathsf{nmSS}(\mu; w)$. $\square$

Thus, we obtain the following theorem, disproving Conjecture 2.

Theorem 6.5. *There is a non-monotone secret-sharing for the family of all access functions* $F : \{0,1\}^n \to \{0,1\}$ *with total share size* $2^{O(\sqrt{n}\log n)}$ *bits.*

6.3   From Non-monotone to Monotone Secret Sharing

We describe the following folklore transformation which transforms a *non-monotone* secret-sharing scheme for any $F : \{0,1\}^n \to \{0,1\}$ into a *monotone* secret-sharing scheme with roughly the same share size for a related function F'.

Lemma 6.6. *For any function* $F : \{0,1\}^n \to \{0,1\}$, *a non-monotone secret sharing scheme for* F *with share size* cc *implies a monotone secret sharing scheme for a monotone access function* $F' : \{0,1\}^{2n} \to \{0,1\}$ *with share size* cc + 2, *where* $F' : \{0,1\}^{2n} \to \{0,1\}$ *is defined as*

$$F'(x_1', \ldots, x_{2n}') = \begin{cases} 1 & \text{if } \exists i \text{ such that } x_{2i-1}' = x_{2i}' = 1 \\ 0 & \text{else if } \exists i \text{ such that } x_{2i-1}' = x_{2i}' = 0 \\ F(x_2', x_4', \ldots, x_{2n}') & \text{otherwise.} \end{cases}$$

Proof. It is easy that F' is monotone. Let nmSS be the non-monotone secret-sharing scheme for F, we construct a monotone secret sharing scheme mSS for F' as follows:

On input $\mu \in \{0,1\}$

1. Sample bits $b_1, \ldots, b_n, r_1, \ldots, r_n$ and $w \in \mathcal{W}$ uniformly at random.
2. Let $(s_{1,0}, s_{1,1}, \ldots, s_{n,0}, s_{n,1}) := \mathsf{nmSS}(\mu \oplus r_1 \oplus \ldots \oplus r_n; w)$.
3. Output $(s'_1, \ldots, s'_{2n})$ where

$$s'_{2i-1} := (s_{i,0}, r_i, b_i) \qquad\qquad s'_{2i} := (s_{i,1}, r_i, b_i \oplus \mu).$$

The reconstruction algorithm for mSS either computes μ from $(b_i, b_i \oplus \mu)$, or runs the one for nmSS to recover $\mu \oplus r_1 \oplus \cdots \oplus r_n$ and thus μ. More precisely, we argue correctness and privacy via a case analysis. For any $(x'_1, \ldots, x'_{2n}) \in \{0,1\}^{2n}$, given the shares $(s_j)_{x'_j = 1}$:

Case 1: $\exists i$ s.t. $(x'_{2i-1}, x'_{2i}) = (1,1)$. Here, $F'(x'_1, \ldots, x'_{2n}) = 1$, and the reconstruction algorithm can recover μ from $b_i, b_i \oplus \mu$ given in s'_{2i-1}, s'_{2i}.

Case 2: $\exists i$ s.t. $(x'_{2i-1}, x'_{2i}) = (0,0)$. Here, $F'(x'_1, \ldots, x'_{2n}) = 0$, and privacy follows from the fact that r_i is perfectly hidden, and therefore μ is also perfectly hidden.

Case 3: $\forall i, (x'_{2i-1}, x'_{2i}) \in \{(0,1), (1,0)\}$. So $F'(x'_1, \ldots, x'_{2n}) = F(x'_2, x'_4, \ldots, x'_{2n})$. First, observe that the shares $(s_j)_{x'_j=1} = (s'_{2i-1+x'_{2i}})_{i \in [n]}$ and reveal exactly

$$s_{1,x'_2}, s_{2,x'_4}, \ldots, s_{n,x'_{2n}}, r_1, \ldots, r_n.$$

Now, if $F(x'_2, x'_4, \ldots, x'_{2n}) = 1$, correctness of nmSS enables Charlie to recover $\mu \oplus r_1 \oplus \ldots \oplus r_n$ and thus μ. Otherwise, privacy of nmSS hides $\mu \oplus r_1 \oplus \ldots \oplus r_n$, thus μ is perfectly hidden.

This completes the proof. $\square$

As for non-monotone secret sharing, there are double exponential different access functions. Under the mapping specified in Lemma 6.6, they are mapped to different monotone functions. Thus, we obtain the following theorem, disproving Conjecture 1.

Theorem 6.7. *There is a collection $\hat{F}_n$ of $2^{2^{n/2}}$ monotone functions over $\{0,1\}^n$, such that there is monotone secret sharing scheme for $\hat{F}_n$ with share size $2^{O(\sqrt{n}\log n)}$ bits.*

To obtain the more general statement in Theorem 1 (informal), we just apply the above construction to the first $\log s$ bits of the input.

7 Linear CDS and Secret Sharing

A CDS protocol is linear if the transmitting functions are linear on the secret and randomness (not necessarily linear on the inputs). A secret sharing scheme is linear if the share generation algorithm is linear on the secret and randomness.

We present a linear non-monotone secret sharing scheme (equivalently, a multiparty CDS protocol) for every access function over $\{0,1\}^n$ with shares of size $O(2^{n/2})$. It is sufficient to construct a linear $(n+1)$-party CDS for $\mathbf{INDEX}_{2^n}^{n+1}$ where each party sends $O(2^{n/2})$ bits. The construction will follow our general framework for building multi-party CDS from 2-party CDS.

In CDS for $\mathbf{INDEX}_{2^n}^{n+1}$, each of the first n parties holds a bit of $i \in \{0,1\}^n$, the last party holds a truth-table $\mathbf{D} \in \{0,1\}^{2^n}$, the secret is disclosed if and only if $\mathbf{D}[i] = 1$. As a warm-up, we recap the 2-party linear CDS for $\mathbf{INDEX}_{2^n}^2$ with total communication $O(2^{n/2})$.

2-party CDS [GKW15]. Bob holds $i \in \{0,1\}^n$ and split it into higher half $j^H \in \{0,1\}^{n/2}$ and lower half $j^L \in \{0,1\}^{n/2}$. The shared randomness is $\mathbf{w}, \mathbf{r} \in \{0,1\}^{2^{n/2}}$. Bob sends

$$\mathbf{m}_B^1 := \mu \mathbf{e}_{j^L} + \mathbf{w} \in \{0,1\}^{2^{n/2}}, \qquad m_B^2 := \mathbf{r}[j^H]$$

to Charlie. Alice holds the truth table $\mathbf{D} \in \{0,1\}^{2^n}$ which can be viewed as a $2^{n/2} \times 2^{n/2}$ matrix, sends

$$\mathbf{m}_A := \mathbf{D}\mathbf{w} + \mathbf{r} \in \{0,1\}^{2^{n/2}}.$$

Charlie computes

$$\begin{aligned}
&\mathbf{e}_{j^H} \cdot \mathbf{D} \cdot \mathbf{m}_B^1 - \mathbf{e}_{j^H} \cdot \mathbf{m}_A + m_B^2 \\
&= \mathbf{e}_{j^H} \cdot \mathbf{D} \cdot (\mu \mathbf{e}_{j^L} + \mathbf{w}) - \mathbf{e}_{j^H} \cdot (\mathbf{D}\mathbf{w} + \mathbf{r}) + \mathbf{r}[j^H] \\
&= \mu \mathbf{e}_{j^H} \cdot \mathbf{D} \cdot \mathbf{e}_{j^L} \\
&= \mu \mathbf{D}[i].
\end{aligned} \tag{7}$$

PSM Building Block. Following our general framework for building multi-party CDS from 2-party CDS, the crux is to construct a "linear" n-party PSM for the following functionality where each party sends $O(2^{n/2})$ bits.

$$(j^H; \mathbf{r}) \qquad \mapsto \mathbf{r}[j^H] \in \{0,1\} \tag{8}$$

$$(j^L; \mathbf{w}, \mu) \mapsto \mu \mathbf{e}_{j^L} + \mathbf{w} \in \{0,1\}^{2^{n/2}} \tag{9}$$

where $\mathbf{w} \in \{0,1\}^{2^{n/2}}$. For this, we should use "partial garbling" [IW14] and exploit the fact that the PSM does not need to protect the privacy of j^H, j^L, which we may treat as "public" inputs. In addition, observe that the computation in (8) and (9) are linear on the "private" inputs $\mathbf{r}, \mathbf{w}, \mu$. Our goal is a protocol where

- the communication of the "partial garbling PSM" is $O(2^{n/2})$.
- each party's PSM message is linear in $\mathbf{r}, \mathbf{w}, \mu$ and the PSM randomness.

As a partial PSM protocol for functionality (8): Sample random vectors $\mathbf{r}_1, \ldots, \mathbf{r}_{n/2} \in \{0,1\}^{2^{n/2}}$ such that $\mathbf{r} = \sum_{t=1}^{n/2} \mathbf{r}_t$. For every $1 \leq t \leq n/2$, the party who holds j_t^H sends $\mathbf{r}_t[j]$ for all $j \in \{0,1\}^{n/2}$ such that $j_t = j_t^H$. Charlie can reconstruct $\mathbf{r}[j^H] = \sum_t \mathbf{r}_t[j^H]$.

As a partial PSM protocol for functionality (9): For every $1 \leq t \leq n/2$, the party who holds j_t^L sends $\mathbf{w}[j]$ for all $j \in \{0,1\}^{n/2}$ such that $j_t \neq j_t^L$. One party sends $\mu + \sum_j \mathbf{w}[j]$ in addition. Charlie can reconstruct vector $\mu \mathbf{e}_{j^L} + \mathbf{w}$ as he receives all the bits of it except the j^L-th bit and the j^L-th bit can be recovered from the sum the vector and the rest of the vector.

CDS for $\mathbf{INDEX}_N^{k+1}$

Input of P_t ($1 \leq t \leq k$): $i_t \in [\sqrt[k]{N}]$ and $\mu \in \{0,1\}$.
Input of P_{k+1} (Alice): $\mathbf{D} \in \{0,1\}^{\sqrt{N} \times \sqrt{N}}$.
Shared Randomness: $\mathbf{w}, \mathbf{r}_{k/2+1}, \ldots, \mathbf{r}_k \in \{0,1\}^{\sqrt{N}}$

The protocol proceeds as follows.

- For $1 \leq t \leq k/2$, the t-th party sends $\mathbf{w}[j]$ for all $j \in [\sqrt[k]{N}]^{k/2}$ such that $j_t \neq i_t$.
- For $k/2 + 1 \leq t \leq k$, the t-th party sends $\mathbf{r}_t[j]$ for all $j \in [\sqrt[k]{N}]^{k/2}$ such that $j_{t-k/2} = i_t$.
- P_{k+1} (Alice) sends $\mathbf{m}_A := \mathbf{D}\mathbf{w} + \sum_t \mathbf{r}_t$.
- In addition, one of parties sends $m_D := \mu + \sum_j \mathbf{w}[j]$.
- Charlie computes $\mathbf{w}' := \mathbf{w} + \mu \mathbf{e}_{i^L}$ by the following:
  For every $j \neq i^L$, $\mathbf{w}'[j] = \mathbf{w}[j]$ and $\mathbf{w}[j]$ is sent by one of the first $k/2$ parties. And $\mathbf{w}'[i^L] := m_D - \sum_{j \neq i^L} \mathbf{w}'[j]$.
- Charlie outputs 1 if $\mathbf{e}_{i^H} \cdot \mathbf{D} \cdot \mathbf{w}' - \mathbf{m}_A[i^H] + \sum_t \mathbf{r}_t[i^H] = 1$, and outputs 0 otherwise.

Fig. 2. $(k+1)$-party linear CDS for $\mathbf{INDEX}_N^{k+1}$, when k is even.

Theorem 7.1. *For even $k \leq \log N$, there is a $(k+1)$-party linear CDS protocol for $\mathbf{INDEX}_N^{k+1}$ (Fig. 2) whose communication complexity is $O(\sqrt{N})$ per party.*

Proof. The argument for correctness is the same as that for 2-party CDS: Let $\mathbf{r} := \sum_t \mathbf{r}_t$. Charlie learns $\mathbf{w}' = \mu \mathbf{e}_{j^L} + \mathbf{w}$ from the first $k/2$ parties, learns $\mathbf{r}[j^H]$ from the following $k/2$ parties, gets $\mathbf{D}\mathbf{w} + \mathbf{r}$ from the last party (Alice), then he can reconstruct $\mu \cdot \mathbf{D}[i]$ using Eq. (7).

Privacy follows from the following observations:

- When $\mathbf{D}[i] = 0$, the joint distribution of $\mathbf{w}', \mathbf{m}_A, \mathbf{r}[i^H]$ can be simulated from $\mathbf{D}, i$ without knowing μ. This is due the security of 2-party CDS for **INDEX**[GKW15].
- The messages from the first $k/2$ parties together with m_D can be determined by $(\mathbf{w}', j^L)$: the messages from the first $k/2$ parties all bits in $\mathbf{w}'$; and $m_D = \sum_j \mathbf{w}'[j]$.
- The messages from the following $k/2$ parties can be simulated from $(\mathbf{r}[j^H], j^H)$: sample $\mathbf{r}_{k/2+1}, \ldots, \mathbf{r}_k$ such that $\sum_t \mathbf{r}_t[j^H] = \mathbf{r}[j^H]$; the t-th party sends $\mathbf{r}_t[j]$ for all $j \in [\sqrt[k]{N}]^{k/2}$ such that $j_{t-k/2} = i_t$.

Theorem 7.2. *There is a linear non-monotone secret-sharing for the family of all access functions $F : \{0,1\}^n \to \{0,1\}$ with share size $O(2^{n/2})$ bits for each party.*

Proof. (Sketch). Use Lemma 6.4 to construct non-monotone secret sharing from multi-party CDS for **ALL**, with the observation that the transformation in Lemma 6.4 preserves linearity.

Acknowledgments. We thank Yuval Ishai for telling us about Conjecture 1. We thank the anonymous EUROCRYPT 2018 reviewers for their insightful comments.

A   Private Simultaneous Message for Branching Program over Commutative Rings

We sketch the PSM schemes from [IK02, CFIK03] for branching programs over commutative rings.

PSM for AFFINE$_m^k$

t-th Party's Input: $\mathbf{x}_t \in \mathcal{R}^n$.
Shared Input: Affine function $f : \mathcal{R}^{nk} \to \mathcal{R}$ such that

$$f_{\mathbf{y}_1, \ldots, \mathbf{y}_k, c}(\mathbf{z}_1, \ldots, \mathbf{z}_k) := \langle \mathbf{z}_1, \mathbf{y}_1 \rangle + \ldots + \langle \mathbf{z}_k, \mathbf{y}_k \rangle + c$$

Shared Randomness: $r_1, \ldots, r_{k-1} \in \mathcal{R}$.

- The t-th party (for $1 \le t < k$) sends $m_t := \langle \mathbf{x}_t, \mathbf{y}_t \rangle + r_t$
- The k-th party sends $m_k := \langle \mathbf{x}_k, \mathbf{y}_k \rangle + c - \sum_{t=1}^{k-1} r_t$,
- Charlie outputs $\sum_{t=1}^k m_t$.

Fig. 3. The $\log |\mathcal{R}|$-bit PSM protocol for **AFFINE**^{k}.

Theorem A.1 (Folklore). *There is a PSM scheme (Fig. 3) for* **AFFINE**^{k} *over commutative ring* $\mathcal{R}$ *such that every party sends one ring element.*

Proof. The correctness is straight-forward,

$$\sum_{t=1}^{k} m_t = \sum_{t=1}^{k-1}(\langle \mathbf{x}_t, \mathbf{y}_t \rangle + r_t) + \langle \mathbf{x}_k, \mathbf{y}_k \rangle + c - \sum_{t=1}^{k-1} r_t = \sum_{t=1}^{k} \langle \mathbf{x}_t, \mathbf{y}_t \rangle + c = f(\mathbf{x}_1, \ldots, \mathbf{x}_k).$$

Privacy follows from the following observations:

- the joint distribution of $m_1, \ldots, m_{k-1}$ is uniformly random, since we are using $(r_1, \ldots, r_{k-1})$ is one-time pads;
- m_k is determined by $m_1, \ldots, m_{k-1}$ and $f(\mathbf{x}_1, \ldots, \mathbf{x}_k)$ as $m_k = f(\mathbf{x}_1, \ldots, \mathbf{x}_k) - \sum_{t=1}^{k-1} m_t$.

Putting the two together, we can simulate Charlie's view given just $f(\mathbf{x}_1, \ldots, \mathbf{x}_k)$. $\square$

Lemma A.2 ([IK02]). *For any matrix* $M \in \mathcal{R}^{m \times m}$ *such that* $M_{i,j} = -1$ *for* $i = j+1$ *and* $M_{i,j} = 0$ *for* $i > j+1$, *there exist matrix* $L, R \in \mathcal{R}^{m \times m}$ *satisfying (10) such that*

$$M = \underbrace{\begin{bmatrix} 1 & L_{1,2} & \cdots & L_{1,m} \\ & 1 & & \\ & & \ddots & \\ & & & 1 \end{bmatrix}}_{L} \cdot \begin{bmatrix} -1 & & & \det M \\ & -1 & & \\ & & \ddots & \\ & & & -1 \end{bmatrix} \cdot \underbrace{\begin{bmatrix} 1 & R_{1,2} & \cdots & R_{1,m} \\ & 1 & \ddots & \vdots \\ & & \ddots & R_{m-1,m} \\ & & & 1 \end{bmatrix}}_{R}$$

Lemma A.3 ([IK02]). *For any matrix* $M \in \mathcal{R}^{m \times m}$ *such that* $M_{i,j} = -1$ *for* $i = j+1$ *and* $M_{i,j} = 0$ *for* $i > j+1$, *the distribution of* $L \cdot M \cdot R$ *is determined by* $\det M$, *where* $L, R \in \mathcal{R}^{m \times m}$ *are random matrices satisfying (10).*

Proof. Let L_M, R_M be the matrices implied by Lemma A.2 such that L_M, R_M are upper triangular matrices with 1's in their diagonal, $L_{i,j}^{(M)} = 0$ for $1 < i < j \leq m$ and

$$M = L_M \cdot \begin{bmatrix} -1 & & & \det M \\ & \ddots & & \\ & & -1 & \end{bmatrix} \cdot R_M$$

$L \cdot M \cdot R$ can be written as

$$L \cdot M \cdot R = \underbrace{L \cdot L_M}_{\text{same distribution as } L} \cdot \begin{bmatrix} -1 & & & \det M \\ & \ddots & & \\ & & -1 & \end{bmatrix} \cdot \underbrace{R_M \cdot R}_{\text{same distribution as } R}$$

The joint distribution of $(L \cdot L_M, R_M \cdot R)$ is the same as (L, R), thus $\det M$ determines the distribution of $L \cdot M \cdot R$. $\square$

Theorem A.4. *There is a PSM scheme (Fig. 4) for* $\mathbf{BP}_m^k$ *over commutative ring* $\mathcal{R}$ *such that every party sends* $\frac{m(m+1)}{2}$ *ring elements.*

Proof. Correctness is straight-forward, as

$$\det C = \det F'(\mathbf{x}_1, \ldots, \mathbf{x}_k) = \det L \cdot \det F(\mathbf{x}_1, \ldots, \mathbf{x}_k) \cdot \det R = \det F(\mathbf{x}_1, \ldots, \mathbf{x}_k).$$

Privacy follows from the following observations:

– The distribution of $C = L \cdot F(\mathbf{x}_1, \ldots, \mathbf{x}_k) \cdot R$ is determined by $\det F(\mathbf{x}_1, \ldots, \mathbf{x}_k)$ (Lemma A.3);
– For each $1 \leq i \leq j \leq m$, Charlie learns $C_{i,j}$ in an independent nested PSM scheme, the corresponding messages can be simulated given C.

Thus Charlie's view can be simulated by first sampling C given $\det F(\mathbf{x}_1, \ldots, \mathbf{x}_k)$, then simulating the messages using C and the simulator of the underlaying PSM. $\qquad\square$

PSM for $\mathbf{BP}^k$

t-**th Party's Input:** $\mathbf{x}_t \in \mathcal{R}^n$.
Shared Input: Affine functions $F_{i,j} : \mathcal{R}^{nk} \to \mathcal{R}$ for $1 \leq i \leq j \leq m$.
Shared Randomness: $\{L_{1,j}\}_{j=2}^m \in \mathcal{R}^{m-1}$, $\{R_{i,j}\}_{1 \leq i < j \leq m} \in \mathcal{R}^{m(m-1)/2}$ and randomness for $\frac{1}{2}m(m+1)$ nested PSM runs for $\mathbf{AFFINE}^k$

– Define matrix L, R such that

$$L_{i,j} = \begin{cases} 1, & \text{if } i = j \\ L_{1,j}, & \text{if } 1 = i < j \\ 0, & \text{otherwise} \end{cases} \qquad R_{i,j} = \begin{cases} 1, & \text{if } i = j \\ R_{i,j}, & \text{if } i < j \\ 0, & \text{if } i > j \end{cases} \qquad (10)$$

Each of the k parties independently computes affine functions $F'_{i,j} : \mathcal{R}^{nk} \to \mathcal{R}$ for $1 \leq i \leq j \leq m$ defined as

$$F'(\mathbf{z}_1, \ldots, \mathbf{z}_n) = L \cdot F(\mathbf{z}_1, \ldots, \mathbf{z}_n) \cdot R$$

where $F' : \mathcal{R}^{nk} \to \mathcal{R}^{m \times m}$ is defined as

$$\left(F'(\mathbf{z}_1, \ldots, \mathbf{z}_k)\right)_{i,j} = \begin{cases} F'_{i,j}(\mathbf{z}_1, \ldots, \mathbf{z}_k), & \text{if } 1 \leq i \leq j \leq m \\ -1, & \text{if } i = j + 1 \\ 0, & \text{if } i > j + 1 \end{cases}$$

– Charlie learns $C = F'(\mathbf{x}_1, \ldots, \mathbf{x}_k)$: For each $1 \leq i \leq j \leq m$, Charlie learns $C_{i,j} = F'_{i,j}(\mathbf{x}_1, \ldots, \mathbf{x}_k)$ via a PSM scheme (Figure 3) for $\mathbf{AFFINE}^k$.
– Charlie outputs $\det C$

Fig. 4. The $(\frac{1}{2}m(m+1)\log|\mathcal{R}|)$-bit PSM protocol for $\mathbf{BP}_m^k$.

References

[BBR94] Barrington, D.A.M., Beigel, R., Rudich, S.: Representing boolean functions as polynomials modulo composite numbers. Comput. Complex. **4**, 367–382 (1994)

[BDL12] Bhowmick, A., Dvir, Z., Lovett, S.: New lower bounds for matching vector codes. CoRR, abs/1204.1367 (2012)

[Bei11] Beimel, A.: Secret-sharing schemes: a survey. In: Chee, Y.M., Guo, Z., Ling, S., Shao, F., Tang, Y., Wang, H., Xing, C. (eds.) IWCC 2011. LNCS, vol. 6639, pp. 11–46. Springer, Heidelberg (2011). https://doi.org/10.1007/978-3-642-20901-7_2

[BF98] Babai, L., Frankl, P.: Linear algebra methods in combinatorics (1998)

[BGP95] Beimel, A., Gál, A., Paterson, M.: Lower bounds for monotone span programs. In: FOCS, pp. 674–681 (1995)

[BGW99] Babai, L., Gál, A., Wigderson, A.: Superpolynomial lower bounds for monotone span programs. Combinatorica **19**(3), 301–319 (1999)

[BI01] Beimel, A., Ishai, Y.: On the power of nonlinear secret-sharing. In: Proceedings of the 16th Annual IEEE Conference on Computational Complexity, Chicago, 18–21 June 2001, pp. 188–202. IEEE Computer Society (2001)

[BIKK14] Beimel, A., Ishai, Y., Kumaresan, R., Kushilevitz, E.: On the cryptographic complexity of the worst functions. In: Lindell, Y. (ed.) TCC 2014. LNCS, vol. 8349, pp. 317–342. Springer, Heidelberg (2014). https://doi.org/10.1007/978-3-642-54242-8_14

[BL88] Benaloh, J.C., Leichter, J.: Generalized secret sharing and monotone functions. In: Proceedings of the 8th Annual International Cryptology Conference on Advances in Cryptology, CRYPTO 1988, Santa Barbara, 21–25 August 1988, pp. 27–35 (1988)

[Bla79] Blakley, G.R.: Safeguarding cryptographic keys. In: Proceedings of AFIPS 1979 National Computer Conference, pp. 313–317 (1979)

[CFIK03] Cramer, R., Fehr, S., Ishai, Y., Kushilevitz, E.: Efficient multi-party computation over rings. In: Biham, E. (ed.) EUROCRYPT 2003. LNCS, vol. 2656, pp. 596–613. Springer, Heidelberg (2003). https://doi.org/10.1007/3-540-39200-9_37

[CKGS98] Chor, B., Kushilevitz, E., Goldreich, O., Sudan, M.: Private information retrieval. J. ACM **45**(6), 965–981 (1998)

[Csi97] Csirmaz, L.: The size of a share must be large. J. Cryptol. **10**(4), 223–231 (1997)

[DG15] Dvir, Z., Gopi, S.: 2-server PIR with sub-polynomial communication. In: STOC, pp. 577–584 (2015)

[DGY11] Dvir, Z., Gopalan, P., Yekhanin, S.: Matching vector codes. SIAM J. Comput. **40**(4), 1154–1178 (2011)

[Efr12] Efremenko, K.: 3-query locally decodable codes of subexponential length, vol. 41, pp. 1694–1703 (2012)

[FKN94] Feige, U., Kilian, J., Naor, M.: A minimal model for secure computation (extended abstract). In: Leighton, F.T., Goodrich, M.T. (eds.) Proceedings of the Twenty-Sixth Annual ACM Symposium on Theory of Computing, 23–25 May 1994, Montréal, pp. 554–563. ACM (1994)

[GIKM00] Gertner, Y., Ishai, Y., Kushilevitz, E., Malkin, T.: Protecting data privacy in private information retrieval schemes. J. Comput. Syst. Sci. **60**(3), 592–629 (2000)

[GKW15] Gay, R., Kerenidis, I., Wee, H.: Communication complexity of conditional disclosure of secrets and attribute-based encryption. In: Gennaro, R., Robshaw, M. (eds.) CRYPTO 2015. LNCS, vol. 9216, pp. 485–502. Springer, Heidelberg (2015). https://doi.org/10.1007/978-3-662-48000-7_24

[GPSW06] Goyal, V., Pandey, O., Sahai, A., Waters, B.: Attribute-based encryption for fine-grained access control of encrypted data. In: ACM Conference on Computer and Communications Security, pp. 89–98 (2006)

[Gro00] Grolmusz, V.: Superpolynomial size set-systems with restricted intersections mod 6 and explicit ramsey graphs. Combinatorica 20(1), 71–86 (2000)

[IK97] Ishai, Y., Kushilevitz, E.: Private simultaneous messages protocols with applications. In: ISTCS, pp. 174–184 (1997)

[IK00] Ishai, Y., Kushilevitz, E.: Randomizing polynomials: a new representation with applications to round-efficient secure computation. In: 41st Annual Symposium on Foundations of Computer Science, FOCS 2000, 12–14 November 2000, Redondo Beach, pp. 294–304. IEEE Computer Society (2000)

[IK02] Ishai, Y., Kushilevitz, E.: Perfect constant-round secure computation via perfect randomizing polynomials. In: Widmayer, P., Eidenbenz, S., Triguero, F., Morales, R., Conejo, R., Hennessy, M. (eds.) ICALP 2002. LNCS, vol. 2380, pp. 244–256. Springer, Heidelberg (2002). https://doi.org/10.1007/3-540-45465-9_22

[ISN89] Ito, M., Saito, A., Nishizeki, T.: Secret sharing scheme realizing general access structure. Electron. Commun. Jpn. (Part III Fundam. Electron. Sci.) 72(9), 56–64 (1989)

[IW14] Ishai, Y., Wee, H.: Partial garbling schemes and their applications. In: Esparza, J., Fraigniaud, P., Husfeldt, T., Koutsoupias, E. (eds.) ICALP 2014. LNCS, vol. 8572, pp. 650–662. Springer, Heidelberg (2014). https://doi.org/10.1007/978-3-662-43948-7_54

[KW93] Karchmer, M., Wigderson, A.: On span programs. In: Structure in Complexity Theory Conference, pp. 102–111. IEEE Computer Society (1993)

[LVW17] Liu, T., Vaikuntanathan, V., Wee, H.: Conditional disclosure of secrets via non-linear reconstruction. In: Katz, J., Shacham, H. (eds.) CRYPTO 2017. LNCS, vol. 10401, pp. 758–790. Springer, Cham (2017). https://doi.org/10.1007/978-3-319-63688-7_25

[OSW07] Ostrovsky, R., Sahai, A., Waters, B.: Attribute-based encryption with non-monotonic access structures. In: ACM Conference on Computer and Communications Security, pp. 195–203 (2007)

[PR17] Pitassi, T., Robere, R.: Lifting nullstellensatz to monotone span programs over any field. Electron. Colloq. Comput. Compl. (ECCC) 24, 165 (2017)

[RPRC16] Robere, R., Pitassi, T., Rossman, B., Cook, S.A.: Exponential lower bounds for monotone span programs. In: FOCS, pp. 406–415 (2016)

[Sha79] Shamir, A.: How to share a secret. Commun. ACM 22(11), 612–613 (1979)

[VV15] Vaikuntanathan, V., Vasudevan, P.N.: Secret sharing and statistical zero knowledge. In: Iwata, T., Cheon, J.H. (eds.) ASIACRYPT 2015. LNCS, vol. 9452, pp. 656–680. Springer, Heidelberg (2015). https://doi.org/10.1007/978-3-662-48797-6_27

[WY05] Woodruff, D.P., Yekhanin, S.: A geometric approach to information-theoretic private information retrieval. In: CCC, pp. 275–284 (2005)

[Yek08] Yekhanin, S.: Towards 3-query locally decodable codes of subexponential length. J. ACM 55(1), 1:1–1:16 (2008)

Improving the Linear Programming Technique in the Search for Lower Bounds in Secret Sharing

Oriol Farràs[1]([✉]), Tarik Kaced[2], Sebastià Martín[3], and Carles Padró[3]

[1] Universitat Rovira i Virgili, Tarragona, Spain
oriol.farras@urv.cat
[2] Sorbonne Université, LIP6, Paris, France
tarik.kaced@ens-lyon.org
[3] Universitat Politècnica de Catalunya, Barcelona, Spain
{sebastia.martin,carles.padro}@upc.edu

Abstract. We present a new improvement in the linear programming technique to derive lower bounds on the information ratio of secret sharing schemes. We obtain non-Shannon-type bounds without using information inequalities explicitly. Our new technique makes it possible to determine the optimal information ratio of linear secret sharing schemes for all access structures on 5 participants and all graph-based access structures on 6 participants. In addition, new lower bounds are presented also for some small matroid ports and, in particular, the optimal information ratios of the linear secret sharing schemes for the ports of the Vamos matroid are determined.

Keywords: Secret sharing · Information inequalities
Rank inequalities · Common information · Linear programming

1 Introduction

Linear programming involving information inequalities has been extensively used in different kinds of information theoretic problems. An early instance is the verification of Shannon information inequalities [63], and we find more examples in secret sharing [15,52], network coding [61,64], and other topics [62].

In this work, we present a new improvement of the linear programming technique in the search for lower bounds on the information ratio of secret

O. Farràs—Supported by the Spanish Government through TIN2014-57364-C2-1-R and by the Catalan government through 2017SGR-705. Tarik Kaced acknowledges the support of the French Agence Nationale de la Recherche (ANR), under grant ANR-16-CE23-0016-01 (project PAMELA). Sebastià Martín and Carles Padró are supported by Spanish Government through MTM2016-77213-R.

J. B. Nielsen and V. Rijmen (Eds.): EUROCRYPT 2018, LNCS 10820, pp. 597–621, 2018.
https://doi.org/10.1007/978-3-319-78381-9_22

sharing schemes. Namely, instead of known non-Shannon information inequalities, we propose to use constraints based on the properties from which those inequalities are deduced.

Secret sharing, which was independently introduced by Shamir [58] and Blakley [9], is a very useful tool that appears as a component in many different kinds of cryptographic protocols. The reader is referred to [4] for a survey on secret sharing and its applications. In a *secret sharing scheme*, a *secret value* is distributed into *shares* among a set of *participants* in such a way that only the *qualified sets* of participants can recover the secret value. This work deals exclusively with *unconditionally secure* and *perfect* secret sharing schemes, in which the shares from any unqualified set do not provide any information on the secret value. In this case, the family of qualified sets of participants is called the *access structure* of the scheme.

In a *linear secret sharing scheme*, the secret and the shares are vectors over some finite field, and both the computation of the shares and the recovering of the secret are performed by linear maps. Because of their homomorphic properties, linear schemes are used in many applications of secret sharing. Moreover, most of the known constructions of secret sharing schemes yield linear schemes.

The *information ratio* of a secret sharing schemes is the ratio between the length of the shares and the length of the secret. The optimization of this parameter, both for linear and general secret sharing schemes, has attracted a lot of attention. This problem has been analyzed for several families of access structures. For example, access structures defined by graphs [5,10,12,16,18,20,30,32], access structures on a small number of participants [20,30–32,37,52,59], bipartite access structures [25,51], the ones having few minimal qualified sets [43,45], or ports of non-representable matroids [7,44,48,52].

That optimization problem is related to the search for asymptotic lower bounds on the length of the shares, which is one of the main open problems in secret sharing. The reader is referred to the survey by Beimel [4] for more information about this topic. For *linear* secret sharing schemes, building up on the superpolynomial lower bounds in [3,6], exponential lower bounds have been proved recently [53,55]. Nevertheless, for the general case, no proof for the existence of access structures requiring shares of superpolynomial size has been found. Moreover, the best of the known lower bounds is the one given by Csirmaz [14,15], who presented a family of access structures on an arbitrary number n of participants whose optimal information ratio is $\Omega(n/\log n)$.

Almost all known lower bounds on the optimal information ratio have been obtained by the same method, which is called here the *linear programming (LP) technique*. In particular, the asymptotic lower bound found by Csirmaz [14,15] and most of the lower bounds for the aforementioned families of access structures. The LP-technique is based on the fact, pointed out by Karnin et al. [39], that a secret sharing scheme can be defined as a collection of random variables such that their joint entropies satisfy certain constraints derived from the access structure.

The technique was first used by Capocelli et al. [12]. In particular, they presented the first examples of access structures with optimal information ratio strictly greater than 1. Csirmaz [15] refined the method by introducing some abstraction revealing its combinatorial nature. This was achieved by using the connection between Shannon entropies and polymatroids discovered by Fujishige [26, 27]. The lower bounds on the optimal information ratio that can be obtained by using that connection between Shannon entropies and polymatroids or, equivalently, by using only Shannon information inequalities are called here *Shannon-type* lower bounds. The known exact values of the optimal information ratio have been determined by finding, for each of the corresponding access structures, both a Shannon-type lower bound and a linear secret sharing scheme whose information ratio equals that bound.

A further improvement, which was first applied in [7], consists in adding to the game constraints that cannot be derived from Shannon information inequalities. Specifically, the so-called *non-Shannon information inequalities* and *non-Shannon rank inequalities*. The former provide lower bounds for the general case, while the bounds derived from the latter apply to linear secret schemes. That addition made it possible to find several new lower bounds [7, 16, 48, 52] and also the first examples of access structures whose optimal information ratios are strictly greater than any Shannon-type lower bound [7], namely the ports of the Vamos matroid.

Finally, Metcalf-Burton [48] and Padró et al. [52] realized that the method consists of finding lower bounds on the solutions of certain linear programming problems, which can be solved if the number of participants is small. In particular, the best Shannon-type lower bound for any given access structure is the optimal value of a certain linear programming problem. Again, new lower bounds for a number of access structures [25, 45, 48, 52] were obtained as a consequence of that improvement.

Some limitations of the LP-technique in the search for asymptotic lower bounds have been found. Namely, the best lower bound that can be obtained by using all information inequalities that were known at the beginning of this decade is linear in the number of participants [8, 15], while at most polynomial lower bounds can be found by using all known or unknown inequalities on a bounded number of variables [46].

Summarizing, while the LP-technique has important limitations when trying to find asymptotic lower bounds, it has been very useful in the search for lower bounds for finite and infinite families of access structures, providing in many cases tight bounds. More details about the LP-technique and its application are discussed in Sect. 2.

Yet another improvement to the LP-technique is presented in this work. Instead of using the known non-Shannon information and rank inequalities, we use the properties from which most of them have been derived. Specifically, most of the known non-Shannon information inequalities are obtained by using the *copy lemma* [22, 66] or the *Ahlswede-Körner lemma* [1, 2, 38, 42]. These two techniques are proved to be equivalent in [38]. All known non-Shannon rank inequalities, which provide lower bounds on the information ratio of *linear* secret sharing schemes, are derived from the *common information property* [23]. We derive from these properties some constraints to be added to the linear programming problems that are used to find lower bounds.

We applied that improvement to several access structures on a small number of players and we find new lower bounds that could not be found before by using the known information and rank inequalities. Specifically, the access structures on five participants, the graph-based access structures on six participants, and some ports of non-representable matroids have been the testbeds for our improvement on the LP-technique.

Jackson and Martin [37] determined the optimal information ratios of most of the access structures on five participants. The use of computers to solve the corresponding linear programming problems provided better Shannon-type lower bounds for some of the unsolved cases [52]. In addition, constructions of linear secret sharing schemes were presented in [31] improving some upper bounds. After those developments, only eight cases remained unsolved. Moreover, the values of the optimal information ratios for all solved cases were determined by a linear secret sharing scheme matching a Shannon-type lower bound. The negative result in [52, Proposition 7.1] clearly indicated that some of the open cases could not be solved in that way. Nevertheless, adding non-Shannon information and rank inequalities to the linear programs did not produce any new lower bound [52]. In contrast, our enhanced LP-technique provides better lower bounds for those unsolved cases, which are tight for *linear* secret sharing schemes. In particular, the optimal information ratio of linear secret sharing schemes is now determined for every access structure on five participants. Even though we present new lower bounds, some values are still unknown for general schemes. So, we partially concluded the project initiated by Jackson and Martin in [37]. Moreover, we found the smallest examples of access structures for which the optimal information ratio does not coincide with the best Shannon-type lower bound.

A similar project was undertaken by van Dijk [20] for graph-based access structures on six participants, that is, access structures whose minimal qualified sets have exactly two participants. Most of the cases were solved in the initial work [20], and several advances were presented subsequently [13,30,32,41,52]. At this point, only nine cases remained unsolved. We have been able to find for them new lower bounds for linear schemes by using our enhanced LP-technique. Once our new lower bounds were made public, Gharahi and Khazaei [33] presented constructions of linear secret sharing schemes proving that they are tight. Therefore, our results made it possible to determine the optimal information rate of linear secret sharing schemes for all graph-based access structures on six participants.

In addition, we present new lower bounds for the ports of four non-representable matroids on eight points and, in particular, we determine the optimal information ratio of linear schemes for the ports of the Vamos matroid and the matroid Q_8.

All the lower bounds that are presented in this paper have been found by solving linear programming problems with conveniently chosen additional constraints derived from the common information property and the Ahlswede-Körner lemma. Since the number of variables and constraints is exponential in the number of participants, this can be done only for access structures on small sets. However, several lower bounds for infinite families of access structures have been obtained by using the LP-technique without solving linear programming problems [10,15,17,18,51]. Nevertheless, a better understanding of those tools is needed to apply our improvement of the LP-technique in a similar way. Since the known limitations of the LP-technique do not imply the contrary, it may be even possible to improve Csirmaz's [14,15] asymptotic lower bound $\Omega(n/\log n)$.

The paper is organized as follows. A detailed discussion on the LP-technique is given in Sect. 2. Our improvement on the method is described in Sect. 3. The new lower bounds that have been obtained by applying our technique are presented in Sect. 4. Constructions of linear secret sharing schemes that are used to prove the tightness of some of those bounds are given in Sect. 5. We conclude the paper in Sect. 6 with some open problems and suggestions for future work.

2   Lower Bounds in Secret Sharing from Linear Programming

We begin by introducing some notation. For a finite set Q, we use $\mathcal{P}(Q)$ to denote its *power set*, that is, the set of all subsets of Q. We use a compact notation for set unions, that is, we write XY for $X \cup Y$ and Xy for $X \cup \{y\}$. In addition, we write $X \smallsetminus Y$ for the set difference and $X \smallsetminus x$ for $X \smallsetminus \{x\}$.

2.1   Entropic and Linear Polymatroids

Only discrete random variables are considered in this paper. For a finite set Q, consider a random vector $(S_x)_{x \in Q}$. For every $X \subseteq Q$, we use S_X to denote the subvector $(S_x)_{x \in X}$, and $H(S_X)$ will denote its Shannon entropy. Given three random variables $(S_i)_{i \in \{1,2,3\}}$, the *entropy of S_1 conditioned on S_2* is

$$H(S_1|S_2) = H(S_{12}) - H(S_2),$$

the *mutual information* of S_1 and S_2 is

$$I(S_1{:}S_2) = H(S_1) - H(S_1|S_2) = H(S_1) + H(S_2) - H(S_{12})$$

and, finally, the *conditional mutual information* is defined by

$$I(S_1{:}S_2|S_3) = H(S_1|S_3) - H(S_1|S_{23}) = H(S_{13}) + H(S_{23}) - H(S_{123}) - H(S_3).$$

A fundamental fact about Shannon entropy is that the conditional mutual information is always nonnegative, and this implies the following connection between Shannon entropy and polymatroids, which was first described by Fujishige [26,27].

Definition 2.1. *A polymatroid is a pair (Q, f) formed by a finite set Q, the ground set, and a rank function $f \colon \mathcal{P}(Q) \to \mathbb{R}$ satisfying the following properties.*

(P1) $f(\emptyset) = 0$.

(P2) f *is* monotone increasing: *if* $X \subseteq Y \subseteq Q$, *then* $f(X) \leq f(Y)$.

(P3) f *is* submodular: $f(X \cup Y) + f(X \cap Y) \leq f(X) + f(Y)$ *for every* $X, Y \subseteq Q$.

A polymatroid is called integer *if its rank function is integer-valued. If* $\mathcal{S} = (Q, f)$ *is a polymatroid and* α *is a positive real number, then* $(Q, \alpha f)$ *is a polymatroid too, which is called a* multiple *of* $\mathcal{S}$.

Theorem 2.2 (Fujishige [26,27]). *Let* $(S_x)_{x \in Q}$ *be a random vector. Consider the mapping* $h \colon \mathcal{P}(Q) \to \mathbb{R}$ *defined by* $h(\emptyset) = 0$ *and* $h(X) = H(S_X)$ *if* $\emptyset \neq X \subseteq Q$. *Then* h *is the rank function of a polymatroid with ground set* Q.

Definition 2.3. *The polymatroids that can be defined from a random vector as in Theorem 2.2 are called* entropic. *Consider a field* $\mathbb{K}$, *a vector space* V *with finite dimension over* $\mathbb{K}$ *and a collection* $(V_x)_{x \in Q}$ *of vector subspaces of* V. *It is clear from basic linear algebra that the map* f *defined by* $f(X) = \dim \sum_{x \in X} V_x$ *for every* $X \subseteq Q$ *is the rank function of a polymatroid. Every such polymatroid is said to be* $\mathbb{K}$-linear.

Because of the connection given in Theorem 2.2, if f is the rank function of a polymatroid, we use the notation $f(A|B) = f(AB) - f(A)$ for every pair of subsets of the ground set.

We discuss in the following the well known connection between entropic and linear polymatroids, as described in [34]. Let $\mathbb{K}$ be a finite field and V a vector space with finite dimension over $\mathbb{K}$. Let S be the random variable determined by the uniform probability distribution on the dual space V^*. For every vector subspace $W \subseteq V$, the restriction of S to W determines a random variable $S|_W$ that is uniformly distributed on its support W^*, and hence $H(S|_W) = \log |\mathbb{K}| \dim W^* = \log |\mathbb{K}| \dim W$. Let $(V_x)_{x \in Q}$ be a collection of subspaces of V. For every $X \subseteq Q$, we notate $V_X = \sum_{x \in X} V_x$. This collection of subspaces determines the $\mathbb{K}$-*linear random vector* $(S_x)_{x \in Q} = (S|_{V_x})_{x \in Q}$. Observe that $S_X = S|_{V_X}$ for every $X \subseteq Q$, and hence

$$H(S_X) = \log |\mathbb{K}| \dim V_X = \log |\mathbb{K}| \dim \sum_{x \in X} V_x.$$

This implies that the $\mathbb{K}$-linear polymatroid determined by the collection of subspaces $(V_x)_{x \in Q}$ is a multiple of the entropic polymatroid defined by the $\mathbb{K}$-linear random vector $(S_x)_{x \in Q} = (S|_{V_x})_{x \in Q}$. By taking also into account that every linear polymatroid admits a linear representation over some finite field [23,54], from this discussion we can conclude the well known fact that every linear polymatroid is the multiple of an entropic polymatroid.

2.2   Secret Sharing

Definition 2.4. *Let* P *be a set of* participants. *An* access structure Γ *on* P *is a* monotone increasing *family of subsets of* P, *that is, if* $A \subseteq B \subseteq P$ *and*

$A \in \Gamma$, then $B \in \Gamma$. The members of Γ are the qualified sets of the structure. An access structure is determined by the family $\min \Gamma$ of its minimal qualified sets. A participant is redundant in an access structure if it is not in any minimal qualified set. All access structures in this paper are assumed to have no redundant participants. The dual Γ^* of an access structure Γ on P is formed by the sets $A \subseteq P$ such that its complement $P \smallsetminus A$ is not in Γ.

Definition 2.5. Let Γ be an access structure on a set of participants P. Consider a special participant $p_o \notin P$, which is usually called dealer, and the set $Q = Pp_o$. A secret sharing scheme on P with access structure Γ is a random vector $\Sigma = (S_x)_{x \in Q}$ such that the following properties are satisfied.

1. $H(S_{p_o}) > 0$.
2. If $A \in \Gamma$, then $H(S_{p_o}|S_A) = 0$.
3. If $A \notin \Gamma$, then $H(S_{p_o}|S_A) = H(S_{p_o})$.

The random variable S_{p_o} corresponds to the secret value, while the shares received by the participants are given by the random variables S_x with $x \in P$. Condition 2 implies that the shares from a qualified set determine the secret value while, by Condition 3, the shares from an unqualified set and the secret value are independent.

Definition 2.6. Let $\mathbb{K}$ be a finite field. A secret sharing scheme $\Sigma = (S_x)_{x \in Q}$ is $\mathbb{K}$-linear if it is a is $\mathbb{K}$-linear random vector.

Definition 2.7. The information ratio $\sigma(\Sigma)$ of the secret sharing scheme Σ is

$$\sigma(\Sigma) = \max_{x \in P} \frac{H(S_x)}{H(S_{p_o})}$$

and its average information ratio $\widetilde{\sigma}(\Sigma)$ is

$$\widetilde{\sigma}(\Sigma) = \frac{1}{n} \sum_{x \in P} \frac{H(S_x)}{H(S_{p_o})}.$$

Definition 2.8. The optimal information ratio $\sigma(\Gamma)$ of an access structure Γ is the infimum of the information ratios of all secret sharing schemes for Γ. The optimal average information ratio $\widetilde{\sigma}(\Gamma)$ is defined analogously. The values $\lambda(\Gamma)$ and $\widetilde{\lambda}(\Gamma)$ are defined by restricting the optimization to linear secret sharing schemes.

2.3   Lower Bounds from Shannon Information Inequalities

We describe next how to find linear programming problems whose optimal values are lower bounds on those parameters. Let Γ be an access structure on a set P and take, as usual, $Q = Pp_o$. Given a secret sharing scheme $\Sigma = (S_x)_{x \in Q}$ with access structure Γ, consider the entropic polymatroid (Q, h) determined by the random vector $(S_x)_{x \in Q}$, that is, $h(X) = H(S_X)$ for every $X \subseteq Q$. Take $\alpha = 1/h(p_o)$ and the polymatroid (Q, f) with $f = \alpha h$. The rank function f can be seen as a vector $(f(X))_{X \subseteq Q} \in \mathbb{R}^{\mathcal{P}(Q)}$ that satisfies the linear constraints

(N) $f(p_o) = 1$,

(Γ1) $f(Xp_o) = f(X)$ for every $X \subseteq P$ with $X \in \Gamma$,

(Γ2) $f(Xp_o) = f(X) + 1$ for every $X \subseteq P$ with $X \notin \Gamma$,

and also the polymatroid axioms (P1)–(P3) in Definition 2.1. Observe that constraints (Γ1), (Γ2) are derived from the chosen access structure Γ. Constraints (P1)–(P3) are equivalent to the so-called *Shannon information inequalities*, that is, the ones implied by the fact that the conditional mutual information is nonnegative. Therefore, the vector f is a feasible solution of Linear Programming Problem 2.9.

Linear Programming Problem 2.9. *The optimal value of this linear programming problem is, by definition, $\widetilde{\kappa}(\Gamma)$:*

$$Minimize \quad (1/n) \sum_{x \in P} f(x)$$

$$subject\ to \quad (N), (\Gamma 1), (\Gamma 2), (P1), (P2), (P3)$$

Since this applies to every secret sharing scheme Σ with access structure Γ and the objective function equals $\widetilde{\sigma}(\Sigma)$, the optimal value $\widetilde{\kappa}(\Gamma)$ of this linear programming problem is a lower bound on $\widetilde{\sigma}(\Gamma)$. Similarly, a lower bound on $\sigma(\Gamma)$ is provided by the optimal value $\kappa(\Gamma)$ of the Linear Programming Problem 2.10.

Linear Programming Problem 2.10. *The optimal value of this linear programming problem is, by definition, $\kappa(\Gamma)$:*

$$Minimize \quad v$$

$$subject\ to \quad v \geq f(x)\ for\ every\ x \in P$$

$$(N), (\Gamma 1), (\Gamma 2), (P1), (P2), (P3)$$

The parameters $\kappa(\Gamma)$ and $\widetilde{\kappa}(\Gamma)$ were first introduced in [44]. They are the best lower bounds on $\sigma(\Gamma)$ and, respectively, $\widetilde{\sigma}(\Gamma)$ that can be obtained by using only Shannon information inequalities, that is, they are the best possible Shannon-type lower bounds. If the number of participants is small, they can be computed by solving the corresponding linear programming problems. This approach has been used in [25, 45, 52]. In more general situations, lower bounds on $\kappa(\Gamma)$ and $\widetilde{\kappa}(\Gamma)$ can be derived from the constraints without solving the linear programming problems, as in [10, 12, 17, 18, 20, 37] and many other works. In particular, the result in the following theorem, which is the best of the known general asymptotic lower bounds, was found in this way.

Theorem 2.11 (Csirmaz [14, 15]). *For every n, there exists an access structure Γ_n on n participants such that $\widetilde{\kappa}(\Gamma_n)$ is $\Omega(n/\log n)$.*

Since not all polymatroids are entropic, the lower bounds $\kappa(\Gamma)$ and $\widetilde{\kappa}(\Gamma)$ are not tight in general. Moreover, Csirmaz [15] proved that $\kappa(\Gamma) \leq n$ for every access structure Γ on n participants, which indicates that those lower bounds may be very far from tight. That result was proved by showing feasible solutions of the linear programming problems with small values of the objective function.

Duality simplifies the search for bounds in secret sharing. Indeed, if Γ^* is the dual of the access structure Γ, then $\lambda(\Gamma^*) = \lambda(\Gamma)$ and $\widetilde{\lambda}(\Gamma^*) = \widetilde{\lambda}(\Gamma)$ [36], and also $\kappa(\Gamma^*) = \kappa(\Gamma)$ and $\widetilde{\kappa}(\Gamma^*) = \widetilde{\kappa}(\Gamma)$ [44]. In contrast, it is not known whether the analogous relation applies to the parameters σ and $\widetilde{\sigma}$ or not.

2.4   Ideal Secret Sharing Schemes and Matroid Ports

The extreme case $\kappa(\Gamma) = 1$ deserves some attention because it is related to ideal secret sharing schemes. Since we are assuming that there are no redundant participants, it is easy to prove that every feasible solution f of the Linear Programming Problems 2.9 and 2.10 satisfies $f(x) \geq 1$ for every $x \in P$. Therefore, $1 \leq \widetilde{\kappa}(\Gamma) \leq \kappa(\Gamma)$ for every access structure Γ, and hence the average information ratio of every secret sharing scheme is at least 1.

Definition 2.12. *A secret sharing scheme $\Sigma = (S_x)_{x \in Q}$ is ideal if its information ratio is equal to 1, which is best possible. Ideal access structures are those that admit an ideal secret sharing scheme.*

Definition 2.13. *A matroid $M = (Q, r)$ is an integer polymatroid such that $r(X) \leq |X|$ for every $X \subseteq Q$. The port of the matroid M at $p_o \in Q$ is the access structure on $P = Q \setminus p_o$ whose qualified sets are the sets $X \subseteq P$ satisfying $r(Xp_o) = r(X)$.*

The following theorem is a consequence of the results by Brickell and Davenport [11], who discovered the connection between ideal secret sharing and matroids.

Theorem 2.14. *Let $\Sigma = (S_x)_{x \in Q}$ be an ideal secret sharing scheme on P with access structure Γ. Then the mapping given by $f(X) = H(S_X)/H(S_{p_o})$ for every $X \subseteq Q$ is the rank function of a matroid M with ground set Q. Moreover, Γ is the port of the matroid M at p_o.*

As a consequence, every ideal access structure is a matroid port. The first counterexample for the converse, the ports of the Vamos matroid, was presented by Seymour [57]. Additional results on matroid ports and ideal secret sharing schemes were proved in [44] by using the forbidden minor characterization of matroid ports by Seymour [56].

Theorem 2.15 ([44]). *Let Γ be an access structure. Then Γ is a matroid port if and only if $\kappa(\Gamma) = 1$. Moreover, $\kappa(\Gamma) \geq 3/2$ if Γ is not a matroid port.*

In particular, there is a gap in the values of the parameter κ. Namely, there is no access structure Γ with $1 < \kappa(\Gamma) < 3/2$. Therefore, the optimal information ratio of an access structure that is not a matroid port is at least $3/2$.

2.5 Lower Bounds from Non-Shannon Information and Rank Inequalities

Better lower bounds can be obtained by adding to the Linear Programming Problems 2.9 and 2.10 new constraints derived from *non-Shannon information inequalities*, which are satisfied by every entropic polymatroid but are not derived from the basic Shannon information inequalities. Zhang and Yeung [66] presented such an inequality for the first time, and many others have been found subsequently [22,24,47,65]. This approach was first applied in [7] to prove that the optimal information ratio of the ports of the Vamos matroid is larger than 1, the first known examples of matroid ports with that property. They are as well the first known examples of access structures with $\kappa(\Gamma) < \sigma(\Gamma)$, and also the first known examples with $1 < \sigma(\Gamma) < 3/2$. Other lower bounds for the ports of the Vamos matroid and other non-linear matroids have been presented [48,52].

When searching for bounds for linear secret sharing schemes, that is, bounds on $\lambda(\Gamma)$ and $\widetilde{\lambda}(\Gamma)$, one can improve the linear program by using *rank inequalities*, which apply to configurations of vector subspaces or, equivalently, to the joint entropies of linear random vectors. It is well-known that every information inequality is also a rank inequality. The first known rank inequality that cannot be derived from the Shannon inequalities was found by Ingleton [35]. Other such rank inequalities have been presented afterwards [23,40]. Better lower bounds on the information ratio of linear secret sharing schemes have been found for some families of access structures by using non-Shannon rank inequalities [7,16,52].

On the negative side, Beimel and Orlov [8] proved that the best lower bound that can be obtained by using all information inequalities on four and five variables, together with all inequalities on more than five variables that were known by then, is at most linear on the number of participants. Specifically, they proved that every linear programming problem that is obtained by using these inequalities admits a feasible solution with a small value of the objective function. That solution is related to the one used by Csirmaz [15] to prove that $\kappa(\Gamma)$ is at most the number of participants. Another negative result about the power of information inequalities to provide asymptotic lower bounds was presented in [46]. Namely, every lower bound that is obtained by using rank inequalities on at most r variables is $O(n^{r-2})$, and hence polynomial on the number n of participants. Since all information inequalities are rank inequalities, this negative result applies to the search for asymptotic lower bounds for both linear and general secret sharing schemes.

3 Improved Linear Programming Technique

Our improvements on the LP-technique are presented in this section. Instead of adding non-Shannon information and rank inequalities to the linear programming problems, which is the strategy described in Sect. 2.5, we add constraints that are obtained by using some properties from which those inequalities are derived.

3.1   Common Information

According to [23], all known non-Shannon rank inequalities are derived from the so-called *common information property*. We say that a random variable S_3 *conveys the common information* of the random variables S_1 and S_2 if $H(S_3|S_2) = H(S_3|S_1) = 0$ and $H(S_3) = I(S_1:S_2)$. In general, given two random variables, it is not possible to find a third one satisfying those conditions [28]. Nevertheless, this is possible for every pair of $\mathbb{K}$-linear random variables. Indeed, if $S_1 = S|_{V_1}$ and $S_2 = S|_{V_2}$ for some vector subspaces V_1, V_2 of a $\mathbb{K}$-vector space V, then $S_3 = S|_{V_1 \cap V_2}$ conveys the common information of S_1 and S_2. The following definition is motivated by the concept of common information of a pair of random variables.

Definition 3.1. *Consider a polymatroid (Q, f) and two sets $A, B \subseteq Q$. Then every subset $X_o \subseteq Q$ such that*

- $f(X_o|A) = f(X_o|B) = 0$, *and*
- $f(X_o) = f(A) + f(B) - f(AB)$

is called a common information *for the pair (A, B). If $X_o = \{x_o\}$, then the element x_o is also called a common information for the pair (A, B).*

Definition 3.2. *An* extension *of a polymatroid (Q, f) is any polymatroid (Q', f') with $Q \subseteq Q'$ and $f'(X) = f(X)$ for every $X \subseteq Q$. Usually, we are going to use the same symbol for the rank function of a polymatroid and that of an extension of it.*

Definition 3.3. *A polymatroid (Q, f) satisfies the* common information property *if, for every pair (A_0, A_1) of subsets of Q, there exists an extension (Qx_o, f) of it such that x_o is a common information for the pair (A_0, A_1).*

Proposition 3.4. *Every linear polymatroid satisfies the common information property. Moreover, given a linear polymatroid (Q, f) and a pair (A_0, A_1) of subsets of Q, it can be extended to a* linear *polymatroid (Qx_o, f) such that x_o is a common information for the pair (A_0, A_1). In particular, the extension also satisfies the common information property.*

Proof. Let $(V_x)_{x \in Q}$ be a collection of vector subspaces representing a $\mathbb{K}$-linear polymatroid (Q, f), and consider two subsets $A_0, A_1 \subseteq Q$. By taking $V_{x_o} = V_{A_0} \cap V_{A_1}$, an extension of our polymatroid to Qx_o is obtained in which x_o is a common information for (A_0, A_1). Obviously, this new polymatroid is $\mathbb{K}$-linear too.

We describe next how to modify the Linear Programming Problems 2.9 and 2.10 by using the common information property in order to obtain better lower bounds on the information ratio of linear secret sharing schemes. Let Γ be an access structure on a set P and $\Sigma = (S_x)_{x \in Q}$ a linear secret sharing scheme for Γ. As usual, associated to Σ consider the polymatroid (Q, f) defined by $f(X) = H(S_X)/H(S_{p_o})$ for every $X \subseteq Q$. Since the scheme Σ is linear,

(Q, f) is the multiple of a linear polymatroid, and hence it satisfies the common information property. Therefore, given any two sets $A_0, A_1 \subseteq Q$, we can find a polymatroid (Qx_o, f), an extension of (Q, f), such that x_o is a common information for the pair (A_0, A_1). Clearly, the vector $(f(X))_{X \subseteq Qx_o} \in \mathbb{R}^{\mathcal{P}(Qx_o)}$ is a feasible solution of the Linear Programming Problem 3.5.

Linear Programming Problem 3.5. *The optimal value of this linear programming problem is a lower bound on $\widetilde{\lambda}(\Gamma)$:*

$$\text{Minimize} \quad (1/n) \sum_{x \in P} f(x)$$

$$\text{subject to} \quad (N), (\Gamma 1), (\Gamma 2)$$
$$f(x_o|A_0) = f(x_o|A_1) = 0$$
$$f(x_o) = f(A_0) + f(A_1) - f(A_0 \, A_1)$$
$$(P1), (P2), (P3) \text{ on the set } Qx_o$$

Since this applies to every linear secret sharing scheme with access structure Γ, the optimal value of that linear programming problem is a lower bound on $\widetilde{\lambda}(\Gamma)$. Of course, we can use the common information for more than one pair of sets. Specifically, given k pairs $(A_{i0}, A_{i1})_{i \in [k]}$ of subsets of Q, the optimal value of the Linear Programming Problem 3.6 is a lower bound on $\widetilde{\lambda}(\Gamma)$. Obviously, analogous modifications on Linear Programming Problem 2.10 provide lower bounds on $\lambda(\Gamma)$.

Linear Programming Problem 3.6. *The optimal value of this linear programming problem is a lower bound on $\widetilde{\lambda}(\Gamma)$:*

$$\text{Minimize} \quad (1/n) \sum_{x \in P} f(x)$$

$$\text{subject to} \quad (N), (\Gamma 1), (\Gamma 2)$$
$$f(x_i|A_{i0}) = f(x_i|A_{i1}) = 0,$$
$$f(x_i) = f(A_{i0}) + f(A_{i1}) - f(A_{i0} \, A_{i1}) \text{ for every } i = 1, \ldots, k$$
$$(P1), (P2), (P3) \text{ on the set } Qx_1 \ldots x_k$$

Remark 3.7. One can also find the common information of a pair of random variables defined from abelian groups. Specifically, given a finite abelian group G and a subgroup $H \subseteq G$, consider the random variables S, uniformly distributed on G, and $S_{/H}$ determined from S by the projection on the quotient group G/H. Given two such random variables $S_1 = S_{/H_1}$ and $S_2 = S_{/H_2}$, the random variable $S_3 = S_{/(H_1 + H_2)}$ conveys the common information of S_1 and S_2. Therefore, the lower bounds obtained from the linear programming problems introduced in this section apply also to secret sharing schemes defined from abelian groups.

3.2 Ahlswede and Körner's Information

In Sect. 3.1, the common information property was used to improve lower bounds on the information ratio of linear secret sharing schemes and, more generally, schemes that are defined from abelian groups. For the general case, we are going to use a similar property motivated by the works of Ahlswede and Körner.

The known non-Shannon-type inequalities can be derived by using two techniques, the so-called Copy lemma [66] and the Ahlswede-Körner lemma as used in [42]. It turns out that the power of these two lemmas is equivalent [38]. In particular, both constructions can be used to derive the same non-Shannon inequalities. Hereafter, we choose to use a version of the Ahlswede and Körner (AK) lemma, as it makes the LP program slightly easier to formulate because the constraints needed for the construction of additional variables are shorter to write down. The original result by Ahlswede and Körner [1,2,19] is a statement about the achievable rate region of a certain communication problem. Here, we use the AK lemma as presented in [38, Lemma 2], a statement that in its part can be derived from the proof of [42, Lemma 5]. That result deals with sequences of random variables, and hence with *almost entropic polymatroids*.

Definition 3.8. *We say that a polymatroid is* almost entropic *if it is the limit of a sequence of entropic polymatroids.*

We introduce next the *AK-information property*, which will play the same role in the general case as the common information for linear schemes.

Definition 3.9. *Consider a polymatroid (Q, f), and subsets $U, V, Z \subseteq Q$. Then every subset $Z_o \subseteq Q$ such that*

- $f(Z_o|UV) = 0,$
- $f(U|Z_o) = f(U|Z),$
- $f(V|Z_o) = f(V|Z),$
- $f(UV|Z_o) = f(UV|Z)$

is called an AK-information *for the triple (U, V, Z). Moreover, we say that a polymatroid (Q, f) satisfies the* AK-information *property, if, for every triple (U, V, Z) of subsets of Q, there exists an extension (Qz_o, f) such that z_o is an AK-information for the triple (U, V, Z).*

The following version of the AK lemma is a straightforward consequence of [38, Lemma 2].

Proposition 3.10 (Ahlswede and Körner lemma). *Let (Q, f) be an entropic polymatroid and consider $U, V, Z \subseteq Q$. Then there exists a sequence $(Qz_o, f_N)_{N>0}$ of entropic polymatroids satisfying the following properties.*

- *The sequence $(Qz_o, (1/N)f_N)_{N>0}$ converges to a polymatroid (Qz_o, f') that is an extension of (Q, f).*
- *The element z_o in (Qz_o, f') is an AK-information for the triple (U, V, Z).*

Loosely speaking, the AK lemma says that given any triple of random variables, we can always construct a new random variable that is as close as we want to their AK-information. The following result is a consequence of Proposition 3.10 and the fact that every multiple of an entropic polymatroid is almost entropic [63].

Proposition 3.11. *Every almost entropic polymatroid satisfies the AK-information property. More specifically, for every almost entropic polymatroid (Q, f) and sets $U, V, Z \subseteq Q$, there exists an almost entropic extension (Qz_o, f) such that z_o is an AK-information for the triple (U, V, Z).*

Of course, this proposition can be repeatedly applied to construct the AK-informations of various triples of subsets. Moreover, entropic polymatroids are trivially almost entropic, therefore we can add any AK-information constraint to the Linear Programming Problems 2.9 and 2.10 in order to obtain lower bounds on $\widetilde{\sigma}(\Gamma)$ and $\sigma(\Gamma)$. For instance, suppose we want to use k such AK-informations, then for $i \in \{1, \ldots, k\}$, let $U_i, V_i, Z_i \subseteq Q$, and let z_i be an AK-information for the triple (U_i, V_i, Z_i). Then the optimal value of the Linear Programming Problem 3.12 is a lower bound on $\widetilde{\sigma}(\Gamma)$. An analogous modification on the Linear Programming Problem 2.9 provides lower bounds on $\sigma(\Gamma)$.

Linear Programming Problem 3.12. *The optimal value of this linear programming problem is a lower bound on $\widetilde{\sigma}(\Gamma)$:*

$$
\begin{aligned}
Minimize \quad & (1/n) \sum_{x \in P} f(x) \\
subject\ to \quad & (N), (\Gamma 1), (\Gamma 2), \\
& f(z_i | U_i V_i) = 0, \\
& f(U_i | z_i) = f(U_i | Z_i), \\
& f(V_i | z_i) = f(V_i | Z_i), \\
& f(U_i V_i | z_i) = f(U_i V_i | Z_i) \ for\ every\ i = 1, \ldots, k \\
& (P1), (P2), (P3)\ on\ the\ set\ Qz_1 \ldots z_k
\end{aligned}
$$

4   New Lower Bounds

We present here the new lower bounds on the optimal information ratio that were obtained by using our improvement on the LP-technique. All of them deal with access structures on small sets of participants and were computed by solving the linear programming problems introduced in Sect. 3.

4.1   Access Structures on Five Participants

Jackson and Martin [37] determined the optimal information ratios of most access structures on five participants. The case of four participants had been previously solved by Stinson [59]. After some additional contributions [21,31,52], both $\sigma(\Gamma)$ and $\widetilde{\sigma}(\Gamma)$ were determined for 172 of the 180 access structures on five participants. All these results were obtained by finding the exact values or lower bounds on $\kappa(\Gamma)$ and $\widetilde{\kappa}(\Gamma)$, and then constructing linear secret sharing schemes whose (average) information ratios equaled the lower bounds. Therefore, $\kappa(\Gamma) = \sigma(\Gamma) = \lambda(\Gamma)$ and $\widetilde{\kappa}(\Gamma) = \widetilde{\sigma}(\Gamma) = \widetilde{\lambda}(\Gamma)$ for each of those 172 access structures. The unsolved cases correspond to the access structures Γ_{30}, Γ_{40}, Γ_{53}, and Γ_{73} (we use the same notation as in [37]) and their duals Γ_{153}, Γ_{150}, Γ_{152}, and Γ_{151}, respectively. Following [37], we take these access structures on the set $\{a, b, c, d, e\}$. The minimal qualified sets of the first four are given in the following.

- $\min \Gamma_{30} = \{ab, ac, bc, ad, bd, ae, cde\}$.
- $\min \Gamma_{40} = \{ab, ac, bc, ad, be, cde\}$.
- $\min \Gamma_{53} = \{ab, ac, ad, bcd, be, ce\}$.
- $\min \Gamma_{73} = \{ab, ac, bd, ce, ade\}$.

We list in the following what is known for them. These results apply also to the corresponding dual access structures.

- $\widetilde{\kappa}(\Gamma) = \widetilde{\sigma}(\Gamma) = \widetilde{\lambda}(\Gamma) = 7/5$ for Γ_{30} and Γ_{40}.
- $\widetilde{\kappa}(\Gamma) = \widetilde{\sigma}(\Gamma) = \widetilde{\lambda}(\Gamma) = 3/2$ for Γ_{53}.
- $3/2 = \widetilde{\kappa}(\Gamma) \leq \widetilde{\sigma}(\Gamma) \leq \widetilde{\lambda}(\Gamma) \leq 8/5$ for Γ_{73}.
- $3/2 = \kappa(\Gamma) \leq \sigma(\Gamma) \leq \lambda(\Gamma) \leq 5/3$ for Γ_{30}, Γ_{53} and Γ_{73}.
- $3/2 = \kappa(\Gamma) \leq \sigma(\Gamma) \leq \lambda(\Gamma) \leq 12/7$ for Γ_{40}.

The values of $\kappa(\Gamma)$ and $\widetilde{\kappa}(\Gamma)$, which coincide with the lower bounds given in [21, 37], were determined in [52] by solving the Linear Programming Problems 2.9 and 2.10. The upper bounds were given in [37], except the one on $\widetilde{\lambda}(\Gamma_{53})$, which was proved in [31].

By [52, Proposition 7.1], there is no linear scheme for Γ_{53} or Γ_{73} with information ratio equal to $3/2$, and there is no linear scheme for Γ_{73} with average information ratio equal to $3/2$. Therefore, it appears that a new technique is required to solve these cases. Our improvement of the LP-technique provided new lower bounds. Namely, by solving problems as the Linear Programming Problems 3.5 and 3.12 with the specified settings, we obtain the bounds in Tables 1 and 2, respectively.

Table 1. Results on five participants using common information.

Access structure	A_0	A_1	New lower bound
$\Gamma_{30}, \Gamma_{40}, \Gamma_{53}, \Gamma_{73}$	a	d	$5/3 \leq \lambda(\Gamma)$
Γ_{73}	a	d	$23/15 \leq \widetilde{\lambda}(\Gamma)$

The values of $\lambda(\Gamma)$ and $\widetilde{\lambda}(\Gamma)$ can be now determined for all access structures on 5 participants by combining the lower bounds in Table 1 with the existing upper bounds and the ones derived from the constructions in Sect. 5. Observe

Table 2. Results on five participants using AK information for the subsets (Z, U, V).

Access structure	Z	U	V	New lower bound
$\Gamma_{30}, \Gamma_{40}, \Gamma_{53}, \Gamma_{73}$	a	d	e	$14/9 \leq \sigma(\Gamma)$
Γ_{73}	a	d	e	$53/35 \leq \widetilde{\sigma}(\Gamma)$

that $\Gamma_{30}, \Gamma_{40}, \Gamma_{53}, \Gamma_{73}$ and their duals are precisely the access structures on least participants satisfying $\kappa(\Gamma) < \lambda(\Gamma)$.

From the bounds in Table 2, we see that $\Gamma_{30}, \Gamma_{40}, \Gamma_{53}, \Gamma_{73}$ are among the smallest access structures with $\kappa(\Gamma) < \sigma(\Gamma)$. Unfortunately, all our attempts to obtain lower bounds on $\sigma(\Gamma)$ for their duals by using AK-informations have been unsuccessful.

4.2 Graph-Based Access Structures on Six Participants

If all minimal qualified sets of an access structure have two participants, it can be represented by a graph whose vertices and edges correspond to the participants and the minimal qualified sets, respectively. Van Dijk [20] determined the optimal information ratio of most graph-based access structures on 6 participants and provided lower and upper bounds for the remaining cases. After several other authors improved those results [13,30,32,41,52], only nine cases remained unsolved. Since the known values of $\sigma(\Gamma)$ have been determined by finding lower bounds on $\kappa(\Gamma)$ and upper bounds on $\lambda(\Gamma)$, we have $\kappa(\Gamma) = \sigma(\Gamma) = \lambda(\Gamma)$ in the solved cases. The unsolved cases correspond to the following graph-based access structures on $P = \{1, 2, 3, 4, 5, 6\}$.

- $\min \Gamma_{55} = \{12, 23, 34, 45, 56, 61, 26, 25\}$
- $\min \Gamma_{59} = \{12, 23, 34, 45, 56, 61, 24, 13\}$
- $\min \Gamma_{70} = \{12, 23, 34, 45, 56, 61, 24, 25, 26\}$
- $\min \Gamma_{71} = \{12, 23, 34, 45, 56, 61, 26, 35, 36\}$
- $\min \Gamma_{75} = \{12, 23, 34, 45, 56, 61, 26, 46, 14\}$
- $\min \Gamma_{77} = \{12, 23, 34, 45, 56, 61, 26, 35, 13\}$
- $\min \Gamma_{84} = \{12, 23, 34, 45, 56, 61, 13, 15, 35, 25\}$
- $\min \Gamma_{91} = \{12, 23, 34, 45, 56, 61, 15, 25, 35, 46\}$
- $\min \Gamma_{93} = \{12, 23, 34, 45, 56, 61, 15, 35, 46, 24\}$

The known lower and upper bounds for those access structures are

- $3/2 = \kappa(\Gamma) \leq \sigma(\Gamma) \leq \lambda(\Gamma) \leq 8/5$ for $\Gamma = \Gamma_{91}$ and $\Gamma = \Gamma_{93}$, and
- $3/2 = \kappa(\Gamma) \leq \sigma(\Gamma) \leq \lambda(\Gamma) \leq 5/3$ for the other seven access structures.

The values of κ were determined by solving the corresponding linear programming problems, and they are equal to the lower bounds in [20]. All upper bounds were presented in [20], except the one for Γ_{93}, which was given in [41].

By using the common information property with the settings specified in Table 3, we found the new lower bound $\lambda(\Gamma) \geq 8/5$ for all those access structures, which is tight for Γ_{91} and Γ_{93}. In particular, those nine graph-based access structures satisfy $\kappa(\Gamma) < \lambda(\Gamma)$. We have to mention here that all our attempts to improve the known lower bounds on $\sigma(\Gamma)$ for those graph-based access structures by using linear programming problems with AK-informations did not give any result.

Table 3. New bounds for graph-based access structures on six participants using common information.

Access structure	A_{00}	A_{01}	A_{10}	A_{11}	New lower bound
$\Gamma_{55}, \Gamma_{70}, \Gamma_{75}, \Gamma_{84}$	3	6			$8/5 \leq \lambda(\Gamma)$
Γ_{71}	5	$p_o 3$			$8/5 \leq \lambda(\Gamma)$
Γ_{91}, Γ_{93}	6	$p_o 5$			$8/5 \leq \lambda(\Gamma)$
Γ_{59}	3	6	5	$p_o 4$	$8/5 \leq \lambda(\Gamma)$
Γ_{77}	4	$p_o 3$	2	$p_o 6$	$8/5 \leq \lambda(\Gamma)$

After a preprint of this work was in circulation, Gharahi and Khazaei [33] proved that all lower bounds on $\lambda(\Gamma)$ in Table 3 are tight by presenting constructions of linear secret sharing schemes for the corresponding graph-based access structures. Therefore, the exact value of $\lambda(\Gamma)$ is now determined for all graph-based access structures on six participants.

4.3   Ports of Non-representable Matroids

Recall from Sect. 2.4 that Γ is a matroid port if and only if $\kappa(\Gamma) = 1$. Moreover, $\kappa(\Gamma) = \sigma(\Gamma) = \lambda(\Gamma) = 1$ if Γ is the port of a linear matroid. In this section, we apply our techniques to find new lower bounds on the optimal information ratio of some ports of non-linear matroids on eight points, which are access structures on seven participants. All matroids on seven points are linear. Hence, the matroids we consider here are amongst the smallest non-linear matroids.

We describe next several matroids (Q, r) on eight points with $r(Q) = 4$ that admit convenient geometric representations on a cube. All of them satisfy that

- $r(X) = |X|$ for every $X \subseteq Q$ with $|X| \leq 3$,
- $r(X) = 4$ for every $X \subseteq Q$ with $|X| \geq 5$, and
- $3 \leq r(X) \leq 4$ for every $X \subseteq Q$ with $|X| = 4$.

In particular, they are *paving* matroids (see [49]). Observe that such a matroid can be described by giving the subsets $X \subseteq Q$ with $|X| = 4$ and $r(X) = 3$, that is, by giving its *4-points planes*.

Consider the 3-dimensional cube with vertices on the points $(x, y, z) \in \{0, 1\}^3$. By using the binary representation, identify each of those vertices to an integer in $\{0, 1, \ldots, 7\}$. For instance, $(0, 1, 0)$ is identified to 2 and $(1, 1, 0)$ to 6. Consider the following 14 sets of vertices.

- The six faces of the cube: $0123, 0145, 0246, 1357, 2367, 4567$,

– the six diagonal planes: $0167, 0257, 0347, 1256, 1346, 2345$, and
– the two twisted planes: $0356, 1247$.

The matroid whose 4-points planes are those fourteen sets is the *binary affine cube* $AG(3,2)$. This matroid is $\mathbb{K}$-linear if and only if the field $\mathbb{K}$ has characteristic 2 [49].

All matroids that are obtained from $AG(3,2)$ by relaxing one of the 4-points planes (that is, by changing the value of its rank to 4) are isomorphic to the matroid $AG(3,2)'$ [49]. We consider here the one obtained by the relaxation of one of the twisted planes, say 1247. The matroid $AG(3,2)'$ is a smallest non-linear matroid [49]. The port of $AG(3,2)'$ at $p_o = 0$ is the access structure $\mathcal{A}$ on the set $\{1,\ldots,7\}$ with minimal qualified sets

$$\min \mathcal{A} = \{123, 145, 167, 246, 257, 347, 356, 1247\}$$

Every port of $AG(3,2)'$ is either isomorphic to $\mathcal{A}$ or to its dual $\mathcal{A}^*$, which has minimal qualified sets

$$\min \mathcal{A}^* = \{123, 145, 167, 246, 257, 347, 1356, 2356, 3456, 3567\}$$

By relaxing the other twisted plane 0356 we obtain from $AG(3,2)'$ the matroid R_8, the *real affine cube*. The 4-points planes of this matroid are the six faces and the six diagonal planes. It is $\mathbb{K}$-linear if and only if $\mathbb{K}$ has characteristic different from 2 [49].

If, instead, the 4-points set 1256 is relaxed in $AG(3,2)'$, one obtains the smallest non-linear matroid F_8 [49]. The port of F_8 at $p_o = 0$ is the access structure $\mathcal{F}$ on $\{1,\ldots,7\}$ with minimal qualified sets

$$\min \mathcal{F} = \{123, 145, 167, 246, 257, 347, 356, 1247, 1256\}$$

The port of F_8 at $p_o = 3$ is isomorphic to $\mathcal{F}$. The ports of F_8 at $p_o = 1$ and $p_o = 2$ are both isomorphic to $\mathcal{F}^*$, whose minimal qualified sets are

$$\min \mathcal{F}^* = \{123, 145, 167, 246, 257, 1356, 2356, 3456, 3567, 1347, 2347, 3457, 3467\}$$

All the other ports of F_8 are isomorphic to the port of F_8 at $p_o = 4$, and hence isomorphic to the access structure $\widehat{\mathcal{F}}$ on $\{1,\ldots,7\}$ with minimal qualified sets

$$\min \widehat{\mathcal{F}} = \{123, 145, 246, 167, 257, 347, 1256, 1356, 2356, 3456, 3567\}$$

Observe that $\widehat{\mathcal{F}}$ is isomorphic to its dual access structure $\widehat{\mathcal{F}}^*$.

The relaxation of one of the diagonal planes of the real affine cube R_8, say 1256, produces the matroid Q_8, again a smallest non-linear matroid [49]. Let $\mathcal{Q}$ be the port of Q_8 at $p_o = 0$. Its minimal qualified sets are

$$\min \mathcal{Q} = \{123, 145, 246, 167, 257, 347, 1256, 1247, 1356, 2356, 3456, 3567\}$$

All ports of Q_8 are isomorphic to $\mathcal{Q}$ or to its dual $\mathcal{Q}^*$. The access structure $\mathcal{Q}^*$ has minimal qualified sets

$$\{123, 145, 246, 167, 257, 1247, 1347, 1356, 2347, 2356, 3456, 3457, 3467, 3567\}$$

Finally, the *Vamos matroid* V_8 is another smallest non-linear matroid [49]. Its 4-points planes are 0123, 0145, 2345, 2367, and 4567. The minimal qualified sets of the port $\mathcal{V}$ of the Vamos matroid V_8 at $p_o = 0$ are the 3-sets $123, 145$ and all 4-sets not containing them, except $2345, 2367, 4567$. Every port of V_8 is isomorphic either to $\mathcal{V}$ or to $\mathcal{V}^*$. The minimal qualified sets of $\mathcal{V}^*$ are the 3-sets $123, 145, 167$ and all 4-sets not containing them, except $2367, 4567$. The known bounds on the optimal information ratio of the ports of those non-linear matroids are summarized as follows.

- $67/59 \leq \sigma(\mathcal{V}) \leq 4/3$.
- $9/8 \leq \sigma(\mathcal{V}^*) \leq 4/3$.
- $5/4 \leq \lambda(\mathcal{V}) = \lambda(\mathcal{V}^*) \leq 4/3$.
- $19/17 \leq \sigma(\Gamma)$ if $\Gamma = \mathcal{A}$ or $\Gamma = \mathcal{Q}$.
- $9/8 \leq \sigma(\Gamma)$ if $\Gamma = \mathcal{A}^*$ or $\Gamma = \mathcal{Q}^*$.
- $5/4 \leq \lambda(\Gamma)$ if Γ is one of the structures $\mathcal{A}, \mathcal{A}^*, \mathcal{Q}, \mathcal{Q}^*$.

The lower bounds were obtained in [7,29,48,52] by using the LP-technique enhanced with the Ingleton inequality or with several non-Shannon information inequalities. The upper bounds for the ports of the Vamos matroid were presented in [44].

By solving the LP Problems 3.6 and 3.12 for those access structures with the given choices, the lower bounds in Tables 4 and 5 are obtained. Except for $\sigma(\mathcal{V}^*)$, they improve all existing lower bounds. In particular, we have determined the exact value of $\lambda(\mathcal{V}) = \lambda(\mathcal{V}^*) = 4/3$. Moreover, the construction we present in Sect. 5 implies $\lambda(\mathcal{Q}) = \lambda(\mathcal{Q}^*) = 4/3$.

Table 4. Results on matroid ports using common information.

Access structure	A_0	A_1	New lower bound
$\mathcal{A}, \mathcal{F}, \widehat{\mathcal{F}}$	06	17	$4/3 \leq \lambda(\Gamma)$
$\mathcal{Q}$	04	15	$4/3 \leq \lambda(\Gamma)$
$\mathcal{V}$	01	23	$4/3 \leq \lambda(\Gamma)$

5   Constructions

We present here linear secret sharing schemes for the access structures Γ_{40} and Γ_{73} on five participants and also for the matroid port $\mathcal{Q}$. These constructions and the lower bounds for linear schemes that have been obtained with our enhancement of the LP-technique determine the exact values of $\lambda(\Gamma_{40})$, $\widetilde{\lambda}(\Gamma_{73})$, and $\lambda(\mathcal{Q})$. As a consequence, the exact values of $\lambda(\Gamma)$ and $\widetilde{\lambda}(\Gamma)$ are now determined for all access structures on five participants.

We present first a linear scheme with information ratio $5/3$ for the access structure Γ_{40} on five participants. For a finite field $\mathbb{K}$ with characteristic larger

Table 5. Results on matroid ports using AK information for the subsets (Z_1, U_1, V_1) and (Z_2, U_2, V_2).

Access structure	Z_1	U_1	V_1	Z_2	U_2	V_2	New lower bound
$\mathcal{A}$	03	12	56				$9/8 \leq \sigma(\Gamma)$
$\mathcal{A}^*$	03	12	47	12	47	56	$33/29 \leq \sigma(\Gamma)$
$\mathcal{F}, \mathcal{Q}$	04	15	37				$9/8 \leq \sigma(\Gamma)$
$\mathcal{F}^*$	04	15	26	14	27	36	$42/37 \leq \sigma(\Gamma)$
$\widehat{\mathcal{F}}$	04	15	37	14	27	36	$42/37 \leq \sigma(\Gamma)$
$\mathcal{Q}^*$	04	15	26	15	26	37	$33/29 \leq \sigma(\Gamma)$
$\mathcal{V}$	01	23	45	23	45	67	$33/29 \leq \sigma(\Gamma)$
$\mathcal{V}^*$	01	23	45				$9/8 \leq \sigma(\Gamma)$

than 5, consider the $\mathbb{K}$-linear secret sharing scheme that is determined by the $\mathbb{K}$-linear code with generator matrix

$$\begin{pmatrix} 1 & 0 & 1 & 0 & 0 & 1 & 0 & 1 \\ 1 & 0 & 0 & 0 & 1 & 2 & 1 & 0 \\ 1 & 1 & 0 & 1 & 0 & 1 & 2 & 0 \end{pmatrix}$$

Namely, every codeword corresponds to a distribution of shares. The vertical bars indicate which positions of the codeword correspond to the secret and to every participant. In this case, a codeword

$$(s_{p_o} \mid s_{a1}, s_{a2} \mid s_{b1}, s_{b2} \mid s_c \mid s_d \mid s_e) \in \mathbb{K}^8$$

corresponds to a distribution of shares in which the secret value is $s_{p_o} \in \mathbb{K}$, the share for a is $(s_{a1}, s_{a2}) \in \mathbb{K}^2$, and so on. The access structure of this linear scheme is Γ_{40}. Another $\mathbb{K}$-linear secret sharing scheme for Γ_{40} is given by the $\mathbb{K}$-linear code with generator matrix

$$\begin{pmatrix} 1 & -1 & 1 & 1 & 0 & 0 & 1 & 0 & 0 & 0 & 0 & 0 & 0 & 1 & 1 & 0 & 1 \\ 2 & 1 & 0 & 0 & 1 & 1 & 0 & 1 & 0 & 0 & 0 & 0 & 1 & 0 & 1 & 0 & 0 & 1 \\ 0 & 3 & 0 & 1 & 1 & 0 & 1 & 1 & 0 & 0 & 0 & 0 & 0 & 0 & 1 & 0 & 0 & 0 \\ 0 & 0 & 0 & 1 & 0 & 0 & 0 & 1 & 1 & 0 & 0 & 0 & 0 & 1 & 0 & 0 & 0 & 1 \\ 1 & 1 & 0 & 0 & 1 & 0 & 0 & 0 & 0 & 1 & 0 & 0 & 0 & 0 & 1 & 0 & 1 & 0 \\ -1 & 2 & 0 & 1 & 0 & 0 & 1 & 0 & 0 & 0 & 1 & 0 & 0 & 0 & 0 & 0 & 0 & 1 \\ 1 & 1 & 0 & 0 & 1 & 0 & 0 & 1 & 0 & 0 & 0 & 1 & 0 & 0 & 1 & 0 & 0 & 0 \\ 0 & 0 & 0 & 0 & 0 & 0 & 0 & 0 & 0 & 0 & 0 & 0 & 0 & 1 & 0 & 0 & 1 \end{pmatrix}$$

By concatenating these two schemes, we obtain a scheme for Γ_{40} with information ratio $5/3$.

If $\mathbb{K}$ is a field with characteristic 2, the $\mathbb{K}$-linear code with generator matrix

$$\begin{pmatrix}
1\,0\,0 & 0\,0\,0\,0 & 1\,0\,0\,0 & 1\,0\,0\,0 & 0\,0\,0\,0 & 1\,0\,0\,0 \\
0\,1\,0 & 0\,0\,0\,0 & 0\,1\,0\,0 & 0\,1\,0\,0 & 0\,0\,0\,0 & 0\,1\,0\,0 \\
0\,0\,1 & 0\,0\,0\,0 & 0\,0\,1\,0 & 0\,0\,1\,0 & 0\,0\,0\,0 & 0\,0\,1\,0 \\
0\,0\,0 & 1\,0\,0\,0 & 0\,0\,0\,1 & 0\,0\,0\,1 & 0\,1\,0\,0 & 0\,1\,0\,0 \\
0\,0\,0 & 0\,1\,0\,0 & 0\,1\,0\,1 & 0\,0\,0\,0\,1 & 0\,0\,0\,0 & 0\,0\,0\,1 \\
0\,0\,0 & 0\,0\,1\,0 & 0\,0\,0\,1 & 0\,0\,1\,0\,0 & 0\,0\,0\,0 & 1\,1\,1\,0 \\
0\,0\,0 & 0\,0\,0\,1 & 0\,0\,0\,0\,0 & 1\,0\,1\,0\,0 & 0\,0\,0\,0 & 0\,1\,1\,1 \\
0\,0\,0 & 0\,0\,0\,0 & 1\,0\,1\,0\,1\,1 & 0\,1\,1\,0\,0 & 0\,0\,0\,0 & 0\,0\,1\,0 \\
0\,0\,0 & 0\,0\,0\,0 & 1\,0\,1\,1\,0 & 1\,1\,1\,0\,1 & 0\,1\,0\,0 & 1\,0\,1\,0 \\
0\,0\,0 & 0\,0\,0\,0 & 1\,0\,0\,1\,1 & 1\,1\,1\,0\,1 & 0\,0\,1\,0 & 1\,1\,0\,0 \\
0\,0\,0 & 0\,0\,0\,0 & 0\,0\,1\,0\,1 & 1\,1\,1\,1\,1 & 0\,0\,0\,1 & 0\,1\,0\,0
\end{pmatrix}$$

defines a $\mathbb{K}$-linear secret sharing scheme with access structure Γ_{73}. Its average information ratio is equal to $23/15$.

Finally, we present a construction of a linear secret sharing scheme with information ratio $4/3$ for the access structure $\mathcal{Q}$. It is obtained by combining four ideal secret sharing schemes in a λ-decomposition with $\lambda = 3$. The reader is referred to [50, 60] for more information about λ-decompositions. Let $\mathbb{K}$ be a finite field with characteristic different from 2. The first scheme is the one given by the $\mathbb{K}$-linear code with generator matrix

$$\begin{pmatrix}
0\,0\,0\,0\,1\,1\,1\,1 \\
0\,0\,1\,1\,0\,0\,1\,1 \\
0\,1\,0\,1\,0\,1\,0\,1 \\
1\,1\,1\,1\,1\,1\,1\,1
\end{pmatrix}$$

Its access structure $\mathcal{R}$ is the port at $p_o = 0$ of the matroid R_8, the real affine cube. One can see that all minimal qualified sets of $\mathcal{Q}$ except 1256 are also qualified sets of $\mathcal{R}$. On the other hand, the unqualified sets of $\mathcal{Q}$ are also unqualified sets of $\mathcal{R}$. The second and third pieces in the decomposition are ideal schemes given by $\mathbb{K}$-linear codes with generator matrices of the form

$$\begin{pmatrix}
0\,0\,0\,0\,1\,1\,1\,1 \\
0\,0\,1\,1\,0\,0\,1\,1 \\
0\,1\,z_2\,1\,z_4\,1\,z_6\,1 \\
1\,1\,1\,1\,1\,1\,1\,1
\end{pmatrix}$$

If $z_2 = 0$ and $z_4 = z_6 = -1$, that linear code represents the matroid that is obtained from R_8 by relaxing the 4-points planes 0347 and 1256. Therefore, we obtain a secret sharing scheme in which 347 is not qualified. If, instead, we take $z_2 = -1$ and $z_4 = z_6 = 0$, the matroid represented by that $\mathbb{K}$-linear code is obtained from R_8 by relaxing the 4-point planes 1256, 0246, and 0257. In the corresponding secret sharing scheme, the sets 246 and 257 are unqualified. The fourth scheme is given by the $\mathbb{K}$-linear code with generator matrix

$$\begin{pmatrix} 0 & 0 & 0\,0\,1\,1\,1\,1 \\ 0 & -1 & 1\,1\,0\,0\,1\,1 \\ 0 & 1 & 0\,1\,0\,1\,0\,1 \\ 1 & 1 & 1\,1\,1\,1\,1\,1 \end{pmatrix}$$

which represents the matroid that is obtained from R_8 by relaxing the 4-points planes 1256, 0145, and 0167. The sets 145 and 167 are not qualified in the corresponding scheme. Observe that every minimal qualified set of $\mathcal{Q}$ appears in at least 3 of those 4 ideal linear secret sharing schemes. Therefore, we get a linear secret sharing scheme for $\mathcal{Q}$ with information ratio 4/3.

6   Open Problems

The first line of future work worth mentioning is to fully conclude the projects initiated by Jackson and Martin [37] and van Dijk [20] by determining the values of $\sigma(\Gamma)$, $\widetilde{\sigma}(\Gamma)$, $\lambda(\Gamma)$, and $\widetilde{\lambda}(\Gamma)$ for all access structures on five participants and all graph-based access structures on six participants. By Remark 3.7, our bounds on $\lambda(\Gamma)$, and $\widetilde{\lambda}(\Gamma)$ apply also to schemes defined by abelian groups.

Many examples of access structures with $\kappa(\Gamma) = \sigma(\Gamma) = \lambda(\Gamma)$ are known, and also examples with $\kappa(\Gamma) < \sigma(\Gamma)$ and $\kappa(\Gamma) < \lambda(\Gamma)$. An open problem is to find the smallest examples with $\sigma(\Gamma) < \lambda(\Gamma)$, and also examples in each of the following situations: $\kappa(\Gamma) = \sigma(\Gamma) < \lambda(\Gamma)$, $\kappa(\Gamma) < \sigma(\Gamma) = \lambda(\Gamma)$, and $\kappa(\Gamma) < \sigma(\Gamma) < \lambda(\Gamma)$. Another interesting problem is to find matroid ports such that $\sigma(\Gamma)$ or $\lambda(\Gamma)$ are greater than 3/2 or even arbitrarily large.

It is worth noticing that, even though we used the common information property to derive lower bounds for linear secret sharing schemes, we could not determine whether that property have a good behavior with respect to duality or not. This may be due to the fact that, by Remark 3.7, those bounds apply to a more general class of schemes. Therefore, when searching for bounds by using common informations, it is worth to apply the method both to an access structure and its dual.

The main direction for future research is to obtain a better understanding of the techniques introduced here in order to improve, if possible, the known asymptotic lower bounds on $\sigma(\Gamma)$. Notice that it is not necessary to solve the corresponding linear programming problem to determine a lower bound. Instead, any feasible solution of the dual linear programming problem provides a lower bound. This strategy, which was suggested by one of the reviewers of this work, has been used, not explicitly, by the authors that have derived lower bounds from the constraints without solving the linear programming problem.

References

1. Ahlswede, R., Körner, J.: On the connection between the entropies of input and output distributions of discrete memoryless channels. In: Proceedings of the 5th Brasov Conference on Probability Theory, Brasov, Editura Academiei, Bucuresti, pp. 13–23 (1977)
2. Ahlswede, R., Körner, J.: Appendix: on common information and related characteristics of correlated information sources. In: Ahlswede, R., Bäumer, L., Cai, N., Aydinian, H., Blinovsky, V., Deppe, C., Mashurian, H. (eds.) General Theory of Information Transfer and Combinatorics. LNCS, vol. 4123, pp. 664–677. Springer, Heidelberg (2006). https://doi.org/10.1007/11889342_41
3. Babai, L., Gál, A., Wigderson, A.: Superpolynomial lower bounds for monotone span programs. Combinatorica **19**, 301–319 (1999)
4. Beimel, A.: Secret-sharing schemes: a survey. In: Chee, Y.M., Guo, Z., Ling, S., Shao, F., Tang, Y., Wang, H., Xing, C. (eds.) IWCC 2011. LNCS, vol. 6639, pp. 11–46. Springer, Heidelberg (2011). https://doi.org/10.1007/978-3-642-20901-7_2
5. Beimel, A., Farràs, O., Mintz, Y.: Secret-sharing schemes for very dense graphs. J. Cryptol. **29**, 336–362 (2016)
6. Beimel, A., Gál, A., Paterson, M.: Lower bounds for monotone span programs. Comput. Complex. **6**, 29–45 (1997)
7. Beimel, A., Livne, N., Padró, C.: Matroids can be far from ideal secret sharing. In: Canetti, R. (ed.) TCC 2008. LNCS, vol. 4948, pp. 194–212. Springer, Heidelberg (2008). https://doi.org/10.1007/978-3-540-78524-8_12
8. Beimel, A., Orlov, I.: Secret sharing and non-Shannon information inequalities. IEEE Trans. Inform. Theory **57**, 5634–5649 (2011)
9. Blakley, G.R.: Safeguarding cryptographic keys. In: AFIPS Conference Proceedings, vol. 48, pp. 313–317 (1979)
10. Blundo, C., De Santis, A., De Simone, R., Vaccaro, U.: Tight bounds on the information rate of secret sharing schemes. Des. Codes Cryptogr. **11**, 107–122 (1997)
11. Brickell, E.F., Davenport, D.M.: On the classification of ideal secret sharing schemes. J. Cryptol. **4**, 123–134 (1991)
12. Capocelli, R.M., De Santis, A., Gargano, L., Vaccaro, U.: On the size of shares for secret sharing schemes. J. Cryptol. **6**, 157–167 (1993)
13. Chen, B.L., Sun, H.M.: Weighted decomposition construction for perfect secret sharing schemes. Comput. Math. Appl. **43**, 877–887 (2002)
14. Csirmaz, L.: The dealer's random bits in perfect secret sharing schemes. Studia Sci. Math. Hungar. **32**, 429–437 (1996)
15. Csirmaz, L.: The size of a share must be large. J. Cryptol. **10**, 223–231 (1997)
16. Csirmaz, L.: An impossibility result on graph secret sharing. Des. Codes Cryptogr. **53**, 195–209 (2009)
17. Csirmaz, L.: Secret sharing on the d-dimensional cube. Des. Codes Cryptogr. **74**, 719–729 (2015)
18. Csirmaz, L., Tardos, G.: Optimal information rate of secret sharing schemes on trees. IEEE Trans. Inf. Theory **59**, 2527–2530 (2013)
19. Csiszar, I., Körner, J.: Information Theory: Coding Theorems for Discrete Memoryless Systems. Academic Press, Akademiai Kiado, New York, Budapest (1981)
20. van Dijk, M.: On the information rate of perfect secret sharing schemes. Des. Codes Cryptogr. **6**, 143–169 (1995)
21. van Dijk, M.: More information theoretical inequalities to be used in secret sharing? Inf. Process. Lett. **63**, 41–44 (1997)

22. Dougherty, R., Freiling, C., Zeger, K.: Six new non-Shannon information inequalities. In: 2006 IEEE International Symposium on Information Theory, pp. 233–236 (2006)
23. Dougherty, R., Freiling, C., Zeger, K.: Linear rank inequalities on five or more variables. arXiv.org, arXiv:0910.0284v3 (2009)
24. Dougherty, R., Freiling, C., Zeger, K.: Non-Shannon information inequalities in four random variables. arXiv.org, arXiv:1104.3602v1 (2011)
25. Farràs, O., Metcalf-Burton, J.R., Padró, C., Vázquez, L.: On the optimization of bipartite secret sharing schemes. Des. Codes Cryptogr. **63**, 255–271 (2012)
26. Fujishige, S.: Polymatroidal dependence structure of a set of random variables. Inf. Control **39**, 55–72 (1978)
27. Fujishige, S.: Entropy functions and polymatroids-combinatorial structures in information theory. Electron. Comm. Japan **61**, 14–18 (1978)
28. Gács, P., Körner, J.: Common information is far less than mutual information. Probl. Control Inf. Theory **2**, 149–162 (1973)
29. Gharahi, M: On the complexity of perfect secret sharing schemes. Ph.D. Thesis, Iran University of Science and Technology (2013) (in Persian)
30. Gharahi, M., Dehkordi, M.H.: The complexity of the graph access structures on six participants. Des. Codes Cryptogr. **67**, 169–173 (2013)
31. Gharahi, M., Dehkordi, M.H.: Average complexities of access structures on five participants. Adv. Math. Commun. **7**, 311–317 (2013)
32. Gharahi, M., Dehkordi, M.H: Perfect secret sharing schemes for graph access structures on six participants. J. Math. Cryptol. **7**, 143–146 (2013)
33. Gharahi, M., Khazaei, S.: Optimal linear secret sharing schemes for graph access structures on six participants. Cryptology ePrint Archive: Report 2017/1232 (2017)
34. Hammer, D., Romashchenko, A.E., Shen, A., Vereshchagin, N.K.: Inequalities for Shannon entropy and Kolmogorov complexity. J. Comput. Syst. Sci. **60**, 442–464 (2000)
35. Ingleton, A.W.: Representation of matroids. In: Welsh, D.J.A. (ed.) Combinatorial Mathematics and its Applications, pp. 149–167. Academic Press, London (1971)
36. Jackson, W.A., Martin, K.M.: Geometric secret sharing schemes and their duals. Des. Codes Cryptogr. **4**, 83–95 (1994)
37. Jackson, W.A., Martin, K.M.: Perfect secret sharing schemes on five participants. Des. Codes Cryptogr. **9**, 267–286 (1996)
38. Kaced, T.: Equivalence of two proof techniques for non-Shannon inequalities. arXiv:1302.2994 (2013)
39. Karnin, E.D., Greene, J.W., Hellman, M.E.: On secret sharing systems. IEEE Trans. Inf. Theory **29**, 35–41 (1983)
40. Kinser, R.J.: New inequalities for subspace arrangements. Combin. Theory Ser. A **118**, 152–161 (2011)
41. Li, Q., Li, X.X., Lai, X.J., Chen, K.F.: Optimal assignment schemes for general access structures based on linear programming. Des. Codes Cryptogr. **74**, 623–644 (2015)
42. Makarychev, K., Makarychev, Y., Romashchenko, A., Vereshchagin, N.: A new class of non-Shannon-type inequalities for entropies. Commun. Inf. Syst. **2**, 147–166 (2002)
43. Martí-Farré, J., Padró, C.: Secret sharing schemes with three or four minimal qualified subsets. Des. Codes Cryptogr. **34**, 17–34 (2005)
44. Martí-Farré, J., Padró, C.: On secret sharing schemes, matroids and polymatroids. J. Math. Cryptol. **4**, 95–120 (2010)

45. Martí-Farré, J., Padró, C., Vázquez, L.: Optimal complexity of secret sharing schemes with four minimal qualified subsets. Des. Codes Cryptogr. **61**, 167–186 (2011)
46. Martín, S., Padró, C., Yang, A.: Secret sharing, rank inequalities, and information inequalities. IEEE Trans. Inform. Theory **62**, 599–609 (2016)
47. Matúš, F.: Infinitely many information inequalities. In: Proceedings of the IEEE International Symposium on Information Theory, (ISIT), pp. 2101–2105 (2007)
48. Metcalf-Burton, J.R.: Improved upper bounds for the information rates of the secret sharing schemes induced by the Vámos matroid. Discret. Math. **311**, 651–662 (2011)
49. Oxley, J.G: Matroid Theory. Oxford Science Publications, The Clarendon Press, Oxford University Press, New York (1992)
50. Padró, C.: Lecture Notes in secret sharing. Cryptology ePrint Archive, Report 2012/674 (2912)
51. Padró, C., Sáez, G.: Secret sharing schemes with bipartite access structure. IEEE Trans. Inform. Theory **46**, 2596–2604 (2000)
52. Padró, C., Vázquez, L., Yang, A.: Finding lower bounds on the complexity of secret sharing schemes by linear programming. Discret. Appl. Math. **161**, 1072–1084 (2013)
53. Pitassi T., Robere R., Lifting Nullstellensatz to Monotone Span Programs over any Field. Electronic Colloquium on Computational Complexity (ECCC), vol. 165 (2017)
54. Rado, R.: Note on independence functions. Proc. Lond. Math. Soc. 3(7), 300–320 (1957)
55. Robere, R., Pitassi, T., Rossman, B., Cook, S.A.: Exponential lower bounds for monotone span programs. In: FOCS 2016, pp. 406–415 (2016)
56. Seymour, P.D.: A forbidden minor characterization of matroid ports. Quart. J. Math. Oxf. Ser. **27**, 407–413 (1976)
57. Seymour, P.D.: On secret-sharing matroids. J. Combin. Theory Ser. B **56**, 69–73 (1992)
58. Shamir, A.: How to share a secret. Commun. ACM **22**, 612–613 (1979)
59. Stinson, D.R.: An explication of secret sharing schemes. Des. Codes Cryptogr. **2**, 357–390 (1992)
60. Stinson, D.R.: Decomposition constructions for secret-sharing schemes. IEEE Trans. Inf. Theory **40**, 118–125 (1994)
61. Thakor, S., Chan, T., Grant, A.: Capacity bounds for networks with correlated sources and characterisation of distributions by entropies. IEEE Trans. Inf. Theory **63**, 3540–3553 (2017)
62. Tian, C.: Characterizing the Rate Region of the $(4, 3, 3)$ Exact-Repair Regenerating Codes. arXiv.org, arXiv:1312.0914 (2013)
63. Yeung, R.W.: A First Course in Information Theory. Kluwer Academic/Plenum Publishers, New York (2002)
64. Yeung, R.W.: Information Theory and Network Coding. Springer, Boston (2008)
65. Zhang, Z.: On a new non-Shannon type information inequality. Commun. Inf. Syst. **3**, 47–60 (2003)
66. Zhang, Z., Yeung, R.W.: On characterization of entropy function via information inequalities. IEEE Trans. Inf. Theory **44**, 1440–1452 (1998)

Author Index

Printed in the United States
By Bookmasters